Lecture Notes in Computer Science 8753

Commenced Publication in 1973
Founding and Former Series Editors:
Gerhard Goos, Juris Hartmanis, and Jan van Leeuwen

More information about this series at http://www.springer.com/series/7412

Organization

Conference Committee

General Chair

Xiaoyi Jiang   University of Münster, Germany

Program Co-chairs

Joachim Hornegger   University of Erlangen-Nuremberg, Germany
Reinhard Koch   University of Kiel, Germany

Program Committee

Horst Bischof   Graz University of Technology, Austria
Thomas Brox   University of Freiburg, Germany
Andres Bruhn   University of Stuttgart, Germany
Joachim Buhmann   ETH Zurich, Switzerland
Daniel Cremers   Technical University of Munich, Germany
Andreas Dengel   Technical University of Kaiserslautern, Germany
Joachim Denzler   University of Jena, Germany
Michael Felsberg   Linköping University, Sweden
Gernot A. Fink   Technical University of Dortmund, Germany
Boris Flach   Czech Technical University in Prague,
                           Czech Republic
Jan-Michael Frahm   University of North Carolina at Chapel Hill, USA
Uwe Franke   Daimler AG, Germany
Ju"rgen Gall   University of Bonn, Germany
Peter Gehler   MPI for Intelligent Systems, Germany
Michael Goesele   Technical University of Darmstadt, Germany
Michal Haindl   UTIA Prague, Czech Republic
Fred A. Hamprecht   University of Heidelberg, Germany
Edwin Hancock   University of York, UK
Matthias Hein   Saarland University, Germany
Olaf Hellwich   Technical University of Berlin, Germany
Vaclav Hlavac   Czech Technical University in Prague, Czech
                           Republic
Thomas Hofmann   ETH Zurich and Google, Switzerland
Reinhard Klette   University of Auckland, New Zealand
Walter G. Kropatsch   Vienna University of Technology, Austria

Christoph H. Lampert	IST Austria, Austria
Bastian Leibe	RWTH Aachen University, Germany
Hendrik Lensch	University of Tübingen, Germany
Marco Loog	Delft University of Technology, The Netherlands
Andreas Maier	University of Erlangen-Nuremberg, Germany
Diana Mateus	Technical University of Munich, Germany
Helmut Mayer	Bundeswehr University Munich, Germany
Rudolf Mester	University of Frankfurt, Germany
Klaus-Robert Müller	Technical University of Berlin, Germany
Sebastian Nowozin	Microsoft Research, UK
Josef Pauli	University of Duisburg-Essen, Germany
Dietrich Paulus	University of Koblenz-Landau, Germany
Nicolai Petkov	University of Groningen, The Netherlands
Thomas Pock Graz	University of Technology, Austria
Christian Riess	University of Erlangen-Nuremberg, Germany
Gerhard Rigoll	Technical University of Munich, Germany
Karl Rohr	University of Heidelberg, Germany
Olaf Ronneberger	University of Freiburg, Germany
Bodo Rosenhahn	University of Hannover, Germany
Stefan Roth	Technical University of Darmstadt, Germany
Volker Roth	University of Basel, Switzerland
Hanno Scharr	Jülich Research Centre, Germany
Daniel Scharstein	Middlebury College, USA
Bernt Schiele	MPI for Informatics and Saarland University, Germany
Christoph Schnörr	University of Heidelberg, Germany
Carsten Steger	MVTec Software GmbH, Germany
Stefan Steidl	University of Erlangen-Nuremberg, Germany
Rainer Stiefelhagen	Karlsruhe Institute of Technology, Germany
Peter Sturm	Inria Grenoble – Rhône-Alpes, France
Christian Theobalt	MPI for Informatics, Saarbrücken, Germany
Klaus Tönnies	University of Magdeburg, Germany
Thomas Vetter	University of Basel, Switzerland
Joachim Weickert	Saarland University, Germany
Martin Welk	UMIT Hall, Austria
Simon Winkelbach	Technical University of Braunschweig, Germany

Additional Reviewers

Ahmed Abdulkadir	David Balduzzi	Johnny Chien
Hanno Ackermann	Thorsten Beier	Vincent Christlein
Sheraz Ahmed	Alexander Binder	Christian Conrad
Ziad Al-Halah	Ghazi Bouabene	Sven Dähne
Freddie Åström	Stefan Breuers	Martin Danelljan
Sebastian Bach	Dirk Buchholz	Edilson de Aguiar

Oliver Demetz	Niko Krasowski	Stephan Richter
Ferran Diego	Michael Krause	Erik Ringaby
Benjamin Drayer	Anna Kreshuk	Saquib Sarfraz
Enrique Dunn	Thorben Kröger	Kevin Schelten
Bernhard Egger	Alina Kuznetsova	Björn Scheuermann
Jakob Engel	Dmitry Laptev	Thorsten Schmidt
Philipp Fischer	Dongwei Liu	Uwe Schmidt
Andreas Forster	Robert Maier	Johannes Schönberger
Thorsten Franzel	Tobias Maier	Christopher Schroers
Faezeh Frouzesh	Muhammad Imran Malik	Peter Schüffler
Qi Gao	Manuel Martinez	Christian Schulze
Pablo Garrido	Nikolaus Mayer	Anke Schwarz
Tobias Gehrig	Brian McWilliams	Bok-Suk Shin
Nico Görnitz	Thomas Möllenhoff	Mohamed Souiai
Miguel Granados	Oliver Müller	Christoph Straehle
Rene Grzeszick	Naveen Shankar Nagaraja	Jan Stühmer
Hendrik Hachmann	Thomas Nestmeyer	Junli Tao
Jared Heinly	Klas Nordberg	Makarand Tapaswi
Alexander Hermans	Peter Ochs	Adnan Ul-Hasan
Laurent Hoeltgen	Kristoffer Öfjäll	Benjamin Ummenhofer
Sebastian Hoffmann	Aljosa Osep	Vladyslav Usenko
Esther Horbert	Sebastian Palacio	Sebastian Volz
Varun Jampani	Danny Panknin	Timo von Marcard
Yong-Chul Ju	Nico Persch	Michael Waechter
Christoph Jud	Axel Plinge	Zhengping Wang
Christian Kerl	Tobias Plötz	Tobias Weyand
Margret Keuper	Sebastian Polsterl	Thomas Windheuser
Fahad Khan	Umer Rafi	Christian Winkens
Martin Kiefel	Mahdi Rezaei	Mehmet Yigitsoy
Hilke Kieritz	Christian Richardt	Jasenko Zivanov

Technical Support, Conference Management System and Proceedings

Sven Grünke	University of Erlangen-Nuremberg, Germany
Daniel Jung	University of Kiel, Germany
Eva Eibenberger	University of Erlangen-Nuremberg, Germany

Local Organizing Committee

Gerlinde Siekaup	University of Münster, Germany
Dimitri Berh	University of Münster, Germany
Klaus Broelemann	University of Münster, Germany

Mohammad Dawood   University of Münster, Germany
Michael Fieseler   University of Münster, Germany
Benjamin Risse   University of Münster, Germany
Michael Schmeing   University of Münster, Germany
Sönke Schmid   University of Münster, Germany
Kathrin Ungru   University of Münster, Germany

Sponsoring Institutions

Cells-in-Motion Cluster of Excellence, Münster, Germany
MVTec Software GmbH, Munich, Germany
Olympus Soft Imaging Solutions GmbH, Münster, Germany
University of Münster, Münster, Germany

Awards 2013

Deutscher Mustererkennungspreis 2013

The "Deutscher Mustererkennungspreis 2013" was awarded to

Thomas Pock

for his outstanding work on "Variational and Optimization Methods for Image Processing and Computer Vision."

DAGM Awards 2013

The main price for the DAGM Best Paper 2013 was awarded to

Timo Scharwächter, Markus Enzweiler, Stefan Roth, and Uwe Franke
"Efficient Multi-Cue Scene Segmentation."

Further DAGM prices were awarded to

Anne Jordt-Sedlazeck, Daniel Jung, and Reinhard Koch
"Refractive Plane Sweep for Underwater Images,"

Christoph Straehle, Sven Peter, Ullrich Köthe, Fred Hamprecht
"K-Smallest Spanning Tree Segmentation."

Contents

Young Researcher Forum

Variational Models for Depth and Flow

Scene Flow Estimation from Light Fields via the Preconditioned Primal-Dual Algorithm

Stefan Heber[1][✉] and Thomas Pock[1,2]

[1] Institute for Computer Graphics and Vision, Graz University of Technology,
Graz, Austria
stefan.heber@icg.tugraz.at
[2] Safety and Security Department, AIT Austrian Institute of Technology,
Seibersdorf, Austria

Abstract. In this paper we present a novel variational model to jointly estimate geometry and motion from a sequence of light fields captured with a plenoptic camera. The proposed model uses the so-called sub-aperture representation of the light field. Sub-aperture images represent images with slightly different viewpoints, which can be extracted from the light field. The sub-aperture representation allows us to formulate a convex global energy functional, which enforces multi-view geometry consistency, and piecewise smoothness assumptions on the scene flow variables. We optimize the proposed scene flow model by using an efficient preconditioned primal-dual algorithm. Finally, we also present synthetic and real world experiments.

1 Introduction

Restricted to geometric optics, the plenoptic function [1] describes the amount of light that travels along rays in 3D space. In this context a ray can be seen as a fundamental carrier of light, where the amount of light traveling along the ray is called radiance. By parameterizing a ray via a position $(x, y, z) \in \mathbb{R}^3$ and a direction $(\xi, \eta) \in \mathbb{R}^2$, one sees that the plenoptic function is five dimensional, and it maps a specific point on a ray to the corresponding radiance. Note that the radiance remains constant along a ray till it hits an object. This observation allows to identify one dimensional redundant information in the plenoptic function, which leads to a reduced 4D function usually denoted as light field in computer vision literature. This 4D light field provides a rich source of information of the captured scene, and thus capturing and processing light fields has become a topic of increased interest in recent years. Whereas a conventional image only provides information about the accumulated radiance of all rays hitting a certain position at the image sensor, a light field provides also the additional directional information about the radiance of the individual light rays.

This research was supported by the FWF-START project *Bilevel optimization for Computer Vision*, No. Y729 and the Vision+ project *Integrating visual information with independent knowledge*, No. 836630.

© Springer International Publishing Switzerland 2014
X. Jiang et al. (Eds.): GCPR 2014, LNCS 8753, pp. 3–14, 2014.
DOI: 10.1007/978-3-319-11752-2_1

There are different devices to capture light fields. The simplest device is a single moving camera, which only allows to capture light fields of static scenes. In order to capture dynamic scenes, one can for example choose the hardware intense solution of a camera array [28], or, in recent years, also plenoptic cameras have become available (e.g. Lytro[1] or Raytrix[2]). In a plenoptic camera a micro-lens array is placed in front of the image sensor, with the effect that incoming light is split up into rays of different directions. Each ray then hits the sensor at a slightly different location, which allows to capture the additional directional information.

The additional information inherent in the light field is beneficial for many image processing applications, like e.g. super-resolution [5,24], image denoising [12], image segmentation [25], or depth estimation [4,14,23], and also led to complete new applications, like e.g. digital refocusing [15,19], extending the depth of field [19], or digital correction of lens aberrations [19].

In this paper we will introduce a further application suitable for light field data, which has not been considered before for this type of data. We will consider the task of scene flow estimation with a single plenoptic camera. Thus, we will show that two consecutive light fields captured with a plenoptic camera can be used to calculate scene flow, in terms of two disparity maps and the optical flow.

2   Related Work

An important characteristic of dynamic scenes is the geometry and motion of objects. Such information can be used in many image processing tasks, including tracking and segmentation. Scene flow is defined by Vedula et al. [22] as a dense 3D motion field of a nonrigid 3D scene. Therefore, scene flow estimation is the challenging problem of calculating geometry and motion at the same time [2]. By considering images from one view point, scene flow estimation is underdetermined. Also a moving camera still creates ambiguities between camera and scene motion. Only after introducing additional cameras, these ambiguities can be resolved. By increasing the number of cameras one can decrease possible ambiguities and increase the robustness.

Most of the existing scene flow approaches decouple the problem of calculating geometry and motion [8,16,22,27,29]. Thus the two problems are solved sequentially, which allows faster computation, but comes with the disadvantage that the spatial-temporal information is not fully exploited. Contrary to these decoupled approaches the proposed method makes use of the original definition of scene flow by Vedula et al. [22], where the problem is defined as jointly estimating motion and disparity. There are also some approaches, which are not limited to two views like e.g. [3,8,10,11,18]. Here the methods proposed by Courchay et al. [10] and Furukawa and Ponce [11] are limited to a fixed mesh topology, and the method proposed by Neumann and Aloimonos [18] was only used for scenes which consist of one connected object. The method most closely related

---

[1] www.lytro.com
[2] www.raytrix.de

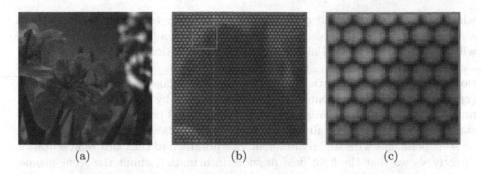

(a)   (b)   (c)

Fig. 1. Illustration of a raw image captured with a plenoptic camera. (a) Shows the complete raw image, (b) and (c) show closeup views of the raw image, where one can clearly see the effect of placing the a micro-lens array in front of the image sensor. This micro-lens array makes it possible to capture the 4D light field.

to ours was proposed by Basha et al. [3]. They use a variational formulation, which enforces smoothness directly on the 3D displacement vectors, whereas our method enforces smoothness on the two disparity maps as wells as on the 2D optical flow. Moreover, contrary to our method they only use a first order regularization, which favors fronto-parallel solutions.

Contribution

In this paper we introduce a novel method for variational scene flow estimation, which is specially designed for light field data captured with a plenoptic camera [17]. We show that the rich structure within a sequence of light fields can be used to calculate scene flow in a multi-view setting. The main idea is to use the multi-view information within the light field to improve the stability of the result and reduces ambiguities. The proposed method is also designed to easily vary between speed and accuracy, by changing the number of involved sub-aperture images. Compared to other scene flow methods, the hardware requirements of the proposed approach are reduced to a single light field camera.

The main contribution of the work is the variational framework, which directly uses the multi-view information provided by the light field to simultaneously calculate both geometry in terms of two disparity maps and motion in terms of the 2D optical flow. To the best of our knowledge, this paper presents the first method, that estimates scene flow from a light field camera setting.

3 Preliminaries

It is common practice to use the so-called two-plane parametrization [13] to mathematically define the 4D light field. Suppose Ω and Π to be the image plane and the lens plane, respectively. Then we can define the light field as

$$\tilde{L} : \Omega \times \Pi \to \mathbb{R}, \qquad (\boldsymbol{p}, \boldsymbol{q}) \mapsto \tilde{L}(\boldsymbol{p}, \boldsymbol{q}), \qquad (1)$$

where $p := (x, y) \in \Omega$ and $q := (\xi, \eta) \in \Pi$. In oder to describe the proposed scene flow algorithm, it is also necessary to introduce a time parameter t, *i.e.* we will describe the light field at time t via the 5D function $L(p, q, t)$.

The light field can be visualized in different ways. The simplest representation (in the case of plenoptic cameras) is the raw image captured at the sensor (*cf.* Fig. 1(a)). Another common visualization goes by the name sub-aperture image. This is a representation where the directional component q is kept constant and one varies over all spatial positions p. Sub-aperture images can also be seen as images with slightly different viewpoints, and thus this representation directly shows that the light field provides information about the scene geometry. Furthermore, this representation also clearly shows the connection between light fields and multi-view systems.

4   Light Field Scene Flow Model

In this section we will describe the proposed light field scene flow model, which can be seen as an extension of the shape from light field model proposed by Heber *et al.* [14] to the task of scene flow estimation. The proposed light field scene flow model enforces multi-view geometry consistency by assuming brightness constancy, and it incorporates global smoothness assumptions on all variables. The method jointly calculates two disparity maps denoted as $d = [d_1, d_2]^T$, and the optical flow $u = [u, v]^T$. Note, that all variables are calculated for the center view $L(p, 0, t)$ of the light field. Our model is based on variational principles and combines a data fidelity term with a suitable regularization term

$$\text{minimize}\quad E_{data}(d, u) + E_{reg}(d, u),\tag{2}$$

where the data fidelity term and the regularization term will be formulated in Sects. 4.1 and 4.2, respectively.

4.1   Data Fidelity Term

The data fidelity term $E_{data}(d, u)$ of the proposed light field scene flow model can be stated in the continuous setting as follows

$$E_{data}(d, u) = \int_\Omega \int_0^R \int_0^{2\pi} \begin{bmatrix} \lambda_1 \\ \lambda_2 \\ \lambda_3 \end{bmatrix}^T \begin{bmatrix} \Psi_{s,r}^1(p, d_1) \\ \Psi_{s,r}^2(p, d_2) \\ \Psi_{s,r}^3(p, d, u) \end{bmatrix} \, \mathrm{d}(s, r, p),\tag{3}$$

with

$$\Psi_{s,r}^1(p, d_1) = \left| L(p, 0, t_1) - L\left(p - \frac{d_1}{R}\varphi_{s,r}, \varphi_{s,r}, t_1\right) \right|,\tag{4}$$

$$\Psi_{s,r}^2(p, d_2) = \left| L(p, 0, t_2) - L\left(p - \frac{d_2}{R}\varphi_{s,r}, \varphi_{s,r}, t_2\right) \right|,\tag{5}$$

$$\Psi_{s,r}^3(p, d, u) = \left| L\left(p - \frac{d_1}{R}\varphi_{s,r}, \varphi_{s,r}, t_1\right) - L\left(p + u - \frac{d_2}{R}\varphi_{s,r}, \varphi_{s,r}, t_2\right) \right|,\tag{6}$$

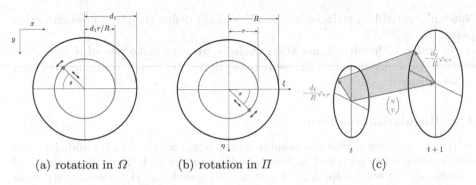

(a) rotation in Ω (b) rotation in Π (c)

Fig. 2. Illustration of the parametrization used in (3). (a) Sketches a scene point's image position and the corresponding rotation circle, (b) shows the according directional sampling position in the lens plane for extracting the sub-aperture image, and (c) sketches the modeled connection between the two light field images at time t and $t+1$.

where $\Omega \subseteq \mathbb{R}^2$ is the image domain, $\lambda_i \in \mathbb{R}_+$ for $1 \leqslant i \leqslant 3$ are positive weighting parameters, $t_2 = t_1 + 1$, and $\varphi_{s,r} = r\,[cos(s), sin(s)]^T$ is a circle parametrization. By taking a closer look at (4) and (5) one sees that those terms denote data fidelity terms for stereo matching at time t_1 and t_2, respectively. Furthermore, (6) denotes a data fidelity term for optical flow calculation between corresponding sub-aperture images at time t_1 and t_2 (*cf.* Fig. 2(c)). Note, similar as in [14] d_1 and d_2 denote the largest scene point's image rotation radii in the image plane at time t_1 and t_2, respectively (*cf.* Fig. 2). Also note that we are using the robust ℓ_1 norm as the loss function.

$E_{data}(\boldsymbol{d}, \boldsymbol{u})$ (*cf.* (3)) is not convex, thus we use first order Taylor approximations to obtain a convex relaxation, *i.e.*

$$L\left(\boldsymbol{p} - \frac{d_1}{R}\varphi_{s,r}, \varphi_{s,r}, t_1\right) \approx \tag{7}$$

$$L\left(\boldsymbol{p} - \frac{\hat{d}_1}{R}\varphi_{s,r}, \varphi_{s,r}, t_1\right) + (d_1 - \hat{d}_1)\frac{r}{R}\nabla_{-\frac{\varphi_{s,r}}{r}}L\left(\boldsymbol{p} - \frac{\hat{d}_1}{R}\varphi_{s,r}, \varphi_{s,r}, t_1\right),$$

where $\nabla_{-\frac{\varphi_{s,r}}{r}}$ is the directional derivative with direction $[-\frac{\varphi_{s,r}}{r}, 0, 0]^T$. In a similar way we approximate $L\left(\boldsymbol{p} - \frac{d_2}{R}\varphi_{s,r}, \varphi_{s,r}, t_2\right)$ of the second stereo term (*cf.* (5)). Finally, the Taylor approximation of the remaining non convex part of the optical flow term (*cf.* (6)) is given as follows

$$L\left(\boldsymbol{p} + \boldsymbol{u} - \frac{d_2}{R}\varphi_{s,r}, \varphi_{s,r}, t_2\right) \approx \tag{8}$$

$$L\left(\boldsymbol{p} + \hat{\boldsymbol{u}} - \frac{\hat{d}_2}{R}\varphi_{s,r}, \varphi_{s,r}, t_2\right) + \begin{bmatrix} u - \hat{u} \\ v - \hat{v} \\ d_2 - \hat{d}_2 \end{bmatrix}^T \begin{bmatrix} \nabla_x L\left(\boldsymbol{p} + \hat{\boldsymbol{u}} - \frac{\hat{d}_2}{R}\varphi_{s,r}, \varphi_{s,r}, t_2\right) \\ \nabla_y L\left(\boldsymbol{p} + \hat{\boldsymbol{u}} - \frac{\hat{d}_2}{R}\varphi_{s,r}, \varphi_{s,r}, t_2\right) \\ \frac{r}{R}\nabla_{-\frac{\varphi_{s,r}}{r}}L\left(\boldsymbol{p} + \hat{\boldsymbol{u}} - \frac{\hat{d}_2}{R}\varphi_{s,r}, \varphi_{s,r}, t_2\right) \end{bmatrix}.$$

Note that variables marked with $\hat{\cdot}$ in (7) and (8) define the given approximation point.

In order to handle illumination changes we also make use of a structure-texture decomposition [26], *i.e.* we remove the low frequency component of each sub-aperture image.

4.2   Regularization Term

In this section we define the regularization term, which will be added to the data-fidelity term proposed in Sect. 4.1. Due to the fact, that the problem of minimizing (3) with respect to d and u is ill-posed, *i.e.* the data fidelity term alone is not sufficient to calculate a reliable solution, an additional smoothness assumption is needed. As in [14] we assume that our solution is piecewise linear, which can be achieved by introducing Total Generalized Variation (TGV) [6] of second order as a regularization term. Moreover, we also use anisotropic diffusion tensors Γ_t, as suggested by Ranftl *et al.* [21]. These diffusion tensors connect the prior with the image content, which leads to solutions with a lower degree of smoothness around depth discontinuities. This image-driven TGV regularization term can be written as

$$\Phi_t(u) = \min_{\boldsymbol{w} \in \mathbb{R}^2} \left\{ \alpha_1 \int_{\Omega} |\Gamma_t (\nabla u - \boldsymbol{w})| \, \mathrm{d}\boldsymbol{x} + \alpha_0 \int_{\Omega} |\nabla \boldsymbol{w}| \, \mathrm{d}\boldsymbol{x} \right\}, \tag{9}$$

with

$$\Gamma_t = \exp(-\gamma |\nabla L(\boldsymbol{p}, \boldsymbol{0}, t)|^\beta) \, \boldsymbol{n}\boldsymbol{n}^T + \boldsymbol{n}^\perp \boldsymbol{n}^{\perp^T}, \tag{10}$$

where $\boldsymbol{n}$ is the normalized image gradient of the center view of the light field at time t, $\boldsymbol{n}^\perp$ is a vector perpendicular to $\boldsymbol{n}$, and α_0, α_1, γ and $\beta \in \mathbb{R}_+$ are predefined scalars. We apply (9) to all involved variables in the following way

$$E_{reg}(\boldsymbol{d}, \boldsymbol{u}) = \Phi_{t_1}(d_1) + \Phi_{t_2}(d_2) + \Phi_{t_1}(u) + \Phi_{t_1}(v). \tag{11}$$

By combining the data term in (3) and the above regularization term (11), we obtain our final variational scene flow model.

4.3   Discretization

In order to handle the discrete set of measurements from the image sensor we define a discrete image domain $\hat{\Omega}$. Moreover we also use a discrete set of circle parametrizations. Therefore, we define $M > 1$ to be the number of different sampling circles, and N_i to be the number of uniform sampling positions of the ith circle. Then the discretized version of the data term (3) is given as

$$\hat{E}_{data}(\boldsymbol{p}, \boldsymbol{u}) = \sum_{\boldsymbol{p} \in \hat{\Omega}} \sum_{i=0}^{M-1} \sum_{j=1}^{N_i} \begin{bmatrix} \lambda_1 \\ \lambda_2 \\ \lambda_3 \end{bmatrix}^T \begin{bmatrix} \Psi^1_{s_{ij}, r_i}(\boldsymbol{p}, d_1) \\ \Psi^2_{s_{ij}, r_i}(\boldsymbol{p}, d_2) \\ \Psi^3_{s_{ij}, r_i}(\boldsymbol{p}, \boldsymbol{d}, \boldsymbol{u}) \end{bmatrix}, \tag{12}$$

with

$$s_{ij} = \frac{2\pi(j-1)}{N_i} \quad \text{and} \quad r_i = \frac{Ri}{M-1}, \tag{13}$$

where r_i represents the radius, and s_{ij} for $1 \leqslant j \leqslant N_i$ represent the discrete circle positions of the ith circle. Note that by choosing $M = 2$ and $N_{0,1} = 1$ the model reduces to the stereo case.

We will denote $\hat{E}_{reg}(d, u)$ as the discrete version of the regularization term (11), where we use finite differences with Neumann boundary conditions to discretize the involved gradient operators.

4.4 Optimization

In this section we show how to optimize the discretized problem

$$\text{minimize} \quad \hat{E}_{data}(d, u) + \hat{E}_{reg}(d, u) \tag{14}$$

with the primal-dual algorithm, proposed by Chambolle *et al.* [9]. Due to the fact, that the linear approximations (7) and (8) are only accurate in a small neighborhood around the current solutions $\hat{d}$ and $\hat{u}$, we will also embed the algorithm into a coarse-to-fine warping scheme [7].

In order to use the primal-dual algorithm [9], we have to rewrite (14) as a generic saddle point problem. To simplify notation we define the following terms:

$$A_{ij}^1 = \frac{r_i}{R} \nabla_{\frac{\varphi_{s_{ij},r_i}}{r_i}} L\left(p - \frac{\hat{d}_1}{R}\varphi_{s_{ij},r_i}, \varphi_{s_{ij},r_i}, t_1\right), \tag{15}$$

$$A_{ij}^2 = L(p, 0, t_1) - L\left(p - \frac{\hat{d}_1}{R}\varphi_{s_{ij},r_i}, \varphi_{s_{ij},r_i}, t_1\right),$$

$$A_{ij}^5 = \nabla_x L\left(p + \hat{u} - \frac{\hat{d}_2}{R}\varphi_{s_{ij},r_i}, \varphi_{s_{ij},r_i}, t_2\right),$$

$$A_{ij}^6 = \nabla_y L\left(p + \hat{u} - \frac{\hat{d}_2}{R}\varphi_{s_{ij},r_i}, \varphi_{s_{ij},r_i}, t_2\right),$$

$$A_{ij}^7 = L\left(p - \frac{\hat{d}_1}{R}\varphi_{s_{ij},r_i}, \varphi_{s_{ij},r_i}, t_1\right) - L\left(p + \hat{u} - \frac{\hat{d}_2}{R}\varphi_{s_{ij},r_i}, \varphi_{s_{ij},r_i}, t_2\right),$$

and the terms A_{ij}^3 and A_{ij}^4 are defined analogously to A_{ij}^1 and A_{ij}^2, where $\hat{d}_1$ and t_1 are replaced with $\hat{d}_2$ and t_2, respectively. If we assume $A_{ij}^* \in \mathbb{R}^{|\hat{\Omega}|}$ to be column vectors, where $|\hat{\Omega}|$ denotes the number of elements of the discrete image domain, then the discretized problem (14) can be rewritten as the following saddle point problem:

$$
\min_{\substack{P}} \max_{\substack{D \\ \|x\|_\infty \leq 1 \\ \forall x \in D}} \left\{ \lambda_1 \sum_{i=0}^{M-1} \sum_{j=1}^{N_i} \left\langle A_{ij}^2 - \mathrm{diag}(A_{ij}^1)(d_1 - \hat{d}_1), dd_{ij}^1 \right\rangle + \right. \tag{16}
$$

$$
\lambda_2 \sum_{i=0}^{M-1} \sum_{j=1}^{N_i} \left\langle A_{ij}^4 - \mathrm{diag}(A_{ij}^3)(d_2 - \hat{d}_2), dd_{ij}^2 \right\rangle +
$$

$$
\lambda_3 \sum_{i=0}^{M-1} \sum_{j=1}^{N_i} \left\langle A_{ij}^7 - I_{|\hat{\Omega}|}^{1,4} \mathrm{diag} \left(\begin{bmatrix} -A_{ij}^1 \\ A_{ij}^3 \\ A_{ij}^5 \\ A_{ij}^6 \end{bmatrix} \right) \begin{bmatrix} d - \hat{d} \\ u - \hat{u} \end{bmatrix}, dd_{ij}^3 \right\rangle +
$$

$$
\alpha_1 \left\langle \begin{bmatrix} \Gamma_{t_1}(\nabla d_1 - p_{d_1}) \\ \Gamma_{t_2}(\nabla d_2 - p_{d_2}) \\ \Gamma_{t_1}(\nabla u - p_u) \\ \Gamma_{t_1}(\nabla v - p_v) \end{bmatrix}, \begin{bmatrix} d_d \\ d_u \end{bmatrix} \right\rangle + \alpha_0 \left\langle \begin{bmatrix} \nabla p_{d_1} \\ \nabla p_{d_2} \\ \nabla p_u \\ \nabla p_v \end{bmatrix}, \begin{bmatrix} d_{p_d} \\ d_{p_u} \end{bmatrix} \right\rangle \right\},
$$

with $I_n^{1,4} = [I_n, I_n, I_n, I_n]$, where I_n is the identity matrix of size $n \times n$. Moreover, P and D represent the set of all primal and dual variables, respectively:

$$
P = \{ d, u, p_d, p_u \}, \tag{17}
$$

$$
D = \left\{ (dd_{ij}^k)_{1 \leq k \leq 3}, d_d, d_u, d_{p_d}, d_{p_u} \right\}. \tag{18}
$$

The saddle point problem (16) can now be solved using the primal-dual algorithm proposed in [9]. An improvement with respect to convergence speed can be obtained by using adequate symmetric and positive definite preconditioning matrices as suggested in [20].

5   Experimental Results

In this section we first evaluate the proposed algorithm on two challenging synthetic data sequences generated with povray[3]. After the synthetic evaluation we will also present some qualitative results for real world data. Here we will use two consecutive raw images captured with a Lytro camera as input for the proposed scene flow model.

5.1   Synthetic Experiments

For the synthetic evaluation we create two datasets denoted as *snails* and *apples*[4]. Both datasets have a spatial resolution of 640×480 micro-lenses, and a directional resolution of 9×9 pixels per micro-lens. In order to create the datasets,

---

[3] www.povray.org
[4] Scenes are taken from www.oyonale.com.

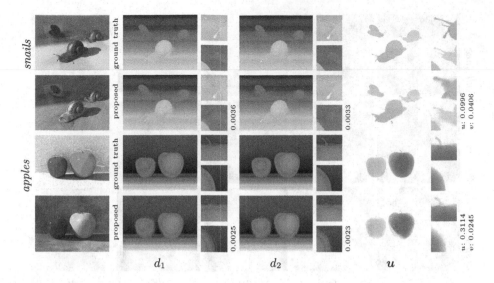

Fig. 3. Qualitative results for the synthetic scenes *snails* and *apples*. The figure shows from left to right, an illustration of the motion and the center view of the light field at time t_1, the two disparity maps d_1 and d_2, and the color coded optical flow u (Middlebury color code). For the variables d_1, d_2 and u we present the result of the proposed model, as well as the corresponding ground truth (Color figure online).

we first render 9×9 sub-aperture images, where the viewpoints are shifted on a regular grid. After rendering we rearrange the light field data to obtain a synthetic raw image similar to a raw image captured with a plenoptic camera. A sequence of such raw images is used as input for the proposed algorithm. Figure 3 presents qualitative results of the proposed model for the two datasets, where $M = 3$, $N_0 = 1$, and $N_{1,2} = 8$ (*cf.* (13)). Furthermore, Fig. 3 also shows the mean squared errors (MSEs) for the different scene flow terms. Although the two datasets are quite challenging, *i.e.* they include specularity, shadow, reflections *etc.*, the proposed model is still capable of estimating a reliable solution for the disparity as well as for the optical flow variables. The results shown in Fig. 3 took about 30 s to compute (17 views). Note, that the computation time drops significantly by reducing the number of involved views.

5.2 Real World Experiments

We now present some qualitative real world results obtained by the proposed light field scene flow model. For capturing the light fields we use a Lytro camera, which is a commercially available plenoptic camera. Such a camera provides a spatial resolution of around 380×330 micro-lenses and a directional resolution of about 10×10 pixels per micro-lens. For the real world experiments we set $M = 2$, $N_0 = 1$ and $N_1 = 16$, and the weighting parameters are tuned for the different scenes. Figure 4 shows some qualitative results of the proposed method

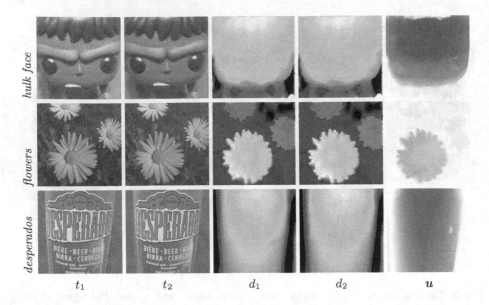

Fig. 4. Qualitative results for real world scenes. The figure shows from left to right, the two center views (800 × 800 pixels) from the light fields captured with a Lytro camera at time t_1 and t_2, the calculated disparity maps d_1 and d_2 and the corresponding 2D optical flow u shown with Middlebury color code.

for different scenes. The captured light fields have a quite low spatial resolution and also include a significant amount of noise, nevertheless the proposed model is able to calculate piecewise linear disparity maps and flow fields.

6   Conclusion

In this paper we proposed an algorithm for calculating scene flow for two given light fields captured at two consecutive times. To this end we formulated a convex variational model, which simultaneously estimates geometry and motion in terms of two disparity maps and the optical flow. We evaluated the model on synthetic data and showed the robustness of the model on real world experiments, where we used a Lytro camera for image capturing. In future work we plan to implement additional occlusion handling strategies to further improve the quality of the results.

References

1. Adelson, E.H., Wang, J.Y.A.: Single lens stereo with a plenoptic camera. IEEE Trans. Pattern Anal. Mach. Intell. **14**(2), 99–106 (1992)
2. Alvertos, P., Patras, I., Alvertos, N., Tziritas, G.: Joint disparity and motion field estimation in stereoscopic image sequences. In: 13th International Conference on Pattern Recognition, pp. 359–362 (1996)

3. Basha, T., Moses, Y., Kiryati, N.: Multi-view scene flow estimation: a view centered variational approach. Int. J. Comput. Vis. **101**(1), 6–21 (2013)
4. Bishop, T., Favaro, P.: Plenoptic depth estimation from multiple aliased views. In: 12th International Conference on Computer Vision Workshops (ICCV Workshops), pp. 1622–1629. IEEE (2009)
5. Bishop, T.E., Favaro, P.: The light field camera: extended depth of field, aliasing, and superresolution. IEEE Trans. Pattern Anal. Mach. Intell. **34**(5), 972–986 (2012)
6. Bredies, K., Kunisch, K., Pock, T.: Total generalized variation. SIAM J. Imaging Sci. **3**(3), 492–526 (2010)
7. Brox, T., Bruhn, A., Papenberg, N., Weickert, J.: High accuracy optical flow estimation based on a theory for warping. In: Pajdla, T., Matas, J.G. (eds.) ECCV 2004. LNCS, vol. 3024, pp. 25–36. Springer, Heidelberg (2004)
8. Carceroni, R.L., Kutulakos, K.N.: Multi-view scene capture by surfel sampling: from video streams to non-rigid 3d motion, shape and reflectance. Int. J. Comput. Vis. **49**, 175–214 (2002)
9. Chambolle, A., Pock, T.: A first-order primal-dual algorithm for convex problems with applications to imaging. J. Math. Imaging Vis. **40**, 120–145 (2011)
10. Courchay, J., Pons, J.-P., Monasse, P., Keriven, R.: Dense and accurate spatio-temporal multi-view stereovision. In: Zha, H., Taniguchi, R., Maybank, S. (eds.) ACCV 2009, Part II. LNCS, vol. 5995, pp. 11–22. Springer, Heidelberg (2010)
11. Furukawa, Y., Ponce, J.: Dense 3d motion capture from synchronized video streams. In: 2008 IEEE Computer Society Conference on Computer Vision and Pattern Recognition (CVPR 2008). IEEE Computer Society, Anchorage, 24–26 June 2008
12. Goldluecke, B., Wanner, S.: The variational structure of disparity and regularization of 4d light fields. In: IEEE Conference on Computer Vision and Pattern Recognition (CVPR) (2013)
13. Gortler, S.J., Grzeszczuk, R., Szeliski, R., Cohen, M.F.: The lumigraph. In: SIGGRAPH, pp. 43–54 (1996)
14. Heber, S., Ranftl, R., Pock, T.: Variational shape from light field. In: International Conference on Energy Minimization Methods in Computer Vision and Pattern Recognition (2013)
15. Isaksen, A., McMillan, L., Gortler, S.J.: Dynamically reparameterized light fields. In: SIGGRAPH, pp. 297–306 (2000)
16. Keriven, R., Faugeras, O.: Multi-view stereo reconstruction and scene flow estimation with a global image-based matching score. Int. J. Comput. Vis. **72**, 2007 (2006)
17. Lumsdaine, A., Georgiev, T.: The focused plenoptic camera. In. Proceedings of IEEE ICCP, pp. 1–8 (2009)
18. Neumann, J., Aloimonos, Y.: Spatio-temporal stereo using multi-resolution subdivision surfaces. Int. J. Comput. Vis. **47**, 2002 (2002)
19. Ng, R.: Digital light field photography. Ph.D. thesis, Stanford University (2006). http://www.lytro.com/renng-thesis.pdf
20. Pock, T., Chambolle, A.: Diagonal preconditioning for first order primal-dual algorithms in convex optimization. In: International Conference on Computer Vision (ICCV), pp. 1762–1769. IEEE (2011)
21. Ranftl, R., Gehrig, S., Pock, T., Bischof, H.: Pushing the limits of stereo using variational stereo estimation. In: Intelligent Vehicles Symposium, pp. 401–407. IEEE (2012)

22. Vedula, S., Baker, S., Rander, P., Collins, R.T., Kanade, T.: Three-dimensional scene flow. In: ICCV, pp. 722–729 (1999)
23. Wanner, S., Goldluecke, B.: Globally consistent depth labeling of 4D lightfields. In: IEEE Conference on Computer Vision and Pattern Recognition (CVPR) (2012)
24. Wanner, S., Goldluecke, B.: Spatial and angular variational super-resolution of 4D light fields. In: Fitzgibbon, A., Lazebnik, S., Perona, P., Sato, Y., Schmid, C. (eds.) ECCV 2012, Part V. LNCS, vol. 7576, pp. 608–621. Springer, Heidelberg (2012)
25. Wanner, S., Straehle, C., Goldluecke, B.: Globally consistent multi-label assignment on the ray space of 4d light fields. In: IEEE Conference on Computer Vision and Pattern Recognition (CVPR) (2013)
26. Wedel, A., Pock, T., Zach, C., Bischof, H., Cremers, D.: An Improved Algorithm for TV-L^1 Optical Flow. In: Cremers, D., Rosenhahn, B., Yuille, A.L., Schmidt, F.R. (eds.) Statistical and Geometrical Approaches to Visual Motion Analysis. LNCS, vol. 5604, pp. 23–45. Springer, Heidelberg (2009)
27. Wedel, A., Rabe, C., Vaudrey, T., Brox, T., Franke, U., Cremers, D.: Efficient dense 3d scene flow from sparse or dense stereo data, Oct 2008
28. Wilburn, B., Joshi, N., Vaish, V., Talvala, E.V., Antunez, E., Barth, A., Adams, A., Horowitz, M., Levoy, M.: High performance imaging using large camera arrays. ACM Trans. Graph. **24**(3), 765–776 (2005)
29. Zhang, Y., Kambhamettu, C., Kambhamettu, R.: On 3d scene flow and structure estimation. In: IEEE Conference on Computer Vision and Pattern Recognition, pp. 778–785 (2001)

Introducing More Physics into Variational Depth–from–Defocus

Nico Persch$^{(\boxtimes)}$, Christopher Schroers, Simon Setzer,
and Joachim Weickert

Mathematical Image Analysis Group, Faculty of Mathematics and Computer Science,
Saarland University, Campus E1.7, 66041 Saarbrücken, Germany
{persch,schroers,setzer,weickert}@mia.uni-saarland.de

Abstract. Given an image stack that captures a static scene with different focus settings, variational depth–from–defocus methods aim at jointly estimating the underlying depth map and the sharp image. We show how one can improve existing approaches by incorporating important physical properties. Most formulations are based on an image formation model (forward operator) that explains the varying amount of blur depending on the depth. We present a novel forward operator: It approximates the thin–lens camera model from physics better than previous ones used for this task, since it preserves the maximum–minimum principle w.r.t. the unknown image intensities. This operator is embedded in a variational model that is minimised with a *multiplicative* variant of the Euler–Lagrange formalism. This offers two advantages: Firstly, it guarantees that the solution remains in the physically plausible positive range. Secondly, it allows a stable gradient descent evolution without the need to adapt the relaxation parameter. Experiments with synthetic and real–world images demonstrate that our model is highly robust under different initialisations. Last but not least, the experiments show that the physical constraints are essential for obtaining more accurate solutions, especially in the presence of strong depth changes.

1 Introduction

The Depth–from–Defocus Problem. Only points with a certain distance to the lens are imaged completely sharp. This distance depends on the focal settings and is described by the *focal plane*. Points with a larger or smaller distance appear blurred, where the amount of blur increases with the object's offset to the focal plane. The range in which points are imaged acceptably sharp is the *depth–of–field* of the camera. In particular macro photography and microscope imaging suffer from a very limited depth–of–field. In these applications, a common remedy is to capture several images by varying the focal settings. Then each of these images differs in the regions that are projected sharply. Given such an image stack, the *depth–from–defocus* problem consists of inferring the underlying topography (depth map) as well as the sharp image as it would

© Springer International Publishing Switzerland 2014
X. Jiang et al. (Eds.): GCPR 2014, LNCS 8753, pp. 15–27, 2014.
DOI: 10.1007/978-3-319-11752-2_2

have been recorded by a pinhole camera. Essentially this corresponds to inverting the imaging process. This inverse problem is ill–posed and much harder to solve than the forward problem that models the image formation. To deal with this ill–posedness, regularisation is required. Variational formulations offer an elegant approach for this task. In our paper we present a novel variational framework that incorporates important physical properties.

Existing Approaches and Related Work. Instead of inverting the physical imaging process, there are approaches that estimate depth using *in–focus* information. They apply a local sharpness criterion, and the depth is assumed to correspond to the slice of the focal stack where the local sharpness achieves its maximum. The *variance method (VM)* [24] for example uses the local variance as a sharpness criterion.

To our knowledge, the first method that estimates depth using *defocus* information goes back to Pentland [19]. He estimates the amount of blurriness of image features or patches and uses this to infer the local depth. While this seminal work requires a completely sharp image as reference, Subbarao [23] describes a possibility to avoid this restriction. Namboodiri and Chaudhuri [16] assume a constant depth and then use the fact that Gaussian blurring can be expressed by linear diffusion. The extension to the more general case, i.e. allowing variations in the depth profile, is treated in many subsequent work [11,17,18,27]. Here the depth–of–field effect is described by means of an isotropic diffusion process with spatially variant diffusivity as it would occur in an inhomogeneous medium. Extensions to anisotropic diffusion processes also exist [9,13]. The depth–from–defocus problem can alternatively be addressed with Markov random fields [3,7]. In the latter work, Bhasin and Chaudhuri consider a scenario restricted to only two different focal planes. They investigate how the *point spread function (PSF)* has to be iteratively corrected in order to represent the energy distribution at depth discontinuities. Also blind deconvolution approaches such as the one by Chan and Wong [6] can be understood as related work in a broader sense.

Most related to our work are the approaches that jointly estimate the sharp image and the depth by minimising a suitable energy. In [10,14] this problem is stated as the minimisation of Csiszár's information divergence between the recorded focal stack and an appropriate model assumption. While the first approach assumes a locally equifocal surface such that the PSF is shift–invariant, the latter one embeds a shift–variant PSF in the imaging model. When regarding a shift–variant PSF as a 4-D function and a shift–invariant as a 2-D one, Aguet et al. [1] propose a compromise between both: They use a shift–invariant 3-D function defined as a family of 2-D Gaussians with varying standard deviation as a PSF. Compared to a 4-D function, this reduces the complexity by incorporating knowledge about how the PSF adapts depending on depth. However, the proposed formulation does not preserve an important physical property, namely the maximum–minimum principle w.r.t. the image intensities. This causes problems, especially at locations where depth changes occur.

Contributions. To address the aforementioned problems, we propose a novel physically motivated forward operator that preserves the maximum–minimum principle w.r.t. the image intensities. This forward operator is derived as an approximation of the thin lens camera model. Given a sharp image and depth information, the thin lens camera model is the established physically based camera model used in computer graphics for generating photorealistic depth–of–field effects. We show how to invert this forward operator within a variational formulation that allows to jointly obtain the unknown depth and intensity values given a focal stack. As it is our goal to preserve important physical properties, we also have to ensure that our solution contains only positive intensity and depth values. To achieve this we employ the multiplicative Euler–Lagrange formalism. Besides enabling us to restrict our solution to physically plausible values, this formalism offers an additional benefit: It allows us to derive an efficient semi–implicit scheme for finding the sought depth and intensity values. This semi–implicit scheme does not require any adaptation of the relaxation parameter.

Organisation of the Paper. In Sect. 2, we discuss image formation models and derive our novel forward operator. Section 3 then explains the variational formulation that effectively allows to invert our forward operator. Experiments show the benefits of our novel model in Sect. 4. We conclude our paper in Sect. 5.

2 Image Formation Models

Let us first obtain a better understanding of forward operators, i.e. image formation models that allow to generate a focal stack given a sharp image and depth information. To this end, we start by briefly discussing the thin lens camera model.

2.1 Image Formation with a Thin Lens

The thin lens imaging model is illustrated in Fig. 1. It uses a thin circular lens with focal length f. This lens is placed in the optical centre ℓ_0 at a distance v

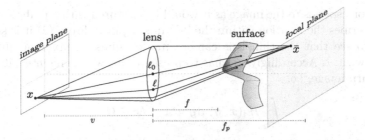

Fig. 1. Thin lens camera model

to the image plane $\Omega_2 \subset \mathbb{R}^2$. Lens and image plane are parallel. The thin lens equation [4]

$$\frac{1}{f_p} = \frac{1}{f} - \frac{1}{v}$$ (1)

characterises the imaging process. It involves a virtual *focal plane* that is parallel to the lens at distance f_p. A point $\bar{x}$ within this plane is sharply focused to a single point x on the image plane. For each image point x the corresponding $\bar{x}$ can be obtained by intersecting the ray from x through the optical centre ℓ_0 with the focal plane.

Generally, the lens focusses a bundle of rays into a single point x. This bundle can be described using $\bar{x}$ and all points on the lens. Following [20], we define the *thin lens operator*:

$$\mathcal{F}_{\mathrm{L}}[t, d](x) := \frac{1}{|\mathcal{A}|} \int_{\mathcal{A}} t\Big(T_d(\ell, \bar{x})\Big)\, \mathrm{d}\ell\,,$$ (2)

where $|\mathcal{A}|$ is the area of the lens and $d : \Omega_2 \to \mathbb{R}^+$ denotes the topography. The function T_d computes the first intersection point of a ray through ℓ and $\bar{x}$ with the topography d and t maps these intersection points to intensity values.

A direct simulation of geometric optics is possible with raytracing methods [8]. However, this is computationally very expensive because a large amount of blur requires processing a huge number of rays per pixel. Therefore, instead of directly considering the thin lens camera model, researchers are interested in finding approximations as alternatives for the simulation of photorealistic depth–of–field effects [2,5,21]. Similarly, we are also interested in finding a good approximation of the thin lens operator, however, with the additional requirement that it well fits into a variational framework. To this end, let us first rewrite the thin lens camera model with a spatially variant convolution.

2.2 Spatially Variant Convolution

Given a topography d, the thin lens camera model can be expressed with a spatially variant point spread function (PSF) $H_d : \Omega_2 \times \Omega_2 \to \mathbb{R}_0^+$:

$$\mathcal{F}_{\mathrm{H}}[u, d](x) := \int_{\Omega_2} H_d(x, y)\, u(y)\, \mathrm{d}y\,,$$ (3)

where u corresponds to the image as it would be captured with a pinhole camera, and x describes the location within the 2-D image plane. From (2) it is straightforward to see that the thin lens camera model fulfils a maximum–minimum principle w.r.t. t. Accordingly, H_d has to preserve this w.r.t. the intensity values of the sharp image, i.e.

$$\int_{\Omega_2} H_d(x, y)\, \mathrm{d}y = 1 \quad \forall x \in \Omega_2\,.$$ (4)

This guarantees that each intensity value of the resulting image lies between the minimum and maximum intensity value of the sharp image. Equation (3) can be understood as a weighted average of the sharp image intensities. To obtain the weights of the PSF, raytracing techniques may be used. However, this is similar to computing the thin lens camera model directly. Thus, let us investigate a different more efficient way to approximate the weights of the PSF.

2.3   Approximation of the PSF

In the thin lens model, a point on the surface spreads its intensity to a *circle of confusion* on the image plane [23]. For the moment, let us assume that the surface is equifocal ($d = $ constant), i.e. aligned parallel to the lens. Then Eq. (3) can be expressed in terms of a convolution with a spatially invariant kernel $h_d : \Omega_2 \subset \mathbb{R}^2 \to \mathbb{R}_0^+$ instead of a spatially variant one H_d. This comes down to reducing the PSF from a 4-D to a 2-D function. More precisely, in case of a circular lens, the kernel corresponds to a pillbox function and its radius is related to the constant topography. However, in practise it can be a better choice to use a Gaussian PSF instead of a pillbox when taking into account the wave character of light [19]. The standard deviation of the Gaussian replaces the radius of the pillbox. In the general case of non–constant topographies, the standard deviation changes with the depth of each surface point. Following this idea, Aguet et al. [1] express the imaging process as

$$\mathcal{F}_{\mathrm{U}}[u, d](\boldsymbol{x}, z) := \int_{\Omega_2} h(\boldsymbol{x} - \boldsymbol{y}, z - d(\boldsymbol{y}))\, u(\boldsymbol{y})\, \mathrm{d}\boldsymbol{y}\,, \qquad (5)$$

where z represents a given focal plane and $h : \Omega_3 \subset \mathbb{R}^3 \to \mathbb{R}_0^+$. Therewith, they lift h_d from a 2-D function to a 3-D one, composed of 2-D PSFs varying in their standard deviation, along the third dimension.

2.4   Our Modification

The formulation above is problematic if partial occlusions occur, which is expected to happen due to depth changes. The forward operator then effectively performs spatially variant 2-D convolutions with unnormalised kernels. This results in a violation of the maximum–minimum principle w.r.t to the images intensities. To avoid this, we propose to replace (5) by the novel forward operator

$$\mathcal{F}_{\mathrm{N}}[u, d](\boldsymbol{x}, z) := \frac{\mathcal{F}_{\mathrm{U}}[u, d](\boldsymbol{x}, z)}{\int_{\Omega_2} h(\boldsymbol{x} - \boldsymbol{x}', z - d(\boldsymbol{x}'))\, \mathrm{d}\boldsymbol{x}'}\,. \qquad (6)$$

The normalisation function guarantees the maximum–minimum principle, and thus handles regions where partial occlusions appear in a more appropriate way. While this normalisation may look like a small modification at first glance, it can have a large impact on the quality of the result: Fig. 2 depicts the behaviour of the different forward operators. We see that in regions where depth changes are present, partial occlusions appear. Applying an unnormalised forward operator

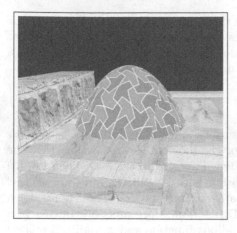 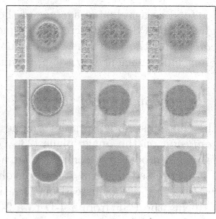

Fig. 2. (a) **Left box:** Our 3-D synthetic test model. **Right box:** Results of different forward operators (from left to right) with changing focal planes (from top to bottom). (b) **Left column:** Unnormalised forward operator of [1]. (c) **Centre column:** Our normalised forward operator (6). (d) **Right column:** Thin lens camera model (2).

results in bright overshoots followed by dark shadows (Fig. 2(b)). These local violations of the maximum–minimum principle lead to wrong model assumptions, which produce results that are not photorealistic. On the other hand, comparing Fig. 2(c, d) shows that our normalised approach comes very close to the physically well–founded thin lens camera model which allows to create realistic depth–of–field effects.

3   Variational Formulation

So far we have discussed image formation models, i.e. operators that can create stacks of blurred images if we know the sharp image and the depth. In this section we are interested in inverting this process, i.e. given an image stack, we wish to jointly estimate the depth map and the sharp image.

3.1   Variational Model

Let $u_R : \Omega_3 \to \mathbb{R}^+$ be the stack of recorded 2-D images that vary in their focal plane. The sought, sharp image $u : \Omega_2 \to \mathbb{R}^+$ in combination with the depth map $d : \Omega_2 \to \mathbb{R}^+$ can then be estimated as a minimiser of the energy

$$E(u,d) = M(u,d) + \alpha \, S(|\boldsymbol{\nabla}d|). \tag{7}$$

The data term M enforces the similarity between the recorded stack and the forward operator applied to the unknown sharp image u and depth d. To penalise deviations from the model assumptions we choose a quadratic cost function which is optimal for Gaussian distributed noise:

$$M(u, d) = \int_{\Omega_3} \underbrace{\left(u_{\mathrm{R}} - \mathcal{F}_{\mathrm{N}}[u, d] \right)}_{=:e[u,d]}^2 \mathrm{d}\boldsymbol{x} \; \mathrm{d}z \; . \tag{8}$$

Exactly like $\mathcal{F}_{\mathrm{U}}$, our forward operator $\mathcal{F}_{\mathrm{N}}$ from (6) is linear in u but nonlinear in d, and the data term is convex in u but nonconvex in d. Especially in homogeneous regions, a minimiser of the data term alone is non–unique. To avoid such ambiguities, we add a regularisation term S that penalises large gradient magnitudes in the depth field:

$$S(|\boldsymbol{\nabla}d|) = \int_{\Omega_2} \Psi(|\boldsymbol{\nabla}d|^2) \, \mathrm{d}\boldsymbol{x} \; , \tag{9}$$

where $\Psi : \mathbb{R} \to \mathbb{R}^+$ is a positive increasing function imposing (piecewise) smoothness. For the results in Sect. 4, we employ the Whittaker–Tikhonov penaliser $\Psi(s^2) = s^2$ which corresponds to homogeneous diffusion [25,29]. Other regularisers such as total variation (TV) [22] are also appropriate. The amount of smoothness can be steered by the regularisation parameter $\alpha > 0$.

3.2 Minimisation

Euler-Lagrange Equations. A minimiser (u, d) of the energy (7) must necessarily fulfill the *Euler–Lagrange* equations

$$\frac{\delta E}{\delta u} = 0 \quad \text{and} \quad \frac{\delta E}{\delta d} = 0 \tag{10}$$

and its corresponding natural boundary conditions. The established approach to find the functional derivatives $\frac{\delta E}{\delta u}$ and $\frac{\delta E}{\delta d}$ is given by applying the classical (additive) Euler–Lagrange formalism [12]. To obtain the functional derivative $\frac{\delta E}{\delta u}$, one uses the definition
$\langle \frac{\delta E}{\delta u}, v \rangle = \delta_v E$, where $\langle \cdot, \cdot \rangle$ denotes the standard inner product, and

$$\delta_v E := \frac{\partial}{\partial \epsilon} E(u + \epsilon v, d)|_{\epsilon=0} \tag{11}$$

acts like a "directional derivative" in the direction of a function v. Then we obtain

$$\frac{\delta E}{\delta u}(\boldsymbol{x}) = -2 \left(\overline{e} * h^* \right)(\boldsymbol{x}, d(\boldsymbol{x})) \tag{12}$$

with E from Eq. (7). Here we have introduced the abbreviation $\overline{e} := N^{-1} \cdot e$, where e is related to the data term (8) and N corresponds to the normalisation function, i.e. the denominator in (6). The operator $*$ expresses a 3-D convolution, and $h^*(x) := h(-x)$. Analogously applying the same formalism w.r.t. the depth d allows to compute the functional derivative

$$\frac{\delta E}{\delta d}(\boldsymbol{x}) = 2 \left(\left(\bar{e} * h_z^*\right)(\boldsymbol{x}, d(\boldsymbol{x})) \cdot u - \left(\left(\bar{e} \cdot \mathcal{F}_N[u, d]\right) * h_z^*\right)(\boldsymbol{x}, d(\boldsymbol{x})) \right.$$
$$\left. - \alpha \cdot \mathrm{div}\left(\Psi'(|\boldsymbol{\nabla} d|^2)\, \boldsymbol{\nabla} d\right) \right). \tag{13}$$

Enforcing Positivity. Since negative intensities as well as negative depth values are not physically plausible, we would like to modify our approach in such a way that both quantities are strictly constrained to be positive. The *multiplicative* Euler–Lagrange formalism offers an interesting and efficient way to achieve this [28]. Here, a multiplicative perturbation is used instead of an additive one. Thus, we consider

$$\delta_v^* E := \frac{\partial}{\partial \epsilon} E(u + \epsilon u \cdot v, d)|_{\epsilon=0} \tag{14}$$

for minimisation w.r.t. u, and an analog expression for d. This gives the following functional derivatives:

$$\frac{\delta^* E}{\delta u} = u \cdot \frac{\delta E}{\delta u} \quad \text{and} \quad \frac{\delta^* E}{\delta d} = d \cdot \frac{\delta E}{\delta d}. \tag{15}$$

There are two different ways to understand why the multiplicative Euler–Lagrange formalism restricts the solution to positive values [28]: The first explanation interprets the multiplicative Euler–Lagrange formalism via the reparametrisations $u = \exp(w)$ and $d = \exp(z)$. Moving unwanted values to infinite distance is the second explanation. To this end, one can show that the multiplicative functional gradients $\frac{\delta^* E}{\delta u}$ and $\frac{\delta^* E}{\delta d}$ occur within the additive formalism when one replaces the Euclidean metric du by a hyperbolic one, i.e. du/u.

3.3 Discretisation and Implementation

The multiplicative approach presented above does not only guarantee the positivity of our solution, it also enables us to introduce an efficient semi–implicit iteration scheme. To this end, we consider a gradient descent scheme with the multiplicative gradient $\frac{\delta^* E}{\delta u}$ from Eq. (15):

$$\frac{u^{k+1} - u^k}{\tau} = 2\, u^{k+1} \left(\bar{e}^k * h^*\right)(\boldsymbol{x}, d), \tag{16}$$

where τ is the relaxation parameter, and the upper index denotes the iteration level. For the multiplicative gradient descent w.r.t. the depth map d, we use the semi-implicit approach

$$\frac{d^{k+1} - d^k}{\tau} = -2 \left(\left(\bar{e}^k * h_z^*\right)(\boldsymbol{x}, d^k)\, u - \left(\left(\bar{e}^k \cdot \mathcal{F}_N[u, d^k]\right) * h_z^*\right)(\boldsymbol{x}, d^k) \right) \cdot d^{k+1}$$
$$+ 2\alpha \cdot \mathrm{div}\left(\Psi'(|\boldsymbol{\nabla} d^k|^2)\, \boldsymbol{\nabla} d^{k+1}\right) \cdot d^k. \tag{17}$$

This semi–implicit scheme is less sensitive sensitive w.r.t. the relaxation parameter such that it can remain fixed during the iterations. Therefore, we can refrain from backtracking line–search or complicated methods such as Brent's algorithm that combines the bisection method, the secant method and inverse quadratic interpolation. Such methods are required when using the standard additive Euler–Lagrange formalism for this problem and are avoided by our approach.

Since we deal with digital images, we replace continuous functions by their discrete counterparts and derivatives by finite differences. The 3-D convolution is implemented in the Fourier domain, using the Fast Fourier Transform and the convolution theorem. While Eq. (16) can be solved directly, Eq. (17) requires to solve a nonsymmetric linear system of equations. We solve the latter one iteratively with a Jacobi algorithm. We use an alternating minimisation scheme [15] as it is commonly used e.g. in blind deconvolution problems [6]: Keeping the solution for one sub–problem fixed (e.g. recovering the sharp image u), the other problem (e.g. the estimation of the depth) is solved with a fixed number of gradient descent steps. After that, roles are exchanged. To account for the nonconvexity, we apply a coarse–to–fine strategy where the solution of the downsampled problem serves as initialisation on the next finer one.

4 Experiments

Synthetic Data. In our first experiment, we generated a stack of images with varying focal planes. This is achieved by rendering the 3-D model from Fig. 2(a) with the thin lens camera model (lens diameter $D = 2.69$ cm, distance to image plane $v = 35$ mm). In total we have rendered 20 images where the distance of the focal plane to the lens changed in equidistant steps from $f_p = 3$ cm to $f_p = 7$ cm. Figure 2(d) shows 3 different slices of this rendered focal stack.

In Fig. 3, we compare the results of different approaches to estimate the topography. For the variance method two undesired hills in front of and behind the hemisphere arise (Fig. 3(a)). Using the forward operator $\mathcal{F}_U$ leads to a violation

Table 1. Quantitative comparison. Error measurement of the estimated topography and the sharp image to its ground–truth. We consider the mean squared error (MSE) as well as the structural similarity (SSIM) [26]. We compare the variance method (VM) with and without Gaussian post–smoothing with variance σ, the operator $\mathcal{F}_U$, and our max–min–preserving imaging model $\mathcal{F}_N$. The latter two are either initialised with a constant depth or an estimation of the VM.

Method		VM		$\mathcal{F}_U$		Ours	
		$\sigma = 0$	$\sigma = 4$	Constant	VM	Constant	VM
Depth	MSE	2.83	1.10	21.66	2.26	0.77	0.58
	SSIM	0.98	0.93	0.94	0.99	1.00	1.00
Image	MSE	48.33	46.38	124.52	52.17	49.09	45.17
	SSIM	0.87	0.87	0.67	0.87	0.90	0.92

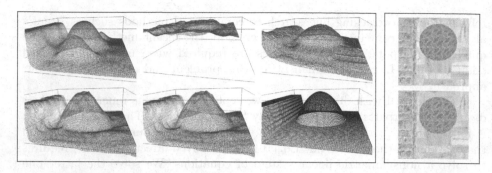

Fig. 3. Visual comparison. **Left box:** *In reading order* **(a)** Variance Method (VM) with Gaussian post–smoothing (patch-size = 6, σ = 4.0). **(b)** $\mathcal{L}_U$ ignoring normalisation, initialised with constant depth ($\alpha = 45$). **(c)** Dito initialised with VM. **(d)** Ours $\mathcal{L}_N$ initialised with constant depth ($\alpha = 150$). **(e)** Dito with VM initialisation. **(f)** Ground–truth of the topography. **Right box:** **(g) Top:** Estimated sharp image. **(h) Bottom:** Ground–truth of the sharp image.

of the model assumptions at large depth changes (Fig. 3(b,c)). It implicitly introduces a regularisation that erroneously prefers only smooth changes of depth. While Fig. 3(b) has been initialised with a constant depth map, Fig. 3(c) has been initialised with the result of the variance method. Apparently the initialisation strongly affects the outcome. Due to wrong assumptions at depth changes, it is not possible to converge to a reasonable solution with a constant initialisation. In contrast to that, the reconstructions with our normalised forward operator $\mathcal{F}_N$ shown in the Fig. 3(d, e) do not suffer from this effect. Our approach is less sensitive to the initialisation, yielding similar error values and reconstructions in both cases. In fact, our model entails a physically plausible behaviour also at depth changes. We obtain reconstructions that match the ground truth in a better way, concerning both the depth and the sharp image; cf. Table 1. Figure 3(g) also shows the estimated sharp image with our approach. It closely resembles the ground truth.

Real–World Data. In our second experiment, we test our approach on a real world focal stack showing a house fly eye (see Fig. 4(a)). We employ a coarse–to–fine strategy with the variance method as initialisation on the coarsest grid. To demonstrate the performance of our approach on this real word data set, we present the reconstructed sharp image along with the estimated depth profile (Fig. 4(b)). For the illustration of the depth profile a grey value coding is used: The brighter the grey value the larger the distance of the object to the lens. In our result, fine structures are clearly visible in the sharp image and the depth profile. This can be seen in Fig. 4(b) when considering the small hair, for example.

In our third experiment, we use an image stack consisting of 22 frames that depict a coffee bean. Figure 5(a) shows 3 different slices of this focal stack. In this experiment we employ a coarse–to–fine strategy again but this time with

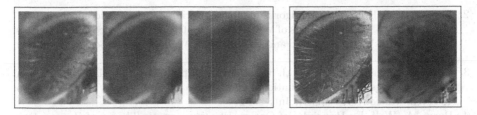

Fig. 4. House fly eye (provided by the Biomedical Imaging Group EPFL, Lausanne, Switzerland). **(a) Left box:** 3 out of 21 images of a focal stack of a house fly eye. **(b) Right box:** Estimated sharp image and topography for $\alpha = 25$.

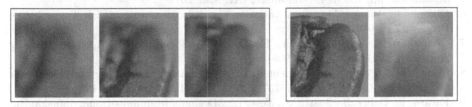

Fig. 5. Coffee bean (provided by the Computer Graphics Group, MPI for Informatics, Saarbrücken, Germany). **(a) Left box:** 3 out of 22 images of the focal stack of a coffee bean. **(b) Right box:** Reconstructed sharp image along with the estimated depth ($\alpha = 160$).

a constant depth value as initialisation on the coarsest grid. Even with such a crude initialisation, accurate results are possible.

5   Conclusions

We have shown the benefits of introducing two physical constraints into variational depth–from–defocus models: a maximum–minimum principle w.r.t. the unknown sharp image, and the positivity of the sought image intensities and the depth values. Our resulting model offers clear advantages especially in the presence of strong depth variations. Moreover, we advocate to replace the traditional Euler–Lagrange formalism by a multiplicative variant, whenever positivity is to be preserved. It is our hope that both physically refined modelling and multiplicative calculi will receive more popularity in future computer vision models.

Acknowledgements. Our research has been partly funded by the Deutsche Forschungs-gemeinschaft (DFG) through a Gottfried Wilhelm Leibniz prize for Joachim Weickert and the Cluster of Excellence *Multimodal Computing and Interaction*.

References

1. Aguet, F., Van De Ville, D., Unser, M.: Model-based 2.5-D deconvolution for extended depth of field in brightfield microscopy. IEEE Trans. Image Process. **17**(7), 1144–1153 (2008)

2. Barsky, B.A., Kosloff, T.J.: Algorithms for rendering depth of field effects in computer graphics. In: Proceedings of WSEAS International Conference on Computers, pp. 999–1010. World Scientific and Engineering Academy and Society, Heraklion, July 2008

3. Bhasin, S., Chaudhuri, S.: Depth from defocus in presence of partial self occlusion. In: Proceedings of IEEE International Conference on Computer Vision, vol. 1, pp. 488–493. Vancouver, Canada, July 2001

4. Born, M., Wolf, E.: Principles of Optics: Electromagnetic Theory of Propagation, Interference and Diffraction of Light, 4th edn. Pergamon Press, Oxford (1970)

5. Cant, R., Langensieoen, C.: Creating depth of field effects without multiple samples. In: Proceeding of IEEE International Conference on Computer Modelling and Simulation, pp. 159–164. Cambridge, UK, Mar 2012

6. Chan, T.F., Wong, C.K.: Total variation blind deconvolution. IEEE Trans. Image Process. **7**, 370–375 (1998)

7. Chaudhuri, S., Rajagopalan, A.: Depth from Defocus: A Real Aperture Imaging Approach. Springer, Berlin (1999)

8. Cook, R.L., Porter, T., Carpenter, L.: Distributed ray tracing. In: Computer Graphics, SIGGRAPH '84, pp. 137–145. ACM, Minneapolis, Jul 1984

9. Favaro, P., Osher, S., Soatto, S., Vese, L.: 3D shape from anisotropic diffusion. In: Proceedings of IEEE Conference on Computer Vision and Pattern Recognition, CVPR, Madison, USA, Jun 2003

10. Favaro, P., Soatto, S.: Shape and radiance estimation from the information divergence of blurred images. In: Vernon, D. (ed.) ECCV 2000. LNCS, vol. 1842, pp. 755–768. Springer, Heidelberg (2000)

11. Favaro, P., Soatto, S., Burger, M., Osher, S.: Shape from defocus via diffusion. IEEE Trans. Pattern Anal. Mach. Intell. **30**(3), 518–531 (2008)

12. Gelfand, I.M., Fomin, S.V.: Calculus of Variations. Dover, New York (2000)

13. Hong, L., Yu, J., Hong, C., Sui, W.: Depth estimation from defocus images based on oriented heat-flows. In: Proceedings of IEEE International Conference on Machine Vision, pp. 212–215. Dubai, UAE (2009)

14. Jin, H., Favaro, P.: A variational approach to shape from defocus. In: Heyden, A., Sparr, G., Nielsen, M., Johansen, P. (eds.) ECCV 2002, Part II. LNCS, vol. 2351, pp. 18–30. Springer, Heidelberg (2002)

15. Luenberger, D., Ye, Y.: Linear and Nonlinear Programming, 3rd edn. Springer, New York (2008)

16. Namboodiri, V.P., Chaudhuri, S.: Use of linear diffusion in depth estimation based on defocus cue. In: Chanda, B., Chandran, S., Davis, L.S. (eds.) Proceedings of Indian Conference on Computer Vision, Graphics and Image Processing, pp. 133–138. Allied Publishers Private Limited, Kolkata (2004)

17. Namboodiri, V.P., Chaudhuri, S.: On defocus, diffusion and depth estimation. Pattern Recogn. Lett. **28**(3), 311–319 (2007)

18. Namboodiri, V., Chaudhuri, S., Hadap, S.: Regularized depth from defocus. In: Proceedings of IEEE International Conference on Image Processing, San Diego, USA, pp. 1520–1523, Oct 2008

19. Pentland, A.P.: A new sense for depth of field. IEEE Trans. Pattern Anal. Mach. Intell. **9**(4), 523–531 (1987)

20. Pharr, M., Humphreys, G.: Physically Based Rendering: From Theory to Implementation. Morgan Kaufmann, San Francisco (2004)

21. Rokita, P.: Fast generation of depth of field effects in computer graphics. Comput. Graph. **17**(5), 593–595 (1993)

22. Rudin, L.I., Osher, S., Fatemi, E.: Nonlinear total variation based noise removal algorithms. Physica D **60**, 259–268 (1992)
23. Subbarao, M.: Parallel depth recovery by changing camera parameters. In: Proceedings of IEEE International Conference on Computer Vision, Washington, USA, pp. 149–155, Dec 1988
24. Sugimoto, S.A., Ichioka, Y.: Digital composition of images with increased depth of focus considering depth information. Appl. Optics **24**(14), 2076–2080 (1985)
25. Tikhonov, A.N.: Solution of incorrectly formulated problems and the regularization method. Sov. Math. Doklady **4**, 1035–1038 (1963)
26. Wang, Z., Bovik, A., Sheikh, H., Simoncelli, E.: Image quality assessment: from error visibility to structural similarity. IEEE Trans. Image Process. **13**(4), 600–612 (2004)
27. Wei, Y., Dong, Z., Wu, C.: Global depth from defocus with fixed camera parameters. In: Proceedings of IEEE International Conference on Mechatronics and Automation, Changchun, China, pp. 1887–1892, Aug 2009
28. Welk, M., Nagy, J.G.: Variational deconvolution of multi-channel images with inequality constraints. In: Martí, J., Benedí, J.M., Mendonça, A.M., Serrat, J. (eds.) IbPRIA 2007. LNCS, vol. 4477, pp. 386–393. Springer, Heidelberg (2007)
29. Whittaker, E.T.: A new method of graduation. Proc. Edinburgh Math. Soc. **41**, 65–75 (1923)

Reconstruction

High-Resolution Stereo Datasets
with Subpixel-Accurate Ground Truth

Daniel Scharstein[1]([⊠]), Heiko Hirschmüller[2], York Kitajima[1], Greg Krathwohl[1],
Nera Nešić[3], Xi Wang[1], and Porter Westling[4]

[1] Middlebury College, Middlebury, VT, USA
schar@middlebury.edu
[2] German Aerospace Center, Oberpfaffenhofen, Germany
[3] Reykjavik University, Reykjavik, Iceland
[4] LiveRamp, San Francisco, USA

Abstract. We present a structured lighting system for creating high-resolution stereo datasets of static indoor scenes with highly accurate ground-truth disparities. The system includes novel techniques for efficient 2D subpixel correspondence search and self-calibration of cameras and projectors with modeling of lens distortion. Combining disparity estimates from multiple projector positions we are able to achieve a disparity accuracy of 0.2 pixels on most observed surfaces, including in half-occluded regions. We contribute 33 new 6-megapixel datasets obtained with our system and demonstrate that they present new challenges for the next generation of stereo algorithms.

1 Introduction

Stereo vision is one of the most heavily researched topics in computer vision [5,17,18,20,28], and much of the progress over the last decade has been driven by the availability of standard test images and benchmarks [7,14,27,28,30,31]. Current datasets, however, are limited in resolution, scene complexity, realism, and accuracy of ground truth. In order to generate challenges for the next generation of stereo algorithms, new datasets are urgently needed.

In this paper we present a new system for generating high-resolution two-view datasets using structured lighting, extending and improving the method by Scharstein and Szeliski [29]. We contribute 33 new 6-megapixel datasets of indoor scenes with subpixel-accurate ground truth. A central insight driving our work is that high-resolution stereo images require a new level of calibration accuracy that is difficult to obtain using standard calibration methods. Our datasets are available at http://vision.middlebury.edu/stereo/data/2014/.

Novel features of our system and our new datasets include the following: (1) a portable stereo rig with two DSLR cameras and two point-and-shoot cameras, allowing capturing of scenes outside the laboratory and simulating the diversity of Internet images; (2) accurate floating-point disparities via robust interpolation of lighting codes and efficient 2D subpixel correspondence search; (3) improved

© Springer International Publishing Switzerland 2014
X. Jiang et al. (Eds.): GCPR 2014, LNCS 8753, pp. 31–42, 2014.
DOI: 10.1007/978-3-319-11752-2_3

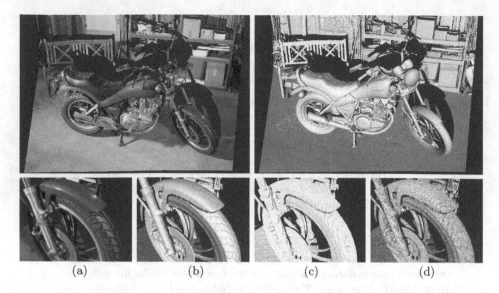

Fig. 1. Color and shaded renderings of a depth map produced by our system; (a), (b) detail views; (c) resulting surface if disparities are rounded to integers; (d) resulting surface without our novel subpixel and self-calibration components.

calibration and rectification accuracy via bundle adjustment; (4) improved self-calibration of the structured light projectors, including lens distortion, via robust model selection; and (5) additional "imperfect" versions of all datasets exhibiting realistic rectification errors with accurate 2D ground-truth disparities. The resulting system is able to produce new stereo datasets with significantly higher quality than existing datasets; see Figs. 1 and 2 for examples.

We contribute our new datasets to the community with the aim of providing a new challenge for stereo vision researchers. Each dataset consists of input images taken under multiple exposures and multiple ambient illuminations with and without a mirror sphere present to capture the lighting conditions. We provide each dataset with both "perfect" and realistic "imperfect" rectification, with accurate 1D and 2D floating-point disparities, respectively.

2   Related Work

Recovery of 3D geometry using structured light dates back more than 40 years [3,4,25,32]; see Salvi et al. [26] for a recent survey. Applications range from cultural heritage [21] to interactive 3D modeling [19]. Generally, 3D acquisition employing active or passive methods is a mature field with companies offering turnkey solutions [1,2]. However, for the goal of producing high-resolution stereo datasets, it is difficult to precisely register 3D models obtained using a separate scanner with the input images. Existing two-view [7] and multiview [30,31] stereo datasets for which the ground truth was obtained with a laser scanner typically

Bicycle2   Playroom   Pipes   Playtable   Adirondack   Piano

Newkuba   Hoops   Classroom2   Staircase   Recycle   Djembe

Fig. 2. Left views and disparity maps for a subset of our new datasets, including a restaging of the Tsukuba "head and lamp" scene [24]. Disparity ranges are between 200 and 800 pixels at a resolution of 6 megapixels.

suffer from (1) limited ground-truth resolution and coverage; and (2) limited precision of the calibration relating ground-truth model and input images. To address the second problem Seitz et al. [30] align each submitted model via ICP with the ground-truth model before the geometry is evaluated, while Geiger et al. [7] recently re-estimated the calibration from the current set of submissions.

Establishing ground-truth disparities from the input views directly avoids the calibration problem and can be done via unstructured light [1,6,34], but only yields disparities for nonoccluded scene points visible in both input images. Scharstein and Szeliski [29] pioneered the idea of self-calibrating the structured light sources from the initial nonoccluded *view disparities*, which yields registered *illumination disparities* in half-occluded regions as well. We extend this idea in this paper and also model projector lens distortion; in addition, we significantly improve the rectification accuracy using the initial correspondences.

Gupta and Nayar [10] achieve subpixel precision using a small number of sinusoidal patterns, but require estimating scene albedo, which is sensitive to noise. In contrast, we use a large number of binary patterns under multiple exposures and achieve subpixel precision via robust interpolation. We employ the maximum min-stripe-width Gray codes by Gupta et al. [9] for improved robustness in the presence of interreflections and defocus.

Overall, we argue that the approach of [29] is still the best method for obtaining highly accurate ground truth for stereo datasets of static scenes. The contribution of this paper is to push this approach to a new level of accuracy. In addition, by providing datasets with both perfect and imperfect rectification, we enable studying the effect of rectification errors on stereo algorithms [13]. In Sect. 4 we show that such errors can strongly affect the performance of

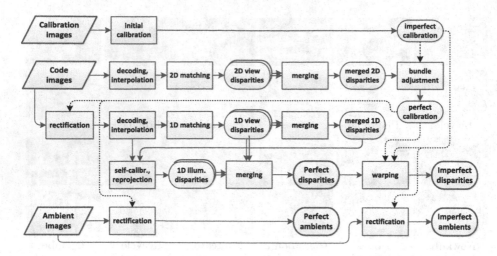

Fig. 3. Diagram of the overall processing pipeline.

high-resolution stereo matching, and we hope that our datasets will inspire novel work on stereo self-calibration [11].

3   Processing Pipeline

The overall workflow of our 3D reconstruction pipeline is illustrated in Fig. 3.

The inputs to our system are (1) calibration images of a standard checkerboard calibration target; (2) code images taken under structured lighting from different projector positions; and (3) "ambient" input images taken under different lighting conditions. The main processing steps (rows 2–4 in Fig. 3) involve the code images taken with the two DSLR cameras.

First, the original (unrectified) code images from each projector are thresholded, decoded, and interpolated, yielding floating-point coordinates of the projector pixel illuminating the scene. These values are used as unique identifiers to establish correspondences between the two input views, resulting in subpixel-accurate *2D view disparities*, which are used in a bundle-adjustment step to refine the initial "imperfect" calibration. The processing then starts over, taking rectified images as input and producing *1D view disparities* (row 3 in the diagram). The merged disparities are used to self-calibrate each projector (row 4), from which *1D illumination disparities* are derived. All sets of view and illumination disparities are merged into the final "perfect" disparities, which are then warped into the imperfect rectification. Corresponding sets of ambient images are produced by rectifying with both calibrations (row 5). We next discuss the individual steps of the processing pipeline in detail.

Image acquisition: To capture natural scenes outside the laboratory we employ a portable stereo rig (Fig. 4) with two Canon DSLR cameras (EOS 450D with 18–55 mm lens) in medium resolution (6 MP) mode. The cameras are mounted

Fig. 4. Top: Portable stereo rig with 2 DSLRs and 2 consumer cameras, and painting of glossy scene. Bottom: Ambient views under different lighting conditions and disparities.

on a horizontal optical rail with variable baseline from 140 mm to 400 mm. We optionally include two point-and-shoot cameras (Canon PowerShot A495, A800, A3000, or SD780), in order to simulate the variability of Internet images. We control the main cameras using DSLR Remote Pro, and trigger the PowerShots with CHDK firmware via their USB power cables.

For each scene we take calibration and ambient images with all cameras, then code images with the two DSLRs only. We typically use 4 different ambient lighting conditions and 8 different exposures. We take a second set of all images with a mirror sphere added to the scene to allow recovering the approximate lighting conditions. We store the original images both in JPG and RAW, and do all subsequent processing in lossless formats.

For the code images we project a series of binary patterns that uniquely encode each pixel location (u, v) of the projector. We use ViewSonic 1024 × 768 DLP projectors, thus 10 patterns are sufficient in each of u and v. Instead of standard Gray codes [29] we use maximum min-stripe-width Gray codes [9] for increased robustness to interreflections and defocus. Following [9,29], we project each pattern and its inverse, and take all code images under 3 exposures (typically 0.1 s, 0.5 s, and 1.0 s) in order to handle a wide range of albedos. We illuminate the scene from $P = 4 \ldots 18$ different projector locations, depending on the complexity of the scene. This allows us to focus the projectors on different scene depths and to minimize shadowed areas.

Our technique allows precise 3D capture of static indoor scenes. If the room can be darkened, we are able to handle very low albedos via long exposures. In the Motorcycle scene shown in Fig. 4 we painted high-gloss surfaces with a clay solution after acquiring the ambient images in order to recover the geometry of even truly reflective surfaces. This is the only scene where we applied paint; most of the other scenes contain semi-glossy surfaces which our system can recover, unless they are close to parallel to the viewing direction.

Decoding and interpolation: Given the 120 code images from one projector as input, we compute the signed difference between each pattern and its inverse

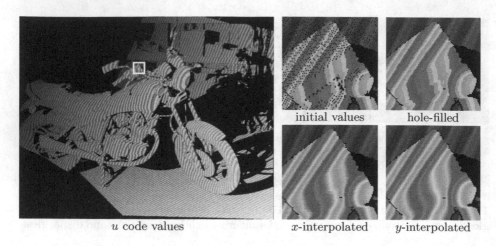

initial values   hole-filled

u code values   *x*-interpolated   *y*-interpolated

Fig. 5. Left: Recovered *u* codes for one of 12 projector positions, visualized using a helix in HSV color space. Right: Zoomed views illustrating the processing of code values.

for the 3 color bands, and select at each pixel the exposure yielding the largest absolute average difference $|d|$. We output "1" if $d \geq t_d$, "0" if $d \leq -t_d$, and "unknown" otherwise (we use $t_d = 16$). We then concatenate the individual bits and decode them into integer code pairs (u, v). Since the cameras have higher resolution than the projectors, a single projector pixel typically spans 2–4 camera pixels. On smooth surfaces, we expect the code values to increase monotonically in the prominent code direction (i.e., rightwards for *u* and downwards for *v*). This knowledge allows smart hole-filling and robust interpolation without blurring across depth boundaries. The process is illustrated in Fig. 5 and involves filling holes less than 6 pixels wide whose border values differ by no more than 2 in the code direction, followed by interpolation by convolution with a bilinear ("tent") kernel with radius 7 in both *x* and *y* directions. Only values that fit the expected monotonic ramp are included, the other values are extrapolated. The resulting floating-point *u* and *v* code images form the input for the subsequent stages.

Fast 2D correspondence search: In order to establish view disparities, we need to find for each pixel the most similar code pair in the other image. Given the initial unrectified images, we have no prior knowledge of the search range. A naive search would take quadratic time in the number of pixels. Instead, we developed a linear algorithm that precomputes the bounding boxes for each integer (u, v) code pair appearing in the target image. Except at depth boundaries, each of these bounding boxes only spans a few pixels. We can find the closest matching code value in expected constant time by restricting the search to the merged bounding boxes of the rounded code values and its 8 neighbors.

Once the pixel with the closest code value has been located, we derive a 2D subpixel disparity estimate by fitting planes to the *u* and *v* code values of the pixels surrounding the target code, respectively. If the fitted planes have poor residuals (e.g., if the planes straddle a depth boundary), no subpixel correction is attempted. Otherwise, the subpixel offset is determined using a least-squares fit of the source values to the u and v planes in the target image.

Filtering and merging: Given $2P$ disparity maps from 2 views and P projectors, we first cross-check each pair of disparity maps by requiring that left and right disparities agree to within 0.5 pixels. During 1D processing we also apply a robust 3×3 smoothing filter, remove isolated disparity components of less than 20 pixels, fill holes of up to 200 pixels if the surrounding disparities fit a planar model, and cross-check again. We then merge each set of P disparity maps, requiring at least two estimates per pixel, average those within 1.0 of the median value, and cross-check the final pair of merged disparities.

Calibration refinement: In our experience, the stereo rectification resulting from a standard calibration procedure (we use OpenCV) has limited accuracy. At a resolution of 6 MP residual vertical disparities of several pixels are common, especially in image corners. Sources of calibration errors include (1) incorrect lens distortion estimation if the calibration target never occupies a large fraction of the image; and (2) errors relating tilt angle and vertical camera position if the calibration target does not adequately sample the depth range of the scene. We have developed a novel method for correcting such errors via bundle adjustment using the precise 2D correspondences recovered from structured lighting.

We first sample these 2D disparities with a stride of 10 pixels and remove outliers by requiring that at least 50 % of the pixels in each 9×9 window fit a planar disparity model. We then reproject the resulting points back into 3D using the initial calibration parameters, and perform sparse bundle adjustment using SBA [23] to minimize the residual y-disparities. Since we are only working on image correspondences and do not have any constraints on the 3D reconstruction, we can only (re-)calibrate the cameras up to an unknown perspective transformation [12]. In order to avoid arbitrary perspective distortions, we keep all parameters of camera 0 fixed and only optimize for lens distortion. For camera 1 we optimize lens distortion parameters, joint focal length $f_{1x} = f_{1y}$, and c_{1y} of the principal point, but keep c_{1x} fixed, because it does not affect y-disparities. We also optimize the extrinsics while keeping the length of the baseline constant. We tried lens distortion models of up to four parameters (radial κ_1, κ_2; tangential p_1, p_2), and found that the 2-parameter model without tangential distortion is sufficiently accurate for the lenses we employ.

Figure 6 plots the maximum and average absolute residual y-disparities for our datasets, rectified with the original "imperfect" and the refined "perfect" calibration. Averaged over all datasets, the maximum absolute y-error is reduced by a factor of 4.1 (from 2.76 to 0.67) while the overall average absolute error is reduced by a factor of 8.1 (from 0.77 to 0.096). Notably, the maximum absolute error in the refined rectification never exceeds 1.0 pixels. This shows that we can achieve excellent rectification even with a simple 2-parameter lens-distortion model, if accurate correspondences are available. Recall from Fig. 3 that we next match the rectified code images; the resulting high-confidence 1D view disparities are used to derive *illumination disparities* as explained next.

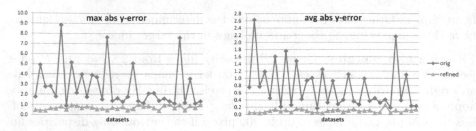

Fig. 6. Reduction of maximum (left) and average (right) absolute vertical rectification errors through our novel calibration refinement technique over all 33 datasets.

Self-calibration of projectors: So far we have only utilized the interpolated code values (u, v) as unique "scene color" allowing us to establish unambiguous subpixel correspondences between the two cameras. But since (u, v) also encodes the pixel location in the projector's illumination pattern, we get additional correspondences for each projector-camera pair. While traditional structured lighting systems require an external calibration of the projector, we follow [29] and self-calibrate the projectors in terms of the already recovered view disparities, which we can interpret as projective depth. This amounts to recovering the projection matrices M_{pc} relating projector p with camera c: $\mathbf{u}_p \cong M_{pc}\mathbf{x}_c$, where $\mathbf{x}_c = [x\,y\,d\,1]^T$ is a (homogeneous) 3D point with disparity d in camera c, and $\mathbf{u}_p = [u\,v\,1]^T$ is the corresponding image location in projector p. As in [29] we can solve for M_{pc} via robust least squares and then derive new d estimates (illumination disparities) from the code values, including in half-occluded regions. We have found, however, that this linear model often yields high residuals on the order of several pixels since projector lens distortion is not modeled.

We therefore employ a novel nonlinear refinement, using the linear solution as start value, where we estimate up to six additional parameters: the distortion center (c_u, c_v), radial distortion parameters κ_1, κ_2, and tangential distortion parameters p_1, p_2. We use the Levenberg-Marquardt (LM) method [22] to minimize the squared reprojection error in u and v, and then invert the estimated nonlinear projection model (also via LM) to translate the code values into illumination disparities. One complication is that for some projector-camera combinations the lens distortion cannot be stably estimated, for instance, if only a small part of the projection pattern is visible. In such cases the disparities derived from the inverted projection model can be dramatically wrong. We solve this problem using a model-selection approach: we compute the illumination disparities for 0, 3, 4, and 6-parameter models, and select the one yielding the smallest residuals. Here 0 means linear; 3 includes c_u, c_v, κ_1; 4 includes κ_2; and 6 includes p_1, p_2. Using this technique we reduce the average absolute error from 0.47 to 0.26; the average reduction per dataset is by a factor of 2.1. More importantly, we significantly reduce the "bad" residuals greater than 1.0 from 7.3 % to 0.75 %.

Merging and accuracy estimates: In the final merging step, we combine all $2P$ view and illumination disparity maps per camera, again using robust averaging. We relax the cross-checking criterion to allow half-occluded regions behind matched surfaces. For each pixel we not only store its final floating-point disparity d, but also the number of disparity samples n and their sample standard deviation s. While we do not have an absolute measure of reconstruction accuracy, we can use s as a quality measure, given that the disparity estimates are derived from separately decoded projection patterns. For the Motorcycle scene with $P = 12$ projector positions, the average number of samples per pixel is $\bar{n} = 5.7$ and the average sample standard deviation at pixels with $n \geq 2$ is $\bar{s} = 0.16$. Averaged over all datasets, the values are $\bar{n} = 7.7$ and $\bar{s} = 0.20$. A qualitative impression of the reconstruction accuracy is provided by the 3D renderings in Fig. 1 and on the dataset webpage. Without our novel subpixel estimation and self-calibration components, the reconstruction is very noisy and exhibits visible aliasing effects (Fig. 1d); the sample standard deviations are typically about twice as large.

We can also assess accuracy in manually selected planar scene regions by fitting a plane to the recovered disparites and measuring the residuals. For 17 planar regions selected from the Motorcycle, Djembe, and Piano datasets, we obtain average absolute residuals of $\bar{r} = 0.032$, an improvement by a factor of 8 over integer disparities ($\bar{r} = 0.252$), and by a factor of 4 over our system *without* the novel subpixel and self-calibration components ($\bar{r} = 0.135$).

Manual cleanup: In some cases we get erroneous reconstructions, mostly due to reflective surfaces close to parallel to the viewing direction. Our pipeline converts disparities to 3D meshes, which can be edited (e.g., using MeshLab) to remove erroneous surfaces. The edited meshes are then used to invalidate the corresponding pixels. Less than half of our scenes required manual editing.

Perfect and imperfect rectification: The last step in our pipeline is to compute 2D disparity maps for the initial "imperfect" calibration. We do this by selecting pairs of corresponding pixel locations and warping them, first using the inverse perfect, then the forward imperfect rectification transform. Finally we crop images and disparity maps to remove any invalid borders due to the rectification transforms. We also introduce a disparity offset to ensure positive disparities over the entire depth range. We provide disparities in PFM format together with camera calibration information for all views. Figure 2 shows a subset of our new datasets available at http://vision.middlebury.edu/stereo/data/2014/.

4   Stereo Experiments

We test our new datasets using 3 state-of-the-art stereo methods that are able to handle high-resolution images: a correlation method employing a 7×7 census transform [33] and aggregation with overlapping 9×9 windows [16]; the fast ELAS method by Geiger et al. [8]; and the semi-global matching (SGM) method by Hirschmüller [15]. We run all three methods on our new datasets, and also on the Middlebury Cones and Teddy benchmark images [29] for comparison. Figure 7 shows the error percentages. The first plot demonstrates that

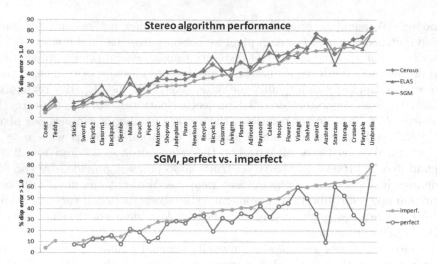

Fig. 7. Stereo algorithm performance on our new datasets. Top: Bad pixel percentages (% error >1.0) for Census, ELAS, and SGM, sorted by SGM error, for imperfect calibration. Bottom: SGM errors for imperfect and perfect calibration. Two existing datasets, Cones and Teddy [29], are included on the left for comparison.

our new datasets provide a range of challenges that significantly exceeds those of existing datasets (plotted on the left). The second plot demonstrates that imperfect rectification can yield significantly higher errors over accurate rectification; visual inspection of the error maps shows that errors are particularly severe in misaligned regions of high-frequency texture. When averaged over multiple datasets, there is also a clear correlation between increase in disparity error and amount of vertical rectification error (see Table 1).

Table 1. Increase in disparity error (from perfect to imperfect) for different amounts of rectification error as measured by average absolute y-disparity.

Avg abs y-disp range	0.0–0.5	0.5–1.0	1.0–1.5	1.5–3.0
Number of datasets	18	6	6	3
Avg disp error increase	9 %	49 %	61 %	154 %

5   Conclusion

We have presented a structured lighting system with novel subpixel and self-calibration components for creating high-resolution stereo datasets of static indoor scenes. We achieve significantly higher disparity and rectification accuracy than those of existing datasets. We contribute 33 new datasets to the community at http://vision.middlebury.edu/stereo/data/2014/, including versions with realistic rectification errors. Experiments demonstrate that our new datasets present a new level of challenge for stereo algorithms, both in terms of resolution and scene complexity. The challenge will be even greater when our images from different exposures, illuminations, or the point-and-shoot cameras are used. We hope that our datasets will inspire research into the next generation of stereo algorithms.

Acknowledgments. This work was supported by NSF awards IIS-0917109 and IIS-1320715 to DS.

References

1. 3dMD: 3D imaging systems. http://www.3dmd.com/
2. Acute 3D: 3D photogrammetry software. http://www.acute3d.com/
3. Batlle, J., Mouaddib, E., Salvi, J.: Recent progress in coded structured light as a technique to solve the correspondence problem: a survey. Pat. Rec. **31**(7), 963–982 (1998)
4. Besl, P.: Active optical range imaging sensors. MV&A **1**(2), 127–152 (1988)
5. Brown, M., Burschka, D., Hager, G.: Advances in computational stereo. TPAMI **25**(8), 993–1008 (2003)
6. Davis, J., Nehab, D., Ramamoorthi, R., Rusinkiewicz, S.: Spacetime stereo: a unifying framework for depth from triangulation. TPAMI **27**(2), 296–302 (2005)
7. Geiger, A., Lenz, P., Urtasun, R.: Are we ready for autonomous driving? The KITTI vision benchmark suite. In: CVPR, pp. 3354–3361 (2012)
8. Geiger, A., Roser, M., Urtasun, R.: Efficient large-scale stereo matching. In: Kimmel, R., Klette, R., Sugimoto, A. (eds.) ACCV 2010, Part I. LNCS, vol. 6492, pp. 25–38. Springer, Heidelberg (2011)
9. Gupta, M., Agrawal, A., Veeraraghavan, A., Narasimhan, S.: A practical approach to 3D scanning in the presence of interreflections, subsurface scattering and defocus. IJCV **102**(1–3), 33–55 (2013)
10. Gupta, M., Nayar, S.: Micro phase shifting. In: CVPR, pp. 813–820 (2012)
11. Hansen, P., Alismail, H., Rander, P., Browning, B.: Online continuous stereo extrinsic parameter estimation. In: CVPR, pp. 1059–1066 (2012)
12. Hartley, R., Zisserman, A.: Multiple View Geometry in Computer Vision. Cambridge University Press, Cambridge (2000)
13. Hirschmüller, H., Gehrig, S.: Stereo matching in the presence of sub-pixel calibration errors. In: CVPR, pp. 437–444 (2009)
14. Hirschmüller, H., Scharstein, D.: Evaluation of stereo matching costs on images with radiometric differences. TPAMI **31**(9), 1582–1599 (2009)
15. Hirschmüller, H.: Stereo processing by semiglobal matching and mutual information. TPAMI **30**(2), 328–341 (2008)
16. Hirschmüller, H., Innocent, P., Garibaldi, J.: Real-time correlation-based stereo vision with reduced border errors. IJCV **47**(1–3), 229–246 (2002)
17. Hosni, A., Bleyer, M., Gelautz, M.: Secrets of adaptive support weight techniques for local stereo matching. CVIU **117**(6), 620–632 (2013)
18. Hu, X., Mordohai, P.: A quantitative evaluation of confidence measures for stereo vision. TPAMI **34**(11), 2121–2133 (2012)

19. Izadi, S., et al.: KinectFusion: real-time 3D reconstruction and interaction using a moving depth camera. In: ACM UIST, pp. 559–568 (2011)
20. Lazaros, N., Sirakoulis, G., Gasteratos, A.: Review of stereo vision algorithms: from software to hardware. J. Optomechatron. **2**(4), 435–462 (2008)
21. Levoy, M., et al.: The digital Michelangelo project: 3D scanning of large statues. In: SIGGRAPH, pp. 131–144 (2000)
22. Lourakis, M.: levmar: Levenberg-Marquardt nonlinear least squares algorithms in C/C++. http://www.ics.forth.gr/~lourakis/levmar/
23. Lourakis, M., Argyros, A.: SBA: a software package for generic sparse bundle adjustment. ACM Trans. Math. Softw. **36**(1), 1–30 (2009). http://www.ics.forth.gr/~lourakis/sba/
24. Nakamura, Y., Matsuura, T., Satoh, K., Ohta, Y.: Occlusion detectable stereo - occlusion patterns in camera matrix. In: CVPR, pp. 371–378 (1996)
25. Posdamer, J., Altschuler, M.: Surface measurement by space-encoded projected beam systems. CGIP **18**(1), 1–17 (1982)
26. Salvi, J., Fernandez, S., Pribanic, T., Llado, X.: A state of the art in structured light patterns for surface profilometry. Pat. Rec. **43**(8), 2666–2680 (2010)
27. Scharstein, D.: Middlebury stereo page. http://vision.middlebury.edu/stereo/
28. Scharstein, D., Szeliski, R.: A taxonomy and evaluation of dense two-frame stereo correspondence algorithms. IJCV **47**(1–3), 7–42 (2002)
29. Scharstein, D., Szeliski, R.: High-accuracy stereo depth maps using structured light. In: CVPR, vol. I, pp. 195–202 (2003)
30. Seitz, S., Curless, B., Diebel, J., Scharstein, D., Szeliski, R.: A comparison and evaluation of multi-view stereo reconstruction algorithms. In: CVPR, vol. 1, pp. 519–528 (2006)
31. Strecha, C., von Hansen, W., Van Gool, L., Fua, P., Thoennessen, U.: On benchmarking camera calibration and multi-view stereo for high resolution imagery. In: CVPR, pp. 1–8 (2008)
32. Will, P., Pennington, K.: Grid coding: A preprocessing technique for robot and machine vision. In: IJCAI, pp. 66–70 (1971)
33. Zabih, R., Woodfill, J.: Non-parametric local transforms for computing visual correspondence. In: ECCV, pp. 151–158 (1994)
34. Zhang, L., Curless, B., Seitz, S.: Spacetime stereo: shape recovery for dynamic scenes. In: CVPR, vol. II, pp. 367–374 (2003)

Semi-Global Matching: A Principled Derivation in Terms of Message Passing

Amnon Drory[1], Carsten Haubold[2]($\boxtimes$), Shai Avidan[1], and Fred A. Hamprecht[2]

[1] Tel Aviv University, Tel Aviv, Israel
carsten.haubold@iwr.uni-heidelberg.de
[2] University of Heidelberg, Heidelberg, Germany

Abstract. Semi-global matching, originally introduced in the context of dense stereo, is a very successful heuristic to minimize the energy of a pairwise multi-label Markov Random Field defined on a grid. We offer the first principled explanation of this empirically successful algorithm, and clarify its exact relation to belief propagation and tree-reweighted message passing. One outcome of this new connection is an uncertainty measure for the MAP label of a variable in a Markov Random Field.

1 Introduction

Markov Random Fields (MRFs) have become a staple of computer vision. Practitioners appreciate the ability to provoke nonlocal effects by specifying local interactions only; and theoreticians like how easy it is to specify valid nontrivial distributions over high-dimensional entities. Unfortunately, *exact* maximum a posteriori (MAP) inference is tractable only for special cases. Important examples include binary[1] MRFs with only submodular potentials (by minimum st-cut [10]), multilabel MRFs with potentials that are convex over a linearly ordered label set (minimum st-cut [7]), tree-shaped MRFs (by dynamic programming), Gaussian MRFs (by solving a linear system) and non-submodular binary pairwise MRFs without unary terms defined over planar graphs (by perfect matching [15]). Many real-world problems have been cast as (and sometimes strong-armed to match) one of these special cases. Inference in more general MRFs, such as non-tree shaped multilabel MRFs, is NP-hard and heuristics such as alpha-expansion [1], damped loopy belief propagation [4] or tree-reweighted message passing [9,17] are often used.

One heuristic that has become very influential, especially in the context of dense disparity from stereo, is Semi-Global Matching (SGM). It is trivial to implement, extremely fast, and has ranked highly in the Middlebury [16], and the KITTI benchmarks [3] for several years. While intuitive, this successful heuristic has not found a theoretical characterization to date.

The present work establishes, for the first time, the precise relation of SGM to non-loopy belief propagation on a subgraph of the full MRF, and to tree-reweighted

---

[1] In the sense that nodes can take one of two possible labels.

© Springer International Publishing Switzerland 2014
X. Jiang et al. (Eds.): GCPR 2014, LNCS 8753, pp. 43–53, 2014.
DOI: 10.1007/978-3-319-11752-2_4

message passing with parallel tree-based updates [17]. This allows SGM to be viewed in the light of the rich literature on these techniques.

Based on these insights, we propose a lower bound based uncertainty measure that may benefit downstream processing.

2   Notation and Background

2.1   Markov Random Fields

Computer Vision problems are often cast in terms of MRFs. Throughout this paper we focus on *pairwise* MRFs, defined on an undirected graph $G=(V, \mathcal{E})$. Each node $\mathbf{p} \in V$ is a random variable which can take values $d_{\mathbf{p}}$ from a finite set of labels Σ. Here we assume that all variables have the same label space. Then a *labeling* $\mathbf{D} \in \Sigma^{|V|}$ is an assignment of one label for each variable. The best labeling $\mathbf{D}^*$ corresponds to the solution with the lowest energy $\text{argmin}_{\mathbf{D} \in \Sigma^{|V|}} E(\mathbf{D})$, with

$$E(\mathbf{D}) = \sum_{\mathbf{p} \in V} \varphi_{\mathbf{p}}(d_{\mathbf{p}}) + \sum_{(\mathbf{p}, \mathbf{q}) \in \mathcal{E}} \varphi_{\mathbf{p}, \mathbf{q}}(d_{\mathbf{p}}, d_{\mathbf{q}}) \qquad (1)$$

where $\varphi_{\mathbf{p}}(\cdot)$ is a *unary term* defined over the node $\mathbf{p}$, and $\varphi_{\mathbf{p}, \mathbf{q}}(\cdot, \cdot)$ is a pairwise term defined over the edge $(\mathbf{p}, \mathbf{q})$.[2]

2.2   Min-Sum Belief Propagation (BP)

Min-sum belief propagation [11,18] is an efficient dynamic programming algorithm for exact energy minimization on MRFs whose underlying graph is a *tree*. It calculates each node's *energy min marginal (EMM)*, or *belief* $\beta_{\mathbf{p}}(d_{\mathbf{p}})$, by sending messages along the edges.

As soon as node $\mathbf{p}$ has received messages from all its neighbors but node $\mathbf{q}$, it can send the following message to $\mathbf{q}$:

$$m_{\mathbf{p} \to \mathbf{q}}(d_{\mathbf{q}}) = \min_{d_{\mathbf{p}} \in \Sigma} \left(\varphi(d_{\mathbf{p}}) + \varphi(d_{\mathbf{p}}, d_{\mathbf{q}}) + \sum_{(\mathbf{k}, \mathbf{p}) \in \mathcal{E}, \mathbf{k} \neq \mathbf{q}} m_{\mathbf{k} \to \mathbf{p}}(d_{\mathbf{p}}) \right) \qquad (2)$$

The belief for each node is calculated from all the messages entering into it:

$$\beta_{\mathbf{p}}(d_{\mathbf{p}}) = \varphi(d_{\mathbf{p}}) + \sum_{(\mathbf{k}, \mathbf{p}) \in \mathcal{E}} m_{\mathbf{k} \to \mathbf{p}}(d_{\mathbf{p}}) \qquad (3)$$

If we consider a certain node $\mathbf{r}$ to be the root of the tree, then the BP algorithm can be divided into 2 parts. In the *inward pass* only the messages that are flowing inwards from the leaves towards $\mathbf{r}$ are calculated. Then, the rest are calculated in the *outward pass*. To calculate $\mathbf{r}$'s own belief, $\beta_{\mathbf{r}}(d_{\mathbf{r}})$, only the inward pass is necessary.

The belief $\beta_{\mathbf{p}}(\cdot)$ is also known as node $\mathbf{p}$'s EMM which means $\beta_{\mathbf{p}}(\ell)$ is the energy of the best labeling where node $\mathbf{p}$ takes label ℓ.

---

[2] We will drop the indices in $\varphi(d_{\mathbf{p}})$ and $\varphi(d_{\mathbf{p}}, d_{\mathbf{q}})$ as they are clear from the function inputs.

2.3   Tree-Reweighted Message-Passing (TRW-T)

Tree-reweighted message-passing with tree-based updates (TRW-T [17]) is an approximate energy minimization algorithm for general MRFs. It works by iteratively decomposing the original MRF into a convex sum of tree MRFs, optimizing each tree separately, and recombining the results. The result of each iteration is a *reparametrization* of the original MRF, which is a different MRF that has the same energy function E, but different unary and pairwise terms, φ.

TRW-T can be applied to an MRF with an underlying graph $G = (V, \mathcal{E})$, potentials φ, and a set of trees $\{T^1, \ldots, T^{N_T}\}$ which are all sub-graphs of G, such that each node and edge in G belongs to at least one tree[3]. Each tree $T^i = (V^i, \mathcal{E}^i)$ has a weight ρ^i so that $\sum_{i=1}^{N_T} \rho^i = 1$. The *node weight* $\rho_{\mathbf{p}}$ is defined as the sum of weights of all trees that contain node $\mathbf{p}$. *Edge weights* $\rho_{\mathbf{pq}}$ are defined similarly.

In each iteration of TRW-T, the following steps are performed:

1. An MRF is defined for each tree by defining tree unary and pairwise terms for each node and edge, by:

$$\varphi^i(d_{\mathbf{p}}) = \frac{1}{\rho_{\mathbf{p}}}\varphi^G(d_{\mathbf{p}}), \qquad \varphi^i(d_{\mathbf{p}}, d_{\mathbf{q}}) = \frac{1}{\rho_{\mathbf{pq}}}\varphi^G(d_{\mathbf{p}}, d_{\mathbf{q}}) \qquad (4)$$

   where by φ^i we denote the unary and pairwise terms for the tree T^i, and by φ^G we denote the terms of the full MRF.

2. The *min-sum belief propagation* algorithm is performed on each tree T^i separately, generating an EMM $\beta^i(\mathbf{p})$ for each node $\mathbf{p}$. A reparametrization of the tree MRF is defined, where the unary terms are: $\widetilde{\varphi}^i_{\mathbf{p}} \triangleq \beta^i(\mathbf{p})$.

3. The reparametrizations of the tree MRFs are aggregated to produce a reparametrization for the graph-MRF. For the unary terms, this is done by:

$$\widetilde{\varphi}^G_{\mathbf{p}} = \sum_{i:\mathbf{p}\in V^i} \rho^i \widetilde{\varphi}^i_{\mathbf{p}} \qquad (5)$$

It is hoped, though not guaranteed, that repeated iterations of this process will flow potential from the entire graph into the global unary terms $\widetilde{\varphi}^G_{\mathbf{p}}$, and turning them into *EMMs*. The final label map can then be found by choosing, for each node, the label that minimizes its unary term:

$$d^*_{\mathbf{p}} = \operatorname*{argmin}_{d_{\mathbf{p}}\in\Sigma} \widetilde{\varphi}^G_{\mathbf{p}}(d_{\mathbf{p}}) \qquad (6)$$

It is important to note that there are no convergence assurances for TRW-T. This means that the energy of the labeling that is produced after each iteration is not guaranteed to be lower, or even equal, to the energy of the previous labeling. Kolmogorov's TRW-S [9] improves upon TRW-T and provides certain convergence guarantees, but can not be related to SGM directly.

---

[3] In the presentation of the algorithm in [17] *spanning* trees are used, but as mentioned in [9], this is not necessary.

Tree Agreement and Labeling Quality Estimation. From the description of the TRW-T process it follows that the graph-MRF's energy function E^G is a convex combination of the tree energy functions E^i, i.e. $E^G(\mathbf{D}) = \sum_{i=1}^{N_T} \rho^i E^i(\mathbf{D})$. Recall we're interested in MAP inference, i.e. finding the labeling with the minimal energy:

$$\min_{\mathbf{D} \in \Sigma^{|V|}} E^G(\mathbf{D}) = \min_{\mathbf{D} \in \Sigma^{|V|}} \sum_{i=1}^{N_T} \rho^i E^i(\mathbf{D}) \geq \underbrace{\sum_{i=1}^{N_T} \min_{\mathbf{D}^i \in \Sigma^{|V|}} \rho^i E^i(\mathbf{D}^i)}_{E_{LB}} \tag{7}$$

We get a *lower bound* on the minimum graph energy by allowing each tree energy to be minimized separately. TRW-T iteratively computes this lower bound, denoted by E_{LB}, by running BP on each tree. Notice that the inequality becomes equality *iff* the best labelings of all trees agree on the label assigned to every node. If this occurs, TRW-T can stop and the labeling that is found is indeed the best labeling $\mathbf{D}^*$. The lower bound can be used for estimating how good a given labeling $\mathbf{D}$ is. For any labeling $\mathbf{D}$ it holds that $E(\mathbf{D}) \geq E(\mathbf{D}^*)$. From this it follows that $E(\mathbf{D}) \geq E_{LB}$. If the relative difference between $E(\mathbf{D})$ and E_{LB} is small, that could indicate that $\mathbf{D}$ is a good labeling. TRW's reparametrizations aim to adjust the unaries such that E_{LB} approaches $E(\mathbf{D}^*)$

2.4   Semi-Global Matching (SGM)

As part of the Semi-Global Matching algorithm [5] for dense stereo matching, Hirschmüller presented an efficient algorithm for approximate energy minimization for a grid-shaped pairwise MRF. It divides the grid-shaped problem into multiple one-dimensional problems defined on *scanlines*, which are straight lines that run through the image in multiple directions. Inference in each scanline is performed separately, and the results are then combined to produce a labeling (in stereo: disparity map) for the image.

We denote by $\mathbf{p}=(x,y)$ the location of a pixel, and by $\mathbf{r}=(dx,dy)$ the direction of a scanline. SGM's processing of a scanline consists of the following recursive calculation:

$$L^{\mathbf{r}}(d_{\mathbf{p}}) = \varphi(d_{\mathbf{p}}) + \min_{d_{\mathbf{p}-\mathbf{r}} \in \Sigma} \left\{ L^{\mathbf{r}}(d_{\mathbf{p}-\mathbf{r}}) + \varphi(d_{\mathbf{p}-\mathbf{r}}, d_{\mathbf{p}}) \right\} \tag{8}$$

We denote the set of all scanline directions by $\mathcal{R}$, and its size by $|\mathcal{R}|$, which is typically 8 or 16. The aggregation of scanline results is accomplished by simple summation:

$$S(d_{\mathbf{p}}) = \sum_{\mathbf{r} \in \mathcal{R}} L^{\mathbf{r}}(d_{\mathbf{p}}) \tag{9}$$

Finally, a pixel labeling is obtained by defining:

$$d_{\mathbf{p}}^{\text{SGM}} = \operatorname*{argmin}_{d_{\mathbf{p}} \in \Sigma} S(d_{\mathbf{p}}) \tag{10}$$

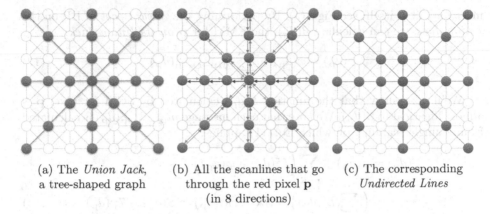

(a) The *Union Jack*, a tree-shaped graph   (b) All the scanlines that go through the red pixel **p** (in 8 directions)   (c) The corresponding *Undirected Lines*

Fig. 1. The *Union Jack*, its 8 scanlines and 4 *Undirected Lines*. When performing inference for the red pixel in the center, SGM ignores all the greyed-out nodes and edges of the original MRF.

3 Synthesis

The qualitative nature of the connection between SGM and BP as well as TRW-T may already have emerged in the foregoing. We now make it quantitative.

3.1 Relation between SGM and BP on Union Jack

Claim 1: $S(d_{\mathbf{p}})$, the effective label costs computed at a pixel **p** by 8-directional SGM are identical to the min-marginals computed on a *Union Jack* (see below) centered at pixel **p**, up to (easily correctable) over-counting of the unary term at **p**.

SGM's treatment of each scanline is similar to a forward pass of *min-sum belief propagation* (BP) on it, flowing messages from the first pixel on the scanline to the last. To be exact, SGM's processing of a scanline amounts to the *Forward Algorithm* [13], a close relative of min-sum BP.[4] The relation between the two algorithms is given by:

$$m_{(\mathbf{p-r})\to\mathbf{p}}(d_{\mathbf{p}}) = L^{\mathbf{r}}(d_{\mathbf{p}}) - \varphi(d_{\mathbf{p}}) \tag{11}$$

In this section we focus on the common case where SGM is used with scanlines in 8 directions. For each pixel **p** we define the *Union Jack of* **p** as the union of all scanlines that pass through **p**. The *Union Jack* is a tree with 8 branches that

---

[4] In terms of message passing on a *factor graph*, one variant represents factor-to-node messages, while the other gives node-to-factor messages.

meet only at **p** itself, see Fig. 1(a). By performing the *inward pass* of BP on the *Union Jack*, we can calculate $\beta(d_\mathbf{p})$, the EMM for **p** *on its Union Jack*:

$$\beta(d_\mathbf{p}) = \varphi(d_\mathbf{p}) + \sum_{r \in \mathcal{R}} m_{(\mathbf{p-r}) \to \mathbf{p}}(d_\mathbf{p}) \tag{12}$$

We'll next show that the function $S(d_\mathbf{p})$ calculated by SGM for pixel **p** is almost identical to $\beta(d_\mathbf{p})$. We note that each message $m_{(\mathbf{p-r}) \to \mathbf{p}}(d_\mathbf{p})$ is calculated from a single scanline, on which Eq. (11) applies:

$$\beta(d_\mathbf{p}) = \varphi(d_\mathbf{p}) + \sum_{r \in \mathcal{R}} \left(L^\mathbf{r}(d_\mathbf{p}) - \varphi(d_\mathbf{p}) \right)$$

$$= \left(\sum_{r \in \mathcal{R}} L^\mathbf{r}(d_\mathbf{p}) \right) - (|\mathcal{R}| - 1)\varphi(d_\mathbf{p}) = S(d_\mathbf{p}) - 7\varphi(d_\mathbf{p}) \tag{13}$$

From the last line we can see that the function $S(d_\mathbf{p})$ that SGM calculates for a pixel **p** is similar to **p**'s EMM calculated on its *Union Jack*. The difference is that SGM *over-counts* the unary term $\varphi(d_\mathbf{p})$ 7 times. This can easily be fixed by adding a final subtraction stage to SGM.

In summary, it turns out that (over-counting corrected) SGM assigns to each pixel the label it would take in the best labeling *of its Union Jack*. This can be seen as an approximation of the label it would take in the optimal labeling *of the entire graph*. Unfortunately, finding that label would require calculating the EMM of each pixel relative to *the entire graph*, which generally is intractable.

Performing BP on each *Union Jack* separately would be inefficient, as many calculations are repeated in different *Union Jacks*. SGM avoids this by reusing the messages L_r computed on each scanline for all *Union Jacks* that contain it.

3.2   Relation between SGM and TRW-T

Claim 2: Performing over-counting corrected SGM on an MRF with energy

$$E = \sum_{\mathbf{p} \in V} \varphi(d_\mathbf{p}) + \sum_{(\mathbf{p},\mathbf{q}) \in \mathcal{E}} \varphi(d_\mathbf{p}, d_\mathbf{q}) \tag{14}$$

is equivalent to performing one iteration of TRW-T on an MRF with energy

$$\widetilde{E} = \sum_{\mathbf{p} \in V} \varphi(d_\mathbf{p}) + C_0 \sum_{(\mathbf{p},\mathbf{q}) \in \mathcal{E}} \varphi(d_\mathbf{p}, d_\mathbf{q}) \tag{15}$$

where $C_0 > 0$ is a constant, e.g. for 8-directional SGM, $C_0 = \frac{1}{4}$

The SGM algorithm closely resembles a single iteration of TRW-T. In both, we perform BP on trees, and then aggregate the results into an approximate belief for each pixel. Next we choose for each pixel the label that minimizes its own approximate belief.

Where SGM differs significantly from TRW-T is in the way the graph MRF's energy terms are used to define the tree MRF terms. SGM simply copies the terms, while TRW-T weighs them as to make the graph terms a convex sum of the tree terms (Eq. (4)). This reveals that SGM is not directly equivalent to a single iteration of TRW-T on the same MRF. Instead, performing SGM on a specific MRF is equivalent to performing one iteration of TRW-T on an MRF whose pairwise terms are multiplied by a constant C_0 (Eq. (15)).

At first glance, the conclusion from claim 2 seems to be that SGM is in fact different from TRW-T, since applying both algorithms to the same MRF will generally produce different results. However, in a real-life setting this is not necessarily the case. In many application scenarios, the energy function is defined as $E = E_{Data} + \lambda E_{Smoothness}$, and λ is learned from a training set. In these cases, the difference between the two algorithms will disappear, since a different value of λ will be learned for each, in such a way that $\lambda_{TRW} = C_0 \lambda_{SGM}$.

To fully define TRW-T we must specify the set of trees $\mathcal{T}$ that we use for TRW-T, and their weights ρ. We define $\mathcal{T}$ to be the set of SGM's *Undirected Lines*, with equal weights ρ_0 for all trees. An *Undirected Line* (UL) is simply a scanline, except no direction of traversal is defined. Each UL corresponds to two opposite-directed scanlines, $\mathbf{r}$ and $\bar{\mathbf{r}}$ (see Fig. 1(b)–(c)). Using Eq. (11), we can show that performing BP on the UL is very closely related to running SGM on both $\mathbf{r}$ and $\bar{\mathbf{r}}$:

$$\beta^{\mathbf{r}}(d_{\mathbf{p}}) + \varphi(d_{\mathbf{p}}) = L^{\mathbf{r}}(d_{\mathbf{p}}) + L^{\bar{\mathbf{r}}}(d_{\mathbf{p}}) \qquad (16)$$

where $\beta^{\mathbf{r}}(d_{\mathbf{p}})$ is the belief calculated for pixel $\mathbf{p}$ by running BP on the UL corresponding to scanline $\mathbf{r}$. SGM's output $S(d_{\mathbf{p}})$ can be explained as a sum of the beliefs calculated for pixel $\mathbf{p}$ on each UL passing through it, with some over-counting of the unary term:

$$S(d_{\mathbf{p}}) = \sum_{\mathbf{r} \in \mathcal{R}} L^{\mathbf{r}}(d_{\mathbf{p}}) = \underbrace{\sum_{\mathbf{r} \in \mathcal{R}^+} \beta^{\mathbf{r}}(d_{\mathbf{p}})} + \frac{|\mathcal{R}|}{2}\varphi(d_{\mathbf{p}}) \qquad (17)$$
$$\tilde{S}(d_{\mathbf{p}})$$

where $\mathcal{R}^+$ is the group of ULs, and $\tilde{S}(d_{\mathbf{p}})$ is the output of an over-counting corrected version of SGM.

Proof of Claim 2: We wish to show that the labeling created by SGM for an MRF with energy E (Eq. (14)) is the same as the labeling created by one iteration of TRW-T on an MRF with energy $\widetilde{E}$ (Eq. (15)). Let us first introduce a scale factor $C_1 > 0$ to define $\widehat{E} = C_1\widetilde{E}$. $\widehat{E}$ and $\widetilde{E}$ have the same labeling after an iteration of TRW-T, as the argmin is unaffected by scaling (Eq. (6)). We will now show how to choose C_0 and C_1 to make the tree MRFs equal. Note that SGM simply copies the potentials $\varphi = \varphi^G$, whereas TRW-T weighs them according to Eq. (4). Here $\rho_{\mathbf{p}} = \rho_0 \frac{|\mathcal{R}|}{2}$, and $\rho_{\mathbf{pq}} = \rho_0 \epsilon$, where ϵ is the number of ULs in which each edge appears (for 8 directional SGM, $\epsilon = 1$).

The unary potential $\hat{\varphi}^i(d_\mathbf{p})$ in a tree of TRW-T applied to $\widehat{E}$ is then given as

$$\hat{\varphi}^i(d_\mathbf{p}) = \frac{1}{\rho_\mathbf{p}}\hat{\varphi}^G(d_\mathbf{p}) = \frac{1}{\rho_\mathbf{p}}C_1\varphi^G(d_\mathbf{p}). \tag{18}$$

Thus, to make $\hat{\varphi}^i(d_\mathbf{p}) = \varphi(d_\mathbf{p})$ hold, we need to choose $C_1 = \rho_\mathbf{p} = \rho_0\frac{|\mathcal{R}|}{2}$. This leaves C_0 to be determined by the pairwise potential:

$$\hat{\varphi}^i(d_\mathbf{p}, d_\mathbf{q}) = \frac{1}{\rho_\mathbf{pq}}C_0\hat{\varphi}^G(d_\mathbf{p}, d_\mathbf{q}) = \frac{1}{\rho_\mathbf{pq}}C_1C_0\varphi^G(d_\mathbf{p}, d_\mathbf{q}) \tag{19}$$

$$= \frac{\rho_0\frac{|\mathcal{R}|}{2}}{\rho_0\epsilon}C_0\varphi^G(d_\mathbf{p}, d_\mathbf{q}) = \frac{|\mathcal{R}|}{2\epsilon}C_0\varphi^G(d_\mathbf{p}, d_\mathbf{q}) \tag{20}$$

Again, if we want the equality to hold, we have to set $C_0 = \frac{2\epsilon}{|\mathcal{R}|}$. For 8 directional SGM this yields $C_0 = \frac{1}{4}$.

We have now shown that, given C_0 and C_1 as above, all tree MRFs are equal for TRW-T ($\widehat{E}$) and SGM (E). The next steps of both algorithms are the same:

1. Perform BP on all trees
2. Sum the beliefs from all scanlines that pass through a pixel (TRW-T Eq. (5), SGM Eq. (17))
3. Choose the label that minimizes each belief (TRW-T Eq. (6), SGM Eq. (10))

Since the same steps are applied to the same trees, the final labeling is also the same for E and $\widehat{E}$, and thus also for $\widetilde{E}$.

Summing up, SGM amounts to the first iteration of TRW-T on a MRF with pairwise energies that have been scaled by a constant and known factor.

4   Implications

Belief propagation was one of the earliest techniques for performing inference in graphical models. Comprehensive literature on its properties exists. Given the formal link that we presented in the previous section, one can now draw from this vast repertoire when examining or improving on SGM. We pointed out that SGM can be treated like a first step of TRW-T, which uses BP at its core.

We now exploit the lower bound computed by TRW-T to construct a per pixel uncertainty measure for the depth map.

We propose a novel uncertainty measure for SGM that is based on SGM's relation to TRW-T. For this we interpret the difference between sum of local optima (lower bound) and global optimum as scanline disagreement.

Many uncertainty measures have been presented in the past. For instance [2,12] model uncertainty based on the image acquisition process. Other evaluations focus on measures that only look at the resulting energy distribution by disparity per pixel [6]. While most of the latter examine the cost curve (of cost vs. disparity) around the optimal disparity, they cannot take into account how this curve was computed as they do not assume any knowledge about the optimization technique. The evaluations in [6] show that none of the different measures is a clear favorite.

Kohli on the other hand derives a confidence measure from the min-marginals of an MRF optimization using dynamic graph cuts [8]. His measure $\sigma_{d_{\mathbf{p}}}$ is given by the ratio of the current max-marginal and the sum of all max-marginals for a certain node. For this ratio, transforming from EMMs to max-marginals is feasible as the partition function cancels out.

$$\sigma_{d_{\mathbf{p}}} = \frac{\exp(-\beta_p(d_{\mathbf{p}}))}{\sum_{\widetilde{d_{\mathbf{p}}} \in \Sigma} \exp(-\beta_p(\widetilde{d_{\mathbf{p}}}))} \tag{21}$$

In contrast, our proposed measure depends on the *Union Jack* optimal solution given by SGM, and its lower bound found by applying TRW-T. For this we decompose the *Union Jack* into eight *half* scanlines that start at the image border and end in $\mathbf{p}$.

Following TRW-T's reparametrization for this decomposition yields equal weights for all trees (lines) $\rho^i = \frac{1}{8}$. Every node and edge appears only once, thus $\rho_{\mathbf{s}} = \rho_{\mathbf{st}} = \frac{1}{8}$, only the root is replicated in all eight scanlines, $\rho_{\mathbf{p}} = 1$. The energy per *half* scanline can then be related to the energy $E^i_{SGM}(D)$ computed by SGM as

$$E^i(D) = \sum_{\mathbf{s} \in V^i} \tfrac{1}{\rho_s} \varphi_{\mathbf{s}}(d_{\mathbf{s}}) + \sum_{(\mathbf{s},\mathbf{t}) \in \mathcal{E}^i} \tfrac{1}{\rho_{st}} \varphi_{\mathbf{st}}(d_{\mathbf{s}}, d_{\mathbf{t}}) \tag{22}$$

$$= \varphi_{\mathbf{p}}(d_{\mathbf{p}}) + \sum_{\mathbf{s} \in V^i \setminus \mathbf{p}} 8\varphi_{\mathbf{s}}(d_{\mathbf{s}}) + \sum_{(\mathbf{s},\mathbf{t}) \in \mathcal{E}^i} 8\varphi_{\mathbf{st}}(d_{\mathbf{s}}, d_{\mathbf{t}}) \tag{23}$$

$$= 8E^i_{SGM}(D) - 7\varphi_{\mathbf{p}}(d_{\mathbf{p}}) \tag{24}$$

According to Eq. (7), the weighted sum of optimal solutions of the subproblems now yields a lower bound to the optimal *Union Jack* energy E_{UJ}:

$$\min_{D \in \Sigma^N} E_{UJ}(D) \geq \sum_{i=1}^{8} \min_{D^i \in \Sigma^N} \rho^i E^i(D^i) = \sum_{i=1}^{8} \min_{D^i \in \Sigma^N} \left(E^i_{SGM}(D^i) - \tfrac{7}{8}\varphi_{\mathbf{p}}(d_{\mathbf{p}}^i) \right) \tag{25}$$

$$\Rightarrow \min_{d_{\mathbf{p}} \in \Sigma} \left(S(d_{\mathbf{p}}) - 7\varphi(d_{\mathbf{p}}) \right) \geq \sum_{i=1}^{8} \min_{d_{\mathbf{p}}^i \in \Sigma} \left(L^i(d_{\mathbf{p}}^i) - \tfrac{7}{8}\varphi_{\mathbf{p}}(d_{\mathbf{p}}^i) \right) \tag{26}$$

An obvious case when this lower bound is tight, is when all directions choose the same label $d_{\mathbf{p}}$. This allows us to interpret the difference between the left and right hand side of (26) as disagreement between the *half* scanlines that end in $\mathbf{p}$. Figure 2 shows a qualitative evaluation of the difference between the left and right hand side for the Tsukuba scene from the Middlebury benchmark [14]. One would expect the disagreement to be large at object borders which impose depth discontinuities, but there are also some other uncertain regions in the background on the cupboard (see the purple frame). There the untextured material causes ambiguous disparities, which is nicely highlighted in the uncertainty heat map.

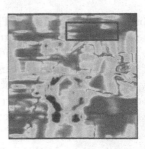

Fig. 2. Left: Cropped Tsukuba image from the Middlebury Benchmark. **Center:** Heat map visualization of our proposed uncertainty measure. **Right:** The uncertainty measure introduced by Kohli [8].

This difference between sum of scanline optima and *Union Jack* optimum can easily be exploited as uncertainty measure. This only incurrs a small overhead of computing the right hand side of (26) by finding the minimum for each direction independently, so searching over $|\Sigma|$ disparities for $|\mathcal{R}|$ directions, and aggregating them. In Fig. 2 we compare Kohli's uncertainty, computed from *Union Jack* EMMs and transformed to the energy domain $1 - \log(\sigma)$, with our proposed uncertainty measure. The difference between confident and less confident regions is much smoother in Kohli's measure, rendering more parts uncertain, but making it harder to e.g. point a user or algorithm towards regions to investigate. This can be explained by the different interpretations of both measures. A low uncertainty in our measure indicates that we cannot find a much better solution in terms of energy, whereas in Kohli's measure this means that the distribution of max-marginals has a high peak at the current label. It is obvious that for SGM our proposed measure is much easier to interpret.

5   Conclusion

In this work we have derived the formal link between SGM and BP, as well as TRW-T, allowing for a new interpretation of SGM's success and affording new insights. As a specific example, we propose an uncertainty measure for the MAP labeling of an MRF. This characteristic is based on the energy difference between the labeling found and a lower bound, and can be computed efficiently alongside SGM. Such uncertainty measures can be useful for downstream processing of a MAP result. We envision that these insights may encourage the application of SGM as a highly efficient, and now also well-motivated, heuristic to new problems beyond dense stereo matching.

Acknowledgments. The research of A.D. and S.A. was partially supported by a Google grant. C.H. and F.A.H. gratefully acknowledge partial financial support by the HGS MathComp Graduate School, the RTG 1653 for probabilistic graphical models and the CellNetworks Excellence Cluster / EcTop.

References

1. Boykov, Y., Veksler, O., Zabih, R.: Fast approximate energy minimization via graph cuts. IEEE Trans. Pattern Anal. Mach. Intell. **23**(11), 1222–1239 (2001)
2. Frank, M., Plaue, M., Hamprecht, F.A.: Denoising of continuous-wave time-of-flight depth images using confidence measures. Opt. Eng. **48**(7), 077003 (2009)
3. Geiger, A., Lenz, P., Urtasun, R.: Are we ready for autonomous driving? the kitti vision benchmark suite. In: 2012 IEEE Conference on Computer Vision and Pattern Recognition (CVPR), pp. 3354–3361. IEEE (2012)
4. Heskes, T., et al.: Stable fixed points of loopy belief propagation are minima of the bethe free energy. In: Advances in Neural Information Processing Systems 15, pp. 359–366 (2003)
5. Hirschmuller, H.: Accurate and efficient stereo processing by semi-global matching and mutual information. In: IEEE Computer Society Conference on Computer Vision and Pattern Recognition, CVPR 2005, vol. 2, pp. 807–814. IEEE (2005)
6. Hu, X., Mordohai, P.: Evaluation of stereo confidence indoors and outdoors. In: 2010 IEEE Conference on Computer Vision and Pattern Recognition (CVPR), pp. 1466–1473. IEEE (2010)
7. Ishikawa, H.: Exact optimization for markov random fields with convex priors. IEEE Trans. Pattern Anal. Mach. Intell. **25**(10), 1333–1336 (2003)
8. Kohli, P., Torr, P.: Measuring uncertainty in graph cut solutions - efficiently computing min-marginal energies using dynamic graph cuts. In: Leonardis, A., Bischof, H., Pinz, A. (eds.) ECCV 2006. LNCS, vol. 3952, pp. 30–43. Springer, Heidelberg (2006)
9. Kolmogorov, V.: Convergent tree-reweighted message passing for energy minimization. IEEE Trans. Pattern Anal. Mach. Intell. **28**(10), 1568–1583 (2006)
10. Kolmogorov, V., Zabin, R.: What energy functions can be minimized via graph cuts? IEEE Trans. Pattern Anal. Mach. Intell. **26**(2), 147–159 (2004)
11. Pearl, J.: Probabllistic Reasoning in Intelligent Systems: Networks of Plausible Inference. Morgan Kaufmann, San Francisco (1988)
12. Perrollaz, M., Spalanzani, A., Aubert, D.: Probabilistic representation of the uncertainty of stereo-vision and application to obstacle detection. In: 2010 IEEE Intelligent Vehicles Symposium (IV), pp. 313–318. IEEE (2010)
13. Russell, S.J., Norvig, P., Candy, J.F., Malik, J.M., Edwards, D.D.: Artificial Intelligence: A Modern Approach. Prentice-Hall Inc, Upper Saddle River (1996)
14. Scharstein, D., Szeliski, R.: A taxonomy and evaluation of dense two-frame stereo correspondence algorithms. Int. J. Comput. Vis. **47**(1–3), 7–42 (2002)
15. Schraudolph, N.N., Kamenetsky, D.: Efficient exact inference in planar ising models. In: NIPS, pp. 1417–1424 (2008)
16. Szeliski, R., Zabih, R., Scharstein, D., Veksler, O., Kolmogorov, V., Agarwala, A., Tappen, M., Rother, C.: A comparative study of energy minimization methods for markov random fields with smoothness-based priors. IEEE Trans. Pattern Anal. Mach. Intell. **30**(6), 1068–1080 (2008)
17. Wainwright, M.J., Jaakkola, T.S., Willsky, A.S.: Map estimation via agreement on trees: message-passing and linear programming. IEEE Trans. Inf. Theor. **51**(11), 3697–3717 (2005)
18. Yedidia, J.S., Freeman, W.T., Weiss, Y.: Understanding belief propagation and its generalizations. In: Exploring Artificial Intelligence in the New Millennium, pp. 239–269. Morgan Kaufmann Publishers Inc., San Francisco (2003). http://dl.acm.org/citation.cfm?id=779343.779352

Submap-Based Bundle Adjustment for 3D Reconstruction from RGB-D Data

Robert Maier[✉], Jürgen Sturm, and Daniel Cremers

TU Munich, Munich, Germany
{maierr,sturmju,cremers}@in.tum.de

Abstract. The key contribution of this paper is a novel submapping technique for RGB-D-based bundle adjustment. Our approach significantly speeds up 3D object reconstruction with respect to full bundle adjustment while generating visually compelling 3D models of high metric accuracy. While submapping has been explored previously for mono and stereo cameras, we are the first to transfer and adapt this concept to RGB-D sensors and to provide a detailed analysis of the resulting gain. In our approach, we partition the input data uniformly into submaps to optimize them individually by minimizing the 3D alignment error. Subsequently, we fix the interior variables and optimize only over the separator variables between the submaps. As we demonstrate in this paper, our method reduces the runtime of full bundle adjustment by 32 % on average while still being able to deal with real-world noise of cheap commodity sensors. We evaluated our method on a large number of benchmark datasets, and found that we outperform several state-of-the-art approaches both in terms of speed and accuracy. Furthermore, we present highly accurate 3D reconstructions of various objects to demonstrate the validity of our approach.

1 Introduction

Low-cost sensors such as the Microsoft Kinect have boosted research in 3D reconstruction and enabled novel applications such as product digitalization, remote inspection and assessment, documentation and reverse-engineering. For example, Fig. 1 shows a colored 3D model of a lawn-mower generated from hand-held RGB-D sensor data with the approach proposed in this paper. The resulting 3D model is highly accurate. It preserves small details such as the plate in the close-up, it allows for metric measurements and can be used to detect dents and deformations.

In all of these applications, it is important that the 3D model is highly accurate, yet the reconstruction process must be fast enough to be truly useful in practice. While volumetric methods such as KinectFusion [4,17] are tailored for real-time use, they are inherently prone to drift because they do not impose any long-range consistency. This may lead to inconsistencies such as the formation of double surfaces and blurry textures. Global optimization techniques such as bundle adjustment [8,10] achieve higher accuracies. Yet, they drastically

© Springer International Publishing Switzerland 2014
X. Jiang et al. (Eds.): GCPR 2014, LNCS 8753, pp. 54–65, 2014.
DOI: 10.1007/978-3-319-11752-2_5

(a) Input im-   (b)   Resulting   3D   (c) Submap visualization with colorized
ages   model and close-up   camera poses and landmarks

Fig. 1. We increase the efficiency of 3D bundle adjustment significantly with a novel submapping approach for RGB-D data (Color figure online).

increase the computation time, scaling poorly with the number of images and 3D points. Submapping techniques offer a remedy to this problem. They divide the large optimization problem into smaller, independent subproblems through marginalization [20,21]. Unfortunately, this strategy only works well when the graph is sparsely connected. For the reconstruction of an object in the center, the challenge considered in this work, many overlaps exist and thus the resulting graph is densely connected. This renders full bundle adjustment slow and existing submapping techniques inefficient.

To cope with these challenges, we developed a novel submapping technique for feature-based 3D reconstruction that significantly speeds up map optimization. Moreover, we provide an experimental study on the trade-off between reconstruction accuracy and computation speed on public benchmark sequences. We finally show several examples of 3D scans demonstrating the accuracy that can be achieved using our approach. To the best of our knowledge, we are the first to introduce submapping into an RGB-D based reconstruction method and to demonstrate its efficiency in the context of 3D object scanning.

2   Related Work

Real-time 3D reconstruction from images has a long history in computer vision [2,5,7,10,13,17]. We focus our review of previous work on approaches using a single, hand-held sensor. In this scenario, called Visual SLAM, both the pose of the camera and the 3D model need to be estimated. Typically, visual features such as SIFT are extracted and matched across the frames. From this, a SLAM graph can be constructed whose vertices correspond to 6-DOF camera poses and 3D landmark positions and whose edges correspond to the observations. Bundle adjustment (BA) [26] optimizes over camera poses and landmarks using non-linear minimization. While impressive reconstructions can be obtained using BA [2], bad initialization and the scale ambiguity can lead to slow convergence, and the computational effort typically grows cubically with the number of cameras and landmarks. Sparse Bundle Adjustment [16] exploits the fact that 3D landmarks are usually not visible in all of the cameras. When instead stereo cameras or RGB-D sensors are used, the scale ambiguity can be resolved and the initialization is simplified [10,11]. Moreover, the absolute camera motion between two frames can be estimated directly. By marginalization over all landmarks, the problem can then be reduced to pose graph SLAM [8,10,14], which is however still cubic in the number of cameras.

Submap-based BA methods [12,15,20,21] aim at partitioning the full BA problem into several smaller submaps, which are connected among each other and optimized independently in an efficient manner. By expressing the camera poses and landmarks relative to a base node for every submap [23], the error per submap is bounded which generally improves convergence. Advanced techniques for graph partitioning such as nested dissection have been proposed [19]. After the individual submaps have been optimized independently, a global optimization over the submap separator variables is carried out and the whole process is iterated. In this work, we employ submapping to pair the high accuracy of full BA with the improved efficiency.

After map optimization, a 3D model can be generated from the data. Common representations for 3D models include point clouds, surfels [10], triangle meshes [27], and signed distance functions [6]. Octrees allow for efficient data storage and fusion at multiple scales [9,24,28]. While KinectFusion [17] demonstrated that impressive 3D models can be acquired by tracking and fusing the depth images directly into a signed distance volume, drift will accumulate in the 3D model and inevitably lead to inconsistencies. Several approaches have been proposed to mitigate these effects in post-processing [27,29,30], however at the cost of having different cost functions at different optimization stages, which is not desirable from a theoretical point of view. Therefore, we propose to use BA with a single cost function to achieve global consistency.

3   3D Object Reconstruction Pipeline

We developed a feature-based 3D reconstruction system similar to [11]. Figure 2 depicts a schematic overview.

We acquire RGB-D data of the object with an off-the-shelf ASUS Xtion Pro Live sensor. To account for increasing sensor noise with distance, we pre-process the acquired depth map with a bilateral filter and cut off large depth values.

Our approach uses frame-to-frame camera tracking to estimate the absolute camera poses for the acquired RGB-D frames by concatenating the relative camera motion. We estimate the relative pose of every frame w.r.t. its predecessor using a robust feature-based 3D alignment algorithm based on the method of Arun [3] and RANSAC.

We use a graph-based map representation. The nodes of the graph correspond to the variables, i.e. camera poses $C_i \in SE(3)$ (with $i \in 1 \ldots M$) and 3D landmarks $\mathbf{X}_j \in \mathbb{R}^3$ (with $j \in 1 \ldots N$). The edges of the graph represent landmark observations $k \in 1 \ldots K$, i.e. $\mathbf{z}_k = (u_k, v_k, d_k)^\top \in \mathbb{R}^3$, where (u_k, v_k) is the observed 2D pixel coordinate and d_k is the observed depth.

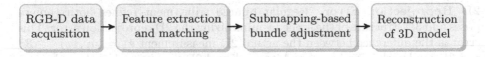

Fig. 2. Processing pipeline of our 3D object reconstruction approach.

To further reduce the drift in the graph, we detect loop closures by performing a 3D alignment of the current frame with 20 uniformly sampled previous frames. After all frames have been processed, we perform bundle adjustment to optimize over the camera poses and landmark positions. More details on this will be provided in Sects. 4 and 5.

After optimization, we use the refined camera poses to generate a dense colored 3D point cloud by fusing the RGB-D frames into an octree-based 3D model representation.

4 Full Bundle Adjustment for RGB-D Sensors

Our approach on 3D bundle adjustment is inspired by the work of Henry et al. [10], that we extended by integrating depth measurements from the RGB-D sensor as additional constraints similar to Scherer et al. [22]. The optimization goal becomes then to minimize the 3D alignment error instead of the 2D reprojection error.

Rigid Body Motion. The camera parameters C_i to be optimized are 3D Euclidean transformations $g = (R, \mathbf{t}) \in SE(3)$, with translation vector $\mathbf{t} \in \mathbb{R}^3$ and rotation $R \in SO(3)$. To have a minimum number of parameters to be optimized, we represent the camera poses C_i with their twist coordinate representations $\boldsymbol{\xi}_i = (\omega_1, \omega_2, \omega_3, v_1, v_2, v_3)^\top \in \mathbb{R}^6$. While this representation is important for the efficiency of solving the NLS problem, we use in the following the notation $C_i \in SE(3)$ to denote the camera poses.

We define the transformations $\mathcal{T}(g, \mathbf{X})$ and $\mathfrak{T}(g, \tilde{g})$ for transforming 3D points $\mathbf{X} \in \mathbb{R}^3$ and transformations $\tilde{g} \in SE(3)$, respectively.

Camera Model. We use the basic pinhole camera model to describe the mapping of 3D points of a three-dimensional scene onto a 2D image plane. The projection function π maps a 3D point $\mathbf{X} = (x, y, z)^\top \in \mathbb{R}^3$ to a 2D image point $\mathbf{x} \in \mathbb{R}^2$. If a depth value d for a 2D image point $\mathbf{x} = (u, v)^\top \in \mathbb{R}^2$ is given, the back-projection of $\mathbf{x}$ to a 3D point $\mathbf{X} \in \mathbb{R}^3$ in the camera coordinate frame is given by:

$$\mathbf{X} = \rho(u, v, d) = \left(\tfrac{(u-c_x)d}{f_x}, \tfrac{(v-c_y)d}{f_y}, d \right)^\top, \tag{1}$$

with focal lengths f_x, f_y and coordinates of the camera center c_x, c_y.

2D Reprojection Error. Regular bundle adjustment refines both camera poses and 3D structure of the scene by minimizing the 2D reprojection error. This error is the difference between the actual 2D measurement $\bar{\mathbf{z}}_k = (u_k, v_k)^\top \in \mathbb{R}^2$ of a 3D landmark in an image and its predicted 2D projection based on the current estimates of the respective camera pose and landmark. The error function minimized in (2D) bundle adjustment is usually defined as

$$\sum_{k=1}^{K} \|h_k(C_{i_k}, \mathbf{X}_{j_k}) - \bar{\mathbf{z}}_k\|^2, \text{ with } h_k(C_{i_k}, \mathbf{X}_{j_k}) = \pi(\mathcal{T}^{-1}(C_{i_k}, \mathbf{X}_{j_k})). \quad (2)$$

where C_{i_k} and $\mathbf{X}_{j_k}$ are the variables to be optimized. We use the subscript k here to indicate that C_{i_k} and $\mathbf{X}_{j_k}$ are related by measurement k.

3D Alignment Error. To utilize the depth data provided by the RGB-D camera, we integrate depth measurements as additional constraints into bundle adjustment to improve accuracy, robustness and convergence behavior. We compute the 3D position of a landmark from its measurement $\mathbf{z}_k = (u_k, v_k, d_k)^\top$ as

$$\mathbf{Z}_k = \rho(u_k, v_k, d_k) \in \mathbb{R}^3. \quad (3)$$

This allows us to directly minimize the 3D alignment error, which is defined as the difference between the measured and the predicted 3D position of a 3D landmark in the local camera coordinate frame. The prediction is accomplished by transforming the landmark $\mathbf{X}_{j_k}$ in global coordinates back into the local camera coordinate frame using the respective absolute camera pose C_{i_k}. With this, we define the error function as

$$\sum_{k=1}^{K} \|\hat{h}_k(C_{i_k}, \mathbf{X}_{j_k}) - \mathbf{Z}_k\|^2, \text{ with } \hat{h}_k(C_{i_k}, \mathbf{X}_{j_k}) = \mathcal{T}^{-1}(C_{i_k}, \mathbf{X}_{j_k}). \quad (4)$$

The solution to this kind of NLS optimization problem is well-investigated and can be computed by applying a sparse LM algorithm. In our implementation, we employ the *Ceres Solver* [1] and *CXSparse* for solving bundle adjustment problems with the above cost functions.

5   Efficient Bundle Adjustment for RGB-D Sensors Using Submapping

With the large amount of data that arises from commodity RGB-D sensors, full bundle adjustment with its high computational complexity quickly becomes intractable due to the large time and memory consumption. In order to make the optimization more efficient, we propose a novel submapping technique tailored to RGB-D sensors, consisting of the following four processing stages:

1. Partition the RGB-D bundle adjustment graph into submaps.
2. Optimize each submap individually.
3. Align the submaps globally.
4. Optimize each submap internally with fixed separator variables.

Note that in contrast to previous work, our approach applies submapping to the 3D (RGB-D) bundle adjustment problem (instead of 2D). In this way, we can achieve high reconstruction accuracies while keeping the computational complexity small. In the following, we present each step in more detail.

Graph Partitioning into Submaps. To facilitate a submap-based optimization approach, the full optimization graph is first partitioned into L submaps. While challenging unordered image sets require advanced graph partitioning techniques [18], we apply a uniform partitioning scheme to achieve spatially coherent partitions, i.e. frames in the same partition have many common features. We split the input trajectory into segments of equal size $\tilde{M} = M/L$. To every submap, we assign a base node $B_l \in SE(3)$ with $l \in 1 \ldots L$ and initialize it to the first contained camera pose such that $B_l = C_{((l-1)\tilde{M}+1)}$. When the submaps are populated, all camera poses and landmarks are parametrized relative to the base node of the respective submap:

$$\tilde{C}_i^l = \mathcal{T}^{-1}(B_l, C_i), \quad \tilde{\mathbf{X}}_j^l = \mathcal{T}^{-1}(B_l, \mathbf{X}_j) \tag{5}$$

We can directly use the measurements as submap measurements $\tilde{\mathbf{Z}}_k^l = \mathbf{Z}_k$.

Submap Optimization. We optimize the submaps independently in the second stage to achieve local consistency. The bundle adjustment problem in each submap consists of its camera poses $\tilde{C}_l$, landmarks $\tilde{\mathbf{X}}_l$ and measurements $\tilde{\mathbf{Z}}_l$. Hence, we have to solve L small optimization problems instead of one large problem. We perform the optimization of submap l by minimizing the 3D alignment error as defined in Eq. 4 using bundle adjustment:

$$\sum_{k=1}^{K_l} ||\hat{h}_k(\tilde{C}_{i_k}^l, \tilde{\mathbf{X}}_{j_k}^l) - \tilde{\mathbf{Z}}_k^l||^2, \tag{6}$$

where K_l is the number of measurements in submap l. After the variables $\tilde{C}_i^l$ and $\tilde{\mathbf{X}}_j^l$ in all submaps have converged, we obtain an optimized local reconstruction for each submap relative to its base node B_l as depicted in Fig. 3.

Global Submaps Alignment. If a landmark of a submap is seen by a camera pose contained in another submap, it is considered a separator landmark and expressed in global world coordinates:

$$\bar{\mathbf{X}}_j^l = \mathcal{T}(B_l, \tilde{\mathbf{X}}_j^l). \tag{7}$$

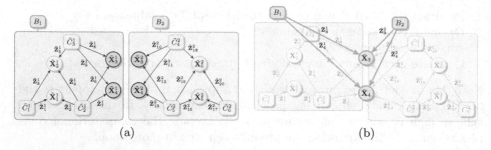

Fig. 3. Example of a SLAM graph consisting of 6 camera poses, 6 landmarks and 20 observations. (a) The map partitioned into two submaps. All camera poses and landmarks are expressed relative to their assigned base node, B_1 and B_2, respectively. (b) The global submaps alignment consists only of the base nodes and separator landmarks. The locations of the separator landmarks relative to their connected submaps' base nodes are used as measurements.

Further, we introduce inter-measurements $\bar{\mathbf{Z}}_k^l = \tilde{\mathbf{X}}_{j_k}^l$ between the submaps, which are the locations of $\bar{\mathbf{X}}_j^l$ relative to B_l. We make this approximation since we assume the relative locations of the submap landmarks to be optimal relative to the base nodes due to the performed local optimization.

After the internal submap alignment, we eliminate the global drift by moving the base nodes. Here, the local coordinate systems in the submaps play an important role, as the landmarks move with their base node and hence keep their locally optimal values from the previous stage. In the global alignment stage, we perform a full optimization of a graph consisting only of the base nodes and the separator landmarks as vertices, connected by the inter-measurements as edges as illustrated in Fig. 3(b). We omit the internal landmarks and camera poses, as they have already been used to determine the optimal relative locations of the separator landmarks in each submap. In particular, we minimize

$$\sum_{k=1}^{\bar{K}} ||\hat{h}_k(B_{l_k}, \bar{\mathbf{X}}_{j_k}) - \bar{\mathbf{Z}}_k^l||^2, \tag{8}$$

over B_l and $\bar{\mathbf{X}}_j$, where $\bar{K}$ is the number of inter-measurements of all submaps.

With the optimization graph consisting only of base nodes and a limited number of separator landmarks, the global optimization itself can be computed very efficiently. After this separator optimization, we obtain the refined poses of the base nodes and locations of the separator landmarks with respect to the global world coordinate system. The effect of moving the base node results in a reduced global drift between the submaps and a globally consistent 3D model.

Internal Submap Update. After the global alignment stage, the locations of the separator landmarks may have changed relative to the base nodes. Therefore, we optimize in the fourth stage the internal landmarks given the refined separator landmarks. To this end, we first update the relative locations of the changed separator landmarks in each submap with the changed global locations:

$$\tilde{\mathbf{X}}_k^l = \mathcal{T}^{-1}(B_l, \bar{\mathbf{X}}_k^l). \tag{9}$$

Subsequently, we perform bundle adjustment with the same input as in stage 1, by minimizing the error function defined in Eq. 6 while keeping the separator landmarks fixed. This is illustrated again in Fig. 3a, where the bold nodes $\tilde{\mathbf{X}}_3^1, \tilde{\mathbf{X}}_4^1, \tilde{\mathbf{X}}_3^2, \tilde{\mathbf{X}}_4^2$ are the separator landmarks. This optimization step allows the internal camera poses and landmarks to optimally fit to the separator while the separator landmarks stay in the same position. This usually also results in quick convergence and small movements of the internal poses and landmarks, since they were already optimized w.r.t. the base node's coordinate system.

In theory, it is reasonable to iterate over the third and the fourth step of the algorithm until convergence. In practice, we found that the results did not change significantly compared to only a single iteration.

To finally get the refined absolute camera poses and 3D landmark locations, we transform the content of each submap back into the global coordinate frame:

$$C_i = \mathfrak{T}(B_l, \tilde{C}_i^l) \quad \text{and} \quad \mathbf{X}_j = \mathcal{T}(B_l, \tilde{\mathbf{X}}_j^l) \tag{10}$$

Finally, we use the computed camera poses C_i to fuse the RGB-D frames into the 3D octree model as explained in Sect. 3.

6 Evaluation and Experimental Results

The following experimental results demonstrate that (1) 3D bundle adjustment leads to a significant improvement in accuracy over 2D bundle adjustment, pose graph optimization and other techniques, (2) submapping strongly reduces the runtime while yielding a comparable accuracy, and (3) accurate 3D models can be acquired of various real agricultural vehicles.

We evaluated our approach on the TUM RGB-D benchmark [25] to allow for a quantitative comparison of its performance with other state-of-the-art methods.

Size of Submaps. First, we evaluated the performance of our algorithm over different submap sizes to find the right trade-off between speed versus accuracy. For this, we computed the Absolute Trajectory Error (ATE) between the estimated and the ground-truth camera trajectory as defined in [25]. We evaluated the ATE over 10 sequences from the TUM RGB-D benchmark and computed its mean and standard deviation.

Figure 4 gives the result. We found that for small submap sizes, the ATE was significantly lower than the ATE of full 3D bundle adjustment. Because bundle adjustment is intrinsically prone to local minima, full (2D and 3D) bundle adjustment sometimes cannot converge from bad initial estimates. As the internal submap optimization in stage 2 already establishes local consistency before

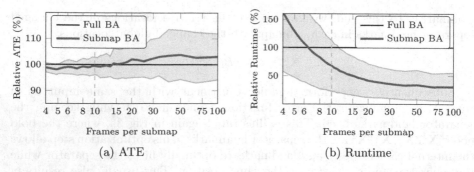

(a) ATE   (b) Runtime

Fig. 4. Performance evaluation of submapping on 10 sequences. (a) Smaller submaps lead to more accurate reconstructions, while (b) larger submaps are more efficient. The black line indicates the average performance of full bundle adjustment (no submapping), while the blue line shows the average performance of our approach w.r.t. to the submap size. The shaded area corresponds to the standard deviation over all benchmark sequences.

the global alignment, submapping can lead to better convergence and helps to find a better solution, in particular for small submap sizes. In contrast, large submap sizes increase the efficiency but decrease accuracy.

As a good compromise between speed and accuracy, we found a submap size of 10 frames to be a reasonable choice that we used in all subsequent experiments. This choice is also indicated by the dashed vertical line in the plot. In values, this submap size leads both to a lower error ($-0.5\,\%$) and faster computation ($-30\,\%$).

Benchmark Sequences. To study in more detail the effect of our cost function, we evaluated the performance of full bundle adjustment using the 2D reprojection error as well as using the 3D alignment error over 10 sequences of the TUM RGB-D benchmark and compared it to our approach. Furthermore, we also compared the performance to the RGB-D SLAM system [8] and a KinectFusion-based implementation [4] using the values reported in the respective publications.

On average, 2D bundle adjustment yields an ATE of 0.066m, while 3D bundle adjustment reduces this error to 0.047m ($-29\,\%$). At the chosen setting of 10 frames per submap, our approach achieves a similar accuracy at a significantly reduced computational cost ($-32\,\%$). Furthermore, the efficiency of bundle adjustment can be improved most for long sequences, with an improvement of up to 84 % of the runtime of full bundle adjustment.

In direct comparison to existing approaches, 3D bundle adjustment outperforms the pose graph RGB-D SLAM method by 13 % (0.047 m vs. 0.054 m) and direct SDF tracking by 17 % (0.047 m vs. 0.058 m). We believe that the higher accuracy of 3D BA in comparison to the RGB-D SLAM system stems from the fact that RGB-D SLAM performs a simplified pose-graph optimization.

KinectFusion-based methods iteratively integrate the new depth image into the signed distance function and therefore inevitably accumulate drift.

Examples of Submap-Based 3D Reconstructions. After we have demonstrated the validity of our framework, we use it to generate 3D reconstructions of three different agricultural machines. The acquired datasets show a soil auger of dimensions $1 \times 1 \times 3$ m, a lawn mower of size $1 \times 1 \times 2$ m and an older model of a Renault farm tractor with dimensions $1 \times 1 \times 3$ m. The reconstructed 3D models are illustrated in Figs. 1 and 5, together with subsets of the acquired RGB images. The generated reconstructions have a compelling visual quality and metric accuracy, which is supported by the fact that no drift is visible in the models. This makes the reconstructions suitable for visual inspection, measuring tasks and reverse-engineering.

Fig. 5. Left: 3D reconstruction of a soil auger, reconstructed from 2349 RGB-D frames. Right: 3D model of a tractor, reconstructed from 2087 RGB-D frames (Color figure online).

7 Conclusion and Future Work

We presented a novel method for 3D object reconstruction from RGB-D data that applies submapping to 3D bundle adjustment. In contrast to prior work, we thereby fully exploit the available depth information during optimization while maintaining efficiency. In an extensive quantitative evaluation on publicly available datasets, we demonstrated that our approach reduces the average runtime by 32 % compared to full bundle adjustment while achieving a similar accuracy. Furthermore, our approach outperforms several state-of-the-art approaches on benchmark datasets with respect to speed and accuracy. The 3D models of various objects reconstructed with our algorithm exhibit a compelling visual quality and metric accuracy, making it suitable for production quality control and reverse-engineering.

References

1. Agarwal, S., Mierle, K.: Ceres Solver: Tutorial & Reference. Google Inc. http://code.google.com/p/ceres-solver/
2. Agarwal, S., Snavely, N., Simon, I., Seitz, S., Szeliski, R.: Building rome in a day. In: ICCV (2009)
3. Arun, K., Huang, T., Blostein, S.: Least-squares fitting of two 3-D point sets. IEEE Trans. Pattern Anal. Mach. Intell. **5**, 698–700 (1987)
4. Bylow, E., Sturm, J., Kerl, C., Kahl, F., Cremers, D.: Real-time camera tracking and 3D reconstruction using signed distance functions. In: RSS (2013)
5. Chiuso, A., Favaro, P., Jin, H., Soatto, S.: 3-D motion and structure from 2-D motion causally integrated over time: implementation. In: Vernon, D. (ed.) ECCV 2000. LNCS, vol. 1843, pp. 734–750. Springer, Heidelberg (2000)
6. Curless, B., Levoy, M.: A volumetric method for building complex models from range images. In: SIGGRAPH (1996)
7. Davison, A.: Real-time simultaneous localisation and mapping with a single camera. In: ICCV (2003)
8. Endres, F., Hess, J., Engelhard, N., Sturm, J., Cremers, D., Burgard, W.: An evaluation of the RGB-D SLAM system. In: ICRA (2012)
9. Fuhrmann, S., Goesele, M.: Fusion of depth maps with multiple scales. ACM Trans. Graph. **30**(6), 148 (2011)
10. Henry, P., Krainin, M., Herbst, E., Ren, X., Fox, D.: RGB-D mapping: Using depth cameras for dense 3D modeling of indoor environments. ISER **20**, 22–25 (2010)
11. Henry, P., Krainin, M., Herbst, E., Ren, X., Fox, D.: RGB-D mapping: using kinect-style depth cameras for dense 3D modeling of indoor environments. Int. J. Rob. Res. (IJRR) **31**(5), 647–663 (2012)
12. Kaess, M., Johannsson, H., Roberts, R., Ila, V., Leonard, J., Dellaert, F.: iSAM2: Incremental smoothing and mapping using the Bayes tree. Int. J. Rob. Res. (IJRR) **31**, 217–236 (2012)
13. Klein, G., Murray, D.: Parallel tracking and mapping for small AR workspaces. In: ISMAR (2007)
14. Konolige, K.: Large-scale map-making. In: AAAI (2004)
15. Lim, J., Frahm, J.M., Pollefeys, M.: Online environment mapping. In: CVPR (2011)
16. Lourakis, M., Argyros, A.: SBA: a software package for generic sparse bundle adjustment. ACM Trans. Math. Softw. (TOMS) **36**(1), 2 (2009)
17. Newcombe, R., Davison, A., Izadi, S., Kohli, P., Hilliges, O., Shotton, J., Molyneaux, D., Hodges, S., Kim, D., Fitzgibbon, A.: KinectFusion: real-time dense surface mapping and tracking. In: ISMAR (2011)
18. Ni, K., Dellaert, F.: Multi-level submap based SLAM using nested dissection. In: IROS (2010)
19. Ni, K., Dellaert, F.: HyperSfM. In: 3DIMPVT (2012)
20. Ni, K., Steedly, D., Dellaert, F.: Out-of-core bundle adjustment for large-scale 3D reconstruction. In: ICCV (2007)

21. Pinies, P., Paz, L., Haner, S., Heyden, A.: Decomposable bundle adjustment using a junction tree. In: ICRA (2012)
22. Scherer, S., Dube, D., Zell, A.: Using depth in visual simultaneous localisation and mapping. In: ICRA (2012)
23. Sibley, G., Mei, C., Reid, I., Newman, P.: Adaptive relative bundle adjustment. In: RSS (2009)
24. Steinbruecker, F., Kerl, C., Sturm, J., Cremers, D.: Large-scale multi-resolution surface reconstruction from RGB-D sequences. In: ICCV (2013)
25. Sturm, J., Engelhard, N., Endres, F., Burgard, W., Cremers, D.: A benchmark for the evaluation of RGB-D SLAM systems. In: IROS (2012)
26. Triggs, B., McLauchlan, P.F., Hartley, R.I., Fitzgibbon, A.W.: Bundle adjustment – a modern synthesis. In: Triggs, B., Zisserman, A., Szeliski, R. (eds.) ICCV-WS 1999. LNCS, vol. 1883, pp. 298–372. Springer, Heidelberg (2000)
27. Whelan, T., Kaess, M., Leonard, J., McDonald, J.: Deformation-based loop closure for large scale dense RGB-D SLAM. In: IROS (2013)
28. Wurm, K., Hornung, A., Bennewitz, M., Stachniss, C., Burgard, W.: OctoMap: a probabilistic, flexible, and compact 3D map representation for robotic systems. In: ICRA Workshop on Best Practice in 3D Perception and Modeling for Mobile Manipulation (2010)
29. Zhou, Q.Y., Koltun, V.: Dense scene reconstruction with points of interest. In: SIGGRAPH (2013)
30. Zhou, Q.Y., Miller, S., Koltun, V.: Elastic fragments for dense scene reconstruction. In: ICCV (2013)

Bio-informatics

A Hierarchical Bayesian Approach for Unsupervised Cell Phenotype Clustering

Mahesh Venkata Krishna[✉] and Joachim Denzler

Computer Vision Group, Friedrich Schiller University Jena,
Ernst-Abbe-Platz 2, 07743 Jena, Germany
{mahesh.vk,joachim.denzler}@uni-jena.de

Abstract. We propose a hierarchical Bayesian model - the *wordless Hierarchical Dirichlet Processes-Hidden Markov Model* (wHDP-HMM), to tackle the problem of unsupervised cell phenotype clustering during the mitosis stages. Our model combines the unsupervised clustering capabilities of the HDP model with the temporal modeling aspect of the HMM. Furthermore, to model cell phenotypes effectively, our model uses a variant of the HDP, giving preference to morphology over co-occurrence. This is then used to model individual cell phenotype time series and cluster them according to the stage of mitosis they are in. We evaluate our method using two publicly available time-lapse microscopy video data-sets and demonstrate that the performance of our approach is generally better than the state-of-the-art.

Keywords: Hierarchical Bayesian methods · Hidden Markov Models · Cell phenotypes · Unsupervised clustering · Mitosis phase modeling · Time-lapse microscopy

1 Introduction

Machine analysis of time-lapse microscopy videos has become a very important application field of computer vision. Presence of a large amount of objects like cells, microbes etc. in these videos often make manual analysis cumbersome and prone to subjective errors. One of the main tasks for machine learning algorithms in this field is the analysis of time-lapse videos of cell culture, where *mitosis* events are happening.

Based on internal cell dynamics and morphology changes, biologists suggest that there are five main stages of mitosis: *prophase, prometaphase, metaphase, anaphase* and *telophase* ([10]: Sect. 18.6, pp. 849–851). The stage between the mitosis events is called the *interphase*. These various stages of cell life cycle are shown in Fig. 1. Given a video, the problem is to classify each cell in it according to the stage of mitosis it is in.

In biology research labs, often, different types of cells are analyzed, using different dyes and illumination methods. This makes the problem challenging, as a learning system trained on one type of cell with certain dye and illumination

© Springer International Publishing Switzerland 2014
X. Jiang et al. (Eds.): GCPR 2014, LNCS 8753, pp. 69–80, 2014.
DOI: 10.1007/978-3-319-11752-2_6

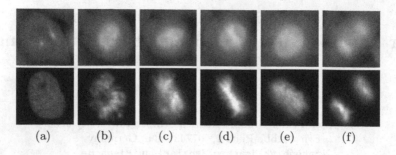

Fig. 1. The six main stages of cell life cycle: (a) *Interphase*, (b) *Prophase*, (c) *Prometaphase*, (d) *Metaphase*, (e) *Anaphase* and (f) *Telophase*. The second row in (a)–(f) show the corresponding phenotypes of the nuclei, shown in orange in the first row.

will not work well in other situations. To retrain the algorithm, an expert has to painstakingly label a new sequence and this is again time consuming, and undesirable.

Using cell population analysis tools such as the one in [2] to perform this task does not often suffice. Studying the temporal dynamics of cell phenotypes at a single cell level across a recording is an important aspect of the state-of-the-art biology research [4].

Thus, from a time-lapse microscopy video analysis system, we desire the following three properties:

1. it should be able to segregate multiple stages of the cell life cycle.
2. it should be unsupervised.
3. it should model the cell phenotypes in tracks of single cells extracted from across the video.

For unsupervised clustering tasks, *Hierarchical Dirichlet Processes* (HDP) and their variants have been used before on tasks ranging from text analysis [13] to Traffic Scene Analysis [15]. Since the traditional HDP models lack the ability to handle temporal information, using HDP to provide prior distribution for a *Hidden Markov Model* (HMM) was proposed in [13]. This HMM will model the temporal changes in cell morphologies, resulting in unsupervised clustering. To model distances in feature spaces better compared to standard topic models, Rematas *et al.* [12], proposed a Kernel Density Estimate (KDE) based scheme for LDA models. Here, the dictionary-of-visual-words representation of standard HDP models is replaced with kernel densities, making the approach *wordless*. We combine the above ideas to formulate a *wordless Hierarchical Dirichlet Processes - Hidden Markov Model* (wHDP-HMM), derive the inference procedure for the model, and develop an unsupervised method based on it to cluster cells in a time-lapse video according to the stage of life cycle they are in.

This paper is arranged as follows. In Sect. 2 we provide a brief overview of the existing literature on the problem. Section 3 discusses our temporal HDP model, and inference procedure. Experiments conducted on two publicly available data-sets are described in Sect. 4, along with results and discussions. Section 5 covers some concluding remarks and ideas for future work.

2   Previous Works

Due to its immense potential in aiding biology research, the problem of cell life cycle modeling has been extensively dealt with in the past. The approach of Yang et al. [16] performs the tasks of segmentation, tracking and mitosis detection. They use level-set methods for segmentation and tracking and use image attributes like circularity, area, average intensity etc. to classify cells under mitosis. While they report good results, it is to be noted that their approach is only effective in detecting late anaphase/telophase. Furthermore, each different sequence has to be separately analyzed and parameters over attributes need to be readjusted.

Supervised cell life stage classification problem has been dealt with previously with considerable success. Online Support Vector Classifiers are used in [14], continuously retraining the model to accommodate changing experimental conditions. But they do not consider temporal information that can be an important influence on the performance. Liu et al. [9] and Huh et al. [6] use Hidden Conditional Random Fields (HCRF) and their variants to perform mitosis detection on four stages of mitosis. Harder et al. [4] performed mitosis stage classification through a finite state machine (FSM), also accounting for abnormal shapes. The authors use cell tracks to construct time series and traverse a FSM for each track. Thus, they prevent biologically impossible results and improve the performance. Further in the time series-based methods, the HMM-based approaches of Held et al. [5] and Gallardo et al. [3] were demonstrated to perform very well. In their approach, a HMM was trained with the mitosis stages as the hidden states of the Markov model. We use this idea in our model (cf. Sect. 3).

The recent Temporally Constrained Combinatorial Clustering (TC3) scheme of [17] tackles a scenario similar to ours, and reports the best results on a publicly available data-set, according to our knowledge. Hence we use this in our performance analysis. The TC3 scheme is a combinatorial clustering scheme with biological causality constraints like "no cell goes back to the previous state of its life". Similarly to the approaches of [4,5], they segment the cells in each frame, extract various features such as shape, intensity etc., and construct multiple time series by tracking each cell using a nearest neighbor tracker. They then use the TC3 stage followed by Gaussian Mixture Model Clustering and then by a HMM stage to improve performance by correcting errors.

Whereas the approach of Zhong et al. [17] performs quite well, the clustering stages consider each individual cell time series and does not take into account the clustering in other time series. However, information regarding the causal progression of life stages in cell time series can be shared among the time series to improve the performance. This leads us to apply HDP models, as they inherently share information among various data groupings.

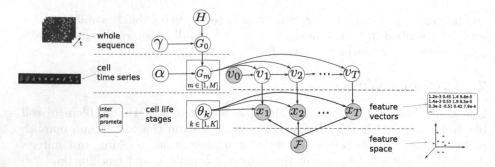

Fig. 2. Our wHDP-HMM model, with correspondences of the model components to data representation for microscopy videos.

Unlike the various methods discussed above, our model has three important aspects, namely, (i) unsupervised clustering, (ii) temporal modeling and (iii) sharing information among different time series.

3   Wordless HDP-HMM

Figure 2 shows our wHDP-HMM model. It is to be noted that, unlike the HDP-HMM models of [8,13], we limit our HDP-HMM model to single HMM. This is due to the fact that biologically, causal ordering in cell life cycle is fixed and one HMM is sufficient to model it.

3.1   Modeling Principles

We now describe the generative process within our model. As shown in Fig. 2, the topmost level is the whole sequence, which is the complete set of cell tracks, represented by a Dirichlet Process (DP), G_0. This DP is parametrized by the concentration parameter γ and the base distribution H. (a more detailed description of HDP is presented in [13]). The base distribution is set to be a Dirichlet distribution, with a hyper-parameter ζ. In the standard HDP terms, this level corresponds to the *data corpus*.

From the DP G_0, M number of cell tracks are sampled. And each cell time series is associated with a DP G_m ($m \in [1, M]$), with α as the concentration parameter and G_0 as base distribution. These correspond to *documents* in HDP models.

And these cell time series DPs provide priors for the HMM hidden states $v_{t,m}$ ($t \in [1, T]$, $m \in [1, M]$). These state variables v_t are indicator variables which denote which of the six states θ_k most likely generated the corresponding feature vectors x_t. The features are seen as originating from the feature space $\mathcal{F}$, which is a deviation from the usual HDP models where the feature words are seen as samples from a dictionary with multinomial distribution for sampling.

In the following, we omit the time series index m for variables v and x for lucidity. In HDP terms, v correspond to *topic mixtures* and x to *words*.

We have the following generative model:

$$G_0 \mid \gamma, H \sim DP(\gamma, H)$$
$$G_m \mid \alpha, G_0 \sim DP(\alpha, G_0) \quad \text{for } m \in [1, M] \tag{1}$$

And for each time series m,

$$v_t \mid v_{t-1}, G_m \sim G_m \quad \text{for } t \in [1, T]$$
$$x_t \mid v_t, \theta_k, \mathcal{F} \sim F(\theta_{v_t}) \quad \text{where } k \in [1, K] \tag{2}$$

Where $F(\cdot)$ is the prior feature distribution given the state. In line with the biological considerations as explained in Sect. 1, we set the number of states K for our problem of mitosis stage modeling to six.

3.2 Inference

Given the observed data x_t for all cell time series, to estimate the hidden states v_t, we perform Bayesian inference using the Markov Chain Monte Carlo procedure, specifically the Gibbs sampling scheme. Here, we iterate over the conditional distribution of the hidden states given the previous state and the feature vectors. This distribution can be factored as follows.

$$P(v_t = k \mid v_{t-1}, x_t) \propto$$
$$P(x_t \mid v_t = k, v_{t-1}, x_{1,\dots,(t-1),(t+1),\dots,T}) P(v_t = k \mid v_{t-1}) \tag{3}$$

Here, the first term in right hand side is the probability of word distribution, and following the idea of [12], we evaluate it through Gibbs iteration as follows.

$$P(x_t \mid v_t = k, v_{t-1}, x_{1,\dots,(t-1),(t+1),\dots,T}) \propto \frac{1}{|\Phi_{k,t' \neq t}|} \sum_{t'}^{T} \Phi_{k,t'} \cdot K(x_t, x_{t'}) \tag{4}$$

where $K(\cdot)$ is the kernel function and Φ is the $|\mathcal{F}| \times K$ matrix of features and their corresponding topic allocations.

The second term in the right hand side of (3), $P(v_t \mid v_{t-1})$ is the transition probability of the HMM. This is evaluated by first initializing it to α and then iteratively evaluating

$$P^i(v_t = k \mid v_{t-1}) \propto \frac{P^{i-1}(v_t = k \mid v_{t-1}) + \alpha \cdot \frac{P^{i-1}(v_0 = k)}{\sum_k P^{i-1}(v_0) + \gamma}}{\sum_{k' \neq k} P^{i-1}(v_t = k' \mid v_{t-1} = k) + \alpha} \tag{5}$$

where i represents the iteration index. The derivation of (5) follows similarly to the one of Infinite HMM discussed in [1].

For the overall inference task, we alternate between the Gibbs sampling and transition probability iteration at each step. The hyper-parameters α and γ are also sampled, whose details are discussed in [1,13].

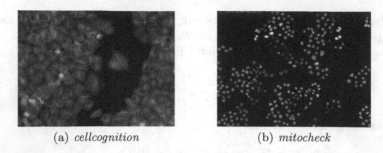

(a) *cellcognition*   (b) *mitocheck*

Fig. 3. Example frames from the two data-sets used for experimental evaluation of our method (Color figure online).

4   Experiments and Results

4.1   Data-Sets

The *cellcognition* data-set. The *cellcognition* data-set from [5,17] is a part of the *Cellcognition*[1] project. An example frame can be seen in Fig. 3(a). The reference data consists of 7 sequences of RNAi treated human HeLa Kyoto cells expressing fluorescent H2B-mCherry (orange) and α-tubulin (green). The frame resolution is 1392×1040 pixels. The data-set, in total, contains 363,120 individual cell objects, segmented and tracked using the *Cellcognition* framework. There are 9,078 cell division events in the whole data-set.

The data-set comes with ground-truth regarding the life cycle stage of each segmented cell. Furthermore, the authors provide 257-dimensional feature vectors they extracted for each cell. These features consist of various intensity, shape, and texture features. We use these features in our experiments to effectively compare our method with respect to theirs.

The *mitocheck* data-set. The *mitocheck* data-set of [11] is a part of the *mitocheck* project[2]. Overall, the database contains more than 129,500 video sequences, each of approximately 32 s, recorded at 2 fps. The frame resolution is 1344×1024 pixels. Since the data-set is extremely large, to restrict the task of evaluation to reasonable limits without giving up significantly on generality, we randomly selected a subset of 100 sequences from them, each containing on an average 64 frames.

In the publicly available data-set videos, frames contain the mCherry expressing nuclear spindles, *i.e.* the microtubule structures formed during mitosis. Thus, the task is to segment the spindles, track and extract them, and to apply our wHDP-HMM to cluster them.

Segmentation for this data-set is performed using the *ilastik*[3] framework and tracking is done using the two-stage graph optimization method of Jiang *et al.* [7].

---

[1] http://www.cellcognition.org
[2] http://www.mitocheck.org
[3] http://www.ilastik.org

Since the ground-truth for the data-set has not been made public, we have manually marked the ground-truth. The ground-truth was marked by two non-experts independently and conflicts in markings were resolved in a second iteration by one of them. Owing to the fact that the ground-truth markings were not done by biology experts, class labels were limited to the clearly defined *telophase*. Thus, evaluation on this data-set is limited to two classes, *telophase* and *non-telophase*. The data-set contains 994,163 cell objects (returned by the segmentation step) and 16,491 cell division instances.

4.2   Experimental Set-Up and Evaluation

We design our experiments in order to demonstrate our various modeling choices such as the use of KDEs as opposed to the standard co-occurrence statistics in HDP models. In the following, we describe the experimental set-up in each case.

Standard HDP. The standard HDP model is used to analyze the video. As the model lacks the capability to handle temporal information, the representation of the data is changed for this case. Here, cells are detected from each frame and features for them are extracted. These features are then used to build a dictionary with 512 visual words, using a Gaussian Mixture Model clustering algorithm. These visual words from each frame form one document for the HDP.

Once we have this grouped representation, we analyze the data through Bayesian inference, with topics representing cell states. For our experiments, we set the three hyper-parameters to 0.5. These hyperparameters are resampled based on the data later and initial values do not affect the results by a large margin.

HDP-HMM. Following the standard HDP, we test the HDP-HMM algorithm presented in [8,13]. Here, we analyze the cell time series data from tracking and make use of the temporal information. In Gibbs sampling, we use the co-occurrence statistics as described in [13]. For this, we construct a dictionary as with the standard HDP model. Thus, we see the effects of not using the cell morphology information explicitly in modeling, and only implicitly through dictionary construction. In this experiment, we set the hyper-parameter values as with the standard HDP.

wHDP-HMM. Here, we use the cell time series information as for the HDP-HMM case. However, unlike before, we now use the kernel matrix of the features extracted in stead of the co-occurrence statistics, as discussed in Sect. 3. For our experiments, we used the Gaussian Radial Basis Function (RBF) kernel, with parameter $\sigma = 1.5$, determined through cross-validation.

Evaluation Metrics. For evaluation, we follow the evaluation schemes of [17]. Thus, if TP represents true positives, TN- true negative, FP- false positive, and FN- false negative, then,

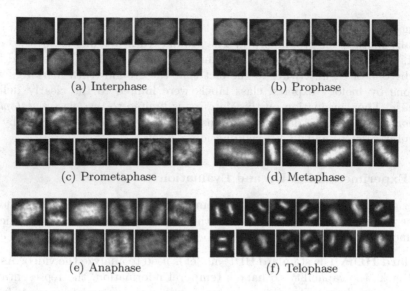

(a) Interphase   (b) Prophase

(c) Prometaphase   (d) Metaphase

(e) Anaphase   (f) Telophase

Fig. 4. Unsupervised clustering results using our wHDP-HMM model for the *cellcognition* data-set.

$$\text{Precision} = \frac{TP}{TP + FP}, \quad \text{Recall} = \frac{TP}{TP + FN}$$

$$\text{F-score} = \frac{2 \cdot (\text{Precision} \times \text{Recall})}{\text{Precision} + \text{Recall}} \tag{6}$$

For both the data-sets, we average and report the results over all the sequences involved, and additionally report the maximum deviation from it.

4.3   Results and Discussion

Results for the *cellcognition* Data-set. For the experiment involving HDP, HDP-HMM and our wHDP-HMM models, we ran 1500 iterations each of the Gibbs sampler. Our unoptimized MATLAB® code for the wHDP-HMM model, the whole process took 20 min per video of approximately 200 frames, with precalculated feature kernels. In comparison, the method of [17] took 15 min with precomputed feature data.

Figure 4 shows some unsupervised clustering results. It shows some nuclear spindle images clustered into each group by our method. As can be seen, the algorithm preforms reasonably well in separating the cell life stages. Due to their closeness in morphology and temporal ordering, *anaphase* and *telophase* are often confused. This is reflected in the quantitative results as well, as we shall see in the following.

Table 1 shows the quantitative results for the *cellcognition* data-set. For [17], we use the results reported by the authors. As can be expected, owing to its

Table 1. Results for various experiments over the *cellcognition* data-set. The numbers represent the mean values over 7 sequences and maximum variations. "[17] TC3+" implies the results of the TC3+GMM+DiscreteHMM approach of Zhong et al., as reported in the cited paper.

Precision

Algorithm	Inter	Pro	Prometa	Meta	Ana	Telo
[17] TC3+	**95.97±0.83**	83.53 ± 2.07	**91.47±2.45**	**96.82±0.92**	80.57 ± 7.67	84.57 ± 5.28
HDP	94.72 ± 1.47	74.03 ± 1.79	69.12 ± 1.05	78.23 ± 2.02	58.02 ± 6.52	51.21 ± 3.49
HDP-HMM	93.45 ± 1.03	79.25 ± 1.93	76.24 ± 1.63	81.02 ± 1.95	72.12 ± 5.67	67.02 ± 2.89
wHDP-HMM	93.11 ± 0.98	**85.43±1.47**	88.94 ± 2.85	92.25 ± 1.37	**81.44±6.73**	**86.74±3.28**

Recall

Algorithm	Inter	Pro	Prometa	Meta	Ana	Telo
[17] TC3+	99.51 ± 0.32	82.75 ± 4.13	84.43 ± 2.96	88.24 ± 3.63	80.22 ± 6.24	79.50 ± 5.09
HDP	96.13 ± 0.75	71.32 ± 2.94	69.03 ± 1.89	64.57 ± 4.31	66.18 ± 4.21	64.14 ± 6.12
HDP-HMM	98.78 ± 1.26	80.37 ± 3.12	81.52 ± 2.41	83.31 ± 4.10	79.05 ± 4.69	69.31 ± 5.64
wHDP-HMM	**99.62±0.94**	**85.01±3.25**	**88.81±1.79**	**90.02±2.72**	**82.33±5.82**	**81.12±4.62**

F-score

Algorithm	Inter	Pro	Prometa	Meta	Ana	Telo
[17] TC3+	**97.69±0.36**	82.84 ± 2.62	87.64 ± 2.35	**92.05±2.00**	80.03 ± 6.79	81.51 ± 4.70
HDP	95.14 ± 0.99	72.36 ± 2.27	68.93 ± 1.35	70.53 ± 2.95	61.71 ± 5.26	58.08 ± 4.72
HDP-HMM	95.86 ± 1.12	79.55 ± 2.35	78.43 ± 1.94	82.04 ± 2.71	75.12 ± 4.93	68.05 ± 4.26
wHDP-HMM	96.03 ± 0.95	**84.89±2.48**	**88.43±2.03**	90.92 ± 1.98	**81.65±6.02**	**83.57±3.63**

being the first stage and morphologically separated from the rest of the stages, performance for the *interphase* is better than the rest. The performances for the following three phases are quite similar to one another. The last two stages, *anaphase* and *telophase* have relatively worse performance, as their morphological closeness results in increased confusion.

Results for the *mitocheck* Data-set. Similar to the previous data-set, for the experiment involving HDP, HDP-HMM and our wHDP-HMM models, we ran 1500 iterations each of the Gibbs sampler, and one round of execution time for our wHDP-HMM was approximately 13 min for a video of 64 frames.

Figure 5 shows the unsupervised clustering for the *mitocheck* data-set. Again, it can be noted that there is greater confusion between *anaphase* and *telophase* than among other phases.

Table 2 shows the quantitative results for the *mitocheck* data-set. For [17], we use the program made public by the authors and use parameter settings used by them. Performance is measured only with respect to the *telophase*, since the ground-truth only involved the this stage. As can be seen, our method performs well compared to other schemes. Nevertheless, the added confusion in the Markov chain shows that low frame rates in video recordings of time-lapse microscopy can have detrimental effects on the performance of cell tracking based approaches like ours, due to gaps in time series.

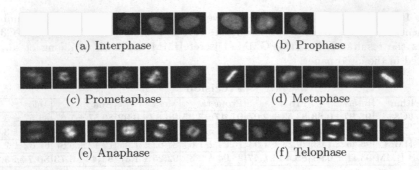

(a) Interphase   (b) Prophase

(c) Prometaphase   (d) Metaphase

(e) Anaphase   (f) Telophase

Fig. 5. Unsupervised clustering results using our wHDP-HMM model for the *mitocheck* data-set.

Table 2. Results for various experiments over the *mitocheck* data-set. The numbers represent the mean values over 100 sequences and maximum variations. "[17] TC3+" implies the results of the TC3+GMM+DiscreteHMM approach of Zhong et al., obtainable from the code made public by the authors.

Method	Precision	Recall	F-Score
[17] TC3+	83.19 ± 4.47	76.29 ± 3.52	79.41 ± 3.83
HDP	51.03 ± 3.63	62.86 ± 5.45	56.11 ± 4.45
HDP-HMM	66.47 ± 5.26	68.13 ± 5.23	67.04 ± 5.24
wHDP-HMM	$\mathbf{84.52 \pm 3.79}$	$\mathbf{79.48 \pm 4.71}$	$\mathbf{81.62 \pm 4.02}$

5   Conclusions

We presented the wHDP-HMM model and as discussed Sect. 1, used it to perform the task of unsupervised mitosis stage modeling in time-lapse microscopy videos, represented in terms of temporal tracks. In our model, HDP provides prior distributions for the HMM, making the system unsupervised and able to handle temporal information. To directly handle cell phenotypes, we replace the co-occurrence based sampling scheme of standard HDP- HMM models with one based on kernel density estimates.

We demonstrated the performance of our method using two publicly available data-sets: *cellcognition* and *mitocheck*. The results compared favorably with the state-of-the-art.

The Bayesian inference step is a time-consuming operation and in deployment scenarios, maybe undesirable. To handle this situation, one can use a two-stage approach, combining our generative model for training and a discriminative classifier for testing. Furthermore, it will be of immense practical use if it is possible to extend the model to jointly perform tracking and clustering. Another interesting idea for future research is to extend the model to allow a certain degree of user interaction, so that an expert user can provide a few inputs to improve system performance.

Acknowledgments. The authors gratefully acknowledge financial support by ZEISS and would like to thank Christian Wojek and Stefan Saur (ZEISS Corporate Research and Technology) for helpful discussions and suggestions.

References

1. Beal, M.J., Ghahramani, Z., Rasmussen, C.E.: The infinite hidden markov model. In: Advances in Neural Information Processing Systems (NIPS), pp. 577–584 (2002)
2. Carpenter, A.E., Jones, T.R., Lamprecht, M.R., Clarke, C., Kang, I.H., Friman, O., Guertin, D.A., Chang, J.H., Lindquist, R.A., Moffat, J., Golland, P., Sabatini, D.M.: Cellprofiler: image analysis software for identifying and quantifying cell phenotypes. Genome Biol. **7**(10), R100 (2006)
3. Gallardo, G.M., Yang, F., Ianzini, F., Mackey, M., Sonka, M.: Mitotic cell recognition with hidden markov models. In: Proceedings of SPIE, vol. 5367, 661–668 (2004)
4. Harder, N., Mora-Bermúdez, F., Godinez, W.J., Wünsche, A., Eils, R., Ellenberg, J., Rohr, K.: Automatic analysis of dividing cells in live cell movies to detect mitotic delays and correlate phenotypes in time. Genome Res. **19**(11), 2113–2124 (2009)
5. Held, M., Schmitz, M.H.A., Fischer, B., Walter, T., Neumann, B., Olma, M.H., Peter, M., Ellenberg, J., Gerlich, D.W.: Cellcognition: time-resolved phenotype annotation in high-throughput live cell imaging. Nat. Methods **7**(9), 747–754 (2010)
6. Huh, S., Chen, M.: Detection of mitosis within a stem cell population of high cell confluence in phase-contrast microscopy images. In: IEEE Conference on Computer Vision and Pattern Recognition (CVPR), pp. 1033–1040 (2011)
7. Jiang, X., Haase, D., Körner, M., Bothe, W., Denzler, J.: Accurate 3D multi-marker tracking in X-ray cardiac sequences using a two-stage graph modeling approach. In: Wilson, R., Hancock, E., Bors, A., Smith, W. (eds.) CAIP 2013, Part II. LNCS, vol. 8048, pp. 117–125. Springer, Heidelberg (2013)
8. Kuettel, D., Breitenstein, M.D., Gool, L.V., Ferrari, V.: What is going on? discovering spatiotemporal dependencies in dynamic scenes. In: Proceedings of the IEEE Conference on Computer Vision and Pattern Recognition (CVPR) (2010)
9. Liu, A.A., Li, K., Kanade, T.: Mitosis sequence detection using hidden conditional random fields. In: IEEE International Symposium on Biomedical Imaging: From Nano to Macro, pp. 580–583, April 2010
10. Lodish, H., Berk, A., Kaiser, C.A., Krieger, M., Bretscher, A., Ploegh, H., Amon, A., Scott, M.P.: Molecular Cell Biology, 7th edn. W.H.Freeman & Co Ltd, New York (2013)
11. Neumann, B., Walter, T., Heriche, J.K., Bulkescher, J., Erfle, H., Conrad, C., Rogers, P., Poser, I., Held, M., Liebel, U., Cetin, C., Sieckmann, F., Pau, G., Kabbe, R., Wuensche, A., Satagopam, V., Schmitz, M.H.A., Chapuis, C., Gerlich, D.W., Schneider, R., Eils, R., Huber, W., Peters, J.M., Hyman, A.A., Durbin, R., Pepperkok, R., Ellenberg, J.: Phenotypic profiling of the human genome by time-lapse microscopy reveals cell division genes. Nature **464**(7289), 721–727 (2010)
12. Rematas, K., Leuven, K., Fritz, M., Tuytelaars, T.: Kernel density topic models: visual topics without visual words. In: Modern Non Parametric Methods in Machine Learning, NIPS Workshop (2012)
13. Teh, Y., Jordan, M., Beal, M., Blei, D.: Hierarchical dirichlet processes. J. Am. Stat. Assoc. **101**, 1566–1581 (2006)

14. Wang, M., Zhou, X., Li, F., Huckins, J., King, R.W., Wong, S.T.: Novel cell segmentation and online svm for cell cycle phase identification in automated microscopy. Bioinformatics **24**(1), 94–101 (2008)
15. Wang, X., Ma, X., Grimson, W.: Unsupervised activity perception in crowded and complicated scenes using hierarchical bayesian models. IEEE Trans. Pattern Anal. Mach. Intell. **31**(3), 539–555 (2009)
16. Yang, F., Mackey, M.A., Ianzini, F., Gallardo, G., Sonka, M.: Cell segmentation, tracking, and mitosis detection using temporal context. In: Duncan, J.S., Gerig, G. (eds.) MICCAI 2005. LNCS, vol. 3749, pp. 302–309. Springer, Heidelberg (2005)
17. Zhong, Q., Busetto, A.G., Fededa, J.P., Buhmann, J.M., Gerlich, D.W.: Unsupervised modeling of cell morphology dynamics for time-lapse microscopy. Nat. Methods **9**(7), 711–713 (2012)

Information Bottleneck for Pathway-Centric Gene Expression Analysis

David Adametz[✉], Mélanie Rey, and Volker Roth

Department of Mathematics and Computer Science, University of Basel,
Basel, Switzerland
{david.adametz,melanie.rey,volker.roth}@unibas.ch

Abstract. While DNA microarrays enable us to conveniently measure expression profiles in the scope of thousands of genes, the subsequent association studies typically suffer from a tremendous imbalance between number of variables (genes) and observations (subjects). Even more so, each gene is heavily perturbed by noise which prevents any meaningful analysis on the single-gene level [6]. Hence, the focus shifted to pathways as groups of functionally related genes [4], in the hope that aggregation potentiates the underlying signal. Technically, this leads to a problem of feature extraction which was previously tackled by principal component analysis [5]. We reformulate the task using an extension of the Meta-Gaussian Information Bottleneck method as a means to compress a gene set while preserving information about a *relevance* variable. This opens up new possibilities, enabling us to make use of clinical side information in order to uncover hidden characteristics in the data.

1 Introduction

Gene expression analysis is concerned with finding individual genes that are connected to a given clinical trait or disease pattern. Typically, however, this task is inherently difficult since (i) a single gene is highly prone to noise and (ii) the number of genes heavily outweighs the number of subjects. Due to these circumstances it is possible to observe genes which perfectly explain an outcome, but cannot be reproduced independently, see Ein-Dor et al. [6].

Therefore, in an attempt to alleviate said shortcomings, the focus was turned to *pathways* [4], which are formally defined as high-level systemic functions in biological processes. These can be categorized into five groups: metabolism, genetic and environmental information processing, cellular processes, organisimal systems and human diseases. For our purposes, a pathway is simply a set of genes given by biological prior knowledge. The underlying idea is that genes will exhibit detectable patterns *only* when grouped with *functionally* related entities. As a result, the conceptual pipeline in gene expression analysis is now extended by an intermediate step to identify common features within such a group. Our focus will be solely on this process of data fusion and feature extraction.

A recognized solution was introduced by Drier et al. [5] and implemented in the software package *Pathifier*. Here, principal component analysis (PCA) is

© Springer International Publishing Switzerland 2014
X. Jiang et al. (Eds.): GCPR 2014, LNCS 8753, pp. 81–91, 2014.
DOI: 10.1007/978-3-319-11752-2_7

$$Z_X = \begin{array}{c} \\ \text{subject 1} \\ \text{subject 2} \\ \text{subject 3} \\ \vdots \\ \text{subject } n \end{array} \begin{bmatrix} \text{gene 1} & \cdots & \text{gene } p \\ -0.25 & \cdots & 0.72 \\ -0.02 & \cdots & -1.49 \\ 0.82 & \cdots & 1.23 \\ \vdots & \ddots & \vdots \\ -0.17 & \cdots & 1.89 \end{bmatrix}, \quad Z_Y = \begin{bmatrix} \text{age} & \text{sex} & \substack{\text{tumor} \\ \text{stage}} & \cdots & \text{trait } q \\ 57 & M & A & \cdots & 1.21 \\ 38 & M & C & \cdots & 0.67 \\ 63 & F & ? & \cdots & ? \\ \vdots & \vdots & \vdots & \ddots & \vdots \\ 48 & M & D & \cdots & ? \end{bmatrix}$$

Fig. 1. Data scheme considered throughout the paper. Matrix Z_X contains the expression values of p genes on a pathway across n patients/subjects. All its entries are continuous. In addition, matrix Z_Y stores the corresponding clinical information over q different traits/features. Typically, this includes discrete, categorical and/or continuous variables. Question marks denote missing values.

used to project all genes of a pathway onto the directions of largest variance. While preserving variance is one possible choice to find a compact representation, it can very well be contradictory to what is of interest in a biological sense. Hence, the question is: What is a substitute for variance and how do we specify such an interest? In that regard, note that we often have access to rich clinical side information, although it is typically not used at this stage of an analysis. This is a prime example for the *Information Bottleneck*, where a variable X (e.g. a gene) is compressed while maintaining information about a second variable Y (a clinical feature, e.g. tumor stage). Hence, it gives rise to the notion of *relevance* in the data.

For a better understanding of the problem, Fig. 1 depicts the structure and composition of our data. Hereby, we think of each gene as a random variable X_j where every subject is an independent realization. Further, we combine these into the p-dimensional random vector X with observations stored in rows of matrix Z_X. The same holds for q clinical variables in random vector Y and the corresponding data matrix Z_Y.

Applying the Information Bottleneck to pathway-centric analysis entails a number of challenges. As can be seen in Fig. 1, gene expression values are continuous and in general follow unknown distributions. Clinical data, however, can exhibit various forms: continuous (viral load), discrete (age), categorical (tumor stage) or binary (sex). Additionally, parts of information could be missing due to practical or ethical reasons. In the following, we introduce the Information Bottleneck and an extension that is capable of handling the above data scheme.

2   Information Bottleneck (IB)

Consider the random vectors X and Y. The framework of the Information Bottleneck [17] is concerned with finding a compression T of X while maintaining information about Y. This leads to an optimization problem

$$\min_{p(t|x)} \mathcal{L} \mid \mathcal{L} \equiv I(X;T) - \beta\, I(T;Y), \tag{1}$$

where $\mathcal{L}$ denotes the likelihood, $I(\cdot)$ is the mutual information and $\beta > 0$ determines the tradeoff between best compression of X and being most informative about Y. For the general problem, there exists no closed-form solution.

2.1   Gaussian Information Bottleneck (GIB)

An interesting special case arises, when X and Y are jointly distributed Gaussian random vectors of dimension p and q respectively (see Chechik et al. [3]):

$$\begin{bmatrix} X \\ Y \end{bmatrix} \sim \mathcal{N}\left(\mathbf{0}_{p+q}, \Sigma\right) \quad \text{with} \quad \Sigma = \begin{bmatrix} \Sigma_x & \Sigma_{xy} \\ \Sigma_{yx} & \Sigma_y \end{bmatrix}. \tag{2}$$

It can be shown that the compression is also Gaussian [8]

$$T = AX + \xi, \tag{3}$$

given by projection matrix $A \in \mathbb{R}^{p \times p}$ and noise component $\xi \sim \mathcal{N}(\mathbf{0}_p, \Sigma_\xi)$. Therefore, rewriting Eq. (1) using Gaussian random vectors leads to the problem of finding the optimal pair (A, Σ_ξ). For $\Sigma_\xi \equiv I_p$[1], we have

$$A = \left[\alpha_1 v_1^t \; ; \; \alpha_2 v_2^t \; ; \; \ldots \; ; \; \alpha_p v_p^t\right], \tag{4}$$

where v_i and $\lambda_i \leq 1$ are the left eigenvectors and -values of $p \times p$ matrix

$$B = \Sigma_{x|y}\Sigma_x^{-1} = I_p - \Sigma_{xy}\Sigma_y^{-1}\Sigma_{yx}\Sigma_x^{-1} \tag{5}$$

in the order of $\lambda_1 < \ldots < \lambda_p$. The scaling of each row in A is given by

$$\alpha_i = \begin{cases} \sqrt{\frac{\beta(1-\lambda_i)-1}{\lambda_i v_i^t \Sigma_x v_i}} & \text{if } \beta > (1 - \lambda_i)^{-1} \\ 0 & \text{else.} \end{cases} \tag{6}$$

In the limit of $\beta \equiv 0$, we have $\alpha_i = 0 \; \forall i$ and thus $A = \mathbf{0}_{p \times p}$. As β increases, a growing number of α_i becomes non-zero and A is filled row-wise from top to bottom. For a sufficiently large $\beta > (1 - \lambda_p)^{-1}$, all v_i are present, meaning T is most informative about Y.

   This concludes the results for Gaussian random vectors. Speaking in terms of gene expression data, we seek the joint covariance between genes and clinical features. Since the assumption of normality is too restrictive, the next section explores a solution.

2.2   Meta-Gaussian Information Bottleneck (MGIB)

The Gaussian formulation of the IB leads to an analytic solution, however, its scope is severely limited in light of our present setting. In general, gene expression

---

[1] (A, Σ_ξ) and (A^*, I_p) are equivalent under $\mathcal{L}$ with a linear transformation between A and A^*.

data—while being continuous—does not necessarily obey normality. Even more critically, clinical information is far from being Gaussian. To that effect, Rey and Roth [13] greatly relaxed the problem as shown in the following.

Using Sklar's theorem [15], every multivariate distribution can be expressed in terms of its univariate margins and their dependence structure. This separation allows us to analyze both properties independently from one another. In general, if random vector (X, Y) follows joint distribution F, it decomposes as

$$F(x_1, \ldots, x_p, y_1, \ldots, y_q) = C\left(F_1(x_1), \ldots, F_p(x_p), F_{p+1}(y_1), \ldots, F_{p+q}(y_q)\right), \quad (7)$$

where $F_j(\cdot) =: u_j \in [0,1]$ is a univariate margin and $C(\cdot)$ is the *copula* — a mapping $[0,1]^{p+q} \mapsto [0,1]$. In the Gaussian case, both margins $\Phi(\cdot)$ and copula $C_P(\cdot)$ with density $c_P(\cdot)$ are Gaussian. Since the Gaussian copula only depends on correlation matrix P, we explicitly denote the subscript $_P$. This is a crucial observation regarding $I(X) = -H(c_{P_x})$, which means the GIB problem can be expressed solely using copula densities. Consequently, as the margins $\Phi(\cdot)$ are not involved anymore, they can be replaced by arbitrary distributions $F(x_i)$ *without changing the optimal solution*. Hereby, we arrive at the family of so-called *Meta-Gaussian* distributions, which all have the Gaussian copula in common, but allow arbitrary margins. Analogous to GIB, the optimal projection A is found by the eigenvectors and -values of $B = I_p - P_{xy} P_y^{-1} P_{yx} P_x^{-1}$.

The goal now is to estimate the joint correlation matrix P of X and Y from their n instances stored in data matrix Z_X and Z_Y. Typically, though, copula models are studied strictly in the context of continuous margins for which there exist robust estimators. Regarding pathway analysis, this approach will fail for discrete or categorical variables, hence, these data types (and mixed data in particular) must be treated differently.

3 Extensions for Pathway Analysis

3.1 Mixed Data and Missing Values

Empirically estimating the margins of gene expression values is a convenient way to derive P, however, such a scheme is only applicable for continuous variables, since it does not consider ties (i.e. non-unique values). We follow the approach of Hoff [9], which avoids to specify the margins altogether. Here, it is assumed that each discrete variable Y_j is a function of a *hidden Gaussian variable* $\bar{Y}_j$. In contrast to a continuous variable, the mapping $\bar{Y}_j \mapsto Y_j$ is surjective and thus leads to ties among the realizations of Y_j. We can use the fact that regardless of the specifics of this mapping (i.e. the margin), $(\bar{Y}_j, Y_j)$ are always linked via the ordering of their instances. In other words, $y_{j1} > y_{j2}$ implies $\bar{y}_{j1} > \bar{y}_{j2}$. Hence, the model is

$$(\bar{X}_1, \ldots, \bar{X}_p, \bar{Y}_1, \ldots, \bar{Y}_q) \sim \mathcal{N}(\mathbf{0}_{p+q}, P) \quad (8)$$

$$\text{with} \quad X_m = F_m^{-1}\left(\Phi(\bar{X}_m)\right) \quad \text{and} \quad Y_j = F_j^{-1}\left(\Phi(\bar{Y}_j)\right). \quad (9)$$

Note that due to surjectivity, the converse $\bar{Y}_j = \Phi^{-1}\left(F_j(Y_j)\right)$ is not unique and therefore violates the requirements of a copula.

In order to make inference about P, Hoff [9] proposed a Gibbs sampling scheme, which can be used to obtain an estimate by averaging. More specifically, data matrix $\bar{Z}$ of latent variables $(\bar{X}, \bar{Y})$ follows truncated Gaussians (due to the constraint of ordering) combined with a conjugate Wishart prior for P. For missing values Z_{ij}, the truncation is replaced by a standard Gaussian without bounds, thereby imputing the latent values. Algorithm 1 shows a computationally more efficient variant of the original sampler that avoids matrix inversion in the innermost loop. Here, subscript $_{-j}$ is a shorthand for set $\{1, \ldots, (p+q)\} \setminus j$.

Algorithm 1. Sampling correlation matrix P and data matrix $\bar{Z}$

Input: $n \times (p+q)$ data matrix $Z = [Z_X, Z_Y]$
Set $B_0 \leftarrow I_{p+q}$, $B \leftarrow I_{p+q}$, $\nu \leftarrow (p+q)+1$
Initialize $\bar{Z} = \left\{ \bar{Z}_{\bullet j} \leftarrow \Phi^{-1}\left(\frac{1}{\#\text{levels}+1}\text{ranks}(Z_{\bullet j})\right) \,\Big|\, j \in \{1, \ldots, (p+q)\} \right\}$
for $k = 1 \ldots N_{\text{samples}}$ **do**
    for variable $j = 1 \ldots (p+q)$ **do**
        Set $\sigma \leftarrow 1/B_{jj}$
        for level r in $Z_{\bullet j}$ **do**
            Find lower bound: $a \leftarrow \max\{\bar{Z}_{\bullet j} | Z_{\bullet j} < r\}$
            Find upper bound: $b \leftarrow \min\{\bar{Z}_{\bullet j} | Z_{\bullet j} > r\}$
            for every $i \in \{1, \ldots, n\}$ where $Z_{ij} = r$ **do**
                Set $\mu_{ij} \leftarrow -\bar{Z}_{i,-j}B_{-j,j}/\sigma$
                Sample from a truncated Gaussian: $\bar{Z}_{ij} \sim \mathcal{TN}(\mu_{ij}, \sigma^2, a, b)$
            end for
        end for
    end for
    Sample $B \sim \mathcal{W}\left(\nu + n, \left[B_0 + \bar{Z}^t\bar{Z}\right]^{-1}\right)$
    Compute $P = \left\{ P_{ij} \leftarrow (B^{-1})_{ij}/\sqrt{(B^{-1})_{ii}(B^{-1})_{jj}} \,\Big|\, (i,j) \in \{1, \ldots, (p+q)\} \right\}$
    Compute $\bar{Z}_{\text{norm}} = \left\{ (\bar{Z}_{\text{norm}})_{\bullet j} \leftarrow \bar{Z}_{\bullet j}/\sqrt{(B^{-1})_{jj}} \,\Big|\, j \in \{1, \ldots, (p+q)\} \right\}$
end for

With regard to the MGIB compression, we are interested in data matrix $\bar{Z}$ of latent variables, which is only a by-product in [9]. This is due to the problem that categorical values (e.g. high, medium, low) cannot be used in conjunction with projection A. Therefore, we estimate the hidden variables $\bar{Z}$ by averaging over samples, analogous to the scheme for P. As the individual $\bar{Z}$ samples are not bound by scale, however, we use an additional normalizing step ($\bar{Z}_{\text{norm}}$) as shown in the algorithm.

Although the sampling scheme was initially motivated by discrete or categorical margins, it also handles continuous variables, and, hence, the setup is suitable for mixed data in general.

3.2 Relation to PCA

Interestingly, (M)GIB is related to principal component analysis (PCA), which is a projection that preserves variance in the data. If data matrix Z_X has zero-mean columns, its PC transformation is given by

$$Z_{PCA} = V^t Z_X \quad \text{with} \quad \Sigma_x = V \Lambda V^t. \tag{10}$$

Here, V and Λ refer to eigenvectors and -values of Σ_x. In (M)GIB, we define Y to be a noisy representation of X, i.e. $Y = X + \epsilon$ with $\epsilon \sim \mathcal{N}(\mathbf{0}_p, I_p)$. This means, we have

$$\Sigma_y = \Sigma_x + I_p \tag{11}$$
$$\Sigma_{xy} = \Sigma_{yx} = \Sigma_x \tag{12}$$

leading to

$$B = I_p - \Sigma_{xy} \Sigma_y^{-1} \Sigma_{yx} \Sigma_x^{-1} \tag{13}$$
$$= I_p - \Sigma_x (\Sigma_x + I_p)^{-1} \tag{14}$$
$$= I_p - V \Lambda V^t (V \Lambda V^t + I_p)^{-1} \tag{15}$$
$$= V(I_p - \Lambda [\Lambda + I_p]^{-1}) V^t \tag{16}$$
$$= V \widetilde{\Lambda} V^t. \tag{17}$$

In this form, it is easy to see that $\widetilde{\Lambda}$ and V correspond to the eigendecomposition of B, therefore the directions coincide with PCA. For a sufficiently large β, A contains all eigenvectors v_i. Note, however, that the compression T will differ from the PC projection in terms of scales α_i (as a function of β).

3.3 Irrelevance Variables

Aside from the IB concept to compress a random variable with respect to a relevance entity, one might also be interested in the converse, meaning a compression that is *explicitly devoid* of some information. In a practical setting, this may include sex and/or age, which often have an undesirable effect on the data.

Conventionally, one would perform multiple regression and control for the variables in question, where categorical inputs require some form of dummy coding. However, as we deal with multiple variables of possibly many levels each, this approach quickly becomes complicated and cumbersome.

In contrast, MGIB offers an alternative which naturally arises from the IB framework. We refer to these features as *irrelevance variables* $\overline{Y} = (\overline{Y}_1, \ldots, \overline{Y}_k)$. In a first step, we compress X with regards to $\overline{Y}$ and hereby obtain an optimal projection A. This specifically captures all unwanted information about X, thus we compute the *nullspace* of A:

$$Q = \text{Null}(A) \in \mathbb{R}^{p \times p} \quad \Leftrightarrow \quad Q^t A = 0_{p \times p}. \tag{18}$$

As a result, compression $T = QX$ cancels all information about $\overline{Y}$. Notice the difference between excluding a variable from the relevance set Y and explicitly treating it as nuisance $\overline{Y}$.

4   Experiments

MGIB is a general tool for data fusion and feature extraction with the flexibility to specify which aspect is of interest to the user. Therefore, it can either be used as preprocessor to reduce dimensionality or as a self-contained module in an existing pipeline.

For the purpose of visualizing, we integrate the method into the previously mentioned Pathifier software package, thereby replacing PCA. The conventional workflow is as follows: Pathifier takes all gene expression values associated with a given pathway and projects them onto their first few principal components. This produces a point cloud, where each point represents a subject living in a reduced set of dimensions. The shape of the cloud carries meaningful information and is captured by the so-called *principal curve*, which is the same as finding the underlying 'skeleton'. When projecting all points onto this path, we can measure the distance between two points (= subjects) *along* the curve. Pathifier assumes two groups of subjects—healthy and diseased—, where the average of healthy subjects serves as a reference point. The distance of a diseased subject to this reference gives the *deregulation score*, the final output of Pathifier. Finally, the subject with maximum distance defines a score of 1.

We use the publicly available colon cancer dataset of Sheffer et al. [14], which consists of $p = 13\,437$ genes measured for $n = 313$ subjects. The latter can be classified into 53 healthy (H), 184 primary tumor (T), 46 polyp (P) and 30 metastasis (M) patients. A separate clinical data table contains the properties age, sex and TNM tumor staging (T = size $\in \{1, 2, 3, 4\}$, N = spread to regional lymph nodes $\in \{0, 1, 2, 3\}$, M = presence of distant metastasis $\in \{0, 1\}$). Since healthy subjects are not affected by cancer, their disease-related values are zero.

As for pathways, we use KEGG[2] gene sets obtained from the curated Molecular Signature Database (MSigDB[3]), version 4.0. In general, these sets contain everything from 5 genes up to 200, thus, for reasons of tractability regarding 313 subjects, we limit the maximum to 60 genes. If a pathway exceeds this number, we cluster the pairwise mutual information using Ward's method and cut the resulting tree on the level of 60 clusters. For each cluster, the gene with highest entropy is selected as a representative.

As we aim for a pathway compression T being most informative about clinical data Y, β is always set to a small value that attains the maximum number of eigenvectors in A. Further increasing β would only alter the scaling, which is not of interest to us.

4.1   Comparison to Classic Pathifier Using PCA

In order to highlight the flexibility of MGIB, we first attempt to recreate the PC projection. This is achieved when Y is a noisy representation of X, i.e. we aim to compress X while maintaining information about itself. The scheme is:

---

[2] http://www.genome.jp/kegg/
[3] http://www.broadinstitute.org/gsea/msigdb/

1. Find all genes corresponding to a pathway, store their expression values in Z_X and a noisy copy in Z_Y.
2. Compress X with regards to Y in order to receive $Z_T = \bar{Z}_X A^t$.
3. Run Pathifier on Z_T to obtain the deregulation score for each subject.

We choose a set of 3 pathways that are directly related to (colon) cancer and run Pathifier on both conventional PC projection and MGIB compression. As can be seen in Fig. 2, this leads to fairly similar deregulation scores, the differences being mainly due to scaling α_j.

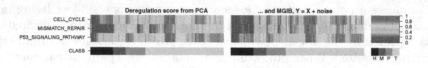

Fig. 2. Pathifier using PCA (left) and MGIB compression T (right). Each row is a pathway and the columns correspond to subjects (same ordering left and right). Subject classes are healthy (H), metastasis (M), polyp (P) and primary tumor (T).

4.2   Analysis with Relevance and Irrelevance Variables

The following experiment shows how to apply MGIB in a practical context to uncover specific characteristics in the data. In particular, the compression is supposed to highlight aspects related to the TNM tumor stage while discarding age and sex of a subject. The required steps are:

1. Find all genes corresponding to a pathway, store expression values in Z_X.
2. Compress X while removing $\overline{Y} = $ (age, sex), calculate $Z_{T_1} = \bar{Z}_X Q_1^t$.
3. Compress T_1 with regards to $Y = $ (T, N, M) and receive $Z_{T_2} = \bar{Z}_{T_1} A_2^t$.
4. Run Pathifier on Z_{T_2} to obtain the deregulation score for each subject.

For comparison purposes, we follow this scheme with and without irrelevance variables. The outcome of both experiments can be interpreted best when sorting the subjects according to age, sex and class. Hence, we report the same deregulation scores 3 times, but only show pathways which are known to be related to the ordered variable.

Irrelevance Variable Age. Figure 3 depicts age-related pathways, which are involved in inflammatory response [10], immune system [2,11] and development of blood vessels (angiogenesis) [1]. The effect ranges from cancelling to amplifying scores of certain age groups.

Irrelevance Variable Sex. Sorting the subjects according to sex results in the score patterns of Fig. 4. Pathways known to be affected by sex cover the categories colorectal cancer [12,16], immune system and cellular processes [7]. As can be seen, subjects are sometimes normalized to a common level (cell cycle) and sometimes put into contrast (chemokine signaling).

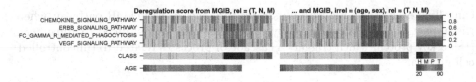

Fig. 3. Impact of age on the deregulation scores: with irrelevance variables (right) and without (left). Subjects (= columns) are sorted by age.

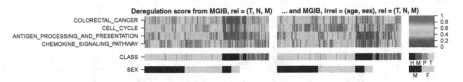

Fig. 4. Impact of sex on the deregulation scores: with irrelevance variables (right) and without (left). Columns are ordered by sex.

Relevance Variables. In a practical analysis, one would evaluate the scores as in Fig. 5. Here, we clearly see two distinct subgroups of primary tumor patients (T), possibly related to stage M = 0. Also, patients of class metastasis (M) appear to be similar to those of primary tumor (T) with stage M = 1.

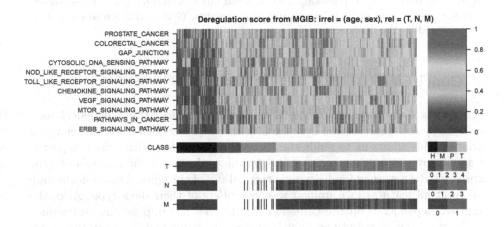

Fig. 5. Deregulation scores for cancer-related pathways, sorted by patient class. The MGIB compression used both relevance and irrelevance variables.

Pathway Map. The MGIB compression can also be analyzed in terms of the projection weights to identify the importance of single genes. Since we used two compression steps (irrelevance and relevance variables), the combined projection of $T_1 = Q_1 X$ and $T_2 = A_2 T_1$ is $T_2 = A_2 Q_1 X$. Therefore, the contribution of gene j is found by $\sum_{i=1}^{p} |(A_2 Q_1)_{ij}|$, where the absolute value prevents weights from cancelling each other. Figure 6 illustrates the map of the mTOR pathway, where the impact of genes is overlayed by color. This pathway is linked to *Rapamycin* [18], a key protein which triggers a cascade of processes related to (uncontrolled) cell growth.

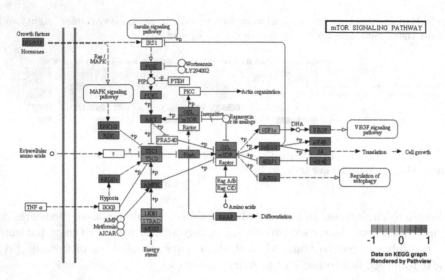

Fig. 6. Map of the mTOR pathway. Each squared box represents a gene and is colored according to the sum of absolute compression weights. Genes with white color were not measured by the microarray.

5   Conclusion

We presented an extension to the Meta-Gaussian Information Bottleneck as a general way to compress a random variable in the presence of a second relevance variable. This allows us to guide the compression and formalize which aspect is of actual interest to the user. Applied to the pathway-centric analysis of gene expression data, we can finally make practical use of the clinical side information and treat all features in a unified way *regardless* of their data type. Also, the exclusive dependence on the Gaussian copula does not impose any restrictions on the marginal distributions, which makes the setup widely applicable, even where data are missing.

Using software package Pathifier, the experiments demonstrate that the approach can easily be incorporated into a given workflow. Not only is it possible to mimic principal component analysis as a special case, but the method can also conveniently remove nuisance variables in an information-theoretic fashion. In summary, we believe this tool greatly enhances gene expression analysis, while at the same time being suitable for a wide class of biological applications.

References

1. Baffert, F., Thurston, G., Rochon-Duck, M., Le, T., Brekken, R., McDonald, D.M.: Age-related changes in vascular endothelial growth factor dependency and angiopoietin-1-induced plasticity of adult blood vessels. Circ. Res. **94**(984), 984–992 (2004)
2. Castle, S.C.: Clinical relevance of age-related immune dysfunction. Clin. Infect. Dis. **31**(2), 578–585 (2000)
3. Chechik, G., Globerson, A., Tishby, N., Weiss, Y.: Information Bottleneck for Gaussian Variables. J. Mach. Learn. Res. **6**, 165–188 (2005)
4. Curtis, R.K., Oresic, M., Vidal-Puig, A.: Pathways to the analysis of microarray data. Trends Biotechnol. **23**(8), 429–435 (2005)
5. Drier, Y., Sheffer, M., Domany, E.: Pathway-based personalized analysis of cancer. Proc. Natl. Acad. Sci. **110**(16), 6388–6393 (2013)
6. Ein-Dor, L., Zuk, O., Domany, E.: Thousands of samples are needed to generate a robust gene list for predicting outcome in cancer. Proc. Natl. Acad. Sci. **103**, 5923–5928 (2006)
7. Elsaleh, H., Joseph, D., Grieu, F., Zeps, N., Spry, N., Iacopetta, B.: Association of tumour site and sex with survival benefit from adjuvant chemotherapy in colorectal cancer. Lancet **355**(9217), 1745–1750 (2000)
8. Globerson, A., Tishby, N.: On the Optimality of the Gaussian Information Bottleneck Curve. The Hebrew University of Jerusalem. Technical report (2004)
9. Hoff, P.D.: Extending the rank likelihood for semiparametric copula estimation. Ann. Appl. Stat. **1**(1), 265–283 (2007)
10. Licastro, F., Candore, G., Lio, D., Porcellini, E., Colonna-Romano, G., Franceschi, C., Caruso, C.: Innate immunity and inflammation in ageing: a key for understanding age-related diseases. Immun. Ageing **2**(1), 8 (2005)
11. Migliore, L., Coppede, F.: Genetic and environmental factors in cancer and neurodegenerative diseases. Mutat. Res. **512**(2–3), 135–153 (2002)
12. Pal, S.K., Hurria, A.: Impact of age, sex, and comorbidity on cancer therapy and disease progression. J. Clin. Oncol. **28**(26), 4086–4093 (2010)
13. Rey, M., Roth, V.: Meta-gaussian information bottleneck. Adv. Neural Inf. Process. Syst. **25**, 1925–1933 (2012)
14. Sheffer, M., Bacolod, M.D., Zuk, O., Giardina, S.F., Pincas, H., Barany, F., Paty, P.B., Gerald, W.L., Notterman, D.A., Domany, E.: Association of survival and disease progression with chromosomal instability: a genomic exploration of colorectal cancer. Proc. Natl. Acad. Sci. **106**(17), 7131–7136 (2009)
15. Sklar, A.: Fonctions de répartition à n dimensions et leurs marges. Université Paris (1959)
16. Söderlund, S., Granath, F., Broström, O., Karlen, P., Löfberg, R., Ekbom, A., Askling, J.: Inflammatory bowel disease confers a lower risk of colorectal cancer to females than to males. Gastroenterology **138**(5), 1697–1703 (2010)
17. Tishby, N., Pereira, F.C., Bialek, W.: The information bottleneck method. In: Proceedings of the 37th Annual Allerton Conference on Communication, Control and Computing, pp. 368–377 (1999)
18. Wullschleger, S., Loewith, R., Hall, M.N.: TOR signaling in growth and metabolism. Cell **124**(3), 471–484 (2006)

Deep Learning and Segmentation

Convolutional Decision Trees for Feature Learning and Segmentation

Dmitry Laptev[✉] and Joachim M. Buhmann

ETH Zurich, 8092 Zurich, Switzerland
`dlaptev@inf.ethz.ch`

Abstract. Most computer vision and especially segmentation tasks require to extract features that represent local appearance of patches. Relevant features can be further processed by learning algorithms to infer posterior probabilities that pixels belong to an object of interest. Deep Convolutional Neural Networks (CNN) define a particularly successful class of learning algorithms for semantic segmentation, although they proved to be very slow to train even when employing special purpose hardware. We propose, for the first time, a general purpose segmentation algorithm to extract the most informative and interpretable features as convolution kernels while simultaneously building a multivariate decision tree. The algorithm trains several orders of magnitude faster than regular CNNs and achieves state of the art results in processing quality on benchmark datasets.

1 Introduction

One of the most important and critical steps for the overwhelming majority of computer vision tasks is feature design. Researchers face a challenging problem of describing local appearance of a patch around the pixel or voxel with a set of features for further processing with different machine learning algorithms.

A very important but by no means the only example of such image processing applications is medical image segmentation. The problem of constructing relevant features arises in this field in the most acute way. Experts that label pixels manually often rely only on local appearance, but are unable to mathematically define the features that appear to be most relevant for them.

In order to overcome the problem of feature design, different methods were proposed to automatically learn discriminative local descriptors [11,13]. Among them, Deep Convolutional Neural Networks (CNN) [13] emerged as probably one of the most attractive methods for supervised feature learning nowadays. This method demonstrated to achieve superior performance for different tasks like face recognition [13], handwritten character recognition [9] and neuronal structure segmentation [7]. On the other hand, CNN suffer from the significant disadvantage that they require very large training data sets and consume an often impractical amount of time to learn the network parameters. Therefore, special hardware cluster architectures have been developed to make CNN applicable for

© Springer International Publishing Switzerland 2014
X. Jiang et al. (Eds.): GCPR 2014, LNCS 8753, pp. 95–106, 2014.
DOI: 10.1007/978-3-319-11752-2_8

real world tasks [7]. These constraints render the process of using CNN for end users very difficult and often even unfeasible.

This paper presents Convolutional Decision Trees (CDT): a significantly accelerated algorithm for adaptive feature learning and segmentation. It belongs to a family of oblique decision tree algorithms [10] adapted for structural data such as spatial structure of the patches in image segmentation. The algorithm builds on the following ideas:

- the method recursively builds multivariate (oblique) decision tree,
- each tree split is represented by a convolution kernel, and therefore encodes a feature of the patch around the pixel,
- convolution kernels are learned in a supervised manner while maximizing the informativeness of the split,
- regularization of kernel gradients produces interpretable and generalizable features,
- regularization parameter adaptively changes from one split to another.

These structured oblique trees significantly differ from non-structural by smoothness regularization of the learned kernels. Complexity control of feature learning renders the optimization problem more robust, regularized learning produces more interpretable features and it largely prevents overfitting. The key advantage is that the features learned adaptively for one task are informative and meaningful and, therefore, can be used for other tasks.

These ideas generate a significant performance increase compared to CNN training procedure (up to several orders of magnitude faster training), while keeping the accuracy at state of the art level. The combination of high accuracy and fast training enables anyone to use this algorithm on general purpose single processor desktop hardware.

The procedure demonstrates convincing result improvements both for medical and for natural image segmentation tasks while it avoids to employ domain-specific prior knowledge. We provide all the details in Sect. 5. In this paper we focus on segmentation task, however, following the method described in [8], the approach can be adapted also for tasks like object detection, tracking and action recognition.

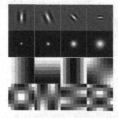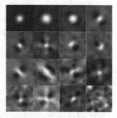

Fig. 1. Examples of different convolutional kernels: commonly used kernels (left) and the kernels obtained with the proposed algorithm (right). The algorithm finds the most informative kernels in a supervised manner, discovering meaningful kernels that look like gaussian blur, edge filters, corner and junction detectors, texture filters, etc.

2   Related Work

2.1   Feature Learning

The problem of feature construction is broadly discussed in the computer vision community. Standard or **commonly used features** are developed to face as many computer vision applications as possible. Different filters [1], SIFT [15] and HoG features [24] are only the few examples of such features. Even though these features work great for some tasks, they are unable to adapt to the specific problems and, therefore, often do not encode all the relevant information.

For some applications experts are able to formalize the desired properties of the object patches based on the local appearance. For such applications **domain-specific features** can be developed. Line filter transform [19] is used for blood vessels segmentation, context cue features [2] – for synapse detection. Domain-specific features proved to be very informative, however the development of these features is time-consuming and expensive while not always possible and it does not generalize to other domains. In contrast, the proposed algorithm learns features in a supervised manner and therefore adapts to the specific application without any prior information.

Unsupervised feature learning overcomes the domain-specificity, as this approach generates features based on the data itself. The "bag of visual words" representation [22] and dictionary learning [11] for sparse coding are procedures that fall in this category together with denoising autoencoders [21]. Even though these methods are powerful for data-representation, compression and image restoration, they exhibit serious limitations when applied to segmentation. This phenomenon happens because neither of the methods rely on the information about the label of the pixels and therefore learns *reconstructive*, not *discriminative* representations.

Supervised feature learning, in contrast, learns the features of the data jointly with learning the classification functions. Sparse coding algorithm can be adapted to this procedure [16], but only with classification functions limited to linear and bilinear models. Convolutional Neural Networks (CNN) [13] are able to learn more complex classification functions and more complex feature representation. CNN, as discussed in the introduction, is a very powerful and flexible technique yet with one major disadvantage: computational time for training.

For example, for neuronal segmentation dataset [6], the authors of [7] use CNN that achieves impressive accuracy, but that trains for almost a week using specially developed GPU cluster. In contrast the proposed method combines the flexibility of arbitrary convolutional kernels with the speed of decision tree training. Depending on the task first reasonable results for smaller trees can be obtained within one hour, while larger trees produce state of the art results in less than 12 h training with one CPU which makes this method feasible for "plug-and-play" experiments. CNNs with the same training time (smaller CNNs or CNNs trained with different strategies) do not achieve comparable results. The description of the method is given in Sect. 3.

2.2 Binary Decision Tree

Learning decision trees pursues the idea to consecutively split the data space into parts according to a predicate ϕ (s.t. $\phi(x) = 0$ for points x in one half-space, and $\phi(x) = 1$ for x in another half-space). The predicate is selected in such a way that it maximizes a task dependent measure of informativeness, e.g. Information Gain or Gini's diversity index [5]. $x \in \mathbb{R}^d$ is a vector of features or attributes of the object: x^j represent j-th feature of the object.

Decision trees most commonly are **univariate** [5]. Formally that means that the form of the predicate is limited to $\phi(x) = [x^j > c]$. Here [*statement*] denotes Iverson brackets which equals to 1 if *statement* is true and zero otherwise. j and c are the parameters of the split. The choice of only univariate splits limits the computational complexity and it allows us to efficiently find the most informative split. However, it has been demonstrated that in many cases univariate trees require many more splits to learn a classifier and lead to results that are difficult to interpret [10].

Therefore **multivariate (oblique) trees** were proposed [5,10,20], that allow the predicate to be more flexible: $\phi(x) = [x^T\beta > c]$. Here $x^T\beta$ is a linear combination of the attributes, $\beta \in \mathbb{R}^d$ and c are the parameters of the split. Depending on the criteria of informativeness, most algorithms only return locally optimal splits.

The proposed algorithm develops the idea of oblique trees for learning convolution kernels in the context of image segmentation problems. Through regularization it incorporates the structural information about the spatial neighborhood of the pixels. Introducing this regularization helps the learned splits to be more interpretable and the optimization problem to be more robust.

3 Proposed Method

Notation. As a training set we assume K pixels $i \in \{1, \ldots, K\}$ with associated binary labels $y_i \in \{-1, 1\}$. Local pixel appearance is described with a patch around it. Let the size of a patch be $w \times w$, then each pixel i is represented with w^2 intensities of the pixels in the patch. In homogeneous coordinates, pixel i is described by a vector $x_i \in \mathbb{R}^{w^2+1}$ with $x_{i,1} \equiv 1$. All the vectors stored row-wise form a data-matrix $X \in \mathbb{R}^{K \times w^2+1}$, $X = [x_1, \ldots, x_K]^T$.

The main idea of the method is to find a smooth convolution kernel that would be informative and discriminative for separating one class from another. A kernel of a convolution is again a $w \times w$ matrix. We also extend a vectorized kernel with a shift parameter b for the predicate. We define vector β to encodes both shift and kernel parameters: $\beta \in \mathbb{R}^{w^2+1}$, $\beta_1 \equiv b$ and $\beta_{2:w^2+1}$ encodes the kernel of the convolution.

The predicate form can be now defined as $\phi(x_i, \beta) = [\beta^T x_i > 0]$. As here we care about the sign of the convolution, we also introduce the constraint $||\beta||_2^2 = 1$ to overcome the disambiguities induced by different scalings.

3.1 Information Gain

We want to estimate the parameter vector β that would maximize the information gain $\mathrm{IG}(\beta)$. Information gain depends on the distribution of positive and negative samples before and after a split with the predicate ϕ. Let's define P as the number of all positive samples: $P = \sum_i [y_i = 1]$, $N = K - P$ — the number of negative samples. After the split, the half-space, where $\phi(x, \beta) = 1$, will contain p positive and n negative samples: $p = \sum_i [\phi(x_i, \beta) = 1, y_i = 1]$, $n = \sum_i [\phi(x_i, \beta) = 1, y_i = -1]$. Both p and n depend on the parameters β. Then

$$\mathrm{IG}(\beta) = \hat{H}\left(P, N\right) - \hat{H}_\beta\left(P, N, p, n\right), \tag{1}$$

where

$$\hat{H}\left(P, N\right) = H\left(\frac{P}{P+N}, \frac{N}{P+N}\right)$$

$$\hat{H}_\beta\left(P, N, p, n\right) = \frac{p+n}{P+N}\hat{H}\left(p, n\right) + \frac{P+N-p-n}{P+N}\hat{H}\left(P-p, N-n\right)$$

And H denotes the entropy: $H(q_0, q_1) = -q_0 \log_2 q_0 - q_1 \log_2 q_1$. Then the problem of finding the most informative split is formalized as follows:

$$\beta \in \arg\max_\beta \mathrm{IG}(\beta) \tag{2}$$

Unfortunately, the maximum of $\mathrm{IG}(\beta)$ cannot be found efficiently because of discontinuity in $\phi(x_i, \beta)$ as it contains an indicator function $[\beta^T x_i > 0]$. To overcome this issue we use an approximation from [17]:

$$\hat{\phi}_\alpha(x_i, \beta) = \frac{1}{1 + \exp(-\alpha\beta^T x_i)} \tag{3}$$

We also introduce $\hat{p}_\alpha = \sum_{i:y_i=1} \hat{\phi}_\alpha(x_i, \beta)$, $\hat{n}_\alpha = \sum_{i:y_i=-1} \hat{\phi}_\alpha(x_i, \beta)$. Then the information gain can be approximated with

$$\hat{\mathrm{IG}}_\alpha(\beta) = \hat{H}\left(P, N\right) - \hat{H}_\beta\left(P, N, \hat{p}_\alpha \hat{n}_\alpha\right)$$

It is easy to see that $\hat{\phi}_\alpha(x_i, \beta)$ converges to $\phi(x_i, \beta)$ in the limit $\alpha \to \infty$. Also $\hat{p}_\alpha$ and $\hat{n}_\alpha$ converges to p and n, respectively. This asymptotics renders it possible to solve the original problem (2) by estimating a limit process, i.e., we investigate the sequence of solutions of a relaxed problem for increasing α:

$$\beta \in \arg\max_\beta \mathrm{IG}(\beta) = \lim_{\alpha \to +\infty} \arg\max_\beta \hat{\mathrm{IG}}_\alpha(\beta)$$

3.2 Regularization

Maximizing the information gain with respect to β usually results in a split that separates the classes, but unfortunately not interpretable (see for example Fig. 2). More than that, when the number of training samples goes to the range

Fig. 2. Examples of convolution kernels learned with different regularization parameters λ. Each row represents a kernel from different levels of CDT (respectively from 1 to 4). Each column stands for different regularization parameters: 0.001, 0.01, 0.1, 0.5, 2, 10. Increasing regularization helps the learned features to be more interpretable (compare first columns with the last ones). However, increasing λ too much results in smoothing out relevant information (for example the orientation of the third kernel disappears in the last two columns).

of $O(w^2)$ (approximately equals to the number of parameters), the linear split model starts to overfit. This problem requires us to introduce a regularization parameter λ that penalizes the complexity of the learned kernel parameters β. We want to assure that the kernel is smooth, and, therefore, we penalize the gradient of the kernel:

$$\beta_\alpha \in \arg\max_\beta L_\alpha(\beta) \quad \text{with} \quad L_\alpha(\beta) = \hat{\text{IG}}_\alpha(\beta) - \lambda||\Gamma\beta||_2^2 \tag{4}$$

Here $\Gamma \in \mathbb{R}^{2w(w-1) \,\times\, (w^2+1)}$ is a matrix of a 2D differentiation operator in a vectorized space, that is a Tikhonov regularization matrix.

Regularization serves two main goals. First of all, it guarantees interpretability of the kernels learned (see Figs. 1 and 2). And second, from an optimization point of view, a strictly concave regularization term steers the gradient descent optimization algorithm out of local minima.

3.3 Optimization

In practice, we need to choose an initial point and the gradient of the functional to effectively find a solution of the problem 4. As initial point, we use the solution to a simple regularized linear regression:

$$\beta_0 = \arg\min_\beta \frac{1}{K}||X\beta - Y||_2^2 + \lambda||\Gamma\beta||_2^2 \tag{5}$$

Here $Y = [y_1, \ldots, y_K]^T$ is a vector of all the responses and Γ denotes a Tikhonov matrix associated with the regularization above. The analytical solution to problem (5) is equal to $\beta_0 = \left(\frac{1}{K}X^T X + \lambda \Gamma^T \Gamma\right)^{-1} X^T Y$.

The derivative of the functional $L_\alpha(\beta)$ in (4) can be also found analytically:

$$\frac{dL_\alpha}{d\beta} = \frac{1}{P+N} \sum_i \frac{\alpha \exp(\alpha\beta^T x_i)}{(1 + \exp(\alpha\beta^T x_i))^2} x_i \left(-\log_2 \frac{p+n}{P+N-p-n} + \right.$$

$$\left. [y_i = 1]\log_2 \frac{p}{P-p} + [y_i = -1]\log_2 \frac{n}{N-n} \right) - 2\lambda\Gamma^T \Gamma\beta \tag{6}$$

As an optimization algorithm, we employ Quasi-Newton Limited-memory BFGS (L-BFGS) [18] that estimates Hessian with low-rank approximation and, therefore, selects optimal step size.

Assuming optimization procedure as a subroutine L-BFGS, we sketch the algorithm that finds one split by learning the most informative convolution kernel in Algorithm 1. We do not directly estimate the limit in Eq. (2), but instead we iteratively increase α and initialize the optimization procedure in the next step with the solution in the previous step.

Algorithm 1. Function findSplit for learning the most informative split

---

Require: training samples $x_i, i = 1, \ldots, K$ with classes y_i; λ

$\beta_0 := \left(\frac{1}{K} X^T X + \lambda^2 \Gamma^T \Gamma \right)^{-1} X^T Y$ ▷ initialize with MSE solution

$\beta_0 := \beta_0 / ||\beta_0||_2^2$ ▷ project on the unit sphere

Set $\alpha := 1$

repeat

     $\beta_\alpha := $ L-BFGS$(L_\alpha(\cdot), \frac{dL_\alpha}{d\beta}(\cdot), \beta_{\alpha-1})$ ▷ find $\arg\max_\beta L_\alpha(\beta)$

     $\beta_\alpha := \beta_\alpha / ||\beta_\alpha||_2^2$ ▷ project on the unit sphere

     $\alpha := \alpha + 1$

until $||\beta_\alpha - \beta_{\alpha-1}||_2^2 < \epsilon$ OR $\alpha >$ MaxIterations

return β_α

---

4 Decision Trees

How can we use the procedure findSplit to build a classifier? A well-known idea is to recursively split the data space into parts. The recursion stops when we achieve certainty about the label of every part. All the sequential splits end up being encoded in a binary decision tree.

The idea is very straightforward, so we do not discuss it in details, except for one important question. So far we defined the regularization parameter λ for only one split, but in principle we can change it from split to split, or from one layer of the tree to the next. Experiments show that just fixing one parameter λ to be the same for every split in a tree often results in kernel overfitting as the tree grows large.

Assume that we want to find two splits in two different parts A and B with volumes respectively $Vol(A)$ and $Vol(B)$. The relation between λ_A and λ_B for this two problems can be established from the following intuition: we want the range of both problems to be the same: for the data part $\frac{1}{K}||X\beta - Y||_2^2$ and for the regularization part $\lambda||\Gamma\beta||_2^2$. With some assumptions and derivations that are out of scope of this paper, we provide the following heuristic rule: $\lambda_B = \lambda_A \left(\frac{Vol(A)}{Vol(B)} \right)^{2/(w^2+1)}$. We approximate this ratio with just the fraction of the data points falling into each of the compacts A and B. The final recursive algorithm for building the convolutional decision tree is sketched in Algorithm 2 (Figs. 3 and 4).

Algorithm 2. Function `buildTree` for decision tree construction

Require: set of indices I, λ, MaxSamples
 $P := \sum_{i \in I}[y_i = 1]$; $N := \sum_{i \in I}[y_i = -1]$
 treeStruct.answer $= \frac{P}{P+N}$
 if $P <$ MaxSamples or $N <$ MaxSamples **then**   ▷ terminal node, stop recursion
   treeStruct.left $= null$; treeStruct.right $= null$
 else   ▷ split recoursively
   $\beta := \texttt{findSplit}(x_i, y_i, \forall i \in I; \lambda)$
   $I_{\text{left}} := \{i \in I : \beta^T x_i > 0\}$; $I_{\text{right}} := \{i \in I : \beta^T x_i <= 0\}$
   $\lambda_{\text{left}} = \lambda \left(\frac{|I|}{|I_{\text{left}}|}\right)^{2/(w^2+1)}$; $\lambda_{\text{right}} = \lambda \left(\frac{|I|}{|I_{\text{right}}|}\right)^{2/(w^2+1)}$ ▷ change lambda
   treeStruct.left $= \texttt{buildTree}(I_{\text{left}}, \lambda_{\text{left}}, \text{MaxSamples})$
   treeStruct.right $= \texttt{buildTree}(I_{\text{right}}, \lambda_{\text{right}}, \text{MaxSamples})$
 end if
 return treeStruct

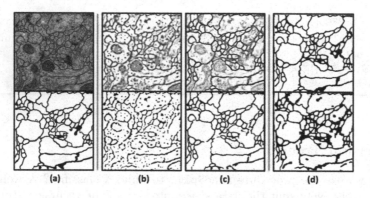

(a)   (b)   (c)   (d)

Fig. 3. The results of the proposed algorithm on Drosophila VNC dataset. Column (a) shows the input image and the ground truth. Columns (b) and (c) demonstrate the qualitative results of the algorithm for the small tree (depth = 3) and the full tree (depth = 17). The results include the probability maps (top) with a Graph Cut segmentation (bottom). The last column (d) shows the results of CNN (top) and RF with predefined features (bottom). Qualitatively the results are comparable with CNN and looks much better than RF results.

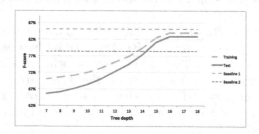

Fig. 4. Quantitative comparison. Baseline 1 is a CNN [7] which produces slightly better results, but is infeasible to train on a single CPU. Baseline 2 is a RF with Graph Cut segmentation [12], which we outperform by 4.5 %.

5   Experiments

5.1   Experimental Settings

We test Convolutional Decision Trees on biological and natural image datasets. First 2/3 of the images are selected for training, and accuracy is reported on the last 1/3. Because in both datasets the classes are imbalanced, we measure the accuracy in F-score: a commonly used metric that combines precision and recall.

We obtain probability maps infered by the proposed algorithm with the following parameters fixed for both datasets: $w = 31$, $\lambda = 0.5$ (initial value for the first call of `buildTree` function) and `MaxSamples` $= 50$.

All the experiments are performed on a single AMD Opteron 6174 CPU. The speed/accuracy tradeoff is controlled by the number of iterations of the `L-BFGS` subroutine. We set it in such a way that all the experiments finish within 12 h (overnight experiment).

To get the final segmentation from the probability maps, we apply simple Graph Cut algorithm [4] with the parameters selected by 5-fold cross-validation.

5.2   Drosophila VNC Dataset

As an example of a biological dataset, we use a publicly available Electron Microscopy dataset of the Drosophila first instar larva ventral nerve cord (VNC) [6]. The dataset consists of 30 images, each image depicts 2×2 microns of tissue with a resolution of 4×4 nm/pixel. The segmentation task is to annotate neuronal structures in tissue as either membranes or the inside volume of neurons.

The best results on this dataset are achieved using Convolutional Neural Networks. In terms of F-score, the accuracy of this algorithm is approximately the same as the accuracy of a human expert [7]. Our results appear to be just 2.2 % worse in absolute values (82.9 % CDT vs 85.1 % CNN), however, the CNN training time for this dataset is around one week using GPU, comparing to just 12 h for the proposed method. Training CNN for 12 h produces results comparable to the other state of the art method that trains in a reasonable time and is described in [12]. A Random Forest (RF) algorithm with specially designed features produces a probability map that is then segmented with a Graph Cut algorithm that uses special potentials. Even though we use a simpler segmentation algorithm, CDT produces better probability maps. Quantitatively that results in 4.5 % increase in F-score (82.9 % CDT vs 78.4 % RF) (Fig. 5).

5.3   Weizmann Horse Dataset

The Weizmann Horse dataset [3] is well-known in the Computer Vision community. It consist of 328 manually labelled images of horses in different environments.

There are many methods that perform well on this dataset [3,8,23]. As a baseline we consider general purpose segmentation method based on superpixel grouping [14]. This method produces the best results on this dataset across

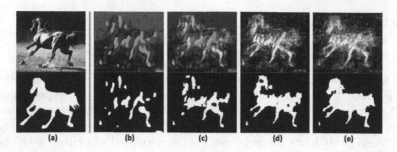

Fig. 5. The results on the Weizmann Horse dataset. Left column (a) shows the input image and the ground truth label. Each of the following columns (b)–(e) demonstrates the qualitative results of the algorithm for different tree sizes (respectively 4, 8, 12, 18). The results include the probability map (top) and the Graph Cut segmentation (bottom). Qualitatively the results improve significantly as the tree grows and more advanced features are learned.

methods that use no prior information: 79.7 %. Quantitatively we achieve 80.4 % and outperform it by 0.7 % (insignificantly). There are also other methods that use domain-specific prior information on the shape of the horse silhouette [23] and achieve superior results of 89.2 % (up to 8.8 % better). However, we do not compare with them as they are limited to specific segmentation tasks where shape information is know a priori and our method is applicable to a much broader class of segmentation tasks.

6   Conclusion

In this paper we propose, for the first time, Convolutional Decision Trees: a general purpose binary segmentation algorithm that is based on learning the most informative features. We represent every feature as a convolution kernel and combine them efficiently in an oblique decision tree. We achieve interpretability and robustness by regularizing the derivative of the kernel.

The method learns features in a supervised manner and adapts to a specific problem. In this sense it works similar to Convolutional Neural Networks (CNN). The key advantage of the proposed algorithm is its run-time; it trains several orders of magnitude faster than regular CNNs which makes it possible to learn features without access to special hardware.

We test the accuracy on two benchmarks: biological (Electron Microscopy) and natural image datasets. For natural images (Weizmann Horse dataset) we achieve state of the art results across general purpose methods that require no domain-specific prior knowledge. For biological imaging (Drosophila VNC dataset) we show the results slightly inferior to CNNs, but outperform the best results that operate within a similar time period by 4.5 %.

References

1. Matlab image processing toolbox. http://www.mathworks.com/help/images/
2. Becker, C., Ali, K., Knott, G., Fua, P.: Learning context cues for synapse segmentation in EM volumes. In: Ayache, N., Delingette, H., Golland, P., Mori, K. (eds.) MICCAI 2012, Part I. LNCS, vol. 7510, pp. 585–592. Springer, Heidelberg (2012)
3. Borenstein, E., Ullman, S.: Class-specific, top-down segmentation. In: Heyden, A., Sparr, G., Nielsen, M., Johansen, P. (eds.) ECCV 2002, Part II. LNCS, vol. 2351, pp. 109–122. Springer, Heidelberg (2002)
4. Boykov, Y., Kolmogorov, V.: An experimental comparison of min-cut/max-flow algorithms for energy minimization in vision. IEEE Trans. Pattern Anal. Mach. Intell. **26**(9), 1124–1137 (2004)
5. Breiman, L., Friedman, J.H., Olshen, R.A., Stone, C.J.: Classification and Regression Trees. Wadsworth & Brooks, Monterey (1984)
6. Cardona, A., Saalfeld, S., Preibisch, S., Schmid, B., Cheng, A., Pulokas, J., Tomancak, P., Hartenstein, V.: An integrated micro-and macroarchitectural analysis of the drosophila brain by computer-assisted serial section electron microscopy. PLoS Biol. **8**(10), e1000502 (2010)
7. Ciresan, D., Giusti, A., Schmidhuber, J., et al.: Deep neural networks segment neuronal membranes in electron microscopy images. Adv. Neural Inf. Process. Syst. **25**, 2852–2860 (2012)
8. Gall, J., Yao, A., Razavi, N., Van Gool, L., Lempitsky, V.: Hough forests for object detection, tracking, and action recognition. IEEE Trans. Pattern Anal. Mach. Intell. **33**(11), 2188–2202 (2011)
9. Graves, A., Schmidhuber, J.: Offline handwriting recognition with multidimensional recurrent neural networks. In: Advances in Neural Information Processing Systems, pp. 545–552 (2008)
10. Heath, D., Kasif, S., Salzberg, S.: Induction of oblique decision trees (1993)
11. Kreutz-Delgado, K., Murray, J.F., Rao, B.D., Engan, K., Lee, T.W., Sejnowski, T.J.: Dictionary learning algorithms for sparse representation. Neural Comput. **15**(2), 349–396 (2003)
12. Laptev, D., Vezhnevets, A., Dwivedi, S., Buhmann, J.M.: Anisotropic ssTEM Image segmentation using dense correspondence across sections. In: Ayache, N., Delingette, H., Golland, P., Mori, K. (eds.) MICCAI 2012, Part I. LNCS, vol. 7510, pp. 323–330. Springer, Heidelberg (2012)
13. Lawrence, S., Giles, C.L., Tsoi, A.C., Back, A.D.: Face recognition: a convolutional neural-network approach. IEEE Trans. Neural Netw. **8**(1), 98–113 (1997)
14. Levinshtein, A., Sminchisescu, C., Dickinson, S.: Optimal image and video closure by superpixel grouping. Int. J. Comput. Vis. **100**(1), 99–119 (2012)
15. Lowe, D.G.: Object recognition from local scale-invariant features. In: The Proceedings of the Seventh IEEE International Conference on Computer Vision, vol. 2, pp. 1150–1157. IEEE (1999)
16. Mairal, J., Bach, F., Ponce, J., Sapiro, G., Zisserman, A.: Supervised dictionary learning (2008). arXiv preprint arXiv:0809.3083
17. Montillo, A., Tu, J., Shotton, J., Winn, J., Iglesias, J., Metaxas, D., Criminisi, A.: Entanglement and differentiable information gain maximization. In: Criminisi, A., Shotton, J. (eds.) Decision Forests for Computer Vision and Medical Image Analysis, pp. 273–293. Springer, Heidelberg (2013)
18. Nocedal, J.: Updating quasi-newton matrices with limited storage. Math. Comput. **35**(151), 773–782 (1980)

19. Sandberg, K., Brega, M.: Segmentation of thin structures in electron micrographs using orientation fields. J. Struct. Biol. **157**(2), 403–415 (2007)
20. Sklansky, J., Michelotti, L.: Locally trained piecewise linear classifiers. IEEE Trans. Pattern Anal. Mach. Intell. **2**, 101–111 (1980)
21. Vincent, P., Larochelle, H., Bengio, Y., Manzagol, P.A.: Extracting and composing robust features with denoising autoencoders. In: Proceedings of the 25th International Conference on Machine Learning, pp. 1096–1103. ACM (2008)
22. Yang, J., Jiang, Y.G., Hauptmann, A.G., Ngo, C.W.: Evaluating bag-of-visual-words representations in scene classification. In: Proceedings of the International Workshop on Multimedia Information Retrieval, pp. 197–206. ACM (2007)
23. Zhu, L., Chen, Y., Yuille, A.: Learning a hierarchical deformable template for rapid deformable object parsing. IEEE Trans. Pattern Anal. Mach. Intell. **32**(6), 1029–1043 (2010)
24. Zhu, Q., Yeh, M.C., Cheng, K.T., Avidan, S.: Fast human detection using a cascade of histograms of oriented gradients. In: 2006 IEEE Computer Society Conference on Computer Vision and Pattern Recognition, vol. 2, pp. 1491–1498. IEEE (2006)

A Deep Variational Model for Image Segmentation

René Ranftl[1]([✉]) and Thomas Pock[1,2]

[1] Institute for Computer Graphics and Vision, Graz University of Technology,
Graz, Austria
ranftl@icg.tugraz.at
[2] Safety and Security Department, AIT Austrian Institute of Technology,
Seibersdorf, Austria

Abstract. In this paper we introduce a novel model that combines Deep Convolutional Neural Networks with a global inference model. Our model is derived from a convex variational relaxation of the minimum s-t cut problem on graphs, which is frequently used for the task of image segmentation. We treat the outputs of Convolutional Neural Networks as the unary and pairwise potentials of a graph and derive a smooth approximation to the minimum s-t cut problem. During training, this approximation facilitates the adaptation of the Convolutional Neural Network to the smoothing that is induced by the global model. The training algorithm can be understood as a modified backpropagation algorithm, that explicitly takes the global inference layer into account.

We illustrate our approach on the task of supervised figure-ground segmentation. In contrast to competing approaches we train directly on the raw pixels of the input images and do not rely on hand-crafted features. Despite its generality, simplicity and complete lack of hand-crafted features, our approach is able to yield competitive performance on the Graz02 and Weizmann Horses datasets.

1 Introduction

Deep Neural Networks have seen an increasing interest in the computer vision community in recent years. While such architectures are interesting due to their biological motivation, the surge of interest can be mainly attributed to practical advancements in training [15] as well as the capability to process huge amounts of training data in reasonable time.

An architecture that has shown promising results is known as the Convolutional Neural Network (CNN) [20]. CNNs consist of multiple layers of convolutions of the input data, followed by an accumulation of intermediate representations and non-linear transformations. Furthermore, specialized layers, such as max-pooling or averaging, can be interleaved with the convolutional layers in order to increase the robustness of the CNN. These networks to some degree

---

The authors acknowledge support from the Austrian science fund (FWF) under the projects No. I1148 and No. Y729.

© Springer International Publishing Switzerland 2014
X. Jiang et al. (Eds.): GCPR 2014, LNCS 8753, pp. 107–118, 2014.
DOI: 10.1007/978-3-319-11752-2_9

emulate biological visual systems and are able to generate mid- and high-level representations of visual data from raw pixels. CNNs constitute highly non-linear and non-convex, but continuously differentiable, functions. This structure makes training of CNNs a hard problem, but it can in many cases be satisfactory solved using gradient-based methods. Despite these problems, CNNs have shown good results for a variety of image processing tasks, such as image classification [18], automatic learning of feature descriptors [12] and object recognition [30].

Since CNNs operate on a patch-level they can be considered, depending on the task, as local classifiers or local regressors and will suffer from clutter and misalignment to object boundaries if they are used for image segmentation. Local classifiers are often augmented with Conditional Random Fields (CRFs) in order to overcome these problems. A typical approach to train such a system is to first train the local classifier (*e.g.* the CNN) and then train the CRF parameters according to the output of the previously trained classifier using for example Structured-Output SVMs [32].

In this work we propose an approach that deviates from the classical two-stage training architecture of segmentation models. Motivated by convex variational relaxations of graph labeling problems, we formulate a global continuous model that tightly couples the pixel-based responses of Convolutional Neural Networks with the global model. Our formulation allows us to jointly train the parameters of the CNNs as well as the parameters of the global model via a modified backpropagation rule which is obtained from a bilevel optimization problem. Similar to classical training of CNNs, the model is trained such that it directly minimizes a quadratic loss function. The advantage of our construction is that the CNN can adapt its output to the global model during the training.

We illustrate our approach on the application of figure-ground segmentation. While training only from raw pixel data and considering only local pairwise interactions between pixels, our model shows competitive performance to state of the art approaches that heavily rely on hand-crafted features and in some instances even on the output of object detectors. Our model offers the following advantages: (i) Fast inference, with computation times below one second for common image sizes. (ii) By training from raw image data, easy adaption to other types of image data than color images, like RGB-D or multispectral images, is possible. (iii) Our approach leverages principles from well-known models, which makes it possible to transfer extensions of those models into our framework.

Related Work. Tappen *et al.* [31] define a Gaussian random field together with a logistic loss function for the task of image segmentation. Their construction, similar to ours, supports the direct optimization of the model parameters under the loss using gradient-based optimization algorithms. They focus on training of linear combinations of features and pairwise model parameters. A similar framework was introduce by Cour *et al.* [10], which shows that gradient-based learning is feasible in the context of spectral graph segmentation. In contrast to this works, we use a smooth approximation to the minimum s-t cut on a graph, which not only enables gradient-based training of all parameters, but also has a strong connection to classical tools like graph cuts [6]. Moreover we train complex functions, which directly extract features from raw input images.

Domke [11] showed how to differentiate the results of optimization algorithms using principles from algorithmic differentiation and used this idea to train the parameters of continuous-valued Markov Random Fields. Brakel *et al.* [7] extended this work in order to train energy-based models that rely on neural networks for the task of time-series imputation. While this approach allows for general optimization-based models to be trained, it relies on truncated steepest-descent methods, due to otherwise excessive memory demand.

Nowozin *et al.* [26] used Decision Trees to learn the interactions in a graphical model. Jancsary *et al.* [17] later expanded on this idea and provided a tractable formulation that can be solved efficiently by fixing the form of the graphical model to a Gaussian random field. Other more general strategies to train global segmentation models include approaches like Structured-Output SVMs [32] and Contrastive Divergence [14].

Global end-to-end training of a system of CNNs and directed acyclic graphs has been used by Bottou *et al.* [5] for the application of automatic check reading. CNNs have previously also been used in the context of image segmentation [2,12]. Alvarez *et al.* [2] derive the unary potential for a CRF from a multi-scale CNN. To make training tractable they use piecewise training. In contrast, we propose to jointly train the parameters of a CNN and a variational global inference model. Turaga *et al.* [33] use a CNN to learn an affinity graph for the segmentation of volumetric electron microscopy data. This work uses normalized cuts and connected component labeling, respectively, in order to extract a labeling from the affinity graph, but does not take this post-processing into account in the learning stage.

2   CNNs and Variational Image Segmentation

In this section we first introduce a variant of a CNN that is easy to integrate within a global segmentation model. We then define a global inference layer, that allows for uncluttered and edge-aligned solutions and also fits to the gradient-based learning paradigm of standard convolutional networks. Figure 1 shows an overview of the complete system. Our construction can be equally understood in two different ways: From the global model perspective, the CNNs act as non-linear response functions to the image content, which define the unary and pairwise potentials, respectively. From the neural network perspective, the global model can be thought of as an additional layer that takes global interactions into account in order to generate an output response.

Convolutional Neural Networks. A CNN is composed of multiple hierarchical layers. Each layer consists of a set of neurons, where a neuron generates its output based on the output of the preceding layer. Let us denote the total number of layers by L, and let $l \in \{1 \ldots L\}$ be the index for each layer. Each neuron of a layer filters the B_{l-1} outputs of its preceding layer, accumulates the results and applies a sigmoidal function to generate its output. The output of

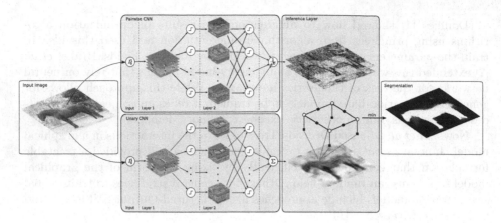

Fig. 1. Schematic overview of the proposed approach. An input image is fed to CNNs, which are defined via an alternating series of convolutions and non-linear transformations. The outputs of the CNNs provide the unary and pairwise potentials for a global inference model. The final segmentation is inferred by minimizing the global model.

the a-th neuron of layer l can be computed as:

$$I_{l,c}^a(I, \beta) = \tanh(\sum_{b=1}^{B_{l-1}} k_{l,c}^{a,b} * I_{l-1,c}^b + t_{l,c}^a), \qquad (1)$$

where $a \in \{1, \ldots, B_l\}$, $b \in \{1, \ldots, B_{l-1}\}$, $k_{l,c}^{a,b}$ is the linear two-dimensional filter that is applied to the b-th output of the previous layer and $t_{l,c}^a$ is the default activation of the neuron. The tanh is applied pixel-wise. The input layer $I_0^1 \in \mathbb{R}^{M \times N \times C}$ of the network is given by the input image I of size $M \times N$ with C channels. In contrast to classical CNN formulations each layer of our net produces a output map of full size, *i.e.* $I_{l,c}^a \in \mathbb{R}^{M \times N}$, where we use mirror-symmetric boundary conditions for the convolutions. The output of the neurons of the last layer as well as their individual channels is averaged to produce the final output:

$$f(I, \beta) = \frac{1}{B_L C} \sum_{c=1}^{C} \sum_{b=1}^{B_L} I_{L,c}^b(I, \beta). \qquad (2)$$

Note that we denote by β the full set of parameters of the CNN, *i.e.* β includes all filters $k_{l,c}^{a,b}$ and thresholds $t_{l,c}^a$. Similar to [16], the network produces a single output value for each input pixel, *i.e.* $f : \mathbb{R}^{M \times N \times C} \times \mathbb{R}^{|\beta|} \to \mathbb{R}^{M \times N}$.

Inference Layer. It is obvious that (2) incorporates local context via the filtering operations in order to generate its output. The extent of this context depends on the sizes of the filters. In order to add global context, we add an additional global inference layer. We model pairwise interactions and unary potentials, which are given by the outputs of CNNs, in a global model. The specific design of this global model will allow us to jointly train the parameters of the model and the parameters of the CNNs using gradient-based optimization algorithms. The major advantage in this approach is that the CNN parameters can be trained to take the global inference layer into account.

Let us first start by defining a global energy. The energy is motivated by graph-based labeling approaches, where to each pixel in an image a graph node is assigned. To allow for uncluttered notation, we assume for all further considerations that quantities of dimensionality $\mathbb{R}^{M \times N}$ are stacked into a vector of size $\mathbb{R}^{MN}$. For a labeling u we define the energy

$$E(u) = \|Au\|_1 + \lambda \langle u, f \rangle, \quad u \in \{0,1\}^{MN}. \tag{3}$$

The matrix A is a weighted node-arc incidence matrix of the graph. More specifically $A = \text{diag}(h(I, \alpha))B$, where the linear operator B of size $m \times MN$ denotes a node-arc incidence matrix, *i.e.* the operator has exactly two entries per-row, where the i-th entry is 1 and the j-th entry is -1, if the nodes i and j are connected by an edge and m is the number of edges in the graph. The matrix $\text{diag}(h(I, \alpha)) \in \mathbb{R}^{m \times m}$ is a diagonal weighting matrix with non-negative entries, which facilitates edge weighting. Note that this matrix depends on the output of a neural network $h(I, \alpha)$ which is defined via the parameter set α. In order to keep the pairwise potentials non-negative, we additionally apply the non-linearity $\exp(-x)$ to the output of the pairwise network. The pixel-wise outputs $f = f(I, \beta)$ of the CNN take the role of the unary potentials in the global model. λ is a regularization parameter that balances the influence of unary and pairwise interactions. It is well-known that energies of this form can be minimized by finding a minimum s-t cut in a graph [8].

In order to allow joint training of the model parameters α, β, λ, we approximate (3) by a convex, and twice differentiable energy. By fulfilling these requirements, we will be able to derive gradients with respect to the parameters using implicit differentiation.

Minimizing (3) constitutes a non-convex optimization problem due to its non-convex domain $\{0,1\}^{MN}$. Fortunately there exists a tight convex relaxation: It was shown in [9] that the domain can be relaxed to the interval $[0,1]^{MN}$. Minimizers of the relaxed problem can be thresholded to recover globally optimal solutions of the original problem. While the relaxation is convex, it is not twice differentiable, which precludes application of implicit differentiation. In order to arrive at a model that can be used in a gradient-based optimization algorithm, we first replace the ℓ^1-norm by a smooth Charbonnier approximation: $\|x\|_\varepsilon = \sum_i (x_i^2 + \varepsilon^2)^{\frac{1}{2}}$. Second, we smooth the box constraints $u \in [0,1]^{MN}$ using log-barrier functions to ensure that minimizers lie in the set $\Omega = (0,1)^{MN}$. The final approximated energy is given by

$$E_{\varepsilon,\mu}(u) = \|Au\|_\varepsilon + \lambda \langle u, f \rangle - \mu \sum_{i=1}^{MN} (\log(u_i) + \log(1 - u_i)), \quad u \in \Omega. \tag{4}$$

Minimizers u^* of (4) will depend on the specific form of the unary potentials and the pairwise potentials, which in turn depend on the image content and the model parameters. We use an optimal first-order method [25] that achieves linear convergence for the optimization of $E_{\varepsilon,\mu}(u)$. We refer the interested reader to the supplementary material for a detailed discussion of this algorithm.

3   Joint Training

The training set consists of K tuples (I_k, v_k) of input images I_k of size $M \times N \times C$ together with their groundtruth segmentations $v_k \in \{0,1\}^{M \times N}$. To simplify notation, we stack the parameters α, β and λ into a single parameter vector $w = (\alpha, \beta, \lambda)$. The goal of our learning algorithm is to find a parameter vector w that is able to reconstruct the groundtruth, where we measure reconstruction error using some loss $L(u, v_k)$. For the model that was introduced in the last section, this can be written as the following bilevel optimization problem:

$$\min_{w \in W} \frac{1}{K} \sum_{k=1}^{K} L(u^*(w, I_k), v_k) + \gamma R(w)$$

$$\text{s.t.} \quad u^*(w, I_k) = \arg\min_{u \in \Omega} E_{\varepsilon,\mu}(u; w, I_k). \tag{5}$$

$R(w)$ is a regularization term for the parameter vector, γ is the regularization parameter, and W is the feasible set of the parameters w, which is in most cases simple (*e.g.* box constraints). $E_{\varepsilon,\mu}$ is called the energy of the lower-level problem, which is given by (4).

While (5) in general poses a non-convex problem even for convex lower-level problems, one can still find a meaningful stationary point using gradient-based optimization algorithms, provided that $L(u, v_k)$ and $R(w)$ are differentiable and the lower-level problem is twice differentiable. Since this is true for (4) by construction, the gradient of the loss function with respect to the parameters w and a single example v_k can be calculated using implicit differentiation [29]:

$$\frac{\partial L(u, v_k)}{\partial w} = -\left(\frac{\partial L}{\partial u} \left(\nabla^2 E_{\varepsilon,\mu} \right)^{-1} \frac{\partial^2 E_{\varepsilon,\mu}}{\partial u \partial w} \right) \Bigg|_{u=u^*}. \tag{6}$$

Equation (6) is evaluated at a minimizer u^*, thus the lower-level problem needs to be solved for given parameters w in order to compute the gradient.

We use a quadratic per-pixel loss for training, which is equivalent to the 0–1 loss for binary u and v. Note that our training scheme can be applied to any differentiable loss function, however.

In order to compute the gradient for a single example, as shown in (6), three quantities have to be computed: The derivative of the loss with respect to a prediction u^*, the Hessian of energy (4) and the derivative of the gradient of (4) with respect to the parameters, all evaluated at the minimizer $u^*(w, I_k)$. The derivative of the loss with respect to u is given by

$$\frac{\partial L(u, v_k)}{\partial u} = u(w, I_k) - v_k \quad \in \mathbb{R}^{MN} \tag{7}$$

The Hessian $\nabla^2 E_{\varepsilon,\mu}$ is given by

$$\nabla^2 E_{\varepsilon,\mu} = A^T D'' A + \nabla^2 S_\mu, \tag{8}$$

where the matrix D is given by $D'' = \text{diag}(\rho_\varepsilon''((Au)_1),\ldots,\rho''((Au)_{MN}))$, with $\rho(x)_\varepsilon'' = \varepsilon^2/(x^2 + \varepsilon^2)^{\frac{3}{2}}$ the second derivative of the Charbonnier penalty and $S_\mu(u) = -\mu \sum_{i=1}^{MN}(\log(u_i) + \log(1 - u_i))$.

The derivative of the loss and the Hessian matrix are the same for all parameters, hence the most interesting quantity that needs to be computed is the derivative of the gradient of the energy $E_{\varepsilon,\mu}$ with respect to the parameters w. Let us focus on the parameter set β of the unary CNN. It is easy to see that

$$\frac{\partial^2 E_{\varepsilon,\mu}}{\partial u \partial \beta} = \lambda \frac{\partial f(I_k,\beta)}{\partial \beta} \quad \in \mathbb{R}^{MN \times |\beta|}. \tag{9}$$

Using (6) we get

$$\frac{\partial L(u,v_k)}{\partial \beta} = -\lambda \left(\frac{\partial f(I_k,\beta)}{\partial \beta}\right)^T ((\nabla^2 E_{\varepsilon,\mu})^{-1}(u^*(w,I_k) - v_k)). \tag{10}$$

By comparing (10) with the derivation of the classical backpropagation rule for CNNs [20], it can be seen that in order to compute the gradient with respect to the parameters, backpropagation with a modified error term can be used: Instead of backpropagating the prediction error $f(I_k,\beta) - v_k$ of the CNN, one needs to backpropagate a regularized version of the prediction error that is determined by the global inference layer, i.e. we backpropagate $e = (\nabla^2 E_{\varepsilon,\mu})^{-1}(u^*(w,I_k) - v_k)$, into the CNN.

The computation of the gradients with respect to the parameters of the pairwise potentials can be done in a similar manner and results in the gradient:

$$\frac{\partial L(u,v_k)}{\partial \alpha} = -(\frac{\partial h}{\partial \alpha})^T(\text{diag}(\rho_\varepsilon'(Au))B + \text{diag}(Bu)D''A)e. \tag{11}$$

While we are able to compute a gradient with respect to the parameters, some care has to be taken for the training. CNNs are known to be computationally expensive to train. Obviously the additional inference layer, which requires the solution of an optimization problem for each gradient computation, adds to the overall training time. Fortunately all computationally expensive parts of the training (convolutions in the CNN and optimization of the global model) allow for massive parallelization. Critical parts were therefore implemented on the GPU, which allows for a speed up of roughly 40 compared to a pure CPU implementation. We use L-BFGS [23] for the optimization of the training problem (5).

4   Experiments

We show the performance of the proposed model on the task of figure-ground segmentation. We fix the pairwise potentials to a 4-neighborhood structure and train three CNNs: One for the unary potentials, and one for horizontal and vertical pairwise potentials, respectively.

The architecture of the CNNs was set as follows: We use a maximum of five layers for the unary potentials. Each layer consists of five output neurons. The filter sizes are chosen as $\{11, 9, 7, 5, 3\}$, with the largest filter size at the input stage. For the pairwise potentials, we use a 4-neighborhood structure and CNNs of depth two, with 5 output neurons and filter sizes $\{11, 9\}$. This architecture results in roughly 30000 parameters which need to be trained.

The filter coefficients were initialized uniformly at random from the interval $[-10^2, 10^2]$. The default activations were initialized to zero. The parameters of the inference layer were set to $\varepsilon = \mu = 10^{-3}$. In order to avoid overfitting, we apply quadratic regularization to the filter coefficients of the CNN, corresponding to classical weight decay. The regularization parameter was set to $\gamma = 50/NC$, where NC is the total number of filter coefficients in the network.

Weizmann Horses. The Weizmann Horses dataset [4] consists of 328 color images. As suggested in [3], we split the data in 3 sets and use 2/3 for training and 1/3 for testing. We report the average score over three runs of the different combinations of training and test sets. The training time for this dataset was 30 h on a NVIDIA Geforce GTX 680 GPU. Inference takes less than 0.5 s per image.

The accuracy in terms of percentage of correctly classified pixels is shown in Table 1. As a baseline we report results of a CNN without the global inference layer. Furthermore we report the results of piecewise training, where we independently train the unary and pairwise CNNs and combine the results in a global inference layer at test time. For the pairwise CNNs we used the edges of the groundtruth segmentations as training targets. In order to show the influence of the depth of the CNNs, we show results for depths three and five respectively.

We observe comparable performance to the state of the art. The inference layer clearly improves the results, regardless of the depth of the underlying CNN. Deeper CNNs result in slightly lower error. The proposed joint training further improves the results by more than 3 % in the case of 5-Layer CNNs, which shows the effectiveness of the proposed approach. Note that this data set is almost saturated, thus an improvement of more than 1 % is significant. Figure 2 shows example results on this dataset together with the unary and pairwise potentials. It can be seen that the model is effective at segmenting the horses, despite appearance variations and different backgrounds.

Table 1. Percentage of correctly classified pixels on the Weizmann Horses dataset.

	CNN	Piecewise	Joint	[19]	[22]	[3]
3 Layers	89.8 %	91.1 %	93.8 %	94.7 %	95.5 %	94.6 %
5 Layers	90.5 %	91.4 %	94.9 %			

(a) Input   (b) Unary   (c) Pairwise   (d) No TH   (e) TH   (f) GT

Fig. 2. Results on the Weizmann Horses dataset. From left to right: (a) Input image, (b) unary potentials $f(I, \beta)$, darker color means higher likelihood of foreground. (c) Pairwise potentials. We show the pixelwise ℓ_2-norm of horizontal and vertical potentials. (d) Result before thresholding. (e) Thresholded result. (f) Groundtruth.

Graz02. The Graz02 dataset [27] is mostly composed of street-side images and features three different subsets of images (*Cars*, *Bike* and *Person*). For each class, binary segmentation masks are provided. This dataset is more challenging since it features objects at many different scales, with significant appearance and pose variations as well as occlusions. We follow the experimental protocol in [24] and use the first 150 *odd*-numbered images for training and the first 150 *even*-numbered images for testing.

Table 2 shows results in terms of F-measure. We observe that the global inference layer as well as the joint training have a positive influence on the performance. Using the inference layer together with piecewise training already gives results that are on-par with most of the compared approaches. Using the joint training procedure additionally yields a huge boost of over 7 %. With an average score of 77.0 %, our model places favorably when compared to the state of the art. Example results for this dataset are shown in Fig. 3.

Table 2. Results in terms of F-measure (%) on the Graz02 dataset. We report the best results of competing methods.

	Cars	Bike	Person	Avg
CNN-5	64.3	75.7	62.7	67.6
CNN-5-Piecewise	66.2	78.4	63.5	69.4
CNN-5-Joint	79.9	80.8	70.3	77.0
Aldavert *et al.* [1]	62.9	58.6	71.9	64.5
Kuettel *et al.* [19]	74.8	63.2	66.4	68.1
Fulkerson *et al.* [13]	72.2	66.1	72.2	70.2
Lempitsky *et al.* [21]	84.2	83.7	81.5	83.1

Fig. 3. Example segmentations for the Graz02 dataset. (Best viewed in color)

5   Conclusion

In this paper we presented a deep variational model for image segmentation. Motivated by graph-based labeling problems, we introduced a global inference layer for Convolutional Neural Networks, which aligns the solution to image edges and adds a global context. We use a smooth approximation to the original labeling problem in order to compute approximate minimum s-t cuts. We showed that the parameters of the network as well as the parameters of the global inference layer can be trained jointly using a modified backpropagation rule. As a result, the parameters of the Convolutional Neural Networks are adapted to the smoothing that is induced by the global inference layer. Our experiments on figure-ground segmentation show that the proposed model together with the joint training procedure is able to achieve performance that compares favorably to the state of the art in this field, while training from raw pixel data only.

Future work includes an extension to multi-label problems, which is straightforward via an approximation to the continuous multi-label Potts model [28].

References

1. Aldavert, D., Ramisa, A., de Mantaras, R.L., Toledo, R.: Fast and robust object segmentation with the integral linear classifier. In: CVPR (2010)
2. Alvarez, J.M., LeCun, Y., Gevers, T., Lopez, A.: Semantic road segmentation via multi-scale ensembles of learned features. In: ECCV Workshops (2012)
3. Bertelli, L., Yu, T., Vu, D., Gokturk, B.: Kernelized structural svm learning for supervised object segmentation. In: CVPR (2011)
4. Borenstein, E., Sharon, E., Ullman, S.: Combining top-down and bottom-up segmentation. In: CVPR (2004)
5. Bottou, L., Le Cun, Y., Bengio, Y.: Global training of document processing systems using graph transformer networks. In: Proceedings of Computer Vision and Pattern Recognition, pp. 489–493. IEEE, Puerto-Rico (1997)

6. Boykov, Y.Y., Jolly, M.P.: Interactive graph cuts for optimal boundary & region segmentation of objects in N-D images. In: ICCV (2001)
7. Brakel, P., Stroobandt, D., Schrauwen, B.: Training energy-based models for time-series imputation. J. of Mach. Learn. Res. **14**, 2771–2797 (2013)
8. Chambolle, A., Darbon, J.: On total variation minimization and surface evolution using parametric maximum flows. IJCV **84**(3), 288–307 (2009)
9. Chan, T.F., Esedoglu, S., Nikolova, M.: Algorithms for finding global minimizers of image segmentation and denoising models. J. App. Math. **66**, 1632–1648 (2004)
10. Cour, T., Gogin, N., Shi, J.: Learning spectral graph segmentation. In: AISTATS (2005)
11. Domke, J.: Generic methods for optimization-based modeling. J. Mach. Learn. Res. **22**, 318–326 (2012)
12. Farabet, C., Couprie, C., Najman, L., LeCun, Y.: Scene parsing with multiscale feature learning, purity trees, and optimal covers. In: ICML (2012)
13. Fulkerson, B., Vedaldi, A., Soatto, S.: Class segmentation and object localization with superpixel neighborhoods. In: ICCV (2009)
14. Hinton, G.: Training products of experts by minimizing contrastive divergence. Neur. Comput. **14**, 1771–1800 (2000)
15. Hinton, G., Osindero, S., Teh, Y.W.: A fast learning algorithm for deep belief nets. Neur. Comput. **18**, 1527–1554 (2006)
16. Jain, V., Seung, H.S.: Natural image denoising with convolutional networks. In: NIPS (2008)
17. Jancsary, J., Nowozin, S., Sharp, T., Rother, C.: Regression tree fields - an efficient, non-parametric approach to image labeling problems. In: CVPR (2012)
18. Krizhevsky, A., Sutskever, I., Hinton, G.E.: ImageNet classification with deep convolutional neural networks. In: NIPS (2012)
19. Kuettel, D., Ferrari, V.: Figure-ground segmentation by transferring window masks. In: CVPR (2012)
20. Lecun, Y., Bottou, L., Bengio, Y., Haffner, P.: Gradient-based learning applied to document recognition. In: Proceedings of the IEEE, pp. 2278–2324 (1998)
21. Lempitsky, V.S., Vedaldi, A., Zisserman, A.: Pylon model for semantic segmentation. In: NIPS (2011)
22. Levin, A., Weiss, Y.: Learning to combine bottom-up and top-down segmentation. IJCV **81**(1), 105–118 (2009)
23. Liu, D.C., Nocedal, J.: On the limited memory BFGS method for large scale optimization. Math. Program. **45**, 503–528 (1989)
24. Marszalek, M., Schmid, C.: Accurate object localization with shape masks. In: CVPR (2007)
25. Nesterov, Y.: Gradient methods for minimizing composite objective function. Math. Program. **140**, 125–161 (2013)
26. Nowozin, S., Rother, C., Bagon, S., Sharp, T., Yao, B., Kohli, P.: Decision tree fields. In: ICCV (2011)
27. Opelt, A., Pinz, A., Fussenegger, M., Auer, P.: Generic object recognition with boosting. PAMI **28**, 416–431 (2004)
28. Pock, T., Chambolle, A., Cremers, D., Bischof, H.: A convex relaxation approach for computing minimal partitions. In: CVPR (2009)
29. Samuel, K.G.G., Tappen, M.F.: Learning optimized map estimates in continuously-valued mrf models. In: CVPR (2009)
30. Sermanet, P., LeCun, Y.: Traffic sign recognition with multi-scale convolutional networks. In: IJCNN, pp. 2809–2813 (2011)

31. Tappen, M.F., Samuel, K.G.G., Dean, C.V., Lyle, D.M.: The logistic random field - a convenient graphical model for learning parameters for mrf-based labeling. In: CVPR (2008)
32. Tsochantaridis, I., Joachims, T., Hofmann, T., Altun, Y.: Large margin methods for structured and interdependent output variables. J. Mach. Learn. Res. 6, 1453–1484 (2005)
33. Turaga, S.C., Murray, J.F., Jain, V., Roth, F., Helmstaedter, M., Briggman, K.L., Denk, W., Seung, H.S.: Convolutional networks can learn to generate affinity graphs for image segmentation. Neural Comput. 22(2), 511–538 (2010)

Feature Computation

Robust PCA: Optimization of the Robust Reconstruction Error Over the Stiefel Manifold

Anastasia Podosinnikova[1]([✉]), Simon Setzer[2], and Matthias Hein[2]

[1] INRIA – Sierra Project-Team, École Normale Supérieure, Paris, France
anastasia.podosinnikova@inria.fr
[2] Computer Science Department, Saarland University, Saarbrücken, Germany

Abstract. It is well known that Principal Component Analysis (PCA) is strongly affected by outliers and a lot of effort has been put into robustification of PCA. In this paper we present a new algorithm for robust PCA minimizing the trimmed reconstruction error. By directly minimizing over the Stiefel manifold, we avoid deflation as often used by projection pursuit methods. In distinction to other methods for robust PCA, our method has no free parameter and is computationally very efficient. We illustrate the performance on various datasets including an application to background modeling and subtraction. Our method performs better or similar to current state-of-the-art methods while being faster.

1 Introduction

PCA is probably the most common tool for exploratory data analysis, dimensionality reduction and clustering, e.g., [9]. It can either be seen as finding the best low-dimensional subspace approximating the data or as finding the subspace of highest variance. However, due to the fact that the variance is not robust, PCA can be strongly influenced by outliers. Indeed, even one outlier can change the principal components (PCs) drastically. This phenomenon motivates the development of robust PCA methods which recover the PCs of the uncontaminated data. This problem received a lot of attention in the statistical community and recently became a problem of high interest in machine learning.

In the statistical community, two main approaches to robust PCA have been proposed. The first one is based on the robust estimation of the covariance matrix, e.g., [4,8]. Indeed, having found a robust covariance matrix one can determine robust PCs by performing the eigenvalue decomposition of this matrix. However, it has been shown that robust covariance matrix estimators with desirable properties, such as positive semidefiniteness and affine equivariance, have a breakdown point[1] upper bounded by the inverse of the dimensionality [4]. The second approach is the so called projection-pursuit [7,11], where one maximizes a robust scale measure, instead of the standard deviation, over all possible

---

[1] The breakdown point [8] of a statistical estimator is informally speaking the fraction of points which can be arbitrarily changed and the estimator is still well defined.

© Springer International Publishing Switzerland 2014
X. Jiang et al. (Eds.): GCPR 2014, LNCS 8753, pp. 121–131, 2014.
DOI: 10.1007/978-3-319-11752-2_10

directions. Although, these methods have the best possible breakdown point of 0.5, they lead to non-convex, typically, non-smooth problems and current state-of-the-art are greedy search algorithms [3], which show poor performance in high dimensions. Another disadvantage is that robust PCs are computed one by one using deflation techniques [12], which often leads to poor results for higher PCs.

In the machine learning and computer vision communities, matrix factorization approaches to robust PCA were mostly considered, where one looks for a decomposition of a data matrix into a low-rank part and a sparse part, e.g., [2,13,14,20]. The sparse part is either assumed to be scattered uniformly [2] or it is assumed to be row-wise sparse corresponding to the model where an entire observation is corrupted and discarded. While some of these methods have strong theoretical guarantees, in practice, they depend on a regularization parameter which is non-trivial to choose as robust PCA is an unsupervised problem and default choices, e.g., [2,14], often do not perform well as we discuss in Sect. 4. Furthermore, most of these methods are slow as they have to compute the SVD of a matrix of the size of the data matrix at each iteration.

As we discuss in Sect. 2, our formulation of robust PCA is based on the minimization of a robust version of the reconstruction error over the Stiefel manifold, which induces orthogonality of robust PCs. This formulation has multiple advantages. First, it has the maximal possible breakdown point of 0.5 and the interpretation of the objective is very simple and requires no parameter tuning in the default setting. In Sect. 3, we propose a new fast TRPCA algorithm for this optimization problem. Our algorithm computes both orthogonal PCs and a robust center, hence, avoiding the deflation procedure and preliminary robust centering of data. While our motivation is similar to the one of [13], our optimization scheme is completely different. In particular, our formulation requires no additional parameter.

2   Robust PCA

Notation. All vectors are column vectors and $I_p \in \mathbb{R}^{p \times p}$ denotes the identity matrix. We are given data $X \in \mathbb{R}^{n \times p}$ with n observations in $\mathbb{R}^p$ (rows correspond to data points). We assume that the data contains t true observations $T \in \mathbb{R}^{t \times p}$ and $n - t$ outliers $O \in \mathbb{R}^{n-t \times p}$ such that $X = T \cup O$ and $T \cap O \neq \varnothing$. To be able to distinguish true data from outliers, we require the standard in robust statistics assumption, that is $t \geq \lceil \frac{n}{2} \rceil$. The Stiefel manifold is denoted as $\mathcal{S}_k = \{U \in \mathbb{R}^{p \times k} \mid U^\top U = I\}$ (the set of orthonormal k-frames in $\mathbb{R}^p$).

PCA. Standard PCA [9] has two main interpretations. One can either see it as finding the k-dimensional subspace of maximum variance in the data or the k-dimensional affine subspace with minimal reconstruction error. In this paper we are focusing on the second interpretation. Given data $X \in \mathbb{R}^{n \times p}$, the goal is to find the offset $m \in \mathbb{R}^p$ and k principal components $(u_1, \ldots, u_k) = U \in \mathcal{S}_k$, which describe $\mathcal{A}(m, U) = \{z \in \mathbb{R}^p \mid z = m + \sum_{j=1}^{k} s_j u_j, \; s_j \in \mathbb{R}\}$, the k-dimensional affine subspace, so that they minimize the reconstruction error

$$\left\{ \hat{m}, \hat{U} \right\} = \underset{m \in \mathbb{R}^p,\, U \in \mathcal{S}_k,\, z_i \in \mathcal{A}(m,U)}{\arg \min} \frac{1}{n} \sum_{i=1}^{n} \| z_i - x_i \|_2^2. \tag{1}$$

It is well known that $\hat{m} = \frac{1}{n} \sum_{i=1}^{n} x_i$, and the optimal matrix $\hat{U} \in \mathcal{S}_k$ is generated by the top k eigenvectors of the empirical covariance matrix. As $U \in \mathcal{S}_k$ is an orthogonal projection, an equivalent formulation of (1) is given by

$$\left\{ \hat{m}, \hat{U} \right\} = \underset{m \in \mathbb{R}^p,\, U \in \mathcal{S}_k}{\arg \min} \frac{1}{n} \sum_{i=1}^{n} \left\| \left(U U^\top - I \right) \left(x_i - m \right) \right\|_2^2. \tag{2}$$

Robust PCA. When the data X does not contain outliers ($X = T$), we refer to the outcome of standard PCA, e.g., (2), computed for the true data T as $\{\hat{m}_T, \hat{U}_T\}$. When there are some outliers in the data X, i.e. $X = T \cup O$, the result $\{\hat{m}, \hat{U}\}$ of PCA can be significantly different from $\{\hat{m}_T, \hat{U}_T\}$ computed for the true data T. The reason is the non-robust squared ℓ_2-norm involved in the formulation, e.g., [4,8]. It is well known that PCA has a breakdown point of zero, that is a single outlier can already distort the components arbitrarily. As outliers are frequently present in applications, robust versions of PCA are crucial for data analysis with the goal of recovering the true PCA solution $\{\hat{m}_T, \hat{U}_T\}$ from the contaminated data X.

As opposed to standard PCA, robust formulations of PCA based on the maximization of the variance (the projection-pursuit approach as extension of (1)), eigenvectors of the empirical covariance matrix (construction of a robust covariance matrix), or the minimization of the reconstruction error (as extension of (2)) are not equivalent. Hence, there is no universal approach to robust PCA and the choice can depend on applications and assumptions on outliers. Moreover, due to the inherited non-convexity of standard PCA, they lead to NP-hard problems. The known approaches for robust PCA either follow to some extent greedy/locally optimal optimization techniques, e.g., [3,11,17,19], or compute convex relaxations, e.g., [2,13,14,20].

In this paper we aim at a method for robust PCA based on the minimization of a robust version of the reconstruction error and adopt the classical outlier model where entire observations (corresponding to rows in the data matrix X) correspond to outliers. In order to introduce the trimmed reconstruction error estimator for robust PCA, we employ the analogy with the least trimmed squares estimator [15] for robust regression. We denote by $r_i(m, U) = \left\| \left(U U^\top - I \right) \left(x_i - m \right) \right\|_2^2$ the reconstruction error of observation x_i for the given affine subspace parameterized by (m, U). Then the trimmed reconstruction error is defined to be the sum of the t-smallest reconstruction errors $r_i(m, U)$,

$$R(m, U) = \frac{1}{t} \sum_{i=1}^{t} r_{(i)}(m, U), \tag{3}$$

where $r_{(1)}(m, U) \leq \cdots \leq r_{(n)}(m, U)$ are in nondecreasing order and t, with $\lceil \frac{n}{2} \rceil \leq t \leq n$, should be a lower bound on the number of true examples T.

If such an estimate is not available as it is common in unsupervised learning, one can set by default $t = \lceil \frac{n}{2} \rceil$. With the latter choice it is straightforward to see that the corresponding PCA estimator has the maximum possible breakdown point of 0.5, that is up to 50 % of the data points can be arbitrarily corrupted. With the default choice our method has no free parameter except the rank k.

The minimization of the trimmed reconstruction error (3) leads then to a simple and intuitive formulation of robust PCA

$$\{m^*, U^*\} = \underset{m \in \mathbb{R}^p, \, U \in \mathcal{S}_k}{\arg\min} \; R(m, U) = \underset{m \in \mathbb{R}^p, \, U \in \mathcal{S}_k}{\arg\min} \; \frac{1}{t} \sum_{i=1}^{t} r_{(i)}(m, U). \tag{4}$$

Note that the estimation of the subspace U and the center m is done jointly. This is in contrast to [2,3,11,14,19,20], where the data has to be centered by a separate robust method which can lead to quite large errors in the estimation of the true PCA components. The same criterion (4) has been proposed by [13], see also [21] for a slightly different version. While both papers state that the direct minimization of (4) would be desirable, [13] solve a relaxation of (4) into a convex problem while [21] smooth the problem and employ deterministic annealing. Both approaches introduce an additional regularization parameter controlling the number of outliers. It is non-trivial to choose this parameter.

3 TRPCA: Minimizing Trimmed Reconstruction Error on the Stiefel Manifold

In this section, we introduce TRPCA, our algorithm for the minimization of the trimmed reconstruction error (4). We first reformulate the objective of (4) as it is neither convex, nor concave, nor smooth, even if m is fixed. While the resulting optimization problem is still non-convex, we propose an efficient optimization scheme on the Stiefel manifold with monotonically decreasing objective. Note that all proofs of this section can be found in the supplementary material [16].

3.1 Reformulation and First Properties

The reformulation of (4) is based on the following simple identity. Let $\tilde{x}_i = x_i - m$ and $U \in \mathcal{S}_k$, then

$$r_i(m, U) = \left\| (UU^\top - I)(x_i - m) \right\|_2^2 = -\left\| U^\top \tilde{x}_i \right\|_2^2 + \left\| \tilde{x}_i \right\|_2^2 := \tilde{r}_i(m, U). \tag{5}$$

The equality holds only on the Stiefel manifold. Let $\tilde{r}_{(1)}(m, U) \leq \ldots \leq \tilde{r}_{(n)}(m, U)$, then we get the alternative formulation of (4),

$$\{m^*, U^*\} = \underset{m \in \mathbb{R}^p, \, U \in \mathcal{S}}{\arg\min} \; \tilde{R}(m, U) = \frac{1}{t} \sum_{i=1}^{t} \tilde{r}_i(m, U). \tag{6}$$

While (6) is still non-convex, we show in the next proposition that for fixed m the function $\tilde{R}(m, U)$ is concave on $\mathbb{R}^{p \times k}$. This will allow us to employ a simple optimization technique based on linearization of this concave function.

Proposition 1. *For fixed $m \in \mathbb{R}^p$ the function $\widetilde{R}(m, U) : \mathbb{R}^{p \times k} \to \mathbb{R}$ defined in (6) is concave in U.*

The iterative scheme uses a linearization of $\widetilde{R}(m, U)$ in U. For that we need to characterize the superdifferential of the concave function $\widetilde{R}(m, U)$.

Proposition 2. *Let m be fixed. The superdifferential $\partial \widetilde{R}(m, U)$ of $\widetilde{R}(m, U)$: $\mathbb{R}^{p \times k} \to \mathbb{R}$ is given as*

$$\partial \widetilde{R}(m, U) = \Big\{ \sum_{i \in I} \alpha_i (x_i - m)(x_i - m)^{\top} U \;\Big|\; \sum_{i=1}^{n} \alpha_i = t, \; 0 \leq \alpha_i \leq 1 \Big\}, \quad (7)$$

where $I = \{i \mid \tilde{r}_i(m, U) \leq \tilde{r}_{(t)}(m, U)\}$ with $\tilde{r}_{(1)}(m, U) \leq \ldots \leq \tilde{r}_{(n)}(m, U)$.

3.2   Minimization Algorithm

Algorithm 1 for the minimization of (6) is based on block-coordinate descent in m and U. For the minimization in U we use that $\widetilde{R}(m, U)$ is concave for fixed m. Let $G \in \partial \widetilde{R}(m, U^k)$, then by definition of the supergradient of a concave function,

$$\widetilde{R}\left(m, U^{k+1}\right) \leq \widetilde{R}\left(m, U^k\right) + \langle G, U^{k+1} - U^k \rangle. \quad (8)$$

The minimization of the linear upper bound on the Stiefel manifold can be done in closed form, see Lemma 1 below. For that we use a modified version of a result of [10]. Before giving the proof, we introduce the polar decomposition of a matrix $G \in \mathbb{R}^{p \times k}$ which is defined to be $G = QP$, where $Q \in \mathcal{S}$ is an orthonormal matrix of size $p \times k$ and P is a symmetric positive semidefinite matrix of size $k \times k$. We denote the factor Q of G by $Polar(G)$. The polar can be computed in $\mathcal{O}(pk^2)$ for $p \geq k$ [10] as $Polar(G) = UV^{\top}$ (see Theorem 7.3.2. in [6]) using the SVD of G, $G = U\Sigma V^{\top}$. However, faster methods have been proposed, see [5], which do not even require the computation of the SVD.

Lemma 1. *Let $G \in \mathbb{R}^{p \times k}$, with $k \leq p$, and denote by $\sigma_i(G)$, $i = 1, \ldots, k$, the singular values of G. Then $\min_{U \in \mathcal{S}_k} \langle G, U \rangle = -\sum_{i=1}^{k} \sigma_i(G)$, with minimizer $U^* = -Polar(G)$. If G is of full rank, then $Polar(G) = G(G^{\top}G)^{-1/2}$.*

Given that U is fixed, the center m can be updated simply as the mean of the points realizing the current objective of (6), that is the points realizing the t-smallest reconstruction error. Finally, although the objective of (6) is neither convex nor concave in m, we prove monotonic descent of Algorithm 1.

Theorem 1. *The following holds for Algorithm 1. At every iteration, either $\widetilde{R}(m^{k+1}, U^{k+1}) < \widetilde{R}(m^k, U^k)$ or the algorithm terminates.*

The objective is non-smooth and neither convex nor concave. The Stiefel manifold is a non-convex constraint set. These facts make the formulation of

Algorithm 1. TRPCA

Input: X, t, d, $U^0 \in \mathcal{S}$, and m^0 median of X, tolerance ε
Output: robust center m^k and robust PCs U^k
repeat for $k = 1, 2, \ldots$
    Center data $\widetilde{X}^k = \{\widetilde{x}_i^k = x_i - m^k, \ i = 1, \ldots, n\}$
    Compute supergradient $\mathcal{G}(U^k)$ of $\widetilde{R}(m^k, U^k)$ for fixed m^k
    Update $U^{k+1} = -Polar\left(\mathcal{G}(U^k)\right)$
    Update $m^{k+1} = \frac{1}{t}\sum_{i \in \mathcal{I}^{k'}} x_i$, where $\mathcal{I}^{k'}$ are the indices of the t smallest
    $\widetilde{r}_i(m^k, U^{k+1})$, $i = 1, \ldots, n$
until relative descent below ε

---

critical points conditions challenging. Thus, while potentially stronger convergence results like convergence to a critical point are appealing, they are currently out of reach. However, as we will see in Sect. 4, Algorithm 1 yields good empirical results, even beating state-of-the-art methods based on convex relaxations or other non-convex formulations.

3.3 Complexity and Discussion

The computational cost of each iteration of Algorithm 1 is dominated by $\mathcal{O}(pk^2)$ for computing the polar and $\mathcal{O}(pkn)$ for a supergradient of $\widetilde{R}(m, U)$ and, thus, has total cost $\mathcal{O}(pk(k+n))$. We compare this to the cost of the proximal method in [2,18] for minimizing $\min_{X=A+E} \|A\|_* + \lambda \|E\|_1$. In each iteration, the dominating cost is $\mathcal{O}(\min\{pn^2, np^2\})$ for the SVD of a matrix of size $p \times n$. If the natural condition $k \ll \min\{p, n\}$ holds, we observe that the computational cost of TRPCA is significantly better. Thus even though we do 10 random restarts with different starting vectors, our TRPCA is still faster than all competing methods, which can also be seen from the runtimes in Table 1.

In [13], a relaxed version of the trimmed reconstruction error is minimized:

$$\min_{m \in \mathbb{R}^p, U \in S_k, s \in \mathbb{R}^k} \left\|X - \mathbf{1}_n m^\top - Us - O\right\|_F^2 + \lambda \|O\|_{2,1}, \tag{9}$$

where $\|O\|_{2,1}$ is added in order to enforce row-wise sparsity of O. The optimization is done via an alternating scheme. However, the disadvantage of this formulation is that it is difficult to adjust the number of outliers via the choice of λ and thus requires multiple runs of the algorithm to find a suitable range, whereas in our formulation the number of outliers $n - t$ can be directly controlled by the user or t can be set to the default value $\lceil \frac{n}{2} \rceil$.

4 Experiments

We compare our TRPCA (the code is available for download at [16]) algorithm with the following robust PCA methods: ORPCA [13], LLD[2] [14], HRPCA [19],

---

[2] Note, that the LLD algorithm [14] and the OPRPCA algorithm [20] are equivalent.

standard PCA, and true PCA on the true data T (ground truth). For background subtraction, we also compare our algorithm with PCP [2] and RPCA [17], although the latter two algorithms are developed for a different outlier model.

To get the best performance of LLD and ORPCA, we run both algorithms with different values of the regularization parameters to set the number of zero rows (observations) in the outlier matrix equal to $\tilde{t}$ (which increases runtime significantly). The HRPCA algorithm has the same parameter t as our method. We write (0.5) in front of an algorithm name if the default value $\tilde{t} = \lceil \frac{n}{2} \rceil$ is used, otherwise, we use the ground truth information $\tilde{t} = |T|$. As performance measure we use the reconstruction error relative to the reconstruction error of the true data (which is achieved by PCA on the true data only):

$$\text{tre}(U, m) = \frac{1}{t} \sum_{\{i \mid x_i \in T\}} r_i(m, U) - r_i(\hat{m}_T, \hat{U}_T), \qquad (10)$$

where $\{\hat{m}_T, \hat{U}_T\}$ is the true PCA of T and it holds that $\text{tre}(U, m) \geq 0$. The smaller $\text{tre}(U, m)$, i.e., the closer the estimates $\{m, U\}$ to $\{\hat{m}_T, \hat{U}_T\}$, the better. We choose datasets which are computationally feasible for all methods.

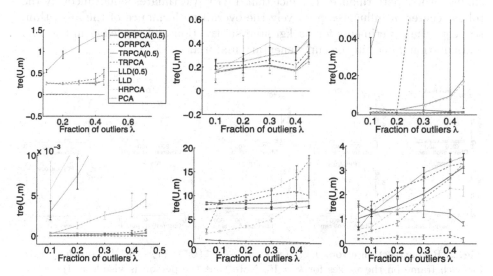

Fig. 1. First row left to right: (1) Data1, $p = 100$, $\sigma_o = 2$; (2) Data1, $p = 20$, $\sigma_o = 2$; (3) Data2, $p = 100$, $\sigma_o = 0.35$; Second row left to right: (1) Data2, $p = 20$, $\sigma_o = 0.35$; (2) USPS10, $k = 1$; (3) USPS10, $k = 10$.

4.1 Synthetic Data Sets

We sample uniformly at random a subspace of dimension k spanned by $U \in S_k$ and generate the true data $T \in \mathbb{R}^{t \times p}$ as $T = AU^\top + E$ where the entries of

$A \in \mathbb{R}^{t \times k}$ are sampled uniformly on $[-1, 1]$ and the noise $E \in \mathbb{R}^{t \times p}$ has Gaussian entries distributed as $\mathcal{N}(0, \sigma_T)$. We consider two types of outliers: (Data1) the outliers $O \in \mathbb{R}^{o \times p}$ are uniform samples from $[0, \sigma_o]^p$, (Data2) the outliers are samples from a random half-space, let w be sampled uniformly at random from the unit sphere and let $x \sim \mathcal{N}(0, \sigma_0 \mathbb{1})$ then an outlier $o_i \in \mathbb{R}^p$ is generated as $o_i = x - \max\{\langle x, w \rangle, 0\} w$. For Data2, we also downscale true data by 0.5 factor. We always set $n = t + o = 200$, $k = 5$, and $\sigma_T = 0.05$ and construct data sets for different fractions of outliers $\lambda = \frac{o}{t+o} \in \{0.1, 0.2, 0.3, 0.4, 0.45\}$. For every λ we sample 5 data sets and report mean and standard deviation of the relative true reconstruction error $\mathrm{tre}(U, m)$.

4.2   Partially Synthetic Data Set

We use USPS, a dataset of 16×16 images of handwritten digits. We use digits 1 as true observations T and digits 0 as outliers O and mix them in different proportions. We refer to this data set as USPS10 and the results can be found in Fig. 1. We notice that TRPCA algorithm with the parameter value $\tilde{t} = t$ (ground truth information) performs almost perfectly and outperforms all other methods, while the default version of TRPCA with parameter $\tilde{t} = \lceil \frac{n}{2} \rceil$ shows slightly worse performance. The fact that TRPCA estimates simultaneously the robust center m influences positively the overall performance of the algorithm, see, e.g., the experiments for background subtraction and modeling in Sect. 4.3 and additional ones in the supplementary material.

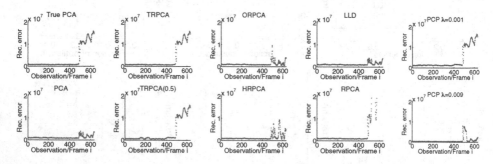

Fig. 2. Reconstruction errors, i.e., $||(x_i - m^*) - U^* (U^*)^\top (x_i - m^*)||_2^2$, on the y-axis, for each frame on the x-axes for $k = 10$. Note that the person is visible in the scene from frame 481 until the end. We consider the background images as true data and, thus, the reconstruction error should be high after frame 481 (when the person enters).

4.3   Background Modeling and Subtraction

In [2,17] robust PCA has been proposed as a method for background modeling and subtraction. While we are not claiming that robust PCA is the best method to do this, it is an interesting test for robust PCA. The data X are the image frames of a video sequence. The idea is that slight change in the background leads to a low-rank variation of the data whereas the foreground changes cannot be modeled by this and can be considered as outliers. Thus with the estimates m^* and U^* of the robust PCA methods, the solution of the background subtraction and modeling problem is given as

$$x_i^b = m^* + U^*(U^*)^\top (x_i - m^*) \tag{11}$$

where x_i^b is the background of frame i and its foreground is simply $x_i^f = x_i - x_i^b$.

We experimentally compare the performance of all robust PCA methods on the water surface data set [1], which has moving water in its background. We choose this dataset of $n = 633$ frames each of size $p = 128 \times 160 = 20480$ as it is computationally feasible for all the methods. In Fig. 3, we show the background

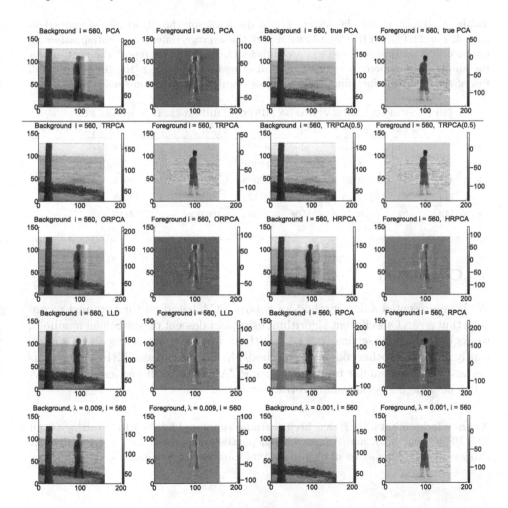

Fig. 3. Backgrounds and foreground for frame $i = 560$ of the water surface data set. The last row corresponds to the PCP algorithm with values of λ set by hand

subtraction results of several robust PCA algorithms. We optimized the value λ for PCP of [2,18] by hand to obtain a good decomposition, see the bottom right pictures of Fig. 3. How crucial the choice of λ is for this method can be seen from the bottom right pictures. Note that the reconstruction error of both the default version of TRPCA and TRPCA(0.5) with ground truth information provide almost perfect reconstruction errors with respect to the true data, cf., Fig. 2. Hence, TRPCA is the only method which recovers the foreground and background without mistakes. We refer to the supplementary material for more explanations regarding this experiment as well as results for another background subtraction data set. The runtimes of all methods for the water surface data set are presented in Table 1, which shows that TRPCA is the fastest of all methods.

Table 1. Runtimes for the water surface data set for the algorithms described in Sect. 4. For TRPCA/TRPCA(0.5) we report the average time of one initialization (in practice, $5 - -10$ random restarts are sufficient). For PCP we report the runtime for the employed parameter $\lambda = 0.001$. For all others methods, it is the time of one full run of the algorithm including the search for regularization parameters.

	TRPCA	TRPCA (.5)	ORPCA	ORPCA (.5)	HRPCA	HRPCA (.5)	LLD	RPCA	PCP ($\lambda = 0.001$)
k=1	7	13	3659	3450	45990	48603	–	1078	–
k=3	99	61	8151	13852	50491	56090	–	730	–
k=5	64	78	2797	3726	72009	77344	232667	3615	875
k=7	114	62	4138	3153	67174	90931	–	4230	–
k=9	119	92	6371	8508	96954	106782	–	4113	–

5   Conclusion

We have presented a new method for robust PCA based on the trimmed reconstruction error. Our efficient algorithm, using fast descent on the Stiefel manifold, works in the default setting $(t = \lceil \frac{n}{2} \rceil)$ without any free parameters and is significantly faster than other competing methods. In all experiments TRPCA performs better or at least similar to other robust PCA methods, in particular, TRPCA solves challenging background subtraction tasks.

Acknowledgements. M.H. has been partially supported by the ERC Starting Grant NOLEPRO and M.H. and S.S. have been partially supported by the DFG Priority Program 1324, "Extraction of quantifiable information from complex systems".

References

1. Bouwmans, T.: Recent advanced statistical background modeling for foreground detection: a systematic survey. Recent Pat. Comput. Sci. **4**(3), 147–176 (2011)
2. Candès, E., Li, X., Ma, Y., Wright, J.: Robust principal component analysis? J. ACM **58**(3), 11:1–11:37 (2011)
3. Croux, C., Pilzmoser, P., Oliveira, M.R.: Algorithms for projection-pursuit robust principal component analysis. Chemometr. Intell. Lab. Syst. **87**, 218–225 (2007)
4. Hampel, F.R., Ronchetti, E.M., Rousseeuw, P.J., Stahel, W.A.: Robust Statistics: The Approach Based on Influence Functions. John Wiley and Sons, New York (1986)
5. Higham, N.J., Schreiber, R.S.: Fast polar decomposition of an arbitrary matrix. SIAM J. Sci. Comput. **11**(4), 648–655 (1990)
6. Horn, R., Johnson, C.: Matrix Analysis. Cambridge University Press, Cambridge (1990)
7. Huber, P.J.: Projection pursuit. Ann. Stat. **13**(2), 435–475 (1985)
8. Huber, P., Ronchetti, E.: Robust Statistics, 2nd edn. John Wiley and Sons, New York (2009)
9. Jolliffe, I.: Principal Component Analysis. Springer Series in Statistics, 2nd edn. Springer, New York (2002)
10. Journée, M., Nesterov, Y., Richtárik, P., Sepulchre, R.: Generalized power method for sparse principal component analysis. J. Mach. Learn. Res. **1**(1), 517–553 (2010)
11. Li, G., Chen, Z.: Projection-pursuit approach to robust dispersion matrices and principal components: primary theory and Monte Carlo. J. Am. Stat. Assoc. **80**(391), 759–766 (1985)
12. Mackey, L.: Deflation methods for sparse PCA. In: 24th Conference on Neural Information Processing Systems, pp. 1017–1024 (2009)
13. Mateos, G., Giannakis, G.: Robust PCA as bilinear decomposition with outlier-sparsity regularization. IEEE Trans. Signal Process. **60**(10), 5176–5190 (2012)
14. McCoy, M., Tropp, J.A.: Two proposals for robust PCA using semidefinite programming. Electron. J. Stat. **5**, 1123–1160 (2011)
15. Rousseeuw, P.J.: Least median of squares regression. J. Am. Stat. Assoc. **79**(388), 871–880 (1984)
16. Supplementary material. http://www.ml.uni-saarland.de/code/trpca/trpca.html
17. De la Torre, F., Black, M.: Robust principal component analysis for computer vision. In: 8th IEEE International Conference on Computer Vision, pp. 362–369 (2001)
18. Wright, J., Peng, Y., Ma, Y., Ganesh, A., Rao, S.: Robust principal component analysis: exact recovery of corrupted low-rank matrices by convex optimization. In: 24th Conference on Neural Information Processing Systems, pp. 2080–2088 (2009)
19. Xu, H., Caramanis, C., Mannor, S.: Outlier-robust PCA: the high dimensional case. IEEE Trans. Inf. Theory **59**(1), 546–572 (2013)
20. Xu, H., Caramanis, C., Sanghavi, S.: Robust PCA via outlier pursuit. IEEE Trans. Inf. Theory **58**(5), 3047–3064 (2012)
21. Xu, L., Yuille, A.L.: Robust principal component analysis by self-organizing rules based on statistical physics approach. IEEE Trans. Neural Networks **6**, 131–143 (1995)

An $\mathcal{O}(n \log n)$ Cutting Plane Algorithm for Structured Output Ranking

Matthew B. Blaschko[1]($\boxtimes$), Arpit Mittal[2], and Esa Rahtu[3]

[1] Inria and École Centrale Paris, Paris, France
matthew.blaschko@inria.fr
[2] Department of Engineering Science, University of Oxford, Oxford, UK
[3] Center for Machine Vision Research, University of Oulu, Oulu, Finland

Abstract. In this work, we consider ranking as a training strategy for structured output prediction. Recent work has begun to explore structured output prediction in the ranking setting, but has mostly focused on the special case of bipartite preference graphs. The bipartite special case is computationally efficient as there exists a linear time cutting plane training strategy for hinge loss bounded regularized risk, but it is unclear how to feasibly extend the approach to complete preference graphs. We develop here a highly parallelizable $\mathcal{O}(n \log n)$ algorithm for cutting plane training with complete preference graphs that is scalable to millions of samples on a single core. We explore theoretically and empirically the relationship between slack rescaling and margin rescaling variants of the hinge loss bound to structured losses, showing that the slack rescaling variant has better stability properties and empirical performance with no additional computational cost per cutting plane iteration. We further show generalization bounds based on uniform convergence. Finally, we demonstrate the effectiveness of the proposed family of approaches on the problem of object detection in computer vision.

1 Introduction

Learning to rank is a core task in machine learning and information retrieval [13]. We consider here a generalization to structured output prediction of the pairwise ranking SVM introduced in [9]. Similar extensions of ranking to the structured output setting [3] have recently been explored in [5,16,22]. In these works, pairwise constraints were introduced between elements in a structured output space, enforcing a margin between a lower ranked item and a higher ranked item proportional to the difference in their structured output losses. These works consider only bipartite preference graphs. Although efficient algorithms exist for cutting plane training in the bipartite special case, no feasible algorithm has previously been proposed for extending this approach to fully connected preference graphs for arbitrary loss functions.

Our work makes feasible structured output ranking with a complete preference graph for arbitrary loss functions. Joachims previously proposed an algorithm for ordinal regression with 0–1 loss and R possible ranks in $\mathcal{O}(nR)$ time for

© Springer International Publishing Switzerland 2014
X. Jiang et al. (Eds.): GCPR 2014, LNCS 8753, pp. 132–143, 2014.
DOI: 10.1007/978-3-319-11752-2_11

n samples [10]. This effectively enables a complete preference graph in this special setting. In practice, however, for structured output prediction with a sufficiently rich output space, the loss values may not be discrete, and may grow linearly with the number of samples. In this case, R is $\mathcal{O}(n)$. Mittal et al. have extended Joachims' $\mathcal{O}(nR)$ result to the structured output ranking setting in the case that there are a discrete set of loss values [15]. A direct extension of these approaches to the structured output setting with a fully connected preference graph and arbitrary loss functions results in a $\mathcal{O}(n^2)$ cutting plane iteration. One of the key contributions of our work is to show that this can be improved to $\mathcal{O}(n \log n)$ time. This enables us to train an objective with 5×10^7 samples on standard hardware (Sect. 5). Furthermore, straightforward parallelization schemes enable e.g. $\mathcal{O}(n)$ computation time on $\mathcal{O}(\log n)$ processors (Sect. 3.1). These results hold not only for the structured output prediction setting, but can be used to improve the computational efficiency of related ranking SVM approaches, e.g. [10].

Analogous to the structured output SVM [17,18], we formulate structured output ranking in slack rescaling and margin rescaling variants. We show uniform convergence bounds for our ranking objective in a unified setting for both variants. Interestingly, the bounds for slack rescaling are dependent on the range of the loss values, while those for margin rescaling are not. Further details are given in Sect. 4. Structured output ranking is a natural strategy for cascade learning, in which an inexpensive feature function, ϕ, is used to filter a set of possible outputs y. We show empirical results in the cascade setting (Sect. 5) supporting the efficiency, accuracy, and generalization of the proposed solution to structured output prediction.

2   Structured Output Ranking

The setting considered here is to learn a compatibility function $g : \mathcal{X} \times \mathcal{Y} \mapsto \mathbb{R}$ that maps an input-output tuple to a real value indicating the prediction of how suitable the input is to a given output. We assume that there is an underlying ground truth prediction for a given input so that every $x_i \in \mathcal{X}$ in a training set is associated with a y_i^* corresponding to the optimal prediction for that input. Additionally, we assume that a loss function $\Delta : \mathcal{Y} \times \mathcal{Y} \mapsto \mathbb{R}$ is provided that measures the similarity of a hypothesized output to the optimal prediction $\Delta(y_i^*, y) \geq 0$. A training set will consist of input-ground truth-output tuples, where the input-ground truth pairs may be repeated, and the outputs are sampled over the output space: $\mathcal{S} = \{(x_i, y_i^*, y_i)\}_{1 \leq i \leq n}$ and (x_i, y_i^*) may equal (x_j, y_j^*) for $j \neq i$ (cf. Sect. 5). We will use the notation Δ_i to denote $\Delta(y_i^*, y_i)$.

In structured output ranking, we minimize with respect to a compatibility function, g, a risk of the form [1]

$$R(g) = \mathbb{E}_{((X_i, Y_i),(X_j, Y_j))} \big[|\Delta_{Y_j} - \Delta_{Y_i}| \cdot \big(\mathbf{1}((\Delta_{Y_j} - \Delta_{Y_i})(g(X_i, Y_i) - g(X_j, Y_j)) < 0)$$

$$+ \frac{1}{2}\mathbf{1}(g(X_i, Y_i) = g(X_j, Y_j)) \big], \tag{1}$$

where $\mathbf{1}(\cdot)$ evaluates to 1 if the argument is true and 0 otherwise, and the term penalizing equality is multiplied by $\frac{1}{2}$ in order to avoid double counting the penalty over the expectation. Here Δ_{Y_i} is the structured output loss associated with an output, Y_i. In contrast to other notions of risk, we take the expectation not with respect to a single sample, but with respect to pairs indexed by the structured output. Given two possible outputs sampled from some prior, the risk determines whether the samples are properly ordered according to the loss associated with predicting that output, and if not pays a penalty proportional to the difference in the losses. This risk penalizes pairs for which sample i has lower loss than sample j and also lower ranking score, i.e. we would like elements with low loss to be ranked higher than elements with high loss.

Two piecewise linear convex upper bounds are commonly used in structured output prediction: a margin rescaled hinge loss, and a slack rescaled hinge loss. The structured output ranking objectives corresponding to regularized risk minimization with these choices are

$$\min_{w\in\mathcal{H},\xi\in\mathbb{R}} \lambda\Omega(w) + \xi \tag{2}$$

$$\text{s.t.} \sum_{(i,j)\in\mathcal{E}} \nu_{ij} \overbrace{\left(\langle w, \phi(x_i, y_i)\rangle - \langle w, \phi(x_j, y_j)\rangle + \Delta_i - \Delta_j\right)}^{\text{margin rescaling}} \geq -\xi \tag{3}$$

$$\text{or} \sum_{(i,j)\in\mathcal{E}} \nu_{ij} \underbrace{\left(\langle w, \phi(x_i, y_i) - \phi(x_j, y_j)\rangle - 1\right)\left(\Delta_j - \Delta_i\right)}_{\text{slack rescaling}} \geq -\xi \tag{4}$$

$$\xi \geq 0 \qquad\qquad \forall \nu \in \{0,1\}^{|\mathcal{E}|} \tag{5}$$

where $\mathcal{E}$ is the edge set associated with a preference graph $\mathcal{G}$,[1] and Ω is a regularizer monotonically increasing in some function norm applied to w [12]. We have presented the one-slack variant here [11]. For a finite sample of (x_i, y_i^*, y_i), such objectives can be solved using a cutting plane approach [10,15,16,20].

The form of $\mathcal{G}$ defines risk variants that encode different preferences in ranking. If an edge exists from node i to node j, this indicates that i should be ranked higher than j. Of particular interest in this work are bipartite graphs, which have efficiencies in computation, and fully connected graphs, which attempt to enforce a total ordering on the samples. Structured output ranking with bipartite preference graphs was previously explored in [16], in which a linear time algorithm was presented for a cutting plane iteration. The algorithm presented in that work shares key similarities with previous strategies for cutting plane training of ranking support vector machines [10], but extends the setting to rescaled structured output losses. A linear time algorithm for fully connected preference graphs was presented in [15] in the special case that the loss values are in a small discrete set. Previous algorithms all degenerate to $\mathcal{O}(n^2)$ when applied to fully connected preference graphs with arbitrary loss values.

---

[1] An edge from i to j in $\mathcal{G}$ indicates that output i should be ranked above output j. It will generally be the case that $\Delta_j \geq \Delta_i$ for all $(i,j) \in \mathcal{E}$.

3 $\mathcal{O}(n \log n)$ Cutting Plane Algorithm

Algorithm 1. Finding maximally violated slack-rescaled constraint for structured output ranking with a complete bipartite preference graph.

Input: Δ, a list of loss values sorted from lowest to highest; s, a vector of the current estimate of compatibility scores $(s_u = \langle w, \phi(x_u, y_u) \rangle_\mathcal{H})$ in the same order as Δ; p, a vector of indices such that $s_{p_v} > s_{p_u}$ whenever $v > u$; t, a threshold such that $(u, v) \in \mathcal{E}$ whenever $u \le t$ and $v > t$

Output: Maximally violated constraint is $\delta - \langle w, \sum_i \alpha_i \phi(x_i, y_i) \rangle \le \xi$

1: $p^+ = p_{\{u | p_u \le t\}}$, $p^- = p_{\{v | p_v > t\}}$
2: $i = 1$, $\delta = \Delta_+ = 0$, $\Delta^{\mathrm{cum}} = \mathbf{0}$, $\alpha = 0$
3: $\Delta^{\mathrm{cum}}_{n-t} = \Delta_{p^-_{n-t}}$
4: **for** $k = n - t - 1$ to 1 descending **do**
5:    $\Delta^{\mathrm{cum}}_{p^-_k} = \Delta_{p^-_k} + \Delta^{\mathrm{cum}}_{p^-_{k+1}}$
6: **end for**
7: **for** $j = 1$ to $n - t$ **do**
8:    **while** $s_{p^-_j} + 1 > s_{p^+_i} \wedge i \le t + 1$ **do**
9:       $\alpha_{p^+_i} = \alpha_{p^+_i} + \Delta^{\mathrm{cum}}_{p^-_j} - (n - t - j + 1)\Delta_{p^+_i}$
10:       $\Delta_+ = \Delta_+ + \Delta_{p^+_i}$, $i = i + 1$
11:    **end while**
12:    $\alpha_{p^-_j} = \alpha_{p^-_j} - ((j-1)\Delta_{p^-_j} - \Delta_+)$
13:    $\delta = \delta + (j-1)\Delta_{p^-_j} - \Delta_+$
14: **end for**
15: **return** (α, δ)

Algorithm 2. An $\mathcal{O}(n \log n)$ recursive algorithm for computing a cutting plane iteration for fully connected ranking preference graphs.

Input: Δ, a list of loss values sorted from lowest to highest; s, a vector of the current estimate of compatibility scores $(s_u = \langle w, \phi(x_u, y_u) \rangle_\mathcal{H})$ in the same order as Δ; p, an index such that $s_{p_v} > s_{p_u}$ whenever $v > u$

Output: Maximally violated constraint is $\delta - \langle w, \sum_i \alpha_i \phi(x_i, y_i) \rangle \le \xi$

1: $n = \mathrm{length}(\Delta)$
2: **if** $\Delta_1 = \Delta_n$ **then**
3:    **return** $(0, 0)$
4: **end if**
5: $t \approx \frac{n}{2}$
6: $p^a = p_{\{u | p_u \le t\}}$
7: $(\alpha_1, \delta_1) = $ Algorithm 2 $(\Delta_{1:t}, s_{1:t}, p^a)$
8: $p^b = p_{\{v | p_v > t\}}$
9: $p^b = p^b - t$ (subtract t from each element of p^b)
10: $(\alpha_2, \delta_2) = $ Algorithm 2 $(\Delta_{t+1:n}, s_{t+1:n}, p^b)$
11: $(\alpha_0, \delta_0) = $ Algorithm 1(Δ, s, p, t)
12: $\alpha = \alpha_0 + \alpha_1 + \alpha_2$, $\delta = \delta_0 + \delta_1 + \delta_2$
13: **return** (α, δ)

Cutting plane optimization of (2)–(5) consists of alternating between optimizing the objective with a finite set of active constraints, finding a maximally violated constraint of the current function estimate and adding it to the active constraint set [11]. Algorithm 1 gives a linear time procedure for finding the maximally violated constraint in the case of a complete bipartite preference graph [10,16] and slack rescaling.[2] This algorithm follows closely the ordinal regression cutting plane algorithm of [10], and works by performing an initial sort on the current estimate of the sample scores. The algorithm subsequently makes use of the transitivity of violated pairwise constraints to sum all violated pairs in a single pass through the sorted list of samples.

---

[2] An analogous algorithm for margin rescaling was given in [16] and has the same computational complexity.

In the case of fully connected preference graphs, Algorithm 2 is a recursive function that ensures that all pairs of samples are considered. Algorithm 2 uses a divide and conquer strategy and works by repeatedly calling Algorithm 1 for various bipartite subgraphs with disjoint edge sets, ensuring that the union of the edge sets of all bipartite subgraphs is the edge set of the preference graph. The set of bipartite subgraphs is constructed by partitioning the set of samples into two roughly equal parts by thresholding the loss function. As the samples are assumed to be sorted by their structured output loss, we simply divide the set by computing the index of the median element. In the event that there are multiple samples with the same loss, the partitioning (Algorithm 2, line 5) may do a linear time search from the median loss value to find a partitioning of the samples such that the first set has strictly lower loss than the second. The notation $p^a = p_{\{u|p_u \leq t\}}$ indicates that p^a contains the elements satisfying the condition in the subscript in the same order that they occured in p. Source code is available for download.[3]

3.1   Complexity

Prior to calling either of the algorithms, the current data sample must be sorted by its structured output loss. Additionally an index vector, p, must be computed that encodes a permutation matrix that sorts the training sample by the current estimate of its compatibility scores, $\langle w, \phi_i \rangle$. Each of these operations has complexity $\mathcal{O}(n \log n)$. The serial complexity of computing the most violated 1-slack constraint is $\mathcal{O}(n \log_2 n)$, matching the complexity of the sorting operation. To show this, we consider the recursion in Algorithm 2. The computational costs of each call consist of (i) the processing needed to find the sorted list of scores for the higher ranked and lower ranked subsets in the bipartite graph, (ii) the cost of calling Algorithm 1, and (iii) the cost of recursion. We will show that items (i) and (ii) can be computed in time linear in the number of samples.

That item (i) is linear in its complexity can be seen by noting that an index p already exists to sort the complete data sample. Rather than pay $\mathcal{O}(n \log n)$ to re-sort the subsets of samples, we may iterate through the elements of p once. As we do so, if $p_i \leq t$, we may add this element to the index that sorts the higher ranked subset. If $p_j > t$, we may add $p_j - t$ to the index that sorts the lower ranked subset. Item (ii) is also linear as the algorithm loops once over each data sample, executing a constant number of operations each time.

We calculate the complexity of Algorithm 2 by a recursive formula $R_n = C_n + 2R_{\frac{n}{2}}$ where C_n is the $\mathcal{O}(n)$ cost of processing items (i) and (ii). It follows that

$$R_n = \sum_{i=0}^{\log_2 n} C_{\frac{n}{2^i}} 2^i. \tag{6}$$

Examining the term $C_{\frac{n}{2^i}} 2^i$, we note that $C_{\frac{n}{2^i}}$ is $\mathcal{O}(\frac{n}{2^i})$ and must be paid 2^i times, resulting in a cost of $\mathcal{O}(n)$ per summand. As there are $\mathcal{O}(\log_2 n)$ summands, the

---

[3] http://pages.saclay.inria.fr/matthew.blaschko/projects/structrank/

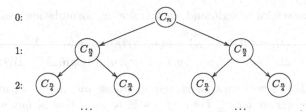

Fig. 1. The recursion tree for Algorithm 2. Each node in the tree corresponds to a set of constraints resulting from a bipartite preference graph. The cost of computing these constraints is labeled in each of the nodes.

total cost is $\mathcal{O}(n \log n)$. Graphically, the recursion tree is a binary tree in which the cost of each node is proportional to $\frac{1}{2^d}$, where d is the depth of the node (Fig. 1). A C implementation of the algorithm takes a fraction second for 10^5 samples on a 2.13 GHz processor.

A straightforward parallelization scheme can be achieved by placing each recursive call in its own thread. Doing so results in $\mathcal{O}(n)$ computation on $\mathcal{O}(\log n)$ processors: each level of a tree at depth i can be computed independently in $C_{\frac{n}{2^i}} 2^i$ instructions, and there are $\mathcal{O}(\log n)$ levels of the tree. Each of $\log n$ processors can be assigned the nodes corresponding to a given level of the tree.

4   Generalization Bounds

In this section, we develop generalization bounds based on the uniform convergence bounds for ranking algorithms presented in [1]. For $\Delta \in [0, 1)$ we have tighter bounds for slack rescaling as compared to margin rescaling. For $\Delta \in [0, \sigma]$ where $\sigma > 1$ bounds are tighter for margin rescaling.

Definition 1 (Uniform loss stability (β)). *A ranking algorithm which is trained on the sample $\mathcal{S}$ of size n has a uniform loss stability β with respect to the ranking loss function ℓ if,*

$$|\ell(\mathcal{S}) - \ell(\mathcal{S}^k)| \leq \beta(n), \ \forall n \in \mathbb{N}, 1 \leq k \leq n \qquad (7)$$

where $\mathcal{S}^k$ is a sample resulting from changing the kth element of $\mathcal{S}$, i.e., changing the input training sample by a single example leads to a difference of at most $\beta(n)$ in the loss incurred by the output ranking function on any pair of examples. Thus, a smaller value of $\beta(n)$ corresponds to a greater loss stability.

Definition 2 Uniform score stability (ν). *A ranking algorithm with an output g_S on the training sample $\mathcal{S}$ of size n, has a uniform score stability ν if*

$$|g_S(x) - g_{S^k}(x)| \leq \nu(n), \ \forall n \in N, 1 \leq k \leq n, \forall x \in \mathcal{X} \qquad (8)$$

i.e., changing an input training sample by a single example leads to a difference of at most $\nu(n)$ in the score assigned by the ranking function to any instance x.

The hinge losses for margin and slack rescaling formulations are given by:

$$\ell_m = (|\Delta_j - \Delta_i| - \langle w, \phi(x_i, y_i) - \phi(x_j, y_j) \rangle \cdot \text{sign}(\Delta_j - \Delta_i))_+, \tag{9}$$

$$\ell_s = (|\Delta_j - \Delta_i| \cdot (1 - \langle w, \phi(x_i, y_i) - \phi(x_j, y_j) \rangle \cdot \text{sign}(\Delta_j - \Delta_i)))_+. \tag{10}$$

Theorem 1. *Let $\mathcal{A}$ be a ranking algorithm whose output on a training sample $\mathcal{S} \in (\mathcal{X}, \mathcal{Y})^n$ we denote by $f_{\mathcal{S}}$. Let $\nu : \mathbb{N} \to \mathbb{R}$ be such that $\mathcal{A}$ has uniform score stability ν. $\mathcal{A}$ has uniform loss stability β with respect to the slack rescaling loss ℓ_s, where for all $n \in \mathbb{N}$*

$$\beta(n) = 2\sigma\nu(n) \tag{11}$$

where $\sigma \geq \Delta$ is an upper bound on the structured output loss function.

Proof. Without loss of generality we assume that $\ell_s(\mathcal{S}) > \ell_s(\mathcal{S}^k)$. There are two non-trivial cases.

Case (i): Margin is violated by both $g_{\mathcal{S}}$ and $g_{\mathcal{S}^k}$.

$$|\ell_s(\mathcal{S}) - \ell_s(\mathcal{S}^k)| = |\Delta_j - \Delta_i| \cdot (1 - (g_{\mathcal{S}}(x_i) - g_{\mathcal{S}}(x_j)) \cdot \text{sign}(\Delta_j - \Delta_i)) - \tag{12}$$
$$|\Delta_j - \Delta_i| \cdot (1 - (g_{\mathcal{S}^k}(x_i) - g_{\mathcal{S}^k}(x_j)) \cdot \text{sign}(\Delta_j - \Delta_i))$$
$$\leq \sigma(|g_{\mathcal{S}}(x_i) - g_{\mathcal{S}^k}(x_i)| + |g_{\mathcal{S}}(x_j) - g_{\mathcal{S}^k}(x_j)|) \leq 2\sigma\nu(n) \tag{13}$$

Case (ii): Margin is violated by either of $g_{\mathcal{S}}$ or $g_{\mathcal{S}^k}$. This is a symmetric case, so we assume that the margin is violated by $g_{\mathcal{S}}$.

$$|\ell_s(\mathcal{S}) - \ell_s(\mathcal{S}^k)| = |\Delta_j - \Delta_i| \cdot (1 - (g_{\mathcal{S}}(x_i) - g_{\mathcal{S}}(x_j)) \cdot \text{sign}(\Delta_j - \Delta_i)) \tag{14}$$
$$\leq |\Delta_j - \Delta_i| \cdot (1 - (g_{\mathcal{S}}(x_i) - g_{\mathcal{S}}(x_j)) \cdot \text{sign}(\Delta_j - \Delta_i)) - \tag{15}$$
$$|\Delta_j - \Delta_i| \cdot (1 - (g_{\mathcal{S}^k}(x_i) - g_{\mathcal{S}^k}(x_j)) \cdot \text{sign}(\Delta_j - \Delta_i))$$
$$\leq \sigma(|g_{\mathcal{S}}(x_i) - g_{\mathcal{S}^k}(x_i)| + |g_{\mathcal{S}}(x_j) - g_{\mathcal{S}^k}(x_j)|) \leq 2\sigma\nu(n) \tag{16}$$

Theorem 2 (Slack Rescaling Generalization Bound). *Let $\mathcal{H}$ be a RKHS with a joint-kernel[4] k such that $\forall (x, y) \in \mathcal{X} \times \mathcal{Y}$, $k((x, y), (x, y)) \leq \kappa^2 < \infty$. Let $\lambda > 0$ and ℓ_r be a rescaled ramp loss. The training algorithm trained on sample $\mathcal{S}$ of size n outputs a ranking function $g_{\mathcal{S}} \in \mathcal{H}$ that satisfies $g_{\mathcal{S}} = \arg\min_{g \in \mathcal{H}} \{\hat{R}_{\ell_s}(g; \mathcal{S}) + \lambda \|g\|_{\mathcal{H}}^2\}$. Then for any $0 < \delta < 1$, with probability at least $1 - \delta$ over the draw of $\mathcal{S}$, the expected ranking error of the function is bounded by:*

$$R(g_{\mathcal{S}}) < \hat{R}_{\ell_r}(g_{\mathcal{S}}; \mathcal{S}) + \frac{32\sigma^2\kappa^2}{\lambda n} + \left(\frac{16\sigma^2\kappa^2}{\lambda} + \sigma\right)\sqrt{\frac{2\ln(1/\delta)}{n}} \tag{17}$$

Proof. From [1, Theorem 11], $\nu(n) = \frac{8\sigma\kappa^2}{\lambda n}$. Substituting this value of $\nu(n)$ in Equation (11) $\beta(n) = \frac{16\sigma^2\kappa^2}{\lambda n}$. Inequality (17) then follows by an application of [1, Theorem 6] which gives the generalization bound as a function of $\beta(n)$.

The proof of [1, Theorem 6] follows closely that of [6] for regression and classification, relying at its core on McDiarmid's inequality [14].

---

[4] We assume a joint kernel map of the form given in [17,18].

Theorem 3 (Margin Rescaling Generalization Bound). *Under the conditions of Theorem 2, and a ranking function $f_{\mathcal{S}} \in \mathcal{H}$ that satisfies $f_{\mathcal{S}} = \arg\min_{f \in \mathcal{H}}\{\hat{R}_{\ell_m}(f;\mathcal{S}) + \lambda\|f\|_{\mathcal{H}}^2\}$. Then for any $0 < \delta < 1$, with probability at least $1 - \delta$ over the draw of $\mathcal{S}$, the expected ranking error of the function is bounded by:*

$$R(f_{\mathcal{S}}) < \hat{R}_{\ell_r}(f_{\mathcal{S}};\mathcal{S}) + \frac{32\kappa^2}{\lambda n} + \left(\frac{16\kappa^2}{\lambda} + \sigma\right)\sqrt{\frac{2\ln(1/\delta)}{n}} \qquad (18)$$

The proof of Theorem 3 follows the outline given in [1, Sect. 5.2.1].

5 Experimental Results

Results are presented as an evaluation of a cascade architecture [21], following the evaluation protocol of Rahtu et al. [16]. The experiments are presented on the VOC 2007 dataset [7]. The images are annotated with ground-truth bounding boxes of objects from 20 classes. VOC 2007 train and validation sets are used only to construct the distribution for the initial window sampling, and the ranking function is learned using the dataset presented in [2]. This is done in order to obtain results comparable to those in [2, 16].

The performance is measured using a recall-overlap curve, which indicates the recall rate of ground truth boxes in the VOC 2007 test set for a given minimum value of the overlap score [19]

$$o(y, \tilde{y}) = \frac{\text{Area}(y \cap \tilde{y})}{\text{Area}(y \cup \tilde{y})}, \qquad (19)$$

where y and $\tilde{y}$ denote the ground truth and predicted bounding box, respectively. We also report the area under the curve (AUC) between overlap scores 0.5 and 1, and normalize its value so that the maximum is 1 for perfect recall. The overlap limit 0.5 is chosen here since less accurately localized boxes have little practical importance.

Our framework for creating the set of predicted bounding boxes broadly follows that of [16]. This setting has three main stages: (i) construction of the initial bounding boxes, (ii) feature extraction, and (iii) window selection. In the first stage we generate a pool of approximately 100,000 initial windows per image using random sampling and superpixel bounding boxes. The random samples are drawn from a distribution learned using the ground truth object boxes in the training and validation sets. The superpixels are computed by a graph based method [8], which is selected for its computational efficiency. At overlap 0.5, the initial windows achieve approximately 98 % recall.

In the second stage, the tentative bounding boxes are scored using several publicly available features. These features are window symmetry (WS), boundary edge distribution (BE), superpixel boundary integral (BI), color contrast (CC), superpixel straddling (SS), and multiscale saliency (MS). The WS, BE, and BI features are described in [16] and SS, CC, and MS are from [2]. The joint

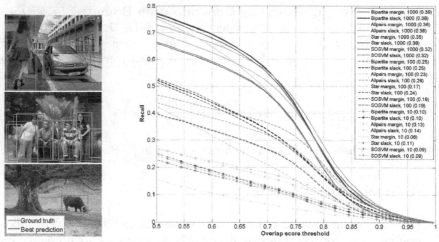

(a) Example detections with (b) Overlap/recall curves. Results are presented for vary-
a complete preference graph ing preference graphs, margin and slack rescaling, and
and slack rescaling. This set- various numbers of returned windows. The AUC score is
ting corresponds to "Allpairs given in parentheses (a higher number at a given number
slack, 1000" in Fig. (b).   of returned windows indicates better performance).

Fig. 2. Example detections and overlap vs. recall for an object detection task. See
Sect. 5 for a complete description of the experimental setting. This figure is best viewed
in color.

feature map, $\phi(x_i, y_i)$, applied in learning is the feature vector corresponding to
the bounding box y_i.

In the last stage, we select the final set of bounding boxes (10, 100 or 1000)
based on the learned score. The feature weights for the linear combination are
learned by using the structured output ranking framework presented in this
paper and the loss function proposed in [4]. This loss is based on the overlap
ratio (19) and is defined as $\Delta_i \equiv 1 - o(y, y_i)$.

In order to run the proposed algorithm, we further need to define the structure of the preference graph $\mathcal{G}$. Three variants were considered: a bipartite graph
in which 1000 best samples per image are ranked higher than all other initial
windows (as in [16]), a fully connected graph (denoted "Allpairs" in the legend of Fig. 2(b)) where full ranking is pursued, or a bipartite graph in which
only ground truth windows are to be ranked higher than all sampled windows
(denoted "Star" in the legend, as the topology of a bipartite graph with one sin-
gleton set is a star graph). Finally, we have trained a standard structured output
SVM (labeled "SOSVM") in the same manor as [4]. To ensure a diverse set of
predictions, we have applied the non-maximal suppression approach described
in [19].

The overlap-recall curves are shown in Fig. 2. The legend in Fig. 2 encodes
the experimental setting for each curve. First, the structure of the preference

graph, $\mathcal{G}$, is specified. The second component of the legend indicates whether slack rescaling or margin rescaling was employed. The third component states the number of top ranked windows used for evaluating the recall. Finally, the fourth component (in parentheses) gives the AUC value.

6 Discussion

The experiments described in Sect. 5 show that structured output ranking is a natural objective to apply to cascade detection models.

On average, a bipartite preference graph performs best if we require 1000 windows as output, which matches the training conditions. The bipartite graph was constructed such that constraints were included between the top 1000 sampled windows, and the remaining 99000 windows. However, when the number of returned windows deviates from 1000, the relative performance of the bipartite ranking decreases and other preference graphs give better performance. The objective is tuned to give the highest performance under a single evaluation setting, at the expense of other settings.

The complete preference graph ranking, labeled "Allpairs" in Fig. 2, gives good performance and tends to have higher performance at high overlap levels. While the difference between slack rescaling and margin rescaling was minimal when using a bipartite preference graph, a much more noticeable difference is present in the case of a complete preference graph. While the bipartite preference graph performs better at certain overlap levels when 1000 windows are returned, the complete preference graph is much more stable across a wide number of windows, and gives the best performance at all overlap levels if 10 windows are returned per image. Finally, the standard structured output SVM (labeled "SOSVM") performs substantially worse than all ranking variants.

7 Conclusions

In this work, we have explored the use of ranking for structured output prediction. We have analyzed both margin and slack rescaling variants of a ranking SVM style approach, showing better empirical results for slack rescaling, and proving generalization bounds for both variants in a unified framework. Furthermore, we have proposed an efficient and parallelizable algorithm for cutting plane training that scales to millions of data points on a single core. We have shown an example application of object detection in computer vision, demonstrating that ranking methods outperform a standard structured output SVM in this setting, and that fully connected preference graphs give excellent performance across a range of settings, particularly at high overlap with the ground truth.

The $\mathcal{O}(n \log n)$ algorithm presented here can be adapted to a wide variety of settings, improving the computational efficiency in a range of ranking approaches and applications. In the setting of [10, 15], the $\mathcal{O}(nR)$ approach for ranking with a complete preference graph and a fixed number, R, of loss values can be improved in an analogous manner to $\mathcal{O}(n \log R)$.

Acknowledgements. This work is partially funded by ERC Grant 259112, and FP7-MC-CIG 334380.

References

1. Agarwal, S., Niyogi, P.: Generalization bounds for ranking algorithms via algorithmic stability. J. Mach. Learn. Res. **10**, 441–474 (2009)
2. Alexe, B., Deselaers, T., Ferrari, V.: What is an object? In: Proceedings of the IEEE Conference on Computer Vision and Pattern Recognition, June 2010
3. Bakır, G.H., Hofmann, T., Schölkopf, B., Smola, A.J., Taskar, B., Vishwanathan, S.V.N.: Predicting Structured Data. MIT Press, Cambridge (2007)
4. Blaschko, M.B., Lampert, C.H.: Learning to localize objects with structured output regression. In: Forsyth, D., Torr, P., Zisserman, A. (eds.) ECCV 2008, Part I. LNCS, vol. 5302, pp. 2–15. Springer, Heidelberg (2008)
5. Blaschko, M.B., Vedaldi, A., Zisserman, A.: Simultaneous object detection and ranking with weak supervision. In: Advances in Neural Information Processing Systems (2010)
6. Bousquet, O., Elisseeff, A.: Stability and generalization. J. Mach. Learn. Res. **2**, 499–526 (2002)
7. Everingham, M., Van Gool, L., Williams, C.K.I., Winn, J., Zisserman, A.: The Pascal visual object classes (VOC) challenge. Int. J. Comput. Vis. **88**(2), 303–338 (2010)
8. Felzenszwalb, P., Huttenlocher, D.: Efficient graph-based image segmentation. Int. J. Comput. Vis. **59**(2), 167–181 (2004)
9. Herbrich, R., Graepel, T., Obermayer, K.: Large margin rank boundaries for ordinal regression. In: Smola, A., Bartlett, P., Schölkopf, B., Schuurmans, D. (eds.) Advances in Large Margin Classifiers, pp. 115–132. MIT Press, Cambridge (2000)
10. Joachims, T.: Training linear SVMs in linear time. In: ACM SIGKDD Conference on Knowledge Discovery and Data Mining, pp. 217–226 (2006)
11. Joachims, T., Finley, T., Yu, C.N.: Cutting-plane training of structural SVMs. Mach. Learn. **77**(1), 27–59 (2009)
12. Lafferty, J., Zhu, X., Liu, Y.: Kernel conditional random fields: representation and clique selection. In: Proceedings of the International Conference on Machine Learning (2004)
13. Manning, C.D., Raghavan, P., Schütze, H.: Introduction to Information Retrieval. Cambridge University Press, New York (2008)
14. McDiarmid, C.: On the method of bounded differences. In: Siemons, J. (ed.) Surveys in Combinatorics, pp. 148–188. Cambridge University Press, Cambridge (1989)
15. Mittal, A., Blaschko, M.B., Zisserman, A., Torr, P.H.S.: Taxonomic multi-class prediction and person layout using efficient structured ranking. In: Fitzgibbon, A., Lazebnik, S., Perona, P., Sato, Y., Schmid, C. (eds.) ECCV 2012, Part II. LNCS, vol. 7573, pp. 245–258. Springer, Heidelberg (2012)
16. Rahtu, E., Kannala, J., Blaschko, M.B.: Learning a category independent object detection cascade. In: Proceedings of the International Conference on Computer Vision (2011)
17. Taskar, B., Guestrin, C., Koller, D.: Max-margin Markov networks. In: Advances in Neural Information Processing Systems (2004)

18. Tsochantaridis, I., Hofmann, T., Joachims, T., Altun, Y.: Support vector machine learning for interdependent and structured output spaces. In: Proceedings of the International Conference on Machine Learning (2004)
19. Vedaldi, A., Gulshan, V., Varma, M., Zisserman, A.: Multiple kernels for object detection. In: Proceedings of the International Conference on Computer Vision (2009)
20. Vedaldi, A., Blaschko, M.B., Zisserman, A.: Learning equivariant structured output SVM regressors. In: Proceedings of the International Conference on Computer Vision, pp. 959–966 (2011)
21. Viola, P., Jones, M.: Robust real-time object detection. Int. J. Comput. Vis. **57**(2), 137–154 (2002)
22. Zhang, Z., Warrell, J., Torr, P.: Proposal generation for object detection using cascaded ranking SVMs. In: Proceedings of the IEEE Conference on Computer Vision and Pattern Recognition (2011)

Exemplar-Specific Patch Features
for Fine-Grained Recognition

Alexander Freytag[1]([⊠]), Erik Rodner[1], Trevor Darrell[2], and Joachim Denzler[1]

[1] Computer Vision Group, Friedrich Schiller University Jena, Jena, Germany
alexander.freytag@uni-jena.de
[2] UC Berkeley ICSI and EECS, Berkeley, USA

Abstract. In this paper, we present a new approach for fine-grained recognition or subordinate categorization, tasks where an algorithm needs to reliably differentiate between visually similar categories, *e.g.*, different bird species. While previous approaches aim at learning a single generic representation and models with increasing complexity, we propose an orthogonal approach that learns patch representations specifically tailored to every single test exemplar. Since we query a constant number of images similar to a given test image, we obtain very compact features and avoid large-scale training with all classes and examples. Our learned mid-level features are built on shape and color detectors estimated from discovered patches reflecting small highly discriminative structures in the queried images. We evaluate our approach for fine-grained recognition on the CUB-2011 birds dataset and show that high recognition rates can be obtained by model combination.

1 Introduction

Nearly all image categorization and object recognition systems are built on the general idea of using a set of patch detectors and their outputs as proper features for classification. This is the case for bag-of-features approaches, *e.g.*, [22], where the set of detectors is usually referred to as codebook or vocabulary of local features, it holds for recent deep convolutional networks, *e.g.*, [21,31], where detectors are convolutional filter masks learned on different levels5, and it also holds for discriminative patch techniques, *e.g.*, [8,18,24]. In all cases, a single set of these detectors is learned and the intra-class variability needs to be tackled by choosing a large number of sparsely coded detectors [22] or by stacking them together into several layers [21].

In contrast, we show how to build patch-based feature representations specifically for each test example. Our approach allows focusing features and the set of patch detectors on the task of differentiating objects with a similar pose and similar global appearance. This ability is especially useful for fine-grained recognition tasks, where finding subtle differences is important. Throughout the paper,

A. Freytag and E. Rodner were supported by a FIT scholarship from the DAAD.

© Springer International Publishing Switzerland 2014
X. Jiang et al. (Eds.): GCPR 2014, LNCS 8753, pp. 144–156, 2014.
DOI: 10.1007/978-3-319-11752-2_12

Fig. 1. Instead of training a global model, we seek for a given test image (*left*) its most similar images among all training samples (*middle left*). We then learn exemplar-specific representations by running our patch discovery on the retrieved image set and thereby figure out parts relevant for differentiation (*middle right*). Learned detectors are used to encode the images from the retrieved set and the query alike (*right*), to train a local model, and to finally classify the query image.

we use the CUB-2011 birds dataset [28] as a running example for fine-grained recognition scenarios. Trying to find suitable patch detectors for the aforementioned differences, within the whole training set, is a very complex task, which we significantly simplify by restricting the patch discovery to the K nearest neighbors leading to very compact feature representations. Figure 1 gives an outline of this idea. In comparison to existing techniques, our approach offers the following main characteristics:

1. Convolution-based bootstrapping without restricting only to initial patches found by heuristic segmentation methods as in [18].
2. Exemplar-specific patch representations instead of global models.
3. A combination of exemplar-specific and semantic patches to improve on previous results on the CUB-2011 birds dataset.

First, we discuss related work in Sect. 2. Our automatic bootstrapping-based patch discovery is presented in Sect. 3. Steps towards exemplar-specific representations are given in Sect. 4 and the results of our experiments in the area of fine-grained recognition are evaluated in Sect. 5.

2   Related Work

There is a large body of literature on the topic of patch discovery, especially when also discriminative clustering and codebook learning methods are included (see [6] and references therein). In the following, we restrict ourselves to patch discovery methods related to our bootstrapping technique. Throughout the paper, we use the term *patch* and the notation x to refer to a small region, window, or block in an image, and the term *patch detector* and its notation w to refer to a template or linear classifier learned with a given set $\mathcal{M}(w)$ of patches.

Patch Discovery. Patch discovery has been an important research field since the early works of [1,27]. Whereas [1] clusters similar patches of training images to obtain a vocabulary in an unsupervised manner, [27] finds class-specific patches by maximizing mutual information. Both papers (and related ones during the same time) use simple detectors based on gray values, which are hardly able to tackle the variability in natural images. The paper of [24] was the first one to present a patch discovery method that made use of recent advances in object localization, such as HOG features and Exemplar-SVM models. The usefulness of patch discovery schemes for fine-grained recognition has been demonstrated recently in [23]. The work of [8] shows how to cast patch discovery as a non-convex optimization problem related to mode seeking. A common drawback is the time-consuming step of model learning with hard negative mining [18,24] presents a time-efficient version using whitened HOG features [15]. Their patch discovery scheme is based on bootstrapping a set of initial seed patches, which are previously derived from an unsupervised segmentation result. In contrast, we perform dense bootstrapping with convolutions, which allows for obtaining useful patch examples after a seeding step considering *every possible position* in the training images. Thus, our approach is more similar to latest techniques within the deep learning field [5,31] where discovery is done in a completely supervised manner and in several layers simultaneously.

Exemplar and Local Models. State-of-the-art categorization techniques are usually based on huge representations and complex models. Local learning approaches aim at an orthogonal solution by learning models on the fly specifically tailored to every test image. Introduced in the early nineties [4], the main focus of those techniques is on a suitable trade-off between model capacity and number of training examples, especially for non-uniform distributions of samples in space. Some rare exceptions followed this idea through the years, *e.g.*, [30] for image categorization, or [16] for recognizing facial expressions. Similar in spirit are [13,14] showing how to transfer image annotations from nearest neighbors in the training set, which have been found by a global matching scheme. Whereas simple, those part detection-by-transferring methods are more flexible compared to global methods that can only tackle a limited number of viewpoints. However, current techniques assume a unique and constant feature space, *i.e.*, all approaches learn local models for fixed representations so far. Similar to our approach, [12] presented how to learn distance functions for every test sample to overcome this issue. In this paper, we even go one step further by learning image representations *and* classification models for every test sample on the fly, which allows focusing on patches important to differentiate quite similar birds already observed.

3   Discovering Mid-Level Patch Representations

Our patch discovery scheme consists of three main parts: (1) finding initial seed patches, (2) learning patch detectors, and (3) convolution-based bootstrapping.

Fig. 2. Seeding results to initialize patches for discovery: seeding is based on unsupervised region segmentation [11] conducted with masked images.

The discovery is followed by a feature extraction step, where we generate features by spatially pooling patch detector outputs.

Finding Initial Seed Patches.   Finding initial seed patches is an important step to guide the following bootstrapping steps in the right direction. The purpose is similar to an interest point detection, which was often used in the earlier works on bag-of-features (see references cited in [19]). Here, we follow the idea of [18] and extract quadratic patches x^k of different sizes centered on the regions found by the region segmentation method of [11]. To further focus seed patches towards bird locations, we mask-out background regions using the pixelwise annotations provided with the dataset, but this is only an optional step. The first sets for the patch detectors w^k are then set to $\mathcal{M}(w^k) = \{x^k\}$. Figure 2 shows some example regions extracted in this manner. Note that we are later on using convolutions to densely bootstrap these initial patches, therefore, the discovered patches are not restricted to initial seed patches as in [18].

Learning Patch Detectors.   Humans usually describe a birds appearance by a mixture of typical color and texture occurrences, *e.g.*, a dotted red belly or feathers with blue stripes. To meet this observation, we represent patches by histogram of oriented gradient features (HOG) and color feature histograms [25] computed for small cells of pixels. For a set of patches, we learn a linear patch detector and detection responses for unseen images are obtained by convolving the weight vector w of the learned model with computed feature planes of the image. As shown by [15], training HOG detectors can be done efficiently using standard Gaussian assumptions. Although only presented for HOG features, their technique can be applied to arbitrary features such as the combined HOG and color features we use in the experiments. Let us consider a single filter w that represents a classifier differentiating between positive (sub-images showing the patch) and negative examples. The paper of [15] assumes that positive and negative examples are Gaussian distributed with the same covariance matrix S_0 and mean vector μ_1 and μ_0, respectively. It can be shown that in this case, the optimal hyperplane separating positives and negatives can be calculated by $w = S_0^{-1}(\mu_0 - \mu_1)$ [15]. Although the underlying assumptions leading to this equation might seem unrealistic, the resulting simple learning step is, at a closer look, a common feature whitening. It implicitly decorrelates all features using statistics of a large set of (negative) examples – an important step to deal with high correlations naturally arising, *e.g.*, between neighboring HOG cells [15].

Fig. 3. Bootstrapping: initial seed patches are indicated with colored frames and the set of positive blocks for resulting detectors is displayed accordingly. While the technique is applicable to supervised and unsupervised scenarios, adding supervision prevents grouping of visually similar blocks from different categories.

Since every detector discriminates a tiny set of positive patches against everything else, the notion of 'everything else' can be shared by all detectors. Thus, the covariance matrix as well as the mean μ_0 of negative examples can be estimated from an arbitrary set of background images and features calculated therein. As a consequence, we can easily pre-compute it and re-use it for every patch model. In summary, the only remaining steps for learning a new patch detector from given examples is to average the features of positive patches and to solve a linear equation system.

Iterative Bootstrapping with Convolutions.   After learning a patch detector for each of the initial seed patches, we proceed with bootstrapping to obtain new useful training examples for each of the patch detectors. Bootstrapping can be performed in a supervised manner by restricting it to images of the category the initial seed patch was extracted from [8,18,24], or unsupervised by using all training images. Both versions are evaluated in our experiments.

During bootstrapping, every patch detector is applied to every corresponding image, such that we obtain scores for every possible patch in all images. Note that this is different from the approach of [18], where only seed patches are considered as potential candidates for bootstrapping. Our method instead allows for discovering important patches not found by the unsupervised segmentation in the beginning.

Given the detection responses, we now seek for possible positive patches to increase the training set size of every detector leading to increased generalization abilities. In contrast to adding a fixed number m of the highest scored examples in each bootstrapping step, we add *at most* m examples. In particular, we do not add training examples that received a detection score worse than a certain percentage γ of the minimum score achieved by examples already used as positive training examples:

$$\mathcal{A}^{t+1}(\boldsymbol{w}^k) = \{\boldsymbol{x} \mid (\boldsymbol{w}^k)^T \boldsymbol{x} > \gamma \cdot \min_{\tilde{\boldsymbol{x}} \in \mathcal{M}^t(\boldsymbol{w}^k)} (\boldsymbol{w}^k)^T \tilde{\boldsymbol{x}}\}. \tag{1}$$

We denoted with $\mathcal{A}^{t+1}(\boldsymbol{w}^k)$ the set of accepted blocks for detector $\boldsymbol{w}^k$ in bootstrapping round $t+1$. If $\mathcal{A}^{t+1}$ is empty, we consider $\boldsymbol{w}^k$ as converged. Otherwise, we select m examples of $\mathcal{A}^{t+1}(\boldsymbol{w}^k)$ with highest score and add them to the set of positive samples:

$$\mathcal{M}^{t+1}(\boldsymbol{w}^k) = \text{top}_m\left(\mathcal{A}^{t+1}\left(\boldsymbol{w}^k\right)\right) \cup \mathcal{M}^t(\boldsymbol{w}^k) . \tag{2}$$

The parameter γ controls the exploration/exploitation trade-off during discovery and we set γ to 0.75 according to preliminary experiments on smaller datasets. Intuitively, larger values for γ prevent patch detectors from being "blurred" during bootstrapping by outliers. Furthermore, this strategy also leads to a convergence during bootstrapping without the necessity of specifying a fixed number of bootstrapping iterations as done in [18,24]. Visualizations for the process of bootstrapping are given in Fig. 3 for supervised and unsupervised scenarios. Initial seeding blocks for every detector are indicated with colored frames.

To finally remove non-discriminative patch detectors, previous approaches introduced supervised feature selection techniques, such as entropy-rank curves used by [18] or mutual information proposed by [24]. We skip this additional selection step and rely on the final classification model for the selection of relevant dimensions. Since bootstrapping is performed independently for each patch, early pruning would also not lead to an advantage in computation time, but would limit the number of features that can be used by a classifier later on drastically.

Extracting Features Based on Discovered Patch Detectors. After discovering a set of patch detectors, it now remains to build a final feature representation for an unseen image. Every detector is trained on just a couple of training patches and fires only on an extremely small number of windows. Thus, the maximum score achieved for an image serves as an indicator whether or not the corresponding patch occurs. Consequently, max-pooling detection responses over the entire image leads to a feature vector with as many dimensions as detectors discovered previously.

Note that since computing detection results can be interpreted as a convolution of images and learned weight vectors of detector models, the overall pipeline shows an interesting parallel to deep learning techniques currently prominent. In direct comparison, we fix the lower layers and instead of learning planes associated with mid-level features by back-propagation, we instead bootstrap patch detectors, which could be also used in unsupervised settings. Given the great results recent approaches obtained by replacing handcrafted representation with rich representations learned in deep architectures [5,31], it would be interesting to see the proposed patch discovery scheme running in a local learning manner on those representations instead of HOG and color names only, i.e., to fine-tune pre-trained deep architectures for *every single test image* and obtaining patches from an additional convolutional layer.

4   Exemplar-Specific Mid-level Features

In order to reliably differentiate between subordinate classes, the identification of relevant features is among the most crucial aspects [7,10,20]. Although pose-alignment techniques [13,14] can almost eliminate effects based on significant pose variations, *e.g.*, of highly deformable objects like birds, they still have to struggle with the identification of discriminative features to encode the yet pose-aligned objects. Early papers in this area used off-the-shelf features such as bag-of-visual-word statistics [20] and more recent approaches further focused extraction on manually defined regions of interest for training samples [10, 14]. Additionally, feature learning techniques have been proposed to distinguish object classes in an offline training step, by asking users [7], train 1-vs-1-features [3], or seek for a subset of useful random patches [9]. Still, all of these methods aim at calculating a unique general representation able to differentiate all training data as good as possible. Coupled with powerful post-processing techniques, *e.g.*, linear embeddings in high dimensional spaces [26], state-of-the-art systems usually work in feature spaces of hundreds of thousands of dimensions, while being trained on orders of magnitude less training samples (denoted with N).

We propose to follow an orthogonal path and find for every unseen image a compact, informative, and image specific feature representation. Thus, we start by querying for a new test image its $K \ll N$ most similar training samples using Euclidean distance and combined HOG and color name (CN) features [25]. Figure 1 displays this first step. All of these images are then used for the patch discovery described in the previous section. This results in a set of patch detectors that are specific for the current test image and especially for the global shape and pose of the object in it. The set of patch detectors is then used to compute feature representations for the test image as well as for the neighbors. Finally, a linear SVM classifier is trained on the neighbor images and used to predict the category for the test image. In terms of asymptotic runtimes, our local learning approach scales with $\mathcal{O}(K^2)$ during testing for patch discovery, and needs no training step. In contrast, discovering patch detectors for a global model demands at least $\mathcal{O}(N^2)$ and $\mathcal{O}(N)$ time during training and testing, respectively.

Combination of Discovered Patches and Semantic Parts. As shown in [14], the performance of fine-grained recognition can be drastically improved when the location of semantic parts can be estimated, such as the head or back position for bird recognition. Therefore, we combine our approach with the exemplar-specific part prediction method proposed by [14]. The combination of both exemplar-specific classification approaches is done by late fusion. In particular, we are combining estimated class probabilities with linear combination $\mathcal{S}(x) = \lambda \cdot \mathcal{S}_{\text{semantic}}(x) + (1 - \lambda)\mathcal{S}_{\text{discovery}}(x)$. We denoted with $\mathcal{S}$ class probabilities obtained via late-fusing probabilities $\mathcal{S}_{\text{semantic}}$ computed with a model using semantic part transfer [14] and probability scores $\mathcal{S}_{\text{discovery}}$ obtained from a model learned on discovered patches. The combination weight $\lambda \in [0, 1]$ serves as trade-off parameter and can be learned with leave-one-out estimation. It is important to note that in this paper, we optimize this parameter on the test set to simply show the potential of a combination.

5 Experiments

We evaluate our approach for fine-grained recognition on the CUB-2011 dataset [28] and use the provided split for training and testing. Following evaluation standards, we use the whole dataset CUB-2011-200 with all classes and the CUB-2011-14 dataset with only 14 classes as done in [10]. In the first part, we are interested in the accuracy using a global patch discovery with all its different flavors, whereas the exemplar-specific extension proposed in Sect. 4 is evaluated in the second part. Finally, we show how combining decisions of models learned on either semantic or discovered parts pays off and results in improved classification performance compared to state-of-the-art results on this dataset. For experimental details, we refer to the supplementary material and our source code, which is available at http://www.inf-cv.uni-jena.de/fine_grained_recognition.

Table 1. Fine-grained recognition results on the CUB-2011 dataset.

Approach	CUB-2011-14	CUB-2011-200
Wah et al. [28]	–	10.25 %
Our approach (global, unsupervised)	54.01 %	-
Our approach (global, supervised)	58.01 %	39.35 %
Our approach ($K = 150$, supervised)	63.95 %	34.16 %
Style-awareness [23]	–	38.31 %
PDL [17]	–	38.91 %
Template learning [29]	–	43.67 %
DPD [32]	–	50.98 %
POOF [3]	70.10 %	56.78 %
Goering et al. [14]	73.39 %	57.99 %
Our approach ($K = 150$) + [14]	**76.64 %**	**58.55 %**
Our approach (global) + [14]	**78.25 %**	**60.81 %**

Note that in [14], reported results are overall recognition rates averaged over all samples, whereas we report average recognition rates (averaged over class accuracies).

5.1 Evaluation of Global Patch Discovery

We ran the patch discovery technique described in Sect. 3 with both supervised and unsupervised bootstrapping (with a maximum of 5 iterations). With the

Table 2. Results on the CUB-2011-14 dataset **without** exemplar-specific discovery.

Approach	CUB-2011-14
Bootstrapping: on seeding blocks only [18]	46.93 %
none	55.64 %
Selection: entropy-rank criterion with merging [18]	26.96 %
representative, without singletons [24]	58.62 %
Our approach (globally discovered parts, convolution-based, no selection)	**58.01** %
Seeding: *human annotated semantic parts*	57.89 %

discovered patch detectors at hand, we encode training images as described previously. Classification accuracies are given in the first rows of Table 1.

First of all, we observe that the supervised bootstrapping clearly outperforms its unsupervised counterpart. Obviously, supervision during bootstrapping avoids rather similar patches with high discrimination abilities being grouped together as can be seen in Fig. 3. Although similar, tiny details still make some patches different from others, *e.g.*, the size of the red dot in the left group of patches. Based on these observations and the fact that unsupervised bootstrapping requires increased computation times compared to the supervised variant, we use supervised bootstrapping in all following evaluations only. Obtained accuracies are in the range of current results from patch discovery techniques such as [23], although not as competitive as latest techniques using ground truth part annotations and additional expert knowledge [3,14].

In Table 2, we analyzed different steps of our approach and compared to alternatives proposed in [18]. It can be seen that our convolution-based bootstrapping outperforms the bootstrapping by [18] which is conducted on seeding blocks only as well as a simple baseline that uses every seeding block as a patch detector without any bootstrapping involved. Furthermore, we can also see that the selection criteria proposed by [18,24] hurt the performance in our case. Interestingly, our part discovery scheme reaches a performance on par with a baseline that uses manually selected semantic parts. We further displayed detection response maps in Fig. 4 obtained by applying our discovered patch detectors on unseen test images.

5.2   Evaluation of Exemplar-Specific Representations

Choosing the Number of Neighbors.   Our fine-grained recognition approach using exemplar-specific patch representations and classifiers is limited by the neighbors found by global matching. When an example of the correct category is not present in the set of neighbors, we are logically not able to predict this category for the current test image. Therefore, we first analyze the quality of the matching scheme and results are given in Fig. 5. We plot the performance

Fig. 4. Detection responses of discovered patch detectors on previously unseen test images. High scores are indicated by warm colors. The very right column displays cases where detectors are distracted by background patterns. Best viewed in color.

of an oracle method, which reflects a perfect classification when at least one example of the correct category is among the neighbors, and the performance of a plain majority vote classification. As can be seen, a small set of neighbors is sufficient for the CUB-2011-14 dataset to provide training examples of the correct class, which is not the case for the larger dataset CUB-2011-200. The majority vote classification is unlikely to already provide proper classification results due to the simple features used for global matching. However, it took us by surprise that this simple kNN-technique already improved over the first baseline ever given for this dataset [28] by more then 4 percent accuracy (first row in Table 1).

Evaluation on CUB-2011. Since the matching accuracy is sufficient to reduce the training images to a reasonable subset, we are now interested in the performance of the whole local learning pipeline. The results for the 14 and 200 class sets are given in Fig. 5a, b, respectively. Interestingly, the local learning approach with $k \geq 80$ neighbors outperforms the global learning approach on the small dataset. This is indeed remarkable, since only a fraction of dimensions is used (an overview of numbers of discovered patch detectors is given in the supplementary material). For the large dataset, however, matching seems to be the limiting factor, which can be already guessed from Fig. 5b. Consequently, non-euclidean distance matching [30] or techniques borrowed from image retrieval [2] would probably overcome current limitations here. Nonetheless, local learning results in quite impressive results given the fact that the dimensionality is about 2.0 % compared to a global model. Thus, we conclude that the discovered patch detectors form a compact and informative representation.

Combining Patch Discovery and Part Transfer. In a final experiment, we combined our approach with the semantic part transfer of [14] and the results are given in the lower part of Table 1. Although our combination technique is a simple weighted combination of class probabilities, it can be seen that we are able to obtain state-of-the-art performance among approaches with fixed feature representations for both the global and the exemplar-specific model, and thus can draw advantage from two complementary encryption techniques. Interestingly, our local variant is inferior to its global counterpart when being

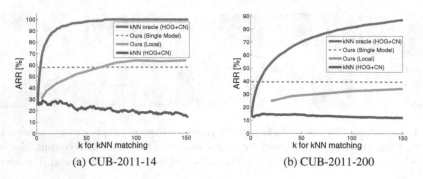

(a) CUB-2011-14   (b) CUB-2011-200

Fig. 5. Accuracy of simple k-NN matching on CUB2011 for a perfect oracle, majority voting, and our approach for different sizes of the query set.

combined with the semantic part transfer. Our intuition is that this is mainly due to the way of transferring part annotations in [14] which also relies on a nearest neighbor search as done in local learning. Furthermore, it should be noted that the results of approaches that learn the underlying feature representations in a supervised manner with convolutional neural networks [5,31] recently obtained higher recognition rates on this dataset. Using these representations together with our discovery scheme is future work.

6   Conclusions

In this paper, we tackled the challenging problem of fine-grained recognition of bird species. Our approach consists of two key ingredients: a novel patch discovery technique and a new variant of local representation learning. For the introduced discovery scheme, we proposed an iterative bootstrapping technique to group re-occurring and informative subimages to patch detectors. In contrast to other papers in this area, our method performs dense bootstrapping without restricting itself to segmentation results and it is suitable both for unsupervised and supervised settings. To overcome computational burdens during learning and to further focus on relevant patches, our second contribution is a novel local representation learning formulation. Thereby, for every test image we learn classification models and image representations jointly by using a subset of training data most similar to the test image. Results on fine-grained recognition tasks have shown that the combination of discovered part and semantic part representations leads to a further boost in performance.

References

1. Agarwal, S., Roth, D.: Learning a sparse representation for object detection. In: Heyden, A., Sparr, G., Nielsen, M., Johansen, P. (eds.) ECCV 2002, Part IV. LNCS, vol. 2353, pp. 113–127. Springer, Heidelberg (2002)

2. Arandjelovic, R., Zisserman, A.: Three things everyone should know to improve object retrieval. In: Conference on Computer Vision and Pattern Recognition (CVPR), pp. 2911–2918 (2012)
3. Berg, T., Belhumeur, P.N.: Poof: part-based one-vs-one features for fine-grained categorization, face verification, and attribute estimation. In: Conference on Computer Vision and Pattern Recognition (CVPR), pp. 955–962 (2013)
4. Bottou, L., Vapnik, V.: Local learning algorithms. Neural Comput. **4**(6), 888–900 (1992)
5. Branson, S., Van Horn, G., Belongie, S., Perona, P.: Improved bird species categorization using pose normalized deep convolutional nets. In: British Machine Vision Conference (BMVC) (2014)
6. Coates, A., Ng, A.Y.: The importance of encoding versus training with sparse coding and vector quantization. In: International Conference on Machine Learning (ICML), pp. 921–928 (2011)
7. Deng, J., Krause, J., Fei-Fei, L.: Fine-grained crowdsourcing for fine-grained recognition. In: Conference on Computer Vision and Pattern Recognition (CVPR), pp. 580–587 (2013)
8. Doersch, C., Gupta, A., Efros, A.A.: Mid-level visual element discovery by discriminative mean-shift. In: Neural Information Processing Systems (NIPS), pp. 1–8 (2013)
9. Duan, K., Parikh, D., Crandall, D., Grauman, K.: Discovering localized attributes for fine-grained recognition. In: Conference on Computer Vision and Pattern Recognition (CVPR), pp. 3474–3481 (2012)
10. Farrell, R., Oza, O., Zhang, N., Morariu, V.I., Darrell, T., Davis, L.S.: Birdlets: Subordinate categorization using volumetric primitives and pose-normalized appearance. In: International Conference on Computer Vision (ICCV), pp. 161–168 (2011)
11. Felzenszwalb, P., Huttenlocher, D.: Efficient graph-based image segmentation. Int. J. Comput. Vis. (IJCV) **59**, 167–181 (2004)
12. Frome, A., Singer, Y., Sha, F., Malik, J.: Learning globally-consistent local distance functions for shape-based image retrieval and classification. In: International Conference on Computer Vision (ICCV), pp. 1–8 (2007)
13. Gavves, E., Fernando, B., Snoek, C., Smeulders, A., Tuytelaars, T.: Fine-grained categorization by alignments. In: International Conference on Computer Vision (ICCV), pp. 1–8 (2013)
14. Göring, C., Rodner, E., Freytag, A., Denzler, J.: Nonparametric part transfer for fine-grained recognition. In: Conference on Computer Vision and Pattern Recognition (CVPR), pp. 1–8 (2014)
15. Hariharan, B., Malik, J., Ramanan, D.: Discriminative decorrelation for clustering and classification. In: Fitzgibbon, A., Lazebnik, S., Perona, P., Sato, Y., Schmid, C. (eds.) ECCV 2012, Part IV. LNCS, vol. 7575, pp. 459–472. Springer, Heidelberg (2012)
16. Ionescu, R., Popescu, M., Grozea, C.: Local learning to improve bag of visual words model for facial expression recognition. In: International Conference on Machine Learning - Workshop on Representation Learning (ICML-WS) (2013)
17. Jia, Y., Vinyals, O., Darrell, T.: Pooling-invariant image feature learning. CoRR abs/1302.5056 (2013)
18. Juneja, M., Vedaldi, A., Jawahar, C., Zisserman, A.: Blocks that shout: Distinctive parts for scene classification. In: Conference on Computer Vision and Pattern Recognition (CVPR), pp. 923–930 (2013)

19. Jurie, F., Triggs, B.: Creating efficient codebooks for visual recognition. In: International Conference on Computer Vision (ICCV), vol. 1, pp. 604–610 (2005)
20. Khan, F.S., Van De Weijer, J., Bagdanov, A.D., Vanrell, M.: Portmanteau vocabularies for multi-cue image representation. In: Neural Information Processing Systems (NIPS), pp. 1323–1331 (2011)
21. Krizhevsky, A., Sutskever, I., Hinton, G.E.: Imagenet classification with deep convolutional neural networks. In: Neural Information Processing Systems (NIPS), vol. 1, p. 4 (2012)
22. Lazebnik, S., Schmid, C., Ponce, J.: Beyond bags of features: Spatial pyramid matching for recognizing natural scene categories. In: Conference on Computer Vision and Pattern Recognition (CVPR), pp. 2169–2178 (2006)
23. Lee, Y.J., Efros, A.A., Hebert, M.: Style-aware mid-level representation for discovering visual connections in space and time. In: International Conference on Computer Vision (ICCV), pp. 1857–1864 (2013)
24. Singh, S., Gupta, A., Efros, A.A.: Unsupervised discovery of mid-level discriminative patches. In: Fitzgibbon, A., Lazebnik, S., Perona, P., Sato, Y., Schmid, C. (eds.) ECCV 2012, Part II. LNCS, vol. 7573, pp. 73–86. Springer, Heidelberg (2012)
25. Van De Weijer, J., Schmid, C.: Applying color names to image description. In: International Conference on Image Processing (ICIP), vol. 3, pp. III-493 (2007)
26. Vedaldi, A., Zisserman, A.: Efficient additive kernels via explicit feature maps. IEEE Trans. Pattern Anal. Mach. Intell. (PAMI) **34**(3), 480–492 (2012)
27. Vidal-Naquet, M., Ullman, S.: Object recognition with informative features and linear classification. In: International Conference on Computer Vision (ICCV), pp. 281–288 (2003)
28. Wah, C., Branson, S., Welinder, P., Perona, P., Belongie, S.: The caltech-ucsd birds-200-2011 dataset. Technical report CNS-TR-2011-001, California Institute of Technology (2011)
29. Yang, S., Bo, L., Wang, J., Shapiro, L.: Unsupervised template learning for fine-grained object recognition. In: Neural Information Processing Systems (NIPS), pp. 3131–3139 (2012)
30. Zhang, H., Berg, A.C., Maire, M., Malik, J.: Svm-knn: discriminative nearest neighbor classification for visual category recognition. In: Conference on Computer Vision and Pattern Recognition (CVPR), pp. 2126–2136 (2006)
31. Zhang, N., Donahue, J., Girshick, R., Darrell, T.: Part-based R-CNNs for fine-grained category detection. In: Fleet, D., Pajdla, T., Schiele, B., Tuytelaars, T. (eds.) ECCV 2014, Part I. LNCS, vol. 8689, pp. 834–849. Springer, Heidelberg (2014)
32. Zhang, N., Farrell, R., Iandola, F., Darrell, T.: Deformable part descriptors for fine-grained recognition and attribute prediction. In: International Conference on Computer Vision (ICCV) (2013)

Video Interpretation

Motion Segmentation with Weak Labeling Priors

Hodjat Rahmati[1]([✉]), Ralf Dragon[2], Ole Morten Aamo[1], Luc van Gool[2],
and Lars Adde[3,4]

[1] Department of Engineering Cybernetics, NTNU, Trondheim, Norway
{hodjat.rahmati,ole.morten.aamo}@itk.ntnu.no
[2] Computer Vision Lab, ETH, Zurich, Switzerland
{dragon,vangool}@vision.ee.ethz.ch
[3] Clinic for Clinical Services, St. Olavs University Hospital, Trondheim, Norway
[4] Department of Laboratory Medicine, Children and Woman's Health,
Faculty of Medicine, NTNU, Trondheim, Norway
lars.adde@ntnu.no

Abstract. Motions of organs or extremities are important features for clinical diagnosis. However, tracking and segmentation of complex, quickly changing motion patterns is challenging, certainly in the presence of occlusions. Neither state-of-the-art tracking nor motion segmentation approaches are able to deal with such cases. Thus far, motion capture systems or the like were needed which are complicated to handle and which impact on the movements. We propose a solution based on a single video camera, that is not only far less intrusive, but also a lot cheaper. The limitation of tracking and motion segmentation are overcome by a new approach to integrate prior knowledge in the form of weak labeling into motion segmentation. Using the example of Cerebral Palsy detection, we segment motion patterns of infants into the different body parts by analyzing body movements. Our experimental results show that our approach outperforms current motion segmentation and tracking approaches.

1 Introduction

We aim at segmenting motion data from monocular videos into underlying body parts. Motion is an important cue for the clinical diagnosis of, for instance, cardiovascular diseases [17], Cerebral Palsy [1], or for gait analysis [9]. In this paper, we focus on the motion of infants to detect Cerebral Palsy (CP) which is a set of chronic conditions affecting body movements, posture and muscle coordination. It is caused by damage to one or more specific areas of the brain, usually occurring during foetal development or infancy. The absence of normal movement between 2 to 4 months post-term age has been shown to be a strong predictor of later Cerebral Palsy [1].

Accurate diagnosis can be achieved by analyzing the infant's bodily motion at the necessary level of detail. Recently, a number of computer-based methods have aimed to quantitatively analyze general movements in order to detect CP. However, they either use extra instruments [13] that are intrusive to the diagnosis task, or they don not provide analytic results [20]. The reciprocal relation

© Springer International Publishing Switzerland 2014
X. Jiang et al. (Eds.): GCPR 2014, LNCS 8753, pp. 159–171, 2014.
DOI: 10.1007/978-3-319-11752-2_13

between body parts is a strong analytic cue for CP detection. E.g., [8] proposed that a high correlation between two limbs might indicate a lack of normal behaviour. Therefore, in order to perform an analytic tool to study the disease we propose a new segmentation method to separate and track different body parts without need for any extra instrument.

The video analysis of infant motion is very challenging. Infants will often cross their limbs, twist arms, move abruptly, etc. As a result during such motions, state-of-the-art trackers, e.g. TLD [7], drift and fail at tracking the body parts at the required level of precision. Similarly, body pose trackers have difficulties due to the high motion variability. Motion segmentation, on the other hand, has no drifting problems if the body parts move distinctly – motion even simplifies the segmentation. However, an initial bounding box or an initial segmentation prior cannot be integrated conveniently and, without it, the segmentation performance is poor. Our key contribution is a new energy-minimization-based formulation of motion segmentation that allows for the integration of prior information. The result is a system that can segment long sequences reliably.

Related Work. Motion segmentation is the task of grouping point trajectories from an image sequence subject to coherence of their motion over time. While earlier work focused on assigning the trajectories to subspaces, e.g. with the generalized PCA [23], subsequent work exploited sparsity [6,12] or nonnegative matrix factorization [4]. Further works exploit temporal smoothness [19] or depth ordering [11]. In most recent works, the pairwise [3–6,12,19] or higher-order [15] relationships between trajectories are aggregated and a final spectral clustering [14] step of an affinity matrix **A** encoding trajectory similarity finds the association of trajectories to motion segments.

In contrast to image segmentation approaches, those for motion segmentation are unsupervised, i.e. it is neither necessary nor possible to specify prior knowledge about the assignment of trajectories to motion segments. Modifying **A** to prevent trajectories with the same prior label to be clustered into different segments has undesired biases which leads to poor results: Since spectral clustering minimizes the inter-segment cut through **A**, weighted by the sum of intra-segment weights [14], tweaking local weights creates a global bias which results in non-intuitive results. In terms of weakly supervised clustering, also called transductive learning, [24] derived a simple iterative framework in which the final k-Means-step in spectral clustering is replaced by an iterative procedure. It alternates between computing a mapping **F** from graph Laplacian to class association and regularizing **F** regarding the initial labels. While they proved that the mapping converges to a reasonable solution respecting the initial labels, our energy-based approach allows specifying additional unary terms for trajectories which are similar to initially-labeled ones. In our work, we overcome the no-prior limitation by formulating the motion clustering task as a multi-label MRF, similar to graph-cut-based image segmentation [2]. We use the generalized Potts model, thus encouraging large motion differences between segments and similar motions within segments. This may seem straightforward, but to the best of our knowledge such a scheme has not been proposed before. MRFs have been used for dense image segmentation, but only based on short-term motion obtained from optical flow, e.g. [21]. In contrast, we segment sparse motion information from trajectories which last over many frames.

The rest of the paper is organized as follows. In Sect. 2, we derive our energy formulation, in Sect. 3, *tracking by segmentation* is explained, in Sect. 4, we present experimental results, and Sect. 5 concludes the paper.

2   Integrating Prior Knowledge into Motion Segmentation

Energy Formulation. This section describes the energy minimization framework used to segment trajectories into the infant's body parts. Let s be a trajectory from the set of all trajectories $\mathcal{S}$, and L be the unknown label vector assigning each trajectory s to a segment L_s. The motion segmentation is obtained through a multi-label graph-cut [10] that minimizes the energy $E(L)$ associated to a segmentation L.

Similar to its use in an already extensive image segmentation literature [2],

$$E(L) = \sum_{s \in \mathcal{S}} D(s, L_s) + \sum_{(s,r) \in \mathcal{N} \times \mathcal{N}} V(s, r, L_s, L_r). \qquad (1)$$

The data term D is a penalty function that encourages high intra-segment motion similarity and V is an interaction potential that enforces low inter-segment similarity for trajectories in a local neighborhood $\mathcal{N}$ which is defined as follows,

$$\mathcal{N} = \{(s, r) \mid (s, r) \in \mathcal{S} \times \mathcal{S} \wedge d_{\mathrm{sp}}(s, r) \leq 10 \wedge (s, r) \text{ have temporal overlap}\}, \qquad (2)$$

where $d_{\mathrm{sp}}(s, r)$ is the average spatial Euclidean distance over the common frames. Due to occlusions and fast motions, trajectories are asynchronous and span different temporal windows. Considering just trajectories that last for the whole shot lower the number of tracked points, leaving us possibly even with an empty set. So, we obtain the energy for all trajectories that have at least one frame in common. Due to transitivity, it can be expected that even trajectories that share no frames can be paired [3].

Data Term. Let $\mathcal{S}_I \subset \mathcal{S}$ be the set of those trajectories s that are initially labeled with I_s. Then the data term in Eq. (1) is defined as

$$D(s, L_s) = \begin{cases} 0 & \text{if } s \in \mathcal{S}_I \wedge L_s = I_s \\ K_s & \text{if } s \in \mathcal{S}_I \wedge L_s \neq I_s \\ g(s, L_s) & \text{if } s \notin \mathcal{S}_I \end{cases} \qquad (3)$$

where $K_s = 1 + \sum_{(r,s) \in \mathcal{N} \times \mathcal{N}} V(s, r, L_s, L_r)$ is a large value that enforces trajectories which are initially labeled to preserve their labels during the optimization process. The first two cases in Eq. (3) enforce the prior knowledge, while $g(s, L_s)$ extends prior knowledge towards initially unlabeled trajectories: $g(s, L_s)$ is a measure of dissimilarity between trajectory s and subset $\mathcal{O}_{L_s}$ – the set of trajectories that are initially assigned label L_s. We define the energy g as negative log-likelihood of the average trajectory similarity $w \in [0,1]$ which will be given in Eq. 8:

$$g(s, L_s) = -log\left(\underset{r \in \mathcal{O}_{L_s}}{mean}((w(s,r))^\gamma)\right). \tag{4}$$

Thus, the average similarity between a trajectory s and the set of initially-labeled trajectories $\mathcal{O}_{L_s}$ is computed using the arithmetic mean, similar to a mixture distribution.

Pairwise Term. We define the pairwise Energy term in Eq. (1) as:

$$V(s, r, L_s, L_r) = (1 - \delta(L_s, L_r))f(w(s,r)), \quad f(w) = -log(1 - w^\phi), \tag{5}$$

where δ is the Kronecker delta function which leads to penalizing neighboring trajectories s and r from different segments by a penalty f that depends on the trajectory similarity $w(s,r)$. The more similar s and r are the higher the penalty of assigning them to different segments will be. f is defined as the negative log-likelihood of the counter probability of w which is weighted by ϕ analog to w in Eq. (4). γ and ϕ non-linearly balances D and V in Eq. (1). They are empirically set to $\gamma = 0.1$ and $\phi = 0.001$.

Trajectory Similarity. The trajectory similarity $w(s,r)$ is a probabilistic measure if the two trajectories s and r belong to the same moving object. Since such trajectories usually move similarly and tend to be spatially closer than trajectories with different associations, our definition contains a motion and a distance term:

$$d_t^2(s, r) = \frac{d_{\mathrm{sp}}(s, r)}{\delta_c(s, r)} \cdot d_{\mathrm{mot}}^2(t, s, r), \quad d_{\mathrm{mot}}^2(t, s, r) = \frac{\|v_t^s - v_t^r\|^2}{5\sigma_t^2(s, r)}, \tag{6}$$

where $d_{\mathrm{sp}}(s, r)$ is the average spatial Euclidean distance over the common frames, d_{mot} is the normalized motion distance at time t, and d_t is the combined distance.

Unlike [3], we scale the spatial distance by the factor $\delta_c(s, r) = log(1 + n_c(s, r))$ where $n_c(s, r)$ is the number of frames that the trajectories s and r have in common. This takes into account that the more frames two trajectory have in common the more reliable the similarity result is. The reasons are twofold: First, the short length of a trajectory indicates severe change in the neighbourhood of that trajectory (trajectories terminate in case of occlusion or fast changes, and the short length shows two of such cases happens in a short period). Thus, the optical flow is likely to be inaccurate in such a situation and that trajectory might be wrongly developed. Second, since similarity is obtained over common frames, two trajectories that show similar motion in their common frames while they have completely different motions in other frames, still get a high similarity.

Therefore, it is reasonable to account less value for similarities obtained over a smaller number of frames. In the definition of d_{mot}, v_t^s is the aggregated motion of a trajectory s over 5 frames ($v_t^s = x_{t+5}^s - x_t^s$). σ_t is an adaptive normalization which enables dealing with both fast and slow motions. In particular, local variations among velocities within a segment should be tolerated more if motions in the segment are changing more rapidly. Therefore, σ_t is defined, as presented in [3]:

$$\sigma_t(s, r) = \min_{a \in \{s, r\}} \sum_{t'=0}^{4} \sigma(x_{t+t'}^a, t + t'), \qquad (7)$$

where $\sigma(x, t)$ is the local variation in the flow field at position x and frame t.

As long as two objects move next to each other, they share similar motions, and it is impossible to separate them as different objects. But as soon as they start to move differently, they can be distinguished. In order to exploit this information, the distance d between two trajectories considers the instance when they start behaving differently. So, $d(s, r) = \max_t d_t(s, r)$. Finally, the edge weight between trajectories is defined as

$$w(s, r) = \exp(-d^2(s, r)), \quad 0 \le w(s, r) \le 1. \qquad (8)$$

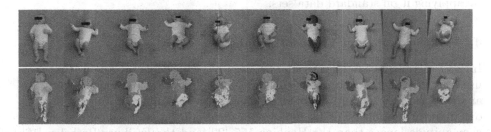

Fig. 1. Sequences 1 (left) to 10 from the experiments (upper row) and segmentation results of our proposed method (lower row) in frame 250.

3 Tracking Based on Segmentation

Along side segmentation, tracking is another important issue. Although there are many tracking algorithms already providing astonishing results on the type of sequences for which they have been designed, we experimentally found none of them to perform sufficiently well on Cerebral Palsy problem. The reasons are manifold: fast motions, high nonrigidity, frequent changes in appearance, etc. For example TLD as proposed in [7], despite being fast and reliable tracking for many applications, fails to track the limbs (upper row of Fig. 9). Therefore, we propose a motion segmentation based tracker. It tracks a specific point x using the motion from segment $\mathcal{O}_i$. We could initialize x manually, or from the center of mass at a labeled frame of all trajectories in $\mathcal{O}_i$. Tracking x using the motion of the center of mass of $\mathcal{O}_i$ would fail due to discontinuity from partial occlusions. Instead, an iterative procedure is used to update the tracking results, as follows. For each segment $\mathcal{O}_i$ and each time step t, we define the subset of all trajectories $s \in \mathcal{O}_i$ that are visible at t and $t+1$ as $\mathcal{S}_t$. Let $\boldsymbol{x}_t^s$ and $\boldsymbol{x}_{t+1}^s$ denote the respective locations of s. Then, $\boldsymbol{x}$ is updated iteratively using the average motion of the trajectories:

$$\boldsymbol{x}_{t+1} = \boldsymbol{x}_t + \frac{1}{|\mathcal{S}_t|} \sum_{s \in \mathcal{S}_t} (\boldsymbol{x}_{t+1}^s - \boldsymbol{x}_t^s). \tag{9}$$

Since Eq. (9) builds the update step by exploiting a large number of trajectories, it can filter out noise and unreliable trajectories, as long as their effects remain small compared to that of the majority of correctly labeled trajectories.

4   Experimental Results and Discussion

In this section the performance of our proposed method is analyzed on two different data sets. First, several videos from infants that have been taken to study Cerebral Palsy are used to analyze the motion of body parts. The second part investigates the generality of the proposed motion segmentation algorithm by applying it on standard data sets.

4.1   Performance on Videos of Infants

In all experiments in this section, we used the experimental set-up that was used in St. Olavs Hospital, Trondheim, Norway. During the experiments, we analyzed the first 1000 frames of 10 sequences showing different infants carrying out different motions (Fig. 1). It is worth mentioning that these sequences are a magnitude longer than the Hopkins 155 [22] and the Freiburg-Berkeley [16] dataset with an average length of 30 and 245 frames, respectively. As ground truth, we manually annotated a dense segmentation of every 250th frame as displayed in Fig. 4. Trajectories are obtained as proposed in [3]. Figure 2 shows the average length of the trajectories for 10 sequences used in this study. As it can be seen, due to occlusions, fast and complicated motion patterns, the trajectories last just for 96.5 frames in average.

Segmentation. Fast and complicated motions render the segmentation and tracking difficult. To illustrate this, consider Fig. 5, that shows the results obtained with the unsupervised segmentation of Brox and Malik [3]. They follow a similar procedure for obtaining trajectories, but with unsupervised spectral clustering on an affinity matrix in order to group trajectories. Compared to the ground truth (Fig. 4), it can be seen that the overall segmentation is not reliable and only very distinct motions could be separated.

   Their approach can hardly be blamed for such failure. Babies exhibit rather erratic limb motions, such that points on the same limb nevertheless move quite

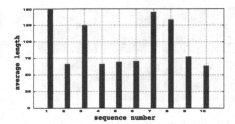

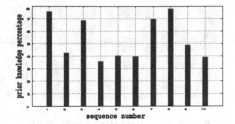

Fig. 2. Average length (in number of frames) of the trajectories in number of frames for different sequences.

Fig. 3. Percentage (%) of trajectories used as prior knowledge with respect to the total trajectories number for different sequences.

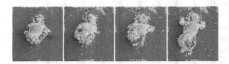

Fig. 4. Seq. 1 ground-truth segmentation for frames 1, 50, 200, 300 from left to right.

Fig. 5. Seq. 1 segmentation results of [3] for frames 1, 50, 200, 300 from left to right.

differently. Motion *per se*, i.e. without further prior knowledge, is hardly strong enough a cue to support a correct segmentation. To supply the segmentation algorithm with such prior information, we label some trajectories for each segment in frames 1 and 500, frames 250 and 750 are used for the evaluation. As it can be observed from Fig. 3, on average only 5 % of the trajectories are initially labeled. Compared to the full annotation needed in current practice, this effort is negligible. As it is visible in Fig. 6 our approach infers the remaining 95 % of the labels. A qualitative comparison between our segmentation result and the ground truth segmentation (Fig. 4) indicates a high precision.

Figure 7 shows a quantitative evaluation where we calculate the ratio of correctly labeled trajectories for three cases: in the first one we manually labeled just the first frame, in the second case two frames (1 and 500). In addition, the result when using the prior labels without transduction are provided as baseline. As it can be observed, there is a substantial gain, which is necessary for our application. With the additional prior knowledge, for most of the sequences the segmentation is very precise, except for Seq. 5 where we deal with very complicated motions, complete occlusions and the body rolling onto the sides. Since we focus on high precision, for the following tracking experiment we used two labeled frames.

Occlusion is a longstanding problem in motion segmentation. Frames 650–800 in Fig. 6 show a case of severe occlusion where the head is occluded by both

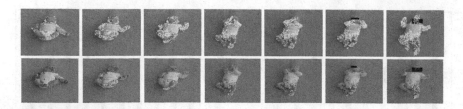

Fig. 6. Seq. 1 segmentation results for frames 1, 50, 200, 300, 650, 700, 800 and 950 from left to right. The upper row shows the results for the baseline where no segmentation method is applied, and the lower row is the results for the proposed method. Frames 800 and 950 have been anonymized after the segmentation.

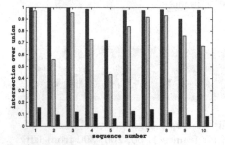

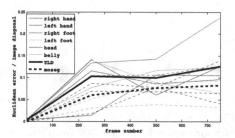

Fig. 7. The GT intersection over union for different sequences. Given are the results for the segmentation of trajectories with one manually labeled frame in green, two manually labeled frames in blue and for the baseline in red.

Fig. 8. The Euclidean tracking error of TLD (straight) and our motion-segmentation approach (dashed) over the frames in a sequence, averaged over all 10 sequences. The colored thin curves denote the errors of the different limbs. The averages over all limbs are displayed in black and bold.

hands. As it can be seen, the segmentation remains correct and in frame 950, new trajectories in the occluded area on the head are labeled correctly. Partial occlusion is less of a problem for our proposed method: there are some trajectories left that can still stand in for the terminated ones. These are joined by novel trajectories upon the reappearance of the previously occluded region. In case of a complete occlusion, trajectories could be linked to each other again by providing further manual labeling or by high dissimilarity to all other motion segments.

Tracking. Although we are tracking body parts and therefore a comparison to pose estimation methods seems reasonable, we compare our tracker performance with a tracking algorithm because pose estimation methods have skeleton constraints as additional prior while our method has the same input as a tracker. In order to study the performance of our tracking method, it is compared with the

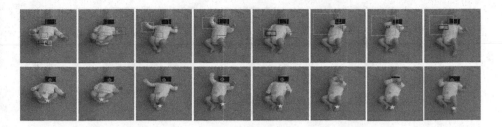

Fig. 9. Seq. 1 tracking results of TLD (upper row) and of proposed method (lower row) for frames 1, 50, 250, 350, 450, 650, 750 and 950 from left to right.

performance of TLD [7] as a representative state-of-the-art tracker. It tracks an object in a 3-step procedure, of tracking, learning, and detection. The tracker follows the object frame by frame, the detector localizes all appearances that have been observed so far, and finally, their *P-N learning* method is applied to estimate the detector error and update it for future use.

Since TLD is a single object tracker, we run it for each body part separately. In some frames TLD falsely does not report a location. To penalize this, we assign these frames the highest error of this sequence. In Fig. 8, the sequence-averaged tracking error of the different body parts is plotted. We evaluate the tracking errors in terms of the Euclidean distance to the ground truth points that are determined manually. As can be observed, our approach is considerably more precise for this task. The main problem with TLD is that it cannot find the object for many frames. The average number of frames without tracking results over all sequences and all body parts for TLD is 12.46 %, while it is 0.38 % for our proposed method which only once looses a body part: in Seq. 5, where the baby rolls onto its right side and the right foot is completely occluded.

The upper row of Fig. 9 shows a qualitative result of TLD. The tracker lost the left foot in frame 50. A bit later, the same happens to the right foot and in frame 450 TLD redetects it wrongly at the right hand. Similar problems regularly occur for the other body parts, hence the performance is insufficient for our task. The results of our proposed tracker are shown in the bottom row of Fig. 9. All body parts are tracked effectively in all the frames, without drifting or part loss. Occlusions (frames 650–750) were dealt with well.

4.2   Comparison with Standard Benchmarks

In this section we challenge our segmentation method with different subjects in order to investigate its generality. To do so, the three video sequences *cats02*, *cats04* and *ducks01* of the *Freiburg-Berkeley* data set [3,16] are considered, in which we deal with occlusion, disocclusion, camera motion, fast motion and low texture objects. The segmentation results of our proposed method (initial labels in frame 50 in *cats02* and *cats04*, and 1 and 200 in *ducks01*) as well as those of Brox and Malik [3] are displayed in Fig. 10. As the results show, [3] only distinguishes very different motions from each other. This is why no object is detected

a)

b)

c)

Fig. 10. Segmentation results for [3] (upper rows) and our proposed method (lower rows) for frames (a) 1, 70, 90, 110 of *cats02*, (b) 1, 40, 50, 70 of *cats04*, and (c) 1, 100, 300, 380 of *ducks01*. For the sake of visibility, the background trajectories are thinned out in *cats04*.

in Seq. *cats04* and *ducks01*. On the other hand, our segmentation method performs reliably: in Seq. *cats02* except very small parts of the legs in a short period of the video, the cat is correctly segmented from the background. Seq. *cats04* shows a case where one of the cats has very low texture as well as fast motions, however the segmentation results are mostly correct. Finally, Seq. *ducks01* shows multiple similar objects that move next to each other, the segmentation task has become even more challenging because of occlusion, disocclusion and exit of one of the ducks. Despite all these, our method managed to segment all the objects correctly through the whole shot.

Figure 11 shows a quantitative comparison between the performance of our proposed method and results of the baseline from the initial labels, where no segmentation algorithm is applied. As it can be seen, the proposed method has managed to segment the objects with high precision (96 % on average).

Although our method performs reliably, it could suffer from some points. First, the segmentation task depends on the distance between trajectories, and since our distance measure (6) only allows for the verification of translational

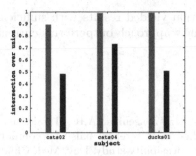

Fig. 11. The GT intersection over union for the proposed method in blue and the baseline in red with no segmentation applied on trajectories.

model, we might have problems with segmenting other models of motion. For example, the legs of the cat in Seq. *cats02* are wrongly segmented because we deal with fast scaling. The second problem could arise from optical flow or trajectory inaccuracy. Although we used one of the promising optical flow methods, we could still have problem in situations with fast motion and low texture. This could be visualized in the last image of Seq. *cats04* where the low texture cat dose a fast jump and trajectories are wrongly developed.

4.3   Performance on Cerebral Palsy Detection

In [18], the proposed method is applied on a set of 82 videos of infants with the same set-up as mentioned in early this section. For each of the infants six trajectories representing motions of different body parts are developed, then a set of features are extracted, and finally, the classification results are compared with those obtained by electromagnetic sensors. The 87 % accuracy of our method in predicting CP shows its advantage.

5   Conclusions

In this paper, we dealt with segmenting and tracking the body limbs of infants in order to provide an analytical tool for clinical diagnosis. The sequences pose multiple problems such as parts having similar appearances, moving in complex ways, and being regularly occluded. Moreover, these parts need to be segmented and tracked with high precision. In our evaluation, the state-of-the-art trackers and motion segmentation algorithms had severe problems with these videos. The manual introduction of prior knowledge appeared to be mandatory at that point, but of course the overhead needed to be kept at a minimum. Therefore, a framework was designed that allows prior knowledge to be integrated into motion segmentation. We proposed a novel energy-minimization-based motion segmentation algorithm. Weak manual annotation came out to suffice to thereupon handle most of the videos automatically. A simple tracker, built on top of the motion segmentation yielded results with sufficient quality. Our experiments showed that our new approach outperforms current tracking and motion segmentation approaches.

References

1. Adde, L., Helbostad, J.L., Jensenius, A.R., Taraldsen, G., Grunewaldt, K.H.: Støen, R.: Early prediction of cerebral palsy by computer-based video analysis of general movements: a feasibility study. Dev. Med. Child Neurol. **52**(8), 773–778 (2010)
2. Boykov, Y., Funka-Lea, G.: Graph cuts and efficient N-D image segmentation. Intl. J. Comput. Vis. **70**(2), 109–131 (2006)
3. Brox, T., Malik, J.: Object segmentation by long term analysis of point trajectories. In: Daniilidis, K., Maragos, P., Paragios, N. (eds.) ECCV 2010, Part V. LNCS, vol. 6315, pp. 282–295. Springer, Heidelberg (2010)
4. Cheriyadat, A.M., Radke, R.J.: Non-negative matrix factorization of partial track data for motion segmentation. In: ICCV, pp. 865–872, Oct 2009
5. Dragon, R., Rosenhahn, B., Ostermann, J.: Multi-scale clustering of frame-to-frame correspondences for motion segmentation. In: Fitzgibbon, A., Lazebnik, S., Perona, P., Sato, Y., Schmid, C. (eds.) ECCV 2012, Part II. LNCS, vol. 7573, pp. 445–458. Springer, Heidelberg (2012)
6. Elhamifar, E., Vidal, R.: Sparse subspace clustering. In: CVPR, pp. 2790–2797 (2009)
7. Kalal, Z., Mikolajczyk, K., Matas, J.: Tracking-learning-detection. TPAMI **34**(7), 1409–1422 (2012)
8. Kanemaru, N., Watanabe, H., Kihara, H., Nakano, H., Takaya, R., Nakamura, T., Nakano, J., Taga, G., Konishi, Y.: Specific characteristics of spontaneous movements in preterm infants at term age are associated with developmental delays at age 3 years. Dev. Med. Child Neurol. **55**, 713–721 (2013)
9. cSen Köktacs, N., Duin, R.P.W.: Statistical analysis of gait data to assist clinical decision making. In: Caputo, B., Müller, H., Syeda-Mahmood, T., Duncan, J.S., Wang, F., Kalpathy-Cramer, J. (eds.) MCBR-CDS 2009. LNCS, vol. 5853, pp. 61–68. Springer, Heidelberg (2010)
10. Kolmogorov, V.: Convergent tree-reweighted message passing for energy minimization. IEEE Trans. Pattern Anal. Mach. Intell. **28**(10), 1568–1583 (2006)
11. Lezama, J., Alahari, K., Sivic, J., Laptev, I.: Track to the future: Spatio-temporal video segmentation with long-range motion cues. In: CVPR, pp. 3369–3376 (2011)
12. Li, Z., Guo, J., Cheong, L.F., Zhou, S.Z.: Perspective motion segmentation via collaborative clustering. In: ICCV (2013)
13. Meinecke, L., Breitbach-Faller, N., Bartz, C., Damen, R., Rau, G., Disselhorst-Klug, C.: Movement analysis in the early detection of newborns at risk for developing spasticity due to infantile cerebral palsy. Hum. Mov. Sci. **25**(2), 125–144 (2006)
14. Ng, A.Y., Jordan, M.I., Weiss, Y.: On spectral clustering: Analysis and an algorithm. NIPS **14**, 849–856 (2002)
15. Ochs, P., Brox, T.: Higher order models and spectral clustering. In: CVPR (2012)
16. Ochs, P., Malik, J., Brox, T.: Segmentation of moving objects by long term video analysis. TPAMI **36**, 1187–1200 (2013)

17. Punithakumar, K., Ayed, I.B., Islam, A., Goela, A., Li, S.: Regional heart motion abnormality detection via multiview fusion. In: Ayache, N., Delingette, H., Golland, P., Mori, K. (eds.) MICCAI 2012, Part II. LNCS, vol. 7511, pp. 527–534. Springer, Heidelberg (2012)
18. Rahmati, H., Aamo, O.M., Stavdahl, Ø., Dragon, R., Adde, L.: Video-based early cerebral palsy prediction using motion segmentation. In: 2014 36th Annual International Conference of the IEEE on Engineering in Medicine and Biology Society (EMBC). IEEE (2014)
19. Shi, F., Zhou, Z., Xiao, J., Wu, W.: Robust trajectory clustering for motion segmentation. In: ICCV (2013)
20. Stahl, A., Schellewald, C., Stavdahl, Ø., Aamo, O.M., Adde, L., Kirkerod, H.: An optical flow-based method to predict infantile cerebral palsy. IEEE Trans. Neural Syst. Rehab. Eng. 20(4), 605–614 (2012)
21. Sun, D., Sudderth, E.B., Black, M.J.: Layered segmentation and optical flow estimation over time. In: CVPR (2012)
22. Tron, R., Vidal, R.: A benchmark for the comparison of 3D motion segmentation algorithms. In: CVPR (2007)
23. Vidal, R., Hartley, R.: Motion segmentation with missing data using powerfactorization and GPCA. In: CVPR, pp. 310–316 (2004)
24. Zhou, D., Bousquet, O., Lal, T., Weston, J., Schölkopf, B.: Learning with local and global consistency. In: Thrun, S., Saul, L., Schölkopf, B. (eds.) Advances in Neural Information Processing Systems. MIT press, Cambridge (2004)

Object-Level Priors for Stixel Generation

Marius Cordts[1,2]([✉]), Lukas Schneider[1], Markus Enzweiler[1], Uwe Franke[1],
and Stefan Roth[2]

[1] Environment Perception, Daimler R&D, Sindelfingen, Germany
[2] Department of Computer Science, TU Darmstadt, Darmstadt, Germany
marius.cordts@daimler.com

Abstract. This paper presents a stereo vision-based scene model for traffic scenarios. Our approach effectively couples bottom-up image segmentation with object-level knowledge in a sound probabilistic fashion. The relevant scene structure, i.e. obstacles and freespace, is encoded using individual Stixels as building blocks that are computed bottom-up from dense disparity images. We present a principled way to additionally integrate top-down prior information about object location and shape that arises from independent system modules, ranging from geometric cues up to highly confident object detections. This results in an efficient exploration of orthogonal image-based cues, such as disparity and gray-level intensity data, combined in a consistent scene representation. The overall segmentation problem is modeled as a Markov Random Field and solved efficiently through Dynamic Programming.

We demonstrate superior segmentation accuracy compared to state-of-the-art superpixel algorithms regarding obstacles and freespace in the scene, evaluated on a large dataset captured in real-world traffic.

1 Introduction

Visual scene understanding is a key problem for autonomous driving and robotics. Especially the knowledge of obstacles that limit the available freespace is crucial for navigation and collision avoidance. This segmentation task was tackled using dense stereo imaging by Badino et al. [4], and resulted in the so-called Stixel world. Subsequently, the model was extended by Pfeiffer et al. [22] to a full image segmentation providing a compact medium-level scene representation accurately capturing multiple depth-layers of objects. These superpixels reduce the complexity for subsequent image processing tasks and are successfully used for numerous applications, such as semantic segmentation [23], object detection [5,12], mapping [21] or segmentation of dynamic objects [13].

However, Stixels are solely based on dense stereo and a strongly simplifying world model with a nearly planar road surface and perpendicular obstacles. Thus, whenever depth measurements are noisy or the world model is violated, Stixels are prone to errors. As can be seen in Fig. 1 top, the car in the center lane, the distant truck and the guard rails are not accurately segmented. In contrast to the bottom-up Stixel segmentation, top-down object detectors do not suffer

© Springer International Publishing Switzerland 2014
X. Jiang et al. (Eds.): GCPR 2014, LNCS 8753, pp. 172–183, 2014.
DOI: 10.1007/978-3-319-11752-2_14

Fig. 1. Original Stixel world (top) and our extension (bottom). Stixel classes are freespace (transp.), obstacle (red), sky (blue), vehicle (green), guard rail (yellow) (Color figure online).

from the mentioned limitations but are specific for one object class, e.g. vehicle, pedestrian or guard rail, and do not provide a generic scene representation.

The main contribution of this work is to show a principled way to incorporate such top-down prior knowledge into the Stixel generation combining the strengths of both methods. We follow a probabilistic approach that allows to find the optimal solution of an extended world model. The additional information not only improves the representation of the detected object classes, but also influences the inference of other parts in the scene, e.g. the freespace. We evaluate our approach in a highway scenario using detectors for vehicles and guard rails, see Fig. 1 for an exemplary output. From a practical point of view, the resulting Stixel world unifies various sources of information and provides a clean, simple and consistent interface for subsequent processing stages.

1.1   Related Work

We see four major categories of publications related to our work. The first contains algorithms for unsupervised image segmentation [1,3,15]. Such methods do not use any semantics and aim for a generic representation of the scene with reduced complexity. Most algorithms are based on appearance only, but there are some that also utilize stereo [26,30]. The Stixel world [22] as introduced above and extended in this publication is naturally closest related to our work.

The second category comprises top-down methods such as detectors for objects [8,10,18,27] or geometric shapes [9,17]. These detectors often show excellent performance, but can only be applied to specific object classes and do not contribute to an understanding of the remaining scene. In this work, we fill this gap and show how to leverage detectors for improving the generic scene model.

The third category covers the task of semantic segmentation, i.e. each segment is associated with a class label. Such methods either operate on a pixel level [19,25] or use superpixels as smallest considered unit [6,7,16]. Although our

174   M. Cordts et al.

proposed algorithm provides an image segmentation with associated class labels,
we do not claim to perform semantic segmentation. Our labels are restricted to
be ground, sky, generic obstacle or those that object detectors provide, which is
not sufficient for a full semantic labeling. However, the proposed Stixel world is
expected to serve well as superpixels for a subsequent labeling step [23].

The fourth and last category uses methods of the second category for semantic segmentation [2,20,29]. From the methodology point of view, we see our work
closest related to publications in this group, using a probabilistic approach for
integrating top-down information in a bottom-up task. However, those methods
either reason on pixel-level [20,29] or use superpixels as the finest element [2].
The first is generally computationally expensive and the latter cannot recover
from errors already present in the superpixels. Thus, in our work we use top-
down knowledge one step earlier, i.e. during the superpixel generation.

2   The Stixel World

In this section, we describe the Stixel computation as introduced in [22]. However, we reformulate the Stixel world using a Markov Random Field (MRF) in
order to integrate object-level prior knowledge in Sect. 3.

The Stixel world S is a segmentation of an image I with size $w \times h$ into
so-called Stixels, where each pixel is assigned to exactly one Stixel. Such a Stixel
$s_{ui} \in S$ can be seen as a superpixel, however restricted to be a vertical line
segment in a certain column $u \in \{1 \ldots w\}$ with bottom and top row $v_{ui}^b \leq v_{ui}^t$. If
the image is horizontally sub-sampled, Stixels become the rectangles visualized in
Fig. 1. The enumeration i of Stixels within a column is such that $v_{ui}^t + 1 = v_{u,i+1}^b$.
According to the assumed world model, a scene is composed of a ground g,
perpendicular objects o and the sky s. Thus, each Stixel is associated with a class
$c_{ui} \in \{g, o, s\}$ and a disparity $d_{ui} \in \mathbb{R}_{\geq 0}$. The latter is discretized, not defined for
the ground, represents the disparity of an object and is zero for the sky. Together,
a Stixel s_{ui} is sufficiently described by the tuple $s_{ui} = (v_{ui}^b, v_{ui}^t, c_{ui}, d_{ui})$.

2.1   Probabilistic Reformulation

Treating all columns independently, searched Stixels $S_{u:}$ in column u are interpreted as random variables $S_{u:} = (S_{u1}, S_{u2}, \ldots, S_{un_u})$ for a fixed number of
Stixels n_u. The measured disparity image in column u is denoted as $d_{u:}$, being
the observations of random variables $D_{u:}$. The posterior $P(S_{u:} \mid D_{u:})$ is defined
using an MRF, depicted as a factor graph in Fig. 2. The graph provides a factorization grouped into a likelihood $\Phi(d_{u:}, s_{u:})$ and a prior $\Psi(s_{u:})$, giving

$$P(S_{u:} = s_{u:} \mid D_{u:} = d_{u:}) = \frac{1}{Z} \Phi(d_{u:}, s_{u:}) \Psi(s_{u:}) , \qquad (1)$$

where Z is the normalizing partition function. The final segmentation s_{ui}^* is
obtained after selecting a model n_u^* and solving the maximum-a-posteriori (MAP)
problem

$$s_{u:}^* = \underset{s_{u:}}{\arg\max} P(S_{u:} = s_{u:} \mid D_{u:} = d_{u:}) . \qquad (2)$$

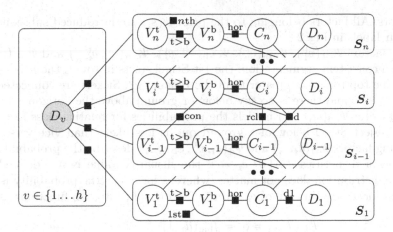

Fig. 2. Stixel world as an MRF. Each Stixel S_i (horizontal boxes) stands for the four random variables V_i^t, V_i^b, C_i, D_i (circles). The prior distribution factorizes according to the labeled factors (black squares, descriptions given in the text). The left part contains exemplary one observed disparity D_v (shaded circle) and the connected factors of the measurement likelihood.

Note that this problem can be solved efficiently using Dynamic Programming (DP), see [22]. Model selection, measurement likelihood and prior are discussed individually in the following, while omitting the column index u for readability.

Model Selection. For selecting a model $n^\star$, first the maximum value $p_n^\star$ of the posterior $P(S_: \mid D_:)$ is determined for each model $n \in \{1 \ldots h\}$. Each value $p_n^\star$ is weighted with a model complexity prior $\exp(-\alpha n_u)$ and the model giving the maximum result is selected. Approximating the partition function Z as being constant for all models, model selection and MAP estimation are performed simultaneously in one DP sweep without explicitly computing the maxima $p_n^\star$.

Likelihood. Assuming conditional independence of the disparities given the segmentation, see Fig. 2, the likelihood factorizes as

$$\Phi(d_:, s_:) = \prod_{v=1}^{h} \prod_{i=1}^{n} \Phi_v(d_v, s_i). \tag{3}$$

The term $\Phi_v(d_v, s_i)$ represents the disparity measurement model that describes the probability of a value d_v for a given Stixel S_i and is non-informative, i.e. uniform, for Stixels not covering row v. For details on the likelihood see [22].

Prior. The prior is modeled using the right part of the Markov random field in Fig. 2. Most important is the first order Markov assumption on Stixel-level, inducing conditional independence of a Stixel S_i and its non-neighbors given its neighbors. All factors belonging to the prior $\Psi(s_{\cdot})$ are introduced subsequently, see their labels in Fig. 2.

The factors $\Psi_{1st}(v_1^b)$, $\Psi_{nth}(v_n^t)$, $\Psi_{t>b}(v_i^b, v_i^t)$, $\Psi_{con}(v_i^b, v_{i-1}^t)$ and $\Psi_{hor}(v_i^b, c_i)$ enforce a consistent segmentation: the 1st Stixel starts in row 1, the nth ends in row h, the top row v_i^t is greater than the bottom v_i^b, Stixels are connected, i.e. $v_i^b = v_{i-1}^t + 1$, and there is no sky below or ground above the horizon.

The factor $\Psi_{rcl}(c_i, c_{i-1})$ models the probabilities for relative class locations, e.g. an object Stixel below of a ground one is less likely than vice versa. The remaining factors $\Psi_{d1}(d_i, c_i)$ and $\Psi_d(d_i, v_i^b, c_i, c_{i-1})$ describe the probability distribution of disparities d_i. For $c_i = $ g the disparity value is not defined and thus the distribution does not matter, whereas for $c_i = $ s the probability is only non-zero for $d_i = 0$. In case of $c_i = $ o the factors split into two functions as

$$\Psi_{d1}(d_i, c_i = \text{o}) = f_{blg}(d_i, 1) \tag{4}$$

$$\Psi_d(d_i, v_i^b, c_i = \text{o}, c_{i-1}) = f_{blg}(d_i, v_i^b)\, f_{grav}(d_i, v_i^b, c_{i-1}) \ . \tag{5}$$

Both functions use the known camera geometry to derive the expected disparity $d_g(v_i^b)$ of the ground in row v_i^b. Then a value $d_i < d_g(v_i^b)$ indicates an object below the ground surface, which is unlikely and captured by f_{blg}. Further, a value $d_i > d_g(v_i^b)$ and a preceding class $c_{i-1} = $ g means that the object is flying above the ground, i.e. no gravity, which is rated with low probability by f_{grav}.

For details on the individual factors, see their corresponding probability distributions in [22].

3   Incorporating Priors

The original Stixel generation is solely based on the disparity image and computed independently for each column. However, other sources of information are often available that take into account the gray value image or multiple columns in the disparity image. This information usually applies only to a specific class and describes its rough location in the image, e.g. using a bounding box, a more complex contour or just a line indicating one end of the object.

3.1   Generic Prior Model

We assume that for each kind of information $j \in \{1 \ldots m\}$, we have a model describing for all pixels the unnormalized probability of being a bottom or top point of the object. The model takes into account the reliability of the source of information, its importance for the Stixel generation, and also a possible uncertainty in the precise location. Such a mapping from a given contour to actual bottom and top point probabilities could either be obtained from training data or by blurring the contour. Thus, each additional input that allows for a meaningful mapping to such probability images can be used for the Stixel generation, see Fig. 3. The resulting images are denoted as $\mathcal{B}_j, \mathcal{T}_j \in \mathbb{R}^{w \times h}$ and are zero where there is no input prior.

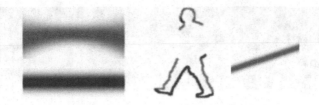

Fig. 3. Probability images derived from the output of three different object detectors: a bounding box, a precise contour and a line. Probabilities for the true contours are encoded as intensities with blue for bottom and red for top points (Color figure online).

3.2 Probabilistic Formulation

In order to use this information for the Stixel generation, we define additional classes a_j that we treat as objects if not stated otherwise. Let $B_{u:}, T_{u:}$ denote the union of all bottom/top probabilities in column u, interpreted as given random variables. Then Eq. (2) is modified to

$$s_{u:}^{\star} = \underset{s_{u:}}{\operatorname{argmax}} P(S_{u:} = s_{u:} \mid D_{u:} = d_{u:}, B_{u:} = b_{u:}, T_{u:} = t_{u:}). \qquad (6)$$

We leave the likelihood unchanged and modify the prior using $b_{u:}$ and $t_{u:}$. First, the factor $\Psi_{\mathrm{rcl}}(c_i, c_{i-1})$ is extended to model the relative location probabilities of the introduced object sub-classes. All additional classes a_j are forced to be on the ground, i.e. $\Psi_{\mathrm{rcl}}(c_i = a_j, c_{i-1} = \mathrm{g}) = 1$ and $\Psi_{\mathrm{rcl}}(c_i = a_j, c_{i-1} \neq \mathrm{g}) = 0$. The case $c_{i-1} = a_j$ is treated as $c_{i-1} = \mathrm{o}$. Second, we introduce an additional factor $\Psi_{\mathrm{pi}}(v_i^{\mathrm{t}}, v_i^{\mathrm{b}}, c_i)$ being the only part where we make use of the bottom and top point probability images. For c_i being one of the standard classes $\mathrm{g}, \mathrm{o}, \mathrm{s}$, the factor Ψ_{pi} is 1. If c_i is one of the additional classes a_j, it holds

$$\Psi_{\mathrm{pi}}(v_i^{\mathrm{t}}, v_i^{\mathrm{b}}, c_i = a_j) = b_j(v_i^{\mathrm{b}}) \, t_j(v_i^{\mathrm{t}}), \qquad (7)$$

where $b_j(v)$ and $t_j(v)$ denote the values of bottom/top point probabilities for class a_j in row v. Where detections are present, b_j and t_j are typically greater than 1, rating the class a_j more likely than the standard classes. If b_j or t_j are 0, i.e. no matching detection, the factor Ψ_{pi} is 0 and hence also the probability for a_j. Third, we add the factor $\Psi_{\mathrm{ht}}(v_i^{\mathrm{t}}, v_i^{\mathrm{b}}, c_i, d_i)$ that captures expectations on the height of an object a_j and evaluates to 1 for valid heights and 0 otherwise. The height in world coordinates can be computed using the given arguments as well as known camera parameters. The described adaptations add only minor computational overhead and still allow for an efficient solution using DP.

4 Experimental Results

In the experimental section, we apply the extended Stixel model to a highway scenario using two external sources of information. Significant improvement is

Fig. 4. Exemplary detections and resulting bottom/top point probabilities.

shown in two key properties: segmentation accuracy (Sect. 4.3) and freespace estimation (Sect. 4.5). Results are compared to the original Stixel formulation [22]. Additionally, the influence of our method on the precision of vehicle Stixels is evaluated (Sect. 4.4). Throughout all experiments, we use identical parameters for our approach and only perform the modifications explained above. Our method increases the runtime of the Stixel computation by 14 %.

4.1   Priors

In this paper we focus on the methodology for incorporating external priors in the Stixel computation. This general idea is independent of the particular information used and can in principle extend to arbitrary object classes. For the experimental evaluation, we utilize detectors for vehicles and guard rails, both highly relevant for autonomous driving on highways. The vehicle detector contributes to all three conducted experiments, see Sects. 4.3–4.5, whereas the guard rails mainly influence the freespace evaluation in Sect. 4.5. Possible detector outputs and the resulting bottom/top point probabilities are visualized in Fig. 4. Note how this knowledge helps to improve the Stixel world, see Fig. 1.

Vehicles. Out of a multitude of proposed vehicle detectors [27], we opted for a two-step system that couples high detection performance at large distances with real-time computational efficiency. In particular, we rely on a very fast vehicle detector, i.e. a Viola-Jones cascade detector [28], to create regions-of-interest for a subsequent strong set of Mixture-of-Experts classifiers using local receptive field features (LRF) [11]. In doing so, the output of the vehicle detector describes the rough location of the vehicle's rear side in the image and its associated confidence value p_{conf}. To incorporate prior knowledge about a vehicle's shape, both the bottom and the top boundary are blurred using a Gaussian truncated at 3σ. The bottom one is centered on the boundary and has a rather small variance. Since the top of a car is less accurately described by a line, we center the Gaussian along a downwards parabola and increase the variance from the box center towards its border, for an example see Fig. 3 left. The maximum value of both probability images is set to $p_{conf}\exp(\alpha_v)$. Eventually, we model the height of vehicles between 0.5 m and 5 m using the factor Ψ_{ht}, see Sect. 3.2.

Fig. 5. Example from dataset containing manual annotations (random colors).

Guard Rails. As guard rail detector, we use parts of [24], based on geometry and appearance. First, the most prominent lines are found using a Hough transform on the gradient image, while restricting the search space to lines matching the expected slope. Only pixels with height values in the range of interest are considered. Second, lines whose disparity does not decrease linearly are discarded. The remaining lines are the guard rail detections and are blurred using a vertical Gaussian to model the top point probabilities with a maximum value of $\exp(\alpha_{gr})$. The height of guard rails is restricted to be between 0.4 m and 1.5 m.

4.2 Dataset

For evaluation, we captured a stereo sequence of 2000 frames on a German highway. Non-occluded vehicles up to very large distances are manually annotated with pixel-accuracy. In addition, the first objects limiting the driving corridor, mainly guard rails, are annotated with pixel-accuracy in every tenth frame. Eventually, occluded and approaching vehicles are annotated as "ignore" with bounding box precision. For an example see Fig. 5.

4.3 Segmentation Accuracy

One key requirement of the Stixel world is to represent the scene and contained objects accurately. To evaluate this aspect, we focus on non-occluded vehicles, since they are the most relevant objects on highways. In the conducted experiment, our method is compared to state-of-the-art superpixel methods with similar runtime, namely SLIC [1] and graph-based image segmentation (GBIS) [15].

To evaluate the algorithms from a semantic point-of-view, we assign to each generated superpixel the majority ground truth label of all covered pixels. Thus, we obtain the upper performance limits of all possible systems for semantic segmentation based on these superpixels. The average number of superpixels that are needed to represent an object serves as a measure of the segmentation's complexity. To evaluate its accuracy, we do not use the PASCAL VOC intersection-over-union (IU) [14], since this measure is dominated by foreground objects and vehicles in larger distances have only little impact. Instead, we answer the question of how well an object can be represented *on average* by computing the IU for each vehicle individually and averaging the results, see Fig. 6 left. Further, we investigate *how many* objects are accurately described via thresholding the

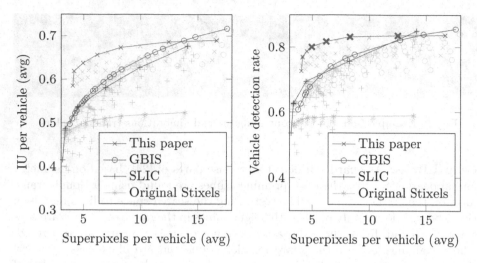

Fig. 6. Our method compared to three baselines in terms of segmentation accuracy over segmentation complexity for the class vehicle. The latter is expressed by the average number of superpixels per object. Accuracy is compared by providing the upper limits for any system based on these superpixels using the average intersection-over-union (IU) per object (left) and the detection rate (right). Each marker stands for one parameter set and solid lines connect the best performing.

IU at 0.5 and determining the detection rate, see Fig. 6 right. To be independent of the parameterization of the algorithms, evaluation is performed for a variation of the most relevant parameters (light markers) taken from [1,15,22] and the upper left part of the convex hull (solid line) is used for comparison of the methods.

The results show that the proposed method significantly improves the performance of the Stixel segmentation and outperforms other state-of-the-art baselines. Especially vehicles in large distances are better segmented due to the information provided by the detector. Note that our method can directly benefit from stronger or additional detectors, whereas an incorporation into SLIC or GBIS is not straightforward.

4.4   Precision

The strength of our approach is the probabilistic integration of detector knowledge into the scene model. In doing so, the whole scene structure in an image column is jointly inferred and unlikely constellations such as vehicles above the ground or inappropriately sized are captured. Thus, erroneous priors due to false positive detections can be suppressed. Further, conflicts due to overlapping detections between objects of the same or different classes are optimally solved, referring to our scene model. These advantages manifest in a high precision of vehicle Stixels. For evaluation, we use the parameter sets belonging to the four red points highlighted in Fig. 6 right. For each setting, we measure the influence

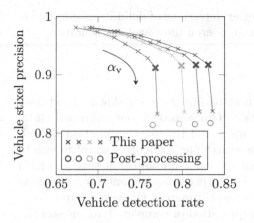 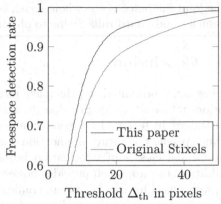

Fig. 7. Influence of detector weight α_v on detection rate (defined above) and precision of vehicle Stixels. A post-processing which forces Stixels to match detections serves as baseline. Highlighted markers correspond to those in Fig. 6.

Fig. 8. Freespace detection rate of original Stixels compared to our proposal. For each Stixel column, the freespace counts as detected, if the deviation of estimation to ground truth is within the range Δ_{th}.

of the scene model's main parameter regarding vehicles, i.e. our confidence in the vehicle detector α_v, see Sect. 4.1. We evaluate the detection rate as defined above and the precision of Stixels with class vehicle. Such a Stixel is considered a true or false positive depending on its covered pixels and their ground truth labels. As baseline serves a more trivial approach for integrating object bounding boxes into the original Stixel world. Here, the obtained segmentation is post-processed to match the given boxes by adding or splitting Stixels where needed and label all Stixels within a box as vehicle.

We significantly improve the precision compared to this baseline without reducing the detection rate, see Fig. 7. Due to modeling the uncertainty of the object's location within a given bounding box, the detection rate is even increased slightly. When weakening the coupling of the detectors by decreasing α_v, the precision improves even more, however at the cost of a lower detection rate.

4.5 Freespace Estimation

For each Stixel column, we extract the row v delimiting the freespace from ground truth. This row is compared to the bottom of the first detected obstacle Stixel from the baseline implementation and our approach, see the transparent areas in Fig. 1. Thresholding this difference at Δ_{th} allows measuring the freespace detection rate, see Fig. 8. Our proposed method outperforms the baseline, even though we do not explicitly influence the freespace estimation. However, due to the joint inference of the whole scene, better detections of delimiting objects, i.e. vehicles and guard rails, helps to obtain an overall improved segmentation.

5   Conclusion

This work presented a principled method to integrate top-down object-level priors into bottom-up Stixel segmentation. Our approach outperformed state-of-the-art in terms of segmentation accuracy and freespace estimation at real-time speeds. For future applications, we expect our model to generalize well to additional classes of information beyond the ones presented in this paper. In addition, our approach provides powerful superpixels for semantic segmentation systems used for rural or urban traffic scenes. Ultimately, this enables to recover a much stronger understanding and interpretation of complex dynamic scenes.

References

1. Achanta, R., Shaji, A., Smith, K., Lucchi, A., Fua, P., Süsstrunk, S.: SLIC superpixels compared to state-of-the-art superpixel methods. IEEE Trans. Pattern Anal. Mach. Intell. **34**, 2274–2282 (2012)
2. Arbeláez, P., Hariharan, B., Gu, C.: Semantic segmentation using regions and parts. In: IEEE Conference on Computer Vision and Pattern Recognition (2012)
3. Arbeláez, P., Maire, M., Fowlkes, C., Malik, J.: Contour detection and hierarchical image segmentation. IEEE Trans. Pattern Anal. Mach. Intell. **33**, 898–916 (2011)
4. Badino, H., Franke, U., Pfeiffer, D.: The stixel world - a compact medium level representation of the 3D-world. In: Denzler, J., Notni, G., Süße, H. (eds.) DAGM 2009. LNCS, vol. 5748, pp. 51–60. Springer, Heidelberg (2009)
5. Benenson, R., Mathias, M., Timofte, R., Van Gool, L.: Pedestrian detection at 100 frames per second. In: IEEE Conference on Computer Vision and Pattern Recognition (2012)
6. Carreira, J., Caseiro, R., Batista, J., Sminchisescu, C.: Semantic segmentation with second-order pooling. In: Fitzgibbon, A., Lazebnik, S., Perona, P., Sato, Y., Schmid, C. (eds.) ECCV 2012, Part VII. LNCS, vol. 7578, pp. 430–443. Springer, Heidelberg (2012)
7. Dann, C., Gehler, P., Roth, S., Nowozin, S.: Pottics – the potts topic model for semantic image segmentation. In: Pinz, A., Pock, T., Bischof, H., Leberl, F. (eds.) DAGM/OAGM 2012. LNCS, vol. 7476, pp. 397–407. Springer, Heidelberg (2012)
8. Dollár, P., Wojek, C., Schiele, B., Perona, P.: Pedestrian detection: an evaluation of the state of the art. IEEE Trans. Pattern Anal. Mach. Intell. **34**, 743–761 (2012)
9. Duda, R., Hart, P.: Use of the Hough transformation to detect lines and curves in pictures. Commun. ACM **15**(1), 11–15 (1972)
10. Enzweiler, M., Gavrila, D.M.: Monocular pedestrian detection: survey and experiments. IEEE Trans. Pattern Anal. Mach. Intell. **31**, 2179–2195 (2009)
11. Enzweiler, M., Gavrila, D.M.: A multi-level mixture-of-experts framework for pedestrian classification. IEEE Trans. Image Process. **20**(10), 2967–2979 (2011)
12. Enzweiler, M., Hummel, M., Pfeiffer, D., Franke, U.: Efficient Stixel-based object recognition. In: IEEE Intelligent Vehicles Symposium (2012)
13. Erbs, F., Schwarz, B., Franke, U.: From Stixels to objects - a conditional random field based approach. In: IEEE Intelligent Vehicles Symposium (2013)

14. Everingham, M., Gool, L.V., Williams, C.K.I., Winn, J., Zisserman, A.: The pascal visual object classes (VOC) challenge. Int. J. Comput. Vis. **88**, 303–338 (2010)
15. Felzenszwalb, P.F., Huttenlocher, D.P.: Efficient graph-based image segmentation. Int. J. Comput. Vis. **59**, 167–181 (2004)
16. Fulkerson, B., Vedaldi, A., Soatto, S.: Class segmentation and object localization with superpixel neighborhoods. In: International Conference on Computer Vision (2009)
17. Gavrila, D.M.: A Bayesian, exemplar-based approach to hierarchical shape matching. IEEE Trans. Pattern Anal. Mach. Intell. **29**, 1408–1421 (2007)
18. Jain, A., Duin, R., Mao, J.: Statistical pattern recognition: a review. IEEE Trans. Pattern Anal. Mach. Intell. **22**(1), 4–37 (2000)
19. Ladický, L., Sturgess, P., Russell, C., Sengupta, S., Bastanlar, Y., Clocksin, W., Torr, P.H.S.: Joint optimisation for object class segmentation and dense stereo reconstruction. In: British Machine Vision Conference (2010)
20. Ladický, L'., Sturgess, P., Alahari, K., Russell, C., Torr, P.H.S.: What, Where and How Many? Combining Object Detectors and CRFs. In: Daniilidis, K., Maragos, P., Paragios, N. (eds.) ECCV 2010, Part IV. LNCS, vol. 6314, pp. 424–437. Springer, Heidelberg (2010)
21. Muffert, M., Schneider, N., Franke, U.: Stix-Fusion: a probabilistic Stixel integration technique. In: Canadian Conference on Computer and Robot Vision (2014)
22. Pfeiffer, D., Franke, U.: Towards a global optimal multi-layer Stixel representation of dense 3D data. In: British Machine Vision Conference (2011)
23. Scharwächter, T., Enzweiler, M., Franke, U., Roth, S.: Efficient multi-cue scene segmentation. In: Weickert, J., Hein, M., Schiele, B. (eds.) GCPR 2013. LNCS, vol. 8142, pp. 435–445. Springer, Heidelberg (2013)
24. Scharwächter, T., Schuler, M., Franke, U.: Visual guard rail detection for advanced highway assistance systems. In: IEEE Intelligent Vehicles Symposium (2014)
25. Shotton, J., Winn, J., Rother, C., Criminisi, A.: TextonBoost for image understanding: multi-class object recognition and segmentation by jointly modeling texture, layout, and context. Int. J. Comput. Vis. **81**(1), 2–23 (2009)
26. Silberman, N., Hoiem, D., Kohli, P., Fergus, R.: Indoor segmentation and support inference from RGBD images. In: Fitzgibbon, A., Lazebnik, S., Perona, P., Sato, Y., Schmid, C. (eds.) ECCV 2012, Part V. LNCS, vol. 7576, pp. 746–760. Springer, Heidelberg (2012)
27. Sun, Z., Bebis, G., Miller, R.: On-road vehicle detection: a review. IEEE Trans. Pattern Anal. Mach. Intell. **28**, 694–711 (2006)
28. Viola, P., Jones, M.J.: Robust real-time object detection. Int. J. Comput. Vis. **4**, 85–107 (2001)
29. Wojek, C., Schiele, B.: A dynamic conditional random field model for joint labeling of object and scene classes. In: Forsyth, D., Torr, P., Zisserman, A. (eds.) ECCV 2008, Part IV. LNCS, vol. 5305, pp. 733–747. Springer, Heidelberg (2008)
30. Zhang, J., Kan, C., Schwing, A.G., Urtasun, R.: Estimating the 3D layout of indoor scenes and its clutter from depth sensors. In: International Conference on Computer Vision (2013)

Coherent Multi-sentence Video Description with Variable Level of Detail

Anna Rohrbach[1]([✉]), Marcus Rohrbach[1,2], Wei Qiu[1,3], Annemarie Friedrich[3], Manfred Pinkal[3], and Bernt Schiele[1]

[1] Max Planck Institute for Informatics, Saarbrücken, Germany
asenina@mpi-inf.mpg.de
[2] UC Berkeley EECS and ICSI, Berkeley, CA, USA
[3] Department of Computational Linguistics, Saarland University, Saarbrücken, Germany

Abstract. Humans can easily describe what they see in a coherent way and at varying level of detail. However, existing approaches for automatic video description focus on generating only single sentences and are not able to vary the descriptions' level of detail. In this paper, we address both of these limitations: for a variable level of detail we produce coherent multi-sentence descriptions of complex videos. To understand the difference between detailed and short descriptions, we collect and analyze a video description corpus of three levels of detail. We follow a two-step approach where we first learn to predict a semantic representation (SR) from video and then generate natural language descriptions from it. For our multi-sentence descriptions we model across-sentence consistency at the level of the SR by enforcing a consistent topic. Human judges rate our descriptions as more readable, correct, and relevant than related work.

1 Introduction

Describing videos or images with natural language sentences is an intriguing but difficult task. Recently it has received increased interest both in the computer vision [1,3,4,10,14] and computational linguistic communities [9,11,22]. The focus of most works on describing videos is to generate single sentences for short video snippets at a fixed level of detail. In contrast, we want to generate coherent multi-sentence descriptions for long videos with multiple activities and allow for producing descriptions at the required levels of detail (see Fig. 1).

Multi-sentence description, our first task, has been explored for videos [1,6, 19], but open challenges remain, e.g. finding a segmentation of appropriate granularity and generating a conceptually and linguistically coherent description. To allow reasoning across sentences we use an intermediate semantic representation (SR) which is inferred from the video. For generating multi-sentence descriptions we ensure that sentences describing the same video are about the same topic (dish in our cooking scenario) and we improve intra-sentence consistency by allowing our language model to choose from a probabilistic SR rather than a single MAP estimate.

© Springer International Publishing Switzerland 2014
X. Jiang et al. (Eds.): GCPR 2014, LNCS 8753, pp. 184–195, 2014.
DOI: 10.1007/978-3-319-11752-2_15

Detailed: A man took a cutting board and knife from the drawer. He took out an orange from the refrigerator. Then, he took a knife from the drawer. He juiced one half of the orange. Next, he opened the refrigerator. He cut the orange with the knife. The man threw away the skin. He got a glass from the cabinet. Then, he poured the juice into the glass. Finally, he placed the orange in the sink.
Short: A man juiced the orange. Next, he cut the orange in half. Finally, he poured the juice into a glass.
One sentence: A man juiced the orange.

Fig. 1. Output of our system for a video, producing coherent multi-sentence descriptions at three levels of detail, using our automatic segmentation.

The second task is generating descriptions with a varying level of detail. While this is a researched problem in natural language generation, e.g. in context of user models [23], we are not aware of any work in computer vision that studies how to select the desired amount of information to be recognized. To understand which information is required for producing a description at a needed level of detail we collected descriptions at three levels of detail for the same video and analyzed which aspects of the video are verbalized in each case.

The first contribution of this paper is to generate coherent multi-sentence descriptions. For this task we (a) propose a model which enforces conceptual consistency across sentences (Sect. 2.2), (b) suggest a simple but effective (and to our knowledge novel) segmentation approach, (c) significantly improve the visual recognition based on the semantic unaries and hand-centric features to provide a consistent description (Sect. 3), (d) couple visual recognition and language generation using a word lattice to improve consistency within each sentence, and (e) improve linguistic cohesiveness/readability (Sect. 4).

Our second contribution is to propose a novel task of describing videos at multiple levels of detail. To approach this task we (a) collected and aligned a corpus of descriptions of three levels of detail, which we provide on our web-page, (b) perform a thorough analysis of the collected data (Sect. 2.1), and (c) propose an approach to handle this new task: namely by selecting the relevant video segments according to the topic and using a language model learned for the right level of detail (Sect. 2.3).

Related Work. To generate descriptions for videos and images rules or templates are a powerful tool but need to be manually defined [4,5,8–10,19]. An alternative is to retrieve sentences from a training corpus [1,3] or to compose novel descriptions based on a language model [10–12,14,18]. We follow [14] which uses an intermediate SR modeled with a CRF. It uses statistical machine translation (SMT) [7] to translate the SR to a sentence. While [14] generates single sentences, the focus of this work is to produce multi-sentence descriptions for an entire video at multiple levels of detail. In contrast to [14] which relies on pre-segmented video snippets, we segment the video automatically.

Multi-sentence generation has been addressed for images by combining descriptions for different detected objects. Reference [10] connects different object detection with prepositions using a CRF and generates a sentence for each pair. Reference [11] models discourse constraints, content planning, and linguistic cohesion using ILP. In contrast we model a global semantic topic to allow descriptions with many sentences while [11] generates in most cases only 1–3 sentences. [6] produces multiple sentences and uses paraphrasing and merging to get the minimum needed number of sentences. In contrast we model consistency across sentences. Using a simple template, [19] generates a sentence every 10 seconds based on concept detection. They recognize a high level event and remove inconsistent concepts. This is similar to our idea of a topic but they work in a much simpler setting of just 3 high level events with manually defined relations to all existing concepts. [1] segments the video based on the similarity of concept detections in neighboring frames and defines verbs manually for all concept pairs, we focus on activity recognition and describing activities with verbs predicted by SMT.

We are not aware of any work in computer vision approaching descriptions at different levels of detail. Closest is [4], which predicts more abstract words if the uncertainty is too high for a more specific prediction. Our approach is complementary, as our goal is to produce descriptions at different levels of detail rather than to decrease uncertainty.

2   Generating Consistent Multi-sentence Video Descriptions at Multiple Levels of Detail

Based on an analysis how humans describe videos we present our approach to generate consistent multi-sentence descriptions at multiple levels on detail.

2.1   Analysis of Human Video Descriptions at Multiple Levels of Detail

We have selected a subset (185 videos) from the *MPII Cooking 2* dataset (update of [15], see our web-page) and collected text descriptions to the videos via Amazon Mechanical Turk (AMT). For each video we asked to describe it in three ways: (1) a detailed description with at most 15 sentences, (2) a short description (3–5 sentences), and (3) a single sentence. We have collected a corpus called *TACoS Multi-Level* with about 20 triples of descriptions for each video.

We draw three conclusions from analyzing the collected descriptions.[1] (1) In detailed descriptions all activities and objects are mentioned, therefore the visual recognition system should identify all of them. (2) Short descriptions could be obtained from detailed descriptions using extractive summarization techniques. However, the various levels show different relative frequency of verbalized concepts, hence it might be beneficial to learn a language model targeted to a desired

---

[1] Details can be found in [17].

level. (3) Single sentence descriptions qualitatively differ from all other types, which suggests that abstractive summarization is required for this level. It is, thus, necessary to recognize the topic (dish that is prepared, in our scenario).

2.2   Multi-sentence Video Descriptions

Assume that a video v can be decomposed into a set of I video snippets represented by video descriptors $\{x_1, ..., x_i, ..., x_I\}$, where each snippet can be described by a single sentence z_i. To reason across sentences we employ an intermediate semantic representation (SR) y_i. We base our approach for a video snippet on the translation approach proposed in [14]. We choose this approach as it allows to learn both the prediction of a semantic representation $x \rightarrow y$ from visual training data (x_i, y_i) and the language generation $y \rightarrow z$ from an aligned sentence corpus (y_i, z_i). While this paper builds on the semantic representation from [14], our idea of consistency is applicable to other semantic representations. The SR y is a tuple of activity and participating objects/locations, e.g. in our case ⟨ACTIVITY, TOOL, OBJECT, SOURCE, TARGET⟩. The relationship is modeled in a CRF where these entities are modeled as nodes $n \in \{1, ..., N\}$ ($N = 5$ in our case) observing the video snippets x_i as unaries. We define s_n as a state of node n, where $s_n \in S$. We use a fully connected graph and linear pairwise (p) and unary (u) terms. In addition, to enable a consistent prediction within a video, we introduce a high level topic node t in the graph, which is also connected to all nodes. In contrast to the other nodes it observes the entire video v rather than a single video snippet. For the topic node t we define a state $s_t \in T$. We then use the following energy formulation for the structured model:

$$E(s_1, ..., s_N, s_t | x_i, v) = \sum_{n=1}^{N} E^u(s_n | x_i) + E^u(s_t | v) + \sum_{\substack{l,m \in \{1,...,N,t\} \\ l \sim m}} E^p(s_l, s_m) \quad (1)$$

with $E^p(s_l, s_m) = w_{l,m}^p$, where $w_{l,m}^p$ are the learned pairwise weights between the CRF node-states s_l and s_m. We discuss the unary features in Sect. 3.

While adding the topic node makes each video snippet aware of the full video, it does not enforce consistency across snippets. Thus, at test time, we compute the conditional probability $p(s_1, ..., s_N | \hat{s}_t)$, setting s_t to the highest scoring state $\hat{s}_t$ over all segments i:

$$(\hat{s}_t, \hat{i}) = \arg \max_{s_t \in T, i \in I} p(s_t | x_i, v). \quad (2)$$

We learn the model by independently training all video descriptors x_i and SR labels $y_i = \langle s_1, ..., s_N, s_t \rangle$ using loopy belief propagation implemented in [16]. The possible states of the CRF nodes are based on the provided video segment labels and topic (dish) labels of the videos from [15].

Segmentation. For the described approach, we have to split the video v into video-snippets x_i. Two aspects are important for this temporal segmentation: it has to find the appropriate granularity so it can be described by a single sentence and it should not contain any unimportant (background) segments which would typically not be described by humans. For the first aspect, we employ agglomerative clustering on a score-vector of semantic attribute classifiers (see Sect. 3). The termination threshold is selected to capture the annotation granularity (number of intervals). The second aspect is achieved by training a background classifier on all unlabeled video segments as negative examples versus all labeled snippets as positive. We evaluate the quality of our segmentation with respect to the final task, namely generating natural language descriptions, in Sect. 5.

2.3   Multi-Level Video Descriptions

Based on the observations discussed in Sect. 2.1, we propose to generate shorter descriptions by extracting a subset of segments from our segmentation. We select relevant segments by scoring how discriminative their predicted SR is for the predicted topic by summing the *tf*idf* scores of the node-states, computed on the training set. For the SR $\langle s_1, \ldots, s_N, s_t \rangle$, its score r equals to:

$$r(s_1, ..., s_N, s_t) = \sum_{n=1}^{N} tf^*idf\,(s_n, s_t) \tag{3}$$

where *tf*idf* is defined as the normalized frequency of the state s_n (i.e. activity or object) in topic s_t times the inverse frequency of its appearance in all topics:

$$tf^*idf\,(s_n, s_t) = \frac{f(s_n, s_t)}{\max_{s'_n \in S} f(s'_n, s_t)} \log \left(\frac{|T|}{\sum_{s'_t \in T} f(s_n, s'_t) > 0} \right) \tag{4}$$

This way we select the K highest scoring segments and use them to produce a short description of the video. One way to produce a description would be to simply extract sentences that correspond to selected segments from the detailed description. However, given that some concepts are not verbalized in shorter descriptions, we additionally explore the approach of learning a translation model targeted to the desired level of detail. For the single sentence descriptions we assume that the predicted topic is sufficient to describe the video. Therefore, we reduce the SR to $\langle$DISH$\rangle$ and learn a translation model to the single sentences.

3   Improving Visual Features

One conclusion drawn in [14] is that the noisy visual recognition is a main limitation. Especially for our problem of multi-sentence generation it is important to recognize the manipulated objects to ensure consistency across sentences. We thus aim to improve the visual recognition by using the semantic unaries and hand-centric features.

Semantic Unaries. The approach of [14] uses visual attributes to obtain the features for CRF unaries. However, this approach ignores the semantic role

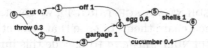

Fig. 2. Encoding probabilistic input for SMT using a word lattice: ⟨CUT OFF,EGG-SHELLS⟩ has the highest confidence but is unlikely according to the language model and other candidate paths, e.g. ⟨CUT OFF, CUCUMBER⟩ can be considered.

of the attributes. E.g. a classifier for a visual attribute *knife* is learned disregarding whether a knife is a TOOL (*cut with a knife*), or an OBJECT (*take out knife*). The CRF unaries use the complete score vectors as features, namely: $E^u(s_n|x_i) = <w_n^u, x_i>$, where w_n^u is a vector of weights between the node-state s_n and the visual attributes' score vector. Unlike the described method, we train SVM classifiers for visual attributes using their semantic role, e.g. we distinguish between *knife*-TOOL and *knife*-OBJECT. This allows us to use a score of each classifier directly as a feature for a corresponding unary: $E^u(s_n|x_i) = w_n^u x_{i,n}$. Here w_n^u is a scalar weight and $x_{i,n}$ is a score of the visual classifier. Thus we get more discriminative unaries and also reduce the number of model parameters (number of connections between node-states and visual features). The topic node unary $E^u(s_t|v)$ is defined similarly, based on the composite activity recognition features [15] as visual descriptors of a video v.

Hand Centric Features for Object Recognition. The visual recognition approach of [14] is based on Dense Trajectory features [21]. In order to improve the object recognition, we propose to focus on hands' regions, in addition to using the holistic features that track all the moving points in the scene. This observation is intuitive, in particular in domains, where people mostly perform hand-related activities. To obtain the hand locations we use our hand detector based on appearance and body pose [17]. We densely sample the points in the hands' neighborhood, extract color Sift features [20] on 4 channels (RGB+grey) and quantize them in a codebook of size 4000. The obtained features are added as another unary to the CRF nodes.

4   Generating Natural Descriptions

Probabilistic Input for SMT. While the translation-based approach can achieve performance comparable with humans on ground truth SRs, this does not hold if the SRs are noisy. The approach of [14] only takes into account the most probable prediction, the uncertainty found in the SR is not used. However, uncertain input is a known problem for SMT as speech based translation is also based on uncertain input. The work of [2] shows that a probabilistic input encoded in a word lattice can improve the performance of translation by decoding alternative hypotheses with lower confidence (see Fig. 2). A *word lattice* is a Directed Acyclic Graph allowing to efficiently decode multiple visual recognition outputs. To construct a word lattice from a set of predicted SRs

$\langle$ACTIVITY,TOOL,INGREDIENT,SOURCE,TARGET$\rangle$, we construct a word lattice for each node and then concatenate them. In case that semantic labels are empty in the SRs, we use a symbol NULL+*node id* to encode this information in the word lattice. SMT combines scores from a phrase-based translation model, a language model, a distortion model and applies word penalties. Word lattice decoding enables us to incorporate confidence scores from the visual recognition.

Creating Cohesive Descriptions. As SMT generates sentences independently for each video segment, the produced descriptions seem more like a 'list of sentences' rather than a 'text' to readers. Hence, we automatically post-process the descriptions such that they are more cohesive using a set of domain-independent rules: we fix punctuation; we combine adjacent sentences if they have the same main verb but different objects, or same object but different verbs; we insert temporal adverbials and replace repeated noun phrases with pronouns (See Footnote 1).

5   Evaluation

For collecting our corpus we rely on the MPII Cooking 2 dataset (update of [15]). This dataset is realistic and typical for assisted daily living or industrial applications which require distinguishing a large number of fine-grained activities and hand-object interaction. Besides, the dataset contains long (average 6 min) videos, allowing to describe them with multiple sentences and at multiple levels.

We evaluate our approach on the TACoS dataset of [13] and on our new corpus TACoS Multi-Level (Sect. 2.1). For TACoS we follow the setup of [14]. For the new corpus we use the training/validation/test split defined for MPII Cooking 2. Comparing to TACoS, our test split is more challenging with more videos (42 vs. 13) and more human subjects (5 vs. 1). We preprocess both corpora by substituting gender specific identifiers with "the person" and transform all sentences to past tense to ensure consistent multi-sentence descriptions.

We evaluate the generated text using BLEU@4, which computes the geometric mean of n-gram word overlaps for n=1,...,4, weighted by a brevity penalty. We also perform human evaluation of the produced descriptions asking human subjects to rate readability (without seeing the video), correctness, and relevance (with respect to the video). Readability is evaluated according to the TAC[2] definition which rates the description's grammaticality, non-redundancy, referential clarity, focus, structure and coherence. Correctness is rated per sentence with respect to the video (independent of completeness), we average the score over all sentences per description. Relevance is rated for the full descriptions and judges if the generated description captures the most important events present in the video. We select all hyperparameters (SVM, CRF, SMT, segmentation) on the validation set and fix them for all experiments; for our segmentation they are the initial segment size (60 frames), the similarity measure (cosine), and the termination threshold (0.982).

---

[2] www.nist.gov/tac/2011/Summarization/Guided-Summ.2011.guidelines.html

Table 1. Visual recognition of SR, accuracy in % (mean over all intervals).

Approach	activity	tool	object	source	target	all	dish
CRF of [14]	59.1	79.6	36.8	71.5	78.2	21.4	-
Our CRF + Semantic unaries	59.2	81.1	39.1	73.8	77.6	23.4	-
+ Hand centric unaries	60.3	82.3	42.6	74.3	78.3	24.2	-
+ Dish unaries	60.4	82.1	48.9	74.3	78.2	26.0	49.3
Number of states	78	53	138	69	49	-	31

5.1 Visual Recognition

We first evaluate the output of our visual recognition (SR) on MPII Cooking 2 dataset. We report accuracy of CRF nodes over all ground truth intervals on the test set in Table 1. The first line shows the results of [14]. We notice that the recognition of the handled object (in many cases the ingredient) is the most difficult, achieving only 36.8 % compared to 59.1 % or more for the other nodes. This lower performance is due to the larger number of states (last line, Table 1) and high intra-class variability of the ingredients. As a first step we add semantic unaries to the CRF. The performance improves for tools by 1.5 % and objects by 2.3 % compared to the first line. Next we add our hand centric color Sift features as second unary to the CRF nodes. This leads to an improvement for each node, especially for objects (+3.5 %). Finally, we add a dish node to the CRF computing unaries with the approach from [15]. This further improves recognition of OBJECT by an impressive 6.3 %. In comparison to [14] we achieve an overall improvement of 1.3 % for ACTIVITY, 2.5 % for TOOL, 12.1 % for OBJECT and 2.8 % for SOURCE (line 1 vs 4). The percentage of segments where the complete SR tuple is correct (column "all") improves on each step and overall increases by 4.6 %. In the next section we show that it leads to more consistent generated descriptions.

5.2 Multi-sentence Generation

We first evaluate the effect of our improved visual recognition and the improvements in natural language sentence generation. We start with the TACoS dataset to allow a direct comparison to [14], using the ground truth intervals provided by TACoS. The first line of Table 2a shows the results using the SR and SMT from [14] (the best version, learning on predictions), which achieves a BLEU score of 23.2 % when evaluated per sentence. This is an increase from 22.1 % reported by [14] due to converting the TACoS corpus to past tense, making it more uniform. The BLEU score evaluated per description is 55.7 %[3] and human judges score these descriptions with 2.5 for readability, 3.3 for correctness, and 2.8 for relevance on a scale from 1–5, where 5 is best. Using our improved SR

---

[3] The BLEU score per description is much higher than per sentence as the n-grams can be matched to the full descriptions.

Table 2. BLEU@4 in % on sentences (Sent) and full descriptions (Desc). Human judgments (Readability, Correctness, Relevance) from 1–5 (5 is best).

Approach	BLEU Sent	BLEU Desc	Read.	Corr.	Relev.
On gt intervals					
[14]	23.2	55.7	2.5	3.3	2.8
Our SR	25.1	63.8	3.3	3.6	3.0
+ prob.	27.5	66.1	3.6	3.7	3.1
Human	36.0[4]	63.6[4]	4.4	4.9	4.8

(a) TACoS

Approach	BLEU Sent	BLEU Desc	Read.	Corr.	Relev.
On gt intervals					
[14]	24.9	60.3	2.8	3.7	3.3
Our	26.9	65.1	3.2/3.4	4.1	3.6
Human	47.8[4]	62.3[4]	4.9	5.0	5.0
On our segmentation					
[14]	-	48.3	2.5	3.5	3.1
Our	-	51.0	2.9/3.2	4.0	3.3

(b) Detailed Descriptions

(line 2 in Table 2a) consistently improves the quality of the descriptions. Judges rate especially the readability much higher (+0.8) which is due to our increased consistency introduced by the dish node. Also correctness (+0.3) and relevance (+0.2) are rated higher, and the BLEU score improves by 1.9 % and 8.1 %.

Next, we evaluate the effect of using probabilistic input for SMT (line 3 in Table 2a). Again all scores increase. Most notably the BLEU by 2.3 % and readability by 0.3. While learning on predictions can recover from systematic errors of the visual recognition, using probabilistic input for SMT allows to recover from errors made at test time by choosing a less likely SR but more likely sentence according to the language model, e.g. *"The person got out a knife and a cutting board from the pot"* is correctly changed to *"The person took out a pot from the drawer"*. While the probabilistic input helps in many cases, we found that it sometimes generates sentences that diverge from the video content.

Now we validate our approach on Detailed Descriptions of the TACoS Multi-Level corpus (Table 2b). The upper part of the Table shows the results on the ground truth intervals provided by the collected descriptions. Here and in the following "Our" denotes the proposed approach with the improved SR and prob-abilistic input. The performance agrees with the results on TACoS. While we make significant improvements over [14], there is still a gap to human descrip-tion, showing the difficulty of the task and the dataset[4]. In the bottom part of Table 2b we evaluate our automatic segmentation and make the following obser-vation: according to human judges, the performance drops only slightly compared to ground truth intervals and it is still higher than the result of [14] on ground truth intervals. This indicates the good quality of our automatic segmentation.

In lines 2 and 5 of Table 2b we evaluate the impact of the linguistic post-processing (Sect. 4) on readability: the score improves from 3.2 to 3.4 and 2.9 to 3.2, respectively (all other reported numbers obtained without post-processing).

---

[4] The BLEU score for human description is not fully comparable due to one reference less, which typically has a strong effect on the BLEU score.

Table 3. BLEU@4 in % on sentences (Sent) and full descriptions (Desc). Human judgments (Readability, Correctness, Relevance) from 1–5 (5 is best).

Approach	BLEU Sent	Desc	Read.	Corr.	Relev.
On gt intervals					
[14]	23.3	52.3	3.6	3.6	3.2
Our	24.7	54.6	3.8/4.0	3.9	3.7
Human	43.9[4]	56.6[4]	4.9	4.9	4.9
On our segmentation					
Our on Det Desc	53.4	-	-	-	
Our on Short Desc	54.3	3.9/4.1	3.7	3.4	

(a) Short Descriptions

Approach	BLEU	Read.	Corr.	Relev.
Upper bound				
Human	53.2[4]	4.9	4.9	4.7
On our segmentation				
Our on Sing Sent Desc	57.7	4.9	3.4	3.3
Our on Det Desc	15.2	-	-	-
Our on Short Desc	21.0	5.0	3.3	2.6

(b) Single Sentence Descriptions

5.3 Multi-Level Generation

On Short Descriptions the results on ground truth intervals (upper part of Table 3a) agree with the previously discussed results. To produce a short description using our segmentation, we select the 3 most relevant segments, as described in Sect. 2. We decide for 3 segments as the average length of short descriptions is 3.5 sentences. In the last two lines of the Table 3a we compare training our system on Detailed vs. Short Descriptions. As expected the language model trained on Short Descriptions performs better (+0.9 % BLEU) supporting our hypothesis that it is beneficial to learn a language model for a desired level of detail.

Table 3b shows the results for the Single Sentence Descriptions. The second line corresponds to our approach of using the dish prediction from the segmentation to translate it into a sentence (Sect. 2.3). We also investigated a retrieval and a template baselines that rely on the dish prediction. They achieve lower BLEU score but nearly identical human judgments, indicating that the dish prediction is the most important aspect for single sentence descriptions. The last two lines compare the extractively produced descriptions, where the single (most relevant) segment was selected. The model trained on the Short Descriptions performs better than the one trained on the Detailed Descriptions, however it is far below Single Sentence Descriptions with respect to relevance (−0.6) and BLEU (−36.7 %), showing the significant difference between these types of descriptions.

6 Conclusion

This work addresses the challenging task of coherent multi-sentence video descriptions. We show that inferring the high level topic helps to ensure consistency across sentences. Using semantic unaries and hand centric features we improve visual recognition, especially for the most challenging semantic category, namely manipulated objects, which consecutively leads to better descriptions.

We also address the so far unexplored task of producing video descriptions at multiple levels of detail with our collected corpus of human descriptions. In an analysis we found that with decreasing length of description, the verbalized information is 'compressed' according to the topic of the video. Based on this we propose a method to extract most relevant segments of the video.

We believe that these results transfer to other domains as our approach is not specific to the kitchen setting. We plan to validate that as part of future work by exploring other domains. While we make a first step to couple visual recognition and language generation by using probabilistic input for SMT on the sentence level, we believe that a direction for future work is to reason jointly about visual recognition and language generation for multi-sentence descriptions.

Acknowledgments. Marcus Rohrbach was supported by a fellowship within the FITweltweit-Program of the German Academic Exchange Service (DAAD).

References

1. Das, P., Xu, C., Doell, R.F., Corso, J.: Thousand frames in just a few words: Lingual description of videos through latent topics and sparse object stitching. In: Proceedings of the IEEE Conference on Computer Vision and Pattern Recognition (CVPR) (2013)
2. Dyer, C., Muresan, S., Resnik, P.: Generalizing word lattice translation. In: Proceedings of the Annual Meeting of the Association for Computational Linguistics (ACL) (2008)
3. Farhadi, A., Hejrati, M., Sadeghi, M.A., Young, P., Rashtchian, C., Hockenmaier, J., Forsyth, D.: Every picture tells a story: generating sentences from images. In: Daniilidis, K., Maragos, P., Paragios, N. (eds.) ECCV 2010, Part IV. LNCS, vol. 6314, pp. 15–29. Springer, Heidelberg (2010)
4. Guadarrama, S., Krishnamoorthy, N., Malkarnenkar, G., Mooney, R., Darrell, T., Saenko, K.: Youtube2text: Recognizing and describing arbitrary activities using semantic hierarchies and zero-shot recognition. In: Proceedings of the IEEE International Conference on Computer Vision (ICCV) (2013)
5. Gupta, A., Srinivasan, P., Shi, J.B., Davis, L.: Understanding videos, constructing plots learning a visually grounded storyline model from annotated videos. In: Proceedings of the IEEE Conference on Computer Vision and Pattern Recognition (CVPR) (2009)
6. Khan, M.U.G., Zhang, L., Gotoh, Y.: Human focused video description. In: Proceedings of the IEEE International Conference on Computer Vision Workshops (ICCV Workshops) (2011)
7. Koehn, P., Hoang, H., Birch, A., Callison-Burch, C., Federico, M., Bertoldi, N., Cowan, B., Shen, W., Moran, C., Zens, R., Dyer, C., Bojar, O., Constantin, A., Herbst, E.: Moses: Open source toolkit for statistical machine translation. In: Annual Meeting of the Association for Computational Linguistics: Human Language Technologies (demo) (2007)
8. Kojima, A., Tamura, T., Fukunaga, K.: Natural language description of human activities from video images based on concept hierarchy of actions. Int. J. Comput. Vis. (IJCV) **50**, 171–184 (2002)

9. Krishnamoorthy, N., Malkarnenkar, G., Mooney, R.J., Saenko, K., Guadarrama, S.: Generating natural-language video descriptions using text-mined knowledge. In: AAAI Conference on Artificial Intelligence (AAAI) (2013)
10. Kulkarni, G., Premraj, V., Dhar, S., Li, S., Choi, Y., Berg, A.C., Berg, T.L.: Baby talk: Understanding and generating simple image descriptions. In: Proceedings of the IEEE Conference on Computer Vision and Pattern Recognition (CVPR) (2011)
11. Kuznetsova, P., Ordonez, V., Berg, A.C., Berg, T.L., Choi, Y.: Collective generation of natural image descriptions. In: Proceedings of the Annual Meeting of the Association for Computational Linguistics (ACL) (2012)
12. Mitchell, M., Dodge, J., Goyal, A., Yamaguchi, K., Stratos, K., Han, X., Mensch, A., Berg, A.C., Berg, T.L., III, H.D.: Midge: Generating image descriptions from computer vision detections. In: Proceedings of the Conference of the European Chapter of the Association for Computational Linguistics (EACL) (2012)
13. Regneri, M., Rohrbach, M., Wetzel, D., Thater, S., Schiele, B., Pinkal, M.: Grounding action descriptions in videos. Trans. Assoc. Comput. Linguist. (TACL) 1, 25–36 (2013)
14. Rohrbach, M., Qiu, W., Titov, I., Thater, S., Pinkal, M., Schiele, B.: Translating video content to natural language descriptions. In: IEEE International Conference on Computer Vision (ICCV) (2013)
15. Rohrbach, M., Regneri, M., Andriluka, M., Amin, S., Pinkal, M., Schiele, B.: Script data for attribute-based recognition of composite activities. In: Fitzgibbon, A., Lazebnik, S., Perona, P., Sato, Y., Schmid, C. (eds.) ECCV 2012, Part I. LNCS, vol. 7572, pp. 144–157. Springer, Heidelberg (2012)
16. Schmidt, M.: UGM: Matlab code for undirected graphical models (2013). http://www.di.ens.fr/~mschmidt/Software/UGM.html
17. Senina, A., Rohrbach, M., Qiu, W., Friedrich, A., Amin, S., Andriluka, M., Pinkal, M., Schiele, B.: Coherent multi-sentence video description with variable level of detail. arXiv:1403.6173 (2014)
18. Siddharth, N., Barbu, A., Siskind, J.M.: Seeing what youre told: Sentence-guided activity recognition in video. In: Proceedings of the IEEE Conference on Computer Vision and Pattern Recognition (CVPR) (2014)
19. Tan, C.C., Jiang, Y.G., Ngo, C.W.: Towards textually describing complex video contents with audio-visual concept classifiers. In: ACM Multimedia (2011)
20. Vedaldi, A., Fulkerson, B.: VLFeat: An open and portable library of computer vision algorithms (2008). http://www.vlfeat.org/
21. Wang, H., Kläser, A., Schmid, C., Liu, C.: Dense trajectories and motion boundary descriptors for action recognition. Int. J. Comput. Vis. (IJCV) 103, 60–79 (2013)
22. Yu, H., Siskind, J.M.: Grounded language learning from videos described with sentences. In: Proceedings of the Annual Meeting of the Association for Computational Linguistics (ACL) (2013)
23. Zukerman, I., Litman, D.: Natural language processing and user modeling: Synergies and limitations. User Model. User-Adap. Inter. 11, 129–158 (2001)

Segmentation and Labeling

Asymmetric Cuts: Joint Image Labeling and Partitioning

Thorben Kroeger[1]([✉]), Jörg H. Kappes[2], Thorsten Beier[1], Ullrich Koethe[1], and Fred A. Hamprecht[1,2]

[1] Multidimensional Image Processing Group, Heidelberg University,
Heidelberg, Germany
`thorben.kroeger@iwr.uni-heidelberg.de`
[2] Heidelberg Collaboratory for Image Processing, Heidelberg University,
Heidelberg, Germany

Abstract. For image segmentation, recent advances in optimization make it possible to combine noisy region appearance terms with pairwise terms which can not only *discourage*, but also *encourage* label transitions, depending on boundary evidence. These models have the potential to overcome problems such as the shrinking bias. However, with the ability to encourage label transitions comes a different problem: strong boundary evidence can overrule weak region appearance terms to create new regions out of nowhere. While some label classes exhibit strong internal boundaries, such as the background class which is the pool of objects. Other label classes, meanwhile, should be modeled as a single region, even if some internal boundaries are visible.

We therefore propose in this work to treat label classes asymmetrically: for some classes, we allow a further partitioning into their constituent objects as supported by boundary evidence; for other classes, further partitioning is forbidden. In our experiments, we show where such a model can be useful for both 2D and 3D segmentation.

1 Introduction

Image segmentation methods typically rely on two complementary sources of information: object appearance and boundary evidence. For example, in semantic labeling tasks [14] a set of object classes of interest is given. Each image can contain one or more of these instances, but might also contain many objects of unknown classes ("background"). One approach for semantic segmentation is to make use of (noisy) local object class probabilities – as obtained from learned appearance models – which can be regularized using local boundary cues.

On the other hand, pure partitioning problems, as in the Berkeley Segmentation Dataset [30], do not specify any object classes but rely on boundary evidence alone [3,5,37].

In this work, we propose a combined semantic labeling *and* partitioning, called *Asymmetric Segmentation*, which can naturally deal with object classes which are known to have strong internal boundaries and jointly optimizes the

© Springer International Publishing Switzerland 2014
X. Jiang et al. (Eds.): GCPR 2014, LNCS 8753, pp. 199–211, 2014.
DOI: 10.1007/978-3-319-11752-2_16

200   T. Kroeger et al.

Fig. 1. Segmentation of image (a) can combine information from both region appearance terms (b) and boundary probabilities (c). We examine different variants of pairwise Conditional Random Field models with Potts potentials. Graph cut (d) uses positive coupling strengths only, which leads to shrinking bias. Multi-way cut (e) uses both negative and positive coupling strengths, such that the creation of boundaries can be actively encouraged. However, this leads to some spurious labelings, induced by strong boundary evidence. Our proposed variant, (f) can yield a better segmentation by allowing boundaries within the background class.

region labeling, the boundaries between classes and the boundaries within classes. Furthermore we present a novel algorithm called Asymmetric Multi-Way Cut (AMWC) for solving those problems.

Many segmentation algorithms, including AMWC, are formulated as second-order Conditional Random Fields [17] over a discrete set of labels, in which the unary potentials transport local evidence for each object class. The pairwise potentials are usually chosen to be Potts functions (2) with varying coupling strengths $w \in \mathbb{R}^+$, which may depend on boundary evidence. The optimal labeling can then be found by minimizing the associated energy function.

Graph cut based algorithms have been extremely influential in the last decade [12,23,33], because they allow to find the optimal solution for binary labeling problems and approximate solutions for multi-label problems with non-negative coupling strengths in polynomial time. They regularize noisy detections by penalizing boundary length. Unfortunately, this leads to "shrinking bias" [36], i.e. thin, elongated objects are cut off (Fig 1d). As a countermeasure, the coupling strength can be chosen as an inverse function of boundary evidence, making label transitions *less* costly when strong boundary evidence exists and *more* costly when boundary evidence is weak. However, the general problem remains: positive coupling strengths cannot actively *encourage* label transitions.

Since negative coupling strengths w *encourage* label transitions and positive w *discourage* label transitions, a model with no restriction on the sign of w may be more expressive: with strong boundary evidence (resulting in $w < 0$) along a thin, elongated object, shrinking bias can be overcome.

Recently, Kappes et al. [18] presented a method that, contrary to others [22,24], is able to find the globally optimal solution for these more general models. This allows us to evaluate the models without any error introduced by approximate optimization.

However, besides their increased computational hardness, models which can *encourage* label transitions also have a major drawback: spurious label transitions are provoked in highly cluttered background (Fig 1e) or within textured objects when strong boundary evidence overrules homogeneous region appearance terms.

In this work, we investigate a new class of models, which do allow intra-category boundaries. The energy function can still be expressed as a pairwise Conditional Random Field. Figure 1f shows the result of our new model, in which we allow internal edges in the "background" class, but disallow internal edges in the "foreground" class.

Contributions. (i) A novel class of segmentation models, where some classes may have internal boundaries and others may not. (ii) An exact solver for AMWC problems based on a formulation using binary edge indicator variables, based on [19]. (iii) Experiments that show when such a formulation is useful and when it is not.
The C++-code of our method will be made available within OpenGM [2].

2   Related Work

Let $G = (\mathcal{V}, \mathcal{E})$ be a given pixel (or superpixel) adjacency graph. Each node $i \in \mathcal{V}$ can be assigned one of k discrete labels: $l_i \in \mathcal{L} = \{0, \ldots, k-1\}$. Many common pixel or superpixel labeling problems [6,12,17,33,35,36] are then formulated as an energy minimization over the sums of unary and pairwise terms:

$$\operatorname*{argmin}_{l \in \mathcal{L}^{|\mathcal{V}|}} \left\{ \sum_{i \in \mathcal{V}} E_i(l_i) + \sum_{(i,j) \in \mathcal{E}} E_{ij}(l_i, l_j) \right\}. \tag{1}$$

The *unary terms* are functions $E_i : \mathcal{L} \to \mathbb{R}$ and indicate the local preference of node i to be assigned a label. The *pairwise terms* are functions $E_{ij} : \mathcal{L} \times \mathcal{L} \to \mathbb{R}$ that express the local joint preferences of two adjacent nodes i and j. A common choice for the binary term is a *Potts* function [12]:

$$E_{ij}^{\boldsymbol{w}}(l_i, l_j) = \begin{cases} 0 & \text{if } l_i = l_j \\ \boldsymbol{w}_{ij} & \text{if } l_i \neq l_j \end{cases}. \tag{2}$$

Depending on the weight $\boldsymbol{w}_{ij}$ a Potts function can either *encourage* ($\boldsymbol{w}_{ij} < 0$) or *discourage* ($\boldsymbol{w}_{ij} > 0$) label transitions.

Binary labeling problems with $\mathcal{L} = \{0, 1\}$ and pairwise potentials with $\forall(i, j) \in \mathcal{E} : \boldsymbol{w}_{ij} \geq 0$ (*graph cut problems*) can be solved in polynomial time with a max-flow algorithm [11,12,23]. Graph cut has been ubiquitous in image segmentation

[33], but penalizes a weighted sum of cut edges which leads to the problem of shrinking bias [36], for which sophisticated countermeasures were developed.

For $k > 2$ labels, (1) becomes a multi-label energy minimization problem that is NP-hard in general, even for non-negative weights $\boldsymbol{w}$. We will refer to the general problem of (1) with $2 \leq k \ll |\mathcal{V}|$ and $\boldsymbol{w} \in \mathbb{R}^{|\mathcal{E}|}$ as the *multi-way cut problem* because the optimal solution can be found as a multicut in a graph with special structure (reviewed in Sect. 3). Approaches to solve this problem approximately are, amongst others, move-making algorithms [6,12] and linear programming [17,19,21,25]. An integer linear program to which violated constraints are added in a cutting-plane fashion [18,19] is able to find the globally optimal solution on many problem instances, see Sect. 3.

A special case of the labeling problem with $E_i(l_i) \equiv 0$, a virtually unlimited set of labels $k = |\mathcal{V}|$ and no restriction on the sign of $\boldsymbol{w}_{ij}$ is called the *correlation clustering* [7] or *multicut problem* [13]. The multicut formulation has recently become popular for unsupervised image segmentation [1,3,4,6,9,20,27,37] where the weights $\boldsymbol{w}$ are either learned in a supervised fashion [3,4,6,20,27], or derived from boundary detectors such as gPb [29] as in [37]. Multicut models have an inherent model-selection ability [6] such that they recover the optimal number of regions needed for an accurate segmentation automatically, based only on boundary evidence. However as region appearance is not taken into account, the resulting segments are not given a class label, which is left as a post-processing step, e.g. done in [20]. Note that we review only unified formulations with a specified objective function here and omit workflows that chain several processing steps.

3   Multi-way Cut Formulation

Before we will define Asymmetric Multi-way Cut in Sect. 4 we first review the Multiway Cut representation of the labeling problem (1) with Potts potentials (2), based on [18,19].

Given a graph $G = (\mathcal{V}, \mathcal{E})$ for (1), we call the nodes $\mathcal{V}$ the *internal nodes* and edges $\mathcal{E}$ the *internal edges*. We then define a new graph $G' = (\mathcal{V}', \mathcal{E}')$, where a set of *terminal nodes* $T = \{t_0, \ldots, t_{k-1}\}$, representing k labels, has been added (Fig. 2, left). Furthermore, *terminal edges* are introduced between each pair of internal and terminal nodes as well as between all pairs of terminal nodes:

$$\mathcal{V}' = \mathcal{V} \cup T \tag{3a}$$

$$\mathcal{E}' = \mathcal{E} \cup \{(t,v) \mid t \in T, v \in \mathcal{V}\} \cup \{(t_i, t_j) \mid 0 \leq i < j < k\}. \tag{3b}$$

Problem (1) with potentials (2) can be written using indicator variables $\boldsymbol{y}$ for the edges $\mathcal{E}'$ and weights $\boldsymbol{w}'$, derived from the potentials $E_i(\cdot)$ and $E_{ij}(\cdot, \cdot)$ [18]:

$$\operatorname*{argmin}_{\boldsymbol{y} \in \{0,1\}^{|\mathcal{E}'|}} \left\{ \sum_{(i,j) \in \mathcal{E}'} \boldsymbol{w}'_{ij} \cdot \boldsymbol{y}_{ij} \right\} \quad \text{s.t. } \boldsymbol{y} \in \mathrm{MWC}_G, \tag{3c}$$

Data:

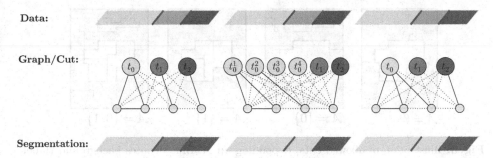

Graph/Cut:

Segmentation:

Fig. 2. Illustration of the graph representation of (1), as given by (3a)–(3b). Shown is a 4×1 pixel image strip ("Data" row). There are three possible classes $\mathcal{L} = \{\text{red}, \text{blue}, \text{yellow}\}$. From the left to the right pixel, the region appearance terms indicate strong preference for the yellow class (pixels 1, 2), a preference for yellow over the blue class (pixel 3, p_3), and a strong preference for the red class for the last pixel. Furthermore, there are strong boundaries to the left and right of p_3. **Left:** Multi-way cut solution. Due to the strong boundary evidence, p_3 is assigned the blue label contrary to its unary potential. **Middle:** The AMWC model with $\mathcal{A} = \{\text{yellow}\}$ is formulated as a MWC by duplicating the terminal nodes for the yellow class $|\mathcal{V}| = 4$ times. **Right:** With the modified constraint (4c'), the number of terminal nodes does not have to be increased.

where MWC_G is the multi-way cut polytope defined by linear constraints [19]:

$$\sum_{(i,j)\in P} \boldsymbol{y}_{ij} \geq \boldsymbol{y}_{uv} \qquad\qquad \forall\, (u,v) \in \mathcal{E}$$

$$P \in \mathrm{Path}(u,v) \subseteq \mathcal{E} \qquad (4a)$$

$$\boldsymbol{y}_{tt'} = 1 \qquad\qquad \forall\, (t,t') \in T, t \neq t' \qquad (4b)$$

$$\boldsymbol{y}_{tu} + \boldsymbol{y}_{tv} \geq \boldsymbol{y}_{uv} \qquad\qquad \forall\, (u,v) \in \mathcal{E}, t \in T \qquad (4c)$$

$$\boldsymbol{y}_{tu} + \boldsymbol{y}_{tv} \geq \boldsymbol{y}_{tv} \qquad\qquad \forall\, (u,v) \in \mathcal{E}, t \in T \qquad (4d)$$

$$\boldsymbol{y}_{tv} + \boldsymbol{y}_{uv} \geq \boldsymbol{y}_{tu} \qquad\qquad \forall\, (u,v) \in \mathcal{E}, t \in T. \qquad (4e)$$

For internal nodes, the *cycle constraint* (4a) [3,13] intuitively forbids dangling boundaries. Constraint (4b) ensures that all terminals are always separated. Finally, (4c)–(4e) constitute cycle constraints for all cycles of three nodes involving one terminal. In particular, (4c) says that, if there is a label transition $(\boldsymbol{y}_{uv} = 1)$, label t cannot belong to both u and v (which would be the case for $\boldsymbol{y}_{tu} = 0$ and $\boldsymbol{y}_{tv} = 0$).

Although a complete description of MWC_G needs a possibly exponential number of constraints, in practice only a small set of active constraints is needed to find the (valid) globally optimal solution. The cutting plane method [18,19] first formulates an unconstrained integer linear program and then identifies violated constraints in the solution, which are subsequently added to the problem. This is repeated until a solution does not violate any of the constraints in (4a)–(4e), yielding the globally optimal solution.

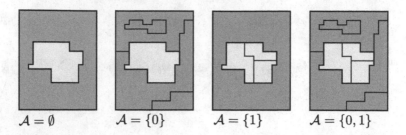

$\mathcal{A} = \emptyset$ $\mathcal{A} = \{0\}$ $\mathcal{A} = \{1\}$ $\mathcal{A} = \{0,1\}$

Fig. 3. Illustration of allowed cuts, depending on $\mathcal{A}$. Blue denotes background (label 0) and yellow foreground (label 1). From left to right we demand no label transitions within foreground and background regions ($\mathcal{A} = \emptyset$), we allow label transitions within the background class ($\mathcal{A} = \{0\}$), or only within the foreground class ($\mathcal{A} = \{1\}$) or in both classes ($\mathcal{A} = \{0,1\}$). In all cases, we admit only closed contours (black boundaries).

4 The Asymmetric Multi-way Cut Formulation

In the proposed asymmetric multiway cut (AMWC) model, we want to allow internal boundaries within regions labeled as $l \in \mathcal{A}$, and disallow internal boundaries in all regions labeled $l \in (\{0, \ldots, k-1\} \setminus \mathcal{A})$, as illustrated in Fig. 3.

4.1 Formulation Within the Binary Edge Labeling Framework

We first give the advantageous formulation of AMWC in terms of the binary labeling of $\mathcal{E}'$ in (3c). One way to formulate the model as a multiway cut is to replace every terminal node $t_a \in \mathcal{A}$ with a set $T_a = \{t_a^0, \ldots, t_a^{|\mathcal{V}|-1}\}$, as shown in Fig. 2, middle and Fig. 4, for which edge weights are copied from the existing terminal edges. These new nodes can represent a partitioning of class a into sub-classes, which are separated by salient boundaries in the image. Similar to the multicut formulation, the label space has to be increased dramatically. By setting $|T_a| = |\mathcal{V}|$, solutions where every node is assigned a different label from the set T_a are made possible.

However, instead of adding $|\mathcal{A}| \cdot (|\mathcal{V}| - 1)$ additional terminal nodes, the same effect can be achieved by simplifying a single constraint in the binary edge labeling formulation of Sect. 3. We relax constraint (4c) to be

$$\boldsymbol{y}_{tu} + \boldsymbol{y}_{tv} \geq \boldsymbol{y}_{uv} \ \forall \, (u, v) \in \mathcal{E}, t \in (T \setminus \mathcal{A}). \tag{4c'}$$

In practice, we can extend the implementation of [19] such that constraint (4c) is only added in the cutting-plane procedure when $t \notin \mathcal{A}$ holds.

With this simple change, we have a method that is able to solve our new AMWC-type models to global optimality.

4.2 Formulation as a Node Labeling Problem

Our AMWC model can still be formulated as a second-order Conditional Random Field (1) with Potts potentials. Starting from the edge labeling formulation

Fig. 4. In this example with $\mathcal{A} = 0$, the background class (label zero, shown in blue) actually consists of several distinct segments t_0^0, t_0^1, t_0^2 and t_0^3 (different shades of blue), all separated by boundaries (black). Another region, $t_1 \notin \mathcal{A}$ cannot be split into sub-segments.

from the previous section, we construct the corresponding model by choosing a new set of labels $\mathcal{L}'$ as well as constructing appropriate unary and pairwise potentials $E_i'(\cdot)$ and $E_{ij}'(\cdot, \cdot)$. Let the original set of labels be $\mathcal{L} = \{t_0, \ldots, t_{k-1}\}$. Then we set

$$\mathcal{L}' = \bigcup_{a \in \mathcal{A}} \left(\{t_a^0, \ldots, t_a^{|\mathcal{V}|-1}\} \right) \cup (\mathcal{L} \setminus \mathcal{A}) \tag{5a}$$

$$E_i'(l_i') = \begin{cases} E_i(l_i') & \text{if } l_i' \in (\mathcal{L} \setminus \mathcal{A}) \\ E_i(a) & \text{if } l_i' = t_a^j \end{cases} \tag{5b}$$

$$E_{ij}'(l_i', l_j') = E_{i,j}(a, b) \quad \text{with } t_a^u = l_i, t_b^v = l_j \qquad \forall u, v. \tag{5c}$$

In (5a), we introduce $|\mathcal{V}| - 1$ additional labels for each label class for which internal boundaries are allowed. These extra labels are assigned the same weights in the new unary terms (5b) as the original label class. Finally, the Potts terms do not change (5c).

The inflated label space makes this formulation unwieldy for practical optimization methods: similar to multicut models, a large number of different labelings, obtained by label permutations, have the same energy.

4.3   Labeling of Regions and Boundaries

At first sight, it may seem that an optimal solution of AMWC can also be obtained by first running the *Multi-way cut* algorithm using the original label set $\mathcal{L} = \{0, \ldots, k-1\}$ to obtain a segmentation into regions $R_1, \ldots, R_n$ and then to run the *multicut* algorithm for each region R_i separately to obtain an internal partitioning. We give two toy examples in the supplementary material which show that this decomposition is not possible.

5   Experiments

Here, we show qualitatively when the AMWC model is useful and when it is not.

2D Segmentation. We consider foreground/background segmentation problems with $\mathcal{L} = \{0, 1\}$ and $\mathcal{A} = \{0\}$. For all images, we first compute an over-segmentation into superpixels using a seeded watershed algorithm on an elevation map combining gradient magnitude and the output of the generalized probability of boundary (gPb) detector from [29]. Then, the superpixel adjacency graph $G = (\mathcal{V}, \mathcal{E})$ defines the structure of the Conditional Random Field (1). For each edge $(i, j) \in \mathcal{E}$ which represents the shared boundary between superpixels i and j, we compute the mean boundary probability, $\mathbf{f}_{ij}$, as given by the gPb detector. Weights $\boldsymbol{w}_{ij} \in \mathbb{R}$ are then obtained as follows

$$\boldsymbol{w}_{ij} = \log \frac{1 - \mathbf{f}_{ij}}{\mathbf{f}_{ij}} + \log \frac{1 - \beta}{\beta}, \tag{6}$$

where $\beta \in [0, 1]$ is a hyper-parameter giving the prior boundary probability. As region appearance terms, we use the output of the object saliency detector [32], or region appearance terms derived from manually placed object bounding boxes. Again, there is a bias hyper-parameter α which gives the prior foreground probability. Finally, we write the energy function (1) as

$$\operatorname*{argmin}_{l \in \mathcal{L}^{|\mathcal{V}|}} \left\{ \gamma \cdot \sum_{i \in \mathcal{V}} E_i(l_i) + \sum_{(i,j) \in \mathcal{E}} E_{ij}^w(l_i, l_j) \right\}, \tag{7}$$

where the hyper-parameter γ weights unary and pairwise terms.

Figure 5 gives examples where the AMWC formulation can help and where it cannot. Column (a) shows the original images, taken from benchmark datasets [10, 30, 31]. Column (b) shows foreground maps either obtained using [32] or given as manual bounding box annotations (rows 1–4). The boundary probability for superpixel edges is visualized in column (c). The Multicut algorithm, column (d) ignores the region appearance terms and gives a decomposition into regions which are shown with random colors. Column (e) shows the visually best solution of a standard graph cut model. Finally, columns (f) and (g) show results for both the Multi-way cut model as well as the AMWC model with $\mathcal{A} = \{\text{background}\}$.

For each row, hyper-parameters α, β, γ – shared among Multicut, Multi-way cut and Asymmetric Multi-way Cutmodels – were chosen to give reasonable and comparable results for these three algorithms.

For the pedestrian detection (Fig. 5, rows 1–4) both the local appearance and edge detection terms are weak. The latter leads to regions which "leak" into the background when using multicut segmentation. Classical graph cut methods suffer severely from the very rough local data terms and show many artifacts caused by shrinking bias. While Multi-way cut generates foreground-artifacts due to strong edges present in the *background*, the proposed AMWC model can handle these by introducing closed contours in the background at these locations. Here, AMWC performs best, even though strong within-foreground contours produce some artifacts (row 2 and 4).

For the examples using saliency detection as the region appearance model (Fig. 5, rows 5–8), both the region and edge terms are more confident. However,

(a) image   (b) region   (c) bound-   (d) MC   (e) GC   (f) MWC   (g) AMWC
             terms   ary terms

Fig. 5. Example segmentations of various pairwise Conditional Random Fields Models. Multicut (MC) is uninformative as regards to category predictions. Graphcut (GC) suffers from shrinking bias. Multi-way cut (MWC) may produce spurious regions induced by strong boundary evidence. AMWC is a joint semantic labeling and partitioning methods that suffers from none of the above. However, it requires specification of the asymmetry (which classes can be partitioned).

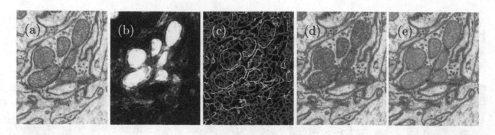

Fig. 6. From the raw volume image (a) a mitochondria versus background probability map (b) was obtained using ilastik [34]. Boundary probabilities are obtained from a random forest classifier using local edge features (c). The multicut algorithm (d) yields a decomposition of the volume image into segments, relying only on the boundary evidence in (c). Finally, Asymmetric Multi-way Cut is the first method that can jointly find a similar segmentation to the one in (d), while at the *same* time labeling regions by their appearance.

graph cut still shows shrinking artifacts and MWC sometimes "hallucinates" foreground-regions in the background, though this is no longer as significant as with the weaker data terms in rows 1–4.

The run times for Multicut, Multi-way cut and AMWC are comparable with less than 10 s per image.

3D Data. To better understand the functioning principles of the brain, researchers use electron microscopy techniques to obtain ultra high-resolution isotropic volumetric images of brain tissue. These exquisite images contain densely packed neurons, which can only be distinguished by their separating membranes. In addition, intra-cellular structures such as mitochondria are visible. In the past, automatic methods to segment mitochondria [28], synapses [8,26], as well as the segmentation of the volume into distinct neurons, e.g. [4,15,16] have been considered separately. The AMWC allows to combine both problems into a single, joint model. We consider two classes, mitochondrion m and cytoplasm c. We set $\mathcal{A} = \{c\}$. Figure 6 shows results on the FIBSEM dataset from [4], where we combine learned pixel-wise mitochondrion probabilities with learned membrane (boundary) probabilities.

This is the first time that a *joint* partitioning into neurons and labeling of intracellular structures becomes possible.

6   Conclusion

We have introduced a new sub-class of non-submodular pairwise multi-label Conditional Random Fields with Potts potentials in which (i) label transitions can be both discouraged as well as encouraged and (ii) some labels, such as background, are allowed to have internal boundaries. As a consequence, strong boundaries within these classes can be naturally accommodated by a further partitioning. The proposed model can be solved exactly using an extension of an existing Multi-way cut solver. We expect this model to be most useful in a regime where regional appearance terms and boundary evidence are both noisy, but (and this is crucial) complementary. In this setting, the present paper offers a principled unified approach to simultaneous labeling and partitioning.

References

1. Alush, A., Goldberger, J.: Break and conquer: efficient correlation clustering for image segmentation. In: 2nd International Workshop on Similarity-Based Pattern Analysis and Recognition (2013)
2. Andres, B., Beier, T., Kappes, J.H.: OpenGM: A C++ library for discrete graphical models, ArXiv e-prints (2012). http://hci.iwr.uni-heidelberg.de/opengm2
3. Andres, B., Kappes, J.H., Beier, T., Kothe, U., Hamprecht, F.A.: Probabilistic image segmentation with closedness constraints. In: ICCV, pp. 2611–2618. IEEE(2011)
4. Andres, B., Kroeger, T., Briggman, K.L., Denk, W., Korogod, N., Knott, G., Koethe, U., Hamprecht, F.A.: Globally optimal closed-surface segmentation for connectomics. In: Fitzgibbon, A., Lazebnik, S., Perona, P., Sato, Y., Schmid, C. (eds.) ECCV 2012, Part III. LNCS, vol. 7574, pp. 778–791. Springer, Heidelberg (2012)
5. Arbelaez, P.: Boundary extraction in natural images using ultrametric contour maps. In: Conference on Computer Vision and Pattern Recognition Workshop, CVPRW'06, pp. 182–182. IEEE (2006)
6. Bagon, S., Galun, M.: Large scale correlation clustering optimization. CoRR abs/1112.2903 (2011)
7. Bansal, N., Blum, A., Chawla, S.: Correlation clustering. Mach. Learn. 56(1–3), 89–113 (2004)
8. Becker, C., Ali, K., Knott, G., Fua, P.: Learning context cues for synapse segmentation in EM volumes. In: Ayache, N., Delingette, H., Golland, P., Mori, K. (eds.) MICCAI 2012, Part I. LNCS, vol. 7510, pp. 585–592. Springer, Heidelberg (2012)
9. Beier, T., Kroeger, T., Kappes, J.H., Koethe, U., Hamprecht, F.: Cut, glue & cut: A fast, approximate solver for multicut partitioning. In: CVPR (2014)
10. Borenstein, E., Ullman, S.: Class-specific, top-down segmentation. In: Heyden, A., Sparr, G., Nielsen, M., Johansen, P. (eds.) ECCV 2002, Part II. LNCS, vol. 2351, pp. 109–122. Springer, Heidelberg (2002)
11. Boykov, Y., Kolmogorov, V.: An experimental comparison of min-cut/max-flow algorithms for energy minimization in vision. Pattern Anal. Mach. Intell. 26(9), 1124–1137 (2004)
12. Boykov, Y., Veksler, O., Zabih, R.: Fast approximate energy minimization via graph cuts. Pattern Anal. Mach. Intell. 23(11), 1222–1239 (2001)
13. Chopra, S., Rao, M.: The partition problem. Math. Program. 59(1–3), 87–115 (1993). http://dx.doi.org/10.1007/BF01581239
14. Everingham, M., Van Gool, L., Williams, C.K.I., Winn, J., Zisserman, A.: The PASCAL Visual Object Classes Challenge 2012 (VOC2012) Results. http://www.pascal-network.org/challenges/VOC/voc2012/workshop/index.html
15. Jain, V., Bollmann, B., Richardson, M., Berger, D.R., Helmstaedter, M.N., Briggman, K.L., Denk, W., Bowden, J.B., Mendenhall, J.M., Abraham, W.C., et al.: Boundary learning by optimization with topological constraints. In: CVPR. IEEE (2010)

16. Jurrus, E., Watanabe, S., Giuly, R.J., Paiva, A.R., Ellisman, M.H., Jorgensen, E.M., Tasdizen, T.: Semi-automated neuron boundary detection and nonbranching process segmentation in electron microscopy images. Neuroinformatics **11**(1), 5–29 (2013)
17. Kappes, J.H., Andres, B., Hamprecht, F.A., Schnörr, C., Nowozin, S., Batra, D., Kim, S., Kausler, B.X., Lellmann, J., Komodakis, N., Rother, C.: A comparative study of modern inference techniques for discrete energy minimization problems. In: CVPR (2013)
18. Kappes, J.H., Speth, M., Andres, B., Reinelt, G., Schn, C.: Globally optimal image partitioning by multicuts. In: Boykov, Y., Kahl, F., Lempitsky, V., Schmidt, F.R. (eds.) EMMCVPR 2011. LNCS, vol. 6819, pp. 31–44. Springer, Heidelberg (2011)
19. Kappes, J.H., Speth, M., Reinelt, G., Schnörr, C.: Higher-order segmentation via multicuts. CoRR abs/1305.6387 (2013)
20. Kim, S., Nowozin, S., Kohli, P., Yoo, C.: Task-specific image partitioning. Transactions on Image Processing (2012)
21. Kohli, P., Shekhovtsov, A., Rother, C., Kolmogorov, V., Torr, P.: On partial optimality in multi-label MRFs. In: ICML, pp. 480–487. ACM (2008)
22. Kolmogorov, V.: Convergent tree-reweighted message passing for energy minimization. IEEE Trans. Pattern Anal. Mach. Intell. **28**(10), 1568–1583 (2006)
23. Kolmogorov, V., Zabin, R.: What energy functions can be minimized via graph cuts? Pattern Anal. Mach. Intell. **26**(2), 147–159 (2004)
24. Komodakis, N., Paragios, N., Tziritas, G.: MRF energy minimization and beyond via dual decomposition. IEEE Trans. Pattern Anal. Mach. Intell. **33**(3), 531–552 (2011)
25. Komodakis, N., Tziritas, G.: Approximate labeling via graph cuts based on linear programming. IEEE Trans. Pattern Anal. Mach. Intell. **29**(8), 1436–1453 (2007)
26. Kreshuk, A., Straehle, C.N., Sommer, C., Koethe, U., Cantoni, M., Knott, G., Hamprecht, F.A.: Automated detection and segmentation of synaptic contacts in nearly isotropic serial electron microscopy images. PloS One **6**(10), e24899 (2011)
27. Kroeger, T., Mikula, S., Denk, W., Koethe, U., Hamprecht, F.A.: Learning to segment neurons with non-local quality measures. In: Mori, K., Sakuma, I., Sato, Y., Barillot, C., Navab, N. (eds.) MICCAI 2013, Part II. LNCS, vol. 8150, pp. 419–427. Springer, Heidelberg (2013)
28. Lucchi, A., Smith, K., Achanta, R., Knott, G., Fua, P.: Supervoxel-based segmentation of mitochondria in EM image stacks with learned shape features. Trans. Med. Imaging **31**(2), 474–486 (2012)
29. Maire, M., Arbeláez, P., Fowlkes, C., Malik, J.: Using contours to detect and localize junctions in natural images. In: CVPR, pp. 1–8. IEEE (2008)
30. Martin, D., Fowlkes, C., Tal, D., Malik, J.: A database of human segmented natural images and its application to evaluating segmentation algorithms and measuring ecological statistics. In: ICCV, vol. 2, pp. 416–423. IEEE (2001)
31. Opelt, A., Pinz, A., Fussenegger, M., Auer, P.: Generic object recognition with boosting. Pattern Anal. Mach. Intell. **28**(3), 416–431 (2006)
32. Perazzi, F., Krähenbühl, P., Pritch, Y., Hornung, A.: Saliency filters: Contrast based filtering for salient region detection. In: CVPR, pp. 733–740. IEEE (2012)
33. Rother, C., Kolmogorov, V., Blake, A.: Grabcut: Interactive foreground extraction using iterated graph cuts. ACM Trans. Graph. **23**, 309–314 (2004)
34. Sommer, C., Straehle, C., Kothe, U., Hamprecht, F.A.: Ilastik: Interactive learning and segmentation toolkit. In: Symposium on Biomedical Imaging: From Nano to Macro, pp. 230–233. IEEE (2011)

35. Szeliski, R., Zabih, R., Scharstein, D., Veksler, O., Kolmogorov, V., Agarwala, A., Tappen, M., Rother, C.: A comparative study of energy minimization methods for Markov random fields. In: Leonardis, A., Bischof, H., Pinz, A. (eds.) ECCV 2006. LNCS, vol. 3952, pp. 16–29. Springer, Heidelberg (2006)
36. Vicente, S., Kolmogorov, V., Rother, C.: Graph cut based image segmentation with connectivity priors. In: CVPR, pp. 1–8. IEEE (2008)
37. Yarkony, J., Ihler, A., Fowlkes, C.C.: Fast planar correlation clustering for image segmentation. In: Fitzgibbon, A., Lazebnik, S., Perona, P., Sato, Y., Schmid, C. (eds.) ECCV 2012, Part VI. LNCS, vol. 7577, pp. 568–581. Springer, Heidelberg (2012)

Mind the Gap: Modeling Local and Global Context in (Road) Networks

Javier A. Montoya-Zegarra[1]([✉]), Jan D. Wegner[1], Ľubor Ladický[2], and Konrad Schindler[1]

[1] Photogrammetry and Remote Sensing, ETH Zürich, Zürich, Switzerland
{javier.montoya,jan.wegner,konrad.schindler}@geod.baug.ethz.ch
[2] Computer Vision Group, ETH Zürich, Zürich, Switzerland
lubor.ladicky@inf.ethz.ch

Abstract. We propose a method to label roads in aerial images and extract a topologically correct road network. Three factors make road extraction difficult: (i) high intra-class variability due to clutter like cars, markings, shadows on the roads; (ii) low inter-class variability, because some non-road structures are made of similar materials; and (iii) most importantly, a complex structural prior: roads form a connected network of thin segments, with slowly changing width and curvature, often bordered by buildings, etc. We model this rich, but complicated contextual information at two levels. Locally, the context and layout of roads is learned implicitly, by including multi-scale appearance information from a large neighborhood in the per-pixel classifier. Globally, the network structure is enforced explicitly: we first detect promising stretches of road via shortest-path search on the per-pixel evidence, and then select pixels on an optimal subset of these paths by energy minimization in a CRF, where each putative path forms a higher-order clique. The model outperforms several baselines on two challenging data sets, both in terms of precision/recall and w.r.t. topological correctness.

1 Introduction

In this paper we deal with automated extraction of the road network from overhead images.[1] The emergence of on-line services like Google Maps, navigation systems, and location-based services has lead to an increased demand for up-to-date maps, particularly in densely populated urban areas. Road extraction is a classical problem which dates back almost 40 years [1] and considerable progress has been achieved, see overviews in [10,24]. Still, no automatic method is robust enough to be employed in practice, and roads are digitized by hand, which is slow and costly. What makes roads (and other linear structures like waterways) special is that topological completeness of the network is often more important than pixel-accurate segmentation. Consider a routing task where the shortest

---

[1] As often done in aerial imaging, when it is available we regard the height-field from dense matching as an additional image channel, and do not separately refer to it.

© Springer International Publishing Switzerland 2014
X. Jiang et al. (Eds.): GCPR 2014, LNCS 8753, pp. 212–223, 2014.
DOI: 10.1007/978-3-319-11752-2_17

connection from A to B is sought. While slightly misplaced road boundaries will not harm the routing, even a very narrow gap may cause a lengthy detour. We thus put an emphasis on network quality, i.e. our main objective is extracting *topologically* complete and correct road networks.

Road extraction in urban environments is challenged by varying road appearance, occlusions as well as heterogeneous background. Unlike highways in the countryside, city streets are frequently occluded (e.g. by trees) or lie in cast shadow. Shape properties like road width, straightness and network density exhibit greater variation. Moreover, many background objects have road-like appearance when viewed from above, e.g. concrete roofs. Thus, classification based on local appearance is unreliable. On the other hand, roads offer a lot of structure and context: *locally*, road pixels form narrow, elongated strips, often bordered by buildings or lined with trees; *globally*, they form a connected network with (mostly) slowly changing segment width. Importantly, these structural properties are quite universal, whereas geometric properties vary from place to place, e.g. American cities have wider roads laid out in a rectangular grid, whereas central European cities have narrower, and more irregular road networks.

We pose road extraction as a pixel-wise labeling task with two classes "road" and "background", and address local context and long-range structure separately. Context is learned directly from data, by training a classifier that uses rich appearance features extracted from a large window, and in this way implicitly includes the local co-occurrence patterns, similar to [17]. The network is modeled explicitly: from the pixel-wise road score, we predict the likelihood that a road of a certain width is present. Based on the resulting $(x, y, width)$-volume of road likelihoods we apply a *recover-and-select* strategy: in the *recover* step many candidates for larger stretches of road are sampled. The *select* step then picks a subset of these candidates that best explains the image evidence (i.e. optimally covers the roads). The selection is formulated as a higher-order CRF, in which the pixels belonging to each road candidate form a large clique, and the clique potential favors consistent labeling of the member pixels. The CRF thus models two preferences: *(i)* pixels should only be labeled as road if they lie on a well-supported long-range connection (or large square) of the network, thus improving precision; and *(ii)* if in a clique the evidence for road outweighs the one for background, then (almost) all of its pixels should be labeled as road, thus improving recall and preventing gaps in the network.

CRF models are currently the standard way of encoding dependencies between pixels in labeling problems, and the search for maximum-evidence (resp. minimum-cost) paths is a classical approach to reconstruct networks – not only roads, but also blood vessels or neurons in medical imaging. Our work is, to our knowledge, the first road extraction framework that attempts to embed minimum-cost paths in a CRF framework, leading on the one hand to a more global solution (because paths can overlap without "double counting"), and on the other hand to a more accurate segmentation (because pixel-to-path membership is a soft constraint and can change during inference).

In the experiments section, we show that together the proposed measures significantly improve the resulting road network. The local context resolves problems due to ambiguous appearance, and yields a significantly higher labeling accuracy than a baseline classifier (16–45 % gain in $F1$-score), which by itself approximately doubles the topological correctness. The long-range network prior on the other hand only brings moderate additional improvement (up to 2 % in $F1$-score) in terms of labeling accuracy, but further increases the topological correctness of the final network by 7 %.

2 Related Work

Since the first early works appeared on road extraction from satellite imagery [1], a large number of methods have been proposed that model road networks with comprehensive sets of ad-hoc rules (*e.g.*, [6,9,25,26,28,35]). Most often they are bottom-up processes that hierarchically stitch together short road segments detected with low-level image processing. The strategy can be successful in rural and suburban areas, where the roads stand out more clearly, there are fewer shadows and occlusions, and the background is relatively homogeneous. Typically many parameters must be tuned empirically. Also, rule-based "expert systems" rely on hard thresholds rather than probabilistic formulations, so they cannot recover from mistakes made at early stages. Some authors have also tried a rule-based approach for more challenging urban environments [11], leading to even more extensive rule sets.

Marked point processes (MPP) offer a probabilistic framework in which road elements (e.g. short line segments) are the basic variables, and allow one to impose high-level topological constraints [16,18,31]. Chai *et al.* [4] recently proposed a comprehensive prior that models both line-segments and junctions of the road network. MPPs are a powerful tool to formulate priors at the level of road elements, but inference is hard and has to rely on all-purpose sampling methods like reversible jump Markov Chain Monte Carlo (RJMCMC), which are computationally expensive and need careful tuning.

A conceptually appealing strategy is to view the road network as a set of minimum cost paths. Such an approach is more flexible in terms of road shape and directly enforces connectivity. Already in early work [7] an A^*-type algorithm is used to iteratively assemble line detector responses to road networks. In medical imaging, minimum-cost paths are widely used to reconstruct vessel trees and neurons (*e.g.*, [2,3,20,34,39]). Recently [33] also tested an algorithm originally developed for medical data on suburban road networks.

Perhaps the most related methods to ours are [32,36]. In [36] the road network is also modeled with the help of a CRF with long-range, higher-order cliques (over super-pixels), using a variant of the P^N-Potts model [14]. The road segments are assumed to be piece-wise straight, and only serve to repair false negatives of the original labeling. Here, we allow for arbitrarily shaped segments, which are found in a data-driven manner. Moreover, our method does not have a foreground bias, but also suppresses false positives in the unaries. Türetken *et al.* [32] (and the earlier [33] for cycle-free networks) compute multi-scale local tubularity scores (reminiscent of our road likelihoods), connect seeds with high foreground scores by minimum cost paths, and prune the over-complete graph to an optimal subgraph with mixed integer programming. They report good performance on aerial images with unoccluded roads, although the method was developed for neurons. The focus is on the center-lines, whereas road pixels are not individually labeled (and the width is not explicitly recovered).

3   Model

We model the road network as the union of elongated segments (termed *paths*) and large compact regions (termed *blobs*). Both *paths* and *blobs* come with an associated scale, *i.e.* we do not only represent road centerlines, but also the local width of the road, respectively the size/diameter of large undirected parts of the network like squares or parking lots. We start with a conventional pixel-wise classification into road and background pixels (Sect. 3.1). In this stage we already include local context via per-pixel feature vectors that encode appearance information over a large spatial neighborhood. In the raw map of road (foreground) scores the local scale (width) of the roads is contained only implicitly. To make it explicit we add a further classification step (Sect. 3.2), which takes as input statistics about the local distribution of the raw scores, and predicts the likelihood that a road *with a specific width* is present.[2] A set of putative *paths* and *blobs* (together referred to as road *candidates*) are then sampled on the basis of the resulting $(x, y, width)$-volume of road likelihoods (Sect. 3.3), by connecting random seed points with paths of maximal cumulative road likelihood, respectively finding large blobs with maximal cumulative likelihood. In order to achieve high recall we follow a recover-and-select strategy: the set of *candidates* is generated such that it is over-complete, but covers as many of the actual road pixels as possible. Finally, a subset of all candidates is selected by energy minimization, resp. MAP estimation, in a CRF (Sect. 3.4). The original road scores form the pixel-wise unaries, and each *candidate* is a higher-order clique with a robust P^N-Potts potential that encourages clique members to take on the same label. Our binary labeling problem allows for globally optimal CRF inference with the min-cut algorithm. The approach is summarized in Fig. 1.

3.1   Context-Aware Road Scores

To obtain pixel-wise road/background scores that take into account the context in the vicinity of a pixel we adopt the multi-feature extension [17] of the TextonBoost algorithm [30]. The pipeline works as follows: first multiple types of features – SIFT [22], textons [23], local ternary patterns [13] and self-similarity [29] – are extracted densely for all pixels. Each feature is soft-quantized to the 8

---

[2] In principle the two-stage classification could potentially be replaced by some form of structured prediction. This would require significantly more training data.

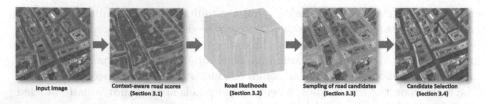

Input Image   Context-aware road scores   Road likelihoods   Sampling of road candidates   Candidate Selection
                (Section 3.1)   (Section 3.2)   (Section 3.3)   (Section 3.4)

Fig. 1. Given an input image, our method first classifies pixels into road and background (Sect. 3.1). Next, the presence of a road with a specific width is predicted (Sect. 3.2). An over-complete set of road candidates is generated (Sect. 3.3), and pruned to an optimal subset (Sect. 3.4).

nearest neighbors in a dictionary of 512 words, using distance-based weighting with the exponential kernel [8]. To include the context, the quantized words are then accumulated into bags-of-words over a (fixed) set of 200 random rectangles that cover a large image region around a pixel. Rectangles range from 4×4 to 80×80 pixels in size, and their locations are sampled in a neighborhood of 160×160 neighborhood, from a Gaussian distribution (*i.e.* their density decreases with distance). The final feature set is the concatenation of all 200 bags-of-words, and thus is aware of the context, in the form of the feature distribution in a large 160-pixel neighborhood around a pixel (compared to an average road width of ≈ 30 pixels in our data).

A classifier is then learned by 5000 rounds of boosting decision stumps on single feature dimensions (also called "shape filters" [30]). Features are not kept in memory, but extracted on the fly using integral images. The boosting output is transformed to pseudo-probabilities S_{road}, S_{bg} in the standard way, by mapping it to the range $[0..1]$ with a sigmoid.

3.2 Road Likelihoods

In order to generate promising candidates in the subsequent step, we need for each pixel not only a single road score, but the likelihood to encounter a piece of road of a particular width w. To estimate that likelihood (for a range of discrete widths), we generate a scale-space representation from the road scores S_{road}, by computing a pyramid of pixel-wise responses to scale-normalized Laplacian-of-Gaussian (LoG) filters of different scales [21]. We found that rather than using the raw LoG responses (or simple transformations of them), it is more robust to resort to a second round of discriminative classification: we feed the mean, median and standard deviation of the pixel-wise LoG responses into a random forest (20 trees, max-depth 15) trained on ground truth road widths, to predict the local likelihood for each possible road width, resulting in a volume $L(x, y, w)$ of likelihoods. In our experience this "learned mapping" from LoG responses to road width likelihoods works better than obvious ad-hoc mappings like rescaling or sigmoid fitting, presumably because it can learn, from the additional information contained in the training labels, to correct typical failures and noise in the raw road scores. At the conceptual level this is in agreement with [27], who also observe an improvement by "cleaning up" raw classifier responses with a second round of classification that looks at their local distribution. In [33] the order is reversed: paths are constructed after the first round, then segments of those paths are reclassified to get more reliable scores.

3.3   Sampling of Road Candidates

The goal of this step is to generate a large set of putative road *candidates*. Candidates are either long-range curvilinear *paths* or large isotropic *blobs* that are likely to belong to the road network. Candidate generation aims for high recall: the union of all *candidates* must contain as much as possible of the road network, even at the cost of low precision. Weak candidates are discarded in the subsequent selection step, but missing candidates cannot be recovered later.

Elongated *paths* are generated by picking two random points with reasonably high road likelihood in $L(x, y, w)$ and connecting them with a path through the volume that maximizes the cumulative likelihood. That path is found with the 3D Fast Marching algorithm [5]. By allowing seed points that are far from each other the paths are a means to impose the long-range network prior: a *path* is always an uninterrupted connection between the seeds and bridges gaps with low road likelihood where necessary. Note that the *path candidates* have an explicit *road width* assigned at each pixel, which changes smoothly (because the path through the volume is continuous also in the w-dimension).

Paths alone are not sufficient to represent the road network. In practice the network also contains large regions without a clear direction like parking lots, squares, roundabouts etc. Paths between different seed points will always traverse such regions along the same routes, where the costs are lowest due unavoidable fluctuations of the likelihood. To nevertheless include such *blob* regions we model them separately. We scan only the top scales w of $L(x, y, w)$, above the maximum road width, for local maxima, and perform non-maxima suppression to obtain a set of *blob candidates*.

3.4   Candidate Selection

The final step is to select a subset of candidates that best covers the road network. Among the given candidates, some paths will pass (partially) through background; different paths will overlap because the fast marching search tends to use the same high-likelihood regions to connect different seeds; and blobs will also overlap smaller blobs as well as many paths.

On the other hand, even the best paths will not always perfectly correspond to roads, especially along the road boundaries. Therefore it is desirable to include a correction step that slightly modifies the candidates where required, but prefers to change them as little as possible in order to maintain coverage and connectivity of the road network. We cast this "selection with correction" as probabilistic inference in a CRF, *i.e.* we minimize an energy $E = \sum_j E_u(x_j) + \sum_i E_p(Q_i)$ over

all pixels x_j of the image.[3] The pixel-wise unaries $E_u(x_j) = -\log(S)$ are negative log-likelihoods of the raw road scores from the original classifier (Sect. 3.1).
The candidates enter in the form of higher-order cliques Q_i, which contain all
pixels that belong to candidate i. Their potentials $E_p(Q_i) = \min(\alpha, N_k \cdot \frac{\alpha-\beta}{\gamma} + \beta)$
are robust P^N-Potts potentials [15], which encourage all member pixels to have
the same label. Here N_k denotes the sum of nodes in the clique that take label
k, and $\{\alpha, \beta, \gamma\}$ are the parameters of a truncated linear function that governs
how the energy increases as more pixels deviate from the dominant label. The
cliques encourage all their member pixels to have the same label. If *sufficient
road evidence* is accumulated inside the clique, pixels are pulled to the road class,
which helps to correct false negatives in the unaries and maintain connectivity
of the network, while still allowing to correct individual pixels that were wrongly
assigned to the clique. If, taken together, the pixels in a clique have *too little
road evidence*, then it is discouraged to label only small, scattered parts of it as
road, which helps to suppress false positives not connected to the road network.

Finally, we still need to encode the model assumption that the candidate set
is over-complete, *i.e.* pixels that are not covered by any path should never be
labeled as roads. In CRF terms this corresponds to a large higher-order clique
Q_{bg} which spans all pixels that are *not member of any candidate* path or blob.
That clique has an asymmetric potential which imposes an infinite penalty if
any of its pixels is labeled as road, and no penalty otherwise. In practice the
same effect can be achieved more efficiently by setting the road likelihoods of
the pixels in Q_{bg} to zero, $\forall x_j \in Q_{bg} : S_{road}(x_j) = 0$, $E_u(x_j) = \infty$. The binary
P^N-Potts model can be solved to global optimality with a graph cut, hence our
inference is guaranteed to find a global minimum of the energy with a single run
of the min-cut algorithm.

4   Experimental Results

We perform experiments on two data sets of urban scenes, GRAZ (Austria) and
VAIHINGEN (Germany).[4] Both data sets are orthophoto mosaics with 3 color
channels plus a normalized height channel computed via dense image matching.
The pixel size is 0.25 m on the ground. In order to enable parallelization and
to reduce the memory footprint, we split up each data set into overlapping tiles
of 1500×1500 pixels. Computations are done on the full tiles to avoid boundary artifacts, while the evaluation is done only for the non-overlapping part of
1000×1000 pixels to avoid double counting. The data sets depict rather different road networks. GRAZ covers the city center of a major city with big building
blocks, inner court yards and parks. There are 67 tiles overall (30 training, 12
validation, 25 testing). Color channels are standard RGB. VAIHINGEN is a small
historic town in hilly countryside, with small buildings, irregular layout, and

---

[3] If desired the P^N-Potts model would also allow for conventional pairwise potentials.
    We did not find them necessary, the context-based unaries are already locally smooth.
[4] GRAZ was kindly provided by Microsoft Photogrammetry. VAIHINGEN is part of the
    ISPRS benchmark http://www.itc.nl/ISPRS_WGIII4/tests_datasets.html.

narrow, winding roads. There are 16 tiles (4 training, 4 validation, 8 testing). Color channels are near infrared, red, and green.

4.1 Evaluation Metrics

As quality measures we report both conventional pixel-based classification scores and topological correctness. Classification accuracy is measured in the standard way with pixel-wise *precision*, *recall*, and *F1-score*. In aerial imaging, variants of the measures called *correctness*, *completeness* and *quality* are popular (*e.g.* [12, 19, 24, 27]) which allow for a few pixels of slack orthogonal to the road centerline to account for geometric uncertainty [37]. We found no significant difference to the standard measures, but nevertheless report both sets. Furthermore, we give the κ-value to assess pixel-wise segmentation accuracy. For a confusion matrix C computed from N pixels, $\kappa = \frac{N \sum_i c_{ii} - \sum_i (\sum_j c_{ij} \cdot \sum_j c_{ji})}{N^2 - \sum_i (\sum_j c_{ij} \cdot \sum_j c_{ji})}$. It quantifies how much the predicted labels differ from a random image with the same label counts.[5]

All these measures are based on pixel area and do not capture the topological correctness of the extracted network. A tiny gap in a road can lead to lengthy detours, but has little impact on recall, and vice versa only few false positive pixels are necessary to produce an inexistent shortcut. We thus additionally report the topological metrics of [36]. These measure what fraction of connecting paths between road points have the correct length within 5 % tolerance, respectively are too short (*2short*), too long (*2long*), or completely infeasible (*noC*). The metrics are computed by randomly sampling paths and counting the occurrence of the four cases until the numbers converge.

4.2 Results

Table 1 shows the results for GRAZ and VAIHINGEN. We report both results of raw classification with the context-aware unaries (*Context*) and results after adding the long-range prior (*CRF*), in order to separate their contributions. Additionally, we add standard baselines for each of the two steps.

As baseline classification we extract per-pixel features with the filter bank of Winn *et al.* [38] and classify them with a random forest (*Winn*). These features consist of multi-scale intensity and derivative responses. They capture the texture properties immediately around the pixel and in our experience work as well as other texture filter banks, but do not capture context in the sense of object-scale shape and co-occurrence patterns. Moreover, as a baseline for a complete system built on top of the features of *Winn* we use our earlier work [36]. That method (*Winn+*) starts from raw road likelihoods obtained by classifying *Winn* features (averaged over superpixels). Straight line segments serve as cliques in a CRF, which is designed to bridge gaps in the road network.

As baseline for the influence of the long-range prior we start from the more powerful *Context* classifier, run the candidate generator in the same way as

---

[5] κ avoids biases due to uneven class distribution. E.g., for an image with 10 % *road* pixels a result without a single *road* pixel has 90 % overall accuracy, but $\kappa=0$ %.

for *CRF*, and discard all candidates whose average unary score is below 0.7 (the threshold which empirically maximizes the *F*1-score). All pixels of the remaining paths and blobs are labeled as road (*RawPath*).

Table 1. Performance of road extraction methods. All numbers are percentages.

	Method	Qual.	Compl.	Corr.	κ	F1	Rec.	Prec.	Corr.	2long	2short	NoC.
GRAZ	Winn	42.8	55.1	65.7	46.9	58.9	54.9	65.8	26.0	10.3	**0.6**	63.1
	Winn+	67.0	84.7	77.1	72.9	80.1	84.8	77.4	74.8	4.3	12.0	8.9
	Context	74.8	88.5	83.1	80.3	85.6	88.5	83.3	77.1	9.1	3.4	10.4
	RawPath	68.2	85.3	78.2	74.0	80.9	85.3	78.4	78.6	**3.2**	16.4	**1.8**
	CRF	**78.0**	**88.5**	**86.9**	**83.1**	**87.6**	**88.6**	**87.1**	**82.8**	5.8	8.4	3.0
VAIH	Winn	56.9	68.2	77.2	62.2	72.4	68.2	77.3	41.3	22.6	**3.4**	32.7
	Winn+	68.7	85.6	78.6	73.5	81.4	85.6	78.7	62.1	5.3	22.8	9.8
	Context	73.0	**89.6**	79.8	77.6	84.4	**89.6**	79.9	72.6	8.0	12.1	7.3
	RawPath	61.0	91.2	64.9	63.8	75.8	91.3	64.9	67.7	**1.8**	30.0	**0.5**
	CRF	**73.3**	88.4	**81.1**	**78.0**	**84.6**	88.4	**81.2**	**77.7**	6.2	12.8	3.3

For both datasets the local context in the unaries drastically improves the pixel-wise performance – see Table 1. The largest contribution comes from increased recall, as the context repairs errors in areas where shadows, trees, unusual surface colour etc. perturb the local appearance – see Fig. 2. Moreover, there is also a significant gain in precision as false positives on concrete roofs, asphalted courtyards etc. are suppressed if they are not supported by the context. Naturally, the greatly improved labeling accuracy is also reflected in much higher topological correctness. *Winn+* does greatly improve the result over the raw labeling of *Winn*, but is still dominated by raw *Context* unaries, which confirms the intuition that one should already include context at the feature level to get stronger unaries.

The proposed *CRF* model further increases per-pixel performance over *Context*, but as expected the effect is comparatively small. Many gaps and false positive patches are cleaned up, but their pixel area is relatively small. Still, these changes significantly increase the topological correctness, mainly by repairing gaps in the network and reducing the number of too long or impossible connections: i.e. the model does what it is designed for, and fills in missing links. The price to pay is that the fraction of too short connections also increases a bit, since some correct gaps are bridged.

On the contrary, heuristically fixing the network (*RawPath*) does not achieve the desired effect. Neither the pixel-wise nor the topological performance of the *Context* unaries is increased, mostly because of false positives. The results suggest that the proposed probabilistic model successfully balances the image evidence against the network prior. It manages to drag concealed roads to the foreground, while at the same time also suppressing false alarms (contrary to *RawPath*, which increases them).

| Winn | Context | RawPath | CRF |

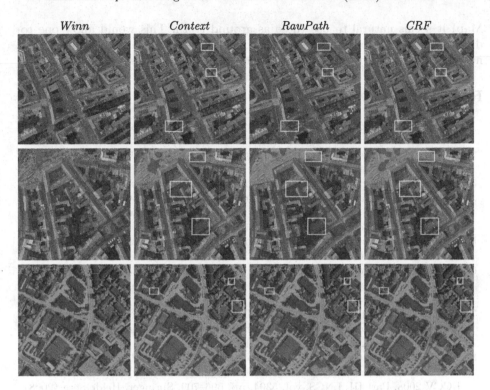

Fig. 2. Road networks extracted in two patches of the orthophoto mosaic of GRAZ (two top rows) and one patch of VAIHINGEN (bottom row). True positives are displayed green, false positives blue, and false negatives red. White boxes highlight improvements by the long-range prior (Color figure online).

5   Conclusions and Future Work

We have proposed methods to exploit context for the semantic segmentation of roads, both at the local and global level. At the local level, expressive features extracted over large neighborhoods implicitly capture the shape and layout of roads and surrounding objects, and lead to much improved classification scores. At the global level the combination of optimal path search and a higher-order CRF models makes it possible to construct an explicit prior about the shape of (pieces of) roads, while still optimizing for pixel-accurate labeling.

Nevertheless, important properties of the road network are still not used. For example, T-junctions and crossings are characteristic network parts that could help to obtain a complete and plausible road network [4,36], and also aspects like a preference for grid layouts and orthogonal intersections are still missing. Moreover, labeling ground truth for training data is time-consuming and costly, and in our scheme must be repeated not only for different sensors or imaging conditions, but also for different building styles, because of the changing context. Since map data is publicly available for many cities (*e.g.*, Open Street

Map) it seems natural to use these as ground truth. This would allow one to directly learn road appearance, shape parameters (*e.g.*, width, straightness), and network topology (*e.g.*, intersection angles at junctions) from big data.

References

1. Bajcsy, R., Tavakoli, M.: Computer recognition of roads from satellite pictures. IEEE Trans. Syst. Man Cybern. **6**(9), 623–637 (1976)
2. Bas, E., Erdogmus, D.: Principal curves as skeletons of tubular objects. Neuroinformatics **9**, 181–191 (2011)
3. Benmansour, F., Cohen, L.D.: Tubular structure segmentation based on minimal path method and anisotropic enhancement. IJCV **92**, 192–210 (2011)
4. Chai, D., Förstner, W., Lafarge, F.: Recovering line-networks in Images by junction-point processes. In: CVPR (2013)
5. Deschamps, T., Cohen, L.D.: Fast extraction of minimal paths in 3d images and applications to virtual endoscopy. Med. Image Anal. **5**(4), 281–299 (2001)
6. Doucette, P., Agouris, P., Stefanidis, A.: Automated road extraction from high resolution multispectral imagery. Photogram. Eng. Remote Sens. **70**(12), 1405–1416 (2004)
7. Fischler, M., Tenenbaum, J., Wolf, H.: Detection of roads and linear structures in low-resolution aerial imagery using a multisource knowledge integration technique. Comput. Graph. Image Process. **15**, 201–223 (1981)
8. van Gemert, J.C., Geusebroek, J.-M., Veenman, C.J., Smeulders, A.W.M.: Kernel codebooks for scene categorization. In: Forsyth, D., Torr, P., Zisserman, A. (eds.) ECCV 2008, Part III. LNCS, vol. 5304, pp. 696–709. Springer, Heidelberg (2008)
9. Grote, A., Heipke, C., Rottensteiner, F.: Road network extraction in suburban areas. Photogram. Rec. **27**(137), 8–28 (2012)
10. Heipke, C., Mayer, H., Wiedemann, C.: Evaluation of automatic road extraction. In: 3D Reconstruction and Modeling of Topographic Objects (1997)
11. Hinz, S., Baumgartner, A.: Automatic extraction of urban road networks from multi-view aerial imagery. ISPRS J. Photogram. Remote Sens. **58**, 83–98 (2003)
12. Hu, J., Razdan, A., Femiani, J.C., Cui, M., Wonka, P.: Road network extraction and intersection detection from aerial images by tracking road footprints. IEEE TGRS **45**(12), 4144–4157 (2007)
13. Hussain, S., Triggs, B.: Visual recognition using local quantized patterns. In: Fitzgibbon, A., Lazebnik, S., Perona, P., Sato, Y., Schmid, C. (eds.) ECCV 2012, Part II. LNCS, vol. 7573, pp. 716–729. Springer, Heidelberg (2012)
14. Kohli, P., Ladicky, L., Torr, P.H.S.: Robust higher order potentials for enforcing label consistency. In: CVPR (2008)
15. Kohli, P., Ladicky, L., Torr, P.H.S.: Robust higher order potentials for enforcing label consistency. IJCV **82**(3), 302–324 (2009)
16. Lacoste, C., Descombes, X., Zerubia, J.: Point processes for unsupervised line network extraction in remote sensing. PAMI **27**(10), 1568–1579 (2005)
17. Ladicky, L., Russell, C., Kohli, P., Torr, P.H.S.: Associative hierarchical CRFs for object class image segmentation. In: ICCV (2009)
18. Lafarge, F., Gimel'farb, G., Descombes, X.: Geometric feature extraction by a multimarked point process. PAMI **32**(9), 1597–1609 (2010)
19. Laptev, I., Mayer, H., Lindeberg, T., Eckstein, W., Steger, C., Baumgartner, A.: Automatic extraction of roads from aerial images based on scale space and snakes. MVA **12**, 23–31 (2000)

20. Li, H., Yezzi, A.: Vessels as 4-D curves: global minimal 4-D paths to extract 3-D tubular surfaces and centerlines. IEEE TMI **26**(9), 1213–1223 (2007)
21. Lindeberg, T.: Scale-space theory: A basic tool for analysing structures at different scales. J. Appl. Stat. **21**(2), 224–270 (1994)
22. Lowe, D.G.: Distinctive image features from scale-invariant keypoints. Int. J. Comput. Vis. **60**(2), 91–110 (2004)
23. Malik, J., Belongie, S., Leung, T., Shi, J.: Contour and texture analysis for image segmentation. Int. J. Comput. Vis. **43**(1), 7–27 (2001)
24. Mayer, H., Hinz, S., Bacher, U., Baltsavias, E.: A test of automatic road extraction approaches. IAPRS **36**(3), 209–214 (2006)
25. Mena, J., Malpica, J.: An automatic method for road extraction in rural and semi-urban areas starting from high resolution satellite imagery. Pattern Recogn. Lett. **26**, 1201–1220 (2005)
26. Miao, Z., Shi, W., Zhang, H., Wang, X.: Road centerline extraction from high-resolution imagery based on shape features and multivariate adaptive regression splines. IEEE GRSL **10**(3), 583–587 (2013)
27. Mnih, V., Hinton, G.E.: Learning to detect roads in high-resolution aerial images. In: Daniilidis, K., Maragos, P., Paragios, N. (eds.) ECCV 2010, Part VI. LNCS, vol. 6316, pp. 210–223. Springer, Heidelberg (2010)
28. Poullis, C., You, S.: Delineation and geometric modeling of road networks. ISPRS J. Photogram. Remote Sens. **65**, 165–181 (2010)
29. Shechtman, E., Irani, M.: Matching local self-similarities across images and videos. In: Conference on Computer Vision and Pattern Recognition (2007)
30. Shotton, J., Winn, J.M., Rother, C., Criminisi, A.: *TextonBoost*: joint appearance, shape and context modeling for multi-class object recognition and segmentation. In: Leonardis, A., Bischof, H., Pinz, A. (eds.) ECCV 2006, Part I. LNCS, vol. 3951, pp. 1–15. Springer, Heidelberg (2006)
31. Stoica, R., Descombes, X., Zerubia, J.: A Gibbs point process for road extraction from remotely sensed images. IJCV **57**(2), 121–136 (2004)
32. Türetken, E., Benmansour, F., Andres, B., Pfister, H., Fua, P.: Reconstructing loopy curvilinear structures using integer programming. In: CVPR (2013)
33. Türetken, E., Benmansour, F., Fua, P.: Automated reconstruction of tree structures using path classifiers and mixed integer programming. In: CVPR (2012)
34. Türetken, E., González, G., Blum, C., Fua, P.: Automated reconstruction of dendritic and axonal trees by global optimization with geometric priors. Neuroinformatics **9**, 279–302 (2011)
35. Ünsalan, C., Sirmacek, B.: Road network detection using probabilistic and graph theoretical methods. IEEE TGRS **50**(11), 4441–4453 (2012)
36. Wegner, J.D., Montoya-Zegarra, J.A., Schindler, K.: A higher-order CRF model for road network extraction. In: CVPR (2013)
37. Wiedemann, C., Heipke, C., Mayer, H., Jamet, O.: Empirical evaluation of automatically extracted road axes. In: CVPR Workshops (1998)
38. Winn, J., Criminisi, A., Minka, T.: Object categorization by learned universal visual dictionary. In: CVPR (2005)
39. Zhao, T., Xie, J., Amat, F., Clack, N., Ahammad, P., Peng, H., Long, F., Myers, E.: Automated reconstruction of neuronal morphology based on local geometrical and global structural models. Neuroinformatics **9**, 247–261 (2011)

Image Processing and Analysis

Guided Image Super-Resolution: A New Technique for Photogeometric Super-Resolution in Hybrid 3-D Range Imaging

Florin C. Ghesu[1], Thomas Köhler[1,2(✉)], Sven Haase[1],
and Joachim Hornegger[1,2]

[1] Pattern Recognition Lab, Friedrich-Alexander-Universität Erlangen-Nürnberg,
Erlangen, Germany
[2] Erlangen Graduate School in Advanced Optical Technologies (SAOT),
Erlangen, Germany
{florin.c.ghesu,thomas.koehler}@fau.de

Abstract. In this paper, we augment multi-frame super-resolution with the concept of guided filtering for simultaneous upsampling of 3-D range data and complementary photometric information in hybrid range imaging. Our guided super-resolution algorithm is formulated as joint maximum a-posteriori estimation to reconstruct high-resolution range and photometric data. In order to exploit local correlations between both modalities, guided filtering is employed for regularization of the proposed joint energy function. For fast and robust image reconstruction, we employ iteratively re-weighted least square minimization embedded into a cyclic coordinate descent scheme. The proposed method was evaluated on synthetic datasets and real range data acquired with Microsoft's Kinect. Our experimental evaluation demonstrates that our approach outperforms state-of-the-art range super-resolution algorithms while it also provides super-resolved photometric data.

1 Introduction

3-D range imaging (RI) based on active sensor technologies such as structured light or Time-of-Flight (ToF) cameras is an emerging field of research. Over the past years, with the development of low-cost devices such as Microsoft's Kinect for the consumer market, RI found its way into various computer vision applications [3,12,22] and most recently also to healthcare [2]. Opposed to passive stereo vision approaches, active RI sensors feature the acquisition of dense range images from dynamic scenes in real-time. In addition to range information, complementary photometric data is often provided by the same device in a hybrid imaging system, e.g. color images in case of the Kinect or amplitude data captured by a ToF camera. However, due to technological or economical restrictions, these sensors suffer from a limited spatial resolution which restricts their use for highly accurate measurements. In particular, this is the case for range sensors that may be distorted by random noise and systematic errors depending on the underlying hardware. In order to overcome a limited sensor resolution,

© Springer International Publishing Switzerland 2014
X. Jiang et al. (Eds.): GCPR 2014, LNCS 8753, pp. 227–238, 2014.
DOI: 10.1007/978-3-319-11752-2_18

image super-resolution has been proposed [14]. A common principle for resolution enhancement is to fuse multiple low-resolution acquisitions with known subpixel displacements into a high-resolution image [7]. The utilized subpixel motion is estimated using image registration methods. For photometric information, the goal of super-resolution is to recover fine structures such as texture barely visible in low-resolution images. In terms of RI, this concept enables accurate 3-D shape scanning [5] which is hard to obtain based on low-resolution range data.

1.1 Related Work

Over the past years, super-resolution has been proposed for a variety of imaging modalities and applications. Traditionally, super-resolution algorithms have been applied to single- or multichannel images encoding photometric information. In many approaches, such as the algorithms proposed by Elad and Feuer [6] or Schultz and Stevenson [18], super-resolution is formulated as maximum a-posteriori (MAP) estimation. In order to consider the presence of outliers, Farsiu et al. [8] introduced a robust extension based on L_1 norm minimization. Babacan et al. [1] have formulated a variational Bayesian approach to estimate high-resolution images and the uncertainty of the underlying model parameters.

In terms of range super-resolution, most prior work adopted techniques originally designed for intensity images. A MAP approach for range super-resolution has been proposed by Schuon et al. [19,20]. Bhavsar and Rajagopalan [4] extended this method by an inpainting scheme to interpolate missing or invalid regions in range data. A Markov Random Field based formulation has been presented by Rajagopalan et al. [17]. Other approaches also exploit complementary photometric data as guidance to reconstruct high-resolution range images. Park et al. [16] utilized adaptive regularization gained from color images to super-resolve range data. A similar technique based on weighted optimization, driven by color images, has been introduced by Schwarz et al. [21]. In the multi-sensor approach proposed by Köhler et al. [10,11], photometric data is utilized as guidance for motion estimation and outlier detection in order to reconstruct reliable high-resolution range images. However, the existence of reliable photometric guidance data required for these methods is not always guaranteed, especially in case of low-cost systems. Furthermore, super-resolution is only applied to a single modality whereas guidance images are not super-resolved.

1.2 Contribution

Opposed to prior work, we propose a novel *guided super-resolution* approach to super-resolve images of two modalities simultaneously. In the context of hybrid RI, we apply our method for photogeometric super-resolution, to reconstruct both, high-resolution range and photometric data. Our algorithm is formulated as joint energy minimization in a MAP framework. In order to exploit correlations between two modalities, we introduce a novel regularizer based on the concept of guided filtering. We employ iteratively re-weighted least square minimization embedded into a cyclic coordinate descent scheme for fast and robust image reconstruction. Our approach is quantitatively and qualitatively evaluated on synthetic data as well as real data captured with Microsoft's Kinect.

2   Photogeometric Super-Resolution Model

Let $\boldsymbol{y}^{(1)}, \ldots, \boldsymbol{y}^{(K)}$ be a sequence of low-resolution input frames, where each frame $\boldsymbol{y}^{(k)}$ is represented by a N_y-dimensional vector. For each input frame $\boldsymbol{y}^{(k)}$, there exists a complementary guidance image $\boldsymbol{p}^{(k)}$ registered to $\boldsymbol{y}^{(k)}$ and denoted as N_p-dimensional vector. In terms of hybrid RI addressed in our work, we use range images as input and corresponding photometric data as guidance. The pair of unknown high-resolution images $(\boldsymbol{x}, \boldsymbol{q})$ with $\boldsymbol{x} \in \mathbb{R}^{N_x}$, $\boldsymbol{q} \in \mathbb{R}^{N_q}$ is related to the low-resolution frames $(\boldsymbol{y}^{(k)}, \boldsymbol{p}^{(k)})$ by a generative model according to:

$$\begin{pmatrix} \boldsymbol{y}^{(k)} \\ \boldsymbol{p}^{(k)} \end{pmatrix} = \begin{pmatrix} \gamma_m^{(k)} \boldsymbol{W}_y^{(k)} & 0 \\ 0 & \eta_m^{(k)} \boldsymbol{W}_p^{(k)} \end{pmatrix} \begin{pmatrix} \boldsymbol{x} \\ \boldsymbol{q} \end{pmatrix} + \begin{pmatrix} \gamma_a^{(k)} \boldsymbol{1} \\ \eta_a^{(k)} \boldsymbol{1} \end{pmatrix}, \tag{1}$$

where the system matrices $\boldsymbol{W}_y^{(k)}$, $\boldsymbol{W}_p^{(k)}$ model geometric displacements between $(\boldsymbol{x}, \boldsymbol{q})$ and $(\boldsymbol{y}^{(k)}, \boldsymbol{p}^{(k)})$, as well as the blur induced by the camera point spread function (PSF) and subsampling with respect to the high-resolution image. The model parameters (γ_m, γ_a) and (η_m, η_a) are used to model out-of-plane motion for range data and photometric differences between different guidance images, respectively [10]. Without loss of generality, we model the PSF as a space invariant Gaussian function, to obtain the matrix elements by:

$$W_{m,n} = \exp\left(-\frac{\|\boldsymbol{v}_n - \boldsymbol{u}_m\|_2^2}{2\sigma^2} \right), \tag{2}$$

where $\boldsymbol{v}_n \in \mathbb{R}^2$ are the coordinates of the nth pixel in the high-resolution image, $\boldsymbol{u}_m \in \mathbb{R}^2$ are the coordinates of the mth pixel in the low-resolution frame mapped to the high-resolution grid and σ denotes the width of the PSF.

In order to reconstruct $(\boldsymbol{x}, \boldsymbol{q})$, we propose a joint energy minimization based on a MAP formulation. The objective function consists of a data fidelity term and two regularization terms ensuring the smoothness of the estimates and exploiting correlations between input and guidance images:

$$(\hat{\boldsymbol{x}}, \hat{\boldsymbol{q}}) = \arg\min_{\boldsymbol{x}, \boldsymbol{q}} \left\{ F_{\text{data}}(\boldsymbol{x}, \boldsymbol{q}) + R_{\text{smooth}}(\boldsymbol{x}, \boldsymbol{q}) + R_{\text{correlate}}(\boldsymbol{x}, \boldsymbol{q}) \right\}. \tag{3}$$

2.1   Data Fidelity Term

The data fidelity term measures the similarity between the back-projected high-resolution images $(\boldsymbol{x}, \boldsymbol{q})$ and all low-resolution frames $(\boldsymbol{y}^{(k)}, \boldsymbol{p}^{(k)})$, $k = 1, \ldots, K$, based on our forward model. In order to account for outliers in low-resolution data, we use a weighted L_2 norm error model [11]:

$$F_{\text{data}}(\boldsymbol{x}, \boldsymbol{q}) = \sum_{i=1}^{KN_y} \beta_{y,i} r_{y,i}(\boldsymbol{x})^2 + \sum_{i=1}^{KN_p} \beta_{p,i} r_{p,i}(\boldsymbol{q})^2, \tag{4}$$

where $r_y : \mathbb{R}^{N_x} \to \mathbb{R}^{KN_y}$ and $r_p : \mathbb{R}^{N_q} \to \mathbb{R}^{KN_p}$ denote the residual terms, while $\beta_y \in \mathbb{R}^{KN_y}$ and $\beta_p \in \mathbb{R}^{KN_p}$ represent confidence maps to weight the residuals element-wise. We concatenate the residual terms for all frames according to $r_y(\boldsymbol{x}) = (\boldsymbol{r}_y^{(1)}, \ldots, \boldsymbol{r}_y^{(K)})^\top$ and $r_p(\boldsymbol{x}) = (\boldsymbol{r}_p^{(1)}, \ldots, \boldsymbol{r}_p^{(K)})^\top$. The residuals of the kth frames are given as:

$$
\begin{aligned}
\boldsymbol{r}_y^{(k)} &= \boldsymbol{y}^{(k)} - \gamma_m^{(k)} \boldsymbol{W}_y^{(k)} \boldsymbol{x} - \gamma_a^{(k)} \mathbf{1} \\
\boldsymbol{r}_p^{(k)} &= \boldsymbol{p}^{(k)} - \eta_m^{(k)} \boldsymbol{W}_p^{(k)} \boldsymbol{q} - \eta_a^{(k)} \mathbf{1}.
\end{aligned}
\tag{5}
$$

In order to set up our model, we employ a variational approach for optical flow estimation [13] to estimate subpixel motion. Following the multi-sensor approach proposed in [10], motion is estimated on photometric data used as guidance. For range images, $\gamma_m^{(k)}$ and $\gamma_a^{(k)}$ are determined using a range correction scheme [10]. In terms of photometric data, $\eta_m^{(k)}$ and $\eta_a^{(k)}$ are obtained by photometric registration, where $\eta_m^{(k)} = 1$ and $\eta_a^{(k)} = 0$ is set to neglect photometric differences.

2.2 Smoothness Regularization

The smoothness regularization is defined as a sum of regularization terms for the high-resolution images weighted by $\lambda_x, \lambda_q \in \mathbb{R}^+$:

$$
R_{\text{smooth}}(\boldsymbol{x}, \boldsymbol{q}) = \lambda_x R(\boldsymbol{x}) + \lambda_q R(\boldsymbol{q}).
\tag{6}
$$

For regularization, we employ the edge-preserving bilateral total variation (BTV) model [8], defined as:

$$
R(\boldsymbol{z}) = \sum_{i=-P}^{P} \sum_{j=-P}^{P} \alpha^{|i|+|j|} \|\boldsymbol{z} - \boldsymbol{S}_v^i \boldsymbol{S}_h^j \boldsymbol{z}\|_1,
\tag{7}
$$

where $\boldsymbol{S}_v^i$ and $\boldsymbol{S}_h^j$ denote a shift of the image $\boldsymbol{z}$ by i pixels in vertical and j pixels in horizontal direction, and P is a local window size. The shift operators act as derivatives across multiple scales, where α $(0 < \alpha \leq 1)$ is used to control the spatial weighting within the window.

2.3 Interdependence Regularization

We propose a novel interdependence regularization term to exploit local correlations between two modalities. In order to include this kind of prior knowledge, we use a linear, pixel-wise correlation model [9] for the high-resolution images $(\boldsymbol{x}, \boldsymbol{q})$. Assuming that $\boldsymbol{x} \in \mathbb{R}^{N_x}$ and $\boldsymbol{q} \in \mathbb{R}^{N_q}$ have the same dimension $N = N_x = N_q$, interdependence regularization is defined as:

$$
R_{\text{correlate}}(\boldsymbol{x}, \boldsymbol{q}) = \lambda_c \|\boldsymbol{x} - \boldsymbol{A}\boldsymbol{q} - \boldsymbol{b}\|_2^2,
\tag{8}
$$

where $\lambda_c \in \mathbb{R}^+$ weights the correlation between $\boldsymbol{x}$ and $\boldsymbol{q}$. The filter coefficients $\boldsymbol{A} \in \mathbb{R}^{N \times N}$ and $\boldsymbol{b} \in \mathbb{R}^N$ are constructed to model this correlation. The higher λ_c is chosen, the higher the correlation between input and guidance images is enforced. In case of $N_x \neq N_q$, we use bicubic interpolation to reshape $\boldsymbol{x}$ and $\boldsymbol{q}$ to the same size in order to compute $R_{\text{correlate}}(\boldsymbol{x}, \boldsymbol{q})$.

3   Numerical Optimization Algorithm

A direct minimization of the joint energy function in Eq. (3) requires a simultaneous estimation of x and q with the associated confidence maps β_y and β_p as well as of the filter coefficients A and b. However, this is a highly ill-posed and computationally demanding inverse problem. For a fast and robust solution, we utilize an iterative re-weighted least square (IRLS) minimization [11] embedded into a cyclic coordinate descent scheme. We decompose the estimation of x and q into a sequence of n least square optimization problems to determine $x^{(t)}$ and $q^{(t)}$ at iteration $t = 1, \ldots, n$. Simultaneously to image reconstruction, the confidence maps $\beta_y^{(t)}$ and $\beta_p^{(t)}$ as well as the filter coefficients $A^{(t)}$ and $b^{(t)}$ are determined analytically and refined at each iteration. In detail, our optimization is performed as follows.

Confidence Map Computation. Let $(x^{(t)}, q^{(t)})$ be the estimates for the high-resolution images at iteration t and $(r_y^{(t)}, r_p^{(t)})$ be the associated residual error computed according to Eq. (5). Then, following [11] the confidence maps $\beta_y^{(t)}$ and $\beta_p^{(t)}$ at iteration t are derived analytically according to:

$$\beta_{y,i}^{(t)} = \begin{cases} 1 & \text{if } |r_{y,i}^{(t)}| \leq \epsilon_y \\ \frac{\epsilon_y}{|r_{y,i}^{(t)}|} & \text{otherwise} \end{cases} \quad \beta_{p,i}^{(t)} = \begin{cases} 1 & \text{if } |r_{p,i}^{(t)}| \leq \epsilon_p \\ \frac{\epsilon_p}{|r_{p,i}^{(t)}|} & \text{otherwise} \end{cases}, \quad (9)$$

where ϵ_y and ϵ_p are initialized by the standard deviations of $r_y^{(t)}$ and $r_p^{(t)}$, respectively. For outlier detection, this scheme assigns a smaller confidence to low-resolution observations that result in higher residual errors.

Guidance Image Super-Resolution. In order to estimate $q^{(t)}$, we solve Eq. (3) with respect to q using the confidence map $\beta_p^{(t)}$ while keeping x fixed. For this step, we employ interdependence regularization in a non-symmetric way, as super-resolution of guidance data would not benefit from the complementary input images. The updated estimate $q^{(t)}$ is obtained according to:

$$q^{(t)} = \arg\min_q \left\{ F_{\text{data}}(x, q) + R_{\text{smooth}}(x, q) \right\}_{x = x^{(t-1)}}. \quad (10)$$

For the solution of this convex optimization problem, scaled conjugate gradient (SCG) iterations [15] are used to determine $q^{(t)}$ using $q^{(t-1)}$ as initial guess.

Guided Filtering. Once the guidance image $q^{(t)}$ and the super-resolved image $x^{(t-1)}$ from the previous iteration are determined, the filter coefficients $A^{(t)}$ and $b^{(t)}$ are updated. In order to exploit the correlation between $q^{(t)}$ and $x^{(t-1)}$, we keep both fixed and determine the filter coefficients analytically using guided filtering [9]. Omitting the iteration index for sake of clarity, we construct A as diagonal matrix and calculate the filter coefficients according to:

$$\tilde{A}_{k,k} = \frac{\frac{1}{|\omega_k|} \sum_{i \in \omega_k} q_i x_i - \mathrm{E}_{\omega_k}(q) \mathrm{E}_{\omega_k}(x)}{\mathrm{Var}_{\omega_k}(q) + \epsilon} \quad (11)$$

$$\tilde{b}_k = \mathrm{E}_{\omega_k}(x) - \tilde{A}_{k,k} \mathrm{E}_{\omega_k}(q), \quad (12)$$

Algorithm 1. Guided image super-resolution algorithm

1: **for** $t = 1 \ldots t_{max}$ **do**   ▷ t_{max}: maximum number of iterations
2:    Update confidence maps $\beta_y^{(t)}$ and $\beta_p^{(t)}$ according to Eq. (9).
3:    Estimate high-resolution photometric data $q^{(t)}$ at step t according to Eq. (10):

$$q^{(t)} = \arg\min_{q} \sum_{i=1}^{KN_p} \beta_{p,i}^{(t)} r_{p,i}^{(t)}(q)^2 + \lambda_q R(q)$$

4:    Determine $A^{(t)}$ and $b^{(t)}$ from $q^{(t)}$ and $x^{(t-1)}$ according to Eqs. (11) and (12).
5:    Estimate high-resolution range image $x^{(t)}$ at step t according to Eq. (13):

$$x^{(t)} = \arg\min_{x} \sum_{i=1}^{KN_y} \beta_{y,i}^{(t)} r_{y,i}^{(t)}(x)^2 + \lambda_x R(x) + \lambda_c \|x - A^{(t)} q^{(t)} - b^{(t)}\|_2^2$$

6:    If not converged proceed with next iteration

---

where ω_k denotes a local neighborhood of size $|\omega_k|$ and radius r centered at the kth pixel in (x, q), ϵ is a regularization factor for guided filtering, and $\mathrm{E}_{\omega_k}(\cdot)$ and $\mathrm{Var}_{\omega_k}(\cdot)$ are the mean and variance in ω_k. The filter coefficients A and b are computed by box filtering of $\tilde{A}$ and $\tilde{b}$ using the window defined by ω_k.

Input Image Super-Resolution. Finally, we solve Eq. (3) with respect to x to obtain a refined estimate $x^{(t)}$ under the guidance of $q^{(t)}$. This is achieved by means of interdependence regularization based on $A^{(t)}$ and $b^{(t)}$ according to:

$$x^{(t)} = \arg\min_{x} \left\{ F_{\text{data}}(x, q) + R_{\text{smooth}}(x, q) + R_{\text{correlate}}(x, q) \right\}_{q=q^{(t)}}. \quad (13)$$

In the same way as for the guidance images, the resulting convex optimization problem is solved employing SCG, where $x^{(t-1)}$ is used as initial guess. The outline of the proposed algorithm is depicted in Table 1.

4   Experiments and Results

We compared the proposed guided super-resolution (GSR) approach to MAP super-resolution using the L_2 norm [6] and L_1 norm model [8] working on a single modality. For photometric data, all methods were directly applied to intensity images using optical flow [13] for motion estimation. For range super-resolution, we employed the multi-sensor super-resolution approach [10] to derive the motion estimate from the corresponding photometric data. Super-resolution was applied in a sliding window scheme using $K = 31$ successive frames, where the central frame was chosen as reference for motion estimation. We used a magnification factor of $f = 4$ for range and photometric data. The PSF width was approximated to $\sigma_p = 0.4$ for photometric data and $\sigma_y = 0.6$ for range data. For BTV regularization, we set $\alpha = 0.7$ and $P = 1$. For guided filtering, we set $\epsilon = 10^{-4}$ and $r = 1$. The regularization weights were optimized using a grid search on a

Table 1. Peak-signal-to-noise ratio (PSNR) and structural similarity (SSIM, in brackets) for synthetic data averaged over ten sequences. We compared bicubic interpolation (second column), MAP super-resolution using the L_1 norm (third column) as well as the L_2 norm (forth column) to the proposed guided super-resolution (GSR, last column).

	Sequence	Interpolation	MAP - L_1	MAP - L_2	GSR
Range	Bunny-1	32.78 (0.96)	34.10 (0.96)	34.05 (0.97)	**35.01 (0.98)**
	Bunny-2	31.29 (0.94)	32.84 (0.95)	33.22 (0.97)	**33.34 (0.98)**
	Dragon-1	24.63 (0.57)	27.68 (0.72)	28.71 (0.84)	**30.00 (0.91)**
	Dragon-2	27.14 (0.75)	29.09 (0.84)	29.76 (0.93)	**30.80 (0.95)**
Photo.	Bunny-1	28.48 (0.79)	29.82 (0.87)	29.79 (0.87)	**29.79 (0.88)**
	Bunny-2	30.05 (0.81)	31.35 (0.86)	31.42 (0.86)	**31.43 (0.86)**
	Dragon-1	23.34 (0.65)	24.25 (0.72)	24.24 (0.71)	**24.27 (0.72)**
	Dragon-2	24.65 (0.66)	25.60 (0.72)	25.51 (0.70)	**25.60 (0.72)**

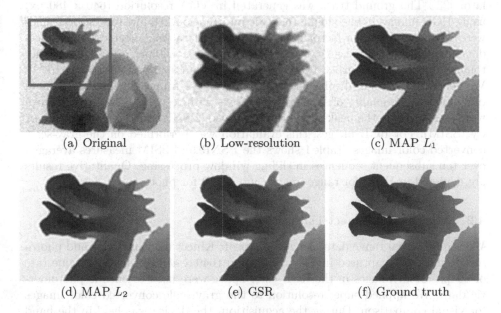

(a) Original   (b) Low-resolution   (c) MAP L_1

(d) MAP L_2 (e) GSR   (f) Ground truth

Fig. 1. Results for simulated *Dragon-2* dataset: Low-resolution range data (a) and selected region of interest (b), results of MAP super-resolution using the L_1 (c) and L_2 norm model (d), and our proposed GSR method (e) compared to the ground truth (f).

training data set with known ground truth and were set to $\lambda_x = 0$, $\lambda_q = 0.002$ and $\lambda_c = 1$ for all experiments. We used SCG with a termination tolerance of 10^{-3} for the pixels of (x, q) and the objective function value. The maximum number of SCG iterations was set to 50 for 15 IRLS iterations[1].

---

[1] Supplementary material is available at http://www5.cs.fau.de/research/data/.

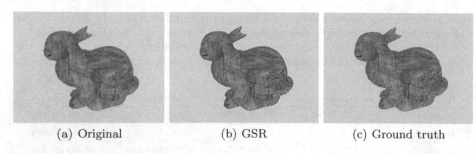

(a) Original   (b) GSR   (c) Ground truth

Fig. 2. Results for simulated *Bunny-1* dataset: Low-resolution photometric data (a) and result of guided super-resolution (GSR) (b) compared to the ground truth (c).

4.1 Synthetic Hybrid Range Data

We simulated synthetic range data with known ground truth using a RI simulator [23]. The ground truth was generated in VGA resolution (640×480 px) using RGB images to encode photometric information. All low-resolution frames were downsampled by a factor of $f = 4$. Range data was affected by distance-dependent Gaussian noise ($\sigma_n = 8\,\text{mm}$) and Gaussian blur ($\sigma_b = 3\,\text{mm}$). Photometric data was disturbed by space invariant Gaussian noise ($\sigma_n = 5 \cdot 10^{-4}$). We generated random displacements of consecutive frames to simulate camera movements. The quality of super-resolved data with respect to the ground truth was assessed using the peak-signal-to-noise ratio (PSNR) and structural similarity (SSIM). For RGB images, this evaluation was performed on the gray-scale converted color images. Table 1 shows the PSNR and SSIM measures averaged over ten subsequent sequences in sliding window processing. Qualitative results are depicted in Fig. 1 for range images and Fig. 2 for photometric data.

4.2 Microsoft's Kinect Datasets

We acquired real range data using a Microsoft Kinect device. Range and photometric data was captured in VGA resolution (640×480 px) using a frame rate of 30 fps. Color images in the RGB color space were used to encode photometric data. We applied super-resolution to the gray-scale converted color images for visual comparison. During the acquisition, the device was held in the hand such that a small shaking ensured the required motion for super-resolution over consecutive frames. We used the same parameter settings as for synthetic data. Qualitative results are shown for different acquisitions in Figs. 3 and 4. Finally, we also rendered 3-D meshes with a texture overlay from super-resolved range data. A comparison of the different approaches is depicted in Fig. 5.

5 Discussion

In this work, we introduce a novel method for photogeometric resolution enhancement based on the concept of guided super-resolution. Unlike prior work, our

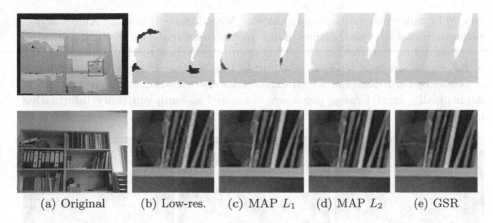

(a) Original   (b) Low-res.   (c) MAP L_1 (d) MAP L_2 (e) GSR

Fig. 3. Real data set example showing range data (first row) and photometric data (second row): Low-resolution range and RGB frame (a), selected low-resolution region of interest (b), results for MAP super-resolution using L_1 norm (c) and L_2 norm (d), and the super-resolved images using the proposed guided super-resolution (GSR) (e).

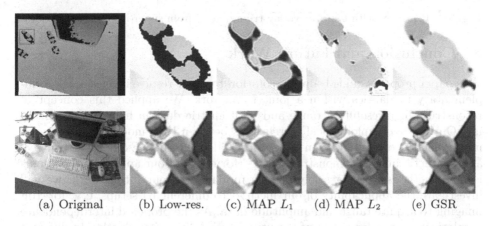

(a) Original   (b) Low-res.   (c) MAP L_1 (d) MAP L_2 (e) GSR

Fig. 4. Real data set example showing range data (first row) and photometric data (second row): Low-resolution range and RGB frame (a), selected low-resolution region of interest (b), results for MAP super-resolution using L_1 norm (c) and L_2 norm (d), and the super-resolved images using the proposed guided super-resolution (GSR) (e).

method super-resolves range and photometric data simultaneously. This allows us to exploit photometric data as guidance for range super-resolution within a joint framework. Our experimental evaluation demonstrates the performance of our method on real as well as synthetic data. In case of range images, we achieved an improvement of ∼1 dB for PSNR and ∼0.04 for the SSIM measure compared to super-resolution applied only on range data (see Table 1). On photometric data, our method achieved similar performance as other state-of-the-art algorithms. However, this behavior was expected since the photometric data is not guided by range images in our formulation. Visual inspection using Kinect acquisitions demonstrates the benefits of our approach for real data. For range images, we observed an improved trade-off between edge reconstruction and smoothing in flat regions as depicted in Figs. 3 and 4. This is caused by the proposed interdependence regularization, which exploits local correlation between modalities. Additionally, invalid range pixels are corrected as complementary information from multiple frames is fused in image reconstruction.

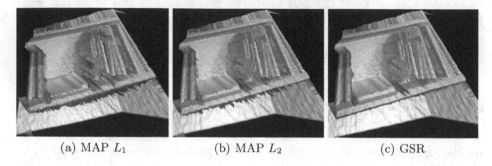

(a) MAP L_1 (b) MAP L_2 (c) GSR

Fig. 5. 3-D meshes with texture overlay triangulated from super-resolved range data.

6 Conclusion and Future Work

This paper proposes guided super-resolution to super-resolve images of two complementary modalities within a joint framework. We applied this concept to reconstruct high-resolution range and photometric data in hybrid range imaging. Our method exploits local correlations between both modalities for a novel interdependence regularization based on guided filtering. Experiments on real and synthetic images demonstrate the performance of our approach in comparison to methods working solely on one modality. In our future work, we will investigate the applicability of our method to different RI setups, e.g. for ToF imaging to acquire range and amplitude data. As the proposed interdependence regularization is independent of the utilized modalities, its adoption to different multi-sensor setups such as RGB or multispectral imaging seems attractive.

Acknowledgments. The authors gratefully acknowledge funding of the Erlangen Graduate School in Advanced Optical Technologies (SAOT) by the German National Science Foundation (DFG) in the framework of the excellence initiative and the support by the DFG under Grant No. HO 1791/7-1.

References

1. Babacan, S.D., Molina, R., Katsaggelos, A.K.: Variational Bayesian super resolution. IEEE Trans. Image Process. **20**(4), 984–999 (2011)
2. Bauer, S., Seitel, A., Hofmann, H., Blum, T., Wasza, J., Balda, M., Meinzer, H.-P., Navab, N., Hornegger, J., Maier-Hein, L.: Real-time range imaging in health care: a survey. In: Grzegorzek, M., Theobalt, C., Koch, R., Kolb, A. (eds.) Time-of-Flight and Depth Imaging. LNCS, vol. 8200, pp. 228–254. Springer, Heidelberg (2013)

3. Beder, C., Bartczak, B., Koch, R.: A comparison of PMD-cameras and stereo-vision for the task of surface reconstruction using patchlets. In: IEEE Conference on Computer Vision and Pattern Recognition, pp. 1–8 (2007)
4. Bhavsar, A.V., Rajagopalan, A.N.: Range map superresolution-inpainting, and reconstruction from sparse data. Comput. Vis. Image Underst. 116(4), 572–591 (2012)
5. Cui, Y., Schuon, S., Chan, D., Thrun, S., Theobalt, C.: 3D shape scanning with a time-of-flight camera. In: IEEE Conference on Computer Vision and Pattern Recognition, pp. 1173–1180 (2010)
6. Elad, M., Feuer, A.: Restoration of a single superresolution image from several blurred, noisy, and undersampled measured images. IEEE Trans. Image Process. 6(12), 1646–1658 (1997)
7. Farsiu, S., Robinson, D., Elad, M., Milanfar, P.: Advances and challenges in super-resolution. Int. J. Imaging Syst. Technol. 14, 47–57 (2004)
8. Farsiu, S., Robinson, D., Elad, M., Milanfar, P.: Fast and robust multiframe super resolution. IEEE Trans. Image Process. 13(10), 1327–1344 (2004)
9. He, K., Sun, J., Tang, X.: Guided image filtering. In: Daniilidis, K., Maragos, P., Paragios, N. (eds.) ECCV 2010, Part I. LNCS, vol. 6311, pp. 1–14. Springer, Heidelberg (2010)
10. Köhler, T., Haase, S., Bauer, S., Wasza, J., Kilgus, T., Maier-Hein, L., Feubner, H., Hornegger, J.: ToF meets RGB: novel multi-sensor super-resolution for hybrid 3-D endoscopy. Med. Image Comput. Comput. Assist. Interv. 16, 139–146 (2013)
11. Köhler, T., Haase, S., Bauer, S., Wasza, J., Kilgus, T., Maier-Hein, L., Feuner, H., Hornegger, J.: Outlier detection for multi-sensor super-resolution in hybrid 3D endoscopy. In: Deserno, T.M., Handels, H., Meinzer, H.-P., Tolxdorff, T. (eds.) Bildverarbeitung für die Medizin 2014. Informatik aktuell, pp. 84–89. Springer, Heidelberg (2014)
12. Kurmankhojayev, D., Hasler, N., Theobalt, C.: Monocular pose capture with a depth camera using a sums-of-gaussians body model. In: Weickert, J., Hein, M., Schiele, B. (eds.) GCPR 2013. LNCS, vol. 8142, pp. 415–424. Springer, Heidelberg (2013)
13. Liu, C.: Beyond pixels: exploring new representations and applications for motion analysis. Ph.D. thesis, Massachusetts Institute of Technology (2009)
14. Milanfar, P.: Super-Resolution Imaging. CRC Press, Boca Raton (2010)
15. Nabney, I.T.: NETLAB: Algorithms for Pattern Recognition. Advances in Pattern Recognition, 1st edn. Springer, Heidelberg (2002)
16. Park, J., Kim, H., Tai, Y., Brown, M., Kweon, I.: High quality depth map upsam-pling for 3D-TOF cameras. In: International Conference on Computer Vision, pp. 1623–1630 (2011)
17. Rajagopalan, A.N., Bhavsar, A., Wallhoff, F., Rigoll, G.: Resolution enhancement of PMD range maps. In: Rigoll, G. (ed.) DAGM 2008. LNCS, vol. 5096, pp. 304–313. Springer, Heidelberg (2008)
18. Schultz, R.R., Stevenson, R.L.: Extraction of high-resolution frames from video sequences. IEEE Trans. Image Process. 5, 996–1011 (1996)
19. Schuon, S., Theobalt, C., Davis, J., Thrun, S.: High-quality scanning using time-of-flight depth superresolution. In: IEEE Conference on Computer Vision and Pattern Recognition, vol. 1, pp. 1–7 (2008)
20. Schuon, S., Theobalt, C., Davis, J., Thrun, S.: LidarBoost: depth superresolution for ToF 3D shape scanning. In: IEEE Conference on Computer Vision and Pattern Recognition, vol. 1, pp. 343–350 (2009)

21. Schwarz, S., Sjostrom, M., Olsson, R.: A weighted optimization approach to time-of-flight sensor fusion. IEEE Trans. Image Process. **23**(1), 214–225 (2014)
22. Shotton, J., Girshick, R., Fitzgibbon, A., Sharp, T., Cook, M., Finocchio, M., Moore, R., Kohli, P., Criminisi, A., Kipman, A., Blake, A.: Efficient human pose estimation from single depth images. Pattern Anal. Mach. Intell. **35**(12), 2821–2840 (2013)
23. Wasza, J., Bauer, S., Haase, S., Schmid, M., Reichert, S., Hornegger, J.: RITK: the range imaging toolkit - a framework for 3-D range image stream processing. In: VMV, pp. 57–64. Eurographics Association (2011)

Image Descriptors Based
on Curvature Histograms

Philipp Fischer[(✉)] and Thomas Brox

Department of Computer Science, University of Freiburg,
Freiburg im Breisgau, Germany
{fischer,brox}@cs.uni-freiburg.de

Abstract. Descriptors based on orientation histograms are widely used in computer vision. The spatial pooling involved in these representations provides important invariance properties, yet it is also responsible for the loss of important details. In this paper, we suggest a way to preserve the details described by the local curvature. We propose a descriptor that comprises the direction and magnitude of curvature and naturally expands classical orientation histograms like SIFT and HOG. We demonstrate the general benefit of the expansion exemplarily for image classification, object detection, and descriptor matching.

1 Introduction

Orientation histograms, such as SIFT [14] or HOG [4], are omnipresent in computer vision. They are the dominant descriptors in all recognition tasks and play a major role in structure from motion and image retrieval. This success is because the descriptor is invariant to local deformations on the one hand, and it is still descriptive due to the spatial grid of multiple histograms (cells) on the other hand. It allows recognition approaches to deal with a large set of natural variations without the need to model them explicitly.

However, the invariance to local deformations also frustrates discrimination of structures. HOG and SIFT locally simplify the image to straight lines. Therefore, the pointed tip of a cat's ear leads to almost the same representation as the roundish ear of a dog (see Fig. 1).

In this paper, we extend the idea of orientation histograms to curvature, i.e., the image is no longer locally simplified to straight lines but curved lines. Consequently, curvature histograms can distinguish local shape at a more detailed level. In contrast to previous work [16], we compute curvature with a per-pixel filter instead of operating on parametric curve segments. Hence, the approach is generally applicable and the descriptor can be computed basically as fast as a SIFT or HOG descriptor.

Moreover, we include the sign of the curvature (convex vs. concave). It is worth noting that the sign of the curvature is different from the sign of the

Supported by a scholarship of the Deutsche Telekom Stiftung.

X. Jiang et al. (Eds.): GCPR 2014, LNCS 8753, pp. 239–249, 2014.
DOI: 10.1007/978-3-319-11752-2_19

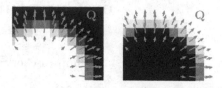

Fig. 1. Curvature is a valuable feature. The image is overlaid with its scalar curvature q, where the opacity is given by the gradient magnitude. The ear tips have high curvature. The average q values are: 62 for the cat's ears, 43 for the dog's ears.

Fig. 2. The two images are considered equivalent when using the proposed curvature descriptors. The vector curvature Q is invariant to the sign of the gradient and always points from the interior to the exterior of the circle. The image shows the product of Q with the local gradient magnitude.

gradient, which often is of little use in classification tasks. In contrast, the sign of the curvature allows the descriptor to get an idea where the interior or exterior of a (convex) object is, independent of the gradient direction; see Fig. 2.

The richer set of features leads to significant improvements in descriptor performance. We demonstrate this with image classification experiments on Caltech 101, detection experiments on Pascal VOC 2007, and descriptor matching experiments on the dataset of Mikolajczyk et al. [15].

2   Related Work

The most closely related work is by Monroy et al. [16], where curvature histograms are used for object detection. In [6] Eigenstetter et al. use these features in combination with self-similarity. Our approach is different in the way how curvature is defined and computed. In [6,16], the *chord-to-point distance accumulation* from [10] is used to measure curvature; see Fig. 3. This requires boundary segments, which Monroy et al. obtain performing a boundary probability computation pB [1]. Due to sparsity of such edge maps, some features in the image are lost for the descriptor. In contrast, we compute curvature densely based on the image gradient, which is significantly faster and preserves all relevant features. While Monroy et al. showed good results of their approach on images with strong line segments, our approach is more generally applicable, as we demonstrate in Sect. 4.

There are many works on curvature computation in general. In the continuous setting, curvature is well defined [17], yet on discrete images there are multiple ways to estimate it. Most works on discrete curvature estimation operate on binary images or discrete contours [3,11,13]. In contrast, the work in [5] also applies to color images. It assumes a height field surface given by the image

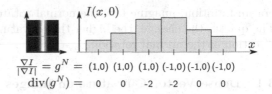

$$\frac{\nabla I}{|\nabla I|} = g^N = \quad (1,0) \ (1,0) \ (1,0) \ (-1,0) \ (-1,0) \ (-1,0)$$

$$\mathrm{div}(g^N) = \quad 0 \quad 0 \quad -2 \quad -2 \quad 0 \quad 0$$

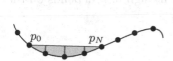

Fig. 3. Curvature computation by chord-to-point distance accumulation. The chord-to-point distances are shown as red lines. These are accumulated to obtain the local curvature at $p_{N/2}$ in [6,16] (Color figure online)

Fig. 4. Profile of a vertical line in a discrete image. The sign of the normalized gradient changes at the center of the line. This leads to a non-zero curvature on straight lines when curvature is calculated via the divergence of the normalized gradient (in this example central differences were used). With the proposed descriptor the straight line does not expose curvature.

intensity, and calculates the principal curvatures on this surface. However, this is fundamentally different from the intuitive understanding of curvature in an image, which is the curvature of isolines. Another classical application of curvature is the use of maxima of the norm of the Hessian to detect interest points. However, in this paper we compute the signed curvature densely on natural color images and curvature is represented by a curvature histogram to increase the expressiveness as a local image descriptor.

In a wider sense, also Carreira et al. [2] is related to our work. They suggested second order pooling of features in the context of semantic segmentation. This work shares the idea that important second order features need to be computed *before* pooling, as they get lost otherwise. While they propose the use of the feature vector covariance, we propose the use of second order derivatives. The two concepts are complementary.

3 Curvature

In general, the curvature of some curve is understood as how much it deviates from a straight line. Quantitatively, for one point on the curve it is defined as $\kappa = 1/r$, where r is the radius of a circle fitting the curve locally; hence straight lines have zero curvature.

As there are different ways to describe a curve, there are also different ways to define curvature, all leading to the same concept as stated above. We focus on the definition from [17] and assume that the curve X is parameterized by the arc-length t, such that equally spaced t map to equally spaced $X(t)$. Let $T(t)$ be the tangent and $N(t)$ the normal at $X(t)$. As in [17], we define curvature as the change of the normal along the tangent:

$$\kappa \cdot T = -\frac{dN}{dt} \tag{1}$$

For an intuition, imagine how the normal rotates when you walk along a curve. The quicker the change, the higher the curvature. The sign of κ depends on the direction of the normal rotation.

3.1   Dense Vector Curvature on Images

When computing curvature in an image, we are limited by its resolution, and hence the underlying curve can never be reproduced accurately. Curvature computed on discrete images can only be an approximation of κ. In the discrete case, the image gradient must be approximated by considering a neighborhood with some spatial extent, for instance, by using finite differences. To distinguish a slightly curved line from a straight line, the observed area must be large enough. Hence, the maximum and minimum curvature that can be detected, are bounded by pixel size and neighborhood size.

Another important issue is illustrated in Fig. 4. While the divergence of the normalized gradient is a standard method to determine the curvature of the isolines, in the discrete case it can yield undesired curvature responses along straight lines.

Following the idea of Eq. 1, the aim is to determine the change of the gradient (or normal) along the tangent. Locally, the tangent coincides with the curve itself. At each point (x, y) we extract the gradient $g = (g_x, g_y)^T = \nabla I$ using standard linear differencing filters. This yields also the tangent $\phi = (-g_y, g_x)^T$ orthogonal to the gradient. Further, let $g^N = g/|g|$ and $\phi^N = \phi/|\phi|$ denote their normalized forms.

The Jacobian matrix of the normalized gradient (i.e. its differential) describes the change of gradient direction:

$$J(g^N) = \begin{pmatrix} (g_x^N)_x & (g_x^N)_y \\ (g_y^N)_x & (g_y^N)_y \end{pmatrix}$$

Thus, $J(g^N) \cdot \phi^N \in \mathbb{R}^2$ gives the gradient change in tangent direction. According to (1), the resulting vector is expected to be $-\kappa \cdot T$ and hence parallel to the tangent. On a discrete image, however, small deviations occur. Thus we project this vector to the tangent to obtain the scalar q approximating κ:

$$q = -\left((\phi^N)^T \cdot J(g^N) \cdot \phi^N \right) \in \mathbb{R} \tag{2}$$

Note that this scalar curvature can be positive or negative. The sign depends on both the convexity and the sign of the gradient.

For deriving a descriptor, we suggest combining the curvature with the orientation of the gradient. In this scope, we are interested in decoupling the sign of the curvature from the sign of the gradient to distinguish convex and concave curves (see Fig. 2). To this end, we multiply q with the gradient vector g^N

$$Q = q \cdot g^N \in \mathbb{R}^2 \tag{3}$$

and obtain a vector Q, which we call *vector curvature*.

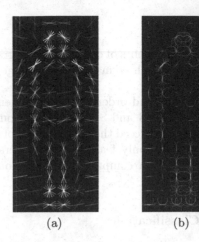

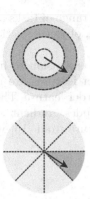

Fig. 5. The curvature vector Q can be binned to marginal histograms for magnitude and direction. This results in 12 bins.

Fig. 6. Positive weights of feature descriptors trained for pedestrian detection using a linear SVM. In (a) HOG features were used, (b) shows a curvature descriptor. High convexity is learned pointing upwards in the head region and downwards in the feet region.

Intuitively, driving along the curved isoline with a car, the vector Q points in the direction of the centrifugal force that acts on the passenger. Its magnitude is proportional to this force, assuming a constant velocity.

3.2 Feature Binning

With the above procedure, we compute the vector curvature Q densely for every pixel in an image and bin it into sparse per-pixel histograms as illustrated in Fig. 5. We quantize the orientations of Q into 8 bins. This histogram is partially redundant to the orientation histogram, but the sign depends on the direction of the curvature rather than the direction of the gradient. Our experiments show that the combination of orientation, sign and magnitude of curvature makes a strong feature.

Following the standard procedures to assemble SIFT and HOG descriptors [4,8,16], we pool these dense features into spatial histograms. The image is divided into a grid of square cells, and for each cell the sparse curvature histograms (one for each pixel) are aggregated by weighting them with their respective gradient magnitude. For detection, we follow the soft binning approach by linearly interpolating between cell centers just as in HOG. Figure 6 shows visualizations of trained HOG and curvature descriptors.

The feature computation on one image of size 300×250px takes around 0.3 s with HOG. Adding our curvature, increases the computation time to 0.7 s. The approach from [16] takes about 35 s due to the costly extraction of boundary segments.

4   Experiments

To demonstrate the benefits of curvature histograms in a wide range of tasks and images, we tested it in three major areas: image classification, object detection, and descriptor matching.

Curvature is a second order feature and hence a natural expansion to first order features like HOG and SIFT. In isolation, second order features are not meaningful enough to exceed the performance of a good first order feature. Thus, we compared gradient-only features to the version that includes gradient and curvature histograms. A comparison to the boundary curvature from [16] was done in the same way.

4.1   Image Classification

For the experiments on image classification we used the Caltech 101 dataset and the VLfeat implementation of spatial pyramid matching [12,18]. The baseline method is based on dense SIFT descriptors. It trains a k-means based bag-of-words classifier with an SVM. By collecting SIFT descriptors from random positions and scales, it builds a dictionary using k-means clustering. Using this dictionary, for every training image a spatial histogram is built with one bin per *word* (i.e. one k-means cluster). Spatial binning is achieved by dividing the image into 2×2 and 4×4 cells and generating histograms for each cell of each division. A χ^2 kernel map is applied before training the SVM on the histograms. For further details, we refer to the public source code.

We compared this baseline to our curvature-augmented approach which was added in the form of a second spatial pyramid (i.e. using a second dictionary). For training the SVM and during testing, both histograms were concatenated.

Results. For every category, we randomly chose 15 training images and 15 test images. We repeated the whole training and testing process for 10 random sets to report statistics on the accuracy in Fig. 8(a). The *accuracy* refers to the share of test images that were classified correctly, i.e., the sum of the confusion matrix diagonal. The average increase in accuracy of our method over the baseline is

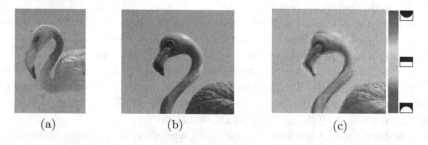

(a)   (b)   (c)

Fig. 7. Two flamingo images from Caltech are shown in (a) and (b). The typical shape of the flamingos' beaks and necks enables reliable recognition when using curvature. The scalar curvature of (b) is shown in (c). Note the high curvature at the beak's tip.

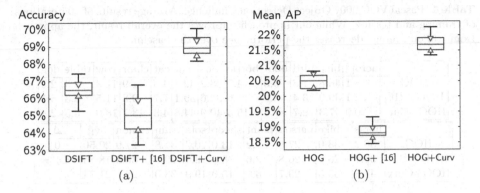

Fig. 8. Statistical results. The blue boxes show the inter-quartile range, the red line is the median. The total extent of the results is given by the black whiskers, while the red triangles enclose the range of significance (Two distributions differ significantly (by 95 %), if the triangle intervals (i.e. notches) do not overlap (cf. MATLAB 2013a boxplot function)). (a) Results of Caltech101 evaluation. The average accuracies are 66.5 % for SIFT only, 65.1 % for SIFT with boundary curvature, and 69.1 % for SIFT and vector curvature. This is an incrase of 2.6 % over the baseline using our approach. There is some variance but the differences are clearly significant. (b) Summarized results of Pascal VOC 2007 evaluation. When changing the random seeds, the mean AP over all classes varies by about 1 %. Still, the boxplots show that the mean AP distributions do not overlap and that they differ significantly. Table 1 shows a per-class performance overview (Color figure online).

2.6 %. While the curvature features from [16] work well on some images, they have problems on other images, where the extraction of good boundary segments is harder. On average, the accuracy is slightly lower than the baseline.

The proposed vector curvature improves performance on almost all classes. For some classes it is particularly useful: on the classes *pyramid, flamingo head* and *electric guitar*, the curvature extension outperforms baseline SIFT by a large margin. Objects belonging to these classes expose sharp corners, which help recognition. Figure 7 shows two sample images. As shown in Fig. 7(c), the flamingo exposes high curvature at the tip of its beak. Also the neck has a typical curvature. This information is largely ignored by orientation histograms.

4.2 Object Detection

For the evaluation in object detection, we trained filter masks for all 20 object classes used in the Pascal VOC 2007 challenge [7]. For every class, we clustered the training examples in 3 aspect ratios, such that in total 60 filter masks were trained. The training was iterated and used random negatives for the first round. Retraining was based on the hard-negatives. Testing was done in a multi-scale sliding window fashion.

As in image classification, we compared gradient-only features (HOG in this case) to the curvature augmented versions, which add either our vector curvature

Table 1. Pascal VOC 2007 Object Detection Challenge. Average results of 10 iterations of training and testing. While our approach improves the overall result, the approach from [16] on average decreases the performance over the baseline.

	aeropl	bike	bird	boat	bottle	bus	car	cat	chair	cow	table	dog
HOG	18.1	46.8	1.8	6.5	14.7	38.2	42.3	4.6	11.9	17.3	8.3	2.5
HOG+ [16]	22.1	39.9	3.4	5.9	11.6	33.6	39.1	7.7	9.8	11.5	8.3	5.1
HOG+Curv	20.9	47.3	2.7	7.4	15.6	40.9	44.8	4.3	13.7	18.0	6.5	2.8

	horse	mbike	person	plant	sheep	sofa	train	tvmon	$\overline{avg}$	Δ
HOG	38.2	33.0	27.6	5.2	13.0	20.2	31.8	27.9	20.50	0
HOG+ [16]	36.7	26.6	26.8	2.6	8.4	18.4	29.1	30.3	18.85	-1.65%
HOG+Curv	41.8	35.5	29.7	6.0	15.0	19.9	33.0	28.9	21.74	+1.24%

or the boundary curvature from [16]. For Pascal we also ran the whole training and testing 10 times with different random seeds for sampling negative examples.

Results. While a combination of HOG and our vector curvature increases the mean AP by 1.24 % over the baseline, the method from [16] combined with HOG decreases the mean AP by 1.65 %. To show statistical significance, the results are summarized using boxplots in Fig. 8(b). A detailed per-class overview can be found in Table 1. The proposed curvature extension of HOG increases the AP for 18 of 20 classes.

As demonstrated in [16], boundary curvature performs well on datasets with clean boundaries, such as the ETHZ shape dataset [9]. The proposed vector curvature descriptor is more generally applicable as it does not fully depend on strong boundaries. Moreover, it includes important cues provided by the curvature direction.

The raw numbers already indicate the benefit of adding vector curvature to the overall descriptor. Some more distinct advantages can be seen when looking at actual samples. Figure 9 shows false-positives of the bus category. The top row lists the top ranked false-positives detected by HOG that did not occur when using curvature. The bottom row analogously shows the top ranked false-positives detected with curvature that did not occur with HOG. Note how the false-positives obtained by adding curvature all show buses and are only due to suboptimal bounding boxes or duplicate detections. This indicates that by adding curvature we obtain an additional performance gain that is not measured by the standard evaluation criterion.

4.3   Descriptor Matching

In this experiment we evaluated the feature description performance in image matching tasks. We use the standard matching dataset by Mikolajczyk et al. [15], which contains different transformations such as zoom, rotation, blur and lighting changes. The dataset has previously been used for two different tasks: local interest point/region detection and local descriptor matching. We are interested in descriptor matching and intentionally disregard the detector performance by

Fig. 9. By adding curvature we obtain an additional performance gain that is not measured by the standard evaluation criterion: The figure shows exclusive top false-positives buses for HOG without (top row) and with curvature (bottom row). Note how the false-positives obtained by adding curvature all show buses and are only due to suboptimal bounding boxes or duplicate detections.

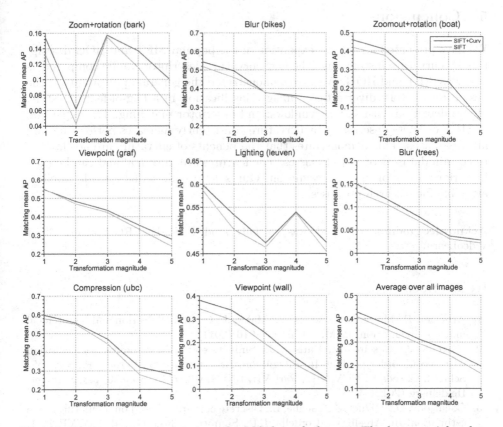

Fig. 10. Mean average precision on the Mikolajczyk dataset. The bottom right plot shows the average over the whole dataset. SIFT features (green) are consistently outperformed by the curvature expanded version (red) (Color figure online).

using the same region detector for both descriptors. We chose the MSER (maximally stable extremal regions) detector, because it is among the best detectors in [15]. The image patches of the detected elliptical regions were normalized to a uniform patch size and rotated to the dominant gradient orientation, which is the standard procedure. Given two images and their regions to be matched, we computed the descriptors of the normalized patches. All possible descriptor pairs were then ranked by their Euclidean distance. As in [15], a pair was considered to be a correct match if the overlap error of the region ellipses is less than 40 % when transformed by the ground truth mapping. Subsequently, for each image pair, we computed a precision-recall graph and its average precision (AP).

Results. The dataset contains 8 image categories with 5 image pairs each. The matching results for all 40 image pairs are given in Fig. 10, while the last graph shows the average performance over all categories. The curvature augmented version performs consistently better than SIFT alone.

5   Conclusions

We have presented an elegant expansion of the popular SIFT and HOG descriptors by curvature histograms. The new curvature descriptor can be computed as easily as SIFT and HOG. We have demonstrated the general applicability of the descriptor on very diverse tasks: image classification in a bag-of-features style, object detection with sliding windows, and descriptor matching. The expansion of SIFT or HOG descriptors by curvature consistently increases performance on all three tasks. This demonstrates that the benefits of curvature are not limited to certain datasets or object classes. The features are universal and can be used in various areas of visual recognition. Our source code will be made publicly available for research purposes.

Acknowledgments. The work was partially funded by the ERC Starting Grant VideoLearn.

References

1. Arbelaez, P., Maire, M., Fowlkes, C., Malik, J.: Contour detection and hierarchical image segmentation. PAMI **33**(5), 898–916 (2011)
2. Carreira, J., Caseiro, R., Batista, J., Sminchisescu, C.: Semantic segmentation with second-order pooling. In: Fitzgibbon, A., Lazebnik, S., Perona, P., Sato, Y., Schmid, C. (eds.) ECCV 2012, Part VII. LNCS, vol. 7578, pp. 430–443. Springer, Heidelberg (2012)
3. Coeurjolly, D., Miguet, S., Tougne, L.: Discrete curvature based on osculating circle estimation. In: Arcelli, C., Cordella, L.P., Sanniti di Baja, G. (eds.) IWVF 2001. LNCS, vol. 2059, pp. 303–312. Springer, Heidelberg (2001)
4. Dalal, N., Triggs, B.: Histograms of oriented gradients for human detection. In: CVPR, pp. 886–893 (2005)

5. Deng, H., Zhang, W., Mortensen, E.N., Dietterich, T.G., Shapiro, L.G.: Principal curvature-based region detector for object recognition. In: CVPR (2007)
6. Eigenstetter, A., Ommer, B.: Visual recognition using embedded feature selection for curvature self-similarity. In: NIPS, pp. 386–394 (2012)
7. Everingham, M., Van Gool, L., Williams, C.K.I., Winn, J., Zisserman, A.: The PASCAL Visual Object Classes Challenge 2007 (VOC2007) Results (2007). http://www.pascal-network.org/challenges/VOC/voc2007/workshop/index.html
8. Felzenszwalb, P.F., Girshick, R.B., McAllester, D., Ramanan, D.: Object detection with discriminatively trained part-based models. PAMI 32(9), 1627–1645 (2010)
9. Ferrari, V., Tuytelaars, T., Van Gool, L.: Object detection by contour segment networks. In: Leonardis, A., Bischof, H., Pinz, A. (eds.) ECCV 2006. LNCS, vol. 3953, pp. 14–28. Springer, Heidelberg (2006)
10. Han, J.H., Poston, T.: Chord-to-point distance accumulation and planar curvature: a new approach to discrete curvature. Pattern Recogn. Lett. 22(10), 1133–1144 (2001)
11. Kerautret, B., Lachaud, J.-O.: Robust estimation of curvature along digital contours with global optimization. In: Coeurjolly, D., Sivignon, I., Tougne, L., Dupont, F. (eds.) DGCI 2008. LNCS, vol. 4992, pp. 334–345. Springer, Heidelberg (2008)
12. Lazebnik, S., Schmid, C., Ponce, J.: Beyond bags of features: Spatial pyramid matching for recognizing natural scene categories. In: CVPR, pp. 2169–2178 (2006)
13. Liu, H., Latecki, L.J., Liu, W., Bai, X.: Visual curvature. In: CVPR (2007)
14. Lowe, D.G.: Distinctive image features from scale-invariant keypoints. IJCV 60(2), 91–110 (2004)
15. Mikolajczyk, K., Tuytelaars, T., Schmid, C., Zisserman, A., Matas, J., Schaffalitzky, F., Kadir, T., Gool, L.J.V.: A comparison of affine region detectors. IJCV 65(1–2), 43–72 (2005)
16. Monroy, A., Eigenstetter, A., Ommer, B.: Beyond straight lines - object detection using curvature. In: ICIP, pp. 3561–3564 (2011)
17. Sapiro, G.: Geometric Partial Differential Equations and Image Analysis. Cambridge University Press (2006). http://books.google.de/books?id=LFzGwdi0KtsC
18. Vedaldi, A., Fulkerson, B.: VLFeat: an open and portable library of computer vision algorithms (2008). http://www.vlfeat.org/

Human Pose and People Tracking

Test-Time Adaptation for 3D Human Pose Estimation

Sikandar Amin[1,2]($\boxtimes$), Philipp Müller[2], Andreas Bulling[2],
and Mykhaylo Andriluka[2,3]

[1] Technische Universität München, Munich, Germany
sikandaramin@gmail.com
[2] Max Planck Institute for Informatics, Saarbrücken, Germany
[3] Stanford University, Stanford, USA

Abstract. In this paper we consider the task of articulated 3D human pose estimation in challenging scenes with dynamic background and multiple people. Initial progress on this task has been achieved building on discriminatively trained part-based models that deliver a set of 2D body pose candidates that are then subsequently refined by reasoning in 3D [1,4,5]. The performance of such methods is limited by the performance of the underlying 2D pose estimation approaches. In this paper we explore a way to boost the performance of 2D pose estimation based on the output of the 3D pose reconstruction process, thus closing the loop in the pose estimation pipeline. We build our approach around a component that is able to identify true positive pose estimation hypotheses with high confidence. We then either retrain 2D pose estimation models using such highly confident hypotheses as additional training examples, or we use similarity to these hypotheses as a cue for 2D pose estimation. We consider a number of features that can be used for assessing the confidence of the pose estimation results. The strongest feature in our comparison corresponds to the ensemble agreement on the 3D pose output. We evaluate our approach on two publicly available datasets improving over state of the art in each case.

1 Introduction and Related Work

In this paper we consider the task of articulated 3D human pose estimation from multiple views. We focus on the setting with uncontrolled environment, dynamic background and multiple people present in the scene, which is more complex and general compared to the motion capture studio environments often considered in the literature [7,13,17]. One of the key challenges in that setting is that the appearance of people is more diverse, and simple means of representing observations based on background subtraction are not applicable due to the presence of multiple people and interactions between people and scene objects. Inspired by recent results in 2D pose estimation [3,18], several approaches have proposed to build upon and adapt these results for pose estimation in 3D [1,4,5,12]. In these approaches 2D detectors are either used to model the likelihood of the 3D pose [5,12], or provide a set of proposals for positions of body

© Springer International Publishing Switzerland 2014
X. Jiang et al. (Eds.): GCPR 2014, LNCS 8753, pp. 253–264, 2014.
DOI: 10.1007/978-3-319-11752-2_20

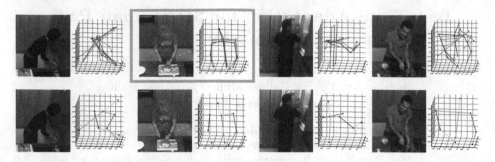

Fig. 1. Overview of our approach. *Top row:* In the first stage we estimate 3D poses of people in each frame with an ensemble of multi-view pictorial structures models. Output of the models from the ensemble is shown in blue, red and black. We select highly confident key-frames either based on (1) agreement between models in the ensemble, or (2) using a classifier trained on features computed from these outputs. The green bounding box indicates a selected key-frame. *Bottom row:* Output of our final model that incorporates evidence from the keyframes (Color figure online).

joints that are subsequently refined by reasoning in 3D [1,4]. Improving the 2D pose estimation performance is thus crucial for each of these methods. Towards this goal we propose an approach to tune the 2D pose estimation component at test time.

Generally, one would expect that the pose estimation results should improve if one is continuously observing the same scene with the same human subjects, as one would be able to learn more specific appearance models than is possible in the general case. However, with a few exceptions [6,14], this idea is rarely explored in the literature, likely because it is unclear how to robustly estimate the person-specific appearance in the presence of noise in the pose estimation. Various approaches for estimating the confidence of the pose prediction have been considered in the literature, ranging from models that are discriminatively trained for both detection and pose estimation [18] to specialized methods that estimate confidence based on the combination of features as a post-processing step [11]. In this paper we follow the direction similar to [11] but also employ features based on the 3D pose reconstruction, which we find to be highly effective for filtering out incorrect pose estimates. Overall we make the following contributions. As a main contribution of this paper we propose a new approach for articulated 3D human pose estimation that builds on the multi-view pictorial structures model [1], and extends it to adapt to observations available at test time. Our approach has an interesting property that it operates on the entire test set, making use of the evidence available in all test images. This is in contrast to prior works [1,4,5,12] that typically operate on single-frames only or are limited to temporal smoothness constraints which are effective only in a small temporal neighborhood of each frame [2,16]. As a second contribution we evaluate two approaches to assess the accuracy of 3D pose estimation. The first is to train a discriminative model based on various pose quality features as in [11], and the second is to consider the agreement of an ensemble of several independently trained models on the 3D pose output. An interesting finding of our evaluation is that pose agreement alone performs on-par or better than the discriminatively trained confidence predictor. The combination of both approaches further improves the results.

Overview of our approach. In this paper we build on the multi-view pictorial structures approach proposed in [1]. This approach first jointly estimates projections of each body joint in each view, and then recovers 3D pose by triangulation. We explore two mechanisms for improving the performance of the multi-view pictorial structures model. Both of them are based on the observation that 3D pose reconstruction provides strong cues for identification of highly confident pose estimation hypotheses (= key-frames) at test time (see Fig. 1 for a few examples). We explore two ways to take advantage of such key-frame hypotheses. We either directly use them as additional training examples in order to adapt the 2D pose estimation model to the scene at hand, or we extend the pictorial structures model with an additional term that measures appearance similarity to the key-frames. As we show in the experiments both mechanisms considerably improve the pose estimation results. In the following we first introduce the multi-view pictorial structures model and then describe our extensions.

2   Multi-view Pictorial Structures

The pictorial structures model represents a body configuration as a collection of rigid parts and a set of pairwise part relationships [8,10]. We denote a part configuration as $L = \{l_i | i = 1, \ldots, N\}$, where $l_i = (x_i, y_i, \theta_i)$ corresponds to the image position and absolute orientation of each part. Assuming that the pairwise part relationships have a tree structure the conditional probability of the part configuration L given the image evidence I factorizes into a product of unary and pairwise terms:

$$p(L|I) = \frac{1}{Z} \prod_{n=1}^{N} f_n(l_n; I) \cdot \prod_{(i,j) \in E} f_{ij}(l_i, l_j). \tag{1}$$

where $f_n(l_n; I)$ is the likelihood term for part n, $f_{ij}(l_i, l_j)$ is the pairwise term for parts i and j and Z is a partition function.

Multi-view model: Recently [1,5] have extended this approach to the case of 3D human pose estimation from multiple views. In the following we include the concise summary of the multiview pictorial structures model that we use in our experiments and refer the reader to the original paper [1] for more details.

The multiview pictorial structures approach proposed by [1] generalizes the single-view case by jointly reasoning about the projections of body parts in each view. Let L_v denote the 2D body configuration and I_v the image observations in view v. Multiview constraints are modeled as additional pairwise factors in the pictorial structures framework that relate locations of the same joint in each view. The resulting multiview pictorial structures model corresponds to the following decomposition of the posterior distribution:

$$p(L_1, ..., L_V | I_1, ..., I_V) = \frac{1}{Z} \prod_v f(L_v; I_v) \prod_{(a,b)} \prod_n f_n^{app}(l_n^a, l_n^b; I_a, I_b) f_n^{cor}(\mathbf{l}_n^a, \mathbf{l}_n^b),$$

(2)

where $\{(a, b)\}$ is the set of all view-pairs, $\mathbf{l}_n^v$ represents the image position of part n in view v, in contrast l_n^v in addition to image position also includes the absolute orientation, $f(L_v; I_v)$ are the single-view factors for view v which decompose into products of unary and pairwise terms according to Eq. 1, and f_n^{app} and f_n^{cor} are multiview appearance and correspondence factors for part n. The inference is done jointly across all views. The 2D pose estimation results are then triangulated to reconstruct the final 3D pose.

3   Test-Time Adaptation

Our approach to test-time adaptation is composed of two stages. In the first stage we mine confident pose estimation examples from the test data. These examples are then used in the second stage to improve the pose estimation model. We now describe these stages in detail.

3.1   Confident Examples Mining

The objective of the first stage is to identify the test examples for which the initial model succeeded in correctly estimating the body pose. To that end, we consider two methods to assess the accuracy of pose estimation.

3D pose agreement. In the first method we proceed by training an ensemble of M multiview pictorial structure models. Each model in the ensemble is trained on a disjoint subset of the training set. In addition we also train a reference model using all training examples. The rationale behind this procedure is that the reference model will typically perform better than ensemble models as it is trained on more examples. Ensemble models will in turn provide sufficient number of independent hypothesis in order to assess the prediction accuracy. At test time we evaluate the agreement between the pose hypotheses estimated by ensemble models and the reference model. Given the estimated 3D poses from all models, we define the pose agreement score s_{pa} as:

$$s_{pa} = \exp\left(-\frac{\sum_n \sum_m \|\mathbf{x}_n^m - \hat{\mathbf{x}}_n\|_2^2}{N}\right),$$

(3)

where $\mathbf{x}_n^m$ represents the 3D position of part n estimated with the ensemble model $m \in \{1, ..., M\}$, and $\hat{\mathbf{x}}_n$ is the location of part n estimated with the reference model.

Pose classification. As a second method we train a discriminative AdaBoost classifier to identify correct 3D pose estimates. The classifier is trained using the following features:

1. *3D pose features.* These features encode the plausible 3D poses and correspond to the torso, head and limb lengths, distance between shoulders, angles between upper and lower limbs, and angles between head and shoulder parts.
2. *Prediction uncertainty features.* We encode the uncertainty in pose estimation by computing the L2 norm of the covariance matrix corresponding to the strongest mode in the marginal posterior distribution of each part in each view. This is the same criteria as used for component selection in [1] and is similar to the features used in [11].
3. *Posterior.* As a separate feature we also include the value of the posterior distribution corresponding to the estimated pose that is given by Eq. 2.

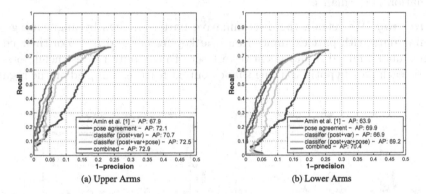

(a) Upper Arms   (b) Lower Arms

Fig. 2. MPIICooking dataset: Key-frame selection using score based on the posterior marginal of the model from [1] (blue) compared to variants of our approach. Best result corresponds to combination of ensemble agreement and pose classification scores given by Eq. 5 (magenta) (Color figure online).

We concatenate these three types of features to produce a combined feature vector of the size $20 + 2VN$ for the upper-body and $26 + 2VN$ for the full-body case, where V is the number of views and N is the number of parts in the pictorial structures model. We rely on a disjoint validation set for training of the pose classifier. We consider 3D poses with all body parts estimated correctly to be positive examples and all other examples as negative. Body part is considered to be correctly estimated if both of its endpoints are within 50 % of the part length from their ground-truth positions. The classifier score corresponding to the mth pictorial structure model $s_{abc,m}$ is given by the weighted sum of the weak single-feature classifiers $h_{m,t}$ with weights $\alpha_{m,t}$ learned using AdaBoost:

$$s_{abc,m} = \frac{\sum_t \alpha_{m,t} h_{m,t}(x_m)}{\sum_t \alpha_{m,t}} \qquad (4)$$

Combined approach. Finally, we consider a weighted combination of agreement and classification scores:

$$s_{comb} = s_{pa} + \sum_m w_m s_{abc,m}, \tag{5}$$

where the weights of the classifier scores are given by $w_m = \frac{\sum_{\hat{m} \neq m} \phi_{\hat{m}}}{(M-1)\sum_{\hat{m}} \phi_{\hat{m}}}$, and $\phi_{\hat{m}}$ are given by the training time mis-classification error. As we demonstrate in Sect. 4 such combination improves results over using each approach individually.

3.2   2D Model Refinement

At test-time we choose 10 % of the highest scoring pose hypotheses according to one of the scoring methods described in Sect. 3.1 and denote them as key-frames. We investigate two different avenues to use key-frames in order to improve pose estimation performance.

Retraining: We retrain the discriminative part classifiers by augmenting the training data with part examples from the key-frames and $n = 5$ of their nearest neighbors, mined from the entire test set. To compute nearest neighbors we encode each part hypothesis using shape context features sampled on the regular

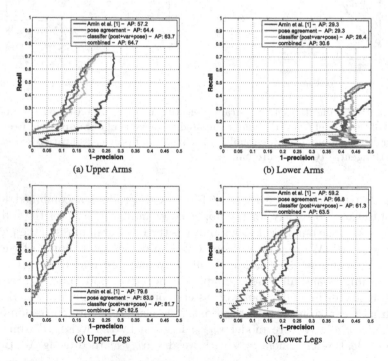

(a) Upper Arms   (b) Lower Arms

(c) Upper Legs   (d) Lower Legs

Fig. 3. Shelf dataset: Key-frame selection using score based on the posterior marginal of the model from [1] (blue) compared to variants of our approach (Color figure online).

grid within the part bounding box and bounding box color histogram. The nearest neighbors are then found using euclidean distance in this feature space. For the rest of the paper, we refer to this approach as **RT**.

Appearance similarity: We introduce additional unary term for each part in the pictorial structures model that encourages similarity to key-frames. The similarity term for part n is given by:

$$f_{SIM}(l_n; I) = \exp\left(-\frac{\min_j \|e(l_n) - e(a_{nj})\|_2^2}{2 * \sigma_n^2}\right), \tag{6}$$

where l_n is image position and absolute orientation of the part hypothesis, a_{nj} is a hypothesis for part n from the j-th key-frame, and $e(l_n)$ corresponds to shape-context and color features extracted at l_n. The variance σ_n^2 is estimated based on the euclidean distances between all feature vectors corresponding to part n in the training set. We refer to this approach as **SIM** later in text.

4   Experiments

Datasets. Our aim is to analyze the performance of our proposed approach in challenging settings with dynamic background and a variety of subjects. Therefore, we evaluate our approach on MPII Cooking [15] and the Shelf [4] datasets. Both of these datasets have been recently introduced for the evaluation of articulated human pose estimation from multiple views.

MPII Cooking: The dataset was originally recorded for the task of fine grained activity recognition of cooking activities, and has been later used in [1] to benchmark the performance of 3D pose estimation. This evaluation dataset consists of 11 subjects with non-continuous images and two camera views. The training set includes 4 subjects and 896 images and the test set includes 7 subjects and 1154 images.

Shelf dataset: This dataset has been introduced in [4] and is focused on the task of multiple human 3D pose estimation. The dataset depicts up to 4 humans interacting with each other while performing an assembly task. The Shelf dataset provides 668 and 367 annotated frames for training and testing respectively. For every frame each fully visible person is annotated in 3 camera views.

In our evaluation we rely on the standard train/test split and evaluation protocols as used by the original publications. As described in Sect. 3, we split the training set in multiple parts to train an ensemble of pose estimation models.

Key-Frames Analysis. We analyze the performance of our key-frame detection procedure using recall-precision curves. The results are shown in Figs. 2 and 3. In this analysis we omit the torso and head body parts as they are almost perfectly localized by all approaches. For all other parts, which are smaller in size hence more susceptible to noise, we observe that directly using the marginal posterior

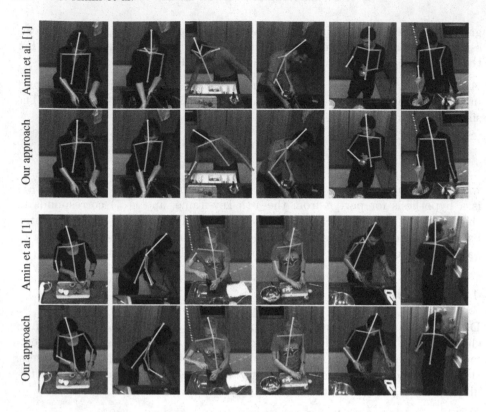

Fig. 4. Examples of pose estimation results obtained with our approach and comparison to the state-of-the-art approach of Amin et al. [1] on the MPII Cooking dataset.

of the PS model as pose confidence leads to poor results (blue curve in Figs. 2 and 3). On the MPII Cooking dataset we get AP of 67.9/63.9 for upper/lower arms respectively using this marginal posterior as confidence measure. On the other hand, training a classifier with posterior and variance features improves the AP to 70.7/66.9. The performance of the classifier increases further to 72.5/69.2 when we extend the feature vector with 3D pose cues. This result underlines the importance of 3D pose features for classification. Interestingly, the results of the pose agreement approach of Eq. 3 i.e. 72.1/69.9, suggest that this measure alone is almost equally effective. Combining both (classifier & pose agreement) scores as in Eq. 5, we step up the average precision of the detected key-frames to 72.9/70.4. These results show the importance of different levels of features in the process of extracting key-frames with high confidence.

The recall-precision curve for the Shelf dataset are shown in Fig. 3. Although, here the detection of lower arms is significantly more difficult, but the AP results for both upper and lower arms are in line with the MPII Cooking AP values. The AP values are slightly worse for the classifier score compared to pose agreement. For the legs, pose agreement alone outperforms the combined score of Eq. 5 i.e.,

Fig. 5. Examples of pose estimation failures on the MPIICooking dataset.

Fig. 6. Examples of pose estimation results of our approach on the Shelf dataset. Last column shows an example of the failure case.

83.0/66.8 as compared to 82.5/63.5 for upper/lower legs. The reason for this behavior is the significantly worse performance of the classifier of Eq. 4. This result suggests the need to learn weights when combining different scores as in Eq. 5. This we will investigate in future work. For the Shelf dataset the training and test splits contain the same subjects. Moreover, the training data splits for ensemble of multiview pictorial structure models also contain the same subjects. This explains the higher performance of the pose agreement cue compared to the classifier output.

Pose Estimation Results. Here we discuss the improvement we achieve in pose estimation performance when we incorporate these key-frames for 2D model refinement as discussed in Sect. 3.2. We use score of the combined approach s_{comb} to select the key-frames. Following [1, 15, 16], we use body-joints instead of limbs as parts in the pictorial structures model. This approach is commonly referred to as flexible pictorial structures model (FPS).

Table 1. MPII Cooking: accuracy measured using percentage of correct parts (PCP) score [9]. We compare the model of Amin et al. [1] with variants of our approach. *RT* stands for model retraining, *SIM* stands for a model augmented with similarity factors.

Model	Torso	Head	Upper arm		Lower arm		All
			r	l	r	l	
Cam-1							
Amin et al. [1]	92.9	89.4	72.6	79.4	68.8	76.8	80.0
Our (RT)	95.2	93.8	**75.3**	**83.4**	73.6	80.9	83.6
Our (SIM)	**95.8**	92.5	74.0	82.6	73.5	82.2	83.4
Our (RT+SIM)	**95.8**	**94.0**	74.8	83.3	**74.2**	**82.3**	**84.0**
Cam-2							
Amin et al. [1]	91.1	92.4	75.4	76.7	72.9	74.7	80.5
Our (RT)	92.1	95.5	79.2	82.8	76.9	78.8	84.2
Our (SIM)	92.4	**96.2**	**79.5**	81.6	**77.1**	79.6	84.4
Our (RT+SIM)	**92.6**	**96.2**	78.9	**83.3**	**77.1**	**79.7**	**84.7**

Table 2. Shelf dataset: accuracy measured using 3D PCP score. We compare the model of Belagiannis et al. [4] with variants of our approach. *RT* stands for model retraining, *SIM* stands for a model augmented with similarity factors.

	Belagiannis et al. [4]	Our (RT)	Our (SIM)	Our (RT+SIM)
Actor1	66	68.5	**72.1**	72.0
Actor2	65	67.2	69.4	**71.3**
Actor3	83	83.9	84.9	**85.7**
Average	71.3	74.4	77.0	**77.3**

MPII Cooking: We use the standard pose configuration, i.e., 10 upper-body parts as introduced in [15]. Amin et al. [1] reports the percentage of correct parts (PCP) for the 2D projections per camera for this dataset. First, we evaluate our proposed retraining approach (RT) to improve overall pose estimation accuracy by adapting the model to test scene specific settings. We show the PCP results for MPII Cooking in Table 1. Our RT approach achieves 83.6/84.2 overall PCP and shows improvement for all individual parts. This improvement can be attributed to the fact that retraining the model including the mined examples can learn the person/scene specific features. The other approach (SIM) which involves adding a new unary term, based on the feature similarity, to the pictorial structures model also achieves competitive results, i.e., 83.4/84.4 overall PCP. The improvement in this case is more pronounced on the lower arms compared to retraining the model. Furthermore, we also evaluate the combination of the two approaches RT+SIM. In this approach, along with retraining the part classifiers using the appearance features from the key-frames we also introduce the appearance similarity based unary term f_{SIM} to the multiview pictorial structures framework. The results in Table 1 show that this works best because it combines the benefits of both approaches and results in stronger unaries for the part hypotheses. We illustrate some example improvements of our approach in Fig. 4, as compared to [1] on MPII Cooking dataset. Figure 5 demonstrates some typical failure cases of our approach.

Shelf dataset: We use a full body model with 14 parts in this case as described in the original paper [4]. The accuracy of the approach from [4] is bounded by the performance of the 2D part detectors. Their model is unable to recover once the 2D part detector fails to fire in the first stage. On the other hand, as our approach is able to utilize the test scene specific information, we achieve far better results in terms of PCP values. Table 2 shows the 3D PCP values in comparison to the recent results of [4]. Our first approach, i.e., retraining the model RT, outperforms the approach from [4] by 3 % PCP. Interestingly, we get 77.0 PCP with our second approach SIM which involves model inference using an extra similarity term and it outperforms our RT approach by further 2.6 %. This result can be explained by the fact that all dimensions of the appearance feature vector are considered equally in this approach. There exist some features which do not perform well during the feature selection process in AdaBoost when learnt together with the training set. Still, they contain similarity information for the test examples when compared against the mined key-frames in terms of euclidean distance for the complete feature vector. Further gain of 0.3 PCP is obtained by combining RT with SIM. Some examples of qualitative results on Shelf dataset are depicted in Fig. 6.

5   Conclusion

In this paper we proposed an approach to 3D pose estimation that adapts to the input data available at test time. Our approach operates by identifying frames in which poses can be predicted with high confidence and then uses them as additional training examples and as evidence for pose estimation in other frames. We analyzed two strategies for finding confident pose estimates: discriminative classification and ensemble agreement. Best results are achieved by combining both strategies. However, ensemble agreement alone already improves considerably over the confidence measure based on the pictorial structures output. We have shown the effectiveness of our approach on two publicly available datasets. In the future we plan to generalize our approach to multiple rounds of confident examples mining, and will explore other approaches for automatic acquisition of training examples from unlabeled images.

Acknowledgements. This work has been supported by the Max Planck Center for Visual Computing and Communication.

References

1. Amin, S., Andriluka, M., Rohrbach, M., Schiele, B.: Multi-view pictorial structures for 3D human pose estimation. In: BMVC (2013)
2. Andriluka, M., Roth, S., Schiele, B.: Monocular 3D pose estimation and tracking by detection. In: CVPR (2010)
3. Andriluka, M., Roth, S., Schiele, B.: Discriminative appearance models for pictorial structures. IJCV 99(3), 259–280 (2012)
4. Belagiannis, V., Amin, S., Andriluka, M., Schiele, B., Navab, N., Ilic, S.: 3D pictorial structures for multiple human pose estimation. In: CVPR (2014)
5. Burenius, M., Sullivan, J., Carlsson, S.: 3D pictorial structures for multiple view articulated pose estimation. In: CVPR (2013)
6. Eichner, M., Ferrari, V.: Appearance sharing for collective human pose estimation. In: Lee, K.M., Matsushita, Y., Rehg, J.M., Hu, Z. (eds.) ACCV 2012, Part I. LNCS, vol. 7724, pp. 138–151. Springer, Heidelberg (2013)
7. El Hayek, A., Stoll, C., Hasler, N., Kim, K.I., Seidel, H.P., Theobalt, C.: Spatio-temporal motion tracking with unsynchronized cameras. In: CVPR (2012)
8. Felzenszwalb, P.F., Huttenlocher, D.P.: Pictorial structures for object recognition. IJCV 61(1), 55–79 (2005)
9. Ferrari, V., Marin, M., Zisserman, A.: Progressive search space reduction for human pose estimation. In: CVPR (2008)
10. Fischler, M., Elschlager, R.: The representation and matching of pictorial structures. IEEE Trans. Comput. C-22(1), 67–92 (1973)
11. Jammalamadaka, N., Zisserman, A., Eichner, M., Ferrari, V., Jawahar, C.V.: Has my algorithm succeeded? an evaluator for human pose estimators. In: Fitzgibbon, A., Lazebnik, S., Perona, P., Sato, Y., Schmid, C. (eds.) ECCV 2012, Part III. LNCS, vol. 7574, pp. 114–128. Springer, Heidelberg (2012)
12. Kazemi, V., Burenius, M., Azizpour, H., Sullivan, J.: Multi-view body part recognition with random forests. In: BMVC (2013)
13. Ofli, F., Chaudhry, R., Kurillo, G., Vidal, R., Bajcsy, R.: Berkeley MHAD: a comprehensive multimodal human action database. In: WACV (2013)
14. Ramanan, D., Forsyth, D.A., Zisserman, A.: Strike a pose: tracking people by finding stylized poses. In: CVPR (2005)
15. Rohrbach, M., Amin, S., Andriluka, M., Schiele, B.: A database for fine grained activity detection of cooking activities. In: CVPR (2012)
16. Sapp, B., Weiss, D., Taskar, B.: Parsing human motion with stretchable models. In: CVPR (2011)
17. Sigal, L., Balan, A., Black, M.J.: Humaneva: synchronized video and motion capture dataset and baseline algorithm for evaluation of articulated human motion. IJCV 87(1–2), 4–27 (2010)
18. Yang, Y., Ramanan, D.: Articulated pose estimation with flexible mixtures-of-parts. In: CVPR (2011)

Efficient Multiple People Tracking Using Minimum Cost Arborescences

Roberto Henschel[1]([✉]), Laura Leal-Taixé[2], and Bodo Rosenhahn[1]

[1] Institut Für Informationsverarbeitung,
Leibniz Universität Hannover, Hannover, Germany
{henschel,rosenhahn}@tnt.uni-hannover.de
[2] Institute of Geodesy and Photogrammetry, ETH Zurich, Zurich, Switzerland
leal@geod.baug.ethz.ch

Abstract. We present a new global optimization approach for multiple people tracking based on a hierarchical tracklet framework. A new type of tracklets is introduced, which we call *tree tracklets*. They contain bifurcations to naturally deal with ambiguous tracking situations. Difficult decisions are postponed to a later iteration of the hierarchical framework, when more information is available. We cast the optimization problem as a minimum cost arborescence problem in an acyclic directed graph, where a tracking solution can be obtained in linear time. Experiments on six publicly available datasets show that the method performs well when compared to state-of-the art tracking algorithms.

1 Introduction

A key challenge in many computer vision domains is to automatically detect objects in video sequences and track them with high accuracy over time. For applications such as surveillance, action recognition, animation or human-computer interaction systems, multiple people tracking has emerged as one of the main tasks to be solved. Algorithms in recent literature have shown great performance in semi-crowded environments. In particular, the hierarchical tracklet approach [11] has evolved as an excellent tracking framework, mainly because of its bootstrapping capabilities, and is hence in the spotlight of current research. While latest improvements have been mainly achieved by exploiting the bootstrapping potential of this approach, little has been done to improve the quality of tracklets. Even though tracklets are the input for many tracking methods, usually these are found in a greedy way, and no appropriate method for finding reliable tracklets has been presented so far. In this paper, we present a global association method for tracklet creation within a hierarchical tracking framework. We propose a generalization of tracklets, which we call *tree tracklets*, that fits better into the hierarchical structure and, finally, a tracking method where tracklet association is formulated as a minimum cost arborescence (MCA) problem. The framework thus performs associations at each iteration with improved time complexity of $\mathcal{O}(n)$ in the number of current tracklets n, and its performance is in the range of less efficient state-of-the-art approaches.

© Springer International Publishing Switzerland 2014
X. Jiang et al. (Eds.): GCPR 2014, LNCS 8753, pp. 265–276, 2014.
DOI: 10.1007/978-3-319-11752-2_21

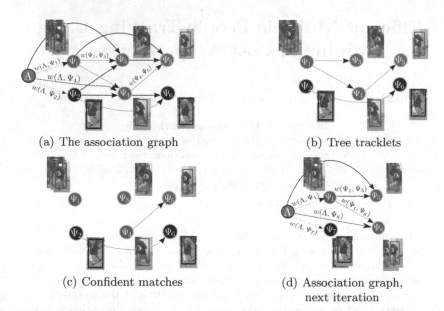

(a) The association graph   (b) Tree tracklets

(c) Confident matches   (d) Association graph,
                                                 next iteration

Fig. 1. In the association graph (a), each detection/tracklet Ψ_i is connected to all possible tracklets in consecutive frames up to some maximum time gap (for simplicity not all edges are drawn). The nodes of all shown figures are ordered from left to right by increasing frame number w.r.t. the tail of the corresponding tracklet. Each incoming edge of minimal weight is colored orange. The virtual start node Λ is used to model a new trajectory start. Detections are then grouped into two sets (b), which we call tree tracklets and cut into confident matches (c). In the next iteration of our algorithm, a new association graph is built up (d) using the confident matches from the last iteration. Ambiguities in the last iteration are then easily resolved (Color figure online).

1.1 Related Work

The problem of multiple people tracking is related to many computer vision applications. A common way to tackle the problem is to divide it into two sub-problems: (i) the detection of all objects and (ii) the correct aggregation of these detections along time to form the final trajectories. People detectors [3,7] achieve high accuracy if the scene is not too crowded. Wrong detections occur especially when people are partially occluded or when they are standing too close to each other, being interpreted as one person.

The data association framework links detections in time, either on a frame-by-frame basis (online systems) or by taking longer sequences into account (offline systems). Recent advances have been made by formulating the problem of finding the trajectories for all objects and all frames. Solutions are obtained by solving a combinatorial optimization problem. If the number of objects is known a-priori, [12] computes trajectories by solving a min-cost flow problem. The formulation of [2] avoids this restriction; a solution is efficiently obtained by solving a k-shortest-path problem. In [17,24], the data association is formulated as a maximum-a-posteriori (MAP) problem; a solution is inferred from a min-cost flow. Instead of computing the complete trajectories at once, [11] introduces a hierarchical approach where, at each iteration, the MAP problem is solved with decreasingly restrictive parameters, using the computed trajectories from the previous iteration to form the current trajectory solution. The Hungarian method [14] is used to compute the associations, therefore, its time complexity is $\mathcal{O}(n^3)$ in the number n of tracklets for each iteration. The hierarchical structure has proven to be a very fruitful approach, since it allows to extract a lot of information out of reliable tracklets. Bootstrapping-like optimization schemes for tracklets have thus become a trend in the tracking field. For instance, the standard MAP formulation of a tracking framework requires probabilities for people entering or leaving the scenery, which are generally not known a-priori. In [11], the distribution of the tails of the tracklets is used to infer these probabilities and [23] constructs an entrance/exit map which is derived from a convex set that is spanned by the tails of long and confident tracklets. Improvements can also be achieved on the motion and appearance affinities. For appearance measurements, [15,18,23] train a classifier based on the computed tracklets. Having tracklets at each iteration, [23] uses a quadratic function to extend the motion model to non-linear movements based on the information of current tracklets. Motion dynamics based on the rank of the Hankel matrix are considered in [4], allowing for identifications of objects by their movement characteristics. If several people walk close to each other, their motion will not be independent of each other. [21] infers group behavior of detected people from the tracklets, while [17] uses a physics-based social force model to predict pedestrians' motion and [16] learns this motion context directly from image features in order to perform tracking in image space. Finally, [10] extends the hierarchical framework to take splits and merges of trajectories into account, which commonly happen when people are walking close to each other and then split, or vice versa.

Nonetheless, few works are dedicated to improving the method to find tracklets or the hierarchical data association method itself. In this paper, we propose an alternative association method with superior time complexity in comparison to the Hungarian method together with the modification of the tracklet model to *tree tracklets* which are constructed especially for a hierarchical framework[1].

1.2   Contributions

The main contribution of this paper is three-fold:

- A tracklet-based hierarchical tracking system where at each iteration the data association is formulated as a minimum cost arborescence problem.
- A better time complexity than current global tracking methods. The time complexity is $\mathcal{O}(n)$ at each iteration, where n is the number of tracklets.
- A new trajectory model, namely *tree tracklets*, for hierarchical tracking frameworks, which handles false alarms and occlusions in a more natural way.

---

[1] The code is publicly available: http://www.tnt.uni-hannover.de/project/MPT/.

2 Tracklet Creation

A hierarchical tracking framework should iteratively connect tracklets by taking decisions at each iteration only in very confident cases, so that difficult selections can be postponed for a later iteration when more information is available thereby reducing error propagation. The modified MAP formulation that we introduce is modeled exactly for that purpose: it returns a generalization of the common tracklets [11] as optimum value, which we call *tree tracklets*. Such tracklets are more robust against error propagation and are obtained by solving a minimum cost arborescence (MCA) problem (see Sect. 2.1).

Having the corresponding ordinary tracklets, false detections are removed and restrictive parameters like the size of the time window are progressively weakened, as associations become more and more confident (see Sect. 3 for more details). Note that such a hierarchical approach has already been formulated in [11] for tracklets using the Hungarian method. However, we reformulate the approach to work on *tree tracklets* and use the MCA formulation instead, resulting in more robust tracklets and faster computation.

Finally, apart from serving as a complete tracking system, the tracklets from any of the intermediate steps can be used as initialization for other tracking frameworks.

2.1 Tracklet Creation with Arborescences

We introduce the notation used in this paper. Let $\mathcal{R} = \{\mathbf{r}_i\}$ be a set of object detections obtained from a detector (*e.g.* [7,8]) on a video sequence, transferred into 3D. Thereby, $\mathbf{r}_i := (\mathbf{p}_i, \mathbf{s}_i, t_i)$, where $\mathbf{p}_i = (x_i, y_i, z_i)$ denotes its position in 3D, $\mathbf{s}_i = (w_i, h_i)$ the size of the bounding box and t_i the time stamp of the detection response, respectively. Let V be an arbitrary set together with time functions $\mathrm{fr}^{\mathrm{head}}, \mathrm{fr}^{\mathrm{tail}} : V \to \mathbb{Z}$. We fix a time window ω of frames that are allowed between two detections to be connected and define the graph $G(V) := G(V, E_\omega)$, where $E_\omega := \{(v, w) \in V \times V \mid \mathrm{fr}^{\mathrm{tail}}(v) < \mathrm{fr}^{\mathrm{head}}(w) \le \mathrm{fr}^{\mathrm{tail}}(v) + \omega\}$.

Now for $V = \mathcal{R}$ and $\mathrm{fr}^{\mathrm{head}}(\mathbf{r}) = \mathrm{fr}^{\mathrm{tail}}(\mathbf{r})$ defined as the timestamp of $\mathbf{r} \in \mathcal{R}$, we call a weakly connected subgraph $\mathbf{I}$ of $G(V)$ a *tracklet*, if $\mathbf{I}$ is a directed path. For example, Fig. 1(c) consists of 4 tracklets (of maximal length). For a tracklet $\mathbf{I}$, we define its head $\mathbf{r}^{\mathrm{head}} = (\mathbf{p}^{\mathrm{head}}, \mathbf{s}^{\mathrm{head}}, t^{\mathrm{head}}) \in V(\mathbf{I})$ to be the unique detection in $\mathbf{I}$ with lowest time stamp. Accordingly, we define the detection $\mathbf{r}^{\mathrm{tail}} \in V(\mathbf{I})$ with highest timestamp to be the tail.

We obtain the (next) tracklets iteratively: We set $\underline{\mathbf{I}}^0 := \mathcal{R}$. Now let $\underline{\mathbf{I}}^n$ denote the current set of computed tracklets in the n-th iteration and let $V := \underline{\mathbf{I}}^n$. For $\mathbf{I} \in V$, we define $\mathrm{fr}^{\mathrm{tail}}(\mathbf{I})$ as the timestamp of $\mathbf{I}$'s tail, and $\mathrm{fr}^{\mathrm{head}}(\mathbf{I})$ of $\mathbf{I}$'s head, respectively, and say a weakly connected subgraph Ψ of $G(\underline{\mathbf{I}}^n)$ is a *tree tracklet*, if each node of Ψ has indegree ≤ 1 in Ψ.

Note that tree tracklets allow a tracklet to be connected to several tracklets in successive frames, maintaining ambiguities until enough information has been collected in following iterations (see Fig. 1(b)). We call a set $\Phi := \{\Psi_1, \ldots, \Psi_s\}$ of (tree) tracklets a (tree) tracklet hypothesis if all (tree) tracklets are pairwise disjoint, and say Φ is covering, if all nodes can be reached by some (tree) tracklet of Φ. By using covering tracklets, we will force the framework to explain all detections as best as possible. Finally, we denote by Φ_ω the set of all covering tree tracklet hypotheses. In particular, we solve multiple people tracking by searching for that tracklet hypothesis that fits best to all detections.

We solve the problem of tracklet association using the MAP method:

$$\Phi^* = \arg\max_{\Phi \in \Phi_\omega} P(\Phi \mid \underline{\mathbf{I}}^n) = \arg\max_{\Phi \in \Phi_\omega} P(\underline{\mathbf{I}}^n \mid \Phi)P(\Phi) = \arg\max_{\Phi \in \Phi_\omega} \prod_{\Psi_k \in \Phi} P(\Psi_k). \quad (1)$$

We have $P(\underline{\mathbf{I}}^n \mid \Phi) = 1$, since Φ is covering. Furthermore, we assume that the tree tracklets Ψ_k are independent of each other. Note that if we restrict (1) to find the maximum only over tracklets, the optimization can be seen as a special case of the ordinary MAP problem used in other frameworks (see [17,24]).

For $\Phi \in \Phi_\omega$ and $\Psi_k \in \Phi$, let $\mathbf{I}_{k_1}$ be the first appearing tracklet of Ψ_k, where $V(\Psi_k) = \{\mathbf{I}_{k_1}, \ldots, \mathbf{I}_{k_m}\}$. We define, modeled as a Markov chain, the probability

$$P(\Psi_k) := P_{\text{init}}(\mathbf{I}_{k_1}) \prod_{s=1}^{|V(\Psi_k)|} \prod_{\mathbf{I} \in \mathbf{I}_{k_s}^+} P_{\text{link}}(\mathbf{I} \mid \mathbf{I}_{k_s}), \quad (2)$$

where $P_{\text{link}}(\mathbf{I}_j|\mathbf{I}_i)$ denotes the probability that the tracklet $\mathbf{I}_i$ belongs to the same object as $\mathbf{I}_j$. Thereby, $\mathbf{I}_{k_s}^+$ is the set of outgoing nodes from $\mathbf{I}_{k_s}$ in Ψ_k. The probability that a new object enters the scenery at the head of $\mathbf{I}_i$ is defined by $P_{\text{init}}(\mathbf{I}_i)$. We provide the concrete calculations of these probabilities in Sect. 3.

Inserting (2) in (1) and applying the negative logarithm, we obtain

$$\Phi^* = \arg\min_{\Phi \in \Phi_\omega} \sum_{\Psi_k \in \Phi} \left(-\log P_{\text{init}}(\mathbf{I}_{k_{i_1}}) + \sum_{s=1}^{|V(\Psi_k)|} \sum_{\mathbf{I} \in \mathbf{I}_{k_s}^+} -\log(P_{\text{link}}(\mathbf{I}|\mathbf{I}_{k_s})) \right). \quad (3)$$

Next we show the relation of Eq. (3) to the MCA problem.

Given a covering tree tracklet hypothesis $\Phi = \{\Psi_1, \cdots, \Psi_s\}$, we construct a graph $G := G(\Phi)$ by adding a virtual node Λ and link it to the first appearing node $\mathbf{I}_{k_1}$ of every $\Psi_k \in \Phi$. Then, G is an arborescence, that is a rooted (in this case at Λ) directed graph such that there is a directed path from Λ to each node [9] (see also Fig. 1(a)). Assigning weights to each edge, the *minimum costs arborescence* (MCA) problem [9,13] is to find the arborescence $\hat{\Phi}$ of G s.t. the total edge weight $w(\hat{\Phi}) := \sum_{e \in E(\hat{\Phi})} w(e)$ is minimal under all possible arborescences of G rooted at Λ (denoted by A_G). If we define the weights $w(\Lambda, \mathbf{I}) := -\log(P_{\text{init}}(\mathbf{I}))$ and $w(\mathbf{I}_i, \mathbf{I}_j) := -\log P_{\text{link}}(\mathbf{I}_j|\mathbf{I}_i)$ for all $\mathbf{I}, \mathbf{I}_i, \mathbf{I}_j \in \underline{\mathbf{I}}^n$ and rewrite (3) in terms of the defined graph G, we obtain

$$\hat{\Phi} = \arg\min_{(V,E) \in A_G} \sum_{(\Lambda, \mathbf{I}_i) \in E} w(\Lambda, \mathbf{I}_i) + \sum_{\substack{(\mathbf{I}_i, \mathbf{I}_j) \in E, \\ \mathbf{I}_i \neq \Lambda}} w(\mathbf{I}_i, \mathbf{I}_j). \quad (4)$$

Hence, a solution of (3) is obtained by solving the MCA problem. Then, Φ^* is the set of weakly connected components of $\hat{\Phi} - \{\Lambda\}$. A solution for the MCA problem can be computed in polynomial time by Edmond's algorithm [5]. In our case though, since G does not have any directed cycle, it can be shown that it is sufficient to select for each node the incoming edge of minimal weight (see for example [13]), so if $(V^*, E^*) := G(\Phi)$, then

$$\hat{\Phi} = \left(V^*, \{(u,v) \in E^* \mid w(u,v) = \min \{w(u',v) \mid (u',v) \in E^*\}\}\right). \qquad (5)$$

Regarding the time complexity of this procedure, we can construct the solution $\hat{\Phi}$ simultaneously while building up the graph G, using (5), and adding only computation costs of $\mathcal{O}(1)$ to the cost of constructing G. From that we obtain Φ^*, having computation costs of $\mathcal{O}(n)$ in the number of current tracklets n. Furthermore, we can update the association information after each frame that has been processed.

2.2 Using Bifurcations to Detect Ambiguities

Since tree tracklets can contain bifurcations, we explain how to deal with them. Note that when we say bifurcation it can be a split into two or more branches. Now bifurcations can be used to spot missing detections caused by splits and merges: If persons walk close together, the detector might create only one box around them, resulting in one trajectory for the group. Once a person leaves the group, the corresponding tracklet will also contain this split in form of a bifurcation. Furthermore, they help to spot and remove false detections. Figure 1 illustrates such a situation. The green box is a false detection and the red box is a correct detection within the same frame. The false detection causes an ambiguity. However, only the most likely detection is assigned to the detection in the next frame (Fig. 1(b)). The tree tracklet structure can thus automatically isolate the false detection and construct the right trajectory in the next iteration. Hence, we handle tree tracklets as explained in Algorithm 1 and obtain the next set of tracklets $\underline{\mathbf{I}}^{n+1}$. Note that a currently computed tracklet Ψ becomes one node in the next iteration. For example in Fig. 1(c), the tracklets Ψ_2 and Ψ_6 are connected. Therefore, they are grouped together to the new node Ψ_7 in the next iteration (Fig. 1(d)). The ambiguity between the nodes Ψ_1, Ψ_3 and Ψ_4 is postponed to the next iteration. However, the tracklet Ψ_4 is connected to another tracklet, resulting in new information that can be used in the next iteration.

3   Implementation Details

We provide the definitions of the functions P_{link} and P_{init} that we use in our experiments. For an arbitrary video sequence, information about the scenery is not available a-priori. Hence, we set the entrance probability to be a constant value $\theta \in [0,1]$, so $P_{\text{init}}(\mathbf{I}_i) := \theta$ for all tracklets $\mathbf{I}_i \in \underline{\mathbf{I}}^n$.

Given tracklets $\mathbf{I}_i, \mathbf{I}_j \in \underline{\mathbf{I}}^n$, we define the time difference $\triangle_{i,j} = |t_i^{\text{tail}} - t_j^{\text{head}}|$ and the forward directed velocity vectors in $\mathbf{v}_i, \mathbf{v}_j$ of $\mathbf{I}_i$ and $\mathbf{I}_j$ at $\mathbf{p}_i^{\text{head}}$ and $\mathbf{p}_j^{\text{tail}}$, respectively. Let $\delta_{i,j}^{\text{f}} := \mathbf{p}_i^{\text{tail}} + \triangle_{i,j}\mathbf{v}_i - \mathbf{p}_j^{\text{head}}$ and $\delta_{i,j}^{\text{b}} := \mathbf{p}_j^{\text{head}} - \triangle_{i,j}\mathbf{v}_j - \mathbf{p}_i^{\text{tail}}$ be the error of a linear extrapolation from $\mathbf{I}_i$ to $\mathbf{I}_j$ and vice versa.

Algorithm 1. Computing tree tracklets

Input: MCA $\hat{\Phi}$, rooted at Λ
Output: Tracklets $\mathbf{I}^{n+1} := \{\Psi_1, \cdots, \Psi_s\}$
  1: $C := \{\Lambda\}$, $c(n) := 0$, for all nodes n of $\hat{\Phi}$.
  2: **for** $n \in C$ **do**
  3:     **if** n has siblings **then**
  4:         $c(n) :=$ new unique number
  5:     **else**
  6:         $c(n) := c(m)$, where m is the parent node
  7:     **end if**
  8:     $C := C \cup C_{\text{new}} \setminus \{n\}$, C_{new} being the set of children of n
  9: **end for**
 10: **for**   each assigned value k of the function c **do**
 11:     Ψ_k is the subgraph of $\hat{\Phi}$ induced by all nodes $n \neq \Lambda$ with $c(n) = k$.
 12: **end for**

We define the link probability as $P_{\text{link}}(\mathbf{I}_j \mid \mathbf{I}_i) = A_t(\mathbf{I}_j \mid \mathbf{I}_i)A_m(\mathbf{I}_j \mid \mathbf{I}_i)$ where the term $A_t(\mathbf{I}_j \mid \mathbf{I}_i)$ (as defined in [11]) is used to avoid too big time jumps, and the motion affinity (based on [11]) is defined as:

$$A_m(\mathbf{I}_i \mid \mathbf{I}_j) = \frac{\mathcal{N}(\delta_{i,j}^{\text{f}}, \mathbf{0}, \triangle_{i,j}\Sigma)\mathcal{N}(\delta_{i,j}^{\text{b}}, \mathbf{0}, \triangle_{i,j}\Sigma)}{\mathcal{N}(\mathbf{0}, \mathbf{0}, \triangle_{i,j}\Sigma)^2}. \tag{6}$$

Here, $\mathcal{N}(d, \mathbf{0}, \Sigma)$ denotes the multivariate normal probability distribution of the linear extrapolation error, with mean $\mathbf{0}$ and covariance Σ. To obtain Σ, we run one iteration of the algorithm using only distance information between detections as in [17]. We then use the resulting tracklets to learn an error distribution of the extrapolation using velocity vectors for each triplet of connected nodes. These tracklets are then discarded since they are only used for initialization.

4   Experiments

We evaluate the proposed tracking framework on several publicly available datasets: Bahnhof, Sunnyday, Jelmoli and Linthescher [6] and TownCenter [1]. Detections for Bahnhof and Sunnyday are obtained from [22], for Linthescher and Jelmoli we use [7]. For TownCenter, starting from the first frame, we take every tenth frame and use the detections provided with the dataset [1]. For the evaluation, we use the following metrics [18]: Recall (correctly matched detections/ ground truth detections), Precision (correctly matched detections/detections of resulting tracks), MT (mostly tracked trajectories, over 80 % tracked), ML (mostly lost trajectories, less than 20 % tracked), PT (partially tracked trajectories, tracked >20 % and <80 %), Frg (track fragments) and Ids (number of identity switches).

4.1   Influence of the Parameters

We first analyze the effect of the parameters of Sect. 3 on the Bahnhof dataset.

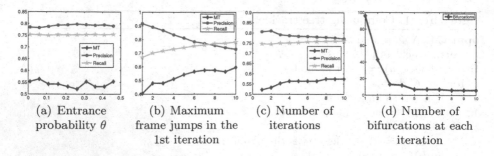

(a) Entrance
probability θ

(b) Maximum
frame jumps in the
1st iteration

(c) Number of
iterations

(d) Number of
bifurcations at each
iteration

Fig. 2. (a,b,c) Parameter analysis on the Bahnhof dataset using MT (mostly tracked), Recall and Precision metrics. (d) Bifurcation evolution at each iteration on the Town-Center sequence.

Entrance probability θ. The decision wether two tracklets are being connected by the algorithm has a dependence on the choice of the entrance probability θ, which controls wether the algorithm connects fewer tracklets with more confidence or more tracklets with less confidence. As we can see in Fig. 2(a), our algorithm is robust against changes of this parameter, but it works better for values <0.1, which is why we use $\theta = 0.09$ at the first iteration and decrease to 0.05 in the last.

Time window ω. On a dataset with a too small time window, detections/ tracklets with a big temporal distance cannot be connected, resulting in shorter tracklets. On the contrary, a too big time window can, depending on the time costs, result in false assignments. Figure 2(b) shows that our algorithm runs on a wide range of time windows producing good MT and Recall values, but as expected, this is at the expense of a decrease in Precision. We find a good compromise with $\omega = 5$, where we are still able to recover reliable tracks in the first iteration and only deal with longer tracks in successive iterations. ω is then increased by one at each iteration.

Number of iterations K. Figure 2(c) plots the MT, Precision and Recall values for the numbers of iterations of our algorithm. As the algorithm proceeds, more and more tracklets are being connected, resulting in an increasing MT value with only small improvements after 4 iterations. For all our experiments, we use $K = 5$ iterations. To remove false detections, we delete tracklets after the second iteration, if their length is smaller than 3 and after the fourth iteration, if their length is smaller than 4. This is why we see a small decrease in Precision at iterations 3 and 5. However, the slightly increasing Recall value shows that our simple tracklet removing approach is already sufficient to remove false detections without producing more missing detections.

Bifurcation handling. Finally, Fig. 2(d) shows the number of bifurcations in the Town center dataset after each iteration. We see how the algorithm improves the results by continuously solving ambiguities. The figure shows nearly an inverse proportional relation between the number of iterations and number of bifurcations, until it approximately converges after 5 iterations.

Fig. 3. Example of tracked pedestrians from the Bahnhof sequence.

4.2 Runtime

Our tracking system is implemented as a non-optimized Matlab code. Using the parameters given in Sect. 3, trajectories on a sequence of 999 frames with 6536 detections and 5 iterations are computed in 8 seconds on a 3.5 GHz machine (see Table 1). Note that we compute the trajectories for the complete sequence at once. Other papers (*e.g.* [17]) separate the sequence into batches to get the results in reasonable time. Our algorithm performs an iteration in linear time, *i.e.* each iteration has a worst-case complexity of $\mathcal{O}(r)$ in the current number of tracklets r, given the cost graph that models the probabilities. In addition to that, the size of our graph decreases after every iteration, since more and more detections are grouped together. Therefore, the total computational complexity is in $\mathcal{O}(Kd)$, where K is the number of iterations and d is the number of detections. In most cases however, the algorithm will have a better runtime complexity, since the number of tracklets in each iteration decreases in most cases exponentially.

Table 1. Runtime in seconds of the algorithm on the Bahnhof dataset (999 frames).

Method	Total runtime	Graph creation	Solver
Simplex [24]	16	9	7
Proposed	**8**	**6**	**2**

Most other globally optimizing tracking frameworks solve a min-cost flow problem [24] on a graph that is similar to the graph used in our model, having entrance/exit, detection and transition edges, respectively. Their algorithm has a worst-case complexity of $\mathcal{O}(n^2 m \log^2(n))$, where n and m denote the number of nodes and edges of the graph, respectively. See also Table 1. Improvements have been made by [20], who solve a k-shortest path problem and obtain a worst-case time complexity of $\mathcal{O}(kN \log N)$, where N denotes the number of frames and k denotes the optimal number of trajectories. Note that typically $K << N$.

4.3 Tracking Performance

Finally, we compare our algorithm with state-of-the-art tracking systems. Since we implemented our tracking system using 3D detections, we compare only to methods that are based on 3D detections, ensuring a fair comparison. Hence, we compare to the following state-of-the-art trackers:

- Zhang *et al.* [24] propose a tracking framework based on linear programming,
- Pellegrini *et al.* [19] take avoidance behavior of humans into account, and
- Leal-Taixé *et al.* [17] combine linear programming with a social force model.

Overall, the results of the proposed method are competitive with state-of-the-art tracking approaches, while being computationally more efficient. Note the results on the TownCenter dataset, where our method achieves 6 % more Recall than the best method and has 13 % more Mostly Tracked (MT) trajectories, while keeping a low number of identity switches. Note that our algorithm neither considers any special social modeling as in [17, 19] nor any type of appearance model. In Fig. 3 we show an example of the trajectories obtained using the proposed method on the Bahnhof sequence (Table 2).

Table 2. Tracking results on different datasets. MT = mostly tracked. PT = partially tracked. ML = mostly lost. Frg = fragmented tracks. Ids = identity switches. Values for [17, 19, 24] are taken from [16].

Dataset	Method	Rec.	Prec.	MT	PT	ML	*Frg*	*Ids*
Bahnhof	[24]	67.6	70.9	43.7	46.8	9.6	176	**39**
	[17]	73.3	75.4	51.1	41.5	7.4	**155**	107
	[19]	71.6	**84.9**	46.8	48.9	**4.3**	173	62
	Proposed	**75.3**	78.4	**56.4**	**38.3**	5.3	166	76
Sunnyday	[24]	76.8	70.6	**73.3**	**16.7**	10.0	39	**9**
	[17]	78.1	75.3	**73.3**	**16.7**	10.0	**29**	24
	[19]	75.5	**80.5**	66.6	23.3	10.1	33	15
	Proposed	**79.4**	75.9	**73.3**	20.0	**6.7**	38	12
Linthescher	[24]	56.7	59.7	18.8	**34.6**	46.7	163	**42**
	[17]	61.1	64.6	**23.1**	37.0	39.9	**149**	107
	[19]	59.7	**75.2**	20.2	40.4	39.4	168	44
	Proposed	**61.7**	65.4	20.1	42.8	**37.0**	223	107
Jelmoli	[24]	53.3	62.3	14.9	**46.8**	38.3	**30**	**4**
	[17]	**55.4**	70.6	14.9	51.1	**34.0**	36	25
	[19]	53.5	**76.7**	**17.0**	**46.8**	36.2	48	15
	Proposed	52.3	68.5	12.8	51.1	36.2	40	17
Town center	[24]	77.2	79.9	53.3	39.6	**7.1**	135	233
	[17]	77.9	**88.7**	53.3	37.8	8.9	56	68
	[19]	74.5	77.5	44.0	48.4	7.6	**53**	42
	Proposed	**83.8**	81.2	**66.7**	**23.1**	10.2	70	**41**

5   Conclusion

In this work we introduced a new type of tracklets, namely *tree tracklets*, which contain bifurcations to naturally deal with ambiguous tracking situations. These are then used in a hierarchical tracking framework, which we formulate as an optimization problem than can be solved in linear time. Experiments show good running time as well as performance compared to state-of-the art tracking systems on six publicly available datasets.

References

1. Benfold, B., Reid, I.: Stable multi-target tracking in real-time surveillance video. In: CVPR (2011)
2. Berclaz, J., Fleuret, F., Türetken, E., Fua, P.: Multiple object tracking using k-shortest paths optimization. TPAMI **33**(9), 1806–1819 (2011)
3. Dalal, N., Triggs, B.: Histograms of oriented gradients for human detection. In: CVPR (2005)
4. Dicle, C., Sznaier, M., Camps, O.: The way they move: tracking targets with similar appearance. In: ICCV (2013)
5. Edmonds, J.: Optimum branchings. J. Res. Natl. Bur. Stan. B. Math. Sci. **71B**(4), 233–240 (1967)
6. Ess, A., Leibe, B., Schindler, K., van Gool, L.: A mobile vision system for robust multi-person tracking. In: CVPR (2008)
7. Felzenszwalb, P., Girshick, R., McAllester, D., Ramanan, D.: Object detection with discriminatively trained part based models. TPAMI **32**(9), 1627–1645 (2010)
8. Gall, J., Yao, A., Razavi, N., van Gool, L., Lempitsky, V.: Hough forests for object detection, tracking and action recognition. TPAMI **33**(11), 2188–2202 (2011)
9. Gibbons, A.: Algorithmic Graph Theory. Cambridge University Press, Cambridge (1985)
10. Henriques, J.F., Caseiro, R., Batista, J.: Globally optimal solution to multi-object tracking with merged measurements. In: ICCV (2011)
11. Huang, C., Wu, B., Nevatia, R.: Robust object tracking by hierarchical association of detection responses. In: Forsyth, D., Torr, P., Zisserman, A. (eds.) ECCV 2008, Part II. LNCS, vol. 5303, pp. 788–801. Springer, Heidelberg (2008)
12. Jiang, H., Fels, S., Little, J.: A linear programming approach for multiple object tracking. In: CVPR (2007)
13. Kamiyama, N.: Arborescence problems in directed graphs: theorems and algorithms. Interdisc. Inf. Sci. **20**(1), 51–70 (2014). Graduate School of Information Sciences, Tohoku University
14. Kuhn, H.W.: The hungarian method for the assignment problem. Naval Res. Logist. Q. **2**(1–2), 83–97 (1955)
15. Kuo, C.H., Huang, C., Nevatia, R.: Multi-target tracking by on-line learned discriminative appearance models. In: CVPR (2010)
16. Leal-Taixé, L., Fenzi, M., Kuznetsova, A., Rosenhahn, B., Savarese, S.: Learning an image-based motion context for multiple people tracking. In: CVPR (2014)
17. Leal-Taixé, L., Pons-Moll, G., Rosenhahn, B.: Everybody needs somebody: modeling social and grouping behavior on a linear programming multiple people tracker. In: ICCV Workshops, 1st Workshop on Modeling, Simulation and Visual Analysis of Large Crowds (2011)

18. Li, Y., Huang, C., Nevatia, R.: Learning to associate: hybrid boosted multi-target tracker for crowded scene. In: CVPR (2009)
19. Pellegrini, S., Ess, A., Schindler, K., van Gool, L.: You'll never walk alone: modeling social behavior for multi-target tracking. In: ICCV (2009)
20. Pirsiavash, H., Ramanan, D., Fowlkes, C.: Globally-optimal greedy algorithms for tracking a variable number of objects. In: CVPR (2011)
21. Qin, Z., Shelton, C.R.: Improving multi-target tracking via social grouping. In: CVPR (2012)
22. Yang, B., Nevatia, R.: An online learned CRF model for multi-target tracking. In: CVPR (2012)
23. Yang, B., Nevatia, R.: Multi-target tracking by online learning of non-linear motion patterns and robust appearance models. In: CVPR (2012)
24. Zhang, L., Li, Y., Nevatia, R.: Global data association for multi-object tracking using network flows. In: CVPR (2008)

Capturing Hand Motion with an RGB-D Sensor, Fusing a Generative Model with Salient Points

Dimitrios Tzionas[1,2]($\boxtimes$), Abhilash Srikantha[1,2], Pablo Aponte[2],
and Juergen Gall[2]

[1] Perceiving Systems Department, MPI for Intelligent Systems, Stuttgart, Germany
{dimitris.tzionas,abhilash.srikantha}@tue.mpg.de
[2] Computer Vision Group, University of Bonn, Bonn, Germany
{aponte,gall}@iai.uni-bonn.de

Abstract. Hand motion capture has been an active research topic, following the success of full-body pose tracking. Despite similarities, hand tracking proves to be more challenging, characterized by a higher dimensionality, severe occlusions and self-similarity between fingers. For this reason, most approaches rely on strong assumptions, like hands in isolation or expensive multi-camera systems, that limit practical use. In this work, we propose a framework for hand tracking that can capture the motion of two interacting hands using only a single, inexpensive RGB-D camera. Our approach combines a generative model with collision detection and discriminatively learned salient points. We quantitatively evaluate our approach on 14 new sequences with challenging interactions.

1 Introduction

Human body tracking has been a popular field of research during the past decades [25], recently gaining more popularity due to the ubiquity of RGB-D sensors. Hand motion capture, a special instance of it, has enjoyed much research interest [11] due to its numerous applications including, but not limited to, computer graphics, human-computer-interaction and robotics.

Despite similarities, robust techniques [40] for full-body tracking are insufficient for hand motion capture, as the latter is more complicated on numerous fronts. Hands are characterized by more degrees of freedom, formulating a higher dimensional optimization problem. Severe occlusions are a usual phenomenon, being either self-occlusions or occlusions from another hand or object. Similarity in shape and appearance causes ambiguities for the differentiation between fingers and hands. Fast motion and lower resolution of hands in images constitute further complicating factors.

Despite theses challenges, there has been substantial progress in hand motion capture in recent years. Ballan et al. [3] have presented a system that successfully captures the motion of two hands strongly interacting with each other and an additional object. Although the approach achieves remarkable accuracy, it is based on an expensive and elaborate multi-camera system. On the other hand,

© Springer International Publishing Switzerland 2014
X. Jiang et al. (Eds.): GCPR 2014, LNCS 8753, pp. 277–289, 2014.
DOI: 10.1007/978-3-319-11752-2_22

Fig. 1. Qualitative results of our pipeline. Each pair shows the aligned RGB and depth input maps after depth thresholding, along with the pose estimate output

Oikonomidis et al. [27–29] have presented a real time hand tracker using just a single off-the-shelf RGB-D camera. Despite their success under challenging scenarios, the exhibited accuracy is not as precise as [3].

Our approach for tracking the pose of two strongly interacting hands is inspired by Ballan et al. [3] and combines a generative model with an occlusion handling method and a discriminatively trained detector for salient points. While [3] relies on an expensive capture setup with 8 synchronized and calibrated RGB cameras recording FullHD footage at 50 fps, we propose an approach that captures hand motion of two interacting hands using a cheap RGB-D camera recording VGA resolution at 30 fps. We evaluate our approach on 14 annotated sequences[1], which include interactions between hands. We further compare our method to the single hand tracker [27] on sequences with one hand, showing that our approach estimates the hand pose with higher accuracy than [27].

2 Related Work

The study of hand motion tracking has its roots in the 90 s [32,33]. Although the problem can be simplified by means of data-gloves [10], color-gloves [48], markers [47] or wearable sensors [21], the ideal solution pursued is the unintrusive, marker-less capture of hand motion. Even until recently the study was mainly confined to the case of a single isolated hand [2,17,24,27,41,42,45]. However, in pursuit of more realistic scenarios, research effort was directed towards the case of a hand interacting with an object [15,16,28], two hands interacting with each other [3,29] and with an additional object [3]. Multiple objects can be tracked by means of hand tracking and physical forces modeling [22].

An analytical review of the field can be found in the work of Erol et al. [11]. In this work a taxonomy is presented, separating the methods met in the literature in two main categories, namely *model-based* and *appearance-based*.

Generative-model approaches [17,31,42] are based on an explicit model used to generate pose hypotheses, which are evaluated against the observed data. The evaluation is based on an objective function which implicitly measures the likelihood by computing the discrepancy between the pose estimate (hypothesis) and the observed data in terms of an error metric. To keep the problem tractable, each iteration is initialized by the pose estimate of the previous step, relying thus heavily on temporal continuity and being prone to accumulative error. The objective function is evaluated in the high-dimensional, continuous parameter space.

---

[1] The annotated dataset sequences and the supplementary material are available at http://files.is.tue.mpg.de/dtzionas/GCPR_2014.html.

Discriminative methods learn a direct mapping from the observed image features to the discrete [2, 34, 35] or continuous [7, 20, 36, 40] target parameter space. Most methods operate on a single frame [2, 7, 36, 40], being thus immune to pose-drifting due to error accumulation. Generalization in terms of capturing illumination, articulation and view-point variation, can be realized only through adequate representative training data. Acquisition and annotation of realistic training data is though a cumbersome and costly procedure. For this reason most approaches rely on synthetic rendered data [20, 35, 40] that has inherent ground-truth. However, the discrepancy between realistic and synthetic data is an important limiting factor, while special care is needed to avoid over-fitting to the training set. The accuracy of discriminative methods heavily depends on the invariance, repeatability and discriminative properties of the features employed and is lower in comparison to generative methods.

A discriminative method can effectively complement a generative method, either in terms of initialization or recovery, driving the optimization framework away from local minima in the search space and aiding convergence to the global minimum. Sridhar et al. [41] combine in a real time system a Sums-of-Gaussians generative model with a discriminatively trained fingertip detector in depth images using a linear SVM classifier. Ballan et al. [3] present an accurate offline tracker that combines in a single framework a generative model with a salient-point (finger-nail) Hough-forest [14] detector in color images. Both approaches [3, 41], however, require an expensive multi-camera hardware setup.

3   Tracking Method

3.1   Hand Model

We resort to the *Linear Blend Skinning* (LBS) model [23], consisting of a triangular mesh, an underlying kinematic skeleton and a set of skinning weights. In our experiments, a triangular mesh of a pair of hands was obtained by a commercial 3D scanning solution and the skeleton structure was manually defined. Additional details are provided in the supplementary material (See Footnote 1). The skinning weight $\alpha_{\mathbf{v},j}$ defines the influence of bone j on 3D vertex $\mathbf{v}$, where $\sum_j \alpha_{\mathbf{v},j} = 1$. The deformation of the mesh is driven by the underlying skeleton with pose parameter vector θ through the skinning weights and is expressed by the LBS operator:

$$\mathbf{v}(\theta) = \sum_j \alpha_{\mathbf{v},j} T_j(\theta) T_j(0)^{-1} \mathbf{v}(0) \qquad (1)$$

where $T_j(0)$ and $\mathbf{v}(0)$ are the bone transformations and vertex positions at the known rigging pose. The skinning weights are computed using [4].

The global rigid motion is represented by a twist [6, 26, 31] in the special Euclidean group $SE(3)$. The articulation of the skeleton is expressed by a kinematic chain of rigid components. For the sake of simplicity, joints with more than 1 degree of freedom (DoF) are modeled by a combination of revolute joints. Using the exponential map operator, the transformation of a bone $T_j(\theta)$ with

k DoF is therefore given by $T_j(\theta) = \prod_{i<j} T_i(\theta) \prod_{k_j} \exp(\theta_{k_j} \hat{\xi}_{k_j})$, where θ is the parameterization of the full pose, $\theta_{k_j} \hat{\xi}_{k_j}$ is the twist representation of a single revolute joint, and $T_{i<j}$ denotes all previous bones in the kinematic chain.

In our experiments, a single hand consists of 31 revolute joints, i.e. 37 DoF. Thus, for sequences with two interacting hands we have to estimate all 74 DoF.

Anatomically inspired joint-angle limits [1] constrain the solution space to the subspace of physically plausible poses as in [3].

3.2 Optimization

Our objective function for pose estimation consists of four terms:

$$E(\theta, D) = \; E_{model \to data}(\theta, D) + E_{data \to model}(\theta, D) +$$
$$E_{salient}(\theta, D) + \gamma_c E_{collision}(\theta) \qquad (2)$$

where θ are the pose parameters of the two hands and D is the current pre-processed depth image. The first two terms minimize the alignment error of the transformed mesh and the depth data. The alignment error is measured by $E_{model \to data}$, which measures how well the model fits the observed depth data, and $E_{data \to model}$, which measures how well the depth data is explained by the model. The last two terms are inspired by [3]. $E_{salient}$ measures the consistency of the generative model with detected salient points in the image. The main purpose of the term in our framework is to recover from tracking errors of the generative model. In our scenario with a single camera of low resolution, the 3D positions of the detected points are less accurate and additional care is needed. $E_{collision}$ penalizes intersections of fingers, ensuring physically plausible poses.

The objective Eq. (2) is minimized by local optimization as described in [31]. In the following, we give details for the terms of the objective function.

3.2.1 Preprocessing

For pose estimation, we first remove irrelevant parts of the RGB-D image by thresholding the depth values and applying skin color segmentation [19] on the RGB image. As a result, we get a masked RGB-D image, which is denoted as D in Eq. (2). The thresholding of the depth image avoids unnecessary processing like normal computation for points far away and the skin color segmentation removes occluding objects from the data.

3.2.2 Fitting the *Model to the Data*

The first term in Eq. (2) aims at fitting the mesh parameterized by pose parameters θ to the preprocessed data D. To this end, the depth values are converted into a 3D point cloud based on the calibration data of the sensor. The point cloud is then smoothed by a bilateral filter [30] and normals are computed [18]. For each vertex of the model $\mathbf{v}_i(\theta)$, with normal $\mathbf{n}_i(\theta)$, we search for the closest point X_i in the point cloud. This gives a 3D-3D correspondence for each vertex.

We discard the correspondence if the angle between the normals of the vertex and the closest point is larger than $45°$ or the distance between the points is larger than $10\,\text{mm}$. We can then write the term $E_{model \rightarrow data}$ as a least squared error of *point-to-point* distances:

$$E_{model \rightarrow data}(\theta, D) = \sum_i \|\mathbf{v}_i(\theta) - X_i\|^2 \qquad (3)$$

An alternative to the *point-to-point* distance is the *point-to-plane* distance, which is commonly used for 3D reconstruction [9,38,39]:

$$E_{model \rightarrow data}(\theta, D) = \|\mathbf{n}_i(\theta)^T(\mathbf{v}_i(\theta) - X_i)\|^2 \qquad (4)$$

The two distance metrics are evaluated in our experiments (Sect. 4.1.3). In general, the *point-to-plane* distance converges faster and is therefore preferred.

3.2.3 Fitting the *Data to the Model*

Only fitting the model to the data is not sufficient as we will show in our experiments. In particular, poses with self-occlusions can have a very low error since the measure only evaluates how well the visible part of the model fits the point cloud. The second term $E_{data \rightarrow model}(\theta, D)$ matches the data to the model to make sure that the solution is not degenerate and explains the data as well as possible. Since matching the data to the model is expensive, we reduce the matching to depth discontinuities [13]. To this end, we extract depth discontinuities from the depth map and the projected depth profile of the model using an edge detector [8]. Correspondences are again established by searching for the closest points, but now in the depth image using a 2D distance transform [12]. Similar to $E_{model \rightarrow data}(\theta, D)$, we discard correspondences with a large distance. The depth values at the depth discontinuities in D, however, are less reliable not only due to the depth ambiguities between foreground and background, but also due to the noise of cheap sensors. The depth of the point in D is therefore computed as average in a local $3x3$ pixels neighborhood and the outlier distance threshold is increased to $30\,\text{mm}$. The approximation is sufficient for discarding outliers, but insufficient for minimization. For each matched point in D we therefore compute the projection ray uniquely expressed as a Plücker line [31,37,43] with direction $\mathbf{d}_i$ and moment $\mathbf{m}_i$ and minimize the least square error between the projection ray and the vertex $\mathbf{v}_i(\theta)$ for each correspondence:

$$E_{data \rightarrow model}(\theta, D) = \sum_i \|\mathbf{v}_i(\theta) \times \mathbf{d}_i - \mathbf{m}_i\|^2 \qquad (5)$$

3.2.4 Collision Detection

Collision detection is based on the observation that two objects cannot share the same space and is of high importance in case of self-penetration, inter-finger penetration or general intensive interaction.

For detecting collisions, we use *bounding volume hierarchies* (BVH) to efficiently determine collisions between meshes [44]. Having found a collision between two triangles f_s and f_t, the amount of penetration can be computed as in [3] using a 3D distance field in the form of a cone. Considering the case where the vertices of f_s are the *intruders* and the triangle f_t is the *receiver* of the penetration (similarly for the opposite case), the cone for computing the 3D distance field Ψ_{f_t} is defined by the circumcenter of the triangle f_t. Letting $\mathbf{n}_{f_t}$ denote the normal of the triangle, $\mathbf{o}_{f_t}$ the circumcenter, and r_{f_t} the radius of the circumcircle, we have

$$\Psi_{f_t}(\mathbf{v}_s) = \begin{cases} |(1 - \Phi(\mathbf{v}_s))\Upsilon(\mathbf{n}_{f_t} \cdot (\mathbf{v}_s - \mathbf{o}_{f_t}))|^2 & \text{when} \quad \Phi(\mathbf{v}_s) < 1 \\ 0 & \text{when} \quad \Phi(\mathbf{v}_s) \geq 1 \end{cases} \quad (6)$$

$$\Phi(\mathbf{v}_s) = \frac{\|(\mathbf{v}_s - \mathbf{o}_{f_t}) - (\mathbf{n}_{f_t} \cdot (\mathbf{v}_s - \mathbf{o}_{f_t}))\mathbf{n}_{f_t}\|}{-\frac{r_{f_t}}{\sigma}(\mathbf{n}_{f_t} \cdot (\mathbf{v}_s - \mathbf{o}_{f_t})) + r} \quad (7)$$

$$\Upsilon(x) = \begin{cases} -x + 1 - \sigma & \text{when} \quad x \leq -\sigma \\ -\frac{1-2\sigma}{4\sigma^2}x^2 - \frac{1}{2\sigma}x + \frac{1}{4}(3 - 2\sigma) & \text{when} \quad -\sigma \leq x \leq +\sigma \\ 0 & \text{when} \quad x \geq +\sigma \end{cases} \quad (8)$$

The parameter σ defines the field of view of the cone and is fixed to 0.5 as in [3].

For each vertex penetrating a triangle, a force can be computed that pushes the vertex back, where the direction is given by the inverse normal direction of the vertex and the strength of the force by Ψ. Using *point-to-point* distances (3), the forces are computed for the set of colliding triangles $\mathcal{C}$:

$$E_{collision}(\theta) = \sum_{(f_s(\theta),f_t(\theta)) \in \mathcal{C}} \left\{ \sum_{\mathbf{v}_s \in f_s} \| - \Psi_{f_t}(\mathbf{v}_s)\mathbf{n}_s\|^2 + \sum_{\mathbf{v}_t \in f_t} \| - \Psi_{f_s}(\mathbf{v}_t)\mathbf{n}_t\|^2 \right\} \quad (9)$$

Though not explicitly denoted, f_s and f_t depend on θ and therefore also Ψ, $\mathbf{v}$ and $\mathbf{n}$. For *point-to-plane* distances (4), the equation gets simplified since $\mathbf{n}^T\mathbf{n} = 1$:

$$E_{collision}(\theta) = \sum_{(f_s(\theta),f_t(\theta)) \in \mathcal{C}} \left\{ \sum_{\mathbf{v}_s \in f_s} \| - \Psi_{f_t}(\mathbf{v}_s)\|^2 + \sum_{\mathbf{v}_t \in f_t} \| - \Psi_{f_s}(\mathbf{v}_t)\|^2 \right\} \quad (10)$$

This term takes part in the objective function (2) regulated by weight γ_c. An evaluation of different γ_c values is presented in our experiments (Sect. 4.1.1).

3.2.5   Salient Point Detection

Our approach is so far based on a generative model, which generally provides accurate solutions, but recovers only slowly from ambiguities and tracking errors. It has been shown in [3] that this can be compensated by integrating a discriminatively trained salient point detector into a generative model. In [3], a finger

Table 1. The graph contains T mesh fingertips ξ_t and S fingertip detections δ_s. The cost of assigning a detection δ_s to a finger ξ_t is given by w_{st} as shown in table (a). The cost of declaring a detection as false positive is λw_s where w_s is the detection confidence. The cost of not assigning any detection to finger ξ_t is given by λ. The binary solution of table (b) is constrained to sum up to 1 for each row and column

(a)	Fingertips ξ_t				V
	ξ_1	ξ_2	$\ldots$	ξ_T	α
Detections δ_s δ_1	w_{11}	w_{12}	$\ldots$	w_{1T}	λw_1
δ_2	w_{21}	w_{22}	$\ldots$	w_{2T}	λw_2
δ_3	w_{31}	w_{32}	$\ldots$	w_{3T}	λw_3
$\vdots$	$\vdots$	$\vdots$	$\ddots$	$\vdots$	$\vdots$
δ_S	w_{S1}	w_{S2}	$\ldots$	w_{ST}	λw_S
V β	λ	λ	$\ldots$	λ	∞

(b)	Fingertips ξ_t				V
	ξ_1	ξ_2	$\ldots$	ξ_T	α
Detections δ_s δ_1	e_{11}	e_{12}	$\ldots$	e_{1T}	α_1
δ_2	e_{21}	e_{22}	$\ldots$	e_{2T}	α_2
δ_3	e_{31}	e_{32}	$\ldots$	e_{3T}	α_3
$\vdots$	$\vdots$	$\vdots$	$\ddots$	$\vdots$	$\vdots$
δ_S	e_{S1}	e_{S2}	$\ldots$	e_{ST}	α_S
V β	β_1	β_2	$\ldots$	β_T	0

nail detector was applied to the high resolution images and due to the multi-camera setup it could be assumed that the nails become visible in some of the cameras. For low-resolution video of a single camera, finger nails cannot be reliably detected. Instead we train a fingertip detector [14] on raw depth data where the training data is not part of the test sequences. More details are given in the supplementary material.

Since we resort to salient points only for additional robustness, it is usually sufficient to have only sparse fingertip detections. We therefore collect detections with a high confidence, choosing a threshold of $c_{thr} = 3.0$ for our experiments. The association between detections and fingertips of the model, as shown in Table 1, is solved by integer programming [3,5]:

$$\text{argmin} \quad \sum_{s,t} e_{st} w_{st} + \lambda \sum_s \alpha_s w_s + \lambda \sum_t \beta_t$$

$$\text{subject to} \quad \sum_s e_{st} + \beta_t = 1 \quad \forall t, \tag{11}$$

$$\sum_t e_{st} + \alpha_s = 1 \quad \forall s$$

$$e_{st}, \alpha_t, \beta_s \in \{0, 1\}$$

The weights w_{st} are given by the 3D distance between the detection δ_s and the finger of the model ξ_t. For each finger ξ_t, a set of vertices are marked in the model. The distance is then computed as the centroid of the visible vertices of ξ_t and the centroid of the detected region δ_s. For the weights w_s, we investigate two approaches. The first approach uses $w_s = 1$ as in [3]. The second approach takes the confidences c_s of the detections into account by setting $w_s = \frac{c_s}{c_{thr}}$. The weighting parameter λ is evaluated in the experimental section (Sect. 4.1.2).

If a detection δ_s has been associated to a finger ξ_t, we have to define correspondences between the set of visible vertices of ξ_t and the point cloud of the detection δ_s. If the finger is already very close and the distance is below 10 mm, we do not compute any correspondences since the localization accuracy of the detector is not higher. In this case the finger is anyway close enough to the data to achieve a good alignment. Otherwise, we compute the closest points between the vertices $\mathbf{v}_i$ and the points X_i of the detection:

$$E_{salient}(\theta, D) = \sum_{s,t} e_{st} \left\{ \sum_{(X_i, \mathbf{v}_i) \in \delta_s \times \xi_t} \|\mathbf{v}_i(\theta) - X_i\|^2 \right\} \tag{12}$$

As in (4), a *point-to-plane* distance metric can replace the *point-to-point* metric. When the overlap of the fingertip and the detection is less than 50 %, replacing the closest points X_i by the centroid of the detection leads to a speed up.

4   Experimental Evaluation

Benchmarking in the context of 3D hand tracking remains an open problem [11] despite recent contributions [41,46]. Related RGB-D methods [27] usually report quantitative results only on synthetic sequences, which inherently include ground-truth, while for realistic conditions they resort to qualitative results.

Although qualitative results are informative, quantitative evaluation based on ground-truth is of high importance. We therefore manually annotate 14 new sequences, 11 of which are used to evaluate the components of our pipeline and 3 for comparison with the state-of-the-art method [27] (details in the supplementary material). The standard deviation for 4 annotators is 1.46 pixels.

The error metric for our experiments is the 2D distance (pixels) between the projection of the 3D joints and the corresponding 2D annotations. Details regarding the joints taken into account in the metric are included in the supplementary material. We report the average error over all frames of all sequences.

4.1   Pipeline Components

Our system is based on an objective function consisting of four terms, described in Sect. 3.2. Two of them minimize the error between the posed mesh and the depth data by fitting the *model to the data* and the *data to the model*. A *salient point* detector further constraints the pose using fingertip detections in the depth image, while a *collision detection* method contributes to realistic pose estimates that are physically plausible. The above terms participate in the objective function in a weighted scheme, which is minimized as in [31].

In the following, we evaluate the parameters used in our components and assess both each component's individual contribution to the overall system performance, as well as of the combination thereof in the objective function (2).

4.1.1   Collision Detection

The collision detection component is regulated in the objective function (2) by the weight γ_c, so that collision and penetration get efficiently penalized. Table 2

Table 2. Evaluation of collision weights γ_c, using a 2D distance error metric (px). Weight 0 corresponds to deactivated collision detection, noted as "$LO + S$" in Table 5. Sequences are grouped (see supplementary material) in 3 categories: "*Severe*" for intense, "*some*" for light and "*no apparent*" for imperceptible collision. Our highlighted decision is based on the union of the "*severe*" and "*some*" sets, noted as "*at least some*"

γ_c	0	1	2	3	5	7.5	10	12.5
All	5.40	5.50	5.63	5.23	5.18	5.19	5.18	5.21
Only Severe	6.00	6.18	6.37	5.73	5.66	5.67	5.65	5.71
Only Some	3.99	3.98	3.98	3.98	3.98	3.99	3.99	3.98
At least some	5.52	5.65	5.80	5.31	5.26	5.27	5.25	5.29

Table 3. Evaluation of the parameter λ of the assignment graph of Eq. (11), using a 2D distance error metric (px). Value $\lambda = 0$ corresponds to deactivation of the detector, noted as "$LO+C$" in Table 5. Both versions of w_s described in Sect. 3.2.5 are evaluated

λ	0	0.3	0.6	0.9	1.2	1.5	1.8
$w_s = 1$	5.24	5.23	5.21	5.21	5.18	5.19	5.30
$w_s = \frac{c_s}{c_{thr}}$		5.21	5.19	5.18	5.18	5.28	5.68

summarizes our evaluation experiments for the values of γ_c. Although we choose $\gamma_c = 10$ for the present dataset, a generally proposed range of values for new sequences would be between 5 and 10.

4.1.2 Salient Point Detection

The salient point detection component depends on the parameters w_s and λ, as described in Sect. 3.2.5. Table 3 summarizes our evaluation of the parameter λ spanning a range of possible values for both cases $w_s = 1$ and $w_s = \frac{c_s}{c_{thr}}$. Although the difference between the two versions of w_s is not very large, the latter performs better for a wide range (0.6 to 1.2) of parameter λ. We therefore choose $w_s = \frac{c_s}{c_{thr}}$ and $\lambda = 1.2$.

4.1.3 Distance Metrics

Table 4 presents an evaluation of the two distance metrics presented in Sect. 3.2.2, namely *point-to-point* (Eq. (3)) and *point-to-plane* (Eq. (4)), along with the number of iterations of the minimization framework. The *point-to-plane* metric leads to adequate minimization with only 10 iterations, providing a significant speed gain, being thus our choice. However, we perform 50 iterations for the first frame in order to ensure an accurate refinement of the manually initialized pose. The runtime for the chosen setup (see supplementary material for benchmark details) is 2.74 and 4.35 s per frame for scenes containing one and two hands respectively.

Table 4. Evaluation of *point-to-point* (*p2p*) and *point-to-plane* (*p2plane*) distance metrics, along with iterations number of the optimization framework, using a 2D distance error metric (px)

Iterations	5	10	15	20	30
p2p	7.39	5.31	5.11	5.04	4.97
p2plane	5.39	5.18	5.14	5.13	5.11

Table 5. Evaluation of the components of our pipeline. "*LO*" stands for local optimization and includes fitting both *data-to-model* (*d2m*) and *model-to-data* (*m2d*), unless otherwise specified. Collision detection is noted as "*C*", while salient point detector is noted as "*S*". The number of sequences where the optimization framework collapses is noted in the last row, while the mean error is reported only for the rest

Components	LO_{m2d}	LO_{d2m}	LO	LO + C	LO + S	LO + CS
Mean Error (px)	15.19	–	5.59	5.24	5.40	5.18
Improvement (%)		–		6.39	3.40	7.36
Failed Sequences	1/11	11/11	0/11	0/11	0/11	0/11

4.1.4   Component Evaluation

Table 5 presents the evaluation of each component and the combination thereof. Simplified versions of the pipeline, fitting either just the *model to the data* or the *data to the model* can lead to a collapse of the pose estimation, due to unconstrained optimization. Our experiments quantitatively show the notable contribution of both the collision detection and the salient point detector components. The best overall system performance is achieved with the combinatorial setup described by the objective function (2). Qualitative results are depicted in Fig. 1.

4.2   Comparison to State-of-the-Art

Recently, Oikonomidis et al. used particle swarm optimization (PSO) for a real-time hand tracker [27–29]. These works constitute the state-of-the-art for single-view RGB-D hand tracking. For comparison we use the software released for tracking one hand [27], with the parameter setups of all the above works. Each setup is evaluated 3 times in order to compensate for the manual initialization and the inherent randomness of PSO. Qualitative results depict the best version, while quantitative results report the average. Figure 2 qualitatively showcases the increased accuracy of our method, along with the decreased accuracy of the FORTH tracker due to the sampling nature of PSO. Quantitative results of Table 6 show that our system outperforms [27] in terms of tracking accuracy. However, it should be noted that the GPU implementation of [27] is real time, in contrast to our CPU implementation. Detailed results for each evaluation of the parameter setups are included in the supplementary material.

Table 6. Comparison of our method against the FORTH tracker. We evaluate the FORTH tracker with 4 parameter-setups met in the referenced literature of the last column

		Mean (px)	St.Dev (px)	Max (px)	Generations	Particles	Reference
FORTH	set 1	8.58	5.74	61.81	25	64	[27]
	set 2	8.32	5.42	57.97	40	64	[28]
	set 3	8.09	5.00	38.90	40	128	[3]
	set 4	8.16	5.18	39.85	45	64	[29]
Proposed		3.76	2.22	19.92			

Fig. 2. Qualitative comparison with [27]. Each image pair corresponds to the pose estimate of the FORTH tracker (left) and our tracker (right)

5   Conclusion

In this work we have presented a system capturing the motion of two highly interacting hands. Inspired by the recent method [3], we propose a combination of a generative model with a discriminatively trained salient point detector and collision modeling to obtain accurate, realistic pose estimates with increased immunity to ambiguities and tracking errors. While [3] depends on expensive, specialized equipment, we achieve accurate tracking results using only a single cheap, off-the-shelf RGB-D camera. We have evaluated our approach on 14 new challenging sequences and shown that our approach achieves a better accuracy than the state-of-the-art single hand tracker [27].

Acknowledgments. The authors acknowledge the help of Javier Romero and Jessica Purmort of MPI-IS regarding the acquisition of the personalized hand model, the assistance of Philipp Rybalov with annotation and the public software release of the FORTH tracker by the CVRL lab of FORTH-ICS, enabling comparison to [27]. Financial support was provided by the DFG Emmy Noether program (GA 1927/1-1).

References

1. Albrecht, I., Haber, J., Seidel, H.P.: Construction and animation of anatomically based human hand models. In: SCA, pp. 98–109 (2003)
2. Athitsos, V., Sclaroff, S.: Estimating 3d hand pose from a cluttered image. In: CVPR, pp. 432–439 (2003)

3. Ballan, L., Taneja, A., Gall, J., Van Gool, L., Pollefeys, M.: Motion capture of hands in action using discriminative salient points. In: Fitzgibbon, A., Lazebnik, S., Perona, P., Sato, Y., Schmid, C. (eds.) ECCV 2012, Part VI. LNCS, vol. 7577, pp. 640–653. Springer, Heidelberg (2012)
4. Baran, I., Popović, J.: Automatic rigging and animation of 3d characters. TOG 26(3), 72 (2007)
5. Belongie, S., Malik, J., Puzicha, J.: Shape matching and object recognition using shape contexts. PAMI 24(4), 509–522 (2002)
6. Bregler, C., Malik, J., Pullen, K.: Twist based acquisition and tracking of animal and human kinematics. IJCV 56(3), 179–194 (2004)
7. de Campos, T., Murray, D.: Regression-based hand pose estimation from multiple cameras. In: CVPR, pp. 782–789 (2006)
8. Canny, J.: A computational approach to edge detection. PAMI 8(6), 679–698 (1986)
9. Chen, Y., Medioni, G.: Object modeling by registration of multiple range images. In: ICRA, pp. 2724–2729 (1991)
10. Ekvall, S., Kragic, D.: Grasp recognition for programming by demonstration. In: ICRA, pp. 748–753 (2005)
11. Erol, A., Bebis, G., Nicolescu, M., Boyle, R.D., Twombly, X.: Vision-based hand pose estimation: a review. CVIU 108(1–2), 52–73 (2007)
12. Felzenszwalb, P.F., Huttenlocher, D.P.: Distance transforms of sampled functions. Technical report. Cornell Computing and Information Science (2004)
13. Gall, J., Fossati, A., Van Gool, L.: Functional categorization of objects using real-time markerless motion capture. In: CVPR, pp. 1969–1976 (2011)
14. Gall, J., Yao, A., Razavi, N., Van Gool, L., Lempitsky, V.: Hough forests for object detection, tracking, and action recognition. PAMI 33(11), 2188–2202 (2011)
15. Hamer, H., Schindler, K., Koller-Meier, E., Van Gool, L.: Tracking a hand manipulating an object. In: ICCV, pp. 1475–1482 (2009)
16. Hamer, H., Gall, J., Weise, T., Van Gool, L.: An object-dependent hand pose prior from sparse training data. In: CVPR, pp. 671–678 (2010)
17. Heap, T., Hogg, D.: Towards 3d hand tracking using a deformable model. In: FG, pp. 140–145 (1996)
18. Holzer, S., Rusu, R., Dixon, M., Gedikli, S., Navab, N.: Adaptive neighborhood selection for real-time surface normal estimation from organized point cloud data using integral images. In: IROS, pp. 2684–2689 (2012)
19. Jones, M.J., Rehg, J.M.: Statistical color models with application to skin detection. IJCV 46(1), 81–96 (2002)
20. Keskin, C., Kıraç, F., Kara, Y.E., Akarun, L.: Hand pose estimation and hand shape classification using multi-layered randomized decision forests. In: Fitzgibbon, A., Lazebnik, S., Perona, P., Sato, Y., Schmid, C. (eds.) ECCV 2012, Part VI. LNCS, vol. 7577, pp. 852–863. Springer, Heidelberg (2012)
21. Kim, D., Hilliges, O., Izadi, S., Butler, A.D., Chen, J., Oikonomidis, I., Olivier, P.: Digits: freehand 3d interactions anywhere using a wrist-worn gloveless sensor. In: UIST, pp. 167–176 (2012)
22. Kyriazis, N., Argyros, A.: Physically plausible 3d scene tracking: the single actor hypothesis. In: CVPR, pp. 9–16 (2013)
23. Lewis, J.P., Cordner, M., Fong, N.: Pose space deformation: a unified approach to shape interpolation and skeleton-driven deformation. In: SIGGRAPH, pp. 165–172 (2000)
24. MacCormick, J., Isard, M.: Partitioned sampling, articulated objects, and interface-quality hand tracking. In: Vernon, D. (ed.) ECCV 2000. LNCS, vol. 1843, pp. 3–19. Springer, Heidelberg (2000)

25. Moeslund, T.B., Hilton, A., Krüger, V.: A survey of advances in vision-based human motion capture and analysis. CVIU **104**(2), 90–126 (2006)
26. Murray, R.M., Sastry, S.S., Zexiang, L.: A Mathematical Introduction to Robotic Manipulation (1994)
27. Oikonomidis, I., Kyriazis, N., Argyros, A.: Efficient model-based 3d tracking of hand articulations using kinect. In: BMVC, pp. 101.1–101.11 (2011)
28. Oikonomidis, I., Kyriazis, N., Argyros, A.: Full dof tracking of a hand interacting with an object by modeling occlusions and physical constraints. In: ICCV, pp. 2088–2095 (2011)
29. Oikonomidis, I., Kyriazis, N., Argyros, A.A.: Tracking the articulated motion of two strongly interacting hands. In: CVPR, pp. 1862–1869 (2012)
30. Paris, S., Durand, F.: A fast approximation of the bilateral filter using a signal processing approach. IJCV **81**(1), 24–52 (2009)
31. Pons-Moll, G., Rosenhahn, B.: Model-Based Pose Estimation, pp. 139–170 (2011)
32. Rehg, J.M., Kanade, T.: Visual tracking of high dof articulated structures: an application to human hand tracking. In: Eklundh, J.-O. (ed.) ECCV 1994. LNCS, vol. 801, pp. 35–46. Springer, Heidelberg (1994)
33. Rehg, J., Kanade, T.: Model-based tracking of self-occluding articulated objects. In: ICCV, pp. 612–617 (1995)
34. Romero, J., Kjellström, H., Kragic, D.: Monocular real-time 3d articulated hand pose estimation. In: HUMANOIDS, pp. 87–92 (2009)
35. Romero, J., Kjellström, H., Kragic, D.: Hands in action: real-time 3d reconstruction of hands in interaction with objects. In: ICRA, pp. 458–463 (2010)
36. Rosales, R., Athitsos, V., Sigal, L., Sclaroff, S.: 3d hand pose reconstruction using specialized mappings. In: ICCV, pp. 378–387 (2001)
37. Rosenhahn, B., Brox, T., Weickert, J.: Three-dimensional shape knowledge for joint image segmentation and pose tracking. IJCV **73**(3), 243–262 (2007)
38. Rusinkiewicz, S., Levoy, M.: Efficient variants of the icp algorithm. In: 3DIM, pp. 145–152 (2001)
39. Rusinkiewicz, S., Hall-Holt, O., Levoy, M.: Real-time 3d model acquisition. TOG **21**(3), 438–446 (2002)
40. Shotton, J., Fitzgibbon, A., Cook, M., Sharp, T., Finocchio, M., Moore, R., Kipman, A., Blake, A.: Real-time human pose recognition in parts from single depth images. In: CVPR, pp. 1297–1304 (2011)
41. Sridhar, S., Oulasvirta, A., Theobalt, C.: Interactive markerless articulated hand motion tracking using rgb and depth data. In: ICCV, pp. 2456–2463 (2013)
42. Stenger, B., Mendonca, P., Cipolla, R.: Model-based 3D tracking of an articulated hand. In: CVPR, pp. 310–315 (2001)
43. Stolfi, J.: Oriented Proj. Geometry: A Framework for Geom. Computation (1991)
44. Teschner, M., Kimmerle, S., Heidelberger, B., Zachmann, G., Raghupathi, L., Fuhrmann, A., Cani, M.P., Faure, F., Magnetat-Thalmann, N., Strasser, W.: Collision detection for deformable objects. In: Eurographics, pp. 119–139 (2004)
45. Thayananthan, A., Stenger, B., Torr, P.H.S., Cipolla, R.: Shape context and chamfer matching in cluttered scenes. In: CVPR, pp. 127–133 (2003)
46. Tzionas, D., Gall, J.: A comparison of directional distances for hand pose estimation. In: Weickert, J., Hein, M., Schiele, B. (eds.) GCPR 2013. LNCS, vol. 8142, pp. 131–141. Springer, Heidelberg (2013)
47. Vaezi, M., Nekouie, M.A.: 3d human hand posture reconstruction using a single 2d image. IJHCI **1**(4), 83–94 (2011)
48. Wang, R.Y., Popović, J.: Real-time hand-tracking with a color glove. TOG **28**(3), 68:1–68:8 (2009)

Interpolation and Inpainting

Flow and Color Inpainting for Video Completion

Michael Strobel$^{(\boxtimes)}$, Julia Diebold, and Daniel Cremers

Technical University of Munich, Munich, Germany
{m.strobel,julia.diebold,cremers}@tum.de

Abstract. We propose a framework for temporally consistent video completion. To this end we generalize the exemplar-based inpainting method of Criminisi et al. [7] to video inpainting. Specifically we address two important issues: Firstly, we propose a color and optical flow inpainting to ensure temporal consistency of inpainting even for complex motion of foreground and background. Secondly, rather than requiring the user to hand-label the inpainting region in every single image, we propose a flow-based propagation of user scribbles from the first to subsequent video frames which drastically reduces the user input. Experimental comparisons to state-of-the-art video completion methods demonstrate the benefits of the proposed approach.

Keywords: Video completion · Video inpainting · Disocclusion · Temporal consistency · Segmentation · Optical flow

1 Introduction

Videos of natural scenes often include disturbing artifacts like undesired walking people or occluding objects. In the past ten years, the technique of replacing disruptive parts with visually pleasing content grew to an active research area in the field of image processing. The technique is known as video inpainting and has its origin in image inpainting. While image inpainting has been researched very active in the past years the problem of video inpainting has received much less attention. Due to the additional temporal dimension in videos, new technical challenges arise and make calculations much more complex and time consuming. At the same time, video completion has a much larger range of applications, including professional post-productions or restoration of damaged film.

In this work, we focus on two central challenges in video completion, namely temporal consistency and efficient mask-definition.

1.1 Related Work

The literature on image inpaiting can be roughly grouped into two complementary approaches, namely inpainting via partial differential equations (PDEs) and exemplar-based inpainting. PDE-based inpainting was first proposed by Masnou and Morel [12,13] and popularized under the name of inpainting by

© Springer International Publishing Switzerland 2014
X. Jiang et al. (Eds.): GCPR 2014, LNCS 8753, pp. 293–304, 2014.
DOI: 10.1007/978-3-319-11752-2_23

Bertalmio *et al.* [3,4]. The key idea is to fill the inpainting region by propagating isolines of constant color from the surrounding region. These techniques provide pleasing results for filling small regions, for example to remove undesired text or scratches from images. For larger regions, however, the propagation of similar colors creates undesired smoothing effects. To account for this shortcoming, texture synthesis techniques were promoted, most importantly exemplar-based techniques [1,8,9] which can fill substantially larger inpainting regions by copy-pasting colors from the surrounding areas based on patch-based similarity. Criminisi *et al.* [6,7] presented an approach which combines the two methods to one efficient image inpainting algorithm. The algorithm works at the image patch level and fills unknown regions effectively by extending texture synthesis with an isophote guided ordering. This automatic priority-based ordering significantly improves the quality of the completion algorithm by preserving crucial image structures.

Patwardhan *et al.* [17,18] and Werlberger [24] extended and adapted Criminisi *et al.*'s [7] method for video inpainting. The approach of Patwardhan *et al.* is using a 5D patch search and takes motion into account. Their approach leads to satisfying results as long camera movement matches some special cases. We are not restricted to specific camera motion.

The idea of using graph cuts for video inpainting was recently introduced by Granados *et al.* [11]. They propose a semi-automatic algorithm which optimizes the spatio-temporal shift map. This algorithm presents impressive results however, the approach only has very limited practicability as the runtime takes between 11 and 90 h for 200 frames.

Newson *et al.* [14,15] provided an important speed-up by extending the PatchMatch algorithm [2] to the spatio-temporal domain thereby drastically accelerating the search for approximate nearest neighbors. Nevertheless, the runtime for high-resolution videos is about 6 h for 82 frames.

1.2 Contributions

We propose a method for video completion which resolves several important challenges:

- We propose a method to interactively determine the inpainting region over multiple frames. Rather than hand-labeling the inpainting region in every single frame, we perform a flow-based propagation of user-scribbles (from the first frame to subsequent frames), followed by an automatic foreground-background segmentation.
- We introduce temporal consistency not by sampling spatio-temporal patches, but rather by a combination of color- and flow-based inpainting. The key idea is to perform an inpainting of the optical flow for the inpainting region and subsequently perform an exemplar-based image inpainting with a constraint on temporal consistency along the inpainted optical flow trajectories - see Fig. 1. As a consequence, the proposed video completion method can handle arbitrary foreground and background motion in a single approach and with substantially reduced computation time.

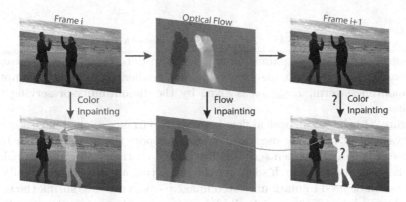

Fig. 1. Sketched approach. We propose an efficient algorithm for semi-automatic video inpainting. In particular, we impose temporal consistency of the inpainting not by a tedious sampling of space-time patches but rather by a strategy of flow- and color inpainting. We inpaint the optical flow and subsequently modify the distance function in an exemplar-based image inpainting such that consistency with corresponding patches in previous frames is imposed.

• The inpainting is computed without any pre- or post-processing steps. An efficient GPU-based implementation provides pleasing video completion results with minimal user input at drastically improved runtimes compared to state-of-the-art methods.

2   Interactive Mask-Definition

In [3,11,25,26] manual labeling of the inpainting region in all frames of the videos is needed. This makes video editing an extremely tedious and somewhat unpleasant process. We present a simple tool for *interactive mask-definition* with minimal user input. The requirements for such a tool include: (i) an intuitive user interface (ii) a robust mask definition and (iii) a real-time capable algorithm.

The method of Nieuwenhuis and Cremers [16] provides a user-guided image segmentation algorithm that generates accurate results even on images with difficult color and lighting conditions. The user input is given by user scribbles drawn on the input image. The algorithm analyzes the spatial variation of the color distributions given by the scribbles. Thanks to their parallel implementation, computation times of around one second per frame can be obtained.

Based on this user input, we (i) automatically relocate the scribbles throughout the video sequence via optical flow and (ii) frame-wise apply the image segmentation method according to Nieuwenhuis and Cremers [16].

2.1   Scribble Relocation via Optical Flow

To transport scribbles over time we use the optical flow method of Brox *et al.* [5] which computes the displacement vector field (u, v) by minimizing an energy functional of the form:

$$E(u, v) = E_{Data} + \alpha \, E_{Smooth} \tag{1}$$

with some regularization parameter $\alpha > 0$. The data term, E_{Data}, measures the global deviations from the grey value and gradient constancy assumption. The smoothness term, E_{Smooth}, is given by the discontinuity-preserving total variation.

Figure 2(b) shows the optical flow between two frames of the image sequence by Newson *et al.* [15]. We use this flow to transport the scribbles from frame to frame (Fig. 2(a,c)). Green scribbles are placed on the region to be inpainted and yellow ones on the search space for the inpainting algorithm. Optionally, red scribbles can be used to mark unrelated image parts in order to shrink the search space. Depending on the user scribbles, a two- or three-region segmentation according to Nieuwenhuis and Cremers [16] is computed.

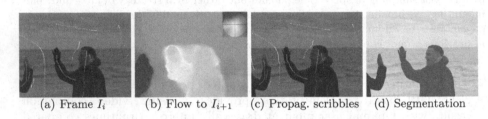

(a) Frame I_i (b) Flow to I_{i+1} (c) Propag. scribbles   (d) Segmentation

Fig. 2. Automatic segmentation by scribble propagation via optical flow. Scribbles are placed on the first frame and propagated to the next frames by optical flow. Segmentation is computed based on the transported scribbles (Color figure online).

2.2 Segmentation According to Nieuwenhuis and Cremers

Let $I : \mathcal{I} \to \mathbb{R}^d$ denote the input frame defined on the domain $\mathcal{I} \subset \mathbb{R}^2$. The task of segmenting the image plane into a set of n pairwise disjoint regions $\mathcal{I}_i$:
$\mathcal{I} = \dot{\bigcup}_{i=1}^{n} \mathcal{I}_i, \, \mathcal{I}_i \cap \mathcal{I}_j = \emptyset$
xmlcommand $\forall \, i \neq j$ can be solved by computing a labeling $u : \mathcal{I} \to \{1, \dots, n\}$, indicating which of the n regions each pixel belongs to: $\mathcal{I}_i = \{x \mid u(x) = i\}$. The segmentation time for a video sequence can be speed-up by initializing the indicator function u with the resulting segmentation of the previous frame.

We compute a segmentation of each video frame by minimizing the following energy [16]:

$$E(\mathcal{I}_1, \dots, \mathcal{I}_n) = \frac{\lambda}{2} \sum_{i=1}^{n} \mathrm{Per}_g(\mathcal{I}_i) + \lambda \sum_{i=1}^{n} \int_{\mathcal{I}_i} f_i(x) \, dx,$$

where $f_i(x) = -\log \hat{\mathcal{P}}(I(x), x \mid u(x) = i)$. $\mathrm{Per}_g(\mathcal{I}_i)$ denotes the perimeter of each set $\mathcal{I}_i$, λ is a weighting parameter. The expression $\hat{\mathcal{P}}(I(x), x \mid u(x) = i)$ denotes the joint probability for observing a color value I at location x given that x is part of region $\mathcal{I}_i$. It can be estimated from the user scribbles. For further details of the segmentation algorithm we refer to [16].

To summarize, our inpainting method brings along a tool which allows the user to quickly define the respective regions on the first video frame, and all the remaining calculations are working automatically. In contrast, state-of-the-art methods require the user to manually draw an exact mask on each single video frame [3,11,25,26] or work with an inflexible bounding box [20].

3   Flow and Color Inpainting for Video Completion

The major challenge in video inpainting is the temporal dimension: The inpainted regions have to be consistent with the color and structure around the hole, and additionally temporal continuity has to be preserved. When applying image inpainting methods frame by frame, the inpainted videos show artifacts, like ghost shadows or flickering [20]. Several investigations have been done in the past years towards a temporally coherent video completion. State-of-the-art methods, however, have some drawbacks: several pre- and post-processing steps are required [14,20], only specific camera motions can be handled [11,14,18,26] and the calculations are extremely time consuming [10,11,14,26].

We propose a novel approach inspired by the exemplar-based image inpainting by Criminisi *et al.* [7] overcoming these problems. We apply inpainting to the optical flow and define a *refined distance function* ensuring temporal consistency in video inpainting. No additional pre- or post-processing steps are required.

3.1   Inpainted Flow for Temporal Coherence

In a temporally consistent video sequence, the inpainted region follows the flow of its surrounding region. Figure 3(a) shows a person who should be removed from the video sequence. The desired patches clearly should not follow the hand of the person, but the flow of the sea. To find the best matching patches, Criminisi *et al.* [7] consider the colors around the hole. We additionally claim a similarity to the patch which naturally flows into this position. This flow is obtained by *inpainting the original flow* - see Fig. 3(d).

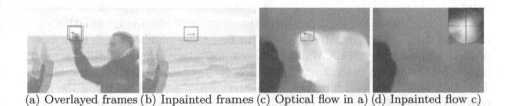

(a) Overlayed frames (b) Inpainted frames (c) Optical flow in a) (d) Inpainted flow c)

Fig. 3. Inpainted flow ensures temporal consistency. In order to ensure temporal consistency, we propose to inpaint the optical flow and additionally request the found patch to be similar to its origin. The inpainted flow (d) should be approximately the flow of the inpainted video sequence.

3.2   Flow Inpainting

For the inpainting of the optical flow we extended the Telea-Inpainting [21] to optical flow. Telea-Inpainting is a fast PDE based approach and hence particularly suited to fill missing parts in optical flow images. Let Ω denote the hole in the optical flow F which has to be replaced, $\delta\Omega$ the contour of the hole and Ω^c the search region (complement of Ω). Telea-Inpainting approximates the value of a pixel p on the boarder of the fill-front $\delta\Omega$ by a first order Taylor expansion combined with a normalized weighting function $w(p, q)$ for $q \in B_\epsilon(p)$ and $\epsilon > 0$:

$$\hat{F}(p) = \frac{\sum_{q \in B_\epsilon(p) \cap \Omega^c} w(p, q)[F(p) - \nabla F(q)(p - q)]}{\sum_{q \in B_\epsilon(p) \cap \Omega^c} w(p, q)}.$$

The pixel values are propagated into the fill region along the isophotes by solving the eikonal equation: $|\nabla T| = 1$ on Ω, $T = 0$ on $\delta\Omega$ using the Tsitsiklis algorithm [19,22]. The solution T of the eikonal equation describes the distance map of the pixels inside Ω to its boundary $\delta\Omega$.

3.3   Exemplar-Based Inpainting

For the general inpainting, we focused on the exemplar-based inpainting method for region filling and object removal by Criminisi *et al.* [7]. This well known *best-first algorithm* uses texture synthesis and successfully propagates continuities of structures along isophotes to the inpainting region.

Computation of the Filling Priorities. Let Ω denote the hole to be replaced and $\delta\Omega$ the contour of the hole. For each pixel p along the contour $\delta\Omega$, a filling priority $P(p)$ is computed. $P(p)$ is defined as the product [7]:

$$P(p) = ((1 - \omega) C(p) + \omega) D(p). \tag{2}$$

$\omega \in \mathbb{R}$ is a weighting factor. $C(p) := \frac{\sum_{q \in \Psi_p \cap (\mathcal{I} - \Omega)} C(q)}{|\Psi_p|}$ is called the *confidence term* and $D(p) := \frac{|\nabla I_p^\perp \cdot n_p|}{\alpha}$ the *data term*. $|\Psi_p|$ denotes the area of the patch Ψ_p, α is a normalization factor and n_p is a unit vector orthogonal to $\delta\Omega$ in the point p.

The confidence term $C(p)$ measures the amount of reliable information surrounding the pixel p. The intention is to fill first those patches which have more of their pixels already filled. Wang *et al.* [23] introduced the weighting factor ω to control the strong descent of $C(p)$ which accumulates along with the filling. The data term $D(p)$ is a function of the strength of isophotes hitting the contour of the hole. This factor is of fundamental importance because it encourages linear structures to be synthesized first. The pixel $\hat{p}$ with the highest priority: $\hat{p} = \arg\max_{p \in \delta\Omega} P(p)$ defines the center of the target patch $\Psi_{\hat{p}}$ which will be inpainted.

Search for the Best Matching Patch. In the next step, the patch $\Psi_{\hat{q}}$ which best matches the target patch $\Psi_{\hat{p}}$ is searched within the source region Φ. Formally [7]:

$$\Psi_{\hat{q}} = \arg\min_{\Psi_q \in \Phi} d\left(\Psi_{\hat{p}}, \Psi_q\right), \tag{3}$$

where the distance $d\left(\cdot, \cdot\right)$ is defined as the sum of squared differences (SSD) of the already filled pixels in the two patches.

This distance, however, is only designed for image inpainting. For the problem of video inpainting the additional temporal dimension is not considered. We present a refined distance function, modeled explicitly to maintain temporal consistency along the video frames. The detailed definition follows in the next Sect. 3.4.

Copy and Refresh. When the search for the best matching patch $\Psi_{\hat{q}}$ is completed, the target region $\Psi_{\hat{p}} \cap \Omega$ is inpainted by copying the pixels from $\Psi_{\hat{q}}$ to the target patch $\Psi_{\hat{p}}$. Besides, the boundary of the target region is updated.

The above steps are done iteratively until the target region is fully inpainted.

3.4 Flow Preserving Distance Function

The main difficulty of generalizing classical exemplar-based inpainting to videos is maintaining temporal consistency. Therefore, we modify the distance function (3) by Criminisi et al. [7]. The key idea of our approach is that scenes do not change vastly and changesets can be determined by optical flow. So we assume to already have a well inpainted frame and for further frames to inpaint we demand similarity to this reference frame. The connection between the reference frame and the current inpainting point is obtained via the inpainted optical flow $\hat{f}$ of the original scene (compare Sect. 3.2).

The corresponding distance function reads as follows:

$$\hat{d}(\Psi_{\hat{p}}, \Psi_q) := d(\Psi_{\hat{p}}, \Psi_q) + \frac{\beta}{|\Psi_{\hat{p}} \cap \Phi|} \, d(\Psi_{\hat{f}^{-1}(\hat{p})}, \Psi_q). \tag{4}$$

The first term ensures local consistency, as proposed by Criminisi et al. The second one enforces similarity to a previous inpainted frame and hence temporal consistency. $\Psi_{\hat{f}^{-1}(\hat{p})}$, using inverse optical flow, points back to the already inpainted image and ensures temporal consistency.

This distance function enables us to reduce complexity of the patch match since we do not have to choose a set of 3D patches. Our algorithm can greedily choose the best patch for the current hole to fill yet can select from all frames to exploit time redundancy. An illustration is shown in Fig. 1.

3.5 Overview of the Algorithm

Interactive Mask Definition. Let $\mathcal{I}[k]$ denote the k'th video frame. The user is asked to roughly scribble (see Sect. 2) the desired regions in the first frame $\mathcal{I}[0]$.

These scribbles are propagated via optical flow (Fig. 2(b)) throughout the video. Depending on the user scribbles a two-region segmentation in object Ω (green) and search space Φ (yellow) or a three-region segmentation with additional region Φ_r (red) for neglecting parts is computed: $\mathcal{I} = \Omega \ \dot{\cup} \ \Phi \ (\ \dot{\cup} \ \Phi_r)$.

This processing gives an accurate mask in an easy and quick manner. State-of-the-art methods do not tackle how to obtain an accurate mask definition.

Video Completion by Flow and Color Inpainting. In the proposed image inpainting algorithm one can choose the number of frames to be inpainted at the same time. This allows to exploit redundancy in the video sequence.

Using the inpainted optical flow $\hat{F}$ of the original video sequence we fill the target region Ω step by step according to Criminisi *et al.* using our new distance function (4). Our distance function ensures, that the chosen patch is both locally consistent and similar to its origin in a previous inpainted frame. This leads to a temporal consistent inpainted video sequence without any flickering.

4   Experiments and Results

In the following we will show results on various datasets and compare our results to state-of-the-art approaches for video inpainting. The evaluations show that we can handle different object and camera motions.

Depending on the video size we choose a patchsize between 8×8 and 12×12 and inpaint 3 to 8 frames at the same time to exploit time redundancy. We choose β around 1.1 to weight local and temporal consistency.

In Fig. 5 we compare two adjacent frames with and without our proposed consistency term. Without the flow consistency term the results have large deviations from one frame to the next one. In the final video such deviations are observed as disruptive flickering. In contrast, the video sequence inpainted with our proposed term shows smooth transitions between the frames. We obtain great results for complex scenes with detailed structures and different types of camera motions at substantially reduced runtime. Figures 4 and 6 compare our results to the results of Patwardhan *et al.* [18] and Newson *et al.* [15]. Table 1 compares the runtime of our method with the state-of-the-art methods [11,14,15,18,26].

4.1   Implementation and Runtime

Runtime is a big challenge to all video inpainting algorithms. Especially on high resolution videos a large amount of data has to be processed. Our parallel implementation takes around 2 to 150 s per frame, depending on the resolution of the input video on a NVIDIA GeForce GTX 560 Ti. This outruns state-of-the-art algorithms, requiring much more computing power (like Granados *et al.* [11] on a mainframe with 64 CPUs) and runtime (compare Table 1).

Table 1. Runtimes. Although our approach includes an interactive mask-definition we outperform state-of-the-art methods up to a factor of five.

	Beach Umbrella	Jumping Girl	Stairs	Young Jaws
	$264 \times 68 \times 98$	$300 \times 100 \times 239$	$320 \times 240 \times 40$	$1280 \times 720 \times 82$
Wexler *et al.* [26]	1h	-	-	-
Patwardhan *et al.* [18]	≈ 30 min	$\approx$ 1h 15min	≈ 15 min	-
Granados *et al.* [11]	11 hours	-	-	-
Newson *et al.* [14]	21 min	62 min	-	-
Newson *et al.* [15]	24 min	40 min	-	5h 48 min
proposed approach	**4.6 min**	**8 min**	**5 min 20 sec**	**3h 20min**

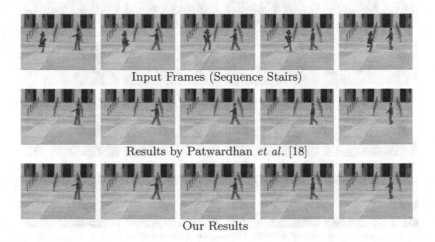

Input Frames (Sequence Stairs)

Results by Patwardhan *et al.* [18]

Our Results

Fig. 4. Comparison to Motion SSD dataset with slight camera movement.

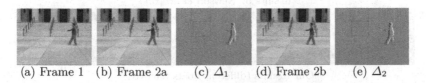

(a) Frame 1   (b) Frame 2a   (c) Δ_1 (d) Frame 2b   (e) Δ_2

Fig. 5. Transition comparison. Δ_1 shows the transition between (a) and (b). The transition is computed without regularization and shows strong video flickering. In contrast, the transition Δ_2 with our approach between (a) and (d) is smooth and does not show disruptive flickering.

Input Frames (Sequence *Fountains*)

Results by Newson *et al.* [15]

Our Results

Input Frames (Sequence *Les Loulous*)

Results by Newson *et al.* [15]

Our Results

Input Frames (Sequence *Young Jaws*)

Results by Newson *et al.* [15]

Our Results (different boats removed)

Fig. 6. Our results compared to state-of-the-art methods. Evaluations on the sequences *Fountains*, *Les Loulous* and *Young Yaws* by [15] show that we obtain the same precision of results, whereas our runtime is much faster. Furthermore, we are not restricted to a static mask and can easily remove different objects - see our results of the *Young Jaws* sequence.

5 Conclusion

We propose an interactive video completion method which integrates two innovations: Firstly, we replace the tedious hand-labeling of inpainting regions in all video frames by a semi-automatic procedure which consists of a flow-based propagation of user scribbles from the first to subsequent frames followed by an automatic foreground-background segmentation. Secondly, we propose a novel solution for assuring temporal consistency of the inpainting. Rather than performing a computationally intense sampling of space-time patches, we perform an optical flow inpainting followed by a flow-constrained image inpainting. An efficient GPU implementation provides a semi-automatic video inpainting method which requires substantially less user input and provides competitive video inpainting results which is around five times faster than competing methods.

References

1. Ashikhmin, M.: Synthesizing natural textures. In: Proceedings of the 2001 Symposium on Interactive 3D Graphics, pp. 217–226. ACM (2001)
2. Barnes, C., Shechtman, E., Finkelstein, A., Goldman, D.B.: Patchmatch: a randomized correspondence algorithm for structural image editing. ACM Trans. Graph. **28**(3), Article 24, 1–11 (2009)
3. Bertalmio, M., Bertozzi, A.L., Sapiro, G.: Navier-stokes, fluid dynamics, and image and video inpainting. In: Proceedings of the IEEE Computer Society Conference on Computer Vision and Pattern Recognition (CVPR), pp. 355–362 (2001)
4. Bertalmio, M., Sapiro, G., Caselles, V., Ballester, C.: Image inpainting. In: Proceedings of the 27th Annual Conference on Computer Graphics and Interactive Techniques, pp. 417–424. ACM Press/Addison-Wesley Publishing Co. (2000)
5. Brox, T., Bruhn, A., Papenberg, N., Weickert, J.: High accuracy optical flow estimation based on a theory for warping. In: Pajdla, T., Matas, J.G. (eds.) ECCV 2004. LNCS, vol. 3024, pp. 25–36. Springer, Heidelberg (2004)
6. Criminisi, A., Perez, P., Toyama, K.: Object removal by exemplar-based inpainting. In: International Conference on Computer Vision and Pattern Recognition, vol. 2, pp. 721–728, June 2003
7. Criminisi, A., Perez, P., Toyama, K.: Region filling and object removal by exemplar-based image inpainting. IEEE Trans. Image Process. **13**(9), 1200–1212 (2004)
8. Efros, A.A., Freeman, W.T.: Image quilting for texture synthesis and transfer. In: Proceedings of the 28th Annual Conference on Computer Graphics and Interactive Techniques, pp. 341–346. ACM (2001)
9. Efros, A.A., Leung, T.K.: Texture synthesis by non-parametric sampling. In: The Proceedings of the Seventh IEEE International Conference on Computer Vision, vol. 2, pp. 1033–1038 (1999)
10. Granados, M., Kim, K.I., Tompkin, J., Kautz, J., Theobalt, C.: Background inpainting for videos with dynamic objects and a free-moving camera. In: Fitzgibbon, A., Lazebnik, S., Perona, P., Sato, Y., Schmid, C. (eds.) ECCV 2012, Part I. LNCS, vol. 7572, pp. 682–695. Springer, Heidelberg (2012)
11. Granados, M., Tompkin, J., Kim, K.I., Grau, O., Kautz, J., Theobalt, C.: How not to be seen - object removal from videos of crowded scenes. Comput. Graph. Forum **31**(2), 219–228 (2012)

12. Masnou, S.: Disocclusion: a variational approach using level lines. IEEE Trans. Image Process. **11**(2), 68–76 (2002)
13. Masnou, S., Morel, J.M.: Level lines based disocclusion. In: International Conference on Image Processing, vol. 3, pp. 259–263 (1998)
14. Newson, A., Almansa, A., Fradet, M., Gousseau, Y., Pérez, P.: Towards fast, generic video inpainting. In: Proceedings of the 10th European Conference on Visual Media Production, CVMP '13, pp. 1–8. ACM, New York (2013)
15. Newson, A., Almansa, A., Fradet, M., Gousseau, Y., Pérez, P.: Video inpainting of complex scenes, January 2014. http://hal.archives-ouvertes.fr/hal-00937795
16. Nieuwenhuis, C., Cremers, D.: Spatially varying color distributions for interactive multi-label segmentation. IEEE Trans. Patt. Anal. Mach. Intell. **35**(5), 1234–1247 (2013)
17. Patwardhan, K., Sapiro, G., Bertalmio, M.: Video inpainting of occluding and occluded objects. In: IEEE International Conference on Image Processing, vol. 2, pp. 69–72 (2005)
18. Patwardhan, K.A., Sapiro, G., Bertalmo, M.: Video inpainting under constrained camera motion. IEEE Trans. Image Process. **16**(2), 545–553 (2007)
19. Sethian, J.A.: A fast marching level set method for monotonically advancing fronts. Proc. Natl. Acad. Sci. **93**(4), 1591–1595 (1996)
20. Shih, T., Tang, N., Hwang, J.N.: Exemplar-based video inpainting without ghost shadow artifacts by maintaining temporal continuity. IEEE Trans. Circuits Syst. Video Technol. **19**(3), 347–360 (2009)
21. Telea, A.: An image inpainting technique based on the fast marching method. J. Graph. Tools **9**(1), 23–34 (2004)
22. Tsitsiklis, J.N.: Efficient algorithms for globally optimal trajectories. IEEE Trans. Autom. Control **40**(9), 1528–1538 (1995)
23. Wang, J., Lu, K., Pan, D., He, N., kun Bao, B.: Robust object removal with an exemplar-based image inpainting approach. Neurocomputing 123, 150–155 (2014), contains Special issue articles: Advances in Pattern Recognition Applications and Methods
24. Werlberger, M.: Convex approaches for high performance video processing. Ph.D. thesis, Institute for Computer Graphics and Vision, Graz University of Technology, Graz, Austria, June 2012
25. Wexler, Y., Shechtman, E., Irani, M.: Space-time video completion. In: International Conference on Computer Vision and Pattern Recognition, vol. 1, pp. 120–127, June 2004
26. Wexler, Y., Shechtman, E., Irani, M.: Space-time completion of video. IEEE Trans. Patt. Anal. Mach. Intell. **29**(3), 463–476 (2007)

Spatial and Temporal Interpolation
of Multi-view Image Sequences

Tobias Gurdan[1,2]($\boxtimes$), Martin R. Oswald[1], Daniel Gurdan[2],
and Daniel Cremers[1]

[1] Department of Computer Science, Technische Universität München,
Munich, Germany
gurdan@in.tum.de
[2] Ascending Technologies GmbH, Krailing, Germany

Abstract. We propose a simple and effective framework for multi-view image sequence interpolation in space and time. For spatial view point interpolation we present a robust feature-based matching algorithm that allows for wide-baseline camera configurations. To this end, we introduce two novel filtering approaches for outlier elimination and a robust approach for match extrapolations at the image boundaries. For small-baseline and temporal interpolations we rely on an established optical flow based approach. We perform a quantitative and qualitative evaluation of our framework and present applications and results. Our method has a low runtime and results can compete with state-of-the-art methods.

1 Introduction

Image interpolation is one of the vital tools in image processing for the creation of visual effects, as can for example be seen in the movie "The Matrix" [3]. While it has been very tedious and costly to create such effects in the past, the rapid increase in computational power has reformed the movie industry substantially. Over the last few years, slow motion and immersive 3d effects had an enormous increase in both quality and quantity. However, there are still many open problems, including for instance free viewpoint video, which becomes more and more popular with the rise of 3d televisions [17]. In this paper, we tackle the challenge of spatiotemporal interpolation of image sequences from multiple, synchronised cameras. The main problem is to robustly estimate pixel correspondences between two given camera frames and subsequently interpolate the pixel motion properly in order to synthesise novel in-between images convincingly. The result is a slow-motion effect by interpolating consecutive frames or a smooth transition from one camera to another as depicted in Fig. 1.

1.1 Related Work

There exist a variety of applications and respective approaches to spatial view point interpolation. Phototourism [21] is a prominent example. Goesele *et al.* [8]

This work was supported by the ERC Starting Grant "Convex Vision".

X. Jiang et al. (Eds.): GCPR 2014, LNCS 8753, pp. 305–316, 2014.
DOI: 10.1007/978-3-319-11752-2_24

Fig. 1. Spatial view point interpolation between two images. Our approach robustly handles wide-baseline view point interpolations. *From left to right*: source image, interpolation at $t = 1/3$, interpolation $t = 2/3$, target image.

used point clouds to register camera positions and transition between views. In the field of sports, a promising technique was introduced by Germann *et al.* [7]. They make use of billboards and foreground-background separation to create a plausible illusion of depth while transitioning from one view to another. Inamato and Saito [10] as well as Replay Technologies Inc. [17] went one step further and applied 3d-registration and projection on top of layer segmentation. For casually captured multi-view video collections, Ballan *et al.* [1] combined a previously reconstructed background geometry and billboards of a foreground object to spatially navigate between camera views. In contrast to our approach, view interpolation is only done for the background geometry and concealed with a strong motion blur. Werlberger *et al.* [26] and Chen and Lorenz [2] addressed temporal image interpolation and suggested the use of optical flow correspondences. Mahajan *et al.* [14] estimated gradient rather than pixel correspondences.

The early years of feature based image morphing, on which our work is based, are covered by Wolberg [27] and more currently by Vlad [24]. In their pioneering work on spatial image interpolation Seitz and Dyer [20] proposed a pre- and postwarping scheme, which was later improved by Fragneto *et al.* [6]. It allows for physically plausible transitions, whereas former methods could lead to unrealistic deformations. One of the first free viewpoint videos filmed with consumer cameras was then presented by Zitnick *et al.* [29], who used layers, matting and a color segmentation-based stereo algorithm to estimate dense depth maps. Vedula *et al.* [23] estimated scene flow on a voxel model for spatiotemporal interpolation. The volume discretisation makes this approach costly to recover fine object details. Lipski *et al.* [11,12] recently proposed a dense matching scheme based on combined edgelet matching and color segmentation that allows free-viewpoint navigation for handheld multi-view video footage in both space and time. Their work is currently considered state of the art for dense image interpolation.

So far, no other work on spatiotemporal image interpolation can simultaneously cope with casual, wide baseline camera setups and preserve image quality in the interpolated results. Or they are not able to navigate in both space and time or need extensive computations. Our framework is designed to combine all aspects in a simple and effective manner.

1.2   Contribution

We present a framework for temporal and spatial multi-view image sequence interpolation. We introduce a novel pipeline for sparse feature matching, filtering and extrapolation that significantly improves correspondence quality compared to common matching techniques. Due to its robustness, our approach is suited even for wide baseline setups, where dense stereo matching approaches reach their limits. To the best of our knowledge feature-based spatiotemporal view point interpolation has not been explored in literature yet. In contrast to works like [1,14,23,26] we do not impose strong assumptions on the scene structure or the camera movement, allowing for arbitrary, uncalibrated input. In several experiments we study the strengths and weaknesses of sparse matching techniques for image interpolation on a variety of sequences.

2   Image Interpolation

In our approach the synthesis of novel views or in-between frames of image-sequences consists of three steps: matching, warping and blending. We consider two cases: spatial interpolation of camera motion between different camera views and temporal interpolation of similar images with dynamic content. The following subsections detail our approach to both cases with respect to the three processing steps, using either sparse or dense correspondences.

2.1   Spatial Interpolation

For spatial interpolation one needs to model camera motion in order to obtain plausible view point interpolations, especially in wide-baseline setups. We rely on established feature matching techniques to get an implicit, locally approximated motion representation. Such sparse matching techniques perform more robustly in wide-baseline setups compared to stereo or optical flow methods.

Instead of exhaustively searching the feature space with subsequent outlier filtering, we follow the ideas of Zhang et al. [28] who use the epipolar constraint to establish matches in the first place, limiting the search space of potential matches from two dimensions to a relaxed one dimensional space. Additionally, we propose two filtering techniques to robustly detect and eliminate false positive matches. We finally propose algorithms to extrapolate missing features and matches on the image borders.

Epipolar Guided Matcher. To get an initial estimate of a scene's epipolar geometry, we rely on SIFT features and descriptors [13]. After descriptor matching using fast approx. nearest neighbours (FLANN) [16], we filter outliers by thresholding descriptor distances, discard non-symmetric matches and finally applying the RANSAC algorithm [5], which simultaneously detects outliers and estimates the fundamental matrix. Figure 2 (top) illustrates the matching pipeline. The so obtained matchings are not suitable for image morphing.

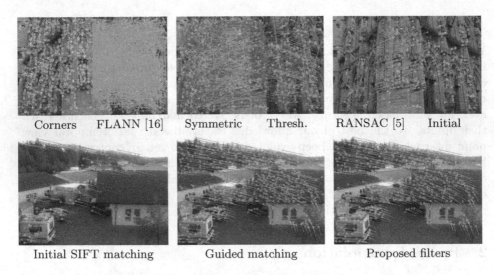

Corners   FLANN [16]   Symmetric   Thresh.   RANSAC [5]   Initial

Initial SIFT matching   Guided matching   Proposed filters

Fig. 2. Our matching pipeline on the Cityhall [22] and KimWest datasets. *Top*: Initial step by step epipolar geometry estimation. *Bottom*: Comparison of SIFT matching, our guided matching and the result after applying our proposed set of outlier filters. Notice the insufficient and poorly distributed keypoints in the initial SIFT matching, whereas our guided matching approach yields good coverage and robust correspondences.

We aim for even *distribution* to avoid degenerate meshes, meaningful *coverage* including most to all relevant image parts and *reliability*, as false positive matches highly impair the final morphing quality. In the following, we evaluated various feature detectors and descriptors and found, that Harris corners with SIFT descriptors fit our requirements best. We use these features as basis for guided matching using the epipolar constraint, effectively reducing the search space beforehand from two dimensions to a relaxed one dimensional space.

We evaluate our approach on different datasets. Figure 2 (bottom) shows exemplary results on the KimWest dataset. As especially on real world scenes false positive matches occur, we present two novel filtering techniques for robust outlier elimination in the following subsections.

Playground Filter. We developed a global filtering procedure, that effectively eliminates obvious false positive matches, i.e. matches with significant displacement deviations. Using k-Means clustering, we group matches into three clusters of keypoint-pairs. They represent small, large and extreme displacements, classifying keypoints based on their relative motion into a foreground, background or an outlier cluster, respectively. To prevent the loss of good matches by discarding a cluster that does not contain actual outliers, we introduce a lower bound for the separation of background and outlier clusters (cf. Fig. 3a). Note that our assumption on scene motion holds especially for camera rotations around a point of interest, but can easily be extended to arbitrary motions by according cluster initialisation. Figure 3a shows some examples of applied playground filter results.

(a) Playground Filter. Matches are globally filtered using clustering. Note, that no outliers are detected in the right most example, as the relative difference between large and extreme displacements is not significant enough.

(b) Council Filter. Matches are locally filtered based on neighbourhood comparison.

Fig. 3. Outlier elimination using our proposed algorithms on the Cityhall [22], Camera and KimWest datasets. *Green lines*: Inlier matches. *Red lines*: Classified outliers (Color figure online).

Council Filter. To address outliers that can not be distinguished by global motion models, we propose a novel, local filtering scheme. We compare subsets of matches in small regions around each keypoint, which we call *council*. We employ a voting scheme to rate the quality of a keypoint. A match is accepted if and only if sufficient nearby matches show similar behaviour. This implies consistency and symmetry among matchings. Table 1 lists the set of free parameters, by which a council votes.

Table 1. Council filter parameters

Table 1: Council Filter Parameters		
Variable	**Value**	**Explanation**
maxDistanceRatio	1/16	neighbourhood radius relative to image width
maxAngle	10	maximal angular difference in degree
minNormRatio	0.8	minimal relative displacement conformance
minNeighbours	3	minimal council size
minVotes	2	minimal number of attained votes
minVoteRatio	0.15	minimal ratio of attained votes

We evaluate the parameters by comparing the number of outliers relative to the total number of matches. Fixing all but one parameters to initial default values and iteratively adapting the free parameter yields the plots in Fig. 4. For

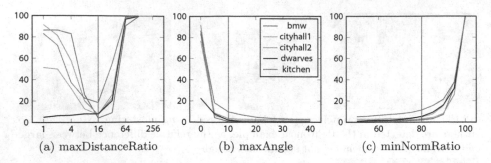

(a) maxDistanceRatio   (b) maxAngle   (c) minNormRatio

Fig. 4. Exemplary council filter parameter evaluation. The plots show the progression of false positive matches for a single, variable parameter. Red lines mark the selected values. The units on the x-axis are in *relative image width* (a), *degree* (b) and *percent* (c) respectively. *Y-axis*: outlier ratio in *percent*. *X-axis*: free parameter value (Color figure online).

each parameter, we choose the value, for which most outliers are identified while simultaneously the amount of misclassifications is minimised. These are optima in either the first or second derivative of the attained measurements. If the results vary among different scenes, we choose conservatively according to the worst performing example, in order to prevent false positive matches. This might lead to discarding true positive matches in other examples, which we tolerate in order to prevent unpleasant distortions in the final interpolations. We make use of quadtrees to allow for efficient search queries in keypoint neighbourhoods.

Affine-SIFT. SIFT descriptors, which are robust against small viewpoint changes, fail to cover all six parameters of affine transformations. Morel *et al.* [15] present an algorithm that compensates for this lack of dimensionality. We extend their approach by discarding inconclusive matches beforehand using relative descriptor distance thresholding and additionally apply our proposed Playground and Council filters.

ASIFT works well for image pairs with high transition tilts, but are often inaccurate in regions of small movements due to the very dense feature space. Thus, we propose to use ASIFT only as an alternative in those situations in which epipolar guided matching fails to establish reliable matchings. Our framework automatically detects these situations by correspondence quality and quantity evaluation and subsequently uses ASIFT matching instead.

Triangulation and Extrapolation. One challenge in sparse image warping is the lack of matches in the boundary regions (cf. Fig. 5a). In the following, we propose a local extrapolation scheme which yields consistent results. New boundary points are attained by projecting outer contour points onto the image boundaries. Let $P \subset \mathbb{R}^2$ be a set of feature points and B be the set of the four image corners. Let further T be the Delaunay triangulation of $P \cup B$. We then introduce the contour of P as the set of points $C \subseteq P$, for which there exists

  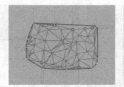 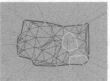

(a) Triangulation of matches for the City-hall dataset. *Left*: Without border extrapolation. *Right*: Proposed extrapolation.

(b) Match extrapolation on the image border using contour projection. Notice our proposed contour on the right (blue).

Fig. 5. Illustration of the proposed mesh extrapolation.

an edge (b, c) in the set of triangles T with $c \in C, b \in B$. In contrast to the convex hull, this definition includes even concave sections (cf. Fig. 5b), which leads to a denser and more accurate extrapolation. We find the contour using neighbourhood cycle closures in undirected, planar graphs. Each of the contour points is then projected onto the image boundary using centric radial projection, yielding an extended set of image boundary points B'. In contrast, orthogonal projection leads to misordered triangle vertices and causes unpleasant artefacts in the final interpolation. The final triangulation is obtained by re-triangulating the set of points $P \cup B'$. In the end, missing displacement information for $B \cup B'$ is attained by assigning each boundary point the motion of its corresponding contour point's match as a local approximation.

2.2 Temporal Interpolation

For temporal interpolation, we incorporate optical flow as presented by Werlberger *et al.* [26] and Chen and Lorenz [2] as a simple and effective tool for generating high quality correspondences between images with small variations (e.g. consecutive video frames or small baseline images). Let $\mathcal{F} : \Omega \subset \mathbb{R}^2 \to \Omega$ be the flow from source image $\mathcal{I}_0 : \Omega \to \mathbb{R}^3$ to target image $\mathcal{I}_1 : \Omega \to \mathbb{R}^3$. A linear dense warping from source to target image is then given by $\mathcal{I}_t(x, y) = \mathcal{I}_1(x + t \cdot \mathcal{F}_x^{-1}(x, y),\ y + t \cdot \mathcal{F}_y^{-1}(x, y))$, $t \in [0, 1]$. We apply this backward-warping to synthesise novel in-between frames. In order to generate spatiotemporal interpolations we first synthesise images in the temporal direction and subsequently perform spatial interpolation as described in the previous section on the temporal in-betweens. For a visual demonstration of interactive results, we refer to the supplemental material, as temporal variations are difficult to show on print.

2.3 Image Warping and Blending

In order to synthesise novel views between two input images $\mathcal{I}_0, \mathcal{I}_1$ at $t \in [0, 1]$, we first warp both images to $\mathcal{I}_0(t)$ and $\mathcal{I}_1(1 - t)$ respectively. In the dense setting, we use pixel-wise backward warping. For sparse matchings, we apply mesh warping and view morphing, where the vertex positions of a textured

Cross-dissolve   DualTV-L1 dense flow   Proposed sparse approach

Fig. 6. Interpolation on the Cityhall dataset [22] at $t = 0.5$. Our approach yields authentic transformations for such affine camera motions. Notice the poor quality of naive image blending (left) and a flow-based approach (middle), where our sparse approach leads to artefact-free interpolations (right).

triangle mesh covering all keypoints are linearly interpolated in rectified image space (cf. [19]). We then blend the intermediate images by adaptive, pixel-wise interpolation, weighted by a logistic sigmoid function $s : [0,1] \mapsto [0,1]$ into the blended image $\mathcal{I}(t) = s(1 - t; a) \cdot \mathcal{I}_0(t) + s(t; a) \cdot \mathcal{I}_1(1 - t)$, where $s(x; a) = \frac{h(-2a(x-0.5))-h(a)}{h(-a)-h(a)}$, with $h(x) = (1 + \exp(x))^{-1}$ and parameter $a \in (0,1)$ steering the steepness of the sigmoid function. For sparse matchings, we backproject the result using homography interpolation as presented in [6].

3   Results

Implementation. We implemented all algorithms in C++ and the GLSL shader language. The software is available for the public on the authors web page[1]. All experiments were run on an Intel Quad-Core i7 3.4 GHz, 16 GB RAM machine with an Intel HD 4000 graphics chip. Establishing and storing dense

Fig. 7. Comparison of dense and sparse image interpolation on the Middlebury "Dwarves" dataset [18]. *Left*: Dense, TV-L1 flow, $e_{RMSE} = 11.19$, Sect. 2.2. *Right*: Sparse, guided matching, $e_{RMSE} = 6.08$, Sect. 2.1. The left two images show interpolations for one skipped frame, which we take as ground truth. The right two images show difference images of the algorithms' results and the ground truth. Despite the large camera motion, our sparse approach still manages to establish reliable correspondences.

---

[1] http://www.tobiasgurdan.de/vision/imageinterpolation/

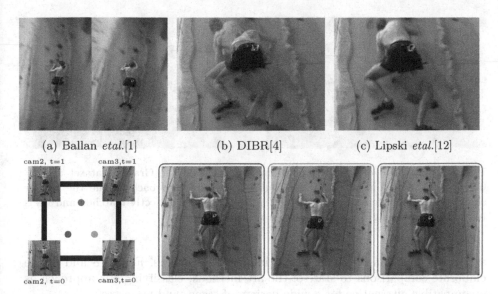

(a) Ballan *etal.*[1]   (b) DIBR[4]   (c) Lipski *etal.*[12]

Fig. 8. *Top*: Comparison to other methods on the Climbing dataset [9]. *Bottom*: Spatiotemporal interpolation using our approach. Notice in (a) the sudden switch of perspective on the climber and the sharpness contrast. (b) shows strong artifacts and ghosting. (c) achieve higher quality results, though there still occur ghosting and warping artifacts. In contrary, notice the high quality of the climber in our approach (bottom). Surrounding artefacts are negligible, as the viewers focus lies on the image centre.

matches between two images of width 480 px using the TV-L1 dense approach takes about one second. Sparse matches for the same input usually take about half a second. Large speed up factors are expected for a GPU-based implementation. After precomputing all matching data, the actual image interpolation can be explored in real time and Full HD, due to GPU accelerated routines.

Evaluation. Our framework can be used for a wide variety of applications, including the synthesis of novel views from small and wide baseline images and videos, the generation of slow-motion effects, virtual avatars, scene exploration and many more (c.f. supplemental material). We evaluated our approach on several challenging data sets. Figure 6 shows a qualitative comparison of simple image blending and the methods provided by our framework on a wide-baseline scene (spatial interpolation). Figure 7 shows a quantitative and qualitative comparison of the flow based approach and our sparse approach on the Middlebury "Dwarves" sequence [18]. We used three consecutive images and interpolated the centre frame from the first and last image and compared the result to the original centre frame as the ground truth by computing the pixel- and channel-wise root mean square error e_{RMSE} for each method. Figure 7 demonstrates that our sparse approach is better suited for larger camera motions. In Fig. 1 we illustrate that our sparse approach also handles wide-baseline scenarios while

Fig. 9. Comparison of spatiotemporal interpolation on the Graffiti dataset at $s = 0.5, t = 0.5$. *Left*: Lipski *et al.* [12]. *Right*: Our sparse approach. While the juggler shows noticeable artefacts, the wall in the background is perfectly matched and leads to an overall pleasing transition between views.

still preserving image quality in the interpolated images. Figures 8 and 9 give elaborate comparisons to state-of-the-art methods, classifying our approach as a compatible alternative for a wide variety of scenes and use-cases.

Runtime. Our framework has low computation times. While Lipski *et al.* need several seconds to compute the correspondence map for an image pair [12], our approach does so in under a second in both spatial and temporal domain. However, our employed subroutines are currently implemented on a single CPU, but well suited for parallelisation on GPUs. Especially, the most computational part, the SIFT feature extraction, could be parallelised, e.g. [25]. This would further decrease computation time significantly, making the approach real-time capable.

Limitations. Although sparse features are robust to large camera movements, with their triangulation we assume a linear distribution of depth values between the features. If the distribution of the feature does not cover depth discontinuities, interpolation artefacts become noticeable as for example can be seen for the person in Fig. 9. In contrast, the piecewise flat background is perfectly interpolated. Similar artefacts are also visible in several frames of the climbing sequence Fig. 8. Moreover, the heuristic interpolation of extrapolated points on the image boundary can lead to distortions, but exact matching information is not available at these locations. Nevertheless, the approach works well in many scenarios and outperforms many related works in terms of speed and simplicity.

4   Conclusion

We presented a framework for spatial and temporal image interpolation. We showed that sparse correspondence based interpolation poses a qualified alternative to state-of-the-art, dense 3d reconstruction or billboard approaches. Establishing robust correspondences among image pairs is still one of the most challenging problems in computer vision. We tackled this problem by introducing a novel matching and outlier filtering pipeline which makes the approach robust even in wide-baseline camera setups. We further proposed an extrapolation method for absent correspondences on image boundaries. For temporal interpolation, a dense correspondence is more suited and can be efficiently solved using established optical flow based methods. We combine the strengths of these approaches in a simple and effective spatiotemporal interpolation method that has a low computation time. We demonstrated competitive results on a variety of sequences with different scene structure and camera baselines.

References

1. Ballan, L., Brostow, G.J., Puwein, J., Pollefeys, M.: Unstructured video-based rendering: interactive exploration of casually captured videos. ACM Trans. Graph. **29**(4) (2010). http://dblp.uni-trier.de/db/journals/tog/tog29.html#BallanBPP10
2. Chen, K., Lorenz, D.A.: Image sequence interpolation based on optical flow, segmentation, and optimal control. IEEE Trans. Image Process. **21**(3), 1020–1030 (2012). http://dblp.uni-trier.de/db/journals/tip/tip21.html#ChenL12
3. Debevec, P.: The campanile movie. In: SIGGRAPH 97 Electronic Theater (1997). http://www.debevec.org/Campanile/ (visited: May 2014)
4. Fehn, C.: Depth-Image-Based Rendering (DIBR), compression and transmission for a new approach on 3D-TV. In: Proceedings of SPIE Stereoscopic Displays and Virtual Reality Systems XI, pp. 93–104 (2004)
5. Fischler, M., Bolles, R.: Random sample consensus: a paradigm for model fitting with applications to image analysis and automated cartography. Commun. ACM **24**(6), 381–395 (1981)
6. Fragneto, P., Fusiello, A., Rossi, B., Magri, L., Ruffini, M.: Uncalibrated view synthesis with homography interpolation. In: 3DIMPVT, pp. 270–277. IEEE (2012). http://dblp.uni-trier.de/db/conf/3dim/3dimpvt2012.html#FragnetoFRMR12
7. Germann, M., Hornung, A., Keiser, R., Ziegler, R., Würmlin, S., Gross, M.: Articulated billboards for video-based rendering. Comput. Graph. Forum (Proc. Eurographics) **29**(2), 585–594 (2010)
8. Goesele, M., Ackermann, J., Fuhrmann, S., Haubold, C., Klowsky, R., Steedly, D., Szeliski, R.: Ambient point clouds for view interpolation. ACM Trans. Graph. **29**(4), 95:1–95:6 (2010). http://doi.acm.org/10.1145/1778765.1778832
9. Hasler, N., Rosenhahn, B., Thormählen, T., Wand, M., Gall, J., Seidel, H.P.: Markerless motion capture with unsynchronized moving cameras. In: CVPR, pp. 224–231 (2009)
10. Inamoto, N., Saito, H.: Free viewpoint video synthesis and presentation from multiple sporting videos. In: ICME, pp. 322–325. IEEE (2005). http://dblp.uni-trier.de/db/conf/icmcs/icme2005.html#InamotoS05
11. Lipski, C.: Virtual video camera: a system for free viewpoint video of arbitrary dynamic scenes. Ph.D. thesis, TU Braunschweig, June 2013
12. Lipski, C., Linz, C., Berger, K., Magnor, M.A.: Virtual video camera: image-based viewpoint navigation through space and time. In: SIGGRAPH Posters. ACM (2009). http://dblp.uni-trier.de/db/conf/siggraph/siggraph2009posters.html#LipskiLBM09
13. Lowe, D.G.: Distinctive image features from scale-invariant keypoints. Int. J. Comput. Vis. **20**, 91–110 (2003)

14. Mahajan, D., Huang, F.C., Matusik, W., Ramamoorthi, R., Belhumeur, P.N.: Moving gradients: a path-based method for plausible image interpolation. ACM Trans. Graph. **28**(3), 42:1–42:11 (2009). doi:10.1145/1531326.1531348

15. Morel, J.M., Yu, G.: Asift: a new framework for fully affine invariant image comparison. SIAM J. Imaging Sci. **2**(2), 438–469 (2009). http://dblp.uni-trier.de/db/journals/siamis/siamis2.html#MorelY09

16. Muja, M., Lowe, D.G.: Fast approximate nearest neighbors with automatic algorithm configuration. In: Ranchordas, A., Arajo, H. (eds.) VISAPP (1), pp. 331–340. INSTICC Press (2009). http://dblp.uni-trier.de/db/conf/visapp/visapp2009-1.html#MujaL09

17. Replay Technologies Inc.: freeD^{TM} technology (2013). http://replay-technologies.com/ (visited: May 2014)

18. Scharstein, D., Pal, C.: Learning conditional random fields for stereo. In: CVPR (2007)

19. Seitz, S.M., Dyer, C.R.: Physically-valid view synthesis by image interpolation. In: Proceedings of the IEEE Workshop on Representations of Visual Scenes, pp. 18–25 (1995)

20. Seitz, S.M., Dyer, C.R.: View morphing. In: SIGGRAPH, pp. 21–30 (1996). http://dblp.uni-trier.de/db/conf/siggraph/siggraph1996.html#SeitzD96

21. Snavely, N., Garg, R., Seitz, S.M., Szeliski, R.: Finding paths through the world's photos. ACM Trans. Graph. (Proceedings of SIGGRAPH 2008) **27**(3), 11–21 (2008)

22. Strecha, C., Tuytelaars, T., Gool, L.J.V.: Dense matching of multiple wide-baseline views. In: ICCV, pp. 1194–1201 (2003)

23. Vedula, S., Baker, S., Kanade, T.: Image-based spatio-temporal modeling and view interpolation of dynamic events. ACM Trans. Graph. **24**(2), 240–261 (2005)

24. Vlad, A.: Image morphing techniques. JIDEG **5**(1) (2010). http://www.sorging.ro/ro/member/serveFile/format/pdf/slug/image-morphing-techniques

25. Warn, S., Apon, A., Cothren, J.: Accelerating sift on hybrid clusters. In: Proceedings of the ACM SIGSPATIAL Second International Workshop on High Performance and Distributed Geographic Information Systems, HPDGIS '11, pp. 2–9. ACM, New York (2011). http://doi.acm.org/10.1145/2070770.2070771

26. Werlberger, M., Pock, T., Unger, M., Bischof, H.: Optical flow guided tv-l1 video interpolation and restoration. In: Energy Minimization Methods in Computer Vision and Pattern Recognition (2011)

27. Wolberg, G.: Image morphing: a survey. Vis. Comput. **14**, 360–372 (1998). http://ci.nii.ac.jp/naid/80010827845/en/

28. Zhang, Z., Deriche, R., Faugeras, O.D., Luong, Q.T.: A robust technique for matching two uncalibrated images through the recovery of the unknown epipolar geometry. Artif. Intell. **78**(1–2), 87–119 (1995). http://dblp.uni-trier.de/db/journals/ai/ai78.html#ZhangDFL95

29. Zitnick, C.L., Kang, S.B., Uyttendaele, M., Winder, S.A.J., Szeliski, R.: High-quality video view interpolation using a layered representation. ACM Trans. Graph. **23**(3), 600–608 (2004). http://dblp.uni-trier.de/db/journals/tog/tog23.html#ZitnickKUWS04

Pose Normalization for Eye Gaze Estimation and Facial Attribute Description from Still Images

Bernhard Egger[✉], Sandro Schönborn,
Andreas Forster, and Thomas Vetter

Department for Mathematics and Computer Science, University of Basel,
Basel, Switzerland
{bernhard.egger,sandro.schoenborn,andreas.forster,
thomas.vetter}@unibas.ch

Abstract. Our goal is to obtain an eye gaze estimation and a face description based on attributes (e.g. glasses, beard or thick lips) from still images. An attribute-based face description reflects human vocabulary and is therefore adequate as face description. Head pose and eye gaze play an important role in human interaction and are a key element to extract interaction information from still images. Pose variation is a major challenge when analyzing them. Most current approaches for facial image analysis are not explicitly pose-invariant. To obtain a pose-invariant representation, we have to account the three dimensional nature of a face. A 3D Morphable Model (3DMM) of faces is used to obtain a dense 3D reconstruction of the face in the image. This Analysis-by-Synthesis approach provides model parameters which contain an explicit face description and a dense model to image correspondence. However, the fit is restricted to the model space and cannot explain all variations. Our model only contains straight gaze directions and lacks high detail textural features. To overcome this limitations, we use the obtained correspondence in a discriminative approach. The dense correspondence is used to extract a pose-normalized version of the input image. The warped image contains all information from the original image and preserves gaze and detailed textural information. On the pose-normalized representation we train a regression function to obtain gaze estimation and attribute description. We provide results for pose-invariant gaze estimation on still images on the UUlm Head Pose and Gaze Database and attribute description on the Multi-PIE database. To the best of our knowledge, this is the first pose-invariant approach to estimate gaze from unconstrained still images.

1 Introduction

Faces play a fundamental role in human interaction. Facial attributes and gaze direction are very important for understanding the plot of a scene. The field of face analysis in still images evolved in the last years and a lot of very powerful

© Springer International Publishing Switzerland 2014
X. Jiang et al. (Eds.): GCPR 2014, LNCS 8753, pp. 317–327, 2014.
DOI: 10.1007/978-3-319-11752-2_25

methods have been developed. However, most of the research is put on the interpretation of a face regardless of its context and ignoring effects of pose variation. We take a step in the direction of facial interaction analysis by estimating the eye gaze. Points of attention can be estimated and faces can be described in their context (e.g. "Person A, male, is looking at Person B, female"). We show an overview of the presented method in Fig. 1.

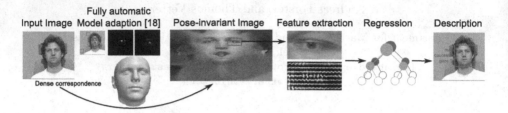

Fig. 1. System overview: The fully automatic 3DMM adaption method of Schönborn [17] is used to obtain a dense correspondence from the input image to the model reference. We extract a pose-invariant face representation preserving the texture from the original image. HOG features and image intensities are used as features for a Random Forest Regression. The output of the system is a gaze estimation and attribute-based image description.

In the broad research field of gaze estimation, most methods focus on tracking. For a single still image there is no pose-invariant method to automatically estimate eye gaze. We propose to use a pose-normalized version of the image and apply simple methods on this pose-invariant representation. Since a face is a three-dimensional object, a 3D Model is the natural way to obtain a pose-normalized representation. We use a generative 3D Morphable Model (3DMM) [3] of faces to solve the pose estimation and registration problem. A facial image is interpreted in an Analysis-by-Synthesis approach. The model is adapted to the face in the image as closely as possible (fitting).

The parameters (Shape, Color, Camera and Light) of the final representation in the model space (fit) contain information on the face and the scene. The eye gaze in the 3DMM is fixed and therefore the gaze estimation cannot be performed on the model parameters directly. The description by the model parameters is limited to what the model is able to reconstruct.

We overcome the model limitations with a discriminative approach. The normalization is based on full and perfect correspondence. We warp the image into a pose-normalized representation by the dense registration of the fit. The warped texture can be seen in Fig. 2. The remaining challenge for features and the classifier are the small correspondence inaccuracies in our fit.

For gaze estimation, we rely on the good registration and work on the histogram-normalized image intensities. For the attribute prediction, we use HOG features [7]. High frequency texture details cannot be encoded in our PPCA model parameters, and therefore the information captured by HOG features is valuable. The HOG features used for attribute estimation are invariant to small

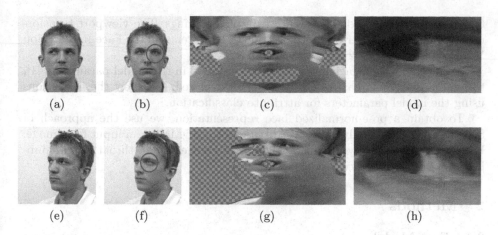

(a)   (b)   (c)   (d)

(e)   (f)   (g)   (h)

Fig. 2. We present a frontal 0° (a) and side view 40° (e) image of the UUlm HPG database. Both contain a relative eye gaze of 40°. The model fits (b) and (f) do not reflect the eye gaze (red circle). As one can see in (c) and (g), the pose-invariant representation obtained by the correspondence of the fit preserves the eye gaze. One can see small misalignments to the background at the border of the face model (green circle). (d) and (h) show the cropped regions we use to estimate the eye gaze (Color figure online).

misalignments due to the binning property. We use Random Forest Regression [6] for the Gaze estimation task and Random Forest Classification for the attribute estimation. To train our forests, we can use all image data which we can fit with the 3DMM.

In a gaze estimation experiment we show that the pose-normalized representation is suitable for gaze analysis. To the best of our knowledge, we present the first results for fully automatic gaze estimation on the UUlm Head Pose Gaze datasets [18] up to yaw angles of 40°. The database was chosen because of their wide range of gaze. The attribute-description experiment and the recognition experiment are based on the Multi-PIE database [9].

1.1 Prior Work

Most works on eye gaze estimation focus on tracking. Hansen et al. [10] give a nice overview of current methods. On single still images there are fewer works and all are limited to frontal pose or need-calibrated settings [8,13].

Kumar et al. [12] give a nice overview on prior work on facial attribute classification and demonstrate the power of attributes for face description and recognition. They classify attributes on affine-aligned face regions from near-frontal images. We extend this idea through a 3D model which adds full pose-invariance. Instead of a single global transformation we use dense local mappings incorporating the full 3D knowledge obtained by the 3DMM.

The power of pose-normalization using a 3D model was demonstrated by Blanz et al. [2]. The 3D Morphable Model is used for preprocessing for various face recognition methods to produce frontal renderings. The viewport transformation improves the performance of 9 out of 10 systems on the Face Recognition Vendor Test FRVT2002.

Even though some facial attributes are encoded in the model parameters [1], the analysis of them has have not yet been explored. We show the limitations using the model parameters for attribute classification.

To obtain a pose-normalized face representation, we use the approach of Schönborn [17] for a fully automatic fitting of the 3DMM to an input face image. We work with a slightly modified version (without ears and throat) of the publicly available Basel Face Model [15].

2   Methods

2.1   Face Model

In this work, we use the 3DMM to extract the scene and face description. Both are obtained through a full adaption of the model to the input image. To achieve full automation, we make use of the probabilistic Data-Driven Markov Chain Monte Carlo integrative fitting algorithm of Schönborn [17], which can handle unreliable detection input. The fitting algorithm recovers the best face description, camera setup and illumination to reconstruct the image. The result of the adaption contains the image location of any point on the face through the correspondence with the model and the obtained camera setup. It also delivers a continuous face representation in terms of the PCA coefficients of the 3D face model.

In contrast to other automatic methods for extracting facial feature points, the 3DMM also results in a fully abstracted face representation which is invariant with respect to pose and illumination. We directly use this representation in terms of model coefficients for face recognition and attribute classification.

We adapted the model likelihood slightly to our needs by using a more general background model. We replace the restrictive original Gaussian background likelihood [17] by an empirical histogram model. Thus, we exchange

$$\mathcal{N}\left(I(p) \mid \mu_{\mathrm{BG}}, \Sigma_{\mathrm{BG}}\right) \quad \text{by} \quad \frac{1}{\delta} \mathrm{h}\left(I(p)\right), \tag{1}$$

where δ is the bin volume and $\mathrm{h}(I(p))$ is the relative frequency of the color value $I(p)$ at location p in input image I. Our histogram consists of 25 bins per RGB color channel.

2.2   Pose Normalization

Our pose-normalized representation is using the full correspondence of the fit for extraction of the image information. The 3D face is textured by the pixel information extracted from the image. The obtained representation corresponds to a texture map known from computer graphics. We use the texture representation proposed by Paysan [14] which builds on a quasi conformal mapping by

Kharevych et al. [11]. This texture map is a warp of the original image and still looks natural. We show examples for the pose-invariant representation in Fig. 2.

2.3 Gaze Estimation

We assume perfect registration and use the histogram-normalized image intensities of the pose-normalized representation. The gaze direction is parametrized relative to the head pose. A Random Forest regression is learned on a training set and used to predict the gaze direction.

2.4 Attribute Classifiers

The attribute classifiers are obtained similarly to the gaze estimation. Histograms of Oriented Gradients (HOG) [7] are used to represent textural details. The edge responses are binned into small spatial regions and can therefore cope with small misalignments.

We train a Random Forest Classifier to predict attributes. The output of the classifier is a certainty of the input image belonging to a class, respectively the face containing a specific attribute. The certainty gives us a more accurate description of the face than a (possibly wrong) binary output (e.g., if the classifier is "0.51 sure to see a male" than just "male").

A classifier is calculated per attribute. The eye, nose, mouth and eyebrow regions are used to predict the attributes. We combine different classifiers for the same attribute by the average prediction of all classifiers to obtain a single global attribute for face description.

2.5 Similarity Measure

A similarity measure in a face space is useful for all applications concerning identity. Different appearances (e.g., through pose or illumination) of the same face should always be similar.

The cosine angle between two face representations f_1 and f_2 is often used as similarity measure for face recognition based on the 3DMM [4]:

$$d = \frac{\langle f_1, f_2 \rangle}{(\|f_1\| \cdot \|f_2\|)} \tag{2}$$

In the classical setting, the vectors f_1 and f_2 are a concatenation of shape and texture parameters. To integrate our attribute predictions into the similarity measure, we concatenate them as a third component into the description vector.

3 Experiments and Results

To evaluate the pose-normalized face representation we performed two different experiments. First, we predict the eye gaze from the eye region cropped out of the normalized face texture. Second, we predict facial attributes from different regions of the texture and evaluate their performance for face description on a face recognition task.

In all experiments we obtain a fit of the 3DMM to the image by a fully automatic DDMCMC method [17]. We draw 10 000 samples and take the best one (maximum posterior probability).

We use the OpenCV 2.4.4 [5] implementation of Random Forests and HOG features. We choose a tree depth of 10, select 10 features per split and trained with 2000 trees for all experiments. For the HOG features we took the preset parameters.

3.1   Gaze Estimation

The gaze estimation experiment was performed on the UUlm Head Pose and Gaze Database. It contains 20 subjects and 111 images per individual. We used the horizontal poses between 0° and 40° and relative gaze direction from −40° to 40°. The fitting is not reliable enough for gaze estimation for yaw angles above 45°. This selection leads to 940 images for our evaluation. We performed leave-one-out cross validation, always excluding all images of one subject. We show variations of the database in Fig. 3. Our gaze estimation is trained on the relative gaze (pose-corrected). The relative gaze is dependent on the estimated head pose and therefore the pose estimation error is propagated to the gaze estimation. The pose estimation error obtained by the model adaption is shown in Fig. 4a. For our gaze estimation experiment we reach a total Mean Approximation Error (MAE) of 9.74°. In Fig. 4b we show the estimation error itemized on each pose seperately. Both plots are reflected in Table 1. Note that due to pose normalization we are able to train a single regression for all poses. The proposed gaze estimator trained on the UUlm HPG database delivers reasonable results on real world images, see Fig. 5.

Fig. 3. These are 5 of 20 subjects of the UUlm HPG database. The variation used in our experiments is shown from left to right. We use yaw angles from 40° to 0° and relative gaze direction from 40° to −40°. The database contains different lighting conditions, glasses and occlusion through hair.

3.2   Attributes

We use 16 attributes to describe a face, see Table 2. For each attribute we learn a regressor on the eyes, nose, mouth or eyebrows. We compare the prediction

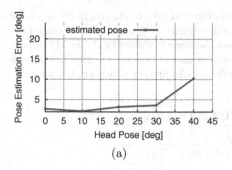
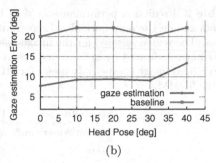

(a)   (b)

Fig. 4. (a) shows the pose estimation performance of the fully automatic fitting method on UUlm HPG database. (b) shows gaze estimation performance per head pose on UUlm HPG database. The baseline results are obtained by always predicting a gaze of 0°.

Table 1. head pose and gaze estimation error (MAE) in degree. The baseline results are obtained by always predicting a gaze of 0°.

	0°	10°	20°	30°	40°
Pose estimation error	2.73	2.16	3.22	3.65	10.26
Gaze estimation error	7.75	9.29	9.41	9.13	13.40
Baseline error	20.00	22.22	22.22	20.00	22.22

based on HOG features and color intensities by a prediction obtained on model parameters. The performances of the particular attribute classifiers are shown in Table 2. The attributes were learned and evaluated on separated subsets of the Multi-PIE database [9]. The database contains over 750 000 images of 337 different persons. We used the 249 identities from the first session for evaluation and the first appearance in session two to four of the other 88 identities for training. The experiment was performed on five poses (15°, 30°, 45°, 60° resp. in Multi-PIE camera names 051, 140, 130, 080, 090) and illumination 16. This leads to 440 images for the training of each classifier. For one attribute we train a single classifier over all poses. There is no tuning to a specific pose.

3.3   Recognition

We use the similarity measure (2) for a face recognition experiment. The performance of attributes estimated on the texture is compared with the recognition rate obtained by the model parameters. We used the output of the classifiers for all 16 attributes on all 4 selected regions (64 attribute estimations). We evaluate our recognition method on the Multi-PIE database. We use the exact same setting for the recognition as Schönborn et al. [17]. The 249 individuals from the first session are used for the recognition task. The results are listed in Table 3. The attribute classifiers are the same as in the attributes section and

324   B. Egger et al.

Table 2. Prediction performance in % of binarized attribute classifiers on pca coefficients and on the pose-normalized representation using HOG features and color intensities. The region selected for the image-based classifier is shown in the fourth column. Attributes indicated by a * are underrepresented in the test set (≤ 20 %). The pose-normalized image is especially useful for attributes like glasses which are not contained in the 3DMM.

Attribute	PCA	HOG	Region
African American*	97.6	98.2	Mouth
Asian	83.4	85.1	Eye
Beard*	96.9	95.5	Eyebrow
Black hair	78.2	75.2	Nose
Blond hair*	87.9	87.4	Eyebrow
Bue eyes	70.9	77.7	Eye
brown eyes	67.8	80.6	Eye
Caucasian	85.4	82.9	Eye
Glasses	71.1	88.5	Nose
Hair on forehead	73.4	67.8	Eyebrow
Indian*	90.3	93.3	Eye
Male	76.2	76.0	Mouth
Mustache	95.3	94.8	Mouth
Nasolabial fold	74.8	75.7	Nose
Thick lips	66.0	69.0	Mouth
Wide nose	59.9	63.9	Nose

Table 3. Rank-1 Identification rates (percent) across pose, obtained by frontal 0° (051_16) images as gallery and the respective pose views as probes.

	15° (140_16)	30° (130_16)	45° (080_16)	60° (090_16)
3DMM shape, texture and attributes	97.6	95.2	80.7	50.6
Attributes only	93.2	82.3	65.5	30.1
3DMM shape only	86.4	63.9	44.2	11.2
3DMM texture only	98.4	94.0	77.5	43.0
3DMM shape and texture	97.6	94.8	79.5	49.0
3DMM shape and texture [17]	-	90.4	74.7	-
3DGEM [16]	97.6	86.7	65.0	44.9

were trained on images from subjects not occurring in the recognition experiment. We compare our results to other fully automatic approaches based on 3D Generic Elastic Models [16] and previous results obtained with a 3DMM [17]. The effect of the added attribute detections is shown in Table 3. As we use an empirical histogram background model, we obtain better recognition results using shape and texture coefficients than previous results by Schönborn et al. [17]. The experiment shows that the description by attributes is powerful and slightly improves the recognition performance.

(a)

(b)

(c)

(d)

Fig. 5. Our gaze estimation approach also works on unconstrained real world images. The automatically extracted gazes relative to head pose are (a): 15°, (b): 27°, (c): 20°, (d): −2°. The images where cropped to the face region after processing. Images: (a) KEYSTONE/AP Photo/Richard Drew, (b) KEYSTONE/EPA/Justin Lane, (c) KEYSTONE/EPA/Dennis M. Sabangan, (d) KEYSTONE/AP Photo/Alastair Grant

4   Conclusion

We proposed to use the registration obtained from the 3DMM fit for gaze estimation and attribute description. A pose-normalized face representation arises
through the dense image correspondence. A regression and classification function
is learned on the region of interest and profits from the pose normalization. The
pose-normalized input image conserves the textural information from the input
image. The information can be extracted by classical image features and lead to
a description not contained in the model parameters. In contrast to the 3DMM,
which needs high resolution 3D scans, our predictors can be learned directly
on image data. By this we overcome model limitations. This approach is fully
automatic, using a fully automatic 3DMM adaption method. In the experiments
we present the first fully automatic and pose-invariant gaze estimation results
on the UUlm HPG database. The gaze estimation is not limited to the database. The learned regression can be applied on real world images, see Fig. 5. In
addition we show the limitation of the model parameters describing attributes
not contained in the model (e.g. glasses, see Table 2). Attribute-based description combined with the 3DMM parameters achieves higher face recognition rates
than other automatic approaches, especially for yaw angles larger than 30°.

Acknowledgment. This work has been partially founded by the Swiss National Science Foundation.

References

1. Amberg, B., Paysan, P., Vetter, T.: Weight, sex, and facial expressions: on the
   manipulation of attributes in generative 3D face models. In: Bebis, G. (ed.) ISVC
   2009, Part I. LNCS, vol. 5875, pp. 875–885. Springer, Heidelberg (2009)
2. Blanz, V., Grother, P., Phillips, P.J., Vetter, T.: Face recognition based on frontal
   views generated from non-frontal images. In: IEEE Computer Society Conference
   on Computer Vision and Pattern Recognition, CVPR 2005, vol. 2, pp. 454–461.
   IEEE (2005)
3. Blanz, V., Vetter, T.: A morphable model for the synthesis of 3D faces. In: SIG-
   GRAPH'99 Proceedings of the 26th Annual Conference on Computer Graphics
   and Interactive Techniques, pp. 187–194. ACM Press (1999)
4. Blanz, V., Vetter, T.: Face recognition based on fitting a 3D morphable model.
   IEEE Trans. Pattern Anal. Mach. Intell. **25**(9), 1063–1074 (2003)
5. Bradski, G.: The opencv library. Dr. Dobb's J. Softw. Tools **25**, 120–126 (2000)
6. Breiman, L.: Random forests. Mach. Learn. **45**(1), 5–32 (2001)
7. Dalal, N., Triggs, B.: Histograms of oriented gradients for human detection. In:
   IEEE Computer Society Conference on Computer Vision and Pattern Recognition,
   CVPR 2005, vol. 1, pp. 886–893. IEEE (2005)
8. Florea, L., Florea, C., Vrânceanu, R., Vertan, C.: Can your eyes tell me how you
   think? a gaze directed estimation of the mental activity (2013)
9. Gross, R., Matthews, I., Cohn, J., Kanade, T., Baker, S.: Multi-pie. Image Vis.
   Comput. **28**(5), 807–813 (2010)

10. Hansen, D.W., Ji, Q.: In the eye of the beholder: a survey of models for eyes and gaze. IEEE Trans. Pattern Anal. Mach. Intell. **32**(3), 478–500 (2010)
11. Kharevych, L., Springborn, B., Schröder, P.: Discrete conformal mappings via circle patterns. ACM Trans. Graph. (TOG) **25**(2), 412–438 (2006)
12. Kumar, N., Berg, A., Belhumeur, P., Nayar, S.: Describable visual attributes for face verification and image search. IEEE Trans. Pattern Anal. Mach. Intell. **33**(10), 1962–1977 (2011)
13. Marku, N., Frljak, M., Pandi, I.S., Ahlberg, J., Forchheimer, R.: Eye pupil localization with an ensemble of randomized trees. Pattern Recogn. **47**(2), 578–587 (2014)
14. Paysan, P.: Statistical modeling of facial aging based on 3D scans. Ph.D. thesis, University of Basel, Switzerland (2010)
15. Paysan, P., Knothe, R., Amberg, B., Romdhani, S., Vetter, T.: A 3D face model for pose and illumination invariant face recognition. In: Proceedings of the 6th IEEE International Conference on Advanced Video and Signal Based Surveillance (AVSS), pp. 296–301. IEEE (2009)
16. Prabhu, U., Heo, J., Savvides, M.: Unconstrained pose-invariant face recognition using 3D generic elastic models. IEEE Trans. Pattern Anal. Mach. Intell. **33**(10), 1952–1961 (2011)
17. Schönborn, S., Forster, A., Egger, B., Vetter, T.: A monte carlo strategy to integrate detection and model-based face analysis. In: Weickert, J., Hein, M., Schiele, B. (eds.) GCPR 2013. LNCS, vol. 8142, pp. 101–110. Springer, Heidelberg (2013)
18. Weidenbacher, U., Layher, G., Strauss, P.M., Neumann, H.: A comprehensive head pose and gaze database (2007)

Posters

Posters

Probabilistic Progress Bars

Martin Kiefel[(✉)], Christian Schuler, and Philipp Hennig

Max Planck Institute for Intelligent Systems, Tübingen, Germany
mkiefel@tue.mpg.de

Abstract. Predicting the time at which the integral over a stochastic process reaches a target level is a value of interest in many applications. Often, such computations have to be made at low cost, in real time. As an intuitive example that captures many features of this problem class, we choose progress bars, a ubiquitous element of computer user interfaces. These predictors are usually based on simple point estimators, with no error modelling. This leads to fluctuating behaviour confusing to the user. It also does not provide a distribution prediction (risk values), which are crucial for many other application areas. We construct and empirically evaluate a fast, constant cost algorithm using a Gauss-Markov process model which provides more information to the user.

1 Introduction

The problem of predicting when the integral over a random rate will reach a certain level, i.e. to solve

$$R = \int_0^T r(t)\, dt \tag{1}$$

for T, where $r(t)$ is a draw from a random process (below we will focus on specific Gaussian processes), is at the heart of a number of applied problems. For motivation, consider the following examples:

- Retailers need to predict the point at which their reserves of individual products run out, an event that incurs a big loss. For products which are sold at sufficiently fine increments such that stocks are essentially a continuous quantity, a stochastic rate of sale is a realistic model.
- The onboard computer of a car needs to predict the remaining range while the fuel tank is slowly emptying. Changes in the drive speed, steepness of the road and wind conditions continuously change the fuel efficiency of the car, making prediction nontrivial.
- A startup company that is trying to attract crowd funding needs to predict the point at which funding goals will likely be reached, to efficiently prepare production (Sect. 4.2).

In this manuscript, we will focus on a fourth example application which shares many of the fundamental properties of the ones above, and adds the complication

© Springer International Publishing Switzerland 2014
X. Jiang et al. (Eds.): GCPR 2014, LNCS 8753, pp. 331–341, 2014.
DOI: 10.1007/978-3-319-11752-2_26

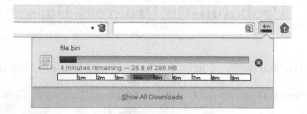

Fig. 1. A graphical representation of uncertainty about the remaining runtime of a download, resulting from the algorithm described below. The top half of the interface shows a classic progress bar, while the bottom half shows, in red, a cloud of uncertainty representing the weighted set of hypotheses over the remaining download time. The shown implementation is available as an open-source plugin. (Note that the plugin currently only works in the pull-down dialogue from the address bar, as shown in the figure, not in the separate download window also provided by Firefox) (Color figure online).

of a particularly restrained computational budget: Plotting a progress bar to predict the completion time of a file download on a desktop computer, under a stochastically varying connection bandwidth.

Although progress bars are ubiquitous elements of user interfaces, they are often treated as 'eye candy', at the margins of both the users' and the designers' attention. Visual design aspects of these pieces of graphical user interface have sometimes been studied [8], but the algorithms behind them are usually simplistic, a combination of simple moving averages, with ad-hoc backstops to avoid pathological values. The resulting predictions are not just unreliable, they are also qualitatively, visually unsatisfying: They give quickly varying predictions, which are confusing to the user, who has to mentally compute the averages.

Instead, we will develop a probabilistic progress bar algorithm, which returns a nonparametric probability distribution over the completion-time of a file download. This distribution can be shown graphically to the user (Fig. 1), providing a less volatile, and more informative interface.

The mathematical 'lever' central to our derivations of a low-cost algorithm is the fact that Gaussian distributions, including nonparametric Gaussian process models, are closed under linear projections, and thus in particular also under integration: The integral over a Gaussian process is itself a Gaussian process. This observation is not new, and has been used repeatedly in the past to establish connections between numerical quadrature rules and Gaussian process regression [1,3,6,7]. The twist in the progress bar setting (and related problems) is that the problem is inverted: Instead of predicting the *value* of the integral over a stochastic process at a particular *time*, one needs to predict the *time* at which the integral over the process reaches a particular *value*. This complicates the computation and typically gives non-Gaussian predictions (Fig. 2).

In addition, our application requires a very low computational cost profile: Progress bars, as elements of the graphical user interface, must not cause serious computational overhead. We thus strive here to develop a algorithm of radically low, constant cost both in memory and computation time. This is achieved

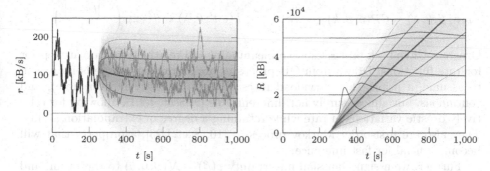

Fig. 2. Gauss-Markov process rate model. Left: an observed download rate $r(t)$ (black) gives rise to a Gauss-Markov process posterior belief about the future $r(t)$ (red, thick mean function, one and two standard deviations in decreasing red intensity, three samples from the belief in faint red). Right: The belief over the integral $R(t) = \int_{t_0}^{t} r(\tilde{t}) d\tilde{t}$ is also a Gaussian process (green, same scheme as left). The belief for when a particular amount R_{target} is reached is a (normalised) cut through this Gaussian belief along a particular output value. In particular, it is *not* a Gaussian distribution (blue lines) (Color figure online).

by using a nonparametric Gauss-Markov process prior for the download rate, with a parametric mean function, which leads to a filtering method [11] that has constant, very low cost, while capturing the important features of the rate distribution. Our algorithm has been implemented as a plugin for the open-source Firefox browser, available on the project page.

To simplify exposition and ease intuition, the remainder of this text will use concepts like the 'download rate' and 'data volume' specific to the example application of web browser's download progress bar. The results translate easily to other applications (e.g. the download rate could be replaced with the rate of products being sold, the rate at which funding pledges arrive, etc.).

2   Model

We assume that the algorithm has access to a rate function $r(t) : \mathbb{R} \to \mathbb{R}$ at recurring (not necessarily regularly spaced) time locations, describing the rate at which data accumulates. Crucially, observations $\boldsymbol{y} = [y(t_1), \ldots, y(t_N)]$ of $r(t)$ at times $T = [t_1, \ldots, t_N]$ can be made without observation noise, i.e. with Dirac likelihood $p(\boldsymbol{y} \,|\, r) = \delta(\boldsymbol{r}(T) - \boldsymbol{y})$. This is a realistic assumption for a web browser, and reasonable for many other applications (e.g. retailers have, of course, noiseless access to individual orders arriving).

We consider a Gaussian process prior over r, with constant mean β and Ornstein-Uhlenbeck [13] covariance function

$$p(r \mid \beta, \theta, \lambda) = \mathcal{GP}(r; \beta, k); \qquad \text{with} \qquad k(t, t'; \theta, \lambda) = \theta^2 \exp\left(-\frac{|t - t'|}{\lambda}\right). \quad (2)$$

Ornstein-Uhlenbeck processes are a meaningful and relatively conservative prior for rate functions: Draws from OU processes are stationary (there is no reason to assume, a priori, that download rates change qualitatively over time), and continuous, but almost surely not differentiable [9, Sect. 4.2.1], allowing for relatively drastic variations in rate while retaining a degree of extrapolation ability. They also are first-order Markov processes [10, Sect. I.23], a property that will become crucial for fast inference.

Further, we assign Gaussian uncertainty $p(\beta) = \mathcal{N}(\beta; 0, b)$ to the mean, and allow for arbitrarily large uncertainty on β, by taking $b \to \infty$. These priors and the Dirac likelihood give a Gaussian process posterior on r with mean and covariance functions [9, Sect. 2.7]

$$\mu(t) = k_{tT} K_{TT}^{-1} y + R_t^\mathsf{T} \bar{\beta} \qquad \text{and} \quad (3)$$
$$\mathbb{V}(t, t') = k_{tt'} - k_{tT} K_{TT}^{-1} k_{Tt'} + R_t^\mathsf{T} A^{-1} R_{t'}. \quad (4)$$

This is using the widely adopted notation $k_{ab} = k(a, b)$ for kernel Gram matrices: if a has n_a elements and b has n_b elements, then k_{ab} is a matrix of size $n_a \times n_b$. We will shorten $K_{TT} \equiv K$ from now on. The other objects in the equation are the residual projection $R_t \equiv 1 - 1^\mathsf{T} K^{-1} k_{Tt}$, the mean of the belief over the empirical mean $\bar{\beta} \equiv A^{-1} 1^\mathsf{T} K^{-1} y$ and its precision $A \equiv 1^\mathsf{T} K^{-1} 1$. The marginal likelihood, the evidence $p(y \mid T, \theta, \lambda)$ for the data, is also Gaussian. Its logarithm is

$$\log p(y \mid T, \theta, \lambda) = -\frac{1}{2} \left[y^\mathsf{T} K^{-1} y - A^{-1} (y^\mathsf{T} K^{-1} 1)^2 \right. \quad (5)$$
$$\left. + \log |K| + \log |A| + N \log 2\pi \right].$$

For efficient inference on the signal variance θ^2, we separate it from the kernel matrices, defining $\tilde{K} \equiv \theta^{-2} K$ and $\tilde{A} \equiv \theta^2 A$. The log evidence, as far as it relates to θ, can then be written up to additive terms as

$$\log p(y \mid T, \theta, \lambda) = -\frac{1}{2\theta^2} \left[y^\mathsf{T} \tilde{K}^{-1} y - \tilde{A}^{-1} (y^\mathsf{T} \tilde{K}^{-1} 1)^2 \right] + \frac{N}{2} \log \theta^{-2} + \text{const.} \quad (6)$$

For hierarchical inference, we introduce a Gamma prior on θ^{-2}:

$$\log p(\theta^{-2} \mid a, b) = a \log b - \log \Gamma(a) + (a - 1) \log \theta^{-2} - \frac{b}{\theta^2}, \quad (7)$$

to get a Gamma posterior on θ^{-2} with

$$a' = a + \frac{N}{2}; \qquad b' = b + \frac{1}{2} \left[y^\mathsf{T} \tilde{K}^{-1} y - \tilde{A}^{-1} (y^\mathsf{T} \tilde{K}^{-1} 1)^2 \right], \quad (8)$$

which has its mean at a/b, its maximum at $(a - 1)/b$, and has variance a/b^2.

2.1   Fast Gauss-Markov Inference

General matrix inversion is computationally expensive if the size of the matrix grows with the number of datapoints. Furtunately, because the Ornstein-Uhlenbeck process is first-order Markov and the input domain (time) is scalar, the Gram matrix K can be inverted analytically, which gives rise to a lightweight, constant cost filtering algorithm.[1] Let there be N observations at sorted locations $t_1 < t_2 < \cdots < t_N$, and let the scaled distance between locations be $\delta_i = (t_{i+1} - t_i)/\lambda$, $i = 1, \ldots, N - 1$. Then a straightforward but tedious inductive argument shows that the inverse of the Gram matrix K with $K_{ij} = \theta^2 e^{-\sum_{k=\min(i,j)}^{\max(i,j)-1} \delta_k}$ is given by the symmetric tri-diagonal matrix

$$K^{-1} = \theta^{-2} \begin{pmatrix} c_1 & b_1 & 0 & \cdots & & 0 \\ b_1 & c_2 & b_2 & & & \vdots \\ 0 & b_2 & c_3 & \ddots & & 0 \\ & & & \ddots & c_{N-1} & b_{N-1} \\ 0 & 0 & & & & \\ 0 & \cdots & 0 & & b_{N-1} & c_N \end{pmatrix} \quad \text{with} \quad \begin{aligned} b_i &= \frac{-e^{-\delta_i}}{1-e^{-2\delta_i}} \\ c_1 &= \frac{1}{1-e^{-2\delta_1}} \\ c_N &= \frac{1}{1-e^{-2\delta_{N-1}}} \\ c_{i \neq 1, N} &= \frac{1-e^{-2(\delta_i+\delta_{i+1})}}{(1-e^{-2\delta_i})(1-e^{-2\delta_{i+1}})}. \end{aligned}$$

$$(9)$$

To simplify notation for the following derivations, we use the shortcut $\Delta_i = \exp(-\delta_i)$. Neatly, the starting case ($N = 0$) can be incorporated by considering an effective additional datapoint at $-\infty$ without changing the results, showing that the matrix K^{-1} is actually circular. We also find that $\log |\tilde{K}^{-1}| = -\sum_{i=1}^{N-1} \log(1 - e^{-2\delta_i})$. From the derivations in the preceding section, we observe that the sufficient statistics for inference are the scalar objects $\mathbf{1}\tilde{K}^{-1}\mathbf{1}$, $\mathbf{1}\tilde{K}^{-1}\mathbf{y}$ and $\mathbf{y}\tilde{K}^{-1}\mathbf{y}$. Using the aforementioned datapoint at $-\infty$, which amounts to $\delta_0 = \delta_N = \infty$ and thus $b_N = 0$, we get the compact form

$$\alpha \equiv \mathbf{1}_i \tilde{K}_{ij}^{-1} y_j = \sum_{i=1}^{N} c_i y_i + b_i(y_i + y_{i+1}) \quad \tilde{A} = \mathbf{1}_i \tilde{K}_{ij}^{-1} \mathbf{1}_j = \sum_{i=1}^{N} c_i + 2b_i \quad (10)$$

$$\gamma \equiv y_i \tilde{K}_{ij}^{-1} y_j = \sum_{i=1}^{N} c_i y_i^2 + 2b_i(y_i y_{i+1}). \tag{11}$$

Up to this point, calculation of the different parts of the inference still is linear in the number of observables. For a constant-cost filtering rule, we treat the update from N to $N+1$ data points explicitly. A few lines of algebra show that updates for the three sufficient statistics, after collecting a new observation y_{new} at time step t_{new}, and given the observation y_{old} and statistics from the previous time step t_{old}, at distance $\Delta \leftarrow \exp[-(t_{\text{new}} - t_{\text{old}})/\lambda]$, are

---

[1] The expositions in this section could also be formulated more generally (for Matérn-class covariances) in the framework of state-space models and associated filters [11]. The derivations here only work for our specific choice of the Ornstein-Uhlenbeck kernel (the first member of the Matérn class), but they allow a more straightforward treatment of the uncertainty on the parametric mean.

$$\tilde{A} \leftarrow \tilde{A} + \frac{1 - \Delta}{1 + \Delta} \qquad\qquad \alpha \leftarrow \alpha + \frac{y_{\text{new}} - y_{\text{old}}\Delta}{1 + \Delta} \qquad (12)$$

$$\gamma \leftarrow \gamma + \frac{(y_{\text{new}} - y_{\text{old}}\Delta)^2}{1 - \Delta^2} \qquad\qquad t_{\text{old}} \leftarrow t_{\text{new}} \qquad y_{\text{old}} \leftarrow y_{\text{new}} \qquad (13)$$

To perform this computation at any point in time, we only need to keep the variables t_{old}, y_{old}, $\tilde{A}$, α_{old}, γ_{old}, a, and b in memory. Due to the first order Markovianity of the Ornstein Uhlenbeck process, mean and covariance of the Gaussian posterior can then be found using the simple forms [11]

$$\Delta_* \equiv \exp[-(t_* - t_{\text{old}})/\lambda] \qquad\qquad R = (1 - \Delta_*) \qquad (14)$$
$$\mu(t_*) = \Delta_* y_{\text{old}} + R^{\mathsf{T}}\bar{\beta} \qquad\qquad \mathbb{V}(t_*) = \theta^2[(1 - \Delta_*^2) + R^{\mathsf{T}}\tilde{A}^{-1}R]. \qquad (15)$$

Using $\bar{\beta} = \frac{\alpha}{A}$. Because of the conjugacy of the Gamma prior to precisions of Gaussians, the posterior over θ is also inverse Gamma, with

$$a = a_0 + \frac{N}{2} \quad \text{and} \quad b = b_0 + \frac{1}{2}\left(\gamma - \frac{\alpha^2}{\tilde{A}}\right). \qquad (16)$$

which can be marginalized to give a Student-t prediction for the rate (see Eq. (23) and following below). We initialise the set of variables to

$$t_{\text{old}} \leftarrow -\infty \qquad y_{\text{old}} \leftarrow 0 \qquad \tilde{A} \leftarrow \tilde{A}_0 \qquad (17)$$
$$\alpha_{\text{old}} \leftarrow \alpha_0 \qquad \gamma_{\text{old}} \leftarrow 0 \qquad a \leftarrow a_0 \qquad b \leftarrow b_0 \qquad (18)$$

with sensible values for $\tilde{A}_0, \alpha_0, a_0, b_0$, which can be used to propagate experience from past runs of the algorithm. For example, if the average download rate in previous runs was $\bar{r}$ with an empirical variance of σ_r^2, we set $\alpha_0 = \bar{r}$, $\tilde{A}_0 = 1$, $b_0 = 1e{-}2$, $a_0 = \sigma_r^2 b_0$ to get a broad Gauss-Gamma prior. For the very first download on a particular network connection, Algorithm 1 below contains sensible standard values for the progress bar setting, with rates measured in kB/s.

3   Constructing Predictions

The Gaussian family is closed under linear operations L:

$$p(t) = \mathcal{N}(t; m, V) \qquad \Rightarrow \qquad p(Lt) = \mathcal{N}(Lt; Lm, LVL^{\mathsf{T}}). \qquad (19)$$

Since integration is a linear operation, a belief over the integral over the rate, the accumulated data at time t_*, $d(t_*) = \int_{t_0}^{t_*} r(t)\, dt$, can be constructed from the Gaussian process posterior mean $\mu(t)$ and covariance $\mathbb{V}(t, t')$ as

$$p(d(t_*)\,|\,\theta) = \mathcal{N}\left(d(t_*); \int_{t_0}^{t_*} \mu(t)\, dt, \iint_{t_0}^{t_*} \mathbb{V}(t, t')\, dt\, dt'\right) \qquad (20)$$

$$= \mathcal{N}\left(d(t_*); \ell_{t_*T} K^{-1} y + \mathfrak{R}_{t_*}^{\mathsf{T}}\bar{\beta}, \kappa_{t_* t_*} - \ell_{t_* T} K^{-1}\ell_{T t_*} + \mathfrak{R}_{t_*}^{\mathsf{T}} A^{-1}\mathfrak{R}_{t_*}\right),$$

with the integrated projection operators (assuming the predictive distribution is only evaluated for future time points $t > t_N$)

$$\ell_{t_*T} = \int_{t_0}^{t_*} k(t,T)\, dt, \quad \Re_{t_*} = \int_{t_0}^{t_*} R_t\, dt, \quad \text{and} \quad \kappa_{t_*t_*} = \iint_{t_0}^{t_*} k(t,t')\, dt\, dt'.$$
(21)

In fact, as pointed out above, due to the first-order Markovianity of the Ornstein Uhlenbeck process, $\mu(t)$ and marginal variance $\mathbb{V}(t,t)$ only depend on the last observed function value, $y_{\text{old}} = r(t_{\text{old}})$. The integral prediction is simply given by

$$p(d(t_*)\,|\,\theta) = \mathcal{N}\left[d(t_*); \ell_{t_*t_{\text{old}}} y_{\text{old}} + \Re_{t_*t_{\text{old}}}\bar{\beta}, \theta^2 [\kappa_{t_*t_*} - \ell^2_{t_*t_{\text{old}}} + \Re^{\mathsf{T}}_{t_*t_{\text{old}}}\tilde{A}^{-1}\Re_{t_*t_{\text{old}}}]\right]$$

$$\equiv \mathcal{N}\left[d(t_*); \mu_f(t_*), \theta^2\sigma_f^2(t_*)\right]$$
(22)

$$\text{with} \qquad \ell_{t_*t_{\text{old}}} = \lambda\left(1 - \exp\left(-\frac{t_* - t_{\text{old}}}{\lambda}\right)\right)$$

$$\text{and} \qquad \Re_{t_*t_{\text{old}}} = (t_* - t_{\text{old}}) - \ell_{t_*f} \qquad \text{and} \qquad \kappa_{t_*t_*} = 2\lambda\left[(t_* - t_{\text{old}}) - \ell_{tt_{\text{old}}}\right]$$

The uncertainty in θ is incorporated by marginalisation, using the Gamma posterior from Eq. (8). The resulting marginal over $d(t_*)$ is a Student t-distribution e.g. [2, Sect. 2.3]

$$p(d(t_*)) = \int_0^\infty p(d(t_*)\,|\,\theta)\mathcal{G}(\theta^{-2}\,|\,a,b)\, d\theta^{-2}$$

$$= \int_0^\infty \frac{b^a \exp(-b/\theta^2)\theta^{-2(a-1)}}{\Gamma(a)\sqrt{2\pi\theta^2\sigma_f^2(t_*)}} \exp\left(-\frac{\theta^{-2}}{2}\left(\frac{d(t_*) - \mu_f(t_*)}{\sigma_f(t_*)}\right)^2\right) d\theta^{-2}$$

$$= \frac{b^a}{\sqrt{2\pi}}\frac{\Gamma\left(a + \frac{1}{2}\right)}{\Gamma(a)}\left[b + \frac{(d(t_*) - \mu_f(t_*))^2}{2\sigma_f^2}\right]^{-a-1/2}$$

$$= \text{St}(d(t_*)/\sigma_f^2; \mu/\sigma_f^2, a/b, 2a).$$
(23)

To construct a density for the probability of target D being reached at time t_*, we interpret the density of Eq. (23) as $p(d(t_*) = D)$ for every t_*, and normalise. Doing so causes a small error: The physical rate is strictly positive $r(t) \geq 0$, so there is one and only one correct D. But our Gaussian model puts a small amount of probability mass on negative rates. Hence interpreting a normalised $p(d(t_*) = D)$ as a density on D puts too much mass on "late" times, which are only possible under negative rates. The exact correction — enforcing strictly positive rates everywhere — involves an intractable integral. One could consider constructing a correction through additional term multiplied with $p(D = d(t_*))$, decaying for large t_*. Another, simpler option is to use a "warped GP" strictly positive prior [12]. Inference in such models can be performed approximately by linearization [5]. However, our empirical studies suggest that the error is small

---

Algorithm 1. Probabilistic progress bar. Every time t_{new} a new rate y_{new} is observed Inference updates the posterior belief. Using its results, Predict(D, t_i, t_*) returns the likelihood that data volume D will accumulate from time t_i to time t_*. The routine Initialise sets prior assumptions.

1: **procedure** INITIALISE
2:　　$t_{\mathrm{old}} \leftarrow -\infty$, $y_{\mathrm{old}} \leftarrow 0$, $A \leftarrow 0$, $\alpha \leftarrow 0$, $\gamma \leftarrow 0$, $\lambda \leftarrow 30[s], a_0 \leftarrow 0.1, b_0 \leftarrow 10^6$
3: **end procedure**
1: **procedure** INFERENCE$(t_{\mathrm{new}}, y_{\mathrm{new}})$
2:　　$\Delta \leftarrow \exp[-(t_{\mathrm{new}} - t_{\mathrm{old}})/\lambda]$　　　　　　　　　　　▷ scaled distance
3:　　$N \leftarrow N + 1$　　　　　　　　　　　　　　　　▷ observation count
4:　　$A \leftarrow A + \frac{1-\Delta}{1+\Delta}$　　　　　　　　▷ residual uncertainty on mean
5:　　$\alpha \leftarrow \alpha + \frac{y_{\mathrm{new}} - y_{\mathrm{old}}\Delta}{1+\Delta}$　　　　　▷ sufficient statistics for signal mean
6:　　$\gamma \leftarrow \gamma + \frac{(y_{\mathrm{new}} - y_{\mathrm{old}}\Delta)^2}{1-\Delta^2}$　　▷ sufficient statistics for signal variance
7:　　$t_{\mathrm{old}} \leftarrow t_{\mathrm{new}}, y_{\mathrm{old}} \leftarrow y_{\mathrm{new}}$
8:　　$\beta \leftarrow \frac{\alpha}{A}$, $a \leftarrow a_0 + \frac{N-1}{2}$, $b \leftarrow b_0 + \frac{1}{2}\left(\gamma - \frac{\alpha^2}{A}\right)$
9: **end procedure**
1: **procedure** PREDICT$(D, t_i, t_*; t_{\mathrm{old}}, y_{\mathrm{old}}, A, \beta, a, b)$
Ensure:　　$t_i > t_{\mathrm{old}} \wedge t_* > t_i$　　　　　　　　　　▷ assumptions valid?
2:　　$\delta_* \leftarrow t_* - t_i$
3:　　$k_* \leftarrow \lambda[e^{-(t_i - t_{\mathrm{old}})/\lambda} - e^{-(t_* - t_{\mathrm{old}})/\lambda}]$
4:　　$\kappa \leftarrow 2\lambda(\delta_* - k_*)$　　　　　　　　▷ prior variance of integral
5:　　$\mu_* \leftarrow k_*(y_{\mathrm{old}} - \beta) + \delta_*\beta$　　　　　　　▷ post. mean
6:　　$\sigma_*^2 \leftarrow \kappa - k_*^2 + (\delta_* - k_*)^2/A$　　　　▷ post. variance
7:　　$\log p_{d(t_*)=D} \leftarrow -(a + \frac{1}{2})\log\left[b + \frac{(d(t_*) - \mu_{t_*})^2}{2\sigma_{t_*}^2}\right] + \mathrm{const.}$
8:　　**return** $\log p_{d(t_*)=D}$　　　　　　　▷ log Student-t likelihood
9: **end procedure**

---

overall, and can simply be ignored. The resulting overall procedure is given in Algorithm 1. It requires the storage of eight floating point numbers, whose update involves two exponential functions, one logarithm, and a handful of sums and products of floats. The procedure Predict returns the (logarithm of) the probability $p_{d(t_*)=D}$ that the download volume D will accumulate form the current time $t_i > t_{\mathrm{new}}$ to the target time t_*. This output can be used in two different ways, depending on the task and setting: To construct a graphical output for a *probabilistic progress bar*, as in Figs. 1 and 2, we evaluate $p_{d(t_*)=D}$ over a set grid of values for t_* (doing so is less expensive than constructing the graphical output itself). If the grid is fine enough, this probability (multiplied with a regulariser, if needed) can also be used to compute mean, variance, and other moments of the distribution over run times. For applications requiring the *most likely* time of completion, this time can be found by an efficient optimization. This can be done very efficiently, because the domain is one-dimensional, and derivatives of $p_{d(t_*)=D}$ can be computed to high order, at diminishing cost. So very efficient numerical optimization techniques, such as Halley's method, are applicable, which we have empirically found to converge within one or two steps in this setting.

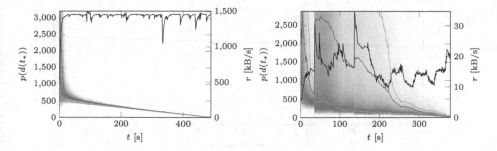

Fig. 3. Behaviour of the probabilistic error bar for two downloads. **Left:** a fast, consistent dial up connection for a 605 Mb file. **Right:** a shaky cell phone connection for a 5 Mb file. Download rates in black, scale on right ordinate. True remaining time of download as linear gray line (scale on left ordinate), probabilistic prediction shaded in green. The dark green line is the mean of the progress bar algorithm's posterior predictive distribution. The edges of the shaded regions mark the 10 % and 90 % quantiles of the predictive distribution (Color figure online).

4   Experiments

We tested our model both in the progress bar setting on several pre-recorded downloads (Sect. 4.1), and in a related task (see Sect. 1), predicting the time of a crowd-funding project to reach its target amount, on the Kickstarter platform (Sect. 4.2).

4.1   Progress Bars for Pre-recorded Downloads

Figure 3 shows results from downloads of two single files in separate settings: A relatively large (605 MB) file over a reliable pipe, and a small (5 MB) file over an unreliable connection. See the figure caption for a description of the plots. A comparison between the predicted completion times (light green quantile shades) and ground truth (gray line) shows how the predictor converges to a good prediction, but also assigns meaningful uncertainty around its prediction. Note the strongly asymmetric form of the prediction, with median and most likely prediction typically close to the true value. The ramp-up phase early in the download is actually a feature of the download itself, and not of the estimation procedure (note that the rates r, plotted as black lines, are initially low).

4.2   Kickstarter

Kickstarter[2] is an online platform on which companies and individuals can ask for financial support for their projects. Every project chooses a deadline and a financial target, community members pledge money during this time window. If the set amount of money is reached within the time period, the project is

---

[2] http://www.kickstarter.com

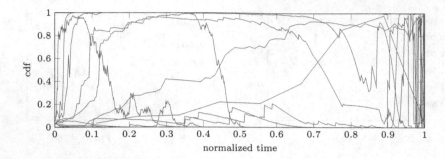

Fig. 4. Trajectories of ten randomly selected kickstarter projects scaled relative to the posterior cumulative distribution function (cdf) of the algorithm. If the model were perfect, these curves should be uniformly distributed across the [0,1] simplex.

considered successfully funded, and the pledged amount is transferred from to the project owners.

To get a sense of the algorithm's probabilistic calibration, we predicted the completion (or failure) time of kickstarter projects with a volume higher or equal to 20.000 USD from a dataset[3] in [4]. 300 projects were set aside as a training set to select the hyperparameters (a_0, b_0) of the prior (7). Independent of the success of the project we fixed the time window to either the point in time when the project got funded, or the original deadline occurred. The funding rate can then be approximated well from finite differences on the 1000 collected funding states.

Figure 4 shows, for 10 randomly selected projects, the position of the true finishing time within the algorithm's cumulative density function over the course of the pledge. The trajectories cover the entire range of the distribution, and often move through it over the course of the window, indicating good coverage of the distribution. At the same time, they are also not truly uniformly distributed, reflecting minor model flaws, the price paid for the low computational cost.

5   Conclusion

We derived a nonparametric algorithm for the probabilistic prediction of the completion time of a stochastically increasing process. Using a Gauss-Markov process prior with a parametric mean function, and analytically integrating over several hyperparameters, we arrived at a filtering algorithm of constant, very low cost, which nevertheless provides a nonparametric probabilistic prediction for the completion time. As pointed out in the introduction, such algorithms have numerous potential applications. One of them is to provide enhanced visual feedback on the progress of a file download in a web browser, a probabilistic error bar. An implementation of this method can be found on the project page[4].

---

[3] Dataset available at http://sidekick.epfl.ch/data.
[4] http://people.tuebingen.mpg.de/mkiefel/projects/mlprogressbar/

References

1. Ajne, B., Daleniua, T.: Några tillämpningar av statistika ideer på numerisk integration. Nordisk Math. Tidskrift **8**, 145–152 (1960)
2. Bishop, C.: Pattern Recognition and Machine Learning. Springer, Berlin (2006)
3. Diaconis, P.: Bayesian numerical analysis. Stat. Decis. Theory Relat. Top. IV **1**, 163–175 (1998)
4. Etter, V., Grossglauser, M., Thiran, P.: Launch hard or go home!: predicting the success of kickstarter campaigns. In: Proceedings of the First ACM Conference on Online Social Networks, COSN '13, pp. 177–182. ACM, New York (2013)
5. Garnett, R., Osborne, M., Hennig, P.: Active learning of linear embeddings for Gaussian processes. In: Uncertainty in Artificial Intelligence (2014)
6. Minka, T.: Deriving quadrature rules from Gaussian processes. Technical report, Statistics Department, Carnegie Mellon University (2000)
7. Osborne, M., Duvenaud, D., Garnett, R., Rasmussen, C., Roberts, S., Ghahramani, Z.: Active learning of model evidence using bayesian quadrature. In: Advances in NIPS, pp. 46–54 (2012)
8. Peres, S., Kortum, P., Stallmann, K.: Auditory progress bars: preference, performance and aesthetics. In: Proceedings of the 13th International Conference on Auditory Display, Montreal, Canada, 26–29 June 2007
9. Rasmussen, C., Williams, C.: Gaussian Processes for Machine Learning. MIT, Cambridge (2006)
10. Rogers, L., Williams, D.: Diffusions, Markov Processes and Martingales, vol. 1: Foundations, 2nd edn. Cambridge University Press, Cambridge (2000)
11. Särkkä, S.: Bayesian Filtering and Smoothing, vol. 3. Cambridge University Press, Cambridge (2013)
12. Snelson, E., Rasmussen, C., Ghahramani, Z.: Warped Gaussian processes. In: Advances in Neural Information Processing Systems (2004)
13. Uhlenbeck, G., Ornstein, L.: On the theory of the brownian motion. Phys. Rev. **36**(5), 823 (1930)

Wide Base Stereo with Fisheye Optics: A Robust Approach for 3D Reconstruction in Driving Assistance

Jose Esparza[1,2]($\boxtimes$), Michael Helmle[2], and Bernd Jähne[1]

[1] Heidelberg Collaboratory for Image Processing, Heidelberg, Germany
[2] Robert Bosch GmbH, Chasis Control Driving Assistance, Leonberg, Germany
jose.esparza@de.bosch.com

Abstract. We propose a new approach to achieve 3D environment reconstruction based on automotive surround view systems with fisheye cameras. In particular, we demonstrate that stereo vision techniques can be applied in overlapping areas of adjacent cameras, which are up to 90 degrees per camera pair in the current setup. Lateral limitations are mainly due to the present system configuration and can be extended. No time accumulation is required, therefore the update rate of the range information is given by the frame rate of the imager. We show by means of experimental results that our approach is capable of delivering 3D information from a pair of images under the described configuration.

1   Introduction

Stereo vision is a topic, which has been largely studied in literature in the last decades [7,8,11]. In recent years, the automotive sector has shown strong interest in bringing this technology into real products, given the high accuracy of the 3D measurements [13,15]. State of the art approaches make use of one or more pairs of identical cameras which are mounted close to each other and have the optical axes aligned in a way to achieve a maximum coverage of a common field of view. The 3D environment reconstruction for such a configuration is normally done via epipolar rectification, feature detection, matching and triangulation of correspondences. Stereo vision based on cameras with strongly disaligned optical axes - close to orthogonality - has been studied only sparsely in literature [3].

Recently, automotive surround view systems have become increasingly popular. These are designed to display the very near range surrounding to the driver in order to prevent collisions during parking and maneuvering [14]. A common layout is to have a camera mounted on each side of the vehicle, usually in the side mirrors, one on front of the vehicle usually hidden in the grill and one in the tail gate or close to it, as in the work of [12]. The mounting position of the cameras is optimized in order to cover close to 360 degrees of the near range of the vehicle surrounding and to be concord with the vehicle exterior design. This is possible with only four cameras thanks to the large fields of view that can be imaged with fisheye optics.

© Springer International Publishing Switzerland 2014
X. Jiang et al. (Eds.): GCPR 2014, LNCS 8753, pp. 342–353, 2014.
DOI: 10.1007/978-3-319-11752-2_27

Normally, there also exist areas in the surrounding of the vehicle where the fields of view of more than one camera overlap. These overlapping areas are of high interest, since stereo measurements can be conducted. This paper proposes the application of stereo vision techniques under the described setup and discusses the main technical challenges of such an approach, as well as its benefits. By means of experimental results it is shown that it is a feasible system for real applications.

Up to our knowledge, there is no previous work addressing stereo vision with fisheye optics considering a similar setup to our proposed one, namely with wide stereo bases and largely disaligned optical axes. The reason why this solution has not been largely studied before is likely to be the low resolution of cameras available for the automotive market. Recently, imagers qualifying for automotive certification have reached megapixel resolutions, creating opportunities for new approaches. A solution to generate ground truth data based on a lidar sensor is also proposed.

This paper is organized in the following way: Sect. 2 introduces a review of existing related work and Sect. 3 describes the different steps in our proposed approach. In Sects. 4 and 5 our experimental setup and ground truth scheme are presented with the obtained results. Discussion about the benefits of our proposal can be found in Sect. 6 and conclusions are presented in Sect. 7.

2   Previous Work

In the context of driving assistance, the work of [12] makes use of a similar camera setup to ours. The aim of the authors is the extrinsic calibration of the camera rig. Although they do not consider the overlap on the fields of view, they perform a time analysis that allows them to match common features accross different camera views based on a local history and motion estimation. With regard to stereo vision using wide stereo bases, the authors of [3] consider stereo matching for widely separated views and propose a feature matching method robust to local affine distortions. The goal of their work is to establish relative camera positions and orientations, since these are not assumed to be known. The work of [9] proposed a stereo vision system with fisheye optics which relies on a pin-hole model rectification prior to the feature detection-matching, thus being limited to camera setups that are placed to the left and right of the rearview mirror.

Epipolar Rectification for Fisheye Optics

Epipolar rectification is a standard step in stereo vision processing, in order to reduce the correspondence search space, thus saving in computation efforts [11]. It can be regarded as a projection of the world into a virtual camera pair, which fulfills the restriction that each common point observed by both cameras is imaged on the same image row. In conventional perspective cameras, a linear rectifying transformation H exists, as described in Eq. 1, where K_C, K_V are the projection matrices of the original and virtual cameras respectively, and R is the rotation that is applied to the original camera to rectify it.

Fig. 1. Example of epipolar lines with fisheye optics corresponding to a front - right camera pair. Left: Front camera. Right: Right camera

$$H = K_V R K_C^{-1} \qquad (1)$$

In the case of fisheye optics, however, K_C has to be substituted by a nonlinear function T_C due to their nonperspectivity. For the same reason, epipolar planes do not project into the image planes as straight lines [1]. An example of the shape of the epipolar lines is depicted in Fig. 1. Several models have been proposed in literature to describe T_C ([16,18]) that also account for lens distortions. The latter, which is used in this work, is a modified version of the unified projection model introduced by [2,10].

Furthermore, K_V accounts for a perspective rectification model, projecting the rectified field of view onto a virtual focal plane. For traditional stereo setups with narrow fields of view this is a valid rectification model. On our proposed setup, however, large field of view overlaps exist between adjacent cameras and perspective rectification models are known to introduce severe image distortions for very large fields of view [9]. Therefore a nonperspective rectification model T_V is more suitable. In the work of [1], a model was proposed for epipolar rectification of fisheye images, which allows a maximum field of view rectification in both vertical and horizontal directions.

3   Proposed Approach

In the following, the steps required to apply stereo vision to surround view systems are introduced. Special attention is paid to aspects which are not relevant in traditional stereo vision, but pose a challenge on the current system configuration.

Camera Calibration

For every camera on our system setup, we assume that previous intrinsic and extrinsic calibration with respect to a common reference frame are available. For the intrinsic calibration we propose using the model described in [16], due to its simplicity and small overall reprojection error. This model accounts both for projection and distortion functions on fisheye cameras. As a global common reference frame for the extrinsic calibration, we consider the vehicle's origin of coordinates as defined by the norm DIN70000, which establishes the origin on the rear axis, at ground level. We use the ground plane as reference, which normal vector is defined as $\hat{e}_z$. This calibration is done by means of special calibration targets as well as additional cameras and bundle adjustment.

Description of the Overlapping Fields of View

On the current system configuration, given the strong disalignment of the optical axes, the fields of view of each camera of a camera pair overlap only partially. Therefore epipolar rectification is only meaningful on the fraction of the images that correspond to this area.

In the following we propose a simple two dimensional model of the camera setup, in order to describe the effective field of view of each camera pair. Let us assume all camera centers to be contained on a single plane parallel to the ground plane, and the maximum field of view of the cameras to be contained on this plane. This is a reasonable assumption since the considered camera setup fulfills the condition that $\|C_{z,L} - C_{z,R}\| \ll \|C_{xy,L} - C_{xy,R}\|$, where C_L and C_R represent the positions of the Left and Right cameras on each stereo pair. Having the camera heights aligned is not a necessary condition for the following algorithmic steps, but the analysis is simplified.

Based on this assumption, and with known intrinsic and extrinsic calibration, Eq. 2 can be defined that expresses the effective field of view of the left camera; i.e. the fraction of its own field of view that may overlap with the field of view of its adjacent right camera.

$$\left[\psi^-, \psi^+\right]_L \approx \left[-\cos^{-1}(\hat{O}_{xy,L}^T R_{F/2} \hat{O}_{xy,R}), F/2\right] \tag{2}$$

In this expression, $R_{F/2}$ represents a 2D rotation matrix over the normal $\hat{e}_z$ of the ground plane, of magnitude equal to half of the field of view of the cameras. $\hat{O}_{xy}$ stands for the normalized projection of the optical axis of each camera over the XY plane. A similar expression can be obtained for the right camera.

This model provides an estimate of the amount of overlap existing for each pair of adjacent cameras. This approximation is only valid as long as the 2D optical axes of the cameras do not intersect in front of both image planes. This situation is not possible on the discussed setup, since the cameras are mounted on each side of the vehicle, looking outwards. From the previous expresion, it can be inferred that the field of view of the resulting virtual cameras will be asymmetric with respect to their principal points. In the following we analyze this effect.

Let us assume $\hat{O}_V$ to be the optical axis of the virtual camera, defined by the principal point. We can describe $\hat{O}_V$ by Eq. 3, where $\hat{t}_{C_R C_L}$ is the normalized vector joining both camera centers.

$$\hat{O}_V = \hat{t}_{C_R C_L} \times \hat{e}_z \tag{3}$$

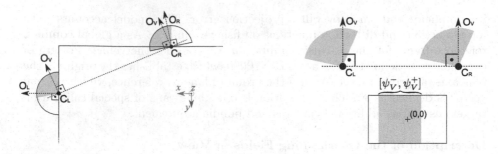

Fig. 2. Overview on the stereo pair overlaps. Left: 2D representation of overlapping fields of view for 2 adjacent cameras. Right: Nonsymmetric rectified fields of view with respect to the principal point

Based on the previous definitions, we can describe the asymmetric field of view $[\psi^-, \psi^+]_V$ of the rectified virtual cameras by means of Eq. 4.

$$[\psi^-, \psi^+]_V = \left[-\cos^{-1}(\hat{\boldsymbol{O}}_{xy,V}^T R_{F/2} \hat{\boldsymbol{O}}_{xy,R}), \cos^{-1}(\hat{\boldsymbol{O}}_{xy,V}^T R_{F/2}^{-1} \hat{\boldsymbol{O}}_{xy,L}) \right] \quad (4)$$

In this expression, $\hat{\boldsymbol{O}}_{xy,V}$ represents the normalized projection of $\hat{\boldsymbol{O}}_V$ over the XY plane. Figure 2 depicts how the principal point is largely displaced with respect to the center of the field of view shared by a camera pair.

Epipolar Rectification

Once the shared field of view of adjacent cameras is defined, the rectification model can be introduced. In particular, considering the use of fisheye cameras, the linear transformation presented in Eq. 1 has to be replaced by a nonlinear transformation. As already discussed earlier in this section, we consider the model proposed in [16] to represent $\boldsymbol{T}_C$, which accounts for both the projection and distortions of the fisheye optics.

For the rectification model $\boldsymbol{T}_V$, we propose using a modified version of the epipolar-equidistance rectification model introduced in [1]. This model allows for epipolar rectification of very large fields of view, providing nonperspective rectification with a well defined center of symmetries (or principal point). In the following we propose a new formulation for the inverse rectification model introduced in [1], where the offset of the center of symmetries with respect to the virtual image center is accounted for.

Let us start by defining a reference epipolar plane that contains the principal point, as introduced in Eq. 3. Over the reference epipolar plane, a horizontal field of view of $\psi_V = [\psi_V^+ - \psi_V^-]$ is covered, which can be computed by means of Eq. 4. On the reference epipolar plane, the view ray that projects onto each pixel position (u, v) can be computed by means of Eq. 5.

$$\hat{\boldsymbol{d}}_0(u) = Rot([t_{C_R C_L} \times \hat{\boldsymbol{e}}_z] \times \hat{\boldsymbol{e}}_z, (\psi_V^- + u \cdot \Delta\psi_V))\hat{\boldsymbol{O}}_V \quad (5)$$

In this expression, $Rot(\boldsymbol{e}, \alpha)$ is a 3×3 matrix that defines a rotation around axis $\boldsymbol{e}$ by an angle α, and $\Delta\psi$ is the angular distance between two consecutive pixels on the same row of the virtual image.

The rest of the epipolar planes can be described as a revolution of the reference one, over the line joining both camera centers. According to this, the inverse projection function $T_V^{-1}(u, v)$ of the virtual cameras can be defined as in Eq. 6.

$$T_V^{-1}(u, v) = Rot(t_{C_R C_L}, (\beta_V^- + v \cdot \Delta\beta_V))\hat{d}_0(u) \tag{6}$$

This inverse function describes the viewing direction for each pixel coordinate $(u, v)_V$ and maps it onto the unit sphere. The angular step $\Delta\beta$ corresponds to the distance between consecutive epipolar planes and $\beta_V = [\beta_V^+ - \beta_V^-]$ can be defined based on ψ_V and on the desired aspect ratio.

Change of Pixel Sizes

We have so far described how the virtual views can be defined so that constraints for epipolar rectification are met. At this point, the next step is to resample the original images at the desired locations, which implies a change in pixel sizes. This effect is especially visible when considering fisheye optics and disaligned optical axes, as depicted in Fig. 1 by means of the epipolar lines. The problem of image interpolation has been largely discussed in literature and several techniques have been proposed to minimize the presence of jaggies and similar image artifacts. In particular, the works of [19, 20] evaluate these effects in detail. In our experiments, we have applied a Lanczos filter [19] with a size parameter $a = 4$ for image interpolation. We have not seen a significant change by applying different size parameters.

After resampling, rastering of the virtual images is possible. In Fig. 3 an example of the results after all the described steps is shown. As can be observed, epipolar rectification is achieved for a large field of view, while avoiding the mentioned image artifacts.

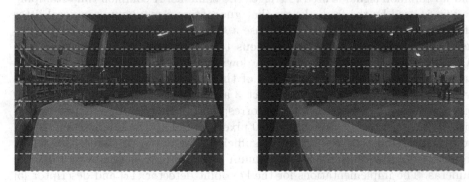

Fig. 3. Result of the rectification process. The image pair corresponds to the original images in Fig. 1. It can be observed how now the epipolar planes project into straight lines on the virtual images

Feature-Based Disparity Estimation

After conducting the epipolar rectification step, detection and matching of features is to take place. In standard stereo vision setups, the vertical disparity of common features after epipolar rectification is expected to be below one pixel. Nonaccurate calibration may lead to failures in fulfilling this requirement. In our surround view setup, due to the large stereo bases, factors like temperature changes, vibrations, etc., introduce a high variance on the relative camera calibration. For this reason, the assumption of a static extrinsic calibration is not strictly valid. In order to deal with this issue, we propose to relax the assumption that the vertical offset between different views of a common features is below one pixel, and accept a larger tolerance on the vertical direction. In this way, vertical offsets larger than a predefined maximum can be discarded. How large this tolerance should be, depends largely on the quality of the camera intrinsic and extrinsic calibration, as well as on the chosen resolution for the epipolar-rectified images. For our setup, the values utilized are described in the Sect. 4. As keypoint detector and descriptor we considered those proposed by [5,17], respectively.

After the matching process, triangulation of correspondences is carried out. For rays not perfectly intersecting, the mid-point of the segment covering the shortest distance between both is considered.

4   Experimental Setup

We consider 4 cameras for our experiments, which offer a resolution of 1280×960 pixels covering a horizontal field of view of approximately 180 degrees per camera. In Table 1, the effective field of view is shown for each adjacent camera pair on the configuration used. The operating frequency is set to 30 frames per second and no common signal is used to trigger the cameras. A common time-stamping system is used by the frame logger that guarantees a maximum temporal jitter of half a frame period. The cameras use a CMOS technology with rolling shutter, and frames are compressed previous to storage using JPEG compression. Although compression is expected to downgrade the performance, the effects are not analyzed here, since it is out of the scope of this paper. The epipolar rectification is done as described in Sect. 3 and an output resolution of 640×480 pixels is used. A search window for corresponding features is set equal to ± 3 pixels on the vertical direction and 200 pixels on the horizontal direction. It can be demonstrated that 200 pixels are sufficient, for our wide stereo base setup, to detect objects which stand a minimum distance of 6 meters away from the cameras. The implementations for the keypoint detector [17] and descriptor [5] are those available within the OpenCV Library [4] and matching is performed based on Hamming distance.

5   Evaluation

In order to evaluate the accuracy of our measurements, we have chosen a Velodyne HDL-64E S2 LiDAR as reference sensor. The lidar was mounted on the

Table 1. Effective fields of view on current setup, calculated by means of Eqs. 2 and 4

Camera pair	$[\psi_L^-, \psi_L^+]$ [deg]	$[\psi_R^-, \psi_R^+]$ [deg]	$[\psi_V^-, \psi_V^+]$ [deg]
Front-Right	$[-4.60, 90.00]$	$[-90.00, 4.60]$	$[-60.76, 33.84]$
Right-Rear	$[-11.20, 90.00]$	$[-90.00, 11.20]$	$[-11.56, 67.23]$
Rear-Left	$[-7.03, 90.00]$	$[-90.00, 7.03]$	$[-70.88, 26.15]$
Left-Front	$[-0.43, 90.00]$	$[-90.00, 0.43]$	$[-29.69, 59.88]$

roof of the vehicle, and its position registered to the vehicle's coordinate system as described in [6], in order to be able to have a common reference for distance measurements. The accuracy of the lidar measurements is in the order of 5 cm, which we consider sufficient for benchmarking our results, since our expected accuracy is at least one order of magnitude lower. The rotation speed was set to 10 Hz, which means that ground truth data is only available for every third frame of the surround view system. We propose using the image plane as a common domain, where visual comparison is possible and an error metric can be defined for the accuracy of the measured distances. The 3D measurements from the lidar and from our stereo setup are projected into the image planes and compared. Figure 4 shows an overview of the complete setup utilized. Since the measurements given by the lidar are very sparse in the vertical direction - it has a vertical resolution of 64 lasers, compared to the 960 pixels of our imager - each measurement is thickened after reprojection, in order to become a denser depth reference. In particular, we use a 20-pixel high mask to achieve this.

As an error metric, we consider the difference between the depth values obtained for each triangulated keypoint and the nearest lidar measurement, projected on the proposed common domain. Let us assume a set C of 3D measurements obtained by our proposed stereo approach, and V the set of 3D points given as a result of the lidar measurements. The distance d^V of each point in V to the camera center can be computed since the lidar has been registered to the vehicle's coordinate system, as in [6]. The distance d^C of each point obtained with the stereo-camera approach to the camera center is trivial. We can now define a L1-norm error metric as in Eq. 7.

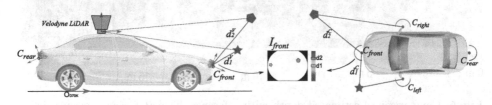

Fig. 4. Setup overview. Left: Configuration for reference measurements. The lidar was registered to the surround-view system, so that 3D measurements can be reprojected onto any image plane for reference. Right: Our proposed approach for 3D measurements

$$e_d = \frac{1}{\dim(C \cap V)} \sum_{i \in (C \cap V)} |d_i^V - d_i^C| \qquad (7)$$

Figure 5 shows an example of the common image domain for evaluation of the stereo depth measurements for the front camera, where color encodes distance to the camera center of each lidar and stereo measurement.

It is clear to the authors that the proposed error metric does not cover the entire measurement space of our system, but it covers the most relevant fraction of it for driving assistance applications. For evaluating this approach, 10 static sequences were considered. Results correspond to single frames, without time accumulation. The error is evaluated by reprojecting all measurements to the front and rear cameras only and results are shown in Table 2.

The average error achieved is approximately between ±1 and ±2 meters on a range which covers distances of up to 20 m, which is reasonable for park & maneuver systems. In most sequences, quartile information shows relative errors lower than 6 % and 20 % for 50 % and 75 % of all measurements, respectively. Although all sequences were recorded on similar conditions, we see a larger level of error on some of them. A deep look into the data shows a high level of confusion on the

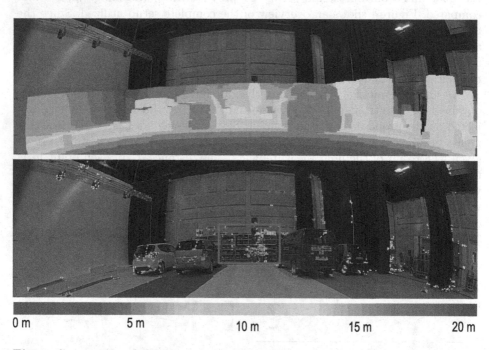

Fig. 5. Comparison of estimated depth with reference. Top: Reference measurements backprojected onto front image. Bottom: Measurements of the camera pairs left-front and front-right backprojected onto front image. Color encodes distances to camera center, on the same scale for both images. Red: High confusion due to repetitive pattern (Color figure online)

Table 2. Absolute and relative error analysis, as presented in Eq. 7. We include quartile information since it is representative for discussion of results

Seq. ID	Measurements	Mean [m]	σ [m]	Q50 [m]	Q75 [m]	Q50 [%]	Q75 [%]
1	5670	1.14	1.69	0.49	1.38	5.16	11.43
2	4280	0.97	1.76	0.38	1.04	4.17	9.23
3	1515	0.98	1.61	0.34	0.96	4.00	10.01
4	5115	1.31	1.85	0.52	1.84	6.51	17.51
5	6205	1.48	2.10	0.58	1.75	7.17	17.35
6	11084	1.10	1.69	0.42	1.31	4.96	12.07
7	4542	0.88	3.04	0.19	0.77	4.43	12.53
8	6487	2.54	6.58	0.25	1.28	4.27	16.54
9	6238	1.93	6.69	0.36	1.32	6.09	17.92
10	22531	0.62	1.36	0.21	0.52	3.34	7.68

feature matching process, due to certain repetitive patterns. This effect has been highlighted on Fig. 5.

6 Discussion

In the present system configuration, the considerable overlap of the field of view of any pair of adjacent cameras is limited to approximately 90 degrees. Therefore only in these regions 3D information could be recovered. Furthermore, objects in the very close vicinity of the ego vehicle show a very large disparity on the rectified images. We restricted our feature search to 200 pixels on the horizontal direction, which allows detection of objects which are, at least, in the order of 6 meters away from the vehicle. In the areas where 3D information could be recovered, the distances compare well with the distances obtained by the lidar, as can been seen in Fig. 5. No time accumulation is required, thus being the 3D information recovered from single pairs of images. We see limitations in using a lidar for benchmarking since the fields of view of both sensors are not completely coincident. In particular, objects too near to the ego vehicle, or too high, remain outside the visibility range of the lidar. The latter is, however, not crucial for our use cases, since for driving assistance such heights usually lack interest. There is a very wide field of potential applications for the proposed system within driving assistance, specially in low speed parking and maneuvering. Future work will focus on optimization of the current camera mounting as well as on performing measurements on a nearer area around the vehicle. The focus of this work was to demonstrate feasibility of our approach using standard open source disparity estimators, although proprietary algorithms with superior performance exist.

7   Conclusion

A robust approach to recover 3D information with the help of surround view cameras has been demonstrated. An overview of the different algorithmic steps required for such an approach to work has been presented, with focus on the restrictions that characterize a surround-view system. A model has been proposed for analysis of the effective stereo field of view on any given camera setup and a scheme for evaluation with respect to ground truth data generated by means of a lidar sensor. An error analysis has been carried out, with discussion on the accuracy of our 3D measurements. More details on the algorithmic part and possible applications will be the subject of forthcoming publications.

References

1. Abraham, S., Förstner, W.: Fish-eye-stereo calibration and epipolar rectification. ISPRS J. Photogramm. Remote Sens. **59**(5), 278–288 (2005)
2. Barreto, J.P., Araujo, H.: Issues on the geometry of central catadioptric image formation. In: Proceedings of the 2001 IEEE Computer Society Conference on Computer Vision and Pattern Recognition, CVPR 2001, vol. 2, pp. II-422. IEEE (2001)
3. Baumberg, A.: Reliable feature matching across widely separated views. In: Proceedings of IEEE Conference on Computer Vision and Pattern Recognition, 2000, vol. 1, pp. 774–781. IEEE (2000)
4. Bradski, G.: The opencv library. Dr. Dobb's J. Softw. Tools **25**, 120–126 (2000)
5. Calonder, M., Lepetit, V., Strecha, C., Fua, P.: BRIEF: binary robust independent elementary features. In: Daniilidis, K., Maragos, P., Paragios, N. (eds.) ECCV 2010, Part IV. LNCS, vol. 6314, pp. 778–792. Springer, Heidelberg (2010)
6. Esparza, J., Vepa, L., Helmle, M., Jaehne, B.: Extrinsic calibration of a 3D laser range finder to a multi-camera system. In: 9. Workshop Fahrerassistenzsysteme, Uni-DAS (2014). ISBN: 3000449558, 9783000449550
7. Faugeras, O.: Three Dimensional Computer Vision: A Geometric Viewpoint. MIT Press, Cambridge (1993)
8. Fusiello, A., Trucco, E., Verri, A.: A compact algorithm for rectification of stereo pairs. Mach. Vis. Appl. **12**(1), 16–22 (2000)
9. Gehrig, S.K.: Large-field-of-view stereo for automotive applications. In: Omnivis 2005, vol. 1 (2005)
10. Geyer, C., Daniilidis, K.: A unifying theory for central panoramic systems and practical implications. In: Vernon, D. (ed.) ECCV 2000. LNCS, vol. 1843, pp. 445–461. Springer, Heidelberg (2000)
11. Hartley, R., Zisserman, A.: Multiple View Geometry in Computer Vision, vol. 2. Cambridge University Press, Cambridge (2000)
12. Heng, L., Li, B., Pollefeys, M.: Camodocal: Automatic intrinsic and extrinsic calibration of a rig with multiple generic cameras and odometry. In: 2013 IEEE/RSJ International Conference on Intelligent Robots and Systems (IROS), pp. 1793–1800. IEEE (2013)
13. Labayrade, R., Aubert, D., Tarel, J.P.: Real time obstacle detection in stereovision on non flat road geometry through. In: IEEE Intelligent Vehicle Symposium, vol. 2, pp. 646–651. IEEE (2002)

14. Liu, Y.-C., Lin, K.-Y., Chen, Y.-S.: Bird's-eye view vision system for vehicle surrounding monitoring. In: Sommer, G., Klette, R. (eds.) RobVis 2008. LNCS, vol. 4931, pp. 207–218. Springer, Heidelberg (2008)
15. Lourakis, M.I., Orphanoudakis, S.C.: Visual detection of obstacles assuming a locally planar ground. In: Chin, R., Pong, T.-C. (eds.) ACCV 1998. LNCS, vol. 1352, pp. 527–534. Springer, Heidelberg (1997)
16. Mei, C., Rives, P.: Single view point omnidirectional camera calibration from planar grids. In: 2007 IEEE International Conference on Robotics and Automation, pp. 3945–3950 (2007)
17. Rosten, E., Drummond, T.: Fusing points and lines for high performance tracking. In: Tenth IEEE International Conference on Computer Vision, ICCV 2005, vol. 2, pp. 1508–1515. IEEE (2005)
18. Scaramuzza, D., Martinelli, A., Siegwart, R.: A toolbox for easily calibrating omnidirectional cameras. In: 2006 IEEE/RSJ International Conference on Intelligent Robots and Systems, pp. 5695–5701 (2006)
19. Turkowski, K.: Filters for common resampling tasks. In: Glassner, A.S. (ed.) Graphics Gems, pp. 147–165. Academic Press Professional, Inc., San Diego (1990)
20. Van Ouwerkerk, J.: Image super-resolution survey. Image Vis. Comput. 24(10), 1039–1052 (2006)

Detection and Segmentation of Clustered Objects by Using Iterative Classification, Segmentation, and Gaussian Mixture Models and Application to Wood Log Detection

Christopher Herbon[1][✉], Klaus Tönnies[2], and Bernd Stock[1]

[1] HAWK Fakultät Naturwissenschaften und Technik,
Von-Ossietzky-Straße 99, 37085 Göttingen, Germany
{herbon,stock}@hawk-hhg.de
[2] Institut für Simulation und Graphik, Otto-von-Guericke-Universität Magdeburg,
Universitätsplatz 2, 39106 Magdeburg, Germany
klaus@isg.cs.uni-magdeburg.de

Abstract. There have recently been advances in the area of fully automatic detection of clustered objects in color images. State of the art methods combine detection with segmentation. In this paper we show that these methods can be significantly improved by introducing a new iterative classification, statistical modeling, and segmentation procedure. The proposed method used a detect-and-merge algorithm, which iteratively finds and validates new objects and subsequently updates the statistical model, while converging in very few iterations.

Our new method does not require any a priori information or user input and works fully automatically on desktop computers and mobile devices, such as smartphones and tablets. We evaluate three different kinds of classifiers, which are used to substantially reduce the number of false positive matches, from which current state of the art methods suffer. Experiments are performed on a challenging database depicting wood log piles, with objects of inhomogeneous sizes and shapes. In all cases our method outperforms the current state of the art algorithms with a detection rate above 99 % and a false positive rate of less than 0.4 %.

1 Introduction

The task of detecting and segmenting objects in large clusters is important, yet very challenging, with applications, e.g. in medical imaging, timber production and other industrial areas. In all these cases the idea is to reduce the manual workload of a user and perform a fully automatic analysis of the input image. While the importance and applicability of this task rises continually, only little research has been done in the area of spatially clustered object detection. Most work addresses the detection, tracking, or segmentation of single objects, where large clusters are not considered. In the timber industry there is a strongly

© Springer International Publishing Switzerland 2014
X. Jiang et al. (Eds.): GCPR 2014, LNCS 8753, pp. 354–364, 2014.
DOI: 10.1007/978-3-319-11752-2_28

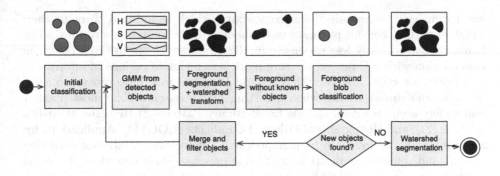

Fig. 1. Method Overview. Green defines an object, which is validated through the classifier; red objects are candidates who are not classified as objects of interest (Color figure online).

increasing demand for such detection systems, that are able to quickly and precisely determine the number of wood logs in large timber stacks. In this paper we apply our novel method to images of such timber stacks, while retaining applicability to similar application cases.

State of the art methods for clustered object detection have recently advanced to become fully automatic with decent detection rates. Still these procedures are insufficient for industrial purposes, especially when applied to challenging data sets. It can be observed that there exists a strong tendency towards high false positive rates, when images are taken outside of a controlled environment. The reason for this behavior is twofold. On one hand there is a strong dependency on the segmentability of background regions. On the other hand the spatial location of the objects is often assumed to meet a certain constraint, such as a dense distribution around the centroid of the cluster.

In this paper we aim to overcome these insufficiencies by proposing a model, which requires no user input, no a priori information, no constraints on image acquisition, and which outperforms state of the art methods, while showing a significantly lower false positive rate. Our main contribution is a detect-and-merge-algorithm, which iteratively finds and validates new objects, until a convergence criterion is met. During the process a Gaussian mixture model is obtained and refined after each iteration. We then utilize the Gaussian mixture model to estimate a probability map, from which regions are estimated, where prospective objects are likely to be found during the next iteration.

2   Related Work

The work, which is relevant to our approach, is quite broad. In this section we will first discuss methods which are generally applicable to our problem and then examine recent advances, which our method builds and improves upon. Graph cuts efficiently solve segmentation problems by employing constrained, iterative energy minimization algorithms [2,14]. While graph cuts show good results, their main drawback is computational complexity, which is induced through their iterative nature. For our purposes watersheds provide comparable results, while being computationally less expensive [6,13]. In our work we use watersheds in combination with a Gaussian mixture models to estimate regions of interest.

Object detection methods, that rely on local features, have proven to show accurate and robust results when detecting high-level cues. The most popular and widely used classifiers include Local Binary Patterns (LBP) [16], Haar-like features [21], and Histograms of Oriented Gradients (HOG) [4]. Applications for these methods are found in human/pedestrian detection [4,12], face detection [22,23], and vehicle classification [3]. In this paper we show how these detection mechanisms may be utilized in detection steps of a larger pipeline.

The method we propose has applications in widely different areas other than wood log segmentation. In biological set ups the classification of plankton in large clusters is a difficult task and some research specifically addresses this issue [1,11]. In [17,18,20] classification of blood cells in medical, microscopic images is performed and the necessity for such methods is explicitly stated. The variety of research in this area underlines the importance of a detection and classification method in general.

Some work has been done which can be regarded as the basis for our method. In [15,19] simple image processing techniques are used to segment and count timber on trucks. Fink [5] shows how histogram based methods, watersheds, and active contours can be used to detect and segment wood log piles in a controlled environment. The results indicate the feasibility of such methods, but all approaches are half automatic at best. Gutzeit et al. [7,8] demonstrate how the methods in [5] can be improved by applying a graph cut segmentation to separate foreground and background. They do however impose significant constraints on the image acquisition process. The location of the wood log pile is constrained to the center of the image and top and bottom of the image must depict background only.

In [9] some of these constraints are loosened by introducing the use of Haar-like features [22]. The method detects a subset of wood log faces through a classifier. From this subset a background and foreground model is built, which still requires the centroid of the object cluster to be located at the image center. The foreground and background model are then used as a probability measure for graph cut segmentation. A flaw of this method is its assumption, that each foreground blob in the image, that is obtained through graph cut, is an object of interest. This leads to a significant number of false positive matches, when background and foreground exhibit similar color distributions, as is often the case in natural images. The approach presented in this paper solves this problem by iteratively applying foreground segmentation and different classifiers. This way, each foreground blob is validated through at least two classifiers and the number of false positives is drastically reduced.

3   Method

In this section we describe our novel clustered object detection method. Our main contribution in this paper can be summarized as follows:

1. The proposed method has a higher rate of correctly detected objects, compared to state of the art methods, achieved through an iterative detect-and-merge algorithm.
2. Our method has a significantly lower false positive detection rate than state of the art methods, by specifically validating each candidate object.

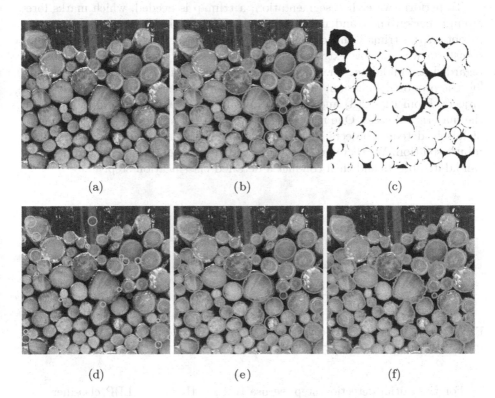

Fig. 2. Detection progress. (a) Enlarged input image, (b) initial detection result, (c) ROI (black) for candidate logs, (d) candidates (blue) and validated objects (green) after first iteration, (e) final detection result after 4 iterations, (f) final segmentation result (Color figure online).

The superior performance of our method is achieved through an iterative detect-and-merge algorithm, which validates each detected object with a dedicated classifier and uses new objects to refine the detection model. Each classifier consists of a cascade of weak classifiers and was trained with training database consisting of 5927 positive and 5161 negative samples. None of them are not part of the evaluation database. The overview of our method in Fig. 1 illustrates the general principle of the proposed pipeline. We first perform an initial detection step. While [9] propose the use of Haar-like features for this step, we found that a combination of LBP and HOG provides a lower false positive rate with a higher detection rate (see Sect. 4.1). We denote the initial detection result by S_{init}, which is the union of the two sets $S_{init} = S_{HOG} \cup S_{LBP}$. S_{init} is then filtered to remove and merge overlapping objects. From S_{init} the Gaussian mixture model representation of the color distribution is obtained [9]. We then utilize this model to calculate a probability map for pixels, which most likely belong to the foreground (objects of interest).

To perform watershed segmentation, a trimap is needed, which marks foreground, background, and unknown pixels. We threshold our probability map to obtain the trimap and calculate the watershed segmentation. Via distance transform we can now extract blobs from this binary image. These blobs are regarded as candidates for objects of interest and will be denoted as S_{blob}. In [9] it is proposed to assume that these candidate objects are correctly detected objects. From Figs. 2(d) and 3 one can easily see, that this assumption fails in the majority of cases. One may certainly apply heuristics to reduce the number of falsely detected objects, but we wish to provide a more general and more robust solution. We therefore propose to extract a region of interest around the candidate objects and then conduct a detailed classification step.

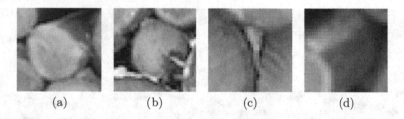

(a)   (b)   (c)   (d)

Fig. 3. Candidate objects for consecutive detection. (a) and (b) are correctly classified, (c) and (d) are correctly rejected.

For the initial detection step we use trained HOG and LBP classifiers. To ensure an accurate detection result, these classifiers include 20 detection stages. Every potentially detected object must pass the evaluation in these 20 stages in order to be validated. For candidate objects, which are blobs that match the Gaussian mixture model, slightly different classifiers are used. Again a combination of HOG and LBP classifiers has shown to provide best results. In the consecutive detection step we reduce the number of stages to 15. This way we can loosen the detection constraints while still being able to accurately dismiss false positive matches. The set of correctly detected objects in the consecutive step is denoted by $S_{cons} = S_{blob,HOG} \cup S_{blob,LBP}$, where $S_{blob,HOG}$ and $S_{blob,LBP}$ contain blobs, which were validated by the respective classifier.

After the consecutive classification step the termination criterion $S_{cons} = \varnothing$ is evaluated. In the case of $S_{cons} \neq \varnothing$ the GMM is updated from the newly detected objects and the consecutive detection procedure is performed again until no further objects are found and we have thus obtained the final detection result.

4   Experimental Results

To the best of our knowledge there is no publicly available test data set for wood log detection. For the statistical evaluation of our method we therefore perform experiments on a data collection of 200 challenging and very different pictures (Fig. 4). All pictures are natural, real world examples, which were provided by the timber production industry in several European countries. The resolution varies from 0.5 to 8.0 megapixels. For comparability only objects are considered, which lie completely within the image and which are more than 75 % visible.

Fig. 4. Test images provided by international timber production organizations

4.1   Initial Detection Rate

The first step of our method is the initial detection step and filtering of overlapping objects. We want to compare the performance of different types of classifiers

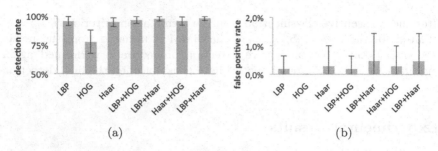

(a)   (b)

	LBP	HOG	Haar	LBP+HOG	LBP+Haar	Haar+HOG	LBP+Haar+HOG
tpr	95.8%	77.9%	95.1%	97.4%	98.0%	96.0%	98.4%
σ_{tpr}	3.7%	10.0%	3.8%	3.1%	1.9%	3.5%	2.1%
fpr	0.24%	0.0%	0.49%	0.24%	0.66%	0.49%	0.66%
σ_{fpr}	0.47%	0.0%	0.71%	0.47%	0.98%	0.71%	0.98%

(c)

Fig. 5. (a) Initial detection rate (tpr), (b) false positive rate (fpr), and (c) numeric
results for different classifiers

and combinations between them. From our experiments it is clearly visible, that
LBP features show the highest number of correctly detected objects. Haar-like
features perform only marginally worse, but also exhibit a higher false detection
rate. HOG performs significantly worse than Haar and LBP, but also features a
0 % false positive rate.

Figure 5 shows the detection and false positive rate. We define the false positive rate fpr as the probability of a falsely detected object for each object,
that is known to exist from ground truth (e.g. a fpr of 10 % would state that
for 100 ground truth objects, there would be 10 falsely detected objects). The
same holds true for the detection rate (true positive rate, tpr). Since we are not
interested in real time performance, we evaluate all different combinations of
classifiers. The union of all three classifiers produces the highest true positive
and highest false positive detection rate, as one would intuitively expect. For
real world applications, the combination of LBP and HOG features promises the
best compromise between high detection rate and low false positive rate.

4.2   Final Detection and False Positive Rate

As the final result of our method we now evaluate the detection and false positive
rate after a given number of iterations. All data sets converged within the first
four iterations. For purposes of comparison we show results for LBP + HOG
and LBP + Haar + HOG classification. The true positive detection rate of our
method exceeds 99 % in both cases. The false positive rate is more than twice as
high for LBP + Haar + HOG compared to LBP + HOG and thus disqualifies it
for practical use. It can be seen that the iterative approach is indeed beneficial,
since the detection rate increases by 1.9 % and 1.2 % respectively. Even more

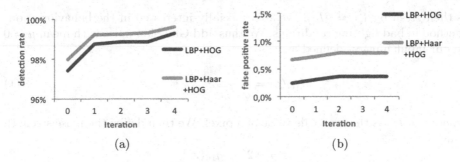

Fig. 6. (a) Detection rate, (b) false positive rate

Table 1. Numeric results for a given number of iterations, classifiers, (a) LBP + HOG (b) Haar + LBP + HOG, and comparison to state of the art method by Gutzeit and Voskamp [9].

	LBP+HOG					Haar+LBP+HOG			
iter.	tpr	σ_{tpr}	fpr	σ_{fpr}	iter.	tpr	σ_{tpr}	fpr	σ_{fpr}
0	97.4%	98.8%	0.24%	0.71%	0	98.4%	2.1%	0.66%	1.5%
1	98.8%	2.6%	0.31%	0.71%	1	99.2%	1.4%	0.71%	1.5%
2	98.8%	1.7%	0.36%	0.75%	2	99.2%	1.4%	0.78%	1.5%
3	98.9%	1.7%	0.36%	0.75%	3	99.3%	1.4%	0.78%	1.5%
4	99.3%	1.3%	0.36%	0.75%	4	99.6%	1.1%	0.78%	1.5%
[9]	99.1%	1.7%	7.01%	13.7%	[9]	99.1%	1.7%	7.01%	13.7%

(a)   (b)

importantly the standard deviation is almost halved for both cases also, while the false positive rate is only slightly increased (Fig. 6).

Table 1(a), (b) compares our method to our implementation of the current state of the art method by Gutzeit and Voskamp [9]. The detection rate of our proposed method is shown to be slightly higher with a slightly lower standard deviation. In terms of false positive rate, the method by Gutzeit and Voskamp shows an exorbitantly high false positive rate, when one does not apply heuristics to remove falsely detected objects. This is caused by test images, where fore- and background are quite similar and thus graph cut methods fail to provide proper segmentation. Our method is a significantly more robust approach to clustered object detection, without the need for applying heuristics of any kind.

4.3 Illumination and Noise

In order to determine the robustness of our method regarding changes in scene illumination, we parametrize a model, which changes the brightness of our input images and adds Gaussian noise. Let β be a multiplicative change in brightness on a logarithmic scale, so that there are no changes for $\beta = 1$ and strong changes for $\beta \ll 1$ and $\beta \gg 1$. The new image intensity for a grayscale pixel $I_{i,j} \in [0,1]$

is then given by $I'_{i,j} = \beta I_{i,j}$. We are especially interested in the behavior of our method in bad lighting conditions. We thus add Gaussian noise with mean $\mu = 0$ to our input images, defined as

$$N_G(z) = \frac{1}{\sigma\sqrt{2\pi}}e^{-\frac{(z-\mu)^2}{2\sigma^2}} \tag{1}$$

where $z = I'_{i,j}$ is the grayscale value of a pixel. We then define the noise strength σ as

$$\sigma = 0.2 \cdot |log_{10}(\beta)| \tag{2}$$

so that the image will exhibit strong noise, when the change in brightness is either very high or very low. From Fig. 7(a) it can be seen that our method is quite robust against moderate changes in lighting conditions. As expected, the detection rate drops, when noise and over-/underexposure are dominant in the image.

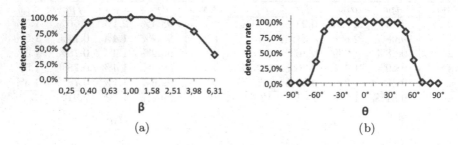

$$(a) \qquad\qquad\qquad\qquad (b)$$

Fig. 7. Detection rate for (a) bad lighting conditions and noise, (b) objects rotated in 3D space by θ around the y-axis.

4.4   Rotation Invariance

In real world scenarios one cannot always assure that the image acquisition process may be constrained to objects, whose surface normal is parallel to the camera's optical axis. Thus the objects in an image may be distorted by a 3D rotation. In our case, where the objects represent locally planar regions, it is especially crucial to analyze the behavior of the proposed method when applied to objects, which are rotated in 3D space.

Since the images satisfy the local planarity constraint, we are able to synthetically rotate the images and apply our object detection method. Figure 7(b) shows the results for a 3D rotation around the y-axis by the angle θ on a scale from $\theta = -90° \ldots 90°$. It can clearly be observed that our algorithm is quite robust between $\theta = -40° \ldots 40°$, while there is a significant drop for $|\theta| > 40°$. The method is therefore adequate for the use in real world scenarios.

5 Conclusions and Future Work

In this paper we have proposed a novel method for clustered object detection with applications to wood log faces, which outperforms the state of the art method in terms of detection and false positive rate. Furthermore we do not rely on heuristics for model estimation. Our method builds on an iterative detect-and-merge algorithm, which finds objects of interest and maintains a Gaussian mixture model of their color distribution for refined detection in the next iteration. Each candidate object is validated individually and our method thus shows a very low false detection rate. We have conducted a series of experiments, which prove the robustness of our method for real world applications. Our method runs well on mobile devices (smartphones/tablets), and may be used independently of desktop computers.

Presently, the data can be made available on request and it is planned to make it publicly accessible as benchmark. In future, we plan to extend our method to panoramic images such as in [10] and we will investigate the suitability of our method to structure from motion data.

References

1. Benfield, M.C., Grosjean, P., Culverhouse, P.F., Irigoien, X., Sieracki, M.E., Lopez-Urrutia, A., Dam, H.G., Hu, Q., Davis, C.S., Hansen, A., et al.: RAPID: research on automated plankton identification. Oceanography **20**, 172–187 (2007)
2. Boykov, Y.Y., Jolly, M.P.: Interactive graph cuts for optimal boundary & region segmentation of objects in N-D images. In: Proceedings. Eighth IEEE International Conference on Computer Vision, 2001. ICCV 2001, vol. 1, pp. 105–112. IEEE (2001)
3. Buch, N., Orwell, J., Velastin, S.A.: 3D extended histogram of oriented gradients (3DHOG) for classification of road users in urban scenes. In: BMVC. British Machine Vision Association (2009)
4. Dalal, N., Triggs, B.: Histograms of oriented gradients for human detection. In: IEEE Computer Society Conference on Computer Vision and Pattern Recognition, 2005. CVPR 2005, vol. 1, pp. 886–893, June 2005
5. Fink, F.: Foto-optische Erfassung der Dimension von Nadelrundholzabschnitten unter Einsatz digitaler, bildverarbeitender Methoden. Ph.D. thesis, Albert-Ludwigs-Universitt (2004)
6. Grau, V., Mewes, A., Alcaniz, M., Kikinis, R., Warfield, S.K.: Improved watershed transform for medical image segmentation using prior information. IEEE Trans. Med. Imag. **23**(4), 447–458 (2004)
7. Gutzeit, E., Ohl, S., Kuijper, A., Voskamp, J., Urban, B.: Setting graph cut weights for automatic foreground extraction in wood log images. In: VISAPP (2), pp. 60–67 (2010)
8. Gutzeit, E., Ohl, S., Voskamp, J., Kuijper, A., Urban, B.: Automatic wood log segmentation using graph cuts. In: Richard, P., Braz, J. (eds.) VISIGRAPP 2010. CCIS, vol. 229, pp. 96–109. Springer, Heidelberg (2011)
9. Gutzeit, E., Voskamp, J.: Automatic segmentation of wood logs by combining detection and segmentation. In: Bebis, G., et al. (eds.) ISVC 2012, Part I. LNCS, vol. 7431, pp. 252–261. Springer, Heidelberg (2012)

10. Herbon, C., Tönnies, K., Stock, B.: Adaptive planar and rotational image stitching for mobile devices. In: Proceedings of the 5th ACM Multimedia Systems Conference, pp. 213–223. ACM (2014)
11. Hu, Q., Davis, C.: Automatic plankton image recognition with co-occurrence matrices and support vector machine. Mar. Ecol. Progr. Ser. **295**, 21–31 (2005)
12. Khan, S.M.: Multi-view approaches to tracking, 3D reconstruction and object class detection. Ph.D. thesis, University of Central Florida (2008)
13. Kim, J.B., Kim, H.J.: Multiresolution-based watersheds for efficient image segmentation. Pattern Recogn. Lett. **24**(1), 473–488 (2003)
14. Lombaert, H., Sun, Y., Grady, L., Xu, C.: A multilevel banded graph cuts method for fast image segmentation. In: Tenth IEEE International Conference on Computer Vision, 2005. ICCV 2005, vol. 1, pp. 259–265. IEEE (2005)
15. Noonpan, V., Chaisricharoen, R.: Wide area estimation of piled logs through image segmentation. In: 2013 13th International Symposium on Communications and Information Technologies (ISCIT), pp. 757–760, Sept 2013
16. Ojala, T., Pietikainen, M., Harwood, D.: Performance evaluation of texture measures with classification based on Kullback discrimination of distributions. In: Proceedings of the 12th IAPR International Conference on Pattern Recognition, 1994, vol. 1 - Conference A: Computer Vision & Image Processing, vol. 1, pp. 582–585, Oct 1994
17. Ongun, G., Halici, U., Leblebicioglu, K., Atalay, V., Beksac, M., Beksaç, S.: An automated differential blood count system. In: Engineering in Medicine and Biology Society, 2001. Proceedings of the 23rd Annual International Conference of the IEEE, vol. 3, pp. 2583–2586. IEEE (2001)
18. Piuri, V., Scotti, F.: Morphological classification of blood leucocytes by microscope images. In: 2004 IEEE International Conference on Computational Intelligence for Measurement Systems and Applications, 2004. CIMSA, pp. 103–108. IEEE (2004)
19. Rahman, S., Yella, S., Dougherty, M.: Image processing technique to count the number of logs in a timber truck. In: Proceedings of the IASTED Conference on Signal and Image Processing, USA (2011)
20. Ross, N.E., Pritchard, C.J., Rubin, D.M., Duse, A.G.: Automated image processing method for the diagnosis and classification of malaria on thin blood smears. Med. Biol. Eng. Comput. **44**(5), 427–436 (2006)
21. Viola, P., Jones, M.: Rapid object detection using a boosted cascade of simple features. In: Proceedings of the 2001 IEEE Computer Society Conference on Computer Vision and Pattern Recognition, 2001. CVPR 2001, vol. 1, pp. I-511–I-518 (2001)
22. Viola, P., Jones, M.J.: Robust real-time face detection. Int. J. Comput. Vis. **57**(2), 137–154 (2004)
23. Zhang, C., Zhang, Z.: A survey of recent advances in face detection. Technical report, Microsoft Research (2010)

Tracking-Based Visibility Estimation

Stephan Lenor[1,2]([✉]), Johannes Martini[2], Bernd Jähne[1], Ulrich Stopper[2],
Stefan Weber[2], and Florian Ohr[2]

[1] Heidelberg Collaboratory for Image Processing, University of Heidelberg,
Speyerer Straße 6, 69115 Heidelberg, Germany
{stephan.lenor,bernd.jaehne}@iwr.uni-heidelberg.de
[2] Robert Bosch GmbH, Daimlerstraße 6, 71229 Leonberg, Germany

Abstract. Assessing atmospheric visibility conditions is a challenging and increasingly important task not only in the context of video-based driver assistance systems. As a commonly used quantity, *meteorological visibility* describes the visual range for observations through scattering and absorbing aerosols such as fog or smog.

We present a novel algorithm for estimating meteorological visibility based on object tracks in camera images. To achieve this, we introduce a likelihood objective function based on Koschmieder's model for horizontal vision to derive the atmospheric *extinction coefficient* from the objects' luminances and distances provided by the tracking. To make this algorithm applicable for real-time purposes, we propose an easy-to-implement and extremely fast minimization method which clearly outperforms classical methods such as Levenberg-Marquardt. Our approach is tested with promising results on real-world sequences recorded with a commercial driver assistance camera as well as on artificial images generated by Monte-Carlo simulations.

1 Introduction

Atmospheric aerosols such as fog or smog scatter and absorb light on its path from an object to an observer. As a consequence, the objects' contrast and in turn the observers' visual range reduces w.r.t. the thickness of the aerosol.

Especially in road traffic scenarios, this can become dangerous (cf. *e.g.* [8]). Although people are able to recognize fog, they mostly can neither quantify their own visual range nor the relative speed at which they are moving towards other road users or static objects. Furthermore, the performance of optical sensors such as multipurpose driver assistance cameras is also compromised by adverse visual conditions.

Therefore, assessing the atmospheric visibility conditions is an increasingly important challenge for visual environment perception (cf. *e.g.* [16,17]). Video-based driver assistance systems could adapt sensor and algorithm parameters according to the current visibility conditions or inform the driver about the appropriate speed; furthermore, the speed and the lighting system (*e.g.* front and rear fog lamps, beam of headlamps) could be adapted automatically. In addition,

© Springer International Publishing Switzerland 2014
X. Jiang et al. (Eds.): GCPR 2014, LNCS 8753, pp. 365–376, 2014.
DOI: 10.1007/978-3-319-11752-2_29

the measurement of the visibility conditions could also be interesting for static cameras, such as traffic surveillance cameras, in order to adjust the speed limit and inform drivers with variable-message signs.

In this work we will focus on the atmospheric parameter $K = K_s + K_a$, which is called the *extinction coefficient*, and represents the sum of the scattering and absorption coefficients. It is directly related to the distance d_{met}, at which it is just possible to distinguish a dark object against the horizon:

$$d_{met} := -\frac{\log(0.05)}{K}. \tag{1}$$

We refer to this distance as *meteorological visibility*. For further details regarding the development and the definition of this term, we recommend the work of Middleton [15, Sect. 6.2.1] and the different definitions of the Verband Deutscher Ingenieure [23] and the International Commission on Illumination [7].

1.1   Related Work

In the literature one can find different approaches for image-based visibility estimation. Most of the methods are based on driver assistance cameras and only provide a heuristic visibility estimation without seriously taking atmospheric models into account.

The first approaches were based on contrast-depth relations. Pomerleau [20] estimated a visibility value between 0 and 1 using the attenuation of the contrast along similar road features such as road markings. Several years later Hautière *et al.* [11] used a stereo disparity map which they combined with the local contrast to estimate the mobilized visibility distance as the maximal distance where the apparent contrast lies above a given threshold. Boussard *et al.* [3] implemented a similar algorithm based on structure from motion.

In 2012, Pavlić *et al.* [19] demonstrated that the visibility range can roughly be obtained from support vector machine classification based on Gabor features.

The first rigorous method to estimate meteorological visibility was that of Hautière *et al.* [12]. They extracted the *road surface luminance curve* (RSLC) from single images taken from a camera pointing along a road. Based on Koschmieder's model they found a correspondence between the RSLCs' inflection point and the extinction coefficient K. Similar to [12], Bronte *et al.* [4] presented another algorithm based on RSLCs. Contrary to Hautière *et al.*, this approach is based on a heuristic relation between the inflection point of the luminance curve and the visibility conditions only. Another optimized implementation of Hautière's RSLC algorithm can be found in [18]. Besides these framework optimizations, Lenor *et al.* [14] presented an improved model (for non-absorbing atmospheres) to describe the relation between the luminance curves' inflection point and the atmospheric extinction coefficient. They showed that the RSLC model should take into account effects caused by non-horizontal vision. All RSLC-based approaches are able to work on just a single frame. However, they require an inflection point model and depend strongly on a specific road scenario that does not allow for objects blocking the view to the horizon.

Other related work which in contrast to our proposed approach consider the case of stationary cameras can be found in Busch and Debes [5]; Babari *et al.* [2]; Geng *et al.* [10]; Song *et al.* [22].

1.2 Setting

The algorithm proposed here is based on luminance observations from objects at different distances to the observer. This requires objects to be moved relative to the camera or multiple cameras observing the same objects. In all cases, one somehow has to identify the objects in each camera image and link these observations to each other (tracking). If the observation angle changes by a relevant degree during the observation, the object surfaces should be approximately Lambertian, so that the light radiated from the objects in the direction of the observer (intrinsic luminance) is comparable across the observation process. Moreover, we rely on a linear camera model, *i.e.* $\exists \alpha_I, \beta_I \in \mathbb{R}$ s.t. an object of luminance L is represented by the image intensity $I = \alpha_I L + \beta_I$ apart from discretization, saturation, and spectral effects (cf. [9]). Since these linear changes have no influence on the estimation results, we do not differentiate between I and L in the following.

For our experiments we use a commercial driver assistance camera, which observes objects in front of a car. Furthermore, we assume the fog density to be homogeneous and the influence of compact light sources such as the own headlights to be negligible compared to the diffuse ambient light. Both constraints are fulfilled in the majority of daylight scenarios.

2 Tracking-Based Algorithm

Due to (1), estimating the meteorological visibility d_{met} is equivalent to an estimation of the atmospheric extinction coefficient K. In this section we introduce a method to estimate K, solving an inverse problem. For this purpose, from Koschmieder's model for horizontal vision, we derive a likelihood functional based on distance and luminance data belonging to tracked objects (cf. Fig. 2). Then, we show that this functional can be minimized remarkably fast.

2.1 Koschmieder's Model for Horizontal Vision

Light transport through scattering and absorbing media can be described by the radiative transfer equation (cf. [6]). By applying it to the line of sight between an observer situated at $p_{\mathrm{obs}} \in \mathbb{R}^3$ and an object of (intrinsic) luminance L_0 at distance d, an initial value problem can be derived

$$\overline{L}(d) = L_0, \quad \frac{d\overline{L}}{dr}(r) = K(r)\overline{L}(r) - K_s(r)\int_{\mathbb{S}^2} \mathcal{L}(p(r),\omega)\psi(\sigma,\omega)dS(\omega), \quad (2)$$

where $\mathcal{L}(x,\omega)$ denotes the luminance at some point $x \in \mathbb{R}^3$ incident from a direction $\omega \in \mathbb{S}^2 := \{x \in \mathbb{R}^3 : |x| = 1\}$. The line of sight is parametrized as $[0,d] \ni r \mapsto p(r) := p_{\mathrm{obs}} - r\sigma$, where $\sigma \in \mathbb{S}^2$ specifies the direction from the object to the observer (cf. Fig. 1), and $\overline{L}(\cdot) := \mathcal{L}(p(\cdot),\sigma)$. In general, the extinction and scattering coefficients K and K_s vary along the line of sight. The phase function $\psi : \mathbb{S}^2 \times \mathbb{S}^2 \to \mathbb{R}_{\geq 0}$ represents the directional scattering distribution.

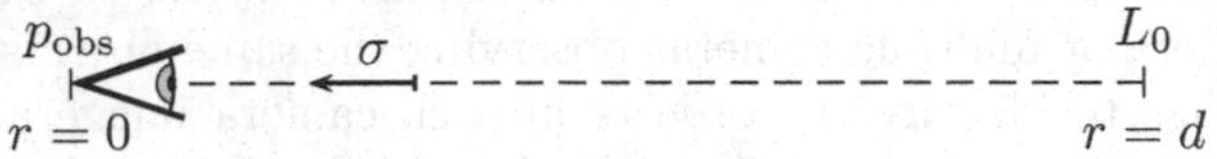

Fig. 1. Line of sight between object of intrinsic luminance L_0 and observer

The derivation in (2) can be removed using standard solving techniques for linear ordinary differential equations (ODEs). Doing this and focusing on the observer position $r = 0$, yields the full radiative transfer model for the luminance $L(d)$ an observer receives from an object at distance d:

$$L(d) = L_0 e^{-\int_0^d K(\tau)d\tau} + \int_0^d K_s(s)e^{-\int_0^s K(\tau)d\tau} \int_{\mathbb{S}^2} \mathcal{L}(p(s),\omega)\psi(\sigma,\omega)dS(\omega)ds. \quad (3)$$

We assume a horizontal line of sight in a plane-layer atmosphere; i.e. K, K_s, and $\mathcal{L}(\cdot,\omega)$ are constant on a plane parallel to the ground. From this reasonable assumption we obtain a highly simplified model

$$L(d) = L_0 e^{-Kd} + L_{\mathrm{air}}\left(1 - e^{-Kd}\right), \quad (4)$$

where L_{air} represents the inscattered ambient light

$$L_{\mathrm{air}} := \frac{K_s}{K} \int_{\mathbb{S}^2} \mathcal{L}(p(0),\omega)\psi(\sigma,\omega)dS(\omega). \quad (5)$$

This model is often referred to as Koschmieder's model (cf. [13]).

2.2 Data Acquisition

For the setting described in Sect. 1.2, we will use the following notation: Let $M \in \mathbb{N}$ be the number of object tracks, where each track has a length $N_m \in \mathbb{N}$, $m \in \{1,\ldots,M\}$. The distance and luminance data shall be given by observation pairs

$$\underbrace{\left(d_1^1, L_1^1\right),\ldots,\left(d_{N_1}^1, L_{N_1}^1\right)}_{\text{1st object}}, \quad \cdots \quad, \underbrace{\left(d_1^M, L_1^M\right),\ldots,\left(d_{N_M}^M, L_{N_M}^M\right)}_{M\text{th object}}. \quad (6)$$

Each object is equipped with its own (unknown) intrinsic luminance L_0^m.

To obtain the luminance values L_n^m, one not only has to segment objects or parts of the objects in the camera images. One also has to carefully select a representative luminance value L_n^m from the given object segment, which can contain a quite heterogeneous luminance distribution. We simply select the mean luminance over the segmented object. It is easy to see that Koschmieder's model (4) is directly passed from single luminance values to their mean. Nevertheless, taking percentile luminances could be a useful alternative as well, especially if they are adapted individually to each object.

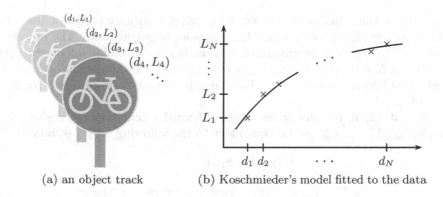

(a) an object track   (b) Koschmieder's model fitted to the data

Fig. 2. The luminance of an observed object changes w.r.t. the distance between the object and the observer. This relation can be described by Koschmieder's model.

2.3   Likelihood Objective Function

For M objects, (4) provides a luminance-distance model with $M+2$ parameters $p = (K, L_{\text{air}}, L_0^1, \dots, L_0^M)$. To infer from some measured data on the underlying parameters, the model can be least-square fitted. We generalize this approach by taking the uncertainties of L_n^m into account, expressed by the standard deviations σ_n^m.

We assume L_n^m to be independently generated from a normally distributed process with an expected value fitting Koschmieder's model in d_n^m and parameters p. This leads to the following likelihood function:

$$p \mapsto \prod_{m=1}^{M} \prod_{n=1}^{N_m} \frac{1}{\sqrt{2\pi}\sigma_n^m} \cdot \exp\left(\frac{1}{2} \left(\frac{L_n^m - L(d_n^m; p)}{\sigma_n^m} \right)^2 \right), \tag{7}$$

which is maximized if and only if the weighted least-square functional

$$p \mapsto \mathcal{F}(p) := \sum_{m=1}^{M} \sum_{n=1}^{N_m} \frac{1}{(\sigma_n^m)^2} \left(\left[L_0^m e^{-Kd_n^m} + L_{\text{air}}(1 - e^{-Kd_n^m}) \right] - L_n^m \right)^2 \tag{8}$$

is minimized.

We select σ_n^m proportional to the statistical error of the mean luminance over a segmented area, *i.e.*

$$\sigma_n^m \sim (\#\text{object pixels at } n\text{th frame of } m\text{th track})^{-\frac{1}{2}}. \tag{9}$$

However, it could be useful to incorporate other uncertainties (*e.g.* from tracking) in σ_n^m as well.

2.4   Fast Minimization

Especially in embedded camera systems for driver assistance purposes, the computational complexity of an algorithm plays an important role. The objective function $\mathcal{F}$ (cf. (8)) can be minimized extremely fast by solving a 1-dimensional equation in K instead of minimizing it over the $(M+2)$-dimensional parameter space. This becomes possible since Koschmieder's model is linear in all parameters but K.

First of all, it requires some straightforward calculations to realize that $\nabla \mathcal{F}(K, L_{\text{air}}, L_0^1, \ldots, L_0^M) = 0$ is equivalent to the following $M+2$ equations:

$$0 = L_{\text{air}}^2 (S_{\text{eed}} - S_{\text{ed}}) + L_{\text{air}} S_{\text{Led}}$$
$$+ \sum_{m=1}^{M} \left[L_{\text{air}} L_0^m (S_{\text{ed}}^m - 2S_{\text{eed}}^m) - L_0^m S_{\text{ed}}^m + L_0^m L_0^m S_{\text{eed}}^m \right], \tag{10}$$

$$\begin{pmatrix} S_{\text{L}} \\ S_{\text{Le}}^1 \\ \vdots \\ S_{\text{Le}}^M \end{pmatrix} = \begin{pmatrix} S_1 - S_e & S_e^1 & \cdots & S_e^M \\ S_e^1 - S_{ee}^1 & S_{ee}^1 & & \\ \vdots & & \ddots & \\ S_e^M - S_{ee}^M & & & S_{ee}^M \end{pmatrix} \begin{pmatrix} L_{\text{air}} \\ L_0^1 \\ \vdots \\ L_0^M \end{pmatrix}, \tag{11}$$

where we introduce the abbreviatory notation

$$S_1^m := \sum_{n=1}^{N_m} \frac{1}{(\sigma_n^m)^2}, \ S_e^m := \sum_{n=1}^{N_m} \frac{e^{-Kd_n^m}}{(\sigma_n^m)^2}, \ S_{ee}^m := \sum_{n=1}^{N_m} \frac{e^{-2Kd_n^m}}{(\sigma_n^m)^2}, \ S_{\text{L}}^m := \sum_{n=1}^{N_m} \frac{L_n^m}{(\sigma_n^m)^2},$$

$$S_{\text{Le}}^m := \sum_{n=1}^{N_m} \frac{L_n^m e^{-Kd_n^m}}{(\sigma_n^m)^2}, \ S_{\text{ed}}^m := \sum_{n=1}^{N_m} \frac{e^{-Kd_n^m} d_n^m}{(\sigma_n^m)^2}, \ S_{\text{eed}}^m := \sum_{n=1}^{N_m} \frac{e^{-2Kd_n^m} d_n^m}{(\sigma_n^m)^2}, \tag{12}$$

$$S_{\text{Led}}^m := \sum_{n=1}^{N_m} \frac{L_n^m e^{-Kd_n^m} d_n^m}{(\sigma_n^m)^2}, \ S_{\Theta} := \sum_{m=1}^{M} S_{\Theta}^m, \ \Theta \in \{1, e, ee, L, Le, ed, eed, Led\}.$$

For $K > 0$, the linear equation system (11) is solved in L_{air} and L_0^m by successively applying

$$L_{\text{air}} = \frac{S_{\text{L}} - \sum_{m=1}^{M} \frac{S_e^m S_{\text{Le}}^m}{S_{ee}^m}}{S_1 - \sum_{m=1}^{M} \frac{S_e^m S_e^m}{S_{ee}^m}}, \quad L_0^m = \frac{S_{\text{Le}}^m + L_{\text{air}}(S_{ee}^m - S_e^m)}{S_{ee}^m}. \tag{13}$$

Therefore, substituting L_0^m and L_{air} in Eq. (10) yields a 1-dimensional equation $f(K) = 0$. It is satisfied if and only if there exist $L_{air}, L_0^1, \ldots, L_0^M$, s.t. $\nabla\mathcal{F}(K, L_{air}, L_0^1, \ldots, L_0^M) = 0$. Numerical experiments suggest that in all reasonable cases this critical point is a unique minimum of $\mathcal{F}$. An analytical discussion of the functionals' properties turns out to be complex and would go beyond the scope of this work.

$f(K) = 0$ can be solved efficiently by applying Newton's method, iterating $K \leftarrow K - f(K)/f'(K)$ (cf. Algorithm 1). In our implementation we start with $K = 10^{-3}$, i.e. $d_{met} \approx 3000\,m$, and understand all meteorological visibilities above $3000\,m$ as unlimited. As the stopping criterion we require the relative change in the estimated d_{met} to be smaller than $1\,\%$:

$$\frac{\left|d_{met} - d_{met}^{old}\right|}{\min\left\{d_{met}, d_{met}^{old}\right\}} \overset{(1)}{=} \frac{\left|K - K_{old}\right|}{\min\{K, K_{old}\}} < 10^{-2}. \tag{14}$$

According to our observations, the algorithm usually stops after 3 iterations. We allow for a maximum number of 10 iterations.

It should also be noted that, in contrast to standard minimization techniques, only one starting value has to be specified, which simplifies both implementation and application.

Algorithm 1. Fast Maximum Likelihood Parameter Estimation

1: **procedure** FASTMLE(d_n^m, L_n^m, σ_n^m)	
2: $K \leftarrow 10^{-3}$	
3: $curIter \leftarrow 1$	▷ current iteration
4: **repeat**	
5: compute $\exp(-Kd_n^m)$	▷ computationally most expensive step
6: compute $S_*^*(K) := \{S_\ominus(K), S_\ominus^m(K)\}$	▷ using $\exp(-Kd_n^m)$, L_n^m, d_n^m, σ_n^m
7: compute $L_{air}(K), L_{air}'(K)$	▷ using $S_*^*(K)$
8: compute $L_0^m(K), L_0^{m\prime}(K)$	▷ using $L_{air}(K)$, $L_{air}'(K)$, $S_*^*(K)$
9: compute $f(K), f'(K)$	▷ using $L_{air}(K)$, $L_{air}'(K)$, $L_0^m(K)$, $L_0^{m\prime}(K)$, $S_*^*(K)$
10: $K_{old} \leftarrow K$	
11: $K \leftarrow K_{old} - f(K)/f'(K)$	
12: **if** $K < 10^{-3}$ **then**	
13: **return** 0	▷ unlimited meteorological visibility
14: **end if**	
15: $curIter \leftarrow curIter + 1$	
16: **until** $curIter > 10 \ \vee\ \|K_{old} - K\| < 10^{-2} \cdot \min\{K, K_{old}\}$	▷ cf. (14)
17: **return** K	
18: **end procedure**	

---

3   Evaluation and Results

To evaluate our approach on real world data, we use images from a monocular driver assistance camera mounted behind the windshield. The tracking, including

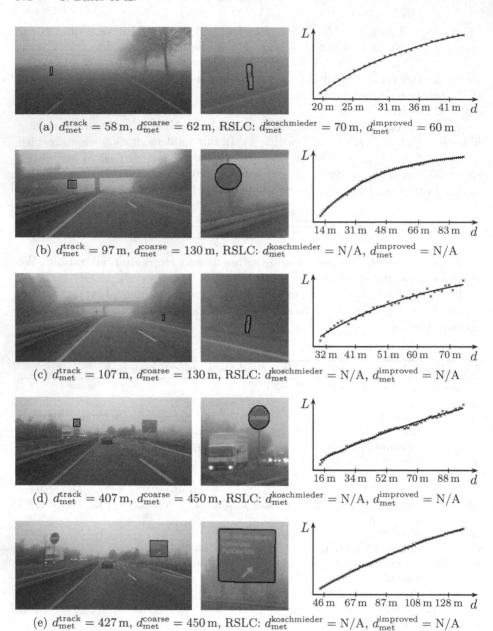

(a) $d_{\mathrm{met}}^{\mathrm{track}} = 58\,\mathrm{m}$, $d_{\mathrm{met}}^{\mathrm{coarse}} = 62\,\mathrm{m}$, RSLC: $d_{\mathrm{met}}^{\mathrm{koschmieder}} = 70\,\mathrm{m}$, $d_{\mathrm{met}}^{\mathrm{improved}} = 60\,\mathrm{m}$

(b) $d_{\mathrm{met}}^{\mathrm{track}} = 97\,\mathrm{m}$, $d_{\mathrm{met}}^{\mathrm{coarse}} = 130\,\mathrm{m}$, RSLC: $d_{\mathrm{met}}^{\mathrm{koschmieder}} = \mathrm{N/A}$, $d_{\mathrm{met}}^{\mathrm{improved}} = \mathrm{N/A}$

(c) $d_{\mathrm{met}}^{\mathrm{track}} = 107\,\mathrm{m}$, $d_{\mathrm{met}}^{\mathrm{coarse}} = 130\,\mathrm{m}$, RSLC: $d_{\mathrm{met}}^{\mathrm{koschmieder}} = \mathrm{N/A}$, $d_{\mathrm{met}}^{\mathrm{improved}} = \mathrm{N/A}$

(d) $d_{\mathrm{met}}^{\mathrm{track}} = 407\,\mathrm{m}$, $d_{\mathrm{met}}^{\mathrm{coarse}} = 450\,\mathrm{m}$, RSLC: $d_{\mathrm{met}}^{\mathrm{koschmieder}} = \mathrm{N/A}$, $d_{\mathrm{met}}^{\mathrm{improved}} = \mathrm{N/A}$

(e) $d_{\mathrm{met}}^{\mathrm{track}} = 427\,\mathrm{m}$, $d_{\mathrm{met}}^{\mathrm{coarse}} = 450\,\mathrm{m}$, RSLC: $d_{\mathrm{met}}^{\mathrm{koschmieder}} = \mathrm{N/A}$, $d_{\mathrm{met}}^{\mathrm{improved}} = \mathrm{N/A}$

Fig. 3. We compare the meteorological visibility estimated by: the new tracking-based algorithm ($d_{\mathrm{met}}^{\mathrm{track}}$), coarse visual inspection ($d_{\mathrm{met}}^{\mathrm{coarse}}$), the RSLC-based algorithm from [12] ($d_{\mathrm{met}}^{\mathrm{koschmieder}}$), the RSLC-based algorithm from [14] ($d_{\mathrm{met}}^{\mathrm{improved}}$). The model fits the data very well, and even large meteorological visibilities are acceptably estimated. Due to obstructed roads, RSLC-based algorithms are often not applicable. Since our approach requires observable objects, tracking-based and RSLC-based algorithms complement each other.

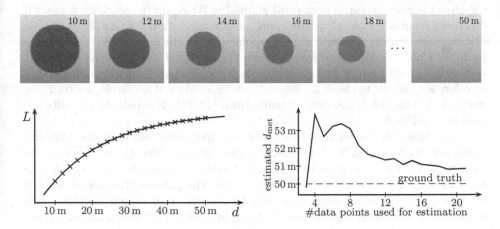

Fig. 4. Synthetic images from the Monte Carlo simulation: A target is situated above the road plane at distances varying from 10 m to 50 m, where the meteorological visibility is given by $d_{\mathrm{met}} = 50$ m. The model perfectly fits the simulated data points (left), and d_{met} is well reconstructed by the tracking-based algorithm (right). The quality of d_{met} estimation increases as more and more data points are successively taken into account (right).

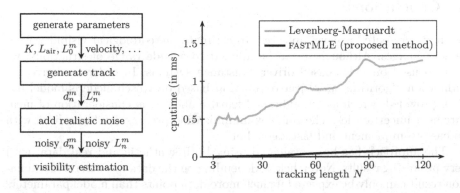

Fig. 5. To compare the fast minimization method proposed in Sect. 2.4 to standard minimization methods, we create artificial tracking data. We generate exact tracking data (of tracking length $N = \sum_{m=1}^{M} N_m$) for simple scenarios based on Koschmieder's model (which is sufficiently exact as shown in Fig. 4). In accordance with our cameras' noise model we disturb the data, and use it for visibility estimation afterwards. The runtime evaluation was performed by counting the basic operations of implementations of Levenberg-Marquardt and fastMLE and multiplying this number with the number of cycles per operation given in the specification of a Cortex-A9 FPU (cf. [1]). To achieve a similar accuracy, the $(M+2)$-dimensional minimization requires a much larger number of iterations than our 1-dimensional approach. Other methods such as Gauß-Newton or gradient descent turn out to be inapplicable due to their strong dependence on the starting values or their extremely slow convergence.

the object segmentation, is performed manually. We equip the tracking data with distances from a simple structure from motion algorithm (cf. *e.g.* [21]). These distances are bundle adjusted and extrapolated with the help of ESC (Electronic Stability Control) data regarding the inter-frame travel distances. As we do not have exact ground truth, we take very coarse estimations from visual inspection as reference values (to check for general practicability). If available, we compare our results with the RSLC-based results from [12, 14]. The real-data results are provided in Fig. 3.

To evaluate the model and the estimation capabilities based on exact ground truth, we implemented a Monte-Carlo simulation (cf. Fig. 4). We assume the road to be an infinitely expanded plane (albedo = 0) and the sun to be a uniform light source at a given height level above the ground (the height level is set to $6d_{met}$). We add homogeneous fog to the atmosphere and simulate each scattering event using the Henyey-Greenstein mean scattering phase function and a forward scattering parameter of 0.85 (cf. *e.g.* [14]). Each photon travels backwards randomly, starting at the camera, and is scattered multiple times (maximal 300 scatterings allowed).

Finally we implemented a simple simulation framework, which synthetically generates noisy tracking data. This enables us to investigate the fast minimization approach from Sect. 2.4 (cf. Fig. 5).

4   Conclusions

We have presented a novel approach to estimate meteorological visibility from camera images. Although it can be applied to a wide range of situations, we have focused on video-based driver assistance systems. In order to derive the estimation algorithm, an objective function based on Koschmieder's model has been provided, which as an additional feature allows the consideration of measurement uncertainties. The complex minimization process can be realized with an easy-to-implement and fast algorithm.

The approach has been evaluated using both synthetic and real data, with very promising results. Nevertheless, depending on the data noise a stable estimation result can only be expected if much more data points than model parameters are involved.

Commercial driver assistance systems already provide numerous useable object tracks (*e.g.* on road signs or vehicles), generic superpixel tracks, or concatenated flow vectors. For the next few years, data quality can be expected to increase considerably.

We therefore expect this algorithm to be integrated easily into existing driver assistance systems where it will be a valuable supplement to RSLC-based algorithms, which on the other hand show their strength for empty road scenarios. Both together will provide strong visibility estimation capabilities and allow for a wide range of visibility-based applications.

References

1. ARM: Specification of Cortex-A9 Floating-Point Unit (1.5 GHz), Revision: r2p2 (2010)
2. Babari, R., Hautière, N., Dumont, É., Brémond, R., Paparoditis, N.: A model-driven approach to estimate atmospheric visibility with ordinary cameras. Atmos. Environ. 45(30), 5316–5324 (2011)
3. Boussard, C., Hautière, N., d'Andréa Novel, B.: Visibility distance estimation based on structure from motion. In: Proceedings of the 11th International Conference on Control Automation, Robotics and Vision, pp. 1416–1421. IEEE, Dec 2010
4. Bronte, S., Bergasa, L.M., Alcantarilla, P.F.: Fog detection system based on computer vision techniques. In: Proceedings of the 12th International Conference on Intelligent Transportation Systems, pp. 1–6. IEEE, Oct 2009
5. Busch, C., Debes, E.: Wavelet transform for analyzing fog visibility. IEEE Intell. Syst. Appl. 13(6), 66–71 (1998)
6. Chandrasekhar, S.: Radiative Transfer. Dover Publications, New York (1960)
7. CIE: Termlist (2012). http://eilv.cie.co.at/
8. Croft, P.J.: Fog. In: Holton, J.R. (ed.) Encyclopedia of Atmospheric Sciences. Academic Press, New York (2003)
9. European Machine Vision Association, et al.: EMVA Standard 1288, Standard for Characterization of Image Sensors and Cameras (2010)
10. Geng, W., Lu, X., Yang, L., Chen, W., Liu, Y.: Detection algorithm of video image distance based on rectangular pattern. In: Proceedings of the 5th International Congress on Image and Signal Processing, pp. 856–860. IEEE, Oct 2012
11. Hautière, N., Labayrade, R., Aubert, D.: Real-time disparity contrast combination for onboard estimation of the visibility distance. IEEE Trans. Intell. Transp. Syst. 7(2), 201–212 (2006)
12. Hautière, N., Tarel, J.P., Lavenant, J., Aubert, D.: Automatic fog detection and estimation of visibility distance through use of an onboard camera. Mach. Vis. Appl. 17(1), 8–20 (2006)
13. Koschmieder, H.: Theorie der Horizontalen Sichtweite. Physik der Freien Atmosphäre 12, 33–55 (1924)
14. Lenor, S., Jähne, B., Weber, S., Stopper, U.: An improved model for estimating the meteorological visibility from a road surface luminance curve. In: Weickert, J., Hein, M., Schiele, B. (eds.) GCPR 2013. LNCS, vol. 8142, pp. 184–193. Springer, Heidelberg (2013)
15. Middleton, W.E.K.: Vision Through the Atmosphere. University of Toronto Press, Toronto (1952)
16. Narasimhan, S.G., Nayar, S.K.: Vision and the atmosphere. Int. J. Comput. Vision 48(3), 233–254 (2002)
17. Narasimhan, S.G., Nayar, S.K.: Shedding light on the weather. In: Proceedings of the Computer Society Conference on Computer Vision and Pattern Recognition, vol. 1, pp. 665–672. IEEE, June 2003
18. Negru, M., Nedevschi, S.: Image based fog detection and visibility estimation for driving assistance systems. In: Proceedings of the International Conference on Intelligent Computer Communication and Processing, pp. 163–168. IEEE, Sept 2013
19. Pavlić, M., Belzner, H., Rigoll, G., Ilić, S.: Image based fog detection in vehicles. In: Proceedings of the Intelligent Vehicles Symposium, pp. 1132–1137. IEEE, June 2012

20. Pomerleau, D.: Visibility estimation from a moving vehicle using the RALPH vision system. In: Proceedings of the Conference on Intelligent Transportation System, pp. 906–911. IEEE, Nov 1997
21. Slama, C.C., Theurer, C., Henriksen, S.W., et al.: Manual of Photogrammetry, 5th edn. American Society of Photogrammetry, Falls Church (1980)
22. Song, H., Chen, Y., Gao, Y.: Homogenous fog condition recognition based on traffic scene. In: Proceedings of the International Conference on Modelling, Identification and Control, pp. 612–617. IEEE, June 2012
23. VDI-Kommission Reinhaltung der Luft: VDI 3786 part 6: Meteorological Measurements of Air Pollution, Turbidity of Ground-Level Atmosphere, Standard Visibility (1983)

Predicting the Influence of Additional Training Data on Classification Performance for Imbalanced Data

Stephen Kockentiedt[1,2]([✉]), Klaus Tönnies[1], and Erhardt Gierke[2]

[1] Department of Simulation and Graphics, Faculty of Computer Science,
University of Magdeburg, Magdeburg, Germany
`stephen@isg.cs.uni-magdeburg.de`
[2] Federal Institute for Occupational Safety and Health, Berlin, Germany

Abstract. It is desirable to predict the influence of additional training data on classification performance because the generation of samples is often costly. Current methods can only predict performance as measured by accuracy, which is not suitable if one class is much rarer than another. We propose an approach which is able to also predict other measures such as G-mean and F-measure, which are used in cases of imbalanced data. We show that our method leads to more correct decisions whether to generate more training samples or not using a highly imbalanced real-world dataset of scanning electron microscopy images of nanoparticles.

1 Introduction

The performance of machine learning classifiers depends on several factors. One important factor is the training set size. A general rule is that a classifier can only produce good results if the training set contains enough samples. However, generating labeled training samples is often expensive. Therefore, a trade-off between classifier performance and cost has to be found. To find a good balance is not an easy task, though. Beforehand, it is not clear, how much the addition of training samples will improve the classifier performance. It is therefore desirable to be able to predict the classifier performance to be expected after a certain amount of labeled samples have been added to the training set. A method to do this should work in a wide range of circumstances including imbalanced data, where instances of one class are much more common than those of another class.

An example of imbalanced is the classification of nanoparticles in scanning electron microscopy images of air samples from nanoparticle production. Here, engineered nanoparticles are much rarer than incidental nanoparticles such as diesel soot. Figure 1 shows a selection of these particles. For a more detailed explanation, see [2]. In such cases, the choice of an appropriate performance measure gains importance. The commonly used accuracy, defined as the percentage of correctly classified samples, is unsuitable. The reason is that instances of rare classes have little impact on the measure. A method to predict the classifier performance should therefore be able to work with measures other than accuracy. Sun et al. [7] suggest using F-measure and G-mean for the two-class case.

© Springer International Publishing Switzerland 2014
X. Jiang et al. (Eds.): GCPR 2014, LNCS 8753, pp. 377–387, 2014.
DOI: 10.1007/978-3-319-11752-2_30

Fig. 1. Nanoparticles. l: Zinc oxide, m: titanium dioxide, r: diesel soot

In some cases, it is possible to specifically add samples of a particular class to the training set. Therefore, it is desirable that the method is able to predict the impact on classifier performance to be expected after the addition of instances of a specific class. This enables the user to sample instances of the class which promises the biggest performance improvement.

In this paper, we propose a method meeting these goals. Section 2 presents related work targeting the same problem while Sect. 3 describes our method. In Sect. 4, we describe how we tested the proposed method, Sect. 5 presents the results of our experiments and Sect. 6 gives some concluding remarks.

2   Related Work

To the best of our knowledge, there are two methods that predict the expected classifier performance after the addition of training samples. Both approaches consist of two phases. In the first phase, the methods estimate the misclassification rate (MR) or components of it for training set sizes smaller than the size of the available dataset. In the second phase, error values for larger training set sizes are predicted using these estimates. Mukherjee et al. [4] use a hold-out method, which repeatedly splits the dataset into a training and a test set to estimate the MR. To predict the error for larger training set sizes, the approach fits a power-law function $c(n) = an^{-\alpha} + b$, where n is the number of samples in the training set and $c(n)$ is the corresponding MR. The same is done for the 25th and 75th quantiles of the error rate vectors to get estimates for the range in which the MR may vary. In our view, the estimation phase has some crucial disadvantages. The fact that they use a hold-out approach means that only a subset of the labeled samples is used to evaluate the trained classifier in each round. Hence, some samples may more often be part of the training set than the test set or vice versa. This can introduce a bias in the error estimates. A more worrisome consequence is that the number of samples used to estimate the MRs can vary greatly between different training set sizes. Using the parameters recommended by the authors, the number of samples used to calculate the MRs can vary from 10 to $|D| - 10$, where $|D|$ is the size of the dataset. Therefore, they may differ by some orders of magnitude. The consequence is that these estimates are much more exact if the training set size is small. Another consequence is that the 25th and 75th quantiles are biased to deviate more from the mean for higher training set sizes, which leads to biased predictions for future training set sizes showing much larger error margins than would be appropriate.

The second approach is proposed by Smith et al. [5,6]. They decompose the MR into bias and variance as proposed by Kohavi and Wolpert [3] under the assumption that the behavior of these components can better be predicted individually. To estimate the values of MR, bias and variance, they use a cross-validation approach by Webb and Conilione [8]. It does not suffer from the disadvantages of the estimation phase of the method by Mukherjee et al. For the second phase, Smith et al. propose that for a given training set size $n \in \{100, 200, \ldots, 1000\}$, future values of the MR, the bias and the variance can be predicted using a simple linear model $V(n+n') = aV(n) + b$, where $V(n)$ stands for the MR, bias or variance, respectively, for a training set size of n. They argue that the values of a and b only depend on the value of n. In other words, they propose that they are independent of the classifier, the data and the value of n'. One consequence is that the error predicted by the model will be the same, no matter whether 10 or 10 000 samples are added to the training set. The authors show that the absolute difference between their predictions and the true MRs is small if the sample count of the training set used for the prediction is large. They do not show how well their method can predict the *change* of the MR.

3   Method

The goal of classification is to assign a class from the set Y to a sample from the set X. In the rest of this paper, we assume that X and Y are countable. If they are not, the sums in the formulas can be replaced by integrals. The true class $f(x) \in Y$ of each sample $x \in X$ is given by a function $f : X \to Y$. Because it is infeasible to estimate the probability distribution from real data, we assume that f is deterministic. A classifier l is trained on a training set $\mathcal{T} \subseteq X \times Y$, which is a random multiset of samples $x \in X$ and the corresponding classes $f(x) \in Y$. The classifier l produces a function $l(\mathcal{T}) : X \to Y$ which assigns a class $l(\mathcal{T})(x) \in Y$ to each sample $x \in X$.

Let $\chi \in X$ be a random variable and $x \in X$ a realization of it. Both approaches described in Sect. 2 predict the MR $\mathrm{P}(l(\mathcal{T})(\chi) \neq f(\chi) \,|\, |\mathcal{T}| = n)$ given the training set size $n \in \mathbb{N}$, which is the same as one minus the accuracy. (To increase readability, we will from now on drop random variables from the probabilities if the context is clear. So '$|\mathcal{T}| = n$' will become 'n'.) However, in many cases, accuracy is a poor choice to measure the classification performance. This is best illustrated using an example such as a cancer screening test for the general public. This is imbalanced data since a high percentage of people, say 99.99 %, will be healthy. An arguably dysfunctional test will always claim that a patient is healthy, even if the person has cancer. While such a test would never detect any case of cancer, it would achieve a very high accuracy of 99.99 % and a very low MR of 0.01 %. Hence, for imbalanced data, a method to predict the classifier performance should be able to work with measures other than accuracy. We will propose such an approach in this chapter.

Sun et al. [7] suggest using the measures F-measure and G-mean in cases of class imbalance. These can be calculated from the true positive rate, the true negative rate and the precision. Given a training set size n, these correspond to the probabilities $P(l(T)(x) = y_+ \,|\, f(x) = y_+, n)$, $P(l(T)(x) = y_- \,|\, f(x) = y_-, n)$ and $P(f(x) = y_+ \,|\, l(T)(x) = y_+, n)$, where y_+ and y_- are the positive and negative class, respectively. The precision can be expressed as

$$
\begin{aligned}
&P(f(x) = y_+ \,|\, l(T)(x) = y_+, n) \\
&= \frac{P(l(T)(x) = y_+ \,|\, y_+, n)\, P(y_+)}{P(l(T)(x) = y_+ \,|\, y_+, n)\, P(y_+) + P(l(T)(x) \neq y_- \,|\, y_-, n)\, P(y_-)},
\end{aligned} \tag{1}
$$

where '$f(x) = y$' was abbreviated as 'y'. Therefore, as $P(f(x) = y)$ can be easily estimated and $P(l(T)(x) = y \,|\, y, n) = 1 - P(l(T)(x) \neq y \,|\, y, n)$, $\forall y \in Y$, for our method, it would be sufficient to be able to predict $P(l(T)(x) \neq f(x) \,|\, y_f, n)$ for all $y_f \in Y$ in order to predict values of F-measure and G-mean.

Similar to [4–6], our method uses an estimation phase and a prediction phase. For the former, we express $P(l(T)(x) \neq f(x) \,|\, f(x) = y_f, n)$ as a sum:

$$
P(l(T)(x) \neq f(x) \,|\, y_f, n) = \sum_{\substack{x \in X \\ f(x) = y_f}} P(x = x \,|\, y_f)\, P(l(T)(x) \neq f(x) \,|\, n) \tag{2}
$$

We use the bias-variance decomposition proposed by Kohavi and Wolpert [3] and used by Smith et al. to test whether it is better to individually predict the behavior of bias and variance than to predict $P(l(T)(x) \neq f(x) \,|\, y_f, n)$ itself. We can decompose $P(l(T)(x) \neq f(x) \,|\, n)$ as follows:

$$
\begin{aligned}
&1 - P(l(T)(x) = f(x) \,|\, n) \\
&= \frac{1}{2}\left[1 - 2P(l(T)(x) = f(x) \,|\, n) + \sum_{y \in Y} P(l(T)(x) = y \,|\, n)^2 \right] \\
&\quad + \frac{1}{2}\left[1 - \sum_{y \in Y} P(l(T)(x) = y \,|\, n)^2 \right] \\
&= \frac{1}{2}\left[(1 - P(l(T)(x) = f(x) \,|\, n))^2 + \sum_{y \neq f(x)} P(l(T)(x) = y \,|\, n)^2 \right] \\
&\quad + \frac{1}{2}\left[1 - \sum_{y \in Y} P(l(T)(x) = y \,|\, n)^2 \right].
\end{aligned} \tag{3}
$$

Along with the definitions

$$
\begin{aligned}
&\text{bias}^2(n, x) := \\
&\frac{1}{2}\left[(1 - P(l(T)(x) = f(x) \,|\, n))^2 + \sum_{y \neq f(x)} P(l(T)(x) = y \,|\, n)^2 \right],
\end{aligned} \tag{4}
$$

$$
\text{variance}(n, x) := \frac{1}{2}\left[1 - \sum_{y \in Y} P(l(T)(x) = y \,|\, n)^2 \right], \tag{5}
$$

we can write $P(l(T)(x) \neq f(x) \mid n) = \text{bias}^2(n, x) + \text{variance}(n, x)$. Together with Eq. (2), we have

$$
\begin{aligned}
& P(l(T)(x) \neq f(x) \mid y_f, n) \\
&= \sum_{\substack{x \in X \\ f(x) = y_f}} P(x \mid y_f)\, \text{bias}^2(n, x) + \sum_{\substack{x \in X \\ f(x) = y_f}} P(x \mid y_f)\, \text{variance}(n, x) \quad (6) \\
&=: \text{bias}^2(y_f, n) + \text{variance}(y_f, n).
\end{aligned}
$$

Kohavi and Wolpert [3] propose unbiased estimators. $P(l(T)(x) \neq f(x) \mid n)$ can simply be estimated as $\widehat{P}(l(T)(x) \neq f(x) \mid n)$, where $\widehat{P}(\cdot)$ is the estimate of $P(\cdot)$ calculated as the observed frequency over multiple repetitions with different training sets. The estimates of $\text{bias}^2(n, x)$ and $\text{variance}(n, x)$ are calculated as

$$
\begin{aligned}
\widehat{\text{bias}}^2(n, x) := \\
\frac{1}{2} \Bigg[& \left(1 - \widehat{P}\left(l(T)(x) = f(x) \mid n\right)\right)^2 + \sum_{y \neq f(x)} \widehat{P}\left(l(T)(x) = y \mid n\right)^2 \quad (7) \\
& - \frac{1}{N-1} \sum_{y \in Y} \widehat{P}\left(l(T)(x) = y \mid n\right) \left(1 - \widehat{P}\left(l(T)(x) = y \mid n\right)\right) \Bigg]
\end{aligned}
$$

and

$$
\widehat{\text{variance}}(n, x) := \widehat{P}\left(l(T)(x) \neq f(x) \mid n\right) - \widehat{\text{bias}}^2(n, x). \quad (8)
$$

Here, N is the number of repetitions used to calculate the estimates. To get estimates of $P(l(T)(x) \neq f(x) \mid y_f, n)$, $\text{bias}^2(y_f, n)$ and $\text{variance}(y_f, n)$, which are independent of x, the mean of the estimates of $P(l(T)(x) \neq f(x) \mid n)$, $\text{bias}^2(n, x)$ and $\text{variance}(n, x)$ over all samples of class y_f are taken. This corresponds to the assumption that the dataset reflects the distribution of x.

To repeat the training and classification using different training sets, we use the sub-sampled cross-validation procedure proposed by Webb and Conilione [8], which is also used by Smith et al. It does not suffer from the disadvantages of the hold-out procedure used by Mukherjee et al. as exactly N training sets are used to classify each sample in the dataset $D \subseteq X \times Y$. Since we want to investigate the influence of the number of samples of a particular class in the training set, we have adapted the algorithm so that one can specify the sample count of each class in the training sets. We assume that there are m classes $Y = \{y_1, \ldots, y_m\}$ and we denote the size of D as $d := |D|$ and the number of samples of class y_j in the dataset D as $d_j := |D \cap (X \times \{y_j\})|, \forall j \in \{1, \ldots, m\}$. The adapted method has the following parameters:

$N \in \mathbb{N}$: The number of training sets used to classify each sample ($N > 0$).
$n_j \in \mathbb{N}, \forall j \in \{1, \ldots, m\}$: The number of samples of class y_j in each training set ($0 \leq n_j < d_j$).

The number of cross-validation folds $k := \left\lceil \max\left(\frac{d}{d_1 - n_1}, \ldots, \frac{d}{d_m - n_m}\right) \right\rceil$ is derived from these parameters so that $k - 1$ folds always contain n_j samples of class y_j. The algorithm has the following steps:

1. For $r = 1, \ldots, N$ do:
   (a) Randomly partition D into k distinct folds $F_1, \ldots, F_k$ so that each fold contains the same number of samples of each class.
   (b) For $i = 1, \ldots, k$ do:
       i. Select a random training set $T \subseteq D - F_i$ so that $|T \cap (X \times \{y_j\})| = n_j, \forall j \in \{1, \ldots, m\}$.
       ii. For each $x \in F_i$: Set $y_{x,r} := l(T)(x)$.
2. For each $x \in D$, compute the estimates of $P(l(T)(x) \neq f(x) \mid n_1, \ldots, n_m)$, $\text{bias}^2(n_1, \ldots, n_m, x)$ and $\text{variance}(n_1, \ldots, n_m, x)$ from $y_{x,1}, \ldots, y_{x,N}$.

In the last step, the $n_1, \ldots, n_m$ indicate that these are the estimates for training sets with n_j samples of class y_j. Note that we opt not to use the inter-training-set variability as a parameter. This has computational reasons as the number of folds k can get quite large in our version of the algorithm, especially if one class is rare, which is common under class imbalance. However, Webb and Conilione [8] show that only the proportion of bias and variance are influenced by this parameter, not the error itself. This means that if we correctly predict the error, it is valid for any inter-training-set variability.

There are two things we want to predict:

1. How the classifier performance will change when adding samples of a particular class to the training set.
2. How the classifier performance will change when adding samples of all classes to the training set. In this case, the added samples are assumed to have approximately the same distribution as the current dataset.

In the first case, we run the estimation algorithm several times where the n_j stay the same for all classes but one. We set $n_j = d_j/2$ for all non-changing classes. This ensures that the inter-training-set variability is relatively high while there are still enough samples to provide a good classifier training. For one class, the number of samples used in each training set is increased for each run of the estimation algorithm, usually in fixed steps. In the second case, the proportion of samples from the dataset used as training samples is increased across estimation procedure runs. However, for all classes, this proportion is equally increased so that it will be the same for each class.

Using these estimates, we follow the prediction approach by Mukherjee et al. [4] by fitting a power-law function to them:

$$c(n) = an^{-\alpha} + b. \tag{9}$$

The reason is that there is theoretical and empirical evidence that the error rate can be described as a function of this kind [4,6]. The $c(n)$ stands either directly for the MR $\widehat{P}(l(T)(x) \neq f(x) \mid y_f, n)$ for class $y_f \in Y$ or for its components $\widehat{\text{bias}}^2(y_f, n)$ and $\widehat{\text{variance}}(y_f, n)$. The meaning of n depends on what we want to predict. If we want to predict the influence of one class, n corresponds to the number of samples from that class in the training set. If we want to predict the classification performance change under the addition of samples of all classes,

n stands for the total number of samples in the training set. To predict the influence of changes to n, its value is simply put into the fitted function. If the bias and variance components are predicted individually, their sum is used to predict the MR for the corresponding class.

Note that our solution is not able to directly predict the error to be expected using a training set which has more samples for a particular class and for all other classes uses all samples from D. To do that, it would be necessary to use all samples of these classes in each training set in the estimation algorithm. However, this would leave no samples for the test sets. While it could be approximated by using high proportions of the samples in each training set, this would lead to high computation times and low inter-training-set variability increasing estimation error. All that said, with our method, it is possible to predict if the addition of samples of a particular class leads to a notable classification performance increase. Thus, it can help the user to make and informed decision if it is worth the additional work or cost.

4   Evaluation

We tested our method on a classification problem which tries to differentiate engineered nanoparticles from other particles in scanning electron microscopy images. The used data is similar to that described in [2]. We looked at three two-class problems: silver (Ag), titanium dioxide (TiO_2) and zinc oxide (ZnO). For all problems, the number of negative samples is 8783. In addition, there are 70 positive samples for Ag, 811 for TiO_2 and 232 for ZnO.

We used the logistic regression classifier from Weka [1] with a ridge value of 1. The features were normalized to have zero mean and unit variance. In addition, the positive samples were weighted 10 times higher than the negative ones for the training of the classifier. This configuration gives relatively good results for all three classification problems.

Because of the nature of the data, our estimation procedure had to be slightly adapted. Particles from the same image may share similar properties. In addition, some images in the dataset show the same particles. Therefore, particles from images taken at the same sample position must end up in the same fold of the cross validation. This has the following consequences:

- The n_j are limited by d_j minus the maximum number of particles from the corresponding class per samples position.
- The number of folds k may have to be chosen higher so that there are always enough samples available for each training set.
- Some folds may contain more particles than others.

For our experiments, we used $N = 50$, the same value as was used by Kohavi and Wolpert [3] and Webb and Conilione [8]. For all three classification problems, we either varied only the positive sample count, the negative sample count or both. The counts were varied in 2.5 % steps from 2.5 % to 75 % of the maximum possible value. The sample counts which were not varied were fixed at

50 %. For every iteration, we recorded $\widehat{\mathrm{P}}\left(l(\mathcal{T})(x) \neq f(x) \mid y, n\right)$, $\widehat{\mathrm{bias}}^2(y, n)$ and $\widehat{\mathrm{variance}}(y, n)$ for $y \in \{y_+, y_-\}$. For brevity, we will from now on refer to them as $\mathrm{error}(y, n)$, $\mathrm{bias}^2(y, n)$ and $\mathrm{variance}(y, n)$. We then provided our prediction method with the values for $n = n_I, n_I + 1 \cdot \Delta_n, \ldots, n_I + (k_I - 1) \cdot \Delta_n$, where $k_I \geq 3$ is the number of input values and n varies from n_I to $n'_I := n_I + (k_I - 1) \cdot \Delta_n$. Its task was to predict $\mathrm{error}(y, n_O)$, where $n_O > n'_I$. Since we are more interested in the prediction of the change of the error instead of the error itself, we evaluated the absolute prediction error relative to $|\mathrm{error}(y, n_O) - \mathrm{error}(y, n'_I)|$.

We compared the variant which uses the decomposition by independently predicting the bias and variance to the variant which directly predicts the error. In addition, we included a baseline predictor as reference which uses the values of $\mathrm{error}(y, n_I + (k_I - 2) \cdot \Delta_n)$ and $\mathrm{error}(y, n_I + (k_I - 1) \cdot \Delta_n)$ to predict $\mathrm{error}(y, n_O)$ assuming a linear relationship between n and $\mathrm{error}(y, n)$.

5   Results

Due to limited space, we are not able to give separate results for each dataset and split by whose class' samples are added to the training set. We present the cumulative results from all experiments. This leads to reliable data based on a large pool of results. Figure 2 shows the performance of our prediction method

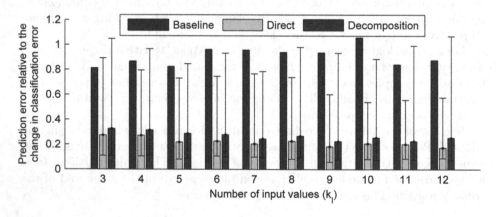

Fig. 2. This box plot shows the performance of our prediction method, with and without bias-variance decomposition, compared to the baseline predictor for different input value counts k_I using all datasets. n_O was set so that $n_O - n'_I = n'_I - n_I$. The bars indicate the median prediction error relative to the change in classification error $|\mathrm{error}(y, n_O) - \mathrm{error}(y, n'_I)|$. The whiskers denote the lower and upper quartiles of the prediction error. For the baseline, we left out the whiskers as they would not have fit inside the diagram. In all cases, our methods show better results than the baseline. In addition, the direct method generally shows a better performance than the one which uses the decomposition. As could be expected, the prediction error decreases as more input values are available.

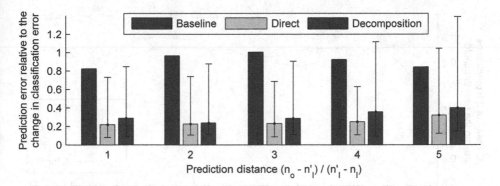

Fig. 3. This box plot shows the performance of our prediction method, with and without bias-variance decomposition, compared to the baseline predictor for $k_I = 5$ and varying prediction distances measures as $(n_O - n'_I)/(n'_I - n_I)$ using all datasets. The notation is the same used in Fig. 2. Again, our methods show better results than the baseline and the direct method generally shows a better performance than the decomposition method. Unsurprisingly, the results get worse as the prediction distance increases. However, the median error of the direct method only goes up from 0.22 to 0.32 showing that in many cases, the approach is capable of predicting the lassification error for much larger training sets.

with varying number k_I of estimated classification errors available to the prediction method. Figure 3 shows the same measures for varying prediction distances $n_O - n'_I$ relative to $n'_I - n_I$, which is the size of the range of sample counts spanned by the estimated classification errors available to the prediction method. Both figures show that the direct method generally produces better results than the approach using the bias-variance decomposition. Unsurprisingly, the prediction performance increases with increasing k_I and decreasing prediction distances. However, the direct method shows good results even for only 3 input estimates and for prediction distances as large as five times the range spanned by the input estimates.

In addition, we wanted to test if predicting the classification accuracy instead of G-mean or F-measure can lead to wrong decisions. To do that, we had to measure performance improvement based on an increase in G-mean or F-measure as these are appropriate measures for our imbalanced data. Figure 4 shows the results of the experiment. As ground truth, we regarded a case as a classification performance improvement if there was a certain relative increase in G-mean/F-measure, ranging from 0.01 to 0.2 (x-axis). We used the predicted F-measure, G-mean and accuracy to predict if an improvement was expected by assessing whether the relative change was higher than a threshold. We varied the values of the thresholds and used the threshold which was most favorable to the respective prediction method. The y-axis shows the G-mean of the predictions compared to the ground truth. Using the predicted accuracy to decide if more samples are beneficial leads to many wrong decisions as measured by an increase in G-mean.

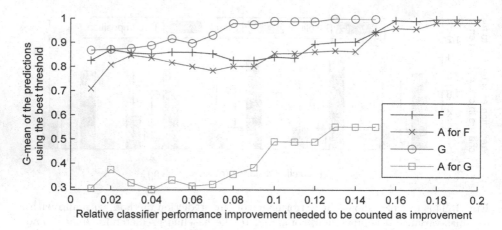

Fig. 4. Using different methods, we predicted if a relative classifier performance improvement (measured by F-measure or G-mean) of a certain magnitude (x-axis) is happening for more training data. The methods were: Using predicted F-measure to predict F-measure improvements (F), using predicted accuracy to predict F-measure improvements (A for F), using predicted G-mean to predict G-mean improvements (G) and using predicted accuracy to predict G-mean improvements (A for G). We varied the threshold for the prediction to be counted as an improvement and calculated the G-mean of the prediction correctness (y-axis) for the best threshold. We used $k_I = 5$ and $(n_O - n'_I)/(n'_I - n_I) = 1$. The results show that the accuracy is not at all appropriate to foretell if it is worthwhile to generate more training samples as measured by G-mean improvement. Measured by F-measure improvement, the accuracy does a good job. However, directly predicting the F-measure is better for all but two values.

If measured by F-measure, the accuracy prediction does better. However, in all but two cases, using the predicted F-measure leads to better decisions.

6   Conclusion

We presented a method to predict the classifier performance under the addition of training samples which is not limited to accuracy as the performance measure. It can predict any measure which can be calculated from the prior probabilities of the classes and the MRs limited to the samples of the respective class. This includes, among others, accuracy, sensitivity/recall, specificity, precision, F-measure, G-mean and all derived measures. Therefore, the method can be applied in cases such as class imbalance. In addition, it is able to predict the performance change under the addition of training samples from particular classes. The method uses a reliable estimation procedure and a prediction approach based a distribution backed by theoretical and empirical evidence. We have shown that the method works well for a varying number of performance estimates as input and that a direct prediction works better than one which uses the bias-variance decomposition of the error. Additionally, we have shown that the approach is able to predict the performance improvement for training sets which are much larger than the dataset used to make the predictions. This enables the user to make informed decisions if it is worth generating more training samples, a process which may be costly and often requires much work. Furthermore, we have demonstrated that, compared to relying on the prediction of classification accuracy, our method leads to a higher rate of correct decisions. Future research could focus on calculating uncertainty measures for the predictions.

References

1. Hall, M., Frank, E., Holmes, G., Pfahringer, B., Reutemann, P., Witten, I.H.: The WEKA data mining software: an update. ACM SIGKDD Explor. Newslett. **11**(1), 10–18 (2009)
2. Kockentiedt, S., Tönnies, K., Gierke, E., Dziurowitz, N., Thim, C., Plitzko, S.: Automatic detection and recognition of engineered nanoparticles in SEM images. In: VMV 2012: Vision, Modeling & Visualization, pp. 23–30. Eurographics Association (2012)
3. Kohavi, R., Wolpert, D.H.: Bias plus variance decomposition for zero-one loss functions. In: Proceedings of the 13th International Conference on Machine Learning, pp. 275–283 (1996)
4. Mukherjee, S., Tamayo, P., Rogers, S., Rifkin, R., Engle, A., Campbell, C., Golub, T.R., Mesirov, J.P.: Estimating dataset size requirements for classifying DNA microarray data. J. Comput. Biol. **10**(2), 119–142 (2003)
5. Smith, J.E., Tahir, M.A.: Stop wasting time: on predicting the success or failure of learning for industrial applications. In: Yin, H., Tino, P., Corchado, E., Byrne, W., Yao, X. (eds.) IDEAL 2007. LNCS, vol. 4881, pp. 673–683. Springer, Heidelberg (2007)
6. Smith, J.E., Tahir, M.A., Sannen, D., Van Brussel, H.: Making early predictions of the accuracy of machine learning classifiers. In: Sayed-Mouchaweh, M., Lughofer, E. (eds.) Learning in Non-stationary Environments, Chap. 6, pp. 125–151. Springer, New York (2012)
7. Sun, Y., Wong, A.K., Kamel, M.S.: Classification of imbalanced data: a review. Int. J. Pattern Recogn. Artif. Intell. **23**(4), 687–719 (2009)
8. Webb, G.I., Conilione, P.: Estimating bias and variance from data. Technical report, Monash University, Melbourne (2003). http://www.csse.monash.edu/~webb/Files/WebbConilione06.pdf

Signal/Background Classification of Time Series for Biological Virus Detection

Dominic Siedhoff[1]([✉]), Hendrik Fichtenberger[2], Pascal Libuschewski[3],
Frank Weichert[1], Christian Sohler[2], and Heinrich Müller[1]

[1] Computer Science VII – Computer Graphics, TU Dortmund, Dortmund, Germany
siedhoff@ls7.cs.tu-dortmund.de
[2] Computer Science II – Efficient Algorithms and Complexity Theory,
TU Dortmund, Dortmund, Germany
[3] Computer Science XII – Embedded Systems, TU Dortmund, Dortmund, Germany

Abstract. This work proposes translation-invariant features based on a wavelet transform that are used to classify time series as containing either relevant signals or noisy background. Due to the translation-invariant property, signals appearing at arbitrary locations in time have similar representations in feature space. Classification is carried out by a condensed k-Nearest-Neighbors classifier trained on these features, i.e. the training set is reduced for faster classification. This reduction is conducted by a k-means clustering of the original training set and using the obtained cluster centers as a new training set. The coreset-technique BICO is employed to accelerate this initial clustering for big datasets. The resulting feature extraction and classification pipeline is applied successfully in the context of biological virus detection. Data from Plasmon Assisted Microscopy of Nano-size Objects (PAMONO) is classified, achieving accuracy 0.999 for the most important classification task.

1 Introduction

Methods for reliable detection of biological viruses by means of inexpensive sensor devices have gained an increasing interest in research. The PAMONO method (Plasmon Assisted Microscopy of Nano-size Objects) [18] enables construction of a biosensor which is capable of indirectly detecting nano-sized particles, including fine dust and biological viruses, using inexpensive components from microscopy. It has potential applications in diagnostics as well as pharmaceutical research because virus-antibody-bindings can be detected, enabling to investigate the ability of antibodies to bind certain viruses.

The sensor produces a time series of 2D images, constituting a large spatio-temporal volume of data (e.g. 4000 images with 1080×145 pixels in one measurement), hence automatic analysis is desirable. One crucial step in this analysis is the separation of pixels that are affected by a nano-particle adhesion (signal) from those that are not (background). This separation can be carried out by examining the time series of intensities measured at the pixels because they exhibit characteristic patterns (cf. Fig. 1): When a particle appears, pixels in the

© Springer International Publishing Switzerland 2014
X. Jiang et al. (Eds.): GCPR 2014, LNCS 8753, pp. 388–398, 2014.
DOI: 10.1007/978-3-319-11752-2_31

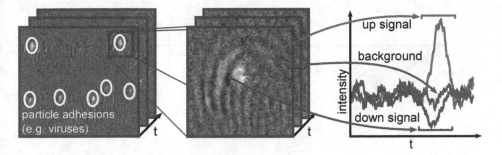

Fig. 1. Time series of PAMONO images (*left*) showing seven nano-particles. Magnification of a single particle (*center*) and per pixel time series (*right*), characterizing pixels as belonging to a particle (up and down signal) or to background noise

center of the particle exhibit an up signal in the time series and pixels around it exhibit a down signal. These two signals indicate a particle adhesion. Regions not affected by particles contain only background noise.

This work proposes wavelet-based features of the time series from which a fast condensed k-Nearest-Neighbors (k-NN) [9] classifier is learned, to separate particle signals from background. Since particles may appear at any point in time, the features need to be translation-invariant, such that signals are similar in feature space, independent of the time the particle appears.

While translation-invariant (TI) wavelets have been developed for denoising in [4], their application for feature computation is limited [14]. In face recognition, non-redundant Discrete Wavelet Transform (DWT) was used as the basis for computing estimates of the power spectrum in local windows. These estimates served as approximately TI features [14]. In contrast, [11] use raw coefficient values as features and ensure the TI property by redundant DWT and signal registration. Our approach combines the use of TI coefficients with feature extraction. This serves the purposes of dimensionality reduction and increases robustness to noise.

Furthermore, to speed up learning and classification, the idea of condensing the training set by removing redundancy and similarities [1,2,7] is explored. We extend an approach from text classification [17] that clusters the training data and uses cluster centers as a condensed training set. In order to reduce the time taken for clustering, we apply the BICO algorithm [6] to reduce the input to a smaller coreset, to which weighted clustering is applied. Finally the cluster centers serve as the training set of a k-NN learner.

The remainder of this work is organized as follows: Sect. 2 describes the extraction of translation-invariant features from wavelet coefficients. Section 3 covers the training procedure for accelerated k-NN by means of a fast coreset-based clustering that condenses the training set. Section 4 demonstrates empirical results of the presented methods as attained for the PAMONO biosensor application. Section 5 provides discussion and future work.

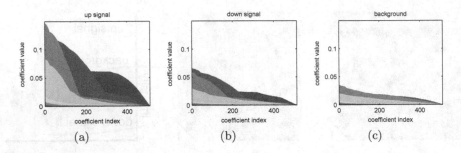

Fig. 2. Examples of **W** for time series from the three classes. Scales (i.e. rows of **W**) are ordered coarsest (back) to finest (front). Positions on the x axis are column indices in **W**. Classes differ in coefficient magnitude and slope, especially for coarser scales, which is exploited by the features in Eqs. (1)–(8)

2   Translation-Invariant Feature Extraction

Translation-invariant (TI) features are desirable in PAMONO time series classification because they make the feature space representation of a series independent of the point in time where the signal (up or down, cf. Fig. 1) appears. PAMONO time series approximately fulfil the circularity condition assumed for the TI wavelets proposed by Coifman and Donoho [4] utilized here: The signal to be detected has finite support at a random, irrelevant location, preceded and followed by uniform noise. Hence series with different locations of the signal can be approximately regarded as circular shifts of each other if we regard the uniform noise as 'basically the same' everywhere in the series.

For extracting TI features from a time series, the following preprocessing is applied: To remove constant offsets in the values, the series mean is subtracted from each value. Let $v = [v_t]_{t=1...T}$ denote the values of a single series after this subtraction. Let $\widehat{\mathbf{W}}$ denote the transform table of the TI wavelet transform from [4], using the maximum level of decomposition and e.g. the Haar basis (used throughout this paper, as argued for in the supplemental material [15]). The $S \times T$ matrix $\widehat{\mathbf{W}}$ is a non-orthogonal decomposition of the series v into $S = \log(T)$ scales with T coefficients per scale. The scales with smaller index s represent coarser structures of v, while those with larger s represent finer details. The notion of translation-invariance that applies to $\widehat{\mathbf{W}}$ is that a circular shift of the series v manifests solely as per-scale permutations in $\widehat{\mathbf{W}}$, i.e. for each of the S rows of $\widehat{\mathbf{W}}$, the T entries are permuted. Note that there are neither permutations between coefficients of *different* scales, nor is there energy-transfer between *any* coefficients. The per-scale permutations are the way in which circular shifts of the input series are encoded by the transform table. Since the goal is computing features that are invariant to shifts of the input series – or synonymously invariant to the location of the finite support of the up or down signal to be detected (cf. Fig. 1) – this locational information must be eliminated. This is done by sorting each scale in $\widehat{\mathbf{W}}$ separately by descending absolute coefficient values. The resulting matrix of sorted absolute values is called $\mathbf{W}$. Figure 2 shows examples of $\mathbf{W}$ for series v from different classes. For any permutation of coefficients in $\widehat{\mathbf{W}}$, the same $\mathbf{W}$ will result, and hence the locational information has been removed. Feature computation is carried out with respect to $\mathbf{W}$ as detailed in the following section.

2.1   Features of Translation-Invariant Wavelet Coefficients

Let $\mathbf{W} = [w_{s,t}]_{s=1...S, t=1...T}$ be the table from Sect. 2 with scale index s and coefficient index t. Furthermore, let $\mu(w_{s,o}) = \frac{1}{T}\sum_{t=1}^{T} w_{s,t}$ yield the mean value over the variable indicated by the wildcard symbol o for a fixed value of the symbol s. The according standard deviation is defined analogously as $\sigma(w_{s,o}) = \sqrt{\frac{1}{T}\sum_{t=1}^{T}(w_{s,t} - \mu(w_{s,o}))^2}$. In the following equations superscript indices are feature names while the subscript s indicates the scale on which a feature is computed. Feature f_s^1 is the mean value of the coefficients on scale s:

$$f_s^1 = \mu(w_{s,o}) . \tag{1}$$

Feature f_s^2 is the ratio of f_s^1 and the mean coefficient value over all scales:

$$f_s^2 = \frac{f_s^1}{\mu(w_{o,o})} . \tag{2}$$

Feature f_s^3 is the standard deviation of the coefficients on scale s,

$$f_s^3 = \sigma(w_{s,o}) , \tag{3}$$

while feature f_s^4 is the ratio of f_s^3 and the coefficient standard deviation accumulated over all scales:

$$f_s^4 = \frac{f_s^3}{\sum_r \sigma(w_{r,o})} . \tag{4}$$

For defining the last four features, let $r_s = [r_{s,t}]_{t=1...T}$ denote T discrete samples of a regression line approximating the coefficients $w_{s,t}$ for a fixed scale s. The line is defined as $a_s x + b_s$, where a_s and b_s are computed as $\mathrm{argmin}_{a_s,b_s} = \sum_t (w_{s,t} - (a_s x_t + b_s))^2$, and the sampling points x_t are chosen at the locations of the coefficients $w_{s,t}$. Hence $r_{s,t} = a_s x_t + b_s$ is the linear approximation of coefficient $w_{s,t}$. Given these prerequisites, f_s^5 is the slope of the regression line,

$$f_s^5 = a_s , \tag{5}$$

and f_s^6 is the linear approximation of the largest coefficient $w_{s,1}$

$$f_s^6 = a_s x_1 + b_s . \tag{6}$$

Feature f_s^7 is the accumulated absolute deviation between coefficients and their linear approximation, normalized by the accumulated coefficient set:

$$f_s^7 = \frac{\sum_t |w_{s,t} - r_{s,t}|}{\sum_t w_{s,t}}.$$ (7)

Feature f_s^8 is defined analogously to f_s^7 but with sums replaced by standard deviations:

$$f_s^8 = \frac{\sigma(w_{s,o} - r_{s,o})}{\sigma(w_{s,o})}.$$ (8)

Note that the differences in the numerator of f_s^8 are taken only between coefficients $w_{s,t}$ and approximations $r_{s,t}$ with the same index t. To prevent numerical issues, a small constant ϵ on the order of computational working precision is added to each denominator.

In order to balance the impact of the different features during the distance computations occurring in subsequent processing stages, the features need to be normalized accordingly. The employed normalization method is shifting and scaling the range of each feature to the unit interval $[0, 1]$.

2.2 Feature Ranking and Selection

The employed feature ranking and selection method is based on computing a figure of merit for each feature. Let $g = f_s^j, s \in \{1 \ldots S\}, j \in \{1 \ldots 8\}$ denote the single scalar feature under consideration. Let $g_i^c, i \in \{1 \ldots N^c\}, c \in \{1 \ldots C\}$ denote the values that feature g attains over the N^c examples of class c in the training dataset, where C is the total number of classes.

The basic measure in computing the figure of merit of a feature g is the mean absolute distance d^{c_1,c_2} between the feature values of class c_1 and the mean feature value of class c_2:

$$d^{c_1,c_2} = \mu\left(|g_o^{c_1} - \mu(g_o^{c_2})|\right).$$ (9)

Using Eq. (9), the figure of merit m_g of feature g is computed as the following ratio:

$$m_g = \frac{\sum_{c_2} \sum_{c_1 \neq c_2} d^{c_1,c_2}}{\sum_c d^{c,c}}.$$ (10)

The numerator of m_g accumulates over all classes c_2, in how far the mean feature value for class c_2 differs from the feature values for all other classes $c_1 \neq c_2$. For features that separate the classes well, this value is large. The denominator accumulates over all classes, in how far feature values vary within a class, hence smaller is better. The features g are then sorted in the order of descending merit m_g, giving a feature ranking assigning rank 1 to the best feature, rank 2 to the second-best and so forth. For determining the final sequence of features, Kuncheva's ranking idea [10] is applied within a 10-fold cross-validation:

The ranks attained by each feature are accumulated over the different folds, and the features are sorted in the order of ascending accumulated ranks. Kuncheva indices are computed during this cross-validation in order to assess feature selection stability. Finally the first F of these features are selected to be used for classification, where F is chosen to maximize the mean classification performance (e.g. accuracy) in a second cross-validation using the rank-sorted features and incrementally increasing the candidate for F.

3 Condensed k-NN Using Fast Coreset Clustering

This section describes how the k-Nearest-Neighbors (k-NN) classifier [9] is accelerated by computing it from cluster centers of the training data. This involves clustering as a preprocessing step, which is accelerated for large input sets by using the coreset-based BICO approach [6]. Section 3.1 explains how coresets can be used for clustering, while Sect. 3.2 depicts the training procedure and the application of a condensed k-NN classifier making use of the clustering results.

3.1 Fast Clustering with Coresets

Clustering is often defined as partitioning a set of objects into groups, such that objects in the same group are similar and objects in different groups are dissimilar. The k-means problem is a well-studied clustering problem defining similarity via Euclidean distance. For two points $p = (p_1, \ldots, p_F), q = (q_1, \ldots, q_F) \in \mathbb{R}^F$, let $||p-q|| := \sqrt{\sum_{i=1}^{F} (p_i - q_i)^2}$ denote their Euclidean distance. Given a matrix $\mathbf{P} = [p_1; \ldots; p_N]$ where each point $p_i \in \mathbb{R}^F$ is a row vector, the k-means problem asks for a matrix of K_m centers $\mathbf{Q} = [q_1; \ldots; q_{K_m}]$ that minimizes the sum of squared distances of all points in $\mathbf{P}$ to their nearest center in $\mathbf{Q}$, i.e.

$$\min_{\mathbf{Q} \in \mathbb{R}^{K_m \times F}} \mathrm{cost}(\mathbf{P}, \mathbf{Q}) := \min_{\mathbf{Q} \in \mathbb{R}^{K_m \times F}} \sum_{p_i \in \mathbf{P}} \min_{q_j \in \mathbf{Q}} ||p_i - q_j||^2 . \tag{11}$$

This is a special case of the weighted k-means problem (with weights all 1), where each point can be weighted by a function $w : \mathbb{R}^F \to \mathbb{R}^+$, i.e. $\mathrm{cost}_w(\mathbf{P}, \mathbf{Q}) = \sum_{p_i \in \mathbf{P}} w(p_i) \min_{q_j \in \mathbf{Q}} ||p_i - q_j||^2$ is minimized. Using k-means as the objective function for condensing the training set is a natural choice when using k-NN as a classifier because both algorithms decide a point's membership to a cluster/class via Euclidean distance. By assigning points to the same cluster, k-means preserves their spatial proximity.

In practice Lloyd's algorithm [13] is frequently used to optimize Eq. (11) and its weighted variant. It is an iterative algorithm converging to a local optimum after a potentially exponential number of steps. The k-means++ algorithm by Arthur and Vassilvitskii [3] is an improvement of Lloyd's algorithm, yielding an $\mathcal{O}(\log K_m)$ approximation guarantee. Its runtime is similar to that of Lloyd's algorithm. Both algorithms do not scale well and are hence time-consuming for large input data sets.

A way to address this problem is to construct a small summary of the input point set first and to cluster that summary instead. Such summaries can be formalized as *coresets*. A (K_m, ϵ)-coreset for K_m sought cluster centers and approximation ratio ϵ is a small weighted set of points $\mathbf{S} \in \mathbb{R}^{N' \times F}$ that ensures that the weighted clustering cost of $\mathbf{S}$ for any set of K_m centers $\mathbf{Q} \in \mathbb{R}^{K_m \times F}$ is a $(1 + \epsilon)$-approximation of the cost of the original input $\mathbf{P} \in \mathbb{R}^{N \times F}$ [8]:

$$|\operatorname{cost}_w(\mathbf{S}, \mathbf{Q}) - \operatorname{cost}(\mathbf{P}, \mathbf{Q})| \leq \epsilon \operatorname{cost}(\mathbf{P}, \mathbf{Q}) . \tag{12}$$

Here $w : \mathbb{R}^F \to \mathbb{R}^+$ is the weight function, and the number N' of points in the coreset may be considerably smaller than the number N of points in the original dataset $\mathbf{P}$. Since large data sets may not fit into main memory, and short construction times are crucial, coresets are often computed in a streaming setting. The BICO algorithm detailed in [6] is a streaming-capable algorithm for computing coresets. It is used in this work to reduce the time taken for clustering large input sets. The following section describes how BICO is integrated into the training procedure of k-NN, yielding a condensed k-NN.

3.2 Training and Application of Condensed k-NN

The input of the training procedure for condensed k-NN is the $N \times F$ matrix $\mathbf{G}$, where rows denote training examples and columns denote normalized features, cf. Sect. 2.1. F is the number of features selected as according to Sect. 2.2. N is the sum $N = \sum_{c=1}^{C} N^c$ of examples belonging to C different classes. In training, class labels are known and can hence be used to partition $\mathbf{G}$ into C per-class matrices $\mathbf{G}^c$. For each class c separately, the BICO approach is used to compute a coreset $\mathbf{S}^c$ of $\mathbf{G}^c$, allowing for very large input sets. Subsequently weighted k-means++ is used to compute a clustering from each coreset $\mathbf{S}^c$, condensing the examples in $\mathbf{G}^c$ to $K_m \ll N_c$ cluster centers $\mathbf{H}^c$. The union $\mathbf{H} = [\mathbf{H}^1; \ldots; \mathbf{H}^C]$ of the per-class cluster centers is then used as the condensed training set for k-NN. The number K_m of cluster centers per class is chosen as to be tractable in a lazy learning approach like k-NN and can be used for class balancing.

Applying the learned classifier to unlabeled input works as follows: The raw input data is preprocessed like the training set, using the same feature normalization and selection. The resulting feature vectors are classified using k-NN with Euclidean distance on the condensed training set $\mathbf{H}$. The number K_n of nearest neighbors in k-NN is not to be confused with the number K_m of cluster centers in k-means. The output of this k-NN are predicted labels for the unlabeled input.

4 Results

Three variants of the PAMONO time series classification task were examined to validate the proposed methods: Task 1 is the three-class separation of the up signal, down signal and background classes (cf. Figs. 1 and 2). Task 2a is the separation of the up signal class from the union of the down signal and background class.

Task 2b is the separation of the up signal from the background class only. The tasks were enumerated in the order of decreasing difficulty and increasing importance: For nano-particle detection it is most important to separate the up signals arising at the centers of particle adhesions from the background arising in regions without particles. Class membership of the small amount of down signals at the fringes of particles can in practice be neglected because down signal pixels can be captured by applying morphological closing to the up signal class mask in image space [5]. This leads to the consideration of task 2b.

Experimental validation was conducted using a total of $N = 315000$ labeled PAMONO time series from $C = 3$ classes as the input. The length was $T = 512$, resulting in 72 features available for selection (8 features on $S = \log(T) = 9$ scales). To obtain 315000 labeled examples, the signal model in [16] was used to create synthetic time series from real background images and particle templates. Three measured datasets were used as the basis for synthesis: 200 nm particles on two differently severe levels of noise and one dataset with 100 nm particles.

The input examples are partitioned into two disjoint subsets: The cross-validation set contains 2/3 of the examples and is used for feature and parameter selection in 10-fold cross-validation and as the basis to train the condensed k-NN classifier. The test set contains the remaining 1/3 of examples and is used solely for performance assessment. Note that for all classification tasks, class distributions were balanced because accuracy was used as the performance metric, which is sensitive to imbalanced class distributions.

Performance and Parameters. The number of per-class cluster centers was fixed at $K_m = 1500$ ($\approx 0.5\%$ of the total number of examples) and the coreset size was fixed at $N' = 5K_m$. The following accuracies were attained on the test set: task 1: 0.870, task 2a: 0.920, task 2b: 0.999. For comparison, matching patterns via cosine similarity to T non-cyclic shifts of C ideal model patterns yields the following accuracies: task 1: 0.894, task 2a: 0.996, task 2b: 0.997. Exhaustive matching against all non-cyclic shifts means that TI conditions hold exactly here, not only approximately. The superior accuracy in tasks 1 and 2a comes at the price of requiring model patterns and increased runtime ($\mathcal{O}(T^2)$ per time series for all shifts and distance computations instead of $\mathcal{O}(T \log^2(T))$ for TI table computation and sorting). Increasing the coreset size N' in the proposed method does not increase accuracy, but only run time: The mean (over all tasks) gain in accuracy when using $N' = 200K_m$, which is equivalent to clustering the full input, is $0.0011, \sigma = 0.0016$. Clustering time, on the other hand, increases from a mean value of 72 s, $\sigma = 23$ s to 2252 s, $\sigma = 811$ s (CPU: Intel(R) Core(TM) i7-2600 at 3.4 GHz). For determining the number F of best features to be selected and the number K_n of neighbors in k-NN, a grid search was conducted within a 10-fold cross-validation. F and K_n were chosen to maximize mean accuracy over the folds. Figure 3 plots mean accuracy over parameters per task, with task 2b achieving values close to 1. The spread of accuracies over the parameter space decreases with decreasing task difficulty. Accuracies saturate with increasing parameter values, meaning that F and K_n just need to be 'large enough'.

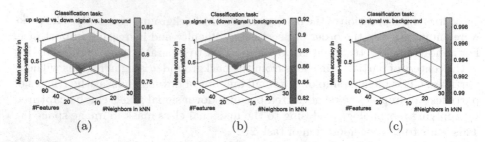

Fig. 3. Mean accuracy in 10-fold cross-validation over the number of best features and the number of neighbors to be used in k-NN for the three classification tasks

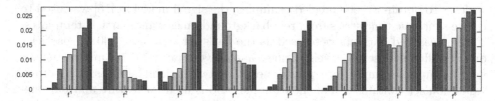

Fig. 4. Kuncheva ranks (lower is more relevant) attained by all features on all scales. Scales are ordered coarse (left) to fine (right) for each feature f^i

Feature Ranking. Feature ranking was carried out as described in Sect. 2.2. Figure 4 shows the Kuncheva ranks (lower means 'more relevant'), normalized by the sum of all ranks over all folds. The three most important features are located on the coarsest scale and exhibit strictly decreasing relevance for finer scales. Two of them (f_s^5 and f_s^6) are based on the regression line. The normalized versions f_s^2 and f_s^4 of f_s^1 and f_s^3 favor finer scales. The approximation-error-based features f_s^7 and f_s^8 are comparatively irrelevant on all scales. The figure shows results for classification task 1; for the other tasks, the results are qualitatively the same. Figure 5(a) plots feature selection stability in terms of the mean Kuncheva indices attained over the cross-validation. Kuncheva indices are above 0.88 for selecting between 10 and 60 features for all classification tasks (task 1: *short dashes*, task 2a: *long dashes*, task 2b: *solid line*). Figure 5(b) illustrates that the feature ranking is meaningful in terms of accuracy on the unseen test set: Each feature was used in isolation to classify the test set. The attained accuracy was plotted over the index that feature had in Kuncheva's ranking. Accuracy decreases approximately linearly with increasing Kuncheva rank.

Robustness to Noise. In order to assess robustness of the features to noise in the input, experiments with artificial noise were conducted. The standard deviation of each input time series was computed, and the mean of these values was used as an estimate m_e of average signal magnitude. Then the interval $[0, 20m_e]$ was equidistantly sampled, and for each sample m_i, zero-mean, unit variance Gaussian noise was scaled by m_i/m_e and added to the original input to give a more noisy input. For each of those noisy inputs a classifier was trained using

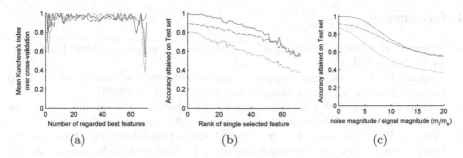

Fig. 5. Kuncheva indices (a), single feature accuracies (b) and robustness to noise (c) for the three classification tasks 'up vs. down vs. background' (*short dashes*), 'up vs. (down ∪ background)' (*long dashes*), 'up vs. background' (*solid line*)

the proposed method, and test set accuracy was measured for each classification task. Figure 5(c) plots these accuracies over m_i/m_e. The values for zero noise provide a baseline for each classification task. Results deteriorate slowly with increasing noise, which is especially true for task 2b (*solid line*), where noise with up to three times the magnitude of m_e causes only minor loss. From there on, the slope of accuracy loss increases. It decreases again, starting at approximately twelve times more noise than signal, and for each task converges in good approximation to the limit of random guessing for the respective task (1/2 for the two-class tasks and 1/3 for the three-class task; balanced class distributions).

5   Discussion and Future Work

A novel set of translation-invariant wavelet-based features for time series classification was proposed and used in a condensed k-NN classifier with fast coreset-based clustering. The efficacy of the approach was demonstrated with respect to nano-particle detection. For the most important classification task of distinguishing the central parts of particle adhesions from noisy background, accuracy close to 1 was achieved. It was demonstrated that the method is robust to increasing noise in the input signal and insensitive to its main parameters (number of features and k in k-NN), as long they are chosen large enough. Feature selection was shown to be stable, and run time was considerably reduced by using the coreset-based BICO approach for clustering. For the future it is planned to port the method into a GPU-based detector [12] and assess its impact on quality, as well as validate it on real data, using quality measures like per-class Precision and Recall, that are suitable for the imbalanced real class distributions. Furthermore an examination of classifiers other than k-NN is planned.

Acknowledgments. Part of the work on this paper has been supported by Deutsche Forschungsgemeinschaft (DFG) within the Collaborative Research Center SFB 876. URL: http://sfb876.tu-dortmund.de/

References

1. Alpaydin, E.: Voting over multiple condensed nearest neighbors. Artif. Intell. Rev. **11**(1–5), 115–132 (1997)
2. Angiulli, F.: Fast condensed nearest neighbor rule. In: Proceedings of the 22nd International Conference on Machine Learning, pp. 25–32 (2005)
3. Arthur, D., Vassilvitskii, S.: k-means++: the advantages of careful seeding. In: Proceedings of the 18th Symposium on Discrete Algorithms (SODA) (2007)
4. Coifman, R.R., Donoho, D.L.: Translation-invariant de-noising. In: Antoniadis, A., Oppenheim, G. (eds.) Wavelets and Statistics. Lecture Notes in Statistics, vol. 103. Springer, New York (1995)
5. Dougherty, E.R.: An Introduction to Morphological Image Processing. SPIE Press, Bellingham (1992)
6. Fichtenberger, H., Gillé, M., Schmidt, M., Schwiegelshohn, C., Sohler, C.: BICO: BIRCH meets coresets for k-means clustering. In: Proceedings of the 21st European Symposium on Algorithms (ESA) (2013)
7. Gowda, K.C., Krishna, G.: The condensed nearest neighbor rule using the concept of mutual nearest neighborhood. IEEE Trans. Inf. Theory **25**(4), 488–490 (1979)
8. Har-Peled, S., Mazumdar, S.: On coresets for k-means and k-median clustering. In: Proceedings of the 36th Symposium on Theory of Computing (STOC), pp. 291–300 (2004)
9. Hastie, T., Tibshirani, R., Friedman, J.: The Elements of Statistical Learning. Springer, New York (2009)
10. Kuncheva, L.I.: A stability index for feature selection. In: Artificial Intelligence and Applications (2007)
11. Li, D., Luo, H., Shi, Z.: Redundant DWT based translation invariant wavelet feature extraction for face recognition. In: ICPR (2008)
12. Libuschewski, P., Siedhoff, D., Timm, C., Gelenberg, A., Weichert, F.: Fuzzy-enhanced, real-time capable detection of biological viruses using a portable biosensor. In: Proceedings of the International Joint Conference on Biomedical Engineering Systems and Technologies (BIOSIGNALS) (2013)
13. Lloyd, S.: Least squares quantization in PCM. IEEE Trans. Inf. Theory **28**(2), 129–137 (1982)
14. Ma, K., Tang, X.: Translation-invariant face feature estimation using discrete wavelet transform. In: Tang, Y.T., Wickershauser, V., Yuen, P.C., Li, C.-H. (eds.) WAA 2001. LNCS, vol. 2251, pp. 200–210. Springer, Heidelberg (2001)
15. Siedhoff, D., Fichtenberger, H., Libuschewski, P., Weichert, F., Sohler, C., Müller, H.: Signal/background classification of time series for biological virus detection - supplemental material. In: Jiang, X., Hornegger, J., Koch, R. (eds.) GCPR 2014. LNCS, vol. 8753, pp. 384–394. Springer, Heidelberg (2014)
16. Siedhoff, D., Libuschewski, P., Weichert, F., Zybin, A., Marwedel, P., Müller, H.: Modellierung und Optimierung eines Biosensors zur Detektion viraler Strukturen. In: Deserno, T.M., et al. (Hrsg.) Bildverarbeitung für die Medizin, pp. 108–113. Springer, Heidelberg (2014)
17. Yong, Z., Youwen, L., Shixiong, X.: An improved KNN text classification algorithm based on clustering. J. Comput. **4**(3), 230–237 (2009)
18. Zybin, A., Kuritsyn, Y.A., Gurevich, E.L., Temchura, V.V., Ueberla, K., Niemax, K.: Real-time detection of single immobilized nanoparticles by surface plasmon resonance imaging. Plasmonics **5**, 31–35 (2010)

Efficient Hierarchical Triplet Merging for Camera Pose Estimation

Helmut Mayer[✉]

Institute of Applied Computer Science, Bundeswehr University Munich,
Neubiberg, Germany
helmut.mayer@unibw.de

Abstract. This paper deals with efficient means for camera pose estimation for difficult scenes. Particularly, we speed up the combination of image triplets to image sets by hierarchical merging and a reduction of the number of merged points. By **image sets** we denote a generalization of image sequences where images can be linked in multiple directions, i.e., they can form a graph. To obtain reliable results for triplets, we use large numbers of corresponding points. For a high-quality and yet efficient merging of the triplets we propose strategies for the reduction of the number of points. The strategies are evaluated based on statistical measures employing the full covariance information for the camera poses from bundle adjustment. We show that to obtain a statistically sound result, intuitively appealing deterministic reduction strategies are problematic and that a simple reduction strategy based on random deletion was evaluated best. We also discuss the benefits of the evaluation measures for finding conceptual and implementation weaknesses. The paper is illustrated with a number of experiments giving standard deviations for all values.

1 Introduction

Camera pose estimation for large image sets is an established field. Agarwal et al. [1] and, particularly, Frahm et al. [6] have demonstrated how to deal with very large image sets. Another direction is presented in Klingner et al. [9] where bundle adjustment is done on a global scale.

Opposed to the above work, we focus on comparably small data sets. While for the experiments we have chosen data sets with 20–40 images, our usual range is from several hundred to a couple of thousand. Yet, also in our case the complexity is high, because we want to find all possible links between images even when no approximate values from GPS and inertial navigation systems (INS) are available and corresponding points look rather different. The latter can be due to changes in perspective, lighting and even the scene itself, i.e., when images are taken at different times, even years. This means, that a reduction of the images to small "stamps" as in [6] is not an option. We found that to obtain reliable results, an extremely robust approach consisting in our case of random sample consensus – RANSAC [3], the geometric information robust information

© Springer International Publishing Switzerland 2014
X. Jiang et al. (Eds.): GCPR 2014, LNCS 8753, pp. 399–409, 2014.
DOI: 10.1007/978-3-319-11752-2_32

criterion – GRIC [15] and robust bundle adjustment for pairs and triplets is a must. We employ triplets, because points in triplets are much more rigorously checked than in pairs [4].

Concerning bundle adjustment, Jian et al. [8] and Jeong et al. [7] have shown how to speed up the solution using, e.g., a conjugate gradient solution (CG). We also use a sparse implementation of bundle adjustment, but because we are interested in the covariance information for quality assurance and also for further processing, we cannot use CG. Instead, this paper focuses on the efficiency improvement possible by the reduction of the number of corresponding points when merging triplets.

Hierarchical merging of image triplets into (sub-)sequences including loop closing has been described already by Fitzgibbon and Zisserman [4] in 1998. Klopschitz et al. [10] have proposed means for merging triplets made up of images generally connected into graphs, i.e., image sets. They identify reliable pairs of image sets and merge the most reliable pairs first. Opposed to this, we focus on the reduction of the points in the image triplets for a more efficient merging. Particularly, we build on Bartelsen et al. [2] as well as Mayer et al. [12], which merge only a single triplet to the image set at each time with a need for robust bundle adjustment after each merge. This leads to an exponential complexity: While extending with the n-th image, a bundle adjustment of all n images is conducted. In this paper this is replaced by a much more efficient hierarchical merging.

The paper is organized as follows: Sect. 2 describes hierarchical merging. Section 3 defines different reduction strategies for a more efficient merging. Based on the statistical evaluation measures introduced in Sect. 4, Sect. 5 presents experiments which show that only random deletion guarantees a high quality solution. Section 6 concludes the paper with a discussion.

2   Hierarchical Merging

Merging starts with image sets initialized by the given triplets. By merging these initial image sets larger and larger image sets are generated. Merging uses a **unique identifier (ID)** for each image point as basis to link corresponding points in different images. This ID is given by

– Image number
– (Sequential) point number in the image.

I.e., we detect points with SIFT [11] or SFOP [5] only once per image. The unique ID is preserved in all stages of matching and geometric reconstruction and is used as the basis for merging. For it, image sets need to have at least two images in common: From the projection matrices of the two common images we derive a rigid 3D transformation (3 translations, 3 rotations, 1 scale) and transform the projection matrices as well as the 3D points of the image set to be merged into the given image set.

Based on the information obtained during matching, corresponding points in corresponding images can be identified in image sets to be merged by means of the unique ID. We assume that corresponding IDs mean corresponding 3D points and, thus, we merge the sets of image points from the two image sets eliminating redundant points. I.e., the set of images a 3D point is linked to is potentially extended. (It is not extended if the sets of image points are equal or one is a subset of the other.) After merging, the image set is robustly bundle adjusted to eliminate outliers and reduce drift.

For the **hierarchical merging**, image sets of ideally equal size and with an overlap of two images are used. Thus, image sets with sizes N and M are merged into a set with $N + M - 2$ images. I.e., merging triplets ($N = M = 3$) leads to quadruples ($N = M = 4$), merging quadruples to sextuples ($N = M = 6$), and merging sextuples to decuples ($N = M = 10$). The merging goes on with sets of 18, 34, 66, 130, 258, 514 and 1026 images, etc.

The hierarchical procedure has a strongly reduced complexity compared to always linking one triplet at a time as in [2, 12], because the robust bundle adjustment is only computed for the few larger and larger image sets. Additionally, the merging of the large number of smaller sets is independent and can, thus, be computed in parallel on different computer cores.

We found that for obtaining reliable results many points are needed for the triplets, much more than are viable and necessary for larger image sets. We, therefore, describe strategies for the reduction of the number of points, leading to a much more efficient merging in the following section.

3   Reduced Hierarchical Merging

For the rest of the discussion we note, that it is possible to conduct the hierarchical merging of the point triplets to 3D points without the geometrical transformations and the bundle adjustment. We call this "simulation" and it only deals with the IDs of the points. The simulation can be done in basically "zero time" compared to the actual geometric merging. From it we know which 3D points are finally produced, which **image point triplets** (image points in the triplets which are merged) are involved, and which images each 3D point links to at maximum (some links will be deleted during matching and robust bundle adjustment). Based on the information in the simulation result we formulate the following principles:

- There is a trade-off between the precision induced by the number of points, i.e., redundancy, and the computational effort.
- 3D points which link as many images as possible are most valuable due to their high redundancy.
- It is important that the points are well distributed over the image to avoid a solution which is valid only for parts of the overlapping area between images.

Based on the principles we defined and tested many strategies. We will focus in the remainder of this paper on a couple of them, which either work, such as our "proposed" strategy, or are intuitively appealing, but not as good as one would expect. They have been tested based on the evaluation measures described in Sect. 4, with results for four data sets reported in Sect. 5.

Proposed: This strategy consists of the random deletion of all image points linked to a 3D point, until on average *avgnr* image points remain per image.

Triplet deletion: This strategy is similar to the "proposed", but is more fine grained by deleting individual image point triplets. For a 3D point this can mean the deletion of none up to three image points, the former if the image points are all also contained in other point triplets, the latter if the 3D point is actually deleted.

Min – Minimum number points in image: Image point triplets are randomly deleted as long as the number of points in the images a point triplet links to does not fall below the threshold *minnr*.

MinMax – Min + maximum number links per 3D point: This extends the "Min" strategy by the second principle, namely that 3D points which link many images are especially advantageous. This is done by keeping all 3D points linking at least a given number of images.

MinMax + grid: As we found that "MinMax" did not work as well as hoped for, we expanded it with the third principle and formulate a grid on the images. For the grid, we keep track of the number of image points in each cell. If the deletion would mean that the number of points in a cell in an image would fall below a given threshold, the deletion is rejected.

Proposed + grid: Here the grid part of the "MinMax + grid" strategy is linked with the random deletion of the "proposed" strategy.

While the image point triplets proposed for deletion are generated randomly, the last four strategies have a deterministic criterion at their core: Image points are kept because of their number of links exceeding a certain value or because their deletion would mean, that there are not enough points in the image or the grid cell any more.

The different strategies are compared based on the evaluation measures described in the following section.

4   Evaluation

The basis for the evaluation are two statistical measures comparing solutions based on the ratio of their variances as well as the relation between the differences of the parameters of the solutions to the covariances for these parameters.

Particularly, we use the covariance matrices Σ_0 for the reference data set 0 and Σ_i for the data set to compare it with i. The first measure is computed from the (real) eigenvalues Λ of the ratio of the covariance matrices

$$\Lambda = Eigenvalues(\Sigma_0 \Sigma_i^{-1}). \tag{1}$$

From Λ we compute the square roots of the maximum $\sqrt{\lambda_{\max}}$ and the average value $\sqrt{\mu_\lambda}$ which give the largest and the average ratio between the standard deviation of a parameter in the reference and the data set to compare it to. This gives an indication of the maximum and the average loss of precision.

The second measure relates the difference between the parameters $\mathbf{x}_i - \mathbf{x}_0$ to the sum of their covariances. Particularly, we compute the test statistic (absolute Mahalanobis distance)

$$T = (\mathbf{x}_i - \mathbf{x}_0)^\mathsf{T}(\Sigma_0 + \Sigma_i)^{-1}(\mathbf{x}_i - \mathbf{x}_0), \tag{2}$$

which is approximately χ_n^2 distributed, with n the number of degrees of freedom (DOF), i.e., the number of parameters.

To use the above measures it is necessary that the parameters $\mathbf{x}$ in the two data sets describe the same influence on the mathematical model. In our case this means that the geometry of the data sets has to be made comparable. For this we use the same image as reference. It is located at the origin, fixing the three parameters of the global translation, and it is not rotated, i.e., the rotation matrix is the unit matrix, fixing the global rotation. The final parameter, namely the scale, is fixed by setting the distance to the image furthest away from the reference image to one.

Our representation of the translations and rotations is based on coordinates and unit quaternions, respectively. For the latter, we keep the largest value fixed and only vary the rest. The same is done for the largest parameter of the image which is furthest away from the reference image for which we normalize the distance.

The above normalization ensures similar margin conditions, yet does not guarantee an optimum solution. For the latter, we determine an overall similarity transformation, i.e., 3D rotation, translation and scale.

5   Experiments

In our experiments we compare the full hierarchical merging (Sect. 2) and the different strategies for reduced hierarchical merging (Sect. 3) concerning computational complexity as well as the achieved precision, the latter based on the measures from the previous section.

Figure 1 gives an overview of the four scenes. B1 and B2 show buildings, E is Ettlingen castle (castle-R20 [14]) and T presents a technical lab. The images were taken from the ground or from small unmanned aerial vehicles (UAVs).

In all scenes the images are either connected in a graph/set structure, or the loop is closed as for E. For each of the one or two cameras used in the experiments (cf. Table 1) we estimate seven parameters, namely five in the calibration matrices as well as two radial distortion parameters. They are added to the $(\#images - 2) * 6 + 5$ parameters of the relative pose to define the DOF, i.e., the n of the χ^2 distribution. The number of images in the sets has been limited to make the large number of experiments more tractable.

By using robust bundle adjustment with computation of the covariance information, we have the full covariance information available. I.e., we use the full $n \times n$ covariance matrices as the basis to compute λ (1) and T (2).

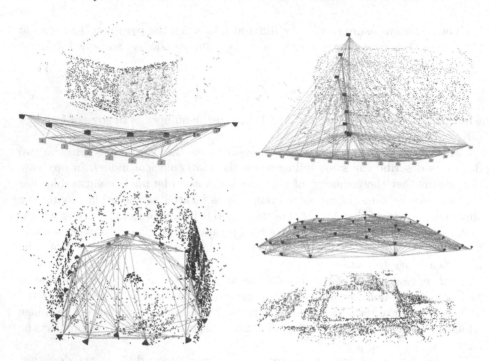

Fig. 1. 3D models – left to right, top to bottom: B1, B2, E (castle-R20 [14]) and T. The camera positions are represented as pyramids (different color means different camera) and red lines connect images with at least ten common 3D points.

We relate the result of the test statistic T to the "3σ" interval $[lb, ub]$ (with lower and upper bound lb and ub) of χ_n^2 (relative Mahalanobis distance):

$$R = \frac{T - lb}{ub - lb}, \tag{3}$$

which ranges from 0 to 1 inside the interval. Some values for R are negative. We regard this as positive as our goal is to maximize the precision and for this small differences and, thus, small T and R are positive. The values for R in Table 1 comparing five different runs of the whole pose estimation procedure using the "proposed" reduction strategy around or a little below 0.5 show that our results can be repeated, though they are estimated slightly too precise. Computation times in Table 1 are given for a standard quad core computer with 8 GB.

Table 2 gives an overview of the full and the reduced merging, the latter according to the "proposed" strategy (cf. Sect. 3). For the reduced merging, we have empirically found that a reduction of the number of points to on average $avgnr = 1000$ per image presents a good compromise between achieved precision and computational complexity. In the experiments below this means a reduction in the number of observations and unknowns of 2.2–3.5. This reduction is mirrored very well by the reduced aggregated computation times for merging and final bundle adjustment. The average standard deviation σ is in the range of a quarter to a little more than two thirds of a pixel.

Table 1. Overview of the four scenes B1, B2, E and T given in Fig. 1. For the end to end "overall" test, run time, test statistic T (2) and relation R (3) to the 3σ bounds of the χ_n^2 distribution are given as averages of five runs, the latter demonstrating differences in or below the bounds ($0 \leq R \leq 1$) and, thus, a high repeatability.

	B1	B2	E	T
Images, triplets, cameras	$20, 19, 2$	$24, 23, 2$	$20, 27, 1$	$39, 57, 1$
DOF (n_{χ^2}), 3σ bounds	$127, [87, 176]$	$151, [107, 204]$	$120, [81, 168]$	$234, [178, 299]$
Overall				
Runtime l	$95.9 \pm 0.7\,\text{s}$	$127.9 \pm 1.2\,\text{s}$	$123.7 \pm 2.6\,\text{s}$	$203.2 \pm 5.5\,\text{s}$
Test statistic T	110.9 ± 38.6	149.1 ± 21	108.8 ± 34.9	193.5 ± 59.7
Relation R bounds/T	0.47 ± 0.26	0.43 ± 0.22	0.32 ± 0.4	0.23 ± 0.37

We also did a comparison with the sequential merging of triplets. E.g., for the scene T we got a speed up of 2.8 for the reduced and 3.4 for the full merging. I.e., comparing the "proposed" reduced with full sequential merging gives a speed up of more than ten. As we merge in parallel on the quad core, we also compared it with serial computation. For this, we got a speedup of 1.4 for the full and 1.1 for the reduced set already for these small image sets.

Tables 3 and 4 present the result of the comparison of the different strategies defined in Sect. 3 based on the evaluation measures defined in Sect. 4. To speed up the tests, we used the same solution for the triplets and only repeated the

Table 2. Full and reduced **merging**: computation times for merging and final bundle adjustment (including estimation of camera and distortion parameters) on a standard quad core computer with 8 GB, estimated average standard deviation σ_0 and the number of observations and unknowns of the final bundle adjustment (to indicate the size of the problem) as average of five runs.

	B1	B2	E	T
Full (all points available in the image triplets)				
Runtime merging	$44.1 \pm 0.5\,\text{s}$	$52.5 \pm 0.5\,\text{s}$	$30.4 \pm 0.2\,\text{s}$	$82.5 \pm 0.1\,\text{s}$
Runtime bundle	$11.6 \pm 0.2\,\text{s}$	$12.7 \pm 0.1\,\text{s}$	$5.6 \pm 0.1\,\text{s}$	$14.5 \pm 0.1\,\text{s}$
σ_0 [pixels]	0.417 ± 0.001	0.267 ± 0.001	0.392 ± 0.004	0.664 ± 0.001
Observations	126538 ± 35	131461 ± 38	84001 ± 88	181450 ± 83
Unknowns	39839 ± 5	39429 ± 15	27907 ± 31	49802 ± 33
Reduced with "proposed" strategy (cf. Sect. 3)				
Runtime merging	$14.3 \pm 0.3\,\text{s}$	$17.3 \pm 0.1\,\text{s}$	$14.1 \pm 0.7\,\text{s}$	$35.1 \pm 1.8\,\text{s}$
Runtime bundle	$3.1 \pm 0.1\,\text{s}$	$4. \pm 0.1\,\text{s}$	$2.4 \pm 0.1\,\text{s}$	$5.1 \pm 0.3\,\text{s}$
σ_0 [pixels]	0.411 ± 0.013	0.265 ± 0.001	0.376 ± 0.007	0.695 ± 0.041
Observations	36525 ± 287	43382 ± 323	37224 ± 81	67546 ± 417
Unknowns	11645 ± 41	13158 ± 29	12426 ± 18	18734 ± 37

406   H. Mayer

Table 3. Average value, standard deviation and maximum value for Test statistic T (2) (top row) and relation R (3) to bounds (bottom rows) of the full versus the reduced merging according to different strategies (left). $0 \leq R \leq 1$ means that the differences of the parameters weighted by the covariances are inside the "3σ" bounds of the χ_n^2 distribution (cf. last row repeated from Table 1). Best values are marked in bold.

T/R	B1	B2	E	T
Proposed	**$106 \pm 28, 145$**	**137 ± 30**$, 207$	$113 \pm 21, 161$	**$125 \pm 26, 155$**
	$0.20 \pm 0.32, 0.65$	**0.30 ± 0.31**$, 1.03$	$0.36 \pm 0.24, 0.92$	**$-0.44 \pm 0.22, -0.19$**
Triplet	$300 \pm 29, 358$	$531 \pm 102, 651$	$111 \pm 16, 132$	$378 \pm 56, 467$
deletion	$2.42 \pm 0.31, 3.05$	$4.38 \pm 1.05, 5.61$	$0.35 \pm 0.18, 0.59$	$1.65 \pm 0.47, 2.39$
Min	$2622 \pm 118, 2762$	$579 \pm 12, 592$	$117 \pm 24, 165$	$656 \pm 29, 695$
	$28.7 \pm 1.2, 30.2$	$4.88 \pm 0.13, 5.01$	$0.41 \pm 0.28, 0.97$	$3.97 \pm 0.24, 4.29$
MinMax	$680 \pm 269, 1161$	$674 \pm 2, 676$	$162 \pm 41, 244$	$616 \pm 88, 723$
	$6.68 \pm 3., 12.1$	$4.12 \pm 0.01, 4.13$	$0.94 \pm 0.47, 1.88$	$3.63 \pm 0.74, 4.52$
MinMax	$558 \pm 348, 1254$	$431 \pm 36, 501$	$91.5 \pm 27, 146$	$278 \pm 36, 333$
+ grid	$5.3 \pm 3.95, 13.2$	$3.35 \pm 0.38, 4.07$	$0.12 \pm 0.31, 0.75$	$0.82 \pm 0.3, 1.28$
Proposed	$205 \pm 66, 339$	$142 \pm 24, $**$180$**	**$77.2 \pm 22, 96.2$**	$323 \pm 190, 608$
+ grid	$1.33 \pm 0.75, 2.83$	$0.39 \pm 0.27, $**$0.75$**	$-0.05 \pm 0.25, $**$0.17$**	$1.45 \pm 1.41, 3.57$
Bounds χ_n^2	$[87, 176]$	$[107, 204]$	$[81, 168]$	$[178, 299]$

merging. To verify the soundness, we made a couple of runs also with other results for triplets, finding in essence the same distribution.

Concerning parameters, we found by many experiments a grid with 6×9 cells and a threshold of at minimum 10 points per grid cell to be a reasonable choice. For the strategies involving the minimum number of points per image, we employed $minnr = 900$ to obtain on average a similar amount of observations and unknowns as for the "proposed" strategy. Concerning the maximum number of points in the "MinMax" strategy, we do not use a fixed threshold, but preserve approximately the $minnr$ points which link as many images as possible. That this usually does not mean that there will be also $minnr$ of points in each image is another reason for using $avgnr = 1000$ and the slightly smaller $minnr = 900$.

The T and corresponding R values in Table. 3 show that only the "proposed" strategy stays (mostly) inside the 3σ bounds for the χ_n^2 distribution. Besides average and standard deviation we also present the maximum value to give an idea about the worst case behavior. The "proposed + grid" strategy is very good for E and partly B2, but produces rather bad results for B1 and T. That similarly "MinMax + grid" works reasonably well for E gives a hint that for E the grid is helpful. It is not, though, for the other data sets. We finally note a "singularity" for "MinMax" for B2: There, the principle for the maximization led to a deterministic selection of most points, leaving nearly no randomness, leading in turn to similar results and low standard deviations. As the results are rather bad also for the other data sets, we did not investigate this further.

Table 4. $\sqrt{\lambda_{\max}}$ (upper row) and $\sqrt{\mu_\lambda}$ (lower row) indicating the maximum and the average loss of precision for the estimated parameters of the full merging versus the reduced merging according to the different merging strategies (left). Given are average, standard deviation and maximum value. Best values are marked in bold.

$\sqrt{\lambda_{\max}}/\sqrt{\mu_\lambda}$	B1	B2	E	T
Proposed	**2.66** ± 0.18, 3.05	**2.36** ± 0.24, 2.91	2.10 ± 0.21, 2.38	**2.07** ± 0.07, **2.25**
	1.94 ± 0.18, 2.23	1.76 ± 0.07, 1.89	1.45 ± 0.06, 1.52	1.67 ± 0.05, 1.81
Triplet	4.55 ± 0.08, 4.68	8.63 ± 0.37, 9.28	3.19 ± 0.24, 3.65	5.2 ± 0.17, 5.37
deletion	2.03 ± 0.02, 2.04	2.17 ± 0.08, 2.3	1.68 ± 0.06, 1.77	2.21 ± 0.03, 2.27
Min	2.9 ± 0.18, 3.17	11.9 ± 0.16, 12.	2.92 ± 0.14, 3.07	2.79 ± 0.02, 2.82
	1.54 ± 0.08, **1.66**	2.36 ± 0.01, 2.37	1.53 ± 0.05, 1.61	**1.43** ± 0.01, **1.45**
MinMax	3.28 ± 0.19, 3.67	2.81 ± 0.01, **2.81**	3.01 ± 0.05, 3.11	2.78 ± 0.03, 2.83
	1.66 ± 0.1, 1.85	**1.43** ± 0.01, **1.43**	1.42 ± 0.04, 1.49	1.48 ± 0.01, 1.50
MinMax	2.74 ± 0.06, **2.83**	3.67 ± 0.1, 3.78	2.19 ± 0.04, 2.26	2.32 ± 0.02, 2.35
+ grid	1.64 ± 0.02, **1.66**	1.92 ± 0.02, 1.95	1.37 ± 0.02, **1.39**	1.45 ± 0.01, 1.46
Proposed	2.69 ± 0.23, 3.08	3.22 ± 1., 5.21	**2.** ± 0.12, **2.14**	2.3 ± 0.15, 2.45
+ grid	2.04 ± 0.13, 2.29	2. ± 0.13, 2.24	1.52 ± 0.06, 1.61	1.84 ± 0.06, 1.92

Table 3 shows that only the "proposed" strategy produces reliable results, Table 4 gives a comparison of the strategies according to the maximum $\sqrt{\lambda_{\max}}$ (upper row) and the mean values $\sqrt{\mu_\lambda}$ (lower row) showing how the precision of the determination of the parameters degrades when reducing the number of 3D points/point triplets at maximum and on average. According to it, the "proposed" strategy is not the best for most scenes concerning the average $\sqrt{\mu_\lambda}$, but it is not far from it and it particularly fits well to the square root of the ratio of the redundancy of the full and the reduced solution (cf. Table 2). Additionally, the "proposed" strategy has in three of four cases the best average of $\sqrt{\lambda_{\max}}$, meaning that it can guarantee a high quality for all parameters.

By means of the principles integrated in the more deterministic strategies (Min and its variants which are based on fixed thresholds) one can reduce the loss of accuracy below the level given by the ratio of the redundancies. Yet, as the values for the relative and absolute Mahalanobis distance R and T in Table 3 demonstrate, this leads to strongly biased results.

6 Conclusion and Discussion

We have presented different strategies for a reduced hierarchical merging leading to faster pose estimation and have evaluated them based on statistical measures for a number of scenes. In summary, the results show that **only a random deletion of points leads to reliable as well as precise results**.

We find most interesting that the intuitively plausible deterministic strategies implementing the second and third principles of Sect. 3 do not work as well as expected. Concerning the second principle preserving points linking many images it is clear that this might reduce the solution to the possibly small parts which are visible in many images making them not representative for the whole image. Thus, while one obtains better average standard deviations σ_0 due to the higher redundancies per point and also good standard deviations for the estimated parameters, the estimated parameters can vary much more than the standard deviations for the parameters predict. I.e., the solutions are estimated as very precise, but still vary considerably and are not reliable.

We had hoped to alleviate the problem by supporting it with the third principle enforcing a good distribution by means of the grid. Yet, as the evaluation shows, this does not help in general which leads us to the insight, that only a random deletion produces reliable results on average.

That the more fine grained deletion of point triplets works less well than the deletion of individual 3D points seems to be an indication that at least in a certain way the preservation of points linking many images (second principle in Sect. 3) is still meaningful. While by deleting point triplets the 3D points linking especially many images are reduced more or less, they are either kept in full or deleted by the "proposed" strategy. Only when all point triplets for a 3D point are deleted at once, or not at all, the result becomes really reliable. It is important to note that the information linking the point triplets to the 3D points of the combined image set is only known after the simulation of the merging (cf. Sect. 3).

Finally, we generally have found the evaluation based on the measures λ (1) and T (2) extremely useful to find conceptual and implementation errors and to decide about different appealing alternatives. Particularly, as recommended, e.g., in the Manual of Photogrammetry [13] as standard solution to begin the iteration of the robust bundle adjustment, we had used the L1–L2 metric in the M-estimator to reweight the covariance matrix by $w = 1/\sqrt{1 + \overline{v}^2/2}$, with $\overline{v} = v/\sigma_v$ the residual divided by its standard deviation. By means of the evaluation we found that this leads to biased results, i.e., T is wide above the 3σ bounds of the χ_n^2 distribution. Opposed to this, the L1 metric with weight $w = 1/|\overline{v}|$ leads to test statistics mostly inside the bounds as presented above.

Acknowledgments. We want to thank Wolfgang Förstner for his invaluable recommendations and clarifications and the reviewers for their helpful comments.

References

1. Agarwal, S., Snavely, N., Simon, I., Seitz, S., Szeliski, R.: Building rome in a Day. In: Twelfth International Conference on Computer Vision, pp. 72–79 (2009)
2. Bartelsen, J., Mayer, H., Hirschmüller, H., Kuhn, A., Michelini, M.: Orientation and dense reconstruction from unordered wide baseline image sets. Photogram. Fernerkun. Geoinformation **4**(12), 4–432 (2012)
3. Fischler, M., Bolles, R.: Random sample consensus: a paradigm for model fitting with applications to image analysis and automated cartography. Commun. ACM **24**(6), 381–395 (1981)

4. Fitzgibbon, A.W., Zisserman, A.: Automatic camera recovery for closed or open image sequences. In: Burkhardt, H.-J., Neumann, B. (eds.) ECCV 1998. LNCS, vol. 1406, pp. 311–326. Springer, Heidelberg (1998)
5. Förstner, W., Dickscheid, T., Schindler, F.: Detecting interpretable and accurate scale-invariant keypoints. In: Twelfth International Conference on Computer Vision, pp. 2256–2263 (2009)
6. Frahm, J.-M., Fite-Georgel, P., Gallup, D., Johnson, T., Raguram, R., Wu, C., Jen, Y.-H., Dunn, E., Clipp, B., Lazebnik, S., Pollefeys, M.: Building rome on a cloudless day. In: Daniilidis, K., Maragos, P., Paragios, N. (eds.) ECCV 2010, Part IV. LNCS, vol. 6314, pp. 368–381. Springer, Heidelberg (2010)
7. Jeong, Y., Nistér, D., Steedly, D., Szeliski, R., Kweon, I.S.: Pushing the envelope of modern methods for bundle adjustment. IEEE Trans. Pattern Anal. Mach. Intell. **34**(8), 1605–1617 (2012)
8. Jian, Y.D., Balcan, D., Dellaert, F.: Generalized subgraph preconditioners for large-scale bundle adjustment. In: Thirteenth International Conference on Computer Vision, pp. 295–302 (2011)
9. Klingner, B., Martin, D., Roseborough, J.: Street view motion-from-structure-from-motion. In: Fourteenth International Conference on Computer Vision, pp. 953–960 (2013)
10. Klopschitz, M., Irschara, A., Reitmayr, G., Schmalstieg, D.: Robust incremental structure from motion. In: Fifth International Symposium on 3D Data Processing, Visualization and Transmission (3DPVT), pp. 1–7 (2010)
11. Lowe, D.: Distinctive image features from scale-invariant keypoints. Int. J. Comput. Vis. **60**(2), 91–110 (2004)
12. Mayer, H., Bartelsen, J., Hirschmüller, H., Kuhn, A.: Dense 3D reconstruction from wide baseline image sets. In: Dellaert, F., Frahm, J.-M., Pollefeys, M., Leal-Taixé, L., Rosenhahn, B. (eds.) Real-World Scene Analysis 2011. LNCS, vol. 7474, pp. 285–304. Springer, Heidelberg (2012)
13. McGlone, J. (ed.): Manual of Photogrammetry, 6th edn. American Society of Photogrammetry and Remote Sensing, Bethesda (2013)
14. Strecha, C., von Hansen, W., Van Gool, L., Fua, P., Thoennessen, U.: On benchmarking camera calibration and multi-view stereo for high resolution imagery. In: Computer Vision and Pattern Recognition, pp. 1–8 (2008)
15. Torr, P., Zisserman, A.: Robust parametrization and computation of the trifocal tensor. Image Vis. Comput. **15**, 591–605 (1997)

Lens-Based Depth Estimation
for Multi-focus Plenoptic Cameras

Oliver Fleischmann[✉] and Reinhard Koch

Institute of Computer Science, University of Kiel, 24118 Kiel, Germany
ofl@informatik.uni-kiel.de, rk@mip.informatik.uni-kiel.de

Abstract. Multi-focus portable plenoptic camera devices provide a reasonable tradeoff between spatial and angular resolution while enlarging the depth of field of a standard camera. Many applications using the data captured by these camera devices require or benefit from correspondences established between the single microlens images. In this work we propose a lens-based depth estimation scheme based on a novel adaptive lens selection strategy. Coarse depth estimates serve as indicators for suitable target lenses. The selection criterion accounts for lens overlap and the amount of defocus blur between the reference and possible target lenses. The depth maps are regularized using a semi-global strategy. For insufficiently textured scenes, we further incorporate a semi-global coarse regularization with respect to the lens-grid. In contrast to algorithms operating on the complete lightfield, our algorithm has a low memory footprint. The resulting per-lens dense depth maps are well suited for volumetric surface reconstruction techniques. We show that our selection strategy achieves similar error rates as selection strategies with a fixed number of lenses, while being computationally less time consuming. Results are presented for synthetic as well as real-world datasets.

1 Introduction

The idea of lightfield imaging, the sampling process of the plenoptic function, dates back to the early 20th century [10]. While it was impossible to build the proposed devices at that time, the technological advances of the last years led to portable lightfield camera devices [1,13]. These devices are basically standard cameras with an additional microlens array (MLA) mounted in front of the camera sensor. The MLA usually consists of thousands of small lenses, sampling the scene from thousands of viewpoints. Within this work we focus on plenoptic images captured by such MLA based camera devices. Most applications using plenoptic image data such as the rendering of novel viewpoints [9], total refocusing [13] or superresolution [3] require or at least benefit from the knowledge of correspondences between the single microlens images. The correspondences (the scene depth) have to be estimated from the plenoptic camera images. Plenoptic depth estimation techniques can roughly be divided into four paradigms:

1. Depth estimation algorithms which operate on very dense four-dimensional lightfields and its epipolar plane images [16,18].

© Springer International Publishing Switzerland 2014
X. Jiang et al. (Eds.): GCPR 2014, LNCS 8753, pp. 410–420, 2014.
DOI: 10.1007/978-3-319-11752-2_33

2. Scene space multi-view stereo algorithms, which directly estimate the scene surface with respect to a regularized photoconsistency cost on a 3D volume computed from a set of views (see [14] for an overview).
3. Image-space multi-view stereo algorithms which estimate a single depth map with respect to a virtual standard camera [2,13].
4. Image-space multi-view stereo depth estimation algorithms which estimate a depth map *per view* (*per microlens image*) which may later on be fused to a 3D surface reconstruction (see e.g. [6,7,15,17]).

In this work we propose a depth estimation pipeline which is governed by the fourth paradim. We argue that the reasons for this choice in contrast to the remaining paradims are:

1. Our algorithm operates on a single image captured by plenoptic camera device while e.g. [18] requires a very dense sampling of the lightfield.
2. Our lens-based depth estimation is well suited for highly parallel processing and integrates in a parallel fashion into volumetric surface reconstruction algorithms such as [4].
3. In contrast to the algorithms operating on the dense lightfield representation such as [18], the memory consumption is rather low.

After a brief review of the basic plenoptic imaging system studied in this article in Sect. 2, we propose a per-lens depth estimation scheme for a single reference lens with respect to a set of target lenses in Sect. 3. The estimated depth maps are regularized using a semi-global strategy [8]. For insufficiently textured scenes, we further use a coarse grained regularization with one depth per lens, which is incorporated in the fine grained depth estimation. In Sect. 4 we introduce a heuristic lens selection strategy. In order to minimize the depth estimation error and the computation time, we describe an adaptive lens selection strategy which estimates a set of coarse depths and selects the final target lenses based on the coarse depth estimates. We evaluate our method with respect to synthetic as well as real-world scenes.

2 Multi-focus Plenoptic Camera

The camera setup we consider within this article is illustrated in Fig. 1(a) and follows the focused plenoptic camera setup described in [11,13]. A standard camera, consisting of a main lens and an image sensor is augmented with a microlens array which is mounted in front of the camera sensor at distance v. The main lens maps surface points to *virtual* surface points behind the image sensor. We refer to the orthogonal distance of the virtual surface points to the MLA plane as the *virtual depth*. The microlenses are modeled as *thin-lenses* focusing on the virtual image, resulting in a finite depth of field. Further, we assume that the microlens array consists of three different lens types with focal lengths f_0, f_1, f_2 which are focused on three different planes with distances u_0, u_1, u_2 behind the image sensor with $\frac{1}{f_i} = \frac{1}{u_i} + \frac{1}{v}$. This model is equivalent to the setup of the Raytrix camera system described in [13].

2.1   Lens Grid

We assume that the microlenses of the MLA are arranged in a regular hexagonal grid aligned with the image sensor. To each microlens we assign a pair of axial coordinates $a = (a_1, a_2)^T$ with respect to the basis

$$B = 2r \begin{bmatrix} \sqrt{3}/2 & 0 \\ 1/2 & 1 \end{bmatrix}$$

where r denotes the radius of the microlenses in pixel units. The center of the lens with axial coordinates $(0,0)^T$ is assumed to coincide with the image center $(h/2, w/2)^T$. Under this assumption, the center of a lens c_a in rectangular pixel coordinates is given by $c_a = Ba + (h/2, w/2)^T$. The full plenoptic image captured by the image sensor is denoted by I_p with domain $dom(I_p) = [0, h] \times [0, w]$. Every microlens image is treated as a single image I_a with local rectangular domain $dom(I_a) = [-r, r]^2$ and values defined in terms of the plenoptic image I_p by $I_a(x) = I_p(c_a + x)$. The focal length, focus distance and f-stop corresponding to the lens a are denoted by f_a, u_a, N_a where the axial coordinates a map to the corresponding lens type by $t(a) = ((-a_1(\mathrm{mod}\ 3)) + a_2)\ (\mathrm{mod}\ 3) \in \{0, 1, 2\}$. Further we assume that neither the main lens nor the microlenses suffer from any lens distortion. The hexagonal grid is illustrated in Fig. 1(b).

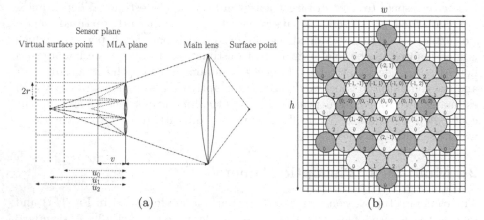

(a)   (b)

Fig. 1. (a) The multi-focus plenoptic camera setup. (b) Hexagonal lens grid over a rectangular pixel grid. The upper indices show the axial coordinates, the lower indices indicate the lens type while the color coding indicates lenses of the same distance with respect to the center.

3   Disparity Estimation

Given a reference lens a and n target lenses $a_1, \ldots, a_n$, we want to estimate the virtual depth at every pixel x within the lens region. For each lens pair

$(a, a_i), i \in \{1, \ldots, n\}$ we follow the classical stereo paradigm and measure the photometric similarity of a local neighborhood $\Omega(x)$ in the reference image I_a and a local neighborhood $\Omega(x - dv)$ in the target image I_{a_i} along the epipolar line for a range of possible disparities $\mathcal{D} = d_0, \ldots, d_m, d_j \in [0, d_{max}], d_{max} < 2r$. For two microlens images I_a, I_{a_i} with centers c_a, c_{a_i}, the epipolar line for a point x in the reference image is $L_i = \{x + tv : t \in \mathbb{R}\}$ with $v = (c_{a_i} - c_a)/2r$. The sum of absolute differences

$$SAD(x, d_j; a, a_i) = \frac{1}{A(x, v, d_j)} \sum_{u \in \Omega(x)} |I_a(u) - I_{a_i}(u - d_j v)| \mathbb{1}(u - d_j v)$$

with

$$A(x, v, d_j) = \sum_{u \in \Omega(x)} \mathbb{1}(u - d_j v) \qquad \mathbb{1}(x) = \begin{cases} 1 & \text{if } ||x|| < r - \delta r \\ 0 & \text{else} \end{cases}$$

serves as a similarity measure with respect to the hypothetical disparity $d_j \in \mathcal{D}$. If the target point $x - d_j v$ is outside the valid image region, a maximum cost C_{max} is assigned resulting in the final cost volume

$$C_i(x, d_j; a) = \begin{cases} SAD(x, d_j; a, a_i) & \text{if } ||x - d_j v|| \leq r - \delta r \\ C_{max} & \text{else.} \end{cases} \qquad (1)$$

3.1 Cost Volume Averaging

The n cost volumes $C_1, \ldots, C_n$ are averaged into a single fine grained cost volume:

$$C^f(x, d_j; a) = \begin{cases} C_{max} & \text{if } \forall i : C_i(x, d_j) = C_{max} \\ \frac{1}{A^f(x, d_j)} \sum_i^n C_i(x, d_j) \mathbb{1}^f(C_i(x, d_j)) & \text{else} \end{cases}$$

where a has been dropped for convenience of notation and $A^f(x, d_j)$ given by

$$A^f(x, d_j) = \sum_{i=1}^n \mathbb{1}^f(C_i(x, d_j; a)) \qquad \mathbb{1}^f(x) = \begin{cases} 1 & \text{if } x < C_{max} \\ 0 & \text{else} \end{cases}$$

serves as a normalization constant with respect to the number of views, which are actually able to see the scene point corresponding to x at the hypothetical disparity d_j.

Fine Regularization. The fine grained per lens cost volumes are regularized using the semi-global strategy introduced in [8]. We favored this variant due to its speed and simplicity. The regularized fine grained cost volume for a single lens a is obtained as

$$C_{reg}^f(x, d_j; a) = \sum_{w \in \mathcal{W}^f} C_w^f(x, d_j; a) \qquad (2)$$

with

$$C_w^f(\boldsymbol{x}, d_j; \boldsymbol{a}) = C^f(\boldsymbol{x}, d_j; \boldsymbol{a}) + \min\{C^f(\boldsymbol{x} - \boldsymbol{w}, d_j; \boldsymbol{a}), C^f(\boldsymbol{x} - \boldsymbol{w}, d_{j+1}; \boldsymbol{a}) + p_1^f,$$
$$C^f(\boldsymbol{x} - \boldsymbol{w}, d_{j-1}; \boldsymbol{a}) + p_1^f,$$
$$\min_d C^f(\boldsymbol{x} - \boldsymbol{w}, d; \boldsymbol{a}) + p_2^f\} \tag{3}$$

where the directions $\mathcal{W}^f$ are the directions induced by the standard rectangular 8-neighborhood, p_1^f is a penalizing constant for deviations of one disparity step and p_2^f a penalizing constant for deviations larger than one disparity step. The fine grained per lens disparity maps are extracted from the cost volume as minima along each disparity slice as

$$\hat{d}^f(\boldsymbol{x}; \boldsymbol{a}) = \arg \min_d C_{reg}^f(\boldsymbol{x}, d; \boldsymbol{a}) \tag{4}$$

The final minima are further refined with an interpolation based on a second order Taylor expansion of the cost slice. In addition to the disparity maps, we calculate a confidence map for the lens according the measure introduced in [12].

3.2   Coarse Regularization

In insufficiently textured scenes, a per lens regularization might not be sufficient to obtain dense depth maps over the complete microlens grid. We therefore propose an optional coarse regularization and calculate a coarse cost volume with a single cost slice per lens and regularize with the semi-global strategy with respect to the lens grid. An averaging of the fine cost volume $C^f(\boldsymbol{x}, d; \boldsymbol{a})$ results in a single coarse cost slice for each lens as

$$C^c(\boldsymbol{a}, d_j) = \frac{1}{A^c(\boldsymbol{a}, d_j)} \sum_{\boldsymbol{x} \in dom(I_a)} C^f(\boldsymbol{x}, d_j; \boldsymbol{a}) \mathbb{1}^c(C^f(\boldsymbol{x}, d_j; \boldsymbol{a})) \tag{5}$$

with

$$A^c(\boldsymbol{x}, d_j) = \sum_{\boldsymbol{x} \in dom(I_a)} \mathbb{1}^c(C^f(\boldsymbol{x}, d_j; \boldsymbol{a})) \qquad \mathbb{1}^c(x) = \begin{cases} 1 & \text{if } x < C_{max} \\ 0 & \text{else.} \end{cases}$$

From the coarse cost volume, a regularized coarse cost volume C_{reg}^c is obtained according to Eqs. (2) and (3) using the directions $\mathcal{W}^c = \{(0,1), (1,0), (1,-1), (0,-1), (-1,0), (-1,1)\}$ with respect to the hexagonal axial coordinates and coarse penalizing constants p_1^c, p_2^c. The coarse disparity estimates are finally extracted from the coarse cost volume as

$$\hat{d}^c(\boldsymbol{a}) = \arg \min_d C_{reg}^c(\boldsymbol{a}, d). \tag{6}$$

Deviations from the coarse disparity estimate are penalized for each cost slice resulting in the fine cost volume

$$C^{c,f}(\boldsymbol{x}, d_j; \boldsymbol{a}) = C^f(\boldsymbol{x}, d_j; \boldsymbol{a}) + \lambda \, |\hat{d}^c(\boldsymbol{a}) - d_j| \, e^{(-\sigma(I_a)^2 / \sigma_{struct}^2)}$$

where $\sigma(I_a)$ denotes the standard deviation of I_a and $\lambda \in \mathbb{R}_+$ is a weighting factor which affects the overall influence of the coarse estimate. The standard deviation of the image is supposed to describe the amount of structure within the lens image where the constant σ_{struct} controls how fast the influence of lens structure should decay. If there is a lot of structure, the weighting will tend to zero since the image content itself already allows good disparity estimates. If there is little structure, the costs are pulled towards the coarse minimum. The cost volume $C^{c,f}$ is afterwards regularized using (2) and (3) resulting in the regularized volume $C^{c,f}_{reg}$. The final disparity estimates are obtained from $C^{c,f}_{reg}$ according to (4).

4   Lens Selection

While we have described the depth estimation for a reference lens and a set of n candidate lenses, it is still an open question how these lenses should be selected. The depth estimation accuracy is influenced by three major factors: (1) The amount of defocus blur influences the localization accuracy during the matching process due to its low-pass behaviour. Consequently, it is desirable to include lenses in the matching process whose amount of defocus blur is similar to the blur amount of the reference lens. (2) The virtual depth error grows quadratically with respect to the virtual depth for a given disparity error [5]. If the baseline of the reference and the target lenses is increased, this error decreases. In the case of scene points close to the main lens it is therefore desirable to choose the baseline of the target lenses as large as possible to reduce the virtual depth error. (3) The number of target lenses which do not actually see the target scene point should be as low as possible. Strategies using a fixed number of target lenses might suffer from this error source for points with high disparity where only the nearby lenses can be used for the matching process. A suitable target lens selection strategy should account for all three potential sources of inaccuracies. In addition the computation time should be kept as low as possible. Since the cost calculation is the most time consuming factor in the depth estimation process, the number of target lenses should be as low as possible.

Adaptive Strategy. Given a reference lens with axial coordinates a, we first choose a maximal baseline s_{max} which determines the size of the neighborhood and therefore the maximal number of candidate lenses $N(a) = \{b \in \mathcal{A} : \|c_b - c_a\| \leq s_{max}\}$. The maximal distance is chosen a priori and depends on the minimal distance of the scene objects to the main lens. The neighborhood is subdivided into $L \in \mathbb{N}$ lens rings $\mathcal{R}_i = \{b \in N(a) : \|c_b - c_a\| = s_i\}, i \in \{0, \dots, L-1\}$ with $s_{max} = s_{L-1}$ whose centers have the same distance and consequently the same baseline s_i with respect to the reference lens. Further we sort them by distance such that $s_0 < s_1 < \cdots < s_{L-1}$. Note that the rings do not necessarily contain lenses of the same type as a. The rings are further subdivided into six sectors spanning a range of $60°$ (cmp. Fig. 2) given by $\mathcal{R}_{i,j} = \{b \in \mathcal{R}_i : \angle(c_a, c_b) < j\,2\pi/6\}$. With respect to lenses from ring

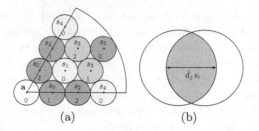

(a)   (b)

Fig. 2. (a) The sectors $\mathcal{R}_{0,0}, \dots, \mathcal{R}_{5,0}$. The lower indices show the lens type while the baselines are indicated by s_0 to s_5. (b) The coverage criterion: if the predicted overlap according to the coarse estimate $\hat{d}_j$ is violated, the lens is discarded.

$\mathcal{R}_1$, which is the innermost ring of the same lens type as $\boldsymbol{a}$, we compute the cost volumes $C_0, \dots, C_5$ and average them into single cost slices according to (1) and (5). Using these six cost slices six coarse disparities $\hat{d}_0, \dots, \hat{d}_5$ are estimated using a simple winner takes all strategy. These six disparities serve as a coarse disparity estimate along the corresponding direction. To maximize the baseline, based on these coarse estimates, appropriate lenses are selected according to the following scheme: For each coarse disparity estimate $\hat{d}_j$, we traverse the sectors $\{\mathcal{R}_{L-1,j}, \dots, \mathcal{R}_{0,j}\} \setminus \{\mathcal{R}_{1,j}\}$ starting with the outermost sector $\mathcal{R}_{L-1,j}$. A sector is a potential candidate sector, if the disparity with respect to lenses from this ring is smaller than $t_{cov} 2r$, i.e. $\hat{d}_j s_i \leq t_{cov} 2r$. The parameter t_{cov} is a coverage factor, which describes the minimal amount of overlap between two lenses, also illustrated in Fig. 2(b). If the overlap is too small, the amount of matching data provided by the target lens does not allow stable depth estimates. If $\hat{d}_j s_i > t_{cov} 2r$ the sector is discarded and the next one is tested. Since we only want to compare lenses with approximately the same amount of defocus blur, we further estimate the circle of confusion for the reference and the potential target lenses in the candidate sector. The circle of confusion of a lens $\boldsymbol{a}$ in pixels with respect to a disparity $\hat{d}_j$ reads

$$c(\boldsymbol{a}) = \frac{|z - u_a|}{z} \frac{f_a^2}{N_a(u_a - f_a) s_p}$$

where the coarse virtual depth is given by $z = (f_a 2r)/(\hat{d}_j)$ and s_p denotes the pixel size of the image sensor. A lens $\boldsymbol{b} \in \mathcal{R}_{i,j}$ is accepted as a target lens, if the absolute difference $|c(\boldsymbol{a}) - c(\boldsymbol{b})|$ is smaller than a threshold t_{coc}. If any lens has been selected from a sector $\mathcal{R}_{i,j}$, we continue with the next sector. Otherwise, if no suitable lenses are found, the lenses from the innermost sector $\mathcal{R}_{0,j}$ are selected, regardless of their circle of confusion. In this case only lenses from the inner lens ring provide enough overlap to cover a sufficient amount of lens pixels. For all accepted lenses the cost volumes are calculated according to (1) and averaged together with the already calculated cost volumes $C_0, \dots, C_5$ using (2). The final disparity results from (4).

5 Results

In the following we assume that the plenoptic input images are grayscale images normalized to $[0, 1]$. For all experiments, we used the following algorithmic parameters: $s_{max} = 8r$, $p_1^f = p_1^c = 0.01$, $p_2^f = p_2^c = 0.03$, $t_{coc} = 1.2$, $t_{cov} = 0.7$, $\sigma_{struct} = 0.01$, $\lambda = 0.01$. The local neighborhood Ω has been chosen as a 3×3 pixel neighborhood and $\mathcal{D} = \{\Delta d, \ldots, k\Delta d\}$, $\Delta d = \frac{2r}{s_{max}}$, $k = \lfloor r/\Delta d \rfloor$.

Synthetic Data. For evaluation purposes, we generated synthetic datasets using the *Cycles* raytracing engine integrated in the 3D modeling tool *Blender*. The setup for the synthetic scenes differs from the real world plenoptic camera setup. Instead of a main lens and a microlens array, we modeled the microlens array as standard cameras following the thin-lens model. The synthetic scenes consist of color camera images as well as ground-truth depth images. In all synthetic scenes, the following camera parameters have been used: Focus distances $(u_0, u_1, u_2) = (0.9\,\text{m}, 1.3\,\text{m}, 2.2\,\text{m})$, lens to sensor distance $v = 0.3\,\text{m}$, pixel size $s_p = 3\,\text{mm}$, f-stop $N = 4$, lens diameter $2r = 25\,\text{px}$, lens border $\delta r = 1\,\text{px}$. All test scenes contain 1203 microlenses. The test scene *Four planes* consists of four fronto-parallel planes at distances $0.9\,\text{m}$, $1.3\,\text{m}$, $2.2\,\text{m}$, $3\,\text{m}$ while the test scenes *Near plane* and *Far plane* consist of single fronto-parallel planes with distance $0.75\,\text{m}$ and $3.5\,\text{m}$ respectively. All planes are textured with a randomly generated procedural texture. In the case of the synthetic data, we did not use the coarse regularization. We measured the mean and the standard deviation of the absolute average error between the estimated and the ground truth disparities in pixels for each lens type. The adaptive strategy is compared to two strategies using a fixed number of target lenses: Strategy $\text{Fixed}_{0,1,4}$ uses the complete rings $\mathcal{R}_0, \mathcal{R}_1, \mathcal{R}_4$ while $\text{Fixed}_{0,\ldots,5}$ uses the rings $\mathcal{R}_0, \ldots, \mathcal{R}_5$. Figure 3 shows an excerpt of the *Four planes* scene. Table 1 shows the corresponding errors in conjunction with the number of used target lenses. The main conclusion is: While the overall errors are comparable to the strategies using a fixed number of target lenses, the number of selected lenses of the adaptive strategy is significantly lower. The number of target lenses has a major influence on the overall runtime, which can be significantly reduced using our adaptive lens selection strategy. The overall errors in the case of the *Near plane* scene are higher due to the lower number of lenses contributing to the correct disparity estimates. Lens type 2 shows the highest standard deviation due to its strong defocus blur. In the *Far plane* Scene, more lenses contribute to the correct estimates yielding more accurate results, while again the lens of the strongest defocus blur, lens type 0, shows the largest standard deviations. The adaptive strategy automatically uses only lenses from the ring $\mathcal{R}_5$ in addition to $\mathcal{R}_1$ yielding a higher localization accuracy. Since the adaptive strategy selects the lenses based on a single coarse estimate, wrong target lenses might be selected at depth edges, resulting in a slightly higher standard deviation in the case of the *Four planes scene*.

418   O. Fleischmann and R. Koch

Table 1. Evaluation results for the synthetical datasets.

	Fixed$_{0,1,4}$		Fixed$_{0,...,5}$		Adaptive	
	Avg.	Std.	Avg.	Std.	Avg.	Std.
Scene: Four planes						
Lens type 0	0.28	0.50	0.30	0.61	0.29	0.56
Lens type 1	0.25	0.37	0.26	0.42	0.26	0.44
Lens type 2	0.25	0.35	0.26	0.86	0.26	0.49
Target lenses	20080		45972		**13546**	
Scene: Near plane						
Lens type 0	0.41	0.09	0.42	0.10	0.41	0.09
Lens type 1	0.41	0.07	0.43	0.08	0.41	0.07
Lens type 2	0.42	0.57	0.45	0.52	0.42	0.50
Target lenses	20080		45972		**13585**	
Scene: Far plane						
Lens type 0	0.08	0.11	0.20	0.73	0.10	0.02
Lens type 1	0.09	0.01	0.10	0.02	0.10	0.01
Lens type 2	0.10	0.02	0.10	0.02	0.10	0.02
Target lenses	20080		45972		**13335**	

Fig. 3. Top row: excerpt from the plenoptic image of the *Four planes* scene along with ground-truth disparity, estimated disparity, and confidence. Bottom rows: Excerpts of the *Watch* and the *Forest* scene along with coarse disparity, fused fine and coarse disparity, and confidence. The brighter the image, the more confident is the estimate.

Real World Data. We also applied our algorithm to publicly available datsets provided by Raytrix[1]. In contrast to our synthetic dataset, the Raytrix datasets do not contain the focal lengths and focus distances of the microlenses but only contain a depth range for each lens which describes the range of acceptable defocus blur for each microlens type. Consequently we modified our defocus criterion in the adaptive strategy such that a lens is accepted if the coarse disparity estimate is within the valid range of the target lens. No ground truth depth images are available for these datasets. Hence, only qualitative visual results are shown in Fig. 3. Due to the large size of the plenoptic input images, any resizing of the complete depth images would result in heavy aliasing effects. Therefore, only excerpts of the scenes are depicted. Detailed images can be found on the authors website[2].

6   Conclusion

We have proposed a depth estimation framework for multi-focus plenoptic images. Instead of a single depth map from a single virtual standard camera, our method estimates a dense regularized depth map *per microlens* which also delivers depth information for surfaces which would be occluded with respect to a single virtual view. Our heuristic adaptive lens selection strategy significantly reduces the computation time with no negative influence on the overall accuracy. Experiments using synthetic datasets support our propositions. Future work includes highly-parallelized integration into volumetric surface reconstruction methods such as [4].

References

1. Press Release: Lytro Redefines Photography with Lightfield Cameras (2011). https://www.lytro.com
2. Adelson, E.H., Wang, J.Y.: Single lens stereo with a plenoptic camera. IEEE Trans. Pattern Anal. Mach. Intell. (PAMI) **14**(2), 99–106 (1992)
3. Bishop, T.E., Zanetti, S., Favaro, P.: Light field superresolution. In: IEEE International Conference on Computational Photography (ICCP), pp. 1–9 (2009)
4. Curless, B., Levoy, M.: A volumetric method for building complex models from range images. In: Proceedings of the 23rd Annual Conference on Computer Graphics and Interactive Techniques, pp. 303–312. ACM (1996)
5. Gallup, D., Frahm, J.M., Mordohai, P., Pollefeys, M.: Variable baseline/resolution stereo. In: IEEE Conference on Computer Vision and Pattern Recognition (CVPR), pp. 1–8 (2008)
6. Goesele, M., Curless, B., Seitz, S.M.: Multi-view stereo revisited. In: IEEE Computer Society Conference on Computer Vision and Pattern Recognition (CVPR), vol. 2, pp. 2402–2409 (2006)

---

[1] http://raytrix.de/index.php/Forschung.html
[2] http://www.informatik.uni-kiel.de/~ofl/

7. Goesele, M., Snavely, N., Curless, B., Hoppe, H., Seitz, S.M.: Multi-view stereo for community photo collections. In: IEEE International Conference on Computer Vision (ICCV), pp. 1–8 (2007)
8. Hirschmüller, H.: Accurate and efficient stereo processing by semi-global matching and mutual information. In: IEEE Conference on Computer Vision and Pattern Recognition (CVPR), vol. 2, pp. 807–814 (2005)
9. Kim, C., Zimmer, H., Pritch, Y., Sorkine-Hornung, A., Gross, M.: Scene reconstruction from high spatio-angular resolution light fields. ACM Trans. Graph. **32**(4), 73:1–73:12 (2013)
10. Lippmann, G.: Epreuves reversibles donnant la sensation du relief. J. Phys. Theor. Appl. **7**(1), 821–825 (1908)
11. Lumsdaine, A., Georgiev, T.: The focused plenoptic camera. In: IEEE International Conference on Computational Photography (ICCP), pp. 1–8 (2009)
12. Merrell, P., Akbarzadeh, A., Wang, L., Mordohai, P., Frahm, J.M., Yang, R., Nistér, D., Pollefeys, M.: Real-time visibility-based fusion of depth maps. In: IEEE International Conference on Computer Vision (ICCV), pp. 1–8 (2007)
13. Perwaß, C., Wietzke, L.: Single lens 3D-camera with extended depth-of-field. In: IS&T/SPIE Electronic Imaging. International Society for Optics and Photonics (2012)
14. Seitz, S.M., Curless, B., Diebel, J., Scharstein, D., Szeliski, R.: A comparison and evaluation of multi-view stereo reconstruction algorithms. In: IEEE Conference on Computer Vision and Pattern Recognition (CVPR), vol. 1, pp. 519–528. IEEE (2006)
15. Strecha, C., Fransens, R., Van Gool, L.: Wide-baseline stereo from multiple views: a probabilistic account. In: IEEE Conference on Computer Vision and Pattern Recognition (CVPR), vol. 1, pp. I-552. IEEE (2004)
16. Tao, M.W., Hadap, S., Malik, J., Ramamoorthi, R.: Depth from combining defocus and correspondence using light-field cameras. In: IEEE International Conference on Computer Vision (ICCV) (2013)
17. Unger, C., Wahl, E., Sturm, P., Ilic, S.: Stereo fusion from multiple viewpoints. In: Pinz, A., Pock, T., Bischof, H., Leberl, F. (eds.) DAGM and OAGM 2012. LNCS, vol. 7476, pp. 468–477. Springer, Heidelberg (2012)
18. Wanner, S., Goldlücke, B.: Globally consistent depth labeling of 4D lightfields. In: IEEE Conference on Computer Vision and Pattern Recognition (CVPR) (2012)

Efficient Metropolis-Hasting Image Analysis for the Location of Vascular Entity

Henrik Skibbe[1]([⊠]), Marco Reisert[2], and Shin Ishii[1]

[1] Graduate School of Informatics, Kyoto University, Kyoto, Japan
skibbe-h@sys.i.kyoto-u.ac.jp, ishii@i.kyoto-u.ac.jp
[2] Department of Radiology, University Hospital Freiburg, Freiburg, Germany
marco.reisert@uniklinik-freiburg.de

Abstract. In this paper we present a novel approach for probabilistically exploring and modeling vascular networks in 3D angiograms. For modeling the vascular morphology and topology a graph-like particle model is used. Each particle represents the intrinsic properties of a small fraction of a vessel including position, orientation and scale. Explicit connections between particles determine the network topology. In evaluation using simulated as well as real X-ray and time-of-flight MRI angiograms the proposed method was able to accurately model the vascular network.

1 Introduction

In our modern society many people suffer from circulatory system diseases. Angiograms acquired with techniques like MR or CT play an indispensable role in diagnostic decision making. Of particular interest are the morphology and the topology of vascular networks having both practical usage in surgery planning.

A fully automated modeling of the vascular network is challenging because the location, shape and the connectivity of vessels are unknown. Furthermore, images are noisy and varying contrast makes the segmentation difficult.

In this paper we propose a new method that automatically models vascular networks in medical images. We use a Metropolis-Hasting (MH) algorithm [8] to fit a graph-like particle system to the image that mimics both vascular morphology and topology. By incorporating prior knowledge about tubular shaped networks into the optimization process the computational effort is kept on a reasonable time scale.

There exists a broad variety of techniques related to computer aided analyses of vascular networks [10]. The majority of the existing approaches extracts the topology of the network by traversing the vessel centerlines in a semi-automated, deterministic manner; see e.g. [1,7]. Fully automated vessel tracking is more challenging because no hint of vessel location, orientation or scale is given [13,14]. A vessel itself is often either implicitly modeled in terms of image derivatives [1,3,6,7,13], or by providing an explicit vessel model [4,14].

We include a model of the vascular network into a global optimization problem. This is a complex and non-convex problem. However, recent work has shown

© Springer International Publishing Switzerland 2014
X. Jiang et al. (Eds.): GCPR 2014, LNCS 8753, pp. 421–431, 2014.
DOI: 10.1007/978-3-319-11752-2_34

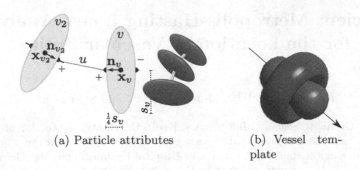

(a) Particle attributes

(b) Vessel template

Fig. 1. The vascular network is modeled by connected, tabled shaped particles. A particle is defined by its attributes including position $\mathbf{x}$, orientation $\mathbf{n}$ and scale s_v; see Sect. 2 for details. In order to measure the similarity of the particle model and the image data a Gaussian derivative based vessel template is used; see Sect. 2.1.

that for similar problems a probabilistic formulation and optimization of the problem leads to remarkable results [2,5,9,12].

The idea presented in this work is partly based on the global fiber tracker proposed in [12]. We use a particle process to model position, orientation and scale of the local vascular structures. Explicit representations of connections determine the graph structure. We use the MH algorithm as in [12] and extent it with novel scale, connections, tracking and insertion proposals. The convergence and behaviour of MH-like algorithm heavily depends on the design of the proposals. In particular, in the case of local minima proposals that do several steps at once can help totunnel through energy barriers [9]. The newly presented tracking and insertion proposals are such kind of joint proposals that speed up the convergence and prevent the algorithm to be trapped in local minima.

It is worth noting that [12] works for DW-MRI data and cannot be used for vessel detection since there exist no scale nor multiple connections for modeling bifurcation as well as a completely different particle model is used.

The vessel appearance in this work is represented in terms of image derivatives; similar to [1,13]. In a comparison to an unsupervised, deterministic vessel tracker [14] we keep the performance while achieving a significantly higher recall.

2   Method

The proposed model $\mathcal{M} = (\mathcal{P}, \mathcal{E})$ is built upon a set $\mathcal{P}$ of tablet shaped particles which can connect via edges $\mathcal{E}$. Figure 1(a) illustrates the attributes of a particle $v := (\mathbf{x}_v, \mathbf{n}_v, s_v)$ which are position $\mathbf{x}_v \in \mathbb{R}^3$, orientation $\mathbf{n}_v \in \mathbb{R}^3, |\mathbf{n}| = 1$ and scale $s_v \in \Phi$, where $\Phi := [\phi_{\min}, \phi_{\max}]$, is a given scale interval. An edge $u \in \mathcal{E}$ is connecting two particles $v_i, v_j \in \mathcal{P}$. This is denoted by $u := (v_i^{\gamma_i}, v_j^{\gamma_j})$, where $\gamma_i, \gamma_j \in \{-, +\}$. With v^+ we distinguish the particle side at position $\mathbf{t}_v^+ := \mathbf{x}_v + (s_v/4)\mathbf{n}_v$ and with v^- the opposite side. Figure 1(a) depicts an edge $u := (v_2^+, v^+)$. Particles cannot overlap with other particles. Example particle models representing real MRI data are depicted in Figs. 7(b) and 9(b).

2.1   The Energy Functional

In order to find a model that represents the tubular network in an image I : $\mathbb{R}^3 \to \mathbb{R}$ we aim at minimizing $E\{I, \mathcal{M}\} = E\{I, \mathcal{M}\}_\text{ext} + E\{\mathcal{M}\}_\text{int}$. The data term $E\{I, \mathcal{M}\}_\text{ext}$ is evaluating the difference of $\mathcal{M}$ and the vascular network within I in a particle-by-particle manner, where

$$E_\text{ext}\{I, \mathcal{M}\} := \sum_{v \in \mathcal{P}} C_\mathcal{P}\{I\}(v). \tag{1}$$

Given a particle v at position $\mathbf{x}_v$, orientation $\mathbf{n}_v$ and scale s_v, the cost $C_\mathcal{P}\{I\}(v)$ turns into a reward if there exists a tube-like structure in I passing $\mathbf{x}_v$ with direction $\mathbf{n}_v$ and scale s_v. Similar to [1] we implicitly model vessels by analyzing the image curvature using both the Hessian matrix and the image gradient leading to

$$C_\mathcal{P}\{I\}(v) := -\alpha_1 V\{I\}(v) + \alpha_2 s_v^2 + \alpha_3 G_V\{I\}(v) + \alpha_4 G_{V\perp}\{I\}(v) + c. \tag{2}$$

The term $V\{I\}$ evaluates how well a particle "fits" into the image. It is worth mentioning that here almost any "vessel-filter" can be used as long as it provides a score for vessels with respect to position, scale and orientation; e.g. [6,7]. For this work we correlate the image with a tubular shaped template as shown in Fig. 1(b). Position, scale and orientation of the template are chosen according to the particle's properties; we shifted further details about $V\{I\}$ to the appendix. The term s_v^2 gives a preference on particles with larger scales to prevent vessels from being modeled by many thin particles.

The functions $G_V\{I\}$ and $G_{V\perp}\{I\}$ are penalty terms based on the image gradients and used to suppress false positive detections. In the ideal case, the image gradients on the centerline of tubular structures are zero. We separately consider the magnitude of the gradient projected in particle direction (with $G_V\{I\}$) and its orthogonal component ($G_{V\perp}\{I\}$). With $G_V\{I\}$ we suppress particles appearing perpendicular to the vessel direction while with $G_{V\perp}\{I\}$ we keep particles on the centerline.

The particle likeliness $c \in \mathbb{R}_{\geq 0}$ gives particles a constant cost and controls the total number of particles (Note that for the free Poisson process, which is the basic underlying process of our simulation in the MH-algorithm, $e^{(-c)}$ is proportional to the expected number of particles). The parameters $\alpha_i \in \mathbb{R}_{\geq 0}$ are weights that must be tuned manually or in combination with a parameter search using an annotated ground truth.

The model term $E\{\mathcal{M}\}_\text{int}$ is used for excluding unlikely models from the optimization process. It is built upon two components:

$$E_\text{int}\{\mathcal{M}\} := C_\text{model}(\mathcal{M}) + \sum_{u \in \mathcal{E}} C_\text{edge}^L(u). \tag{3}$$

The term C_model ensures that particles do not intersect with other particles. It further suppresses loops for which we use a depth-limited search where the maximum depth is set as a number of particles. Note that it prevents self-connections, too. We prevent models from violating these constraints by setting the costs C_{model} to infinity in case of violation and otherwise setting them to zero.

The cost function C_{edge}^L controls the interaction between connected particles. It favors connected particles with short edges and having the same orientation and the same scale. With the edge likeliness $L \in \mathbb{R}$ we control the reward for edges. We shifted the definition of C_{edge}^L to the appendix.

2.2 Optimization

Minimizing $E\{I, \mathcal{M}\}$ is equivalent to maximizing the a-posteriori probability $P(\mathcal{M}|I)$ which can be split up into the maximization of the image likelihood $P(I|\mathcal{M}) = e^{-E_{\text{ext}}\{I,\mathcal{M}\}}$ and a model prior $P(\mathcal{M}) = e^{-E_{\text{int}}\{\mathcal{M}\}}$. For the optimization we use the iterative MH algorithm [8]: starting with the empty model $\mathcal{M} = (\emptyset, \emptyset)$ we provide several reversible transitions (proposals) from one state $\mathcal{M}$ into another $\mathcal{M}_\bullet$. At each iteration a transition is randomly proposed with probability $p^{\text{prop}}(\mathcal{M}_\bullet|\mathcal{M})$ and accepted when Green's ratio $R = e^{-(\Delta E)/T}\frac{p^{\text{prop}}(\mathcal{M}|\mathcal{M}_\bullet)}{p^{\text{prop}}(\mathcal{M}_\bullet|\mathcal{M})}$ is greater than 1, otherwise it is accepted with probability R. With $\Delta E = E\{I, \mathcal{M}_\bullet\} - E\{I, \mathcal{M}\}$ we denote the energy difference (the reward) of the transition. We use simulated annealing to explore unfavorable states during the early iterations for which the temperature $T \in \mathbb{R}_{>0}$ is used. Hence we always accept transitions decreasing the energy $(R \geq 1)$, and accept proposals increasing the energy with probability R (which becomes less likely with decreasing temperature T).

The fundamental proposals are the birth/death and connect/disconnect proposals. Similar to [12] we complement them with move, rotate and connect proposals. Bifurcations can only be proposed by the connect proposal. They are implicitly modeled by two edges connecting three particles. Novel are the scale, track and insert proposals:

Scale Proposal. A particle $v_\circ$ is uniformly sampled from $\mathcal{P}$. A new scale $s_{v_\bullet}$ is uniformly chosen from a scale interval $[\phi_0^{s_{v_\circ}}, \phi_1^{s_{v_\circ}}]$ with $\phi_0^s := \max(\phi_{\min}, s/(\sqrt{T}+1))$ and $\phi_1^s := \min(\phi_{\max}, s(\sqrt{T}+1))$. That is, with a decreasing temperature the boundaries of the interval are getting closer to $s_{v_\circ}$. Green's ratio becomes $R = e^{-(\Delta E)}(\phi_1^{s_{v_\circ}} - \phi_0^{s_{v_\circ}})/(\phi_1^{s_{v_\bullet}} - \phi_0^{s_{v_\bullet}})$.

Track-Birth and Track-Death Proposal. With the tracking proposal we make use of a common property of vessels: they continue for a longer distance. Since vessels are modeled by trajectories of connected particles they can be continued by generating and connecting a particle in front of one of their endpoints.

Track-birth: We uniformly choose an existing particle v out of the subset of terminal particles $\mathcal{P}_{\text{terminal}} \subseteq \mathcal{P}$. Particles in $\mathcal{P}_{\text{terminal}}$ have exactly one connection. Let $\gamma_v \in \{-, +\}$ denote the side that is not connected. We create a new particle $v_\bullet$. A good position $\mathbf{x}_{\text{opt}}$ for $v_\bullet$ would be directly in front of v, in our implementation $\mathbf{x}_{\text{opt}} = \mathbf{x}_v + (s_v/2 + \epsilon)\mathbf{n}_{\text{opt}}$ with $\mathbf{n}_{\text{opt}} := \gamma_v \mathbf{n}_v$. This would minimize the edge cost between v and the new particle $v_\bullet$ in case of identical orientation, scale and $\epsilon = 0$. We experimentally found that slightly fostering particles farther away from the optimal point by choosing $\epsilon = s_v/2$ improves tracking performance. A good orientation and scale for $v_\bullet$ are the values $\mathbf{n}_v$ and s_v of v. We randomly obtain a new position $\mathbf{x}_{v_\bullet}$ by adding normally distributed noise with standard deviation $\sigma_\mathbf{x}$ to $\mathbf{x}_{\mathrm{opt}}$. Similar for the orientation where we add Gaussian noise with $\sigma_\mathbf{n}$ to $\mathbf{n}_v$ and normalize the result. We heuristically choose $\sigma_\mathbf{x} = \sqrt{T}\frac{s_v}{2}$ and $\sigma_\mathbf{n} = \sqrt{T}$ such to take the particle size and temperature into account. Scale is uniformly chosen in the same way as for the scale proposal but with respect to s_v. A new edge $u_\bullet = (v^{\gamma_v}, v_\bullet{}^{(-\gamma_v)})$ is created connecting the previous terminal v with the new terminal node $v_\bullet$. The new model is $M_\bullet = (\mathcal{P} \cup v_\bullet, \mathcal{E} \cup u_\bullet)$. Green's ratio for the track-birth proposal is $R = e^{-(\Delta E_\bullet)/T} R^{\mathrm{track}}(v_\bullet|v)$. With the ratio

$$R^{\mathrm{track}}(v_\bullet|v) := \frac{(\phi_1^{s_v} - \phi_0^{s_v})}{G_{\sigma_\mathbf{x}}(\mathbf{x}_{\mathrm{opt}} - \mathbf{x}_\bullet)G_{\sigma_\mathbf{n}}(\mathbf{n}_\bullet, \mathbf{n}_{\mathrm{opt}})(\phi_{\max} - \phi_{\min})} \qquad (4)$$

we account for the probability for choosing the extension $v_\bullet$ given the terminal node v, where $G_{\sigma_\mathbf{x}}$ is a normalized 3D Gaussian with standard deviation $\sigma_\mathbf{x}$ and the function $G_{\sigma_\mathbf{n}}$ refers to the probability of $\mathbf{n}_\bullet$ given $\mathbf{n}_{\mathrm{opt}}$ (see appendix).

Track-death: We uniformly draw v_0 out of $\mathcal{P}_{\mathrm{terminal}}$. If v_0 has been the result of the track-birth proposal it is connected to a previous terminal v by an edge u_0. The particle v must have two edges, exactly one at each side so that after removing v_0 and u_0 the particle v itself in turn inherits the role as a terminal particle again. If this is not the case we continue with the next iteration. Green's ratio is given by $R = e^{-(\Delta E_0)/T}/R^{\mathrm{track}}(v_0|v)$.

Insert-Birth and Insert-Death Proposal. It happens that particles are connected with long edges. Instead of moving particles together a new particle can be added between them by breaking up a connection, inserting a particle and adding two new connections to make the gap smaller:

Insert-birth: For the insert operation we uniformly pick an existing edge $u_0 \in \mathcal{E}$ with $u_0 = (v_i^{\gamma_i}, v_j^{\gamma_j})$, $v_i, v_j \in \mathcal{P}$. We uniformly pick one of the particles as a reference point; w.l.o.g v_i. Similar to the tracking proposal we create a new particle by randomly choosing the properties based on a good position, orientation and scale. To keep it simple we choose the orientation in the same way as in the tracking proposal with v_i as reference. For the position we chose $\mathbf{x}_{\mathrm{opt}} := (\mathbf{x}_{v_i} + \mathbf{x}_{v_j})/2$ and for the scale $s_{\mathrm{opt}} = (s_{v_i} + s_{v_j})/2$ as reference values. Differently to the tracking proposal we account for the distance between v_i and v_j by choosing $\sigma_\mathbf{x} = \sqrt{T}\|\mathbf{x}_{v_i} - \mathbf{x}_{v_j}\|/4$. Finally, the two edges $u_{\bullet_i} = (v_i^{\gamma_i}, v_\bullet^{-\gamma_i})$ and $u_{\bullet_j} = (v_j^{\gamma_j}, v_\bullet^{\gamma_i})$ are added and u_0 is removed so that $M_\bullet = (\mathcal{P} \cup v_\bullet, \{\mathcal{E} \setminus u_0\} \cup \{u_{\bullet_i}, u_{\bullet_j}\})$. Green's ratio for the insert-birth proposal is $R = e^{-(\Delta E_\bullet)/T} \frac{R^{\mathrm{insert}}(v_\bullet|v_i, v_j)2|\mathcal{E}|}{(|\mathcal{P}|+1)}$ where $R^{\mathrm{insert}}(v_\bullet|v_i, v_j)$ coincides with (4).

Insert-death: For the inverse operation we uniformly pick one particle v_0 out of $\mathcal{P}$. The particle v_0 must have exactly one connection per side

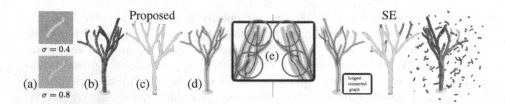

Fig. 2. Experiment: 20 trees, 265 bifurcations in total (image sizes about 160^3). (a) The two noise levels used in the experiment. (b) A model found by the proposed method. (c) The model has been rendered into a voxel-grid for evaluation; see Fig. 3(a). (d) The extracted centerline together with the bifurcation points. (e) The displacement of leaf nodes with respect to a given ground truth is used to for evaluating how correctly a tree has been explored; see Fig. 4 for results.

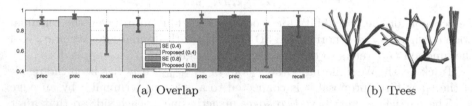

Fig. 3. We evaluated the overlap of the longest graph with the ground truth by rendering the graph into a voxel grid and counting (see Fig. 2(c)) the correctly (yellow), incorrectly (green) and missed voxels (red). The results for both noise levels in terms of precision and recall are represented in the bar plot where the error bars are representing the standard deviation. On the right you see examples of the simulated dataset (Color figure online).

connecting to a particle v_i and v_j, respectively. Otherwise we continue with the next iteration. We remove the two edges u_{o_i}, u_{o_j} connecting to v_o from $\mathcal{E}$, adding a new edge $u_\bullet$ directly connection v_i with v_j and remove v_o so that $R = e^{-(\triangle E_o)/T}|\mathcal{P}|/(R^{\mathrm{insert}}(v_o|v_i, v_j)2(|\mathcal{E}| - 1))$.

3   Experiments

We performed experiments using simulated data as well as X-ray and time-of-flight MRI angiograms.

Simulated Data. We generated twenty vessel tree images like shown in Fig. 3(b) covering a broad variety of bifurcation angles and with intensities in $[0, 1]$. The scale of the vessels was in the range of $[1.25, 6]$. We added Gaussian noise to the data with $\sigma = 0.4$ and $\sigma = 0.8$, see Fig. 2(a) for two example patches from those images. For the proposed method we set $\Phi := [1.25, 8]$. We used a tree like in Fig. 3(b) with $\sigma = 0.4$ for parameter tuning. This is done by first manually

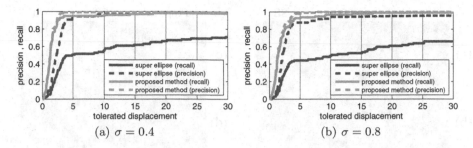

Fig. 4. Displacement of leave positions: ground truth with respect to detected graph.

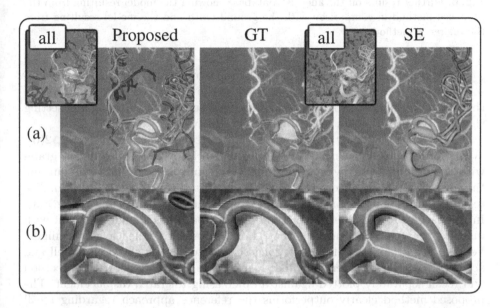

Fig. 5. Results for an X-ray angiogram (256^3 voxels). We show the results for the proposed methods together with a (partially) labeled ground truth available together with the datasets. On the right the results for the reference approach are shown. For the proposed and the reference approach the longest connected graph has been selected from within all graphs shown in the small images labeled "all". In the second row (b) we show a closeup of the aneurism which was challenging for both methods. An overlay of the centerline and bifurcation points that in contrast to SE most of the bifurcations have been correctly detected with the proposed method.

finding some good parameters for (2) and (5). The parameters have a reasonable, directly observable effect so that a good set of parameters can be found quickly based on visually inspection after running the proposed method for one million iterations ($\sim$1 s). For fine-tuning we use a parameter search and optimize with respect to the quality measures described in the experiment below.

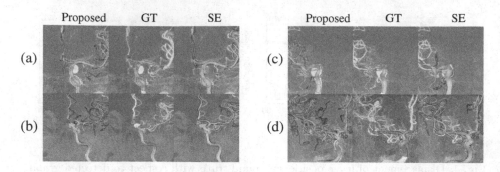

Fig. 6. Further results on the aneurisk database showing the model resulting from the proposed method in comparison with the ground truth and the model resulting from the reference method.

For comparison we used the method proposed by [14] (SE for "superellipsoids") using the implementation provided by the FARSIGHT project[1]. We get the best performance for SE when enabling vessel enhancing filtering and setting the density of seed points to the highest possible value.

For evaluation we only considered the longest connected graph; see Fig. 2(d). We performed two experiments: (1) We evaluated the overlap of the longest graph with the ground truth by rendering the graph into a voxel grid and counting the correctly (yellow), incorrectly (green) and missed voxels (red); Fig. 2(c). The results in terms of precision and recall are represented in the bar plot in Fig. 3(a). In comparison to the reference method our approach has similar precision with much higher recall. (2) The displacement of the position of the leaves (terminals) between the ground truth and the longest detected graph shows how well and accurately a tree has been explored; Fig. 2(e). In Fig. 4 we show the precision and recall separately plotted against an increasing tolerated displacement. The proposed method clearly outperforms the reference approach regarding recall and precision.

X-Ray Angiograms. We further did experiments on real world clinical X-ray angiograms publicly available from the aneurisk project[2]. In Fig. 5(a) we exemplarily show the tracking results for the C0036 X-ray dataset of the aneurisk database for which a partially annotated ground truth (GT) is provided. We show the isosurfaces of the largest connected graphs. Figure 5(b): Our method models the aneurysm as an ordinary vessel which terminates just behind the vessel coming from the upper right part, however, in contrast to the reference approach most of the bifurcations have been detected and modeled correctly. The tracking results for four further X-ray angiograms of the database are shown in Fig. 6 (a–d).

---

[1] FARSIGHT Project: http://www.farsight-toolkit.org/
[2] AneuriskWeb project http://ecm2.mathcs.emory.edu/aneuriskweb

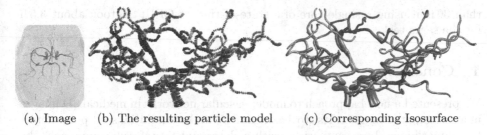

(a) Image   (b) The resulting particle model   (c) Corresponding Isosurface

Fig. 7. MRI angiogram ($384 \times 348 \times 136$). Particle colors: red particles: ordinary birth; green: tracking; blue: insertion (Color figure online).

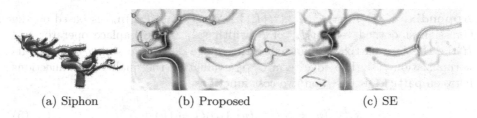

(a) Siphon   (b) Proposed   (c) SE

Fig. 8. The siphon (circle in Fig. 7(a)) is particularly challenging because of its high curvature and varying intensities within the vessel. The isosurface rendering is based on the proposed method. We show the three longest extracted graphs only. The reference approach is not tracking the vessel correctly; see (c).

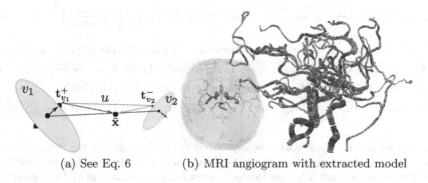

(a) See Eq. 6   (b) MRI angiogram with extracted model

Fig. 9. (a) See Eq. 6 and (b) MRI angiogram with extracted model.

MRI Angiograms. In Fig. 7(a) we show the minimum intensity projection (inverted intensities for better visibility) of an MRI angiogram. In (b) we show the corresponding model of the two largest graphs and in (c) the corresponding isosurface. In Fig. 8(a) we see that the proposed method correctly modeled the siphon (red circle in Fig. 7(a)) with its high curvature and touching vessels (three largest, connected graphs are shown). The resulting model of a second MRI angiogram is shown in Fig. 9(b).

Computation time: starting with a temperature $T = 1$ we successively decreased the temperature to $T = 0.01$ within 400 million iterations which took less than 30 min using a single core of a state-of-the art CPU; SE took about 3.5 h for the same task.

4   Conclusion

We presented a novel approach to model vascular networks in medical 3D images in an automated manner. Its stochastic nature makes an initial seed-point detection superfluous. In a comparison with a deterministic reference approach, the proposed method showed superior performance in accurately modeling the topology and morphology of vascular networks in simulated and real angiograms.

Appendix. $V\{I\}(v) := s_v^3(\mathbf{n}_v{}^T(\mathcal{H}\{I_{s_v}\} - \mathrm{diag}(\triangle I_{s_v}))(\mathbf{x}_v)\mathbf{n}_v)$ is based on the Gaussian smoothed version I_{s_v} of I (width s_v), $\triangle$ the Laplace operator and $\mathcal{H}$ the Hessian matrix. We normalize the image derivatives using a local soft-normalization [11]. The edge term C^L_{edge} evaluating the quality of connections between particles is based on two cost functions

$$C^L_{\mathrm{edge}}(u) = \beta_1 C_{\mathrm{con}}(u) + \beta_2 C_{\mathrm{scale}}(u) + L, \tag{5}$$

where $\beta_1, \beta_2 \in \mathbb{R}$ are weights and the connection likeliness $L \in \mathbb{R}$ controls the reward for an edge. Let $u = (v_1^{\gamma_1}, v_2^{\gamma_2})$ be an edge. The components

$$C_{\mathrm{con}}(u) = \frac{\|\bar{\mathbf{x}} - \mathbf{t}_{v_1}^{\gamma_1}\|^2}{s_{v_1}^2} + \frac{\|\bar{\mathbf{x}} - \mathbf{t}_{v_2}^{\gamma_2}\|^2}{s_{v_2}^2} \quad \text{and} \quad C_{\mathrm{scale}}(u) = \left(\frac{\max(s_{v_1}, s_{v_2})}{\min(s_{v_1}, s_{v_2})}\right)^2 - 1 \tag{6}$$

evaluate the distance and relative orientation of particles on the one hand (similar to [12]) and C_{scale} smooth scale along vessels. With $\bar{\mathbf{x}} = \frac{(s_{v_2}\mathbf{x}_{v_1} + s_{v_1}\mathbf{x}_{v_2})}{(s_{v_1} + s_{v_2})}$ we denote the scale weighted center between both particles; see Fig. 1(c).

Needed in (4): $G_\sigma(\mathbf{n}_1, \mathbf{n}_2) = \sigma\sqrt{\frac{\pi}{2}}(1 + \mathrm{erf}(\langle\mathbf{n}_1, \mathbf{n}_2\rangle/\sqrt{2}\sigma)) \exp\frac{-(1-\langle\mathbf{n}_1, \mathbf{n}_2\rangle^2)}{(2\sigma^2)}$.

Acknowledgment. This work was supported by Bioinformatics for Brain Sciences under the Strategic Research Program for Brain Sciences, MEXT (Japan). The work of M. Reisert is supported by the Deutsche Forschungsgemeinschaft (DFG), grant RE 3286/2-1.

References

1. Aylward, S.R., Bullitt, E.: Initialization, noise, singularities, and scale in height ridge traversal for tubular object centerline extraction. IEEE Trans. Med. Imaging **21**(2), 61–75 (2002)
2. Basu, S., Kulikova, M., Zhizhina, E., Ooi, W.T., Racoceanu, D.: A stochastic model for automatic extraction of 3D neuronal morphology. In: Mori, K., Sakuma, I., Sato, Y., Barillot, C., Navab, N. (eds.) MICCAI 2013, Part I. LNCS, vol. 8149, pp. 396–403. Springer, Heidelberg (2013)

3. Bauer, C., Bischof, H.: A novel approach for detection of tubular objects and its application to medical image analysis. In: Rigoll, G. (ed.) DAGM 2008. LNCS, vol. 5096, pp. 163–172. Springer, Heidelberg (2008)
4. Bogunović, H., Pozo, J.M., Villa-Uriol, M.C., Majoie, C.B., van den Berg, R., van Andel, H.A.G., Macho, J.M., Blasco, J., Román, L.S., Frangi, A.F.: Automated segmentation of cerebral vasculature with aneurysms in 3DRA and TOF-MRA using geodesic active regions: an evaluation study. Med. Phys. **38**, 210 (2011)
5. Chai, D., Forstner, W., Lafarge, F.: Recovering line-networks in images by junction-point processes. In: Proceedings of the CVPR, IEEE (2013)
6. Frangi, A.F., Niessen, W.J., Vincken, K.L., Viergever, M.A.: Multiscale vessel enhancement filtering. In: Wells, W.M., Colchester, A.C.F., Delp, S.L. (eds.) MIC-CAI 1998. LNCS, vol. 1496, pp. 130–137. Springer, Heidelberg (1998)
7. Gülsün, M.A., Tek, H.: Robust vessel tree modeling. In: Metaxas, D., Axel, L., Fichtinger, G., Székely, G. (eds.) MICCAI 2008, Part I. LNCS, vol. 5241, pp. 602–611. Springer, Heidelberg (2008)
8. Hastings, W.K.: Monte Carlo sampling methods using Markov chains and their applications. Biometrika **57**(1), 97–109 (1970)
9. Lacoste, C., Descombes, X., Zerubia, J.: Point processes for unsupervised line network extraction in remote sensing. IEEE Trans. Pattern Anal. Mach. Intell. **27**(10), 1568–1579 (2005)
10. Lesage, D., Angelini, E.D., Bloch, I., Funka-Lea, G.: A review of 3D vessel lumen segmentation techniques: models, features and extraction schemes. Med. Image Anal. **13**(6), 819–845 (2009)
11. Reisert, M., Burkhardt, H.: Harmonic filters for generic feature detection in 3D. In: Denzler, J., Notni, G., Süße, H. (eds.) Pattern Recognition. LNCS, vol. 5748, pp. 131–140. Springer, Heidelberg (2009)
12. Reisert, M., Mader, I., Anastasopoulos, C., Weigel, M., Schnell, S., Kiselev, V.: Global fiber reconstruction becomes practical. NeuroImage **54**(2), 955–962 (2011)
13. Shikata, H., McLennan, G., Hoffman, E.A., Sonka, M.: Segmentation of pulmonary vascular trees from thoracic 3D CT images. J. Biomed. Imaging **2009**, 24 (2009)
14. Tyrrell, J.A., di Tomaso, E., Fuja, D., Tong, R., Kozak, K., Jain, R.K., Roysam, B.: Robust 3-D modeling of vasculature imagery using superellipsoids. IEEE Trans. Med. Imaging **26**(2), 223–237 (2007)

Automatic Determination of Anatomical Correspondences for Multimodal Field of View Correction

Hima Patel, Karthik Gurumoorthy, and Seshadri Thiruvenkadam$^{(\boxtimes)}$

Medical Image Analysis Lab, GE Global Research, Bangalore, India
sheshadri.thiruvenkadam@ge.com

Abstract. In spite of a huge body of work in medical image registration, there seems to be very little effort in Field of View (FOV) correction or anatomical overlap estimation especially for multi-modal studies. This is a key step for most registration algorithms to work on image volumes of different coverages. In this work, we consider the FOV correction problem between Computed Tomography (CT) and Magnetic Resonance Imaging (MRI) image volumes for the same patient. A novel algorithm composed of a cascade of (a) symmetry based gross rotation/translation correction (b) multi-modal feature descriptor and (c) matching scheme using dynamic programming is presented. The above combination deals with the challenges of multi-modal studies namely intensity differences, inhomogeneity, and gross patient movement. Validation and comparisons of the proposed algorithm is quantitatively shown on **73** CT-MRI pairs and has yielded promising results.

1 Introduction

3D multi-modal volume registration is a necessary feature requirement for diagnosis, therapy, and treatment follow-up. Depending on the nature of the disease in a clinical work-flow, different modalities are often used in unison or sequentially to provide complementary information to the radiologist. Handling of multi-modal medical data within registration is fairly well studied with Mutual information (MI) and extensions [1,2] being quite successful. However, the success of standard registration algorithms is based on the key assumption that *the image volumes are scans of the same anatomical region* which may not hold for many studies especially follow up scans, where, large field-of-view (FOV) mismatch occurs between multi-modal or multi-time point volumes. More often than not, the earlier scan in the sequence is a full body scan and once the affected region/anatomy is grossly identified, the follow up scan acquires only the identified anatomy to save time, cost or induced radiation dose. In FOV correction, we try to *identify*—rather than register—the best portion of a full body scan that is compatible with an anatomic specific second volume. Once identified,

---

Equal contributions by the first and second authors. Thanks to GE HCS Tech. Eng. team, BUC, for their technical and application inputs.

© Springer International Publishing Switzerland 2014
X. Jiang et al. (Eds.): GCPR 2014, LNCS 8753, pp. 432–442, 2014.
DOI: 10.1007/978-3-319-11752-2_35

the matching portions can then be registered using any well known registration algorithm. In this sense, our algorithm should be employed as a precursor and *not* as a replacement for registration algorithms.

We wish to point out that, in spite of a huge body of work in medical image registration, there seems to be very little effort in FOV initialization or anatomical overlap estimation especially for multi-modal studies. Many clinical applications require either manual intervention or reliable header information to align the FOV's which may not be feasible in many scenario. In the absence of reliable header information, automated FOV correction is a key step that has to be addressed before registration.

In this work, we consider the FOV correction problem for volumes acquired using two commonly used imaging modalities - Magnetic Resonance Imaging (MRI) and Computed Tomography (CT) for the *same patient* with different coverages. This is a challenging task due to intrinsic differences between these scanner systems which leads to high amounts of intensity variation. Due to inhomogeneity issues, intensity variation can be found even within the MRI scans across patients for the same anatomical region. Any translation and rotations induced by gross patient motion or acquisition protocols further complicates the task. This problem has received some attention for mono-modality cases for CT. In [3] an all-pair cost matrix between slices in the two volumes is computed using Minkowski distance and a deformable displacement between pre-segmented ROI's in slices. The work by Subramanian et al. [4] takes into account bone voxel count in both image volumes and generates 1-D profiles that are subsequently matched with a model 1-D profile. In [5] a cost-matrix is generated between all pairs of slices using the CSSD metric [6].

The above methods though fairly robust to handle challenges of CT multi time-point studies are hard to adapt for multi-modal studies due to non-standard intensity values, MRI in-homogeneity, and presence of gross mis-matches between scans. To the best of our knowledge we are not cognizant of any prior art focused on multi-modal estimation of anatomical correspondence. We propose an algorithm that tackles the above challenges for MRI and CT acquisitions. The **two main contributions** of our work are as follows. First, we propose an elegant way of solving the translation and rotation correction errors by embedding symmetry information in Fast Fourier Transform (FFT) framework. Second, we propose a multi-modal feature descriptor that captures anatomy specific signatures while being relatively insensitive to multi-modal acquisition challenges. To effectively detect anatomical correspondences across multi-modality scans, we break down our problem from finding corresponding anatomies to matching axial slices across the two multi-modal volumes. Say, we have N slices corresponding to whole body CT volume and M slices from the shoulder MRI station. Our goal is to identify the best matching subset M of the N axial slices such that the selected subset from CT corresponds to MRI shoulder scan. In Fig. 1 we show an illustration to explain the problem. The yellow lines on the whole body CT scan mark the best matching subset to MRI shoulder scan. The reader should bear in mind that the volumes are three dimensional and a two dimensional coronal slice view

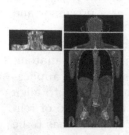

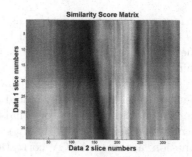

Fig. 1. The yellow lines in the figure show the best matching subset in whole body CT (right) with MRI shoulder scan(left).

Fig. 2. The above figure shows a sample Similarity Score Matrix represented as an image. Darker pixel values indicate a low matching score - thereby implying a better match. Note the dark line in the center which represents the matching subset from Data 2 that best matches Data 1.

serves only to pinpoint the identified matching anatomical portion. In order to determine the best subset, we compare every slice from the first volume with every slice of the second volume and compute an associated *matching score*. This score is *low* for similar slices and *high* for disparate ones. This information is then recorded in the form of a $M \times N$ *Similarity score matrix*, refer Fig. 5. We test our algorithm on 73 test pairs and report our validations and comparisons in Sect. 3.

2   Methods

2.1   Feature Description

This section describes our first contribution. We generate anatomy specific features for every axial slice—planar 2D slices obtained by sectioning the 3D human anatomy along the top-down z axis—such that they are discriminative within the modality and similar across same anatomies in the other modality. This is made challenging as MRI and CT modality scans differ to a great extent in terms of intensity values even for the same anatomical region. For example, bones are highly pronounced in CT as compared to MRI. Due to inhomogeneity issues in MRI systems [7] intensity variations are observed within the image itself. We overcome the above challenges by (a) concentrating on non-local gradient-magnitude changes instead of actual intensity or gradient magnitude values and (b) aggregating several non-local features over entire image to ensure robustness to small artifacts or pathology. Next, we describe the mechanism for our feature description technique.

The feature for every slice is built by concatenating features on non-local patches that describe the characteristics of a specific portion of the anatomy.

We shall call the feature that describes the entire slice as slice descriptor and the feature that describes a specific portion of the slice as patch descriptor. The patch descriptors are described at each $(p,q)^{th}$ pixel for every axial slice i in both the volumes. We do so by considering a large cubical neighborhood of size $m \times n \times 3$ around each $(p,q)^{th}$ location. This neighborhood that also spills over the neighboring slices $i-1$ and $i+1$ is then partitioned into 27 rectangular blocks of equal size, 9 blocks per slice. The mean gradient magnitude of each block is compared with the mean of the center block within the same axial slice. Each block is then tagged with a label of 0 or 1 based on the whether the mean gradient value of that block is smaller or greater than the center block's mean. This results in a 24 bit binary vector, 8 per each slice. Considering the mean gradient value of a large neighborhood is particularly helpful for MRI images as it makes the patch descriptor resistant to small levels of noise. Furthermore, as the human anatomy is contiguous in the axial direction, patch descriptor generated by considering a larger 3D neighborhood—where the neighboring $i-1$ and $i+1$ slices are also included—adds more information when compared to 2D neighborhood that is confined to the same axial slice i.

Expressed mathematically, the 24 bit patch descriptor $\gamma_i(p,q)$ at the $(p,q)^{th}$ pixel in slice i is defined as:

$$\gamma_i(p,q) \equiv\ <\beta_{i-1}(p,q), \beta_i(p,q), \beta_{i+1}(p,q)> \tag{1}$$

where the computation of the slice-wise 8 bit binary feature vector $\beta_i(p,q)$ proceeds as follows. Let $R_i(p,q)$ denote the rectangular region within the same slice i centered at pixel (p,q) of size $\delta_y = m/3$ and $\delta_x = n/3$ consisting of the gradient magnitude values and let $\alpha_i(p,q)$ represent the mean gradient value within the region $R_i(p,q)$. We define $\beta_i(p,q)$ as:

$$\beta_i(p,q) \equiv\ <f(\alpha_i(p,q), \alpha_i(p-\delta_y, q-\delta_x)), \ldots, f(\alpha_i(p,q), \alpha_i(p+\delta_y, q+\delta_x))> \tag{2}$$

where

$$f(\alpha_i(p,q), \alpha_i(p',q')) = \begin{cases} 1, & \text{if } \alpha_i(p,q) \geq \alpha_i(p',q'), \\ 0, & \text{otherwise.} \end{cases} \tag{3}$$

The slice descriptor for every axial slice—obtained by concatenating each patch descriptor $\gamma_i(p,q)$—is an aggregation of changes in the gradient directions in a neighborhood for all local neighborhoods in the image.

As discussed in the previous section, our goal is to determine the best subset of slices from one volume that matches the other volume. This is determined by computing a matching score between every pair of slices between the two volumes. Thus, for every pair (i,j) of slices between the first and the second volume, we calculate a matching score $S(i,j)$ by computing the Hamming distance between their respective slice descriptors, i.e. $S(i,j) = HD(\chi_i, \chi_j)$ where χ_i is the slice descriptor for the ith slice and HD is the Hamming distance and compactly record the information in the Similarity score matrix S. We chose Hamming distance based measure as it is computationally efficient and requires fewer operations [8]. An example of the Similarity score matrix S is shown in

Fig. 2 where the darker pixel values indicate a lower matching score implying a better match. As we can observe, there is a prominent dark band in Fig. 2. This represents the matching subset from one volume that best matches the other volume.

It is important to underscore that while computing the matching score, we are comparing the gradient-magnitude changes across the *same locations*—gradient changes around the pixel (p, q) in the ith axial slice of the first volume is likened to gradient changes around the same pixel location (p, q) in the jth slice of the second volume—in the corresponding two axial slices, each belonging to a different modality. Thus, it is imperative for two axial slices to be corrected for changes due to translation and rotation and we discuss these in the subsequent sections. Since we consider the two volumes from the same patient, we do not explicitly address scale issues arising out of patient variations.

2.2   Symmetry Based Translation Correction

We describe the second contribution of our paper here. Traditionally, translation correction is attempted using two sets of images of same anatomy using one as fixed and the other as moving. However in our case we cannot utilize this approach as we do not know beforehand the matching slices resulting in an chicken and egg problem. We propose to correct for changes in translation utilizing a single image by exploiting the symmetry of human body in the axial direction. The symmetry-based method is also *not* modality specific and hence both the MRI and the CT volumes can be subjected to same preprocessing treatment. An approach availing symmetry was first proposed in [9] for aligning functional brain images to a standard vertical orientation. Image rotation and centering are performed by determining the angle of rotation and centerline coordinate that result in maximal left-right correlation. This method was refined in [10] wherein the authors introduced a stochastic sign change metric for rotational correction and centering of 3D- functional brain images. The symmetry idea was further exploited in [11] for detecting and computing the mid-sagittal in 3D brain images. We propose to extend the symmetry approach throughout the human anatomy as explained below.

Consider correcting for translation about the anterior-posterior y-axis which will be a vertical line when the volume is sliced axially. As the image may be translated, it is symmetrical about an *unknown* y^* axis parallel to the standard y-axis. Let Δx denote the perpendicular distance between y and y^*. We treat the given image as the target (I_T) and exploit the symmetry to create the moving image (I_M) by *flipping* the fixed image about the central y-axis. The moving image I_M will now be symmetrical about the flipped y^*-axis, y_F^*, which is at a distance of $-\Delta x$ from the y-axis. It is important to emphasize that y^* and y_F^* respectively are *equidistant* from y and in *opposite* sides of it. We then see that translating I_M by $2\Delta x$ will align the left region of I_M (right region of I_T) with the left region of I_T. As I_T is symmetrical, this alignment will result in maximum correlation between the two. A global search for translation of I_M and the comparison with I_T can be performed efficiently by realizing that

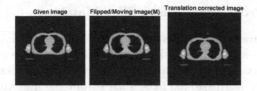

Fig. 3. Given image I_T, the Flipped/Moving image I_M and the final translation corrected image.

the flipping and shifting steps are tantamount to *convolving* the image with itself which is efficiently computable using Fast Fourier Transforms [12]. Since the human anatomy, to some extent, also exhibits symmetry about the x-axis, similar approach can be used to reckon the optimal y-translation value Δy. An example of the translation corrected image is shown in Fig. 3.

2.3 Symmetry Based Rotation Correction

Rotations can be handled in the similar fashion as translation by representing the image in the polar co-ordinate system [13]. Since translation correction requires that the image is rotationally adjusted first, any translation effects needs to nullified prior to correcting for rotations. This is achieved by taking the Fourier transform of the image and considering its magnitude value. Translation effects are relegated to the exponent of the Fourier transform which are annulled when we consider the magnitude alone [13]. By representing the Fourier magnitude image in the polar-representation (r-θ), the translation correction for θ based on the procedure described above will result in an image that is symmetrical about an axis parallel to the y-axis.

Figure 4 gives an example of a rotated MRI image that is centered about the anterior-posterior y axis.

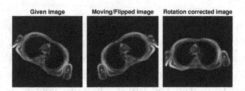

Fig. 4. Given image I_T, the Flipped/Moving image I_M and the final rotation corrected image.

Note that rotation correction is far more complicated than translation correction as it involves: (i) Considering the Fourier transform, (ii) Computing the magnitude square, (iii) Representation in the polar coordinate and (iv) Translation correction for θ. As rotation correction subsumes the steps involved in translation correction, we show experimental results on rotation correction alone in Sect. 3.

2.4   Dynamic Programing Based Slice Matching

Correcting for rotation and translation errors ensures location wise correspondence between two axial slices which is necessary for an accurate matching of two slices. Recall from 2.1 that the $(i, j)^{th}$ entry of the Similarity score matrix informs how similar (or dissimilar) the slice i from the first volume is to slice j from the other volume. Our goal is to find the best subset of slices from the larger volume that correspond with the slices in the smaller volume. We find the best subset using dynamic programming (DP) [14] as it provides an optimal solution. The red dots in Fig. 5 correspond to the best-matching slice pairs output by DP. As human anatomy is contiguous we could expect the optimal set of slice pairs to fall on a straight line. However, when there are deformations/non-rigid motions (for e.g. due to breathing), the generated slice pairs may not stick to a straight line. We counter this effect by constructing the *best-fit* line that passes close through most of the slice pairs, denoted by the yellow line in Fig. 5. The slices that lie on the yellow line are the best matches between the two volumes.

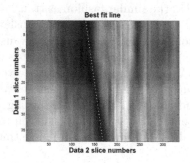

Fig. 5. This figure shows the output of Dynamic Programming on Similarity Score Matrix S. The yellow line shows the output of fitting a line on output of DP which is epresented by red dots.

We wish to point out here that our feature extraction strategy described in Sect. 2.1 differs from [15,16] as we give equal weight to all spatial directions as opposed to [15]. Also, since our broad aim is to compare shape information across slices, we directly match the binary feature vector as opposed to generating a histogram for matching. The use of Hamming distance for comparison is motivated from [8] for efficient computations. However, as opposed to [8] we extract a single binary vector per image. This is attainable as we correct for translations and rotations using our symmetry based approach. Thus, we do not need to search for spatial correspondences by extracting features at certain key locations. It also allows us to consider a simpler neighborhood as compared to [8], but in 3D.

3   Experimental Results

We experimented with **73** test cases each consisting of whole body CT scan and anatomic specific MRI scan. The CT/MRI volumes were first down-sampled in the z-direction using the slice-thickness information. This ensures that any K number of contiguous slices covers the same *extent* (albeit different portions) of anatomical region in both the volumes. The slope of the best-fit line computed using the DP output will therefore be 1. The anatomies in the MRI scans can be broadly classified into head, neck, lung, abdomen, pelvic and the leg regions. Additionally, the MRI volume can either be an In-Phase, Out-Phase or Water data. We have also introduced artificial axial rotations $(-38°, -18°, 3°, 17°, 32°)$ in the MR scans to simulate gross patient movement. For quantitative comparisons, we requested an experienced radiologist to manually mark the ground truth by specifying the start and end matching slices in the CT volume for each of the test cases. We compared the output of our algorithm with the radiologist marked ground truth by estimating the *slice error (err)* defined as:

$$err \equiv \min\left\{|c_s - t_s|, |c_e - t_e|\right\} \tag{4}$$

where c_s and t_s are respectively the computed and the true *start* CT slices and c_e and t_e are the calculated and the true *end* CT slices respectively. c_s and c_e are set to the smallest and the largest matching CT axial slices—numbered incrementally from head to toe–lying on the best-fit line.

3.1   Evaluating Symmetry Based Rotation Correction

To start with, we ascertain the accuracy of our rotation correction step. We considered **1875** MRI slices and induced artificial rotations of varying degrees $(-38°, -18°, 3°, 17°, 32°)$ about the y-axis. The slices were picked from all MRI volumes in our test set. We then estimated the angle of rotation using our technique and constructed a histogram on the computed angles, see Fig. 6. We notice that most errors are within $10°$ of the ground truth value. These were primarily noticed mostly in the head section where the anatomy is circular and symmetrical about more than one axis. It is important to note here that both rotation correction and feature description steps assume a translation corrected image. Though we don't directly evaluate translation correction, the efficiency of this step is indirectly determined by the accuracy of feature description and rotation correction steps.

3.2   Comparison with Normalized Mutual Information Metric

To assess the robustness of our proposed slice descriptor (χ_i) we computed an equivalent Similarity score matrix S_{NMI} by:

$$S_{NMI}(i, j) \equiv NMI(a_i, b_j) \tag{5}$$

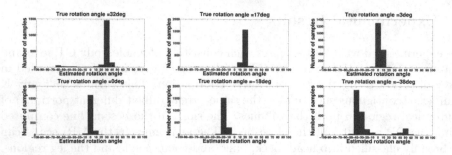

Fig. 6. Histograms of the estimated rotation angle for various degrees of rotations.

where a_i and b_j are the ith and the jth axial slices in the first and second volume, and NMI is the well-known normalized mutual information metric [1].

To keep the comparison fair, we retained our translation correction, rotation correction and DP steps for both our slice descriptor and NMI while computing their respective slice errors. The entries in Table 1 represent the number of MRI volumes (out of 73) where the estimated start and end slices numbers derived by running DP on both S and S_{NMI} lie within a certain range from the ground truth marked on CT cases. It is seen from Table 1 that our slice descriptor and Hamming distance based matching does significantly better than NMI based matching across different rotations.

Table 1. The table below shows a comparative result between our proposed descriptor (PD) and normalized mutual information (NMI). The table entries correspond to number of test cases out of 73 having slice errors less than a specified value.

Angle (→)	32°		17°		3°		0°		−18°		−38°	
err (↓)	PD	NMI	PD	NMI	PD	NMI	PD	NMI	PD	NMI	PD	NMI
≤2	**62**	46	**67**	46	**67**	47	**65**	49	**67**	49	**61**	41
≤3	**68**	51	**69**	52	**68**	54	**70**	56	**69**	54	**64**	49
≤7	**69**	63	**71**	64	**71**	66	**71**	66	**71**	63	**65**	60
≤10	**69**	65	**71**	66	**71**	67	**72**	67	**71**	66	**68**	61

To reason out the difference in results between our approach and NMI, we quantify the robustness of our proposed descriptor to small rotation differences. We artificially introduced rotations of angles 3°, 5° and 10° to MRI scans and deliberately sidestepped our rotation correction step. Table 2 report the results obtained by the generating the similarity score matrices using both our feature descriptor and NMI. The entries denote the number of test cases (out of 73) having slice errors less than a given value for each angle of rotation. We notice our proposed descriptor to be tolerant to small mismatches and fare better than NMI. We note that our algorithm pre-supposes presence of axial rotations only

Table 2. Represents the robustness of our proposed descriptor (PD) and normalized mutual information (NMI) to rotation errors. The table entries denote the number of volumes out of 73 with slice errors less than specified value.

Angle ($\rightarrow$)	3°		5°		10°	
err ($\downarrow$)	PD	NMI	PD	NMI	PD	NMI
≤ 2	**66**	52	**61**	45	**45**	26
≤ 3	**67**	57	**65**	51	**50**	29
≤ 7	**69**	64	**68**	58	**58**	44
≤ 10	**70**	67	**68**	61	**60**	46

which is a reasonable assumption for body cases. However, for neuro cases, 3D rotations are possible (e.g. about X/Y axis) which could potentially affect our rotation/translation correction and slice matching costs. In our robustness experiments, we noticed that our results are accurate up to 10 degrees of non-axial rotation.

4 Computational Complexity

All the three major steps of our algorithm are relatively computationally inexpensive. Consider two volumes consisting of M and N axial slices respectively with each slice containing P pixels. As the symmetry based correction steps using FFT for each slice can be performed in $O(P \log P)$ [17], the preprocessing steps can be completed in $O((M + N)P \log P)$. Using integral image [18], the mean gradient value for each region R can be obtained in $O(P)$ operations and direct comparison of these mean values produces the binary feature vector for each slice. Since MN slices pairs need to be compared, the overall complexity for obtaining the similarity score matrix S is $O(PMN)$. The best-matching slice pairs produced by running DP on S can be recovered in $O(MN)$ [14].

The implementation of our algorithm takes $\sim$ **3.5 s** for volumes of sizes $512 \times 512 \times 335$ and $512 \times 512 \times 104$ on a Linux workstation with 16 GB RAM and 8 core CPU without any parallel processing.

5 Conclusion

We present a method for estimating anatomical overlap to address multi modal FOV correction. Here we show an elegant mechanism for rotation and translation corrections using symmetry as the guiding principle. We also present a multi-modal feature descriptor that is robust to acquisition differences across modalities. We compare the effectiveness of our method to compute slice costs with well-known multi-modal metric NMI and observe that our proposed solution fares better than NMI. Future work consists of extending the technique to other orthogonal and oblique acquisitions.

References

1. Pluim, J.P.W., et al.: Mutual-information-based registration of medical images: a survey. IEEE Trans. Med. Imaging **22**(8), 986–1004 (2003)
2. Wells III, W.M., et al.: Multi-modal volume registration by maximization of mutual information. Med. Image Anal. **1**(1), 35–51 (1996)
3. Fiorin, D., et al.: On the registrability of two CT volumes. In: ISBI, pp. 1079–1082 (2008)
4. Subramanian, N., et al.: Method for registration and navigation of volumetric scans using energy profiles. US 7907761 B2, 15 March 2011
5. Thiruvenkadam, S., et al.: Automatic non-rigid mismatch correction algorithm for CT-CT longitudnal oncology studies. In: ISBI, pp. 1079–1082 (2008)
6. Bunyak, F., Palaniappan, K.: Efficient Segmentation Using Feature-based Graph Partitioning Active Contours. In: ICCV (2009)
7. McVeigh, E.R., et al.: Phase and sensitivity of receiver coils in magnetic resonance imaging. Med. Phys. **13**(6), 806–814 (1986)
8. Alahi, A., et al.: Freak: fast retina keypoint. In: CVPR, pp. 510–517 (2012)
9. Junck, L., et al.: Correlation methods for the centering, rotation, and alignment of functional brain images. J. Nucl. Med. **31**(7), 1220–1226 (1990)
10. Minoshima, S., et al.: An automated method for rotational correction and centering of 3D functional brain images. J. Nucl. Med. **33**(8), 1579–1585 (1992)
11. Ardekani, B.A., et al.: Automatic detection of the mid-sagittal plane in 3D brain images. IEEE Trans. Med. Imaging **16**(6), 947–952 (1997)
12. Bracewell, R.: The Fourier transform and its applications, 3rd edn. McGraw-Hill, New York (1999)
13. Reddy, B.S., Chatterji, B.N.: An FFT-based technique for translation, rotation, and scale invariant image registration. IEEE Trans. Image Process. **5**(8), 1266–1271 (1996)
14. Cormen, T.H., et al.: Introduction to Algorithms, 2nd edn. MIT Press and McGraw-Hill, New York (2001)
15. Ojala, T., et al.: Multiresolution gray-scale and rotation invariant texture classification with local binary patterns. IEEE Trans. Pattern Anal. Mach. Intell. **24**(7), 971–987 (2002)
16. Zhao, G., Pietikäinen, M.: Dynamic texture recognition using local binary patterns with an application to facial expressions. IEEE Trans. Pattern Anal. Mach. Intell. **29**(6), 915–928 (2007)
17. Cooley, J.W., Tukey, J.W.: An algorithm for the machine calculation of complex Fourier series. Math. Comp. **19**(90), 297–301 (1965)
18. Franklin, C.: Summed area tables for texture mapping. In: SIGGRAPH, pp. 207–212 (1984)

Encoding Spatial Arrangements of Visual Words for Rotation-Invariant Image Classification

Hafeez Anwar$^{(\boxtimes)}$, Sebastian Zambanini, and Martin Kampel

Computer Vision Lab, Vienna University of Technology, Vienna, Austria
hafeezcse@gmail.com

Abstract. Incorporating the spatial information of visual words enhances the performance of the well-known bag-of-visual words (BoVWs) model for problems like object category recognition. However, object images can undergo various in-plane rotations due to which the spatial information must be added to the BoVWs model in rotation-invariant manner. We present a novel approach to integrate the spatial information to BoVWs model in a rotation-invariant way by encoding the triangular relationship among the positions of identical visual words in the $2D$ image space. Our proposed BoVWs model is based on densely sampled local features for which the dominant orientations are calculated. Thus we achieve rotation-invariance both globally and locally. We validate our proposed method for rotation-invariance on datasets of ancient coins and butterflies and achieve better performance than the conventional BoVWs model.

1 Introduction

Bag-of-visual words (BoVWs) is a simple yet efficient technique used for a multitude of computer vision problems like scene classification [6,7], large-scale image retrieval [10,11] and object category recognition [2,16]. In this technique, local features extracted from a set of images are clustered to form a visual vocabulary which is comprised of the visual words. From a given image, features are extracted and assigned to visual words from the vocabulary based on a similarity measure, such as the Euclidean distance. The image is then represented as the histogram of visual words. However, such a representation ignores the spatial information of visual words in the $2D$ image space. For problems like object category recognition, spatial information is important for better performance as objects have specific geometric structures. Moreover, object images can undergo various in-plane rotations. Therefore the spatial information must be added to the BoVWs model in a manner that is robust to rotations.

Related Work
The most notable work for incorporating spatial information to the BoVWs model is the spatial pyramid matching (SPM) proposed by Lazebnik et al. [6]. Statistics from the spatial pyramids made in the image space are combined to

© Springer International Publishing Switzerland 2014
X. Jiang et al. (Eds.): GCPR 2014, LNCS 8753, pp. 443–452, 2014.
DOI: 10.1007/978-3-319-11752-2_36

achieve improved performance. Zhang et al. [15] proposed to use the log-polar tiling to add spatial information to BoVWs. They use single, multiple and multi-scale log-polar tiling and report better results than SPM on three benchmark datasets. Recently another method for encoding the spatial arrangement of visual words is proposed by Penatti et al. [9] which they call *word spatial arrangement* (WSA). They capture the relative positions of visual words by partitioning the image space into four quadrants making the position of a given word as the origin. Afterward they aggregate the statistics of all the words of the vocabulary from each quadrant. Their proposed method showed comparative results against SPM on several benchmark datasets. Khan et al. [5] proposed to use the angles made by pair-wise identical visual words (PIWs) and then aggregating them using the pair-wise identical angles histogram (*PIWAH*). Their concept is inspired by the works that model similar image cues [3]. Their method combined with SPM achieves increase in performance on four benchmark datasets. However all these methods add spatial information to the BoVWs model in a manner that is not robust to image rotations.

Anwar et al. [1] propose circular tiling for symbol based ancient coin classification which is robust to rotations. However their method is used for coin images that can be automatically segmented from the background. We propose to add spatial information to the BoVWs model in a rotation-invariant manner. Our work is inspired by the proposed method of Khan et al. [5] but we differ from them in the following points.

- We use pairs of identical words for angles calculation where each pair consists of three identical words. Khan et al. [5] use pairs of two identical words.
- Our method is based on dense sampling of SIFT [8] features where the dominant orientation is calculated for each feature.
- We utilize the segmentation masks for vocabulary construction.

With the above points, we achieve the following advantages.

- Using three identical words in a pair results in global rotation-invariant triangular relationship.
- Rotation invariance is also achieved locally.
- Vocabulary constructed with features sampled from the object area is more discriminating.

The rest of the paper is organized as follows. Section 2 describes our proposed method to incorporate spatial information to BoVWs. Experimental settings and their results are discussed in Sect. 3. Finally, Section 4 concludes the paper and gives future directions for the currently proposed framework.

2   Rotation-Invariant Angles Histogram of Pair-Wise Identical Visual Words (*rPIWAH*)

Discriminative information about the image contents can be obtained by explicitly modeling similar image cues as shown by [3]. Identical image patches are represented by identical visual words in the BoVWs model. Recently Khan et al. [5] proposed a method for integrating the spatial information to BoVWs by using identical visual words. They calculate angles made by pair-wise identical visual words (PIW) with respect to x-axis in the 2D image space. An angles histogram is then built from these angles to represent the image which they call pair-wise identical words histogram ($PIWAH$). Since a pair consists of two identical words we name their pair-wise identical visual words as 2-PIWs. Their proposed method is not rotation invariant as they calculate angles with respect to x-axis. Therefore we modify their proposed method by considering three identical words in a given PIW to calculate angles. This will make a rotation-invariant triangular relationship among the words of a given PIW. After calculating the angles, we produce the angles histogram in a similar manner as proposed by Khan et al. [5] and call it rotation-invariant pair-wise identical words histogram ($rPIWAH$). In our proposed method, a pair of PIW consists of three identical words therefore we name it 3-PIWS. In this Section we give the details about $rPIWAH$.

In the BoVWs model, a visual vocabulary $voc = \{v_1, v_2, v_3, \ldots, v_M\}$ comprises of M visual words. A given image is first represented as a set of descriptors

$$I = \{d_1, d_2, d_3, \ldots, d_N\} \tag{1}$$

where N is the total number of descriptors. A given descriptor d_k is then mapped to a visual word v_i using some similarity measures like the Euclidean distance as follows.

$$v(d_k) = \arg \min_{v \in voc} \text{Dist}(v, d_k) \tag{2}$$

where d_k is the kth descriptor in the image, $v(d_k)$ is the visual word assigned to this descriptor based on the distance $\text{Dist}(v, d_k)$. The given image is then represented as the histogram of visual words where the number of bins of this histogram is equal to the size of the visual vocabulary which is M. The value of the bin b_i of this histogram gives the number of occurrences of a visual word v_i in an image. Let D_i be the set of all descriptors mapped to a visual word v_i, then the ith bin of the histogram of visual words b_i, is the cardinality of the set D_i.

$$b_i = Card(D_i) \text{ where } D_i = \{d_k, k \in [1, \ldots, M] \mid v(d_k) = v_i\} \tag{3}$$

From set D_i, all the distinct pairs of three descriptors are considered to calculate angles between the spatial positions of the descriptors as shown in Fig. 1. The spatial position of a descriptor is given by its position on the dense sampling grid. The set of all 3-PIWs related to a visual word v_i is defined as:

$$3 - PIW_i = \{(P_a, P_b, P_c) \mid (d_a, d_b, d_c) \in D_i^3, d_a \neq d_b \neq d_c\} \tag{4}$$

where P_a, P_b and P_c are the spatial positions of the descriptors d_a, d_b and d_c respectively. The value of the ith bin of the histogram shows the frequency of the visual word v_i. Therefore the cardinality of 3-PIW$_i$ is $^{b_i}C_3$ which is the

Pair of PIWs with 2 visual words   Pair of PIWs with 3 visual words

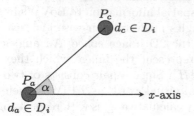 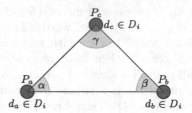

Fig. 1. 2-PIWs and 3-PIWs

number of all possible subsets of three distinct elements among the elements of D_i. However, we ignore those pairs of 3-PIWs in which all the three words are collinear. In order to suppress very small angles, we also ignore triangles made by pairs of non-collinear 3-PIWs where one of the three angles is less than $5°$. Therefore for a word v_i, the number of candidate 3-PIWs for angles computation is most likely less than $^{b_i}C_3$. The three angles shown in Fig. 1 made by 3-PIWs are calculated according to the law of cosines. The angles histogram is built from these angles for which the bins are empirically chosen between $0°$ and $180°$. The angles histogram for a specific word v_i is named as $rPIWAH_i$. Khan et al. [5] proposed a 'bin replacement' technique to combine the $rPIWAH_i$ from all the visual words. The bin b_i of the histogram of visual words associated with visual word v_i is replaced with $rPIWAH_i$ in such a way that the spatial information is added without altering the frequency information of v_i. Finally $rPIWAH_i$ of all the visual words are combined to represent a given image.

$$rPIWAH = (\psi_1 rPIWAH_1, \psi_2 rPIWAH_2, \ldots, \psi_M rPIWAH_M)$$

$$\text{where} \quad \psi_i = \frac{b_i}{\|rPIWAH_i\|} \tag{5}$$

where ψ_i is the normalization coefficient. For a visual vocabulary of size M, if the number of bins in angles histogram is θ, then the size of the $rPIWAH$ is $M\theta$.

3   Experiments and Results

In this section we give a brief detail of the datasets that we used to evaluate our proposed method, present the effect of segmentation used for vocabulary construction on the classification rate and finally experiments for rotation invariance.

3.1   Datasets

We performed our experiments on two different datasets. The purpose of choosing these datasets is that we want to validate our method on both symmetric and non-symmetric objects. The two datasets also differ in the amount of texture on the objects as the ancient coins have less texture than the butterflies.

Fig. 2. Symbols of the coin dataset

Fig. 3. Butterfly classes

Ancient Coins Dataset: The ancient coins dataset is provided by the ILAC project [4]. We performed symbol recognition based classification of ancient coins using our proposed framework. Symbols are minted on the reverse of ancient coins. We considered 11 classes i.e. 11 types of symbols which are shown in Fig. 2. The total number of images is 800. We use the coins dataset because coins can undergo various in-plane rotations.

Butterflies Dataset: Our second dataset consists 694 butterfly images that belong to 8 different classes, i.e. 8 species of butterflies. We obtained images of 6 classes from the Leed's butterfly dataset [13]. These 6 classes are Grand surprise, Orange puppy, Small white, Painted lady, Monarch and Zebra. For images of the other 2 classes i.e. Machaon and Peacock, we used Google image search. Exemplar images of these butterflies are shown in Fig. 3. Due to the regular and symmetric texture on butterflies, we utilize their images for the validation of our

| Given Image | Densely Sampled Features | Segmentation Mask | Features from foreground only |

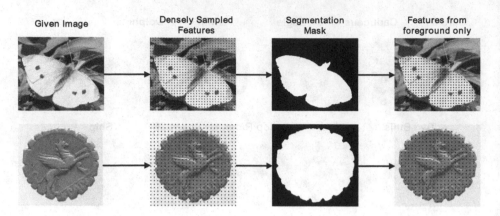

Fig. 4. Features extracted from foreground for vocabulary construction using segmentation masks

proposed method. Apart from that, like coins, butterfly images can also undergo in-plane rotations.

3.2   Segmentation for Vocabulary Construction

We use PIWs for angles calculation and then use these angles to form the angles histogram for image representation. Similar or identical image patches should be assigned to identical visual words from the vocabulary. Therefore the visual vocabulary should be discriminative enough to assign identical words to identical image patches. For vocabulary construction, we use dense sampling where features are extracted from the foreground as well as the background. These features are quantized by the k-means clustering to form the visual vocabulary. Due to the unsupervised nature of k-means, the visual vocabulary is more likely to be non-discriminative and noisy. Therefore we propose to utilize segmentation at the stage of vocabulary construction as shown in Fig. 4. Segmentation will help to select features for vocabulary construction from proper areas of the images. In case of ancient coins, we use the automatic segmentation method proposed by [14] to segment the coin from the homogeneous background as shown in Fig. 4. The results for the coins dataset are shown in Table 1 where *PIWAH* refers to

Table 1. Classification rates of various settings for segmented and non-segmented images of the coins dataset

	Ancient coins dataset	
	Non-segmented images	Segmented images
BoVWs	77.24	77.59
PIWAH	88.97	87.93
rPIWAH	81.03	80.00

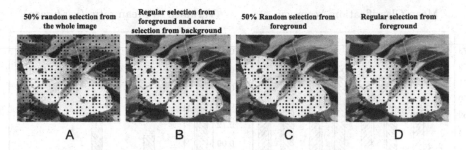

Fig. 5. Various settings of feature selection from images with and without the use of segmentation masks

the angles histogram made from pairs of two identical words and *rPIWAH* refers to the angles histogram made from pairs of three identical words. Due to the homogeneous background of the coin images in our dataset, segmentation used for vocabulary construction has minimal effect on the classification rates of all the three methods.

We use the segmentation masks for butterfly images which are generated by the a semi-automatic segmentation method proposed by [12]. The backgrounds of images in the butterfly dataset consists of trees, grass, stones, sky and flowers as shown in Fig. 3. Therefore, unlike coins the use of segmentation will have a significant effect on the discriminating nature of the visual vocabulary. We extracted features from butterfly images for vocabulary construction using various settings which are shown in Fig. 5. Here we briefly explain these settings.

(A) **50 % random selection from the whole image:** From each image, 50 % features are randomly selected among the densely extracted features.

(B) **Regular selection from the foreground and coarse selection from the background:** From each image, features are regularly selected from the foreground and coarsely extracted from the background.

(C) **50 % random selection from the foreground:** 50 % features are randomly selected among the densely extracted features from the foreground only.

(D) **Regular selection from the foreground:** From each image, features are regularly selected from the foreground only. In each row, every second feature is considered for vocabulary construction.

The features for vocabulary construction are quantized using k-means clustering where the cluster centers are initialized randomly. Apart from that, in settings 'A' and 'C', features are randomly selected for vocabulary construction. Due to these reasons, for each of the above settings, we perform our experiments ten times and report the mean classification accuracies of all the three methods i.e. BoVWs, *PIWAH* and *rPIWAH* in Fig. 6. We also report the variance to mean ratio of each method for all the settings to show their robustness to the randomness of the k-means and features selection. Each time, a visual vocabulary is constructed

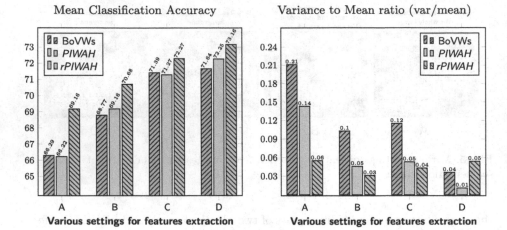

Fig. 6. Mean classification accuracy and variance to mean ratio

for image classification. For the image representation of training and test sets, we do not use segmentation. For 'C' and 'D' in Fig. 6, the classification rates are higher because the vocabularies are only constructed from the foreground i.e. the object area. Therefore the visual words are more discriminating resulting in better performance than 'A' and 'B'. It is noteworthy that our proposed method not only outperforms the other methods but also it has the least variance to mean ratios on settings 'A', 'B' and 'C'. In case of 'A', features are randomly selected from the whole image while in case of 'C', they are randomly selected from the foreground only. Our proposed method has the least variance to mean ratio on both 'A' and 'C' which shows that it is robust to the randomness of feature selection. In addition, our proposed method outperforms the other methods on setting 'A'. This shows that *rPIWAH* performs better even if the segmentation masks are not used for vocabulary construction. In method 'B', features are coarsely selected from the background as well. Due the variance in the background, the variance to mean ratios of BoVWs and *PIWAH* are higher. Our proposed method has the least variance to mean ratio in 'B' which shows that it is also robust to the variations in the background. Therefore we conclude that the use of segmentation masks during vocabulary construction results in higher classification rates. Over all our proposed method outperformed BoVWs and *PIWAH* on both segmented and non-segmented images and it is more robust to the randomness of both the k-means and the feature selection.

3.3   Experiments and Results for Rotation Invariance

We evaluate our proposed method for robustness to rotations. For the butterflies dataset, the training set consists of 289 images while the test set consists of 405 images. For ancient coins dataset, the training set consists of 500 images while the test set consists of 300 images. For both of our datasets, we bring the

Table 2. Mean classification accuracies of various methods on rotated and non-rotated images

	Butterflies dataset		Ancient coins dataset	
	Non-rotated images	Rotated images	Non-rotated images	Rotated images
BoVWs	68.62	67.44	77.58	74.95
PIWAH	68.62	63.03	**87.93**	61.28
rPIWAH	**71.56**	**70.92**	80.00	**77.27**

training images roughly to same orientation i.e. in the training set all the images have identical orientations. The test set is also brought to same orientation as the training set and then it is rotated through a predefined set of angles which is $\{30°, 60°, 90°, 120°, 150°, 180°, 210°, 240°, 270°, 300°, 330°\}$. Thus our test set for rotation-invariance evaluation consists of 12 subsets where 11 subsets consist of rotated images and one subset is comprised of non-rotated images. For classification we use one-vs-all approach of SVM with Helinger kernel. We also evaluate the method proposed by Khan et al. [5] on our rotated and non-rotated datasets. For each method, the mean classification rate of all the rotated subsets is reported in Table 2. *PIWAH* refers to the method of Khan et al. [5] while *rPIWAH* refers to our proposed method. In case of butterflies dataset, *rPIWAH* outperforms the other methods on both rotated and non-rotated images. It has the highest classification rate for the rotated images which shows that it is more robust to image rotations. *PIWAH* performs better than our method on non-rotated coin images. However, *rPIWAH* shows more robustness to rotations in coin images than *PIWAH* and BoVWs. Therefore we conclude that on both the datasets, our proposed method is more robust to image rotations than BoVWs and *PIWAH*.

4   Conclusion

A rotation-invariant encoding scheme of the spatial arrangements of visual words is presented. The global encoding is achieved by calculating angles between the positions of three identical visual words resulting in a rotation-invariant triangular relationship. Local rotation-invariance is achieved by calculating dominant orientation of each SIFT keypoint. In addition, segmentation masks of objects are used at the stage of vocabulary construction, which resulted in better performance. The proposed method is evaluated for rotation-invariance where it outperformed the conventional BoVWs model. In future we plan to explore the use of other rotation-invariant descriptors with our proposed method.

References

1. Anwar, H., Zambanini, S., Kampel, M.: Supporting ancient coin classification by image-based reverse side symbol recognition. In: Wilson, R., Hancock, E., Bors, A., Smith, W. (eds.) CAIP 2013, Part II. LNCS, vol. 8048, pp. 17–25. Springer, Heidelberg (2013)
2. Csurka, G., Dance, C.R., Fan, L., Willamowski, J., Bray, C.: Visual categorization with bags of keypoints. In: ECCV, pp. 1–22 (2004)
3. Deselaers, T., Ferrari, V.: Global and efficient self-similarity for object classification and detection. In: CVPR, pp. 1633–1640 (2010)
4. Kavelar, A., Zambanini, S., Kampel, M., Vondrovec, K., Siegl, K.: The ILAC-project: supporting ancient coin classification by means of image analysis. In: XXIV International CIPA Symposium (2013)
5. Khan, R., Barat, C., Muselet, D., Ducottet, C.: Spatial orientation of visual word pairs to improve bag-of-visual-words model. In: BMVC, pp. 1–11 (2012)
6. Lazebnik, S., Schmid, C., Ponce, J.: Beyond bags of features: spatial pyramid matching for recognizing natural scene categories. In: CVPR, pp. 2169–2178 (2006)
7. Li, F.F., Perona, P.: A bayesian hierarchical model for learning natural scene categories. In: CVPR, pp. 524–531 (2005)
8. Lowe, D.G.: Distinctive image features from scale-invariant keypoints. Int. J. Comput. Vision **60**, 91–110 (2004)
9. Penatti, O.A.B., Silva, F.B., Valle, E., Gouet-Brunet, V., da Silva Torres, R.: Visual word spatial arrangement for image retrieval and classification. Pattern Recogn. **47**(2), 705–720 (2014)
10. Perdoch, M., Chum, O., Matas, J.: Efficient representation of local geometry for large scale object retrieval. In: CVPR, pp. 9–16 (2009)
11. Philbin, J., Chum, O., Isard, M., Sivic, J., Zisserman, A.: Lost in quantization: Improving particular object retrieval in large scale image databases. In: CVPR (2008)
12. Veksler, O.: Star shape prior for graph-cut image segmentation. In: Forsyth, D., Torr, P., Zisserman, A. (eds.) ECCV 2008, Part III. LNCS, vol. 5304, pp. 454–467. Springer, Heidelberg (2008)
13. Wang, J., Markert, K., Everingham, M.: Learning models for object recognition from natural language descriptions. In: BMVC, pp. 2.1–2.11 (2009)
14. Zambanini, S., Kampel, M.: Robust automatic segmentation of ancient coins. In: VISAPP, pp. 273–276 (2009)
15. Zhang, E., Mayo, M.: Enhanced spatial pyramid matching using log-polar-based image subdivision and representation. In: DICTA, pp. 208–213 (2010)
16. Zhang, J., Marszałek, M., Lazebnik, S., Schmid, C.: Local features and kernels for classification of texture and object categories: a comprehensive study. Int. J. Comput. Vision **73**(2), 213–238 (2007)

Accurate Detection in Volumetric Images Using Elastic Registration Based Validation

Dominic Mai[1,2]([✉]), Jasmin Dürr[3], Klaus Palme[2,3], and Olaf Ronneberger[1,2]

[1] Computer Science Department, University of Freiburg, Freiburg, Germany
maid@informatik.uni-freiburg.de
[2] BIOSS Centre of Biological Signalling Studies, University of Freiburg, Freiburg, Germany
[3] Institute for Biologie II, University of Freiburg, Freiburg, Germany

Abstract. In this paper, we propose a method for accurate detection and segmentation of cells in dense plant tissue of *Arabidopsis Thaliana*. We build upon a system that uses a top down approach to yield the cell segmentations: A discriminative detection is followed by an elastic alignment of a cell template. While this works well for cells with a distinct appearance, it fails once the detection step cannot produce reliable initializations for the alignment. We propose a validation method for the aligned cell templates and show that we can thereby increase the average precision substantially.

1 Introduction

Multi class segmentation is an important task in biomedical image analysis. It enables statistically meaningful analysis of signals by relating them to the underlying structures. There are basically two approaches to this problem: 1. In the bottom up approach, one generates a set of region hypotheses that are later classified and merged to obtain the class label, e.g. [3,9]. 2. In the top down approach one uses a detector to obtain coarse object localizations that are refined by a finer grained alignment of a model to the data, e.g. [2,10].

In [10] we presented a paper that deals with detection and alignment of plant cells in volumetric data in a top down approach. The goal of this paper was to detect single cells of a certain layer from an Arabidopsis root and to reconstruct this cell layer. *Arabidopsis thaliana* is a model organism widely used in plant biology [7,11]. We use a rigid cell detector based on 3D HOG features to coarsely localize the cells, similar to Dalal and Trigg's approach for 2D human detection [5]. Then we align a template image (*sharp mean image*) for the respective cell type to the data using elastic registration. Finally we reconstruct the root in a greedy fashion by assembling the aligned cell templates iteratively, beginning with the aligned detection whose associated detection filter received the highest score.

This approach works well for the cell layer 3 (Fig. 3), as the rigid detection filter produces reliable hypotheses for the alignment. Unfortunately it fails to

© Springer International Publishing Switzerland 2014
X. Jiang et al. (Eds.): GCPR 2014, LNCS 8753, pp. 453–463, 2014.
DOI: 10.1007/978-3-319-11752-2_37

produce satisfactory results for cell layer 4 as is illustrated in Fig. 2(a). The reason for this is the greedy reconstruction based on the scores obtained by the HOG based rigid detector. This coarse localization step is needed as it would be computationally impossible to perform alignments of all cell models in all image locations. The detection system is hence optimized to deliver a high recall. This, however, leads to many false positive detections, as the image data often allows for multiple different interpretations with similar scores. For example, it happens frequent that two adjacent small cells are interpreted as a bigger cell that encompasses both (or vice versa).

Therefore, the score from the rigid detector is a bad foundation to decide whether the suggested model describes the recorded data correctly. While the alignment step can correct for coarse localizations of the right cell type, it cannot correct the error if the wrong cell type has been chosen. This is especially bad with the greedy reconstruction approach: once a false alignment is accepted, it is likely to also prevent valid alignments in its direct neighborhood.

Bourdev et al. [1] use a linear support vector machine to rescore detections of persons based on mutually consistent poselet activations. Their framework, however, is more directed to deal with appearance changes due to the camera viewpoint and the articulated nature of their objects, opposed to the deformable nature of the plant cells that we consider.

Contribution. We propose an effective validation step that makes use of the finer grained localization of the elastic alignment. For every detector, we train a discriminative classifier that verifies whether the alignment of the sharp mean image is valid for a certain location. This validation step results in a much better greedy reconstruction (Fig. 2(b)).

2   Detection and Alignment

The foundation of the approach presented here is our detection and alignment pipeline from [10]. We will outline the pipeline and introduce our formal notation along the way. For an overview, have a look at Fig. 1.

We define a 3D volumetric image I of an Arabidopsis root as a function $I : \Omega \mapsto \mathbb{R}, \Omega \subset \mathbb{R}^3$. It comes with a set of ground truth cell segmentations $S_i : \Omega \mapsto \{0, 1\}$. Attached to the root is a root coordinate system (RCS) [12] consisting of the direction of the main axis of rotation of the root and an arbitrary but fixed "up" component perpendicular to this axis.

The root has a cylindrical structure with cells organized in concentric rings around its core (Fig. 3). The RCS is used to normalize for the orientation and the location of the cells with a rigid transformation $\mathbf{H}_i$. The normalized cells are clustered into K clusters with a k-means clustering based on their cuboidal bounding volumes. For every cluster, a discriminative detection filter D_k is trained. It is based on 3D HOG features that bin the image gradients into 20 orientation directions (vertices of a Dodecahedron). The soft binning and spatial pooling is realized with a convolution by a triangular filter with a radius r_{HOG}. The HOG features are sampled on a regular grid at a distance of s_{HOG}. The D_k are realized

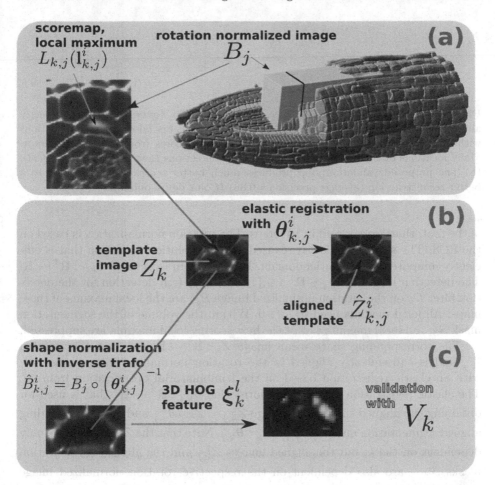

Fig. 1. Overview of our processing pipeline. (a) We run a sliding window detector D_k on the rotation normalized image B_j. On the left you see a slice of B_j and a local maximum of the produced scoremap $L_{k,j}$. (b) The template image Z_k is aligned to the data with the transformation $\theta^i_{k,j}$. (c) We use the inverse transformation $(\theta^i_{k,j})^{-1}$ to shape normalize the root image at this location. Finally, we compute a 3D HOG feature and validate it with the proposed classifier V_k.

as linear support vector machines, with the orientation normalized cell images of a cluster being the positive training examples and randomly sampled orientation normalized images from other parts of the root as negative examples. Along with the detector a *sharp mean image* Z_k [13] is generated from the positive training examples. It represents the centroid of the cell cluster with respect to appearance and shape. The sharp mean image Z_k comes with a segmentation mask S_{Z_k}.

At test time, all detection filters D_k are tested in a sliding window fashion on overlapping rotation normalized cuboid shaped image regions $B_j, j \in \{1, \ldots, N_B\}$

(a)   (b)

Fig. 2. Greedy reconstructions of layer 4 of the root r06. **(green)** cells are correctly aligned detections of cells with an IOU ≥ 0.5, **(red)** cells are falsely aligned detections with an IOU <0.5. (a) Reconstruction based on the scores from the rigid detector. Many locations of the root are occupied by false detections (average precision $= 0.64$). (b) The proposed validation step produces much better scores and thus leads to a better reconstruction (average precision $= 0.88$) (Color figure online).

of the root, that are sampled in $10°$ steps. The rotation normalization is based on the RCS. The sliding window is realized as a convolution operation that is efficiently computed in the Fourier domain. This results in score maps $L_{k,j} : \mathbb{R}^3 \mapsto \mathbb{R}$. The detection locations $\mathbf{l}_{k,j}^i \in \mathbb{R}^3, i \in \{1, \ldots, N_{k,j}\}$ (ith detection for the detection filter D_k on the rotation normalized image B_j) are the local maxima of these maps. All local maxima need to be >0. Within the volume of the segmentation mask S_{Z_k}, all local maxima except the best scoring local maxima are suppressed.

The corresponding sharp mean image $Z_k : \mathbb{R}^3 \mapsto \mathbb{R}$ is put at the location $\mathbf{l}_{k,j}^i$ and is subsequently aligned to the rotation normalized image region B_j with an elastic registration based on the combinatorial optimization from [8]. The elastic registration yields a transformation $\boldsymbol{\theta}_{k,j}^i : \mathbb{R}^3 \mapsto \mathbb{R}^3$ that is used for obtaining the aligned sharp mean image $\hat{Z}_{k,j}^i = Z_k \circ \boldsymbol{\theta}_{k,j}^i$ and the corresponding aligned segmentation mask $\hat{S}_{Z_k,j}^i = S_{Z_k} \circ \boldsymbol{\theta}_{k,j}^i$. Note that the Z_k and S_{Z_k} are only dependent on the k, but the aligned images $\hat{Z}_{k,j}^i$ and the aligned segmentation masks $\hat{S}_{Z_k,j}^i$ are also dependent on the respective rotation normalized image region B_j and the implied transformation $\boldsymbol{\theta}_{k,j}^i$.

As last step, the aligned sharp mean images are transformed back into the coordinate system of the original root to obtain a reconstruction. This is done in an iterative greedy fashion, beginning with the aligned sharp mean image $\hat{Z}_{k,j}^i$ corresponding to the *highest scoring* detection location $\mathbf{l}_{k,j}^i$. The indices $\{k, j, i\}$ are formally given by

$$\{k, j, i\} = \arg \max_{\substack{k \in \{1, \ldots, K\} \\ j \in \{1, \ldots, N_B\} \\ i \in \{1, \ldots, N_{k,j}\}}} L_{k,j}(\mathbf{l}_{k,j}^i). \tag{1}$$

The aligned sharp mean image $\hat{Z}_{k,j}^i$ is transformed back to the original root image, then it is removed from the pool of available detection hypotheses and the detection hypotheses with the next best score is processed. Note that once a location in the original root image is occupied, it is not possible to put other aligned images at this location. This gives a crucial importance to the ordering of the aligned candidates implied by Eq. 1: Due to the continuous nature of the cells wrt. deformation, we usually have multiple competing aligned candidates per ground truth location. It is crucial to pick the well aligned candidates first during this greedy iterative reconstruction, as a badly aligned candidate can not be corrected and will probably also prevent subsequent well aligned candidates in its direct neighborhood.

If the scores delivered by the rigid detector fail to provide a good sorting of the aligned candidates prior to the greedy reconstruction, the results for layer 4 are not satisfactory (Fig. 2(a)). We propose a validation of the aligned sharp mean images that makes use of the finer grained information that is available due to the alignment. This results in much better reconstruction results as we will show in the experiments section (Fig. 2(b)).

3   Training of the Alignment Classifiers

In order to validate a candidate alignment of the sharp mean template we propose to use the metric induced by a discriminative classifier. To this end we will use a support vector machine, as it gives a normalized score around zero. Values >1 indicate a confident decision for the positive class (well aligned candidate), values < -1 indicate a confident decision for the negative class (badly aligned candidate). As SVMs have good generalization properties, decision values in the interval $[-1, 1]$ mark a gradual change between the classes.

In our case the data used for training and testing is the 3D HOG representation of the root image within the support of the aligned cell template. The support vector machine, however, needs input data of a fixed size. This means that we cannot use the image data "below" the aligned template directly, as its volume is variable due to the elastic alignment. Therefore we will use the inverse transformation θ^{-1} to warp the image data onto the cell template. This assures that all training and test data for a cluster k will have the same number of features.

We need to mine positive $(+)$ and negative $(-)$ training examples from a training and validation root to train the validation classifier V_k. We compare the aligned segmentation masks $\hat{S}^i_{Z_k,j}$ with the ground truth segmentations. The *Intersection over Union* is the measure M_{IOU} used in the PASCAL VOC [6] challenge to assess the quality of a detection:

$$M_{\mathrm{IOU}}(S_1, S_2) = \frac{\int_\Omega S_1(\mathbf{x}) \cdot S_2(\mathbf{x}) \quad d\mathbf{x}}{\int_\Omega \max\left(S_1(\mathbf{x}), S_2(\mathbf{x})\right) \quad d\mathbf{x}}. \tag{2}$$

This area based measure is well suited to evaluate the degree of alignment in a detection setting, especially when it is based on 3D segmentation masks. We assign the class $\{+, -\}$ based on the rule that is also used for the evaluation of the complete pipeline. An aligned candidate is accepted, iff the intersection over union of the corresponding aligned segmentation mask with ground truth segmentation S_l is greater than 0.5:

$$c(\hat{S}^i_{Z_k,j}) = \begin{cases} +, & M_{\mathrm{IOU}}(\hat{S}^i_{Z_k,j}, S_l) \geq 0.5 \\ -, & M_{\mathrm{IOU}}(\hat{S}^i_{Z_k,j}, S_l) < 0.5 \end{cases}. \tag{3}$$

We shape normalize the corresponding root image by transforming it with the inverse transformation:

$$\hat{B}^i_{k,j} = B_j \circ \left(\boldsymbol{\theta}^i_{k,j}\right)^{-1}. \tag{4}$$

After the this transformation, we extract the 3D HOG feature that will be used in the training for the validation classifiers V_k.

$$\boldsymbol{\xi}^l_k = \mathbf{f}(\hat{B}^i_{k,j}), \quad \mathbf{f} : (\Omega \mapsto \mathbb{R}) \mapsto \mathbb{R}^{N^f_k} \tag{5}$$

The function $\mathbf{f}$ transforms the image $\hat{B}^i_{k,j}$ into a vectorial feature representation, i.e. the 3D HOG feature, and crops it along the support of the sharp mean image Z_k. For simplicity of notation we replace the indices j, k with $l \in \{1, \ldots, N_k\}$, as its no longer important, from which B_j the $\boldsymbol{\xi}^l_k$ originates. After the classification of the training examples into $(+)$ and $(-)$ with Eq. 3 we end up with a set $S^+_k = \{l^+_1, \ldots, l^+_{N^+}\}$ for positive examples and a set $S^-_k = \{l^-_1, \ldots, l^-_{N^-}\}$ for the negative training examples for every cluster k.

As wish to investigate the effect of the model complexity of the classifier, train a *linear* support vector machine V^{lin}_k a *RBF kernel* support vector machine V^{RBF}_k. We use 5-fold cross validation to estimate suitable parameters for the outlier penalty c and the radius γ of the radial basis function for V^{RBF}_k. The training is done with *libsvm* [4].

3.1   Validating Aligned Templates

The setting at test time is identical to the mining of the training examples for the V_k, except that we run the greedy iterative reconstruction of the root image after the detection and alignment phase. For each detection location $\mathbf{l}^i_{k,j}$, we perform the elastic registration of the corresponding sharp mean image Z_k to the data and yield the transformation $\boldsymbol{\theta}^i_{k,j}$. Then we shape normalize the root image at this location by warping it with the inverse transformation $(\boldsymbol{\theta}^i_{k,j})^{-1}$ and compute the 3D HOG features $\boldsymbol{\xi}^l_k$ (Eq. 5). Thus we end up with a set of aligned template image candidates and the corresponding 3D HOG feature representations of the locally shape normalized root image:

$$\left(\hat{Z}^l_k, \boldsymbol{\xi}^l_k\right) \text{ with } k \in \{1, \ldots, K\} \text{ and } l \in \{1, \ldots, N_k\}. \tag{6}$$

We perform the iterative greedy reconstruction, but replace the sorting induced by the rigid detector scores (Eq. 1) with a sorting based on the proposed validation classifier. We begin with the best scoring candidate image

$$\hat{Z}^l_k \text{ with } \{k, l\} = \arg \max_{\substack{k \in \{1, \ldots, K\} \\ l \in \{1, \ldots, N_k\}}} V^{\{\text{lin}, \text{RBF}\}}_k (\boldsymbol{\xi}^l_k). \tag{7}$$

Note that we run the reconstruction either with the scores from the linear SVM V_{lin} or with the scores from the RBF SVM V_{RBF}.

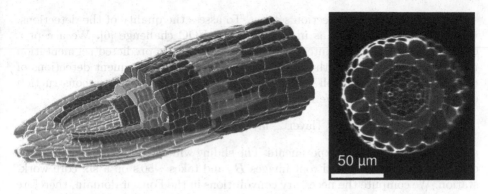

Fig. 3. Volume rendering and slice of the raw data. The root has a cylindrical structure and is made up of concentric layers of different cell types. In this paper, we consider cells from **layer 2 (blue)**, **layer 3 (green)**, and **layer 4 (red)** for the detection task (Color figure online).

4 Experiments

In this section we show a quantitative and a qualitative evaluation of the effectiveness of the proposed validation approach. We had three Arabidopsis roots with ground truth segmentations available: r06, r14, and pi005. The generation of the ground truth is very time consuming, as each root contains ~2500 cells. The ground truth segmentations are obtained by manually checking segmentations from a watershed algorithm on enhanced data [9]. For each root, we trained rigid detectors D_k and validation classifiers V_k for the cell layers 2, 3, and 4 (Fig. 3). We used a *round robin* scheme (Table 1), such that each root takes every role (training, validation, test) once. For the training of the rigid detectors we only used the *training* root. We always split the data into $k = 15$ clusters, as this value has proven to be good for layer 3 [10]. We do not perform a mining of hard negative examples for the detectors, as it did not improve the average precision of the results.

Table 1. Round robin scheme for training and testing.

Training	Validation	Test
r06	pi005	r14
r14	r06	pi005
pi005	r14	r06

As the rigid detector returns virtually no false positives when tested on the training root, we mine the training examples for the validation classifier V_k on the *training* and the *validation* root. We train a linear support vector machine V_k^{lin} and a RBF support vector machine V_k^{RBF} using *libsvm* [4].

Our test setup is a detection setting. To assess the quality of the detections, we use the same method as in the PASCAL VOC challenge [6]. We accept a detection as valid, iff the intersection over union of the predicted segmentation mask with the ground truth segmentation is ≥ 0.5. All subsequent detections of the same ground truth cell that are not suppressed during the reconstruction count as false positives. We investigate all combinations

$$\{r06, r14, pi005\} \times \{layer\ 2, layer\ 3, layer\ 4\} \times \{D_k, V_k^{lin}, V_k^{RBF}\}$$

and thus end up with 27 experiments. The sliding window detection is performed on the rotation normalized root images B_j and takes ~50 s on a six core workstation. We compute the necessary convolutions in the Fourier domain, therefore the runtime is not dependent on the size of the detection filter D_k. The alignment of a cell template to the image data is computed with the combinatorial registration from our previous work [10], using a gradient orientation based data term. The computation of one alignment takes ~1.5 s. The scoring of the aligned cell templates with the validation classifier takes <0.1 s. These steps are nearly perfectly parallelizable. When executed on a computing cluster with 5×32 cores the detection for a whole root takes ~5 min, the alignment of the cell templates in average ~20 min, depending on the number of cell hypotheses. The limiting factor with our setup was the hard disk IO.

The iterative greedy reconstruction is more difficult to parallelize and takes in average ~30 min, also dependent on the number of aligned cell candidates. For some more statistics of the roots, have a look at Table 2.

Table 2. Statistics for the roots ("GT" = ground truth).

	r06	r14	pi005
Size (voxels)	$1030 \times 433 \times 384$	$944 \times 413 \times 360$	$855 \times 458 \times 329$
Layer 2 #GT cells	542	487	554
Layer 3 #GT cells	216	188	208
Layer 4 #GT cells	266	211	222
B_j arrangement	3×36	3×36	2×36
B_j size (voxels)	$301 \times 101 \times 131$	$301 \times 101 \times 131$	$301 \times 101 \times 131$

Our findings are summarized in Fig. 4 as precision-recall graphs and in Table 3 as the mean average precision (øAP) per cell layer. The average precision is computed as the area under the precision-recall curve. Our original processing pipeline (cyan curve) works reasonably well for layer 2 and layer 3 with øAP = 0.71 and øAP = 0.82 respectively. It fails for layer 4 with øAP = 0.52. The reason for this can be found in the less distinctive cell shapes of this layer and in its location within the root. For the volumetric recording of layer 4, the light has to pass layers 1, 2, and 3 during the recording with a confocal microscope, which results in a more distorted signal.

Table 3. Mean average precision (øAP) for scoring strategy and cell layer.

	Layer 2	Layer 3	Layer 4
Raw detector score	0.71	0.82	0.52
Linear SVM	0.86	**0.87**	0.80
RBF SVM	**0.88**	0.86	**0.83**

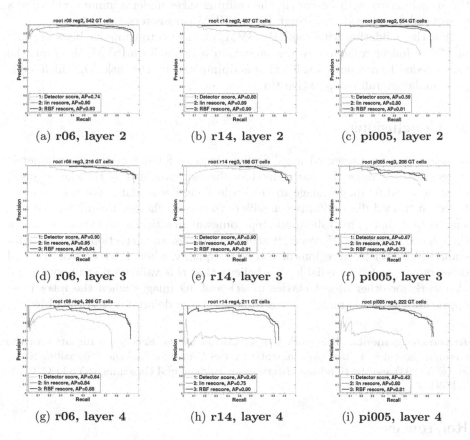

(a) **r06, layer 2** (b) **r14, layer 2** (c) **pi005, layer 2**

(d) **r06, layer 3** (e) **r14, layer 3** (f) **pi005, layer 3**

(g) **r06, layer 4** (h) **r14, layer 4** (i) **pi005, layer 4**

Fig. 4. Precision-Recall curves for cell layer reconstructions of the roots r06, r14, and pi005 organized in columns, e.g. root r06: (a), (d), (g) and cell layers 2, 3, and 4 organized in rows, e.g. layer 2: (a), (b), (c). **(cyan curve)** Reconstruction based on the detector scores. **(black curve)** Reconstruction based on the scores after validating the aligned cell templates with a linear SVM. **(magenta curve)** Reconstruction based on the scores after validating with an RBF SVM. All settings benefit from the better scores produced by the validation. The benefit is the biggest for layer 4, as the cells in this layer have the least distinctive cell shape and the data quality is worst for this layer. (d) is the configuration examined in our previous work [10].

When performing the greedy reconstruction based on the scores of the proposed validation approach, we yield substantially better results for the difficult layer 4. For an illustration see Fig. 2. For every other root and layer combination we also achieve better results through the validation scores. The linear SVM based scores (black curve) achieve the best reconstructions on layer 3. With the RBF SVM based scores (red curve), we achieve the best reconstructions for layer 2 and layer 4. The performance of the linear scoring and the RBF scoring is very similar, maybe with a slight edge for the RBF based rescoring. The training times of the SVMs are practically identical. When training directly on the kernel matrix with *libsvm* [4], the training takes under a minute including a cross validation based grid search for the SVM parameters γ and C.

For the validation with the RBF SVM, one needs to compute between $50\times$ and $200\times$ longer compared to the validation with the linear SVM. However, the time needed to compute a validation is dominated by the disk IO, which leads to a similar overall computation time.

5 Conclusions

In this paper we presented a validation strategy for detections in volumetric images that leverages the fine grained localization provided by the elastic alignment of a cell template image to the underlying image data. We use a metric based on trained discriminative classifiers to decide whether this alignment was successful or not. This validation step comes at practically no extra cost given the aligned detections. However, it achieves to boost the detection accuracy substantially, especially in regions of lower data quality, where the scores of the rigid detector are no longer reliable. We believe that this validation strategy should also work for other object classes in 2D and 3D images when the intra class appearance variation mainly stems from an elastic deformation of the objects.

Acknowledgements. This work was supported by the Excellence Initiative of the German Federal and State Governments: BIOSS Centre for Biological Signalling Studies (EXC 294) and the Bundesministerium für Bildung und Forschung (SYSTEC, 0101-31P5914).

References

1. Bourdev, L., Maji, S., Brox, T., Malik, J.: Detecting people using mutually consistent poselet activations. In: Daniilidis, K., Maragos, P., Paragios, N. (eds.) ECCV 2010, Part VI. LNCS, vol. 6316, pp. 168–181. Springer, Heidelberg (2010)
2. Brox, T., Bourdev, L., Maji, S., Malik, J.: Object segmentation by alignment of poselet activations to image contours. In: IEEE International Conference on Computer Vision and Pattern Recognition (CVPR) (2011)
3. Carreira, J., Sminchisescu, C.: CPMC: automatic object segmentation using constrained parametric min-cuts. IEEE Trans. Pattern Anal. Mach. Intell. **34**(7), 1312 (2012)

4. Chang, C.C., Lin, C.J.: LIBSVM: a library for support vector machines. ACM Trans. Intell. Syst. Technol. **2**, 27:1–27:27 (2011)
5. Dalal, N., Triggs, B.: Histograms of oriented gradients for human detection. In: Schmid, C., Soatto, S., Tomasi, C. (eds.) International Conference on Computer Vision & Pattern Recognition (2005)
6. Everingham, M., Van Gool, L., Williams, C.K.I., Winn, J., Zisserman, A.: The pascal visual object classes (voc) challenge. Int. J. Comput. Vis. **88**(2), 303–338 (2010)
7. Fernandez, R., Das, P., Mirabet, V., Moscardi, E., Traas, J., Verdeil, J., Malandain, G., Godin, C.: Imaging plant growth in 4D: robust tissue reconstruction and lineaging at cell resolution. Nat. Methods **7**(7), 547–553 (2010)
8. Komodakis, N., Tziritas, G., Paragios, N.: Performance vs computational efficiency for optimizing single and dynamic MRFS: setting the state of the art with primal-dual strategies. Comput. Vis. Image Underst. **112**(1), 14–29 (2008)
9. Liu, K., Schmidt, T., T.Blein, Dürr, J., Palme, K., Ronneberger, O.: Joint 3D cell segmentation and classification in the Arabidopsis root using energy minimization and shape priors. In: IEEE International Symposium on Biomedical Imaging (ISBI) (2013)
10. Mai, D., Fischer, P., Blein, T., Dürr, J., Palme, K., Brox, T., Ronneberger, O.: Discriminative detection and alignment in volumetric data. In: Weickert, J., Hein, M., Schiele, B. (eds.) GCPR 2013. LNCS, vol. 8142, pp. 205–214. Springer, Heidelberg (2013)
11. Marcuzzo, M., Quelhas, P., Campilho, A., Mendonça, A.M., Campilho, A.: Automated arabidopsis plant root cell segmentation based on SVM classification and region merging. Comput. Biol. Med. **39**(9), 785–793 (2009)
12. Schmidt, T., Pasternak, T., Liu, K., Blein, T., Aubry-Hivet, D., Dovzhenko, A., Dürr, J., Teale, W., Ditengou, F.A., Burkhardt, H., Ronneberger, O., Palme, K.: The irocs toolbox - 3D analysis of the plant root apical meristem at cellular resolution. Plant J. **77**(5), 806–814 (2014). http://lmb.informatik.uni-freiburg.de//Publications/2014/SLBR14
13. Wu, G., Jia, H., Wang, Q., Shen, D.: Sharpmean: groupwise registration guided by sharp mean image and tree-based registration. NeuroImage **56**(4), 968–1981 (2011)

A Human Factors Study of Graphical Passwords Using Biometrics

Benjamin S. Riggan[1]([✉]), Wesley E. Snyder[1], Xiaogang Wang[2], and Jing Feng[1]

[1] North Carolina State University, Raleigh, USA
{bsriggan,wes,jing_feng}@ncsu.edu
[2] US Army Research Office, Durham, USA
cliff.x.wang.civ@mail.mil

Abstract. One mode of authentication used in modern computing systems is graphical passwords. Graphical passwords are becoming more popular because touch-sensitive and pen-sensitive technologies are becoming ubiquitous. In this paper, we construct the "BioSketch" database, which is a general database of sketch-based passwords (SkPWs) with pressure information used as a biometric property. The BioSketch database is created so that recognition approaches may be commensurable with the benchmark performances. Using this database, we are also able to study the human-computer interaction (HCI) process for SkPWs. In this paper, we compare a generalized SKS recognition algorithm with the Fréchet distance in terms of the intra/inter-class variations and performances. The results show that the SKS-based approach achieves as much as a 7 % and 17 % reduction in equal error rate (EER) for random and skilled forgeries respectively.

1 Introduction

In recent years, the explosion of the smart-device market has motivated the development of new security measures. Improved methods of security are necessary because of increasing phishing, spoofing, and brute force attacks on new and existing technologies. One alternative is graphical passwords, specifically sketch-based passwords (SkPWs). SkPWs are well-suited for accessing secured information on smart-devices, web-based systems, and local computing systems, many of which already include hardware for capturing raw sketch data.

There is some debate over the reliability of SkPWs, which primarily stems from the absence of standardized testing of such systems. Biddel et al. [4] claim that graphical passwords alone do not overcome the problems of text-based passwords, but evidence in [4,11,13] also suggests that next generation systems are capable of making graphical passwords more practical. Therefore, we provide a baseline for future performance.

In this paper, our goals are to construct a comprehensive database of sketches with biometric information and to provide a human factors analysis for recognition of SkPWs. In particular, the database and experiments in this work are used

© Springer International Publishing Switzerland 2014
X. Jiang et al. (Eds.): GCPR 2014, LNCS 8753, pp. 464–475, 2014.
DOI: 10.1007/978-3-319-11752-2_38

to further extend the novel recognition approach in [13] (discussed in Sect. 4.1). To our knowledge, there is no existing database of sketches containing biometric information such as pressure. Using this database, we also demonstrate that SkPWs are a secure, yet usable, means of protecting critical information. We also aim to experimentally show the human potential for reproducing a SkPW with sufficient accuracy and with a reduced impact on security.

The remainder of the paper is organized as follows. First, alternative approaches for recognizing sketches are discussed in Sect. 2. Section 3 presents the construction of the database, including the experimental design and information collected from participants. Section 4 provides a detailed analysis of the human-computer interaction (HCI) for SkPWs, including dissimilarity/similarity measures, drawing variations, and performance results. Lastly, conclusions from a human factors study are given in Sect. 5.

2   Graphical Passwords

Graphical passwords fall into one of three categories: drawmetric, cognometric, or locimetric, as suggested in [5]. Drawmetric, or recall-based, passwords require a user to remember a drawing as a password. Cognometric systems are recognition based, meaning that a user must recognize and select a correct set of images which are displayed among random images. Locimetric, or cued-recall, passwords require users to select a correct set of locations within one or more images. In this paper, we focus on understanding the HCI problem for SkPWs using biometric pressure. Therefore, we briefly review other drawmetric passwords.

Drawmetric passwords involve the combination of multiple complex processes, including:

- **Synthesis.** The initial creation (or visualization) of the "password."
- **Recall.** The process of remembering the drawing.
- **(Re)production.** The physical drawing process during enrollment or login.

Each of these processes are greatly influenced by the HCI, i.e. *human factor*.

One drawmetric system, called Draw-A-Secret (DAS) [9], uses a grid to map the drawing to a string that is matched like a text-based password. Each cell is represented by a number, and the string is the sequence of numbers corresponding to the grid cells that the drawing enters. If any single number in a sequence generated at login time does not match that in the string stored during password enrollment, then access is denied. Therefore, the only difference between text-based passwords and DAS is the fact that the string is generated from a sketch.

There have been some improvements to DAS using background images [6], a finer grid [7], an invisible grid, stroke color [8,17], and grid intersections [16]. However, the basis of the approach remains the same.

Another drawmetric system [11,12] uses Dynamic Time Warping (DTW), an approach used in signature recognition, and reports equal error rates (EERs) as low as 2.7 % and 28.0 % for random and skilled forgeries of doodles (or sketches). DTW is a generalized approach for finding the optimal set correspondences between two signals, e.g. signatures or drawings, using dynamic programming [3]. Using a dissimilarity measure (a distance) to quantify the differences between local features, e.g. relative position, velocity, acceleration, etc., DTW aligns the two drawings by finding the constrained alignment path resulting in the minimum cumulative dissimilarity. The constraints usually include boundary conditions and monotonic constraints. One disadvantage of DTW is the bias due to the constraints, which potentially degrade performance.

3 Biometric Sketch Database

In this section, the Biometric Sketch (BioSketch) database[1] is discussed, including the purpose of the database, the type of data collected, and how it was collected.

The BioSketch database provides a collection of SkPWs with biometric pressure and temporal features. The problem with SkPWs is the lack of a sufficiently large and complete dataset to adequately compare performance between methods. The DooDB database [11,12] comes close, but does not measure biometric pressure because the developers used a resistive[2] touchscreen.

The BioSketch database is designed to capture the following local properties:

- Spatial Coordinates—(x, y)
- Pressure—b
- Time—t.

Other local features may be derived directly from these properties, e.g. relative distances, curvature, velocity, acceleration, etc. Therefore, the BioSketch database is designed to be more general than the DooDB database.

The database was constructed using a Samsung Galaxy Note 10.1, which used a software application that was developed for the sole purpose of collecting sketches from participants.

3.1 Participant Study

The study was designed in collaboration with Jing Feng, a professional cognitive psychologist, and *Institutional Review Board* (IRB) approval was obtained.

The overall experimental design for collecting the data is shown in Fig. 1. The experiment is designed to account for comfort, fatigue, stability, and short/long term recall. The experiment consists of two nearly identical sessions, each composed of practice and data collection phases.

During session 1, the user is required to type his/her username for purposes of identification. The practice phase allows users to become comfortable drawing on the tablet. Every user is allowed the same amount of practice time; writing initials (5×) and numbers 0–5 (5× each). Sufficient practice also allows users

---

[1] http://www.ece.ncsu.edu/imaging/biosketch_form.html
[2] Resistive touchscreens cannot measure pressure.

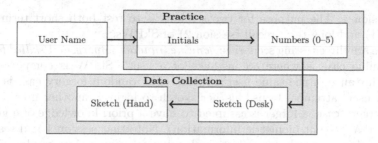

Fig. 1. Block diagram of data collection experiment. The experiment consists of two phases: practice and data collection.

that have not used a Samsung Galaxy Note 10.1 (or similar device) to become familiar with it for the purposes of this experiment. The data collection phase is where the sketches for the BioSketch database are obtained. Users are first asked to consider a SkPW.

For the purposes of this study, we asked that a sketch: *(1) only consists of a single uninterrupted stroke, (2) consists of more than a single line segment or circle, (3) not be a signature,* and *(4) be perceptively reproducible (to the user).* These conditions ensure that all sketches have similar complexity.

Although the matching algorithm we use is capable of extension to multiple strokes, reproducibility decreases with the number of strokes. Therefore, we allow the user to draw *almost* anything as long as they can draw it in a single stroke. Then, we asked that the sketch be more complex than a single line segment or a circle because these are similar to using "password" as a text-based password or "1234" as a PIN. We also asked users not to use their signature because signatures are known to have a higher intra-class variation than simple sketches. For example, consider a simple shape (e.g. a fish) and a signature. The complexity of each may be described by the number of self-intersections. The number of intersections for a fish is low (probably only one intersection) and for a signature the number of intersections is much higher. Lastly, we simply ask that every user be sufficiently confident in their ability to reproduce the drawing. For example, we prefer that participants do not draw a Chinese character or symbol unless they already know Chinese. These restrictions only attempt to control and limit the scope of the experiment.

A total of 20 sketches are collected from every user; 10 drawn with the tablet lying flat on the desk (more stable), and 10 drawn with the tablet in hand (less stable). In order to account for the effects of fatigue, users are asked to draw at most 5 consecutive sketches.

Session 2 is practically identical to the first session. Users are required to refamiliarize themselves with the tablet using practice time, and then 20 more sketches are collected; 10 more stable and 10 less stable. However, the only difference is that they are asked to *recall* the sketch they produced during the first session. Users were asked to return for the second session between 2–10 days after session 1. The purpose for two sessions is to test both short term recall (session 1) and long term recall (session 2) of SkPWs.

There are three testing scenarios: *genuine*, *random forgery*, and *skilled forgery*. The genuine case is where an instance of a user's SkPW is compared with another instance of the same user's SkPW. The random forgery case is where a second users attempts to use his/her sketch to login as another user. For the skilled forgery case, a forger is assumed to have a priori knowledge of a genuine user's SkPW (except biometric information). Note that session 1 and session 2 described above are only for collecting sketches for genuine and random forgery cases (see examples in Fig. 2); skilled forgeries are collected during a separate session from a different set of users.

In order to obtain skilled forgeries, selected forgers were presented with 10 examples of a particular user's SkPW (Fig. 3). Each example provides the forger with the sketch, start point, and end point. While forging a sketch, skilled forgers were allowed to re-draw any attempts that were poor replications, ensuring the sketches are actually "skilled."

Fig. 2. Subset of sketches

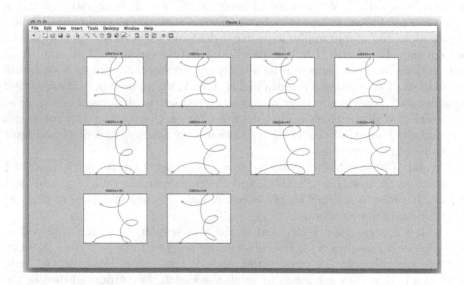

Fig. 3. Example window shown to a forger for skilled forgeries.

3.2 Participants

The participants consisted of different genders, handedness, ages, and backgrounds. In total, there were 35 participants in session 1; 23 in session 2. Table 2 summarizes the number of male/female and left/right handed participants from each session. Additionally, another 15 participants were selected to provide skilled forgeries of sketches from session 1.

4 Human Factor Analysis

In this section, the reproducibility of a sketch by the average person is studied. The inter-class and intra-class variations provide the most insight regarding the human reproducibility of sketches. However, the problem is determining how to measure these differences. Ideally, we want to study these variations independent of a method, but this is difficult because a model must be used to quantify the differences between sketches. There are many measures, such as the Fréchet distance and *Simple K-space* (SKS) similarity measure, which have been used to quantify the differences between parameterized curves. These two measures are discussed in detail, and then we provide the variability analysis and benchmark performances below.

4.1 Similarity Measures

The Fréchet and SKS measures (a.k.a. scores) are computed using a specific set of features, which describe certain properties such as shape, trajectory, biometrics, and parameterization. which may be computed from the measurements collected in Sect. 3. The features for a sketch α are denoted by the vector $v_\alpha(s) = (\rho(s), \kappa(s), \theta(s), b(s), s)^T$, where $\rho(s) = \|\alpha(s) - x_0\|$ (a distance)[3]; $\kappa(s) = |\alpha''(s)|$ (local curvature); $\theta(s)$ is angle of the tangent vector with respect to $+x$ axis (drawing direction); $b(s)$ is the biometric feature (pressure); and s is arc length (order). These specific features are used because SkPWs are more than just shape, they are distinguishable by shape (e.g. distance and curvature) and how they are drawn (e.g. direction and order). Then, the addition of the biometric feature provides a way to verify whether or not the SkPW is drawn in a similar manner, indicating that the same person provided it. For more details see [13].

The *Fréchet distance* [1,2] is used for measuring the dissimilarity between parameterized curves, or in this case sketches. This distance is commonly described using the analogy of person walking a dog with a leash. The Fréchet distance represents the shortest leash required when the person walks forward along a designated path at some speed, and the dog walks forward along another path at another speed. The Fréchet distance is defined as

$$d_F(\alpha, \beta) = \inf_{f,g} \max_s d(f \circ v_\alpha(s), g \circ v_\beta(s)) \tag{1}$$

---

[3] The distance is with respect to x_0, which represents an arbitrary reference point (usually the centroid).

where f and g represent any monotonic functions, which are used to reparameterize sketches α and β respectively; and $\circ$ denotes the function composition operator.

SKS [10,15] was originally developed to match 2-dimensional shapes and images. It is well suited to that application because it is invariant to similarity transforms. Fundamentally, the motivating idea driving SKS is accumulation of similarity evidence, where the term *similarity* implies some measure of consistency between a model and a SkPW. The extension of SKS to SkPWs [13] is briefly explained below.

Given a template SkPW, β, SKS first constructs a model representation, given by:

$$m(v) = \int_0^1 \delta(v - v_\beta(s)) \, ds, \qquad (2)$$

where

$$\delta(z) = \exp\left(-\frac{1}{2}z^T \Sigma^{-1} z\right). \qquad (3)$$

Equation 2 may be considered similar to a density estimation [14] of the properties from the SkPW.

Then, the consistency between another SkPW, α, and the density-like model comes from the accumulator:

$$A(x) = \int_0^1 m(v_\alpha(s; x)) \, ds, \qquad (4)$$

where $v_\alpha(s; x) = (\|\alpha(s) - x\|, \ldots, s)^T$. The accumulator in Eq. 4 represents a path integral over the model for β, where the path is defined by the properties of the SkPW β. If the "passwords" are indeed consistent, then a peak should occur in the accumulator. Otherwise, no significant peaks occur. The amplitude of this peak represents the level of consistency. If sufficiently consistent, then similarity between the SkPWs is inferred.

Therefore, SKS is summarized by the following equation:

$$L(\alpha, \beta) = \max_x \left\{ \int_0^1 \int_0^1 \delta(v_\alpha(s; x) - v_\beta(\hat{s})) \, ds \, d\hat{s} \right\}. \qquad (5)$$

The likelihood L, in Eq. 5, by definition is not symmetric. However, a new *symmetric* SKS measure is defined as follows:

$$L_{sym}(\alpha, \beta) = L(\alpha, \beta) + L(\beta, \alpha), \qquad (6)$$

which can be shown to be a semi-metric.

4.2   Variability Analysis

Next, we study inter-class and intra-class variations to better understand the HCI for SkPWs. These variations are analyzed using both Fréchet and SKS measures. The variations are intended to show the differences between a "matching" and "non-matching" sketch. The differences between quantitative similarity and human perception of similarity motivate us to study theses variations.

The BioSketch database is composed of sketches from two sessions, each composed of two sets of sketches: session 1 with tablet flat on the desk (D1), session 1 with tablet in hand (H1), session 2 with tablet flat on the desk (D2), and session 2 with tablet in hand (H2). For each of these sets, we histogram the scores for genuine sketches, random forgeries, and skilled forgeries.

The histograms (Fig. 4) are used to illustrate the differences between genuine sketches, random forgeries, and skilled forgeries for D1; histograms for H1, D2, H2 are not shown, but are very similar to those shown above. Genuine scores are computed using $d_F(\alpha, \beta)$ and $L_{sym}(\alpha, \beta)$ respectively, where α and β are two distinct sketches drawn by the same person. Random forgery scores are also computed using Eqs. 1 and 6, but α and β are sketches drawn by different users (from session 1 and session 2). α and β are also drawn by different individuals for skilled forgeries, however, skilled forgers attempt to accurately replicate a genuine user's sketch using a priori knowledge of the SkPW.

In Fig. 4, we are interested in the random forgery overlap, skilled forgery overlap, and histogram features (e.g. tails, peaks, and variance). The amount of overlap between the genuine scores and random forgery scores is very small for both Fréchet and SKS, and the overlap between the genuine and skilled forgeries is greater. When comparing the Fréchet distance and SKS for D1 (and other sets) it is evident that the amount of overlap is generally less for SKS. Now, consider the features of the histograms. The histogram of the genuine scores for d_F exhibits a tail, which indicates the presence of outliers in the data. The peaks in the histograms indicate the most probable scores for genuine sketches and forgeries. Notice that peaks for the genuine sketches and skilled forgeries occur nearly at the same location using d_F, and they are further apart using L_{sym}. Finally, compare the variances of the genuine and forgery histograms: the

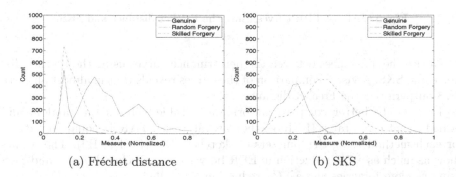

(a) Fréchet distance   (b) SKS

Fig. 4. The plots show the genuine, random forgery, and skilled forgery distributions for D1. Recall that the Fréchet distance (SKS) is a dissimilarity (similarity) measure, which means that genuine sketches have lower (higher) scores.

genuine variance for the Fréchet distance is smaller than that for SKS, which indicates that Fréchet has a more restricted[4] definition of similar than SKS.

A fundamental problem with comparing and analyzing SkPWs is knowing the ground truth. In this paper, we assume that if a person actually drew the sketch, then it is similar to all other sketches he/she provided. However, people are not perfect when it comes to drawing. In fact, from the histograms we observed the presence of outliers in the set of presumably genuine sketches. Therefore, the assumption made is not ideal, but it is the method most commonly used in various pattern recognition problems.

4.3   Benchmark Performance

In most systems, performance is measured by the false acceptance rate (FAR)—percentage of forgeries wrongly considered to be genuine—versus the false rejection rate (FRR)—percentage of genuine sketches considered to be a forgery. Therefore, the performance curves are shown for both Fréchet and SKS methods (Fig. 5).

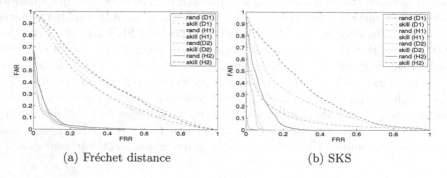

(a) Fréchet distance   (b) SKS

Fig. 5. FAR vs. FRR curves using the Fréchet distance and SKS.

Notice the differences between the performance curves using the Fréchet distance and SKS. A visual comparison of the curves reveals that in almost all cases SKS outperforms the Fréchet distance.

In order to better compare performance, Table 1 shows the random and skilled equal error rates (EERs)—the rate at which FAR is equal to FRR—for each method using the four sets of sketches (D1, H1, D2, H2). The results show as much as a 7 % reduction in EER between the Fréchet distance and SKS using random forgeries and a 17 % reduction for skilled.

---

[4] SKS may be more robust than the Fréchet distance because it has broad definition of *similar* and a more narrow definition of *different*.

Table 1. Summary of EER (%) performances using the Fréchet distance and SKS.

Session	Method	EER (Random)	EER (Skilled)
D1	Fréchet	9.05	34.36
	SKS	**1.74**	**16.75**
H1	Fréchet	9.76	34.73
	SKS	**6.85**	**19.10**
D2	Fréchet	9.97	39.55
	SKS	**5.72**	**27.86**
H2	Fréchet	**11.22**	39.96
	SKS	14.57	**34.52**

Table 2. Participant demographics.

Session	Males	Females	Right	Left
1	20	15	32	3
2	13	10	20	3

Table 3. Survey statistics.

Question	Min	Max	Mean	Median
Comfort	5.0	10.0	9.2	10.0
Consistency	3.0	10.0	7.6	8.0
Usability	1.0	10.0	8.2	9.0

4.4 Survey

After collecting the sketches from each user, participants filled out a survey regarding the proposed graphical password system and human perceptions. The survey included questions regarding user comfort (when drawing on a tablet), consistency (as perceived by the user), and usability. The user indicated a number between one and ten, where one represents very little (if any at all) comfort, consistency, or usability and ten represents the most comfort, consistency, or usability. Table 3 provides the minimum, maximum, mean, and median of all user responses. The responses show that most users are extremely comfortable in drawing on the device. There is less user confidence in terms of their own consistencies. Some believe they are fairly consistent, while others believe they are not very consistent. Naturally there is some variation between users in terms of usability, however, on average more people tend to like the idea using a sketch as a password. These results indicates that a majority users are willing to accept the use of drawings for the purposes of security.

5 Conclusions

Availability of the BioSketch database now allows other researchers to make adequate and meaningful comparisons of alternative methods, features (e.g. velocity, acceleration, etc.), and computing devices (e.g. resistive or capacitive touch screens). The variations between genuine sketches and forgeries (random and skilled) provide substantial evidence that people are highly capable of reproducing a sketch with local biometric pressure, and possibly other biometrics too.

Lastly, benchmark performance demonstrated that the Fréchet distance and SKS measures are both useful measures for matching SkPWs, however, SKS appears to perform slightly better in most cases.

SKS can also be extended to higher spatial dimensions, such as 3D curves, 3D objects, and video sequences. The most obvious application for 3D contours is gesture recognition using a device like Microsoft's Kinect. The main idea is to use a person's hand(s) to generate a complex gesture in 3-space, which is represented using a parameterized curve. Similarly, SKS could be extended to detect not only images, but 3D representations of objects. For example, we could construct a model where the amplitude at each point represent the likelihood that there exists a point some distance from the reference and has some mean curvature (opposed to curvature in the plane).

Acknowledgements. The information in this paper is based on work partially funded by the United States Army Research Office (ARO) grant W911NF-04-D-0003-0019.

References

1. Agarwal, P.K., Avraham, H.K., Kaplan, H., Sharir, M.: Computing the discrete freéchet distance in subquadratic time. In: Proceedings of the 24th Annual ACM-SIAM Symposium on Discrete Algorithms, pp. 156–167 (2013)
2. Alt, H., Godau, H.: Computing the fréchet distance between two polygonal curves. Int. J. Comput. Geom. Appl. **5**(1–2), 75–91 (1995)
3. Bellman, R.: Dynamic Programming. Princeton University Press, Princeton (1957)
4. Biddle, R., Chiasson, S., Van Oorschot, P.C.: Graphical passwords: learning from the first twelve years. ACM Comput. Surv. **44**(4), 1–25 (2012)
5. De Angeli, A., Coventry, L., Johnson, G., Renaud, K.: Is a picture really worth a thousand words? Exploring the feasibility of graphical authentication systems. Int. J. Hum. Comput. Stud. **63**(1–2), 128–152 (2005)
6. Dunphy, P., Yan, J.: Do background images improve "draw a secret" graphical passwords? In: Proceedings of the 14th ACM Conference on Computer and Communications Security, pp. 36–47 (2007)
7. Gao, H., Guo, X., Chen, X., Wang, L., Liu, X.: Yet another graphical password strategy. In: Proceedings of the Annual Computer Security Applications Conference, pp. 121–129 (2008)
8. Goldberg, J., Hagman, J.: Doodling our way to better authentication. In: Proceedings of the ACM Conference on Human Factors in Computing Systems, pp. 868–869 (2002)
9. Jermyn, I., Mayer, A., Monrose, F., Reiter, M., Rubin, A.: The design and analysis of graphical passwords. In: Proceedings of the 8th USENIX Security Symposium, pp. 1–14 (1999)
10. Krish, K., Heinrich, S., Snyder, W.E., Cakir, H., Khorram, S.: Global registration of overlapping images using accumulative image features. Pattern Recogn. Lett. **31**, 112–118 (2010)
11. Martinez-Diaz, M., Fierrez, J., Galbally, J.: The DooDB graphical password database: data analysis and benchmark results. IEEE Access **1**, 596–605 (2013)

12. Martinez-Diaz, M., Fierrez, J., Martin-Diaz, C., Ortega-Garcia, J.: DooDB: a graphical password database containing doodles and pseudo-signatures. In: 12th International Conference on Frontiers in Handwriting Recognition, pp. 339–344 (2010)
13. Riggan, B.S.: Recognition of sketch-based passwords with biometric information using a generalized simple K-space model. Ph.D thesis/Dissertation, North Carolina State University (2014)
14. Rousson, M., Cremers, D.: Efficient kernel density estimation of shape and intensity priors for level set segmentation. In: Duncan, J.S., Gerig, G. (eds.) MICCAI 2005. LNCS, vol. 3750, pp. 757–764. Springer, Heidelberg (2005)
15. Snyder, W.E.: A strategy for shape recognition. In: Srivastava, A. (ed.) Workshop on Challenges and Opportunities in Image Understanding, College Park, MD, January 2007
16. Tao, H., Adams, C.: Pass-go: a proposal to improve the usability of graphical passwords. Int. J. Netw. Secur. 7(2), 273–292 (2008)
17. Varenhorst, C.: Passdoodles: a lightweight authentication method. MIT Research Science Institute, July 2004

Pedestrian Orientation Estimation

Joe Lallemand[1,2]([✉]), Alexandra Ronge[1], Magdalena Szczot[1],
and Slobodan Ilic[2,3]

[1] BMW Group, Munich, Germany
{Joe.Lallemand,Alexandra.Ronge,Magdalena.Szczot}@bmw.de
[2] Computer Aided Medical Procedures,
Technische Universität München, Munich, Germany
Slobodan.Ilic@in.tum.de
[3] Siemens AG, Munich, Germany

Abstract. This paper addresses the task of estimating the orientation of
pedestrians from monocular images provided by an automotive camera.
From an initial detection of a pedestrian, we analyze the area within
their bounding box and give an estimation of the orientation. Using
ground truth mocap data, we define the orientations as a direction and
a rough human pose. A random forest classifier trained on this data
using HOG features assigns each detected pedestrian to their orientation
cluster. Evaluation of the method is performed on a new dataset and on
a publicly available dataset showing improved results.

1 Introduction

Road accidents involving pedestrians still account for the highest fatality to
injury rate throughout the world. In 2012, 3 % of all accidents in the United
States involved pedestrians while 13 % of all road fatalities consisted of pedestrians[1]. In Europe, the pedestrians constitute 20 % of all road fatalities[2] and
even 26 % in China [20]. For several years now, automotive companies have been
developing driver assistance systems addressing pedestrian safety. These systems
identify dangerous situations for pedestrians and support the driver in avoiding accidents, or decreasing the speed with which an impact occurs. Pedestrian
detection is the most crucial task for enabling these driver assistance systems.
Most pedestrian detectors, however, only contain information about the relative
position of pedestrians and by tracking additionally estimate the relative velocity. In order to develop better systems, which can react earlier or even dodge the
pedestrians instead of only reducing the impact speed, more information about
the pedestrians is needed. A crucial information for those systems is the orientation of the pedestrians. In [8], Enzweiler et al. have proposed a pedestrian
detector which gives an orientation estimation using 4 directions, front, back,
left and right.

---

[1] http://www.nhtsa.gov/Pedestrians
[2] http://ec.europa.eu/transport/road_safety/index_en.htm

© Springer International Publishing Switzerland 2014
X. Jiang et al. (Eds.): GCPR 2014, LNCS 8753, pp. 476–487, 2014.
DOI: 10.1007/978-3-319-11752-2_39

In this work, a new method is proposed which goes one step further, not only predicting the orientation but also delivering a rough pose of the pedestrian. Using a dataset of pedestrians with ground truth body joint annotations acquired by a motion capture system, so called orientation clusters, see Fig. 1, are created which contain information about the orientation of a pedestrian as well as a rough estimation of their pose, e.g. walking or standing still or moving his arms. Apart from retrieving the orientation, this also serves as an important preprocessing step for determining the body pose of pedestrians [1,11]. The problem is addressed using classification random forests. Unlike previous works which rely mainly on SVMs [8,9,14], random forests can inherently handle multi-class problems and due to the iterative training scheme, at each node determine the best-sized HOG feature for the training samples. The first approach, called holistic classification consists of classifying the image cutout of each pedestrian as a whole by training a random classification forest based on HOG features computed at specific locations within the cutout. The second approach called pixel-based classification trains a random forest based on a subset of pixels from the detected pedestrian cutout. The contributions of this work are as follows: The formulation of the orientation clusters contains not only the orientations but also pose information about the pedestrian and can serve as a preprocessing step for human pose estimation. For the holistic approach, we simultaneously learn the best cutout locations and HOG feature sizes at each node and show a huge improvement over a fixed HOG size. The pixel-based formulation has not been used in other works and shows improved accuracy over the holistic approach but suffers from higher computational expanse during prediction.

2   Related Works

Much work has been conducted on the detection of pedestrians. In two recent surveys, an overview of pedestrian detection approaches is given [5,7]. Several approaches for orientation estimation depend on multiple cameras [10,13,18] and are thus not applicable in an automotive setup. Human pose estimation from depth sensors has become popular in the last years [6,15,16]. Here the extraction of a continuous orientation is trivial if the more complex problem of pose estimation has been solved. Unfortunately, all of these approaches rely on depth information from infrared sensors, which have a very limited operational range and do not work outdoors. Shimizu and Poggio [14] have proposed to estimate the direction of pedestrians in steps of $22,5°$ using Haar-features and training several SVMs using a *one vs. rest* scheme. Gandhi et al. [9] have proposed a system based on HOG features and *one vs. one* SVM training scheme to differentiate between 8 pedestrian orientations then use HMMs to predict future orientations. Unlike both previous methods, this work uses a random forest classifier which can inherently handle multi-class problems. In [8], Enzweiler and Gavrila have introduced a method for jointly detecting pedestrians and estimating their orientation using a probabilistic framework. The authors use four view-dependent clusters relating to the four directions front, back, left and right. Unlike Enzweiler and Gavrila, the orientation-related clusters are defined differently in this work, containing both orientation and pose dependent information which leads to a finer discrimination of pedestrian posture. In [19], Weinrich et al. have proposed an orientation estimation framework for detecting people and estimating their upper body orientation in a unified framework relying on SVM decision trees trained on monocular images. At each decision tree node, a linear SVM is trained to split the data. The upper body orientation is labeled from $-180°$ to $180°$ in steps of $45°$. Unlike Weinrich et al. which rely on predefined HOG features and extensive feature selection schemes, the algorithm in this work autonomously learns the optimal feature to split the training data at each node and dispenses the need to train an svm at each node.

In [2], Benenson et al. reevaluated standard procedures regarding feature computation for pedestrian detection. They propose to abandon the dense block computation within the HOG window, instead using rectangular blocks at sparsely sampled positions within the HOG windows. The training algorithm presented in this paper incorporates the use of sparsely samples HOG windows. Additionally unlike in [2] we can automatically learn the best positioning and size of those windows.

In [17], Tosato et al. have introduced a method for estimating human body and head orientation using Riemannian Manifolds, introducing a new feature called WARCO (weighted array of covariances). They have published a dataset for human orientation classification on which the method of this paper is evaluated.

3   Method

The objective of this work is to analyze the detected bounding box around the pedestrian to estimate his orientation. This section gives a detailed overview of the method. In Sect. 3.1 we shortly describe the training process for classification random forests [3]. In Sect. 3.3 the pixel-based approach is presented, which aims to learn a mapping from given pixels to orientations by analyzing the pixel's environment. In Sect. 3.4, the holistic approach aiming to learn the mapping to orientations by analyzing differences between images at specific locations is detailed. In the last part of the section, the differences between both approaches will be analyzed.

3.1   Random Forests

A random forest $\mathcal{T} = \{t_i\}_{i=1}^{T}$ is an ensemble of T decorrelated binary decision trees t_i. At each node of the decision tree, a splitting function $g_{\nu,\tau} \equiv f(\nu) > \tau$ divides the incoming samples Δ, consisting of image information and labels, into left and right subsets Δ_L and Δ_R. The function is defined using the feature ν and a threshold τ which achieve the highest information gain at the current node. The information gain $IG(\nu, \tau) = \frac{w_L}{w} H(\Delta_L) + \frac{w_R}{w} H(\Delta_R)$ is based on Shannon's entropy H computed for the histogram of class distributions of each subset:

Fig. 1. Training labels produced by running k-means on the ground truth data. It can be seen that the clusters represent one orientation as well as similar poses. From left to right, the clusters are defined as: forward, backward, left walking, right walking, forward both hands, left - arm movement, right - arm movement, forward - left arm movement, forward - right arm movement.

$$H\left(\Delta\right) = -\sum_{c=1}^{C} p_c * \log\left(p_c\right),\qquad(1)$$

where C denotes the number of classes and p_c the probability of class c. This is iteratively repeated at each node until either the maximum depth has been reached or the node contains less than a fixed number of samples or the information gain is lower than a given threshold. Once a leaf node is reached, the posterior distribution, which is a class distribution histogram, is stored. For further information on random forests, we refer the reader to [4] from Criminisi et al.

3.2   Ground Truth Class Generation

The orientation classes, which are used as ground truth for the forest training, are generated using mocap skeleton data. To determine the classes, the skeleton data for all frames of all people is translated into the same coordinate system centered at the pelvis joint. In this way, the orientation with regard to the camera is preserved, but it allows to use the k-means clustering algorithm [12] on the euclidean distance between the joint positions. As can be seen in Fig. 1 this results in clusters which are separated both with regard to orientations and body poses. K-means algorithm was executed for a number of clusters ranging from 5 to 50 clusters. The data has been analyzed and the 8 cluster separation was chosen by experimental evaluation. We manually post-processed 1 of the clusters which contained images of similar orientations but a high variation in poses, so k-means was reapplied to this cluster to separate the poses within it. Finally each image is assigned it's class based on the cluster it belongs to.

3.3   Pixel-Based Approach

As a first step for pixel-based training, we rescale all pedestrian cut-outs to a fixed height, which ensures that the algorithm learns the same features regardless of the position or distance of a pedestrian with regard to camera.

For pixel-based classification, the training dataset is defined as $\mathcal{S} = \{p_k, l_k\}_{k=1}^{M}$ with M the number of cutouts. For each cutout k, we further defined $p_k = \{x_i, y_i, c_k\}_{i=1}^{N}$ with N the number of pixels. (x_i, y_i) denotes the pixel location of each sample in the cutout c_k. l_k is the label associated to the pedestrian in the cutout c_k as described in Sect. 3.2. For each tree, a subset of pedestrian cutouts c_k from the full training dataset is chosen. For each of these cutouts, a fixed number of randomly sampled pixels (x_i, y_i) constitute the final samples p_k. The size of the training set is thus M cutouts $\times N$ pixels.

Training: At each node, the forest separates the data by testing a number of different features. Each feature is described by a HOG window height and width and is randomly sampled, allowing a maximum width and height equal to the cutout size. The HOG features are then computed for the different pixels, centered at their respective location. For each feature, 10 thresholds are tested to split the data, distributed equally between the minimum and maximum value of the feature dimension. The split which maximizes the information gain is chosen and the corresponding HOG width and height as well as threshold are stored at the node. This is done iteratively for the child nodes. Once a leaf node is reached, the normalized histogram of orientations of all incoming samples is stored.

Testing: To determine the orientation of an unknown pedestrian, N pixels are sampled from the cutout and sent through all T trained trees. At each node, the stored HOG feature is computed at the pixel location and sent left or right based on the stored threshold until a leaf node is reached and the normalized histogram of orientations is extracted. This means that in total $N \times T \times Depth$ features have to be evaluated. Finally all histograms are summed up and the maximum bin determines the orientation.

3.4   Holistic Approach

For holistic classification, a stricter notion of the pedestrian cutout is introduced. In addition to scaling the height of the cutout, the width is fixed and is centered to the middle of the detected bounding box as well. The training dataset is defined as $\mathcal{S} = \{c_k, l_k\}_{k=1}^{M}$. Each pedestrian cutout c_k is treated as a training sample with associated label l_k as described in Sect. 3.2, so in total there are M samples in the training set.

Training: At each node, the forest separates the samples by testing a number of different features. Each feature is described by a *HOG window height and width and a position within the pedestrian cutout.* For each sample, each HOG feature is computed at its given position, which is possible due to the fully fixed size of the cutout and the information gain is computed for 10 different thresholds. Having determined the maximum information gain, in each node the HOG size, the position in the cutout and the threshold are stored in the node. This process is iteratively repeated for the child nodes until a leaf node is reached which stores the normalized histogram of orientations.

Testing: To determine the orientation of an unknown pedestrian, the pedestrian cutout is sent through all trained trees T. At each node, the stored HOG feature is computed at the stored pedestrian location and sent left or right based on the threshold until a leaf node is reached and the normalized histogram of orientations is extracted. This means that in total $T \times Depth$ features have to be evaluated. Finally all histograms are summed up and the maximum bin determines the orientation.

3.5 Pixel-Based vs. Holistic Approach

Figure 2 illustrates the differences between both the pixel-wise and holistic approach. In the pixel-wise approach, the classification forest learns a mapping from pixels to orientation clusters, by analyzing the surrounding area of a pixel. In image (a) and (b) of Fig. 2 the black dots denote different training pixels. Each different HOG feature is centered at these pixels and analyzes a part of their surrounding. While traversing a tree, different parts around a pixel are analyzed to give an estimated orientation.

In the holistic approach, the forest learns a direct mapping from pedestrian cutouts to the orientation clusters. During training HOG features are computed at the same locations in all images and the forest learns how images belonging to different orientation clusters differ from one another in different regions of the cutout, as visualized in image (c) and (d) of Fig. 2. While traversing a tree,

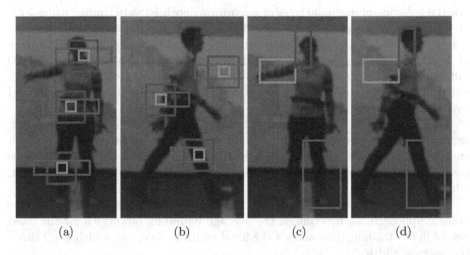

(a)   (b)   (c)   (d)

Fig. 2. The difference between pixel-wise and holistic approach. The red, green and blue rectangles represent the windows within which the HOG is computed. (a) and (b) show the pixel-based approach where all HOGs of different size are computed at the same location. This location is given by one training sample. Figures (c) and (d) show the holistic approach. The position within the image as well as the HOG window for testing are determined within the current node (Color figure online).

482 J. Lallemand et al.

different learned image locations are analyzed using a learned HOG feature to give an estimate of the orientation.

In terms of run-time, the holistic approach is much more efficient as detailed in the testing paragraphs of Sects. 3.3 and 3.4 because fewer features have to be computed. On the other hand, misclassifications from single trees have a much higher influence on the final prediction.

4 Results

4.1 Dataset

For this work, we propose a dataset recorded with a standard automotive camera synchronized to the ART motion capture system[3]. Twenty people both male and female were recorded where the actor is walking, running and taking sharp turns the same as pedestrians which can be encountered in everyday scenarios. For the recordings, the probands were wearing motion capture markers used to generate ground truth skeleton data. The dataset consists of nearly 250000 images after running a pruning step as described in [15], which means that all poses in the dataset are differing by at least 5 cm, for example static poses have been discarded.

4.2 Evaluation on the Proposed Dataset

For the holistic approach, 30 trees of maximum depth 10 were trained using 20 % of the training images. At each node 100 random positions and HOG window sizes were tested to find the optimal split. For the pixel-based approach, 10 trees of depth 18 were trained using 25 % of the training images. At each node 20 different HOG window sizes have been tested. The number of trees, tree depth and number of features evaluated per node have experimentally been determined by performing a grid search.

Evaluation on the proposed dataset is done in two different manners. First, we analyze how well our method is able to separate the proposed clusters which encode an orientation and similar poses. In the second step, we follow the approach proposed by many related works only using four directions: front, back, left and right.

All pedestrian bounding boxes are re-sized to a height of 96 pixels. For the holistic approach, we determined the average bounding box width for the fixed height in our training data and set the fixed width to 58 pixels, which is 1.2 times the average width.

Figures 3 and 4 show the confusion matrices of the cluster classification evaluation for both the pixel-based approach (a) and the holistic approach (b). In general the pixel-based approach performs better than the holistic approach, because misclassification of single pixels extracted from a pedestrian cutout do not directly lead to falsely classified images, while for holistic prediction, a wrong

---

[3] www.ar-tracking.com

Real \ Prediction	Forward	Backward	Left Walking	Right Walking	Forward Both Hands	Left Arm Movement	Right Arm Movement	Forward - Left Arm Movement	Forward - Right Arm Movement
Forward	55,7	30,5	1,3	1,3	0,3	0,5	1,0	5,5	3,9
Backward	8,2	81,6	2,7	2,9	0,0	0,4	0,7	2,4	1,0
Left Walking	4,1	2,2	67,9	0,8	0,2	19,9	0,1	2,5	2,4
Right Walking	5,9	1,2	0,6	80,9	0,1	0,1	6,0	4,3	0,8
Forward Both Hands	1,5	3,8	6,0	0,0	62,8	0,0	4,4	4,1	17,5
Left Arm Movement	0,8	0,7	14,9	1,7	0,2	74,4	0,0	4,6	2,7
Right Arm Movement	0,8	0,4	2,3	17,7	0,1	0,0	73,9	1,6	3,2
Forward - Left Arm Movement	11,9	9,1	2,2	11,9	1,4	2,1	0,9	59,3	1,1
Forward - Right Arm Movement	12,3	9,3	2,4	1,6	1,6	1,0	8,6	2,1	61,2
Total Images:	3798	1128	4220	4249	685	6633	4932	3333	2250

Overall Accuracy: 69,5%

Fig. 3. Evaluation of clusters for the holistic. The rows show the ground truth clusters, the columns show predicted values. The diagonal shows the correct predictions.

classification has a much higher impact on the final outcome. Both approaches have most problems discerning between the forward and backward clusters which are in general very hard to separate. In fact the most discriminate part of the pedestrian for separating both is the face which only constitutes a small portion of the image. Classification into left and right clusters show less confusion as both are more discriminative. Most errors stem from two sources. The ground truth is generated using k-means on poses from continuous scenes, and cluster borders are quite fuzzy which poses problems especially when switching between two orientation clusters of the same direction. For example the transition from "left - walking" to "left - arm movement" cluster is continuous in image space, and although each pose is attributed to one cluster, in terms of image information it is fuzzy. The second problem are self-occlusions. For example, moving your left arm while facing right is often not visible in the images because the torso conceals the arm movement from the camera. The forest classifies the image into "right - walking", but due to the motion capture ground truth which captures the correct pose, it's ground truth cluster is "right - arm movement".

Evaluating only on the four main directions shows improved results as can be seen in Fig. 5. Overall, we achieve an accuracy of almost 83 % for holistic

Real \ Prediction	Forward	Backward	Left Walking	Right Walking	Forward Both Hands	Left Arm Movement	Right Arm Movement	Forward - Left Arm Movement	Forward - Right Arm Movement
Forward	93,5	1,5	0,9	0,4	0,1	0,0	0,0	0,8	2,7
Backward	38,8	53,4	2,7	3,3	0,0	0,0	1,8	0,0	0,0
Left Walking	7,6	0,5	75,8	0,5	0,1	13,4	0,0	0,0	2,0
Right Walking	12,0	0,2	1,2	75,3	0,1	0,0	5,8	5,2	0,0
Forward Both Hands	1,6	0,0	0,1	0,3	95,0	0,0	1,5	1,0	0,4
Left Arm Movement	1,9	0,0	15,2	0,5	0,2	77,7	0,0	2,5	2,0
Right Arm Movement	1,9	0,0	0,0	13,9	0,0	0,0	71,0	5,9	7,2
Forward - Left Arm Movement	23,1	0,0	0,9	1,3	0,1	2,0	0,7	69,8	2,0
Forward - Right Arm Movement	30,2	0,0	0,0	5,9	0,8	0,3	2,7	1,0	59,0
Total Images:	3798	1128	4220	4249	685	6633	4932	3333	2250
Overall Accuracy:					76,1%				

Fig. 4. Evaluation of clusters for the pixel-based approach. Apart from a higher accuracy in general, it is noticeable that misclassifications tend to stay within the same orientation, e.g. when comparing lines 8 and 9.

and 89 % for pixel-based evaluation. As it has been observed in Figs. 3 and 4 and explained in the previous paragraph, many misclassifications within the orientation clusters keep the correct general direction which allows a very good separation into the four general directions.

4.3 Evaluation on the Human Orientation Classification Dataset

Tosato et al. [17] published the HOC Dataset, which is separated into a training set containing 6860 images and a test set containing 5021 images. All images are labeled into 4 directions, front, back, left and right. We trained our system using their training dataset. For the holistic approach, 30 trees of maximum depth 10 were trained using 20 % of the training images equally samples from all 4 directions. Given the much smaller size of the images compared to our training data, at each node 50 random positions and HOG window sizes were tested to find the optimal split. For the pixel-based approach, 10 trees of depth 18 were trained using 25 % of the training images. At each node, 20 different HOG window sizes were tested.

	Front	Back	Left	Right
Front	72,1	16,8	3,2	7,7
Back	9,7	81,6	3,1	3,6
Left	8,3	1,3	87,9	1,3
Right	7,9	0,8	1,6	89,0
Total:	10066	1128	10853	9181
Accuracy : 82,9%				

(a)

	Front	Back	Left	Right
Front	95,2	0,6	1,9	2,9
Back	38,8	53,4	2,7	5,1
Left	7,8	0,2	91,4	0,5
Right	16,1	0,1	0,6	83,1
Total:	10066	1128	10853	9181
Accuracy : 89%				

(b)

Fig. 5. Evaluation of clusters for (a) the holistic approach and (b) the pixel-based approach. The rows show the percentage of images with a given ground truth label that were classified into the different orientations

	Front	Back	Left	Right
Front	77,5	9,5	5,6	7,4
Back	3,0	84,6	6,7	5,7
Left	5,9	10,1	77,3	6,6
Right	6,1	7,2	6,1	80,5
Total:	1014	1211	1557	1239
Accuracy : 79,9%				

(a)

	Front	Back	Left	Right
Front	72,6	12,4	6,8	8,2
Back	3,0	76,4	11,9	8,8
Left	5,9	18,5	65,8	9,8
Right	6,1	11,2	10,2	72,5
Total:	1014	1211	1557	1239
Accuracy : 71,4%				

(b)

	Front	Back	Left	Right
Front	77,0	4,0	9,0	10,0
Back	0,0	88,0	8,0	4,0
Left	4,0	15,0	73,0	8,0
Right	3,0	14,0	6,0	77,0
Total:	1014	1211	1557	1239
Accuracy : 78,7%				

(c)

Fig. 6. Evaluation of clusters for (a) the pixel-based approach and (b) the holistic approach. (c) shows the results provided by [17]

Figure 6 shows the results for both pixel-based (a) and holistic (b) evaluation. Analogously to our dataset, using forests trained on a pixel level shows a higher accuracy by 8.5 %. For the pixel-based approach we achieve a slightly higher accuracy then Tosato et al. which report an accuracy of 78.7 %.

5 Conclusion

This paper presents a new method for human orientation classification using random classification forests in combination with HOG features [2] which are learned during training. Two different approaches were proposed: The first classifying pixels extracted from each pedestrian bounding box to determine the orientation cluster which is very accurate, the second classifying the whole pedestrian bounding box in one shot, which is slightly less accurate but much faster. We classify pedestrians not only with regard to orientation but also predict a rough pose of the pedestrian. Future work will concern integrating temporal information into the prediction pipeline which should greatly enhance the accuracy. Starting from the resulting clusters of this method, the next step consists of determining the body pose of the pedestrian.

References

1. Belagiannis, V., Amann, C., Navab, N., Ilic, S.: Holistic human pose estimation with regression forests. In: Perales, F.J., Santos-Victor, J. (eds.) AMDO 2014. LNCS, vol. 8563, pp. 20–30. Springer, Heidelberg (2014)

2. Benenson, R., Mathias, M., Tuytelaars, T., Van Gool, L.: Seeking the strongest rigid detector. In: 2013 IEEE Conference on Computer Vision and Pattern Recognition (CVPR), pp. 3666–3673. IEEE (2013)
3. Breiman, L.: Random forests. Mach. Learn. **45**(1), 5–32 (2001)
4. Criminisi, A., Shotton, J., Konukoglu, E.: Decision forests: a unified framework for classification, regression, density estimation, manifold learning and semi-supervised learning. Found. Trends. Comput. Graph. Vis. **7**(2–3), 81–227 (2012)
5. Dollar, P., Wojek, C., Schiele, B., Perona, P.: Pedestrian detection: an evaluation of the state of the art. IEEE Trans. Pattern Anal. Mach. Intell. **34**(4), 743–761 (2012)
6. Droeschel, D., Behnke, S.: 3D body pose estimation using an adaptive person model for articulated ICP. In: Jeschke, S., Liu, H., Schilberg, D. (eds.) ICIRA 2011, Part II. LNCS, vol. 7102, pp. 157–167. Springer, Heidelberg (2011)
7. Enzweiler, M., Gavrila, D.M.: Monocular pedestrian detection: survey and experiments. IEEE Trans. Pattern Anal. Mach. Intell. **31**(12), 2179–2195 (2009)
8. Enzweiler, M., Gavrila, D.M.: Integrated pedestrian classification and orientation estimation. In: 2010 IEEE Conference on Computer Vision and Pattern Recognition (CVPR), pp. 982–989. IEEE (2010)
9. Gandhi, T., Trivedi, M.M.: Image based estimation of pedestrian orientation for improving path prediction. In: 2008 IEEE Intelligent Vehicles Symposium, pp. 506–511. IEEE (2008)
10. Hofmann, M., Gavrila, D.: Multi-view 3D human pose estimation in complex environment. Int. J. Comput. Vision **96**(1), 103–124 (2012)
11. Lallemand, J., Szczot, M., Ilic, S.: Human pose estimation in stereo images. In: Perales, F.J., Santos-Victor, J. (eds.) AMDO 2014. LNCS, vol. 8563, pp. 10–19. Springer, Heidelberg (2014)
12. Lloyd, S.: Least squares quantization in PCM. IEEE Trans. Inf. Theory **28**(2), 129–137 (1982)
13. Rybok, L., Voit, M., Ekenel, H.K., Stiefelhagen, R.: Multi-view based estimation of human upper-body orientation. In: 2010 20th International Conference on Pattern Recognition (ICPR), pp. 1558–1561. IEEE (2010)
14. Shimizu, H., Poggio, T.: Direction estimation of pedestrian from multiple still images. In: 2004 IEEE Intelligent Vehicles Symposium, pp. 596–600. IEEE (2004)
15. Shotton, J., Girshick, R., Fitzgibbon, A., Sharp, T., Cook, M., Finocchio, M., Moore, R., Kohli, P., Criminisi, A., Kipman, A., et al.: Efficient human pose estimation from single depth images. IEEE Trans. Pattern Anal. Mach. Intell. **35**(12), 2821–2840 (2013)
16. Taylor, J., Shotton, J., Sharp, T., Fitzgibbon, A.: The vitruvian manifold: inferring dense correspondences for one-shot human pose estimation. In: 2012 IEEE Conference on Computer Vision and Pattern Recognition (CVPR), pp. 103–110. IEEE (2012)
17. Tosato, D., Spera, M., Cristani, M., Murino, V.: Characterizing humans on riemannian manifolds. IEEE Trans. Pattern Anal. Mach. Intell. **35**(8), 1972–1984 (2013)
18. Voit, M., Stiefelhagen, R.: A system for probabilistic joint 3D head tracking and pose estimation in low-resolution, multi-view environments. In: Fritz, M., Schiele, B., Piater, J.H. (eds.) ICVS 2009. LNCS, vol. 5815, pp. 415–424. Springer, Heidelberg (2009)

19. Weinrich, C., Vollmer, C., Gross, H.M.: Estimation of human upper body orientation for mobile robotics using an svm decision tree on monocular images. In: 2012 IEEE/RSJ International Conference on Intelligent Robots and Systems (IROS), pp. 2147–2152. IEEE (2012)
20. Zhang, X., Hongyan, Y., Guoqing, H., Mengjing, C., Yue, G., Xiang, H.: Basic characteristics of road traffic deaths in China. Iran. J. Pub. Health **42**(1), 7 (2013)

Distance-Based Descriptors and Their Application in the Task of Object Detection

Radovan Fusek[✉] and Eduard Sojka

Department of Computer Science, Technical University of Ostrava, FEECS,
17. Listopadu 15, 708 33 Ostrava-Poruba, Czech Republic
{radovan.fusek,eduard.sojka}@vsb.cz

Abstract. In this paper, we propose an efficient and interesting way how to encode the shape of the objects. A lot of state-of-the art descriptors (e.g. HOG, Haar, LBP) are based on the fact that the shape of the objects can be described by brightness differences inside the image. It means that the descriptors encode the gradient or intensity differences inside the image (i.e. edges). In the cases that the edges are very thin, the edge information can be difficult to obtain and the dimensionally of feature vector (without the method for reduction) is typically large and contains redundant information. These ills are motivation for the proposed method in that the edges need not be hit directly; the input brightness function is transformed using the appropriate image distance function. After this transformation, the values of distance function inside objects and backgrounds are different and the values can be used for description of object appearance. We demonstrate the properties of the method for the case of solving the problem of face detection using the classical sliding window technique.

1 Introduction

The detectors that are based on the sliding window technique showed a great performance in the last decade. The main idea behind the sliding window detection technique is based on the fact that the input image is scanned by a rectangular window in different scales. Inside the sliding window the appropriate image descriptors are calculated and composed to the final feature vector. The feature vector is then used as an input for the trainable classifiers (e.g. support vector machine, neural network, random forest). After the classification process, each window is marked as the background or object of interest.

In this area, the three types of features and their modifications that can be used in the sliding window technique became dominant in recent years; HOG, Haar, and LBP features. In [22], the detection framework based on the Haar-like features was presented by Viola and Jones. The framework consists of the image representation called the integral image combined with the rectangular Haar-like features, and AdaBoost algorithm [10]. In [7], the authors proposed the method in that the histograms of oriented gradients (HOG) are used to encode the appearance of the object. Ojala et al. [19] proposed the Local Binary Patterns

© Springer International Publishing Switzerland 2014
X. Jiang et al. (Eds.): GCPR 2014, LNCS 8753, pp. 488–498, 2014.
DOI: 10.1007/978-3-319-11752-2_40

(LBP) in that the local image structures (e.g. lines, edges, spots, and flat areas) can be efficiently encoded by comparing every pixel with its neighboring pixels (more details can be found in Sect. 2).

All mentioned features are based on the fact that the appearance of the objects is described by the image edge information (intensity differences). In general, the features based on the edge information (e.g. length, magnitude, orientation, localization) require large training sets due to their high dimensionality. Additionally, in the cases that the edges are very thin, it is obvious that the edges information is difficult to hit (by the samples). Therefore, the proposed method is based on the distance function in that the information about its changes is not so important. In essence, we divide the image inside the sliding window into the blocks and cells (similarly as in HOG), but instead of the histograms of gradients we encode the values of distance function inside each cell. This leads to the reasonable dimensionality of the feature vector; furthermore, the values of distance function can be easily obtained by sampling. The feature vector that contains the distance function values is then used as an input for the SVM classifier.

2   Related Work

As was mentioned in the previous section, the three types of features are considered as the state-of-the-art in the area of feature based detectors; HOG, Haar, and LBP. In essence, the HOG descriptors can be considered as the dense version of SIFT [17,18]. The sliding window is divided into the cells in that the histograms of oriented gradients are calculated. The cells are normalized across the large blocks. The vector of features that is obtained from each sliding window is then used as input for the SVM classifier. In recent years, many modifications and applications of classical HOG were presented. In [26], the authors proposed the fast way of calculating the HOG features with the use of the integral image. The authors also integrated the HOG features into the Viola and Jones cascade framework. In [5], the authors presented the pyramid of histogram of orientation gradients (PHOG) descriptors in that the HOG descriptors are combined with the image pyramid representation of Lazebnik et al. [13]. In [12], the authors applied the Principal Component Analysis (PCA) to the HOG feature vector to obtain the PCA-HOG vector. The part-based detector based on HOG was proposed by Felzenszwalb et al. in [9].

The Haar-like features which are similar to Haar basis function were proposed by Papageorgiou and Poggio [20] and popularized by Viola and Jones [22]. Viola and Jones combined the Haar-like features with the integral image representation, AdaBoost algorithm, and cascade of classifiers. The extension of the Haar feature set has been presented by Lienhart et al. [16]. For example, the multi-view face detection system was presented by Wu et al. [23], the front-view car and bus detector based on the Haar-like features was proposed in [24].

The LBP operator was proposed by Ojala et al. [19] for the texture analysis, hoverer, the operator was successfully used in many detection and recognition tasks. In the basic form of LBP operator, every pixel is compared with its neighbors to encode the local image structures such as lines, edges, spots, and flat areas. In [11], LBP were used for face detection problem in low-resolution images. The face recognition problem was solved using LBP in [1,2]. Multi-block Local Binary Patterns (MB-LBP) for face detection were proposed in [25].

Since the geodesic distance is used in the paper, it is important to mention works in that this distance was used in the area of image processing. Image segmentation and object detection methods based on geodesic distance were presented in [3,6,8,21].

3   Proposed Method

The proposed method is based on the fact that the properties of the image (especially the properties of the objects) can effectively be described by the distance function. The goal is to obtain more meaningful values for recognition than the classical state-of-the-art method. The usefulness of this function can be described in the following way.

Suppose the simple theoretical image that contains one object of constant brightness (Fig. 1(a)). The appearance (shape) of this object can be described using the gradient of this object (Fig. 1(b)). In the classical sliding window methods (HOG, Harr, LBP), the samples (e.g. blocks, rectangular features) must hit the places with the intensity differences (edges) to obtain the information about the object. In the situation that edges can be very thin (theoretically infinity thin), it is difficult to hit the places with the edge information, and many samples contain the redundant information without the gradient (edges) information.

Suppose the case that the samples (e.g. blocks, rectangular features) are placed inside the image in the way as is depicted in Fig. 1(c). In such a case, the samples do not detect any important information; the values of gradient sizes and directions are null (HOG principle), as well as the intensity differences inside the samples (Haar principle). This situation was motivation to use another way how to encode the appearance of the objects inside images.

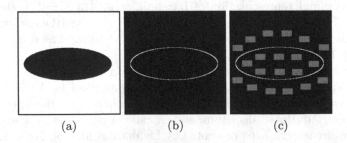

(a)   (b)   (c)

Fig. 1. The image with one object with constant brightness (a). The gradient of the image (b). The samples (red color blocks) in that the information about the object is encoded (c) (Color figure online).

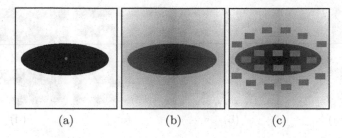

(a)   (b)   (c)

Fig. 2. The image with one object with constant brightness that contains the centroid point (red color) (a). The visualization of the distance function (b). The values of distance function are depicted by the level of brightness. The samples (red color blocks) in that the information about the object is encoded (c) (Color figure online).

Suppose an arbitrary point that is placed inside the previously mentioned object. Say that in the gravity center of the object (Fig. 2(a)); this point can be called as the centroid of the object c_i. Let us compute the geodesic distance function d from the centroid c_i to all other points inside the image. The visualization of the distance function values is shown in Fig. 2(b). In this particular case, we use the geodesic distance, nevertheless, it is important to note that any appropriately distance function can be used in the proposed detection framework (e.g. resistance distance, diffusion distance). In general, the geodesic distance $d(c_1, c_2)$ between two points c_1, c_2 computes the shortest curve that connects booth points along the image manifold; the geodetic distance reflects the topology of the image.

Suppose the same distribution of the samples as in the previous case (Fig. 2(c)). The main contribution of using the distance function is that the values of this function are different inside and outside the object of interest. In essence, the values of distance function reliably reflect the image information and the appearance of objects, and the meaningful values can be reliably obtained by sampling. Even the simple samples in Fig. 2(c) can be used to describe the properties of the image; the sample values can be used to encode properties (shape) of the objects without a large number of redundant information.

It is clear that the situation is more complicated in the real images and one centroid will not be enough to cover more complicated image structures. Therefore, we divide the whole image into the cells. The gravity centers of each cell are defined as the centroid points c_i; the distance is computed from these points to all other points inside each cell.

The visualizations of geodesic distance values inside the cells of different sizes are shown in Fig. 3. Based on the cell sizes, information with various levels of details is obtained. To compress the information contained in the distance function in to a reasonable number of values, we use four values from each cell only. These values take into account the distance in four different directions (Fig. 4) and the values are then used in the feature vector. Each of four neighboring cells create a block in which into the large blocks (Fig. 4) in that the distance

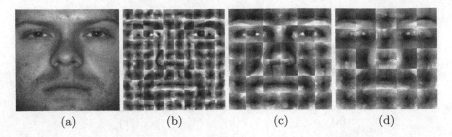

| (a) | (b) | (c) | (d) |

Fig. 3. The visualization of the distance function values inside each cell. The example of face image (a). The sizes of cell 15×15 (b), 25×25 (c), 35×35 (d).

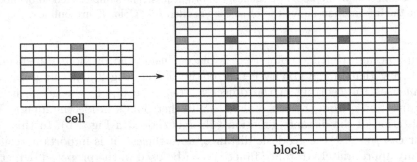

Fig. 4. An example of 9×9 cells that are grouped into one block. In this particular case, from each cell, four values (depicted by green color) are used in the feature vector (Color figure online).

values are normalized. In our experiment, we use the overlapping blocks; the second half of one block correspond with the first half of the next block. The final feature vector is then used as an input for the SVM classifier.

4   Experiments

In this section, we demonstrate the properties of the proposed method for the case of solving the problem of face detection using the classical sliding window technique. For this task we collected 2300 faces and 4300 non-faces. The faces were obtained from the BIOID database combined with the Extended Yale Face Database B [14]. The negative set consists of 3000 images that were obtained from the MIT-CBCL database combined with the 1300 hard negative examples. In the detection process, the sliding window scanned 10 different resolutions of input image. We experimented with many sizes of sliding windows, cells, blocks of the proposed method and we suggested the following configurations.

The configuration $Dist_1$ is designed with the size of sliding window $= 70 \times 70$, size of blocks $= 14 \times 14$, size of cells $= 7 \times 7$, horizontal block step size $= 7$. This configurations consists of 1296 descriptors for one position of sliding window; each window consists of 81 overlapping blocks and each block consists of 4 cells,

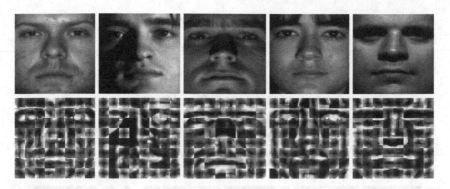

Fig. 5. The visualization of the distance function values inside each cell of $Dist_2$ configuration.

i.e. 324 cells are defined in each window. Finally, each cell is described using 4 distance values, i.e. 1296 descriptors are used (324×4).

The configuration $Dist_2$ is designed with the size of sliding window = 72×72, size of blocks = 18×18, size of cells = 9×9, horizontal block step size = 9. This configurations consists of 784 descriptors for one position of sliding window; each window consists of 49 overlapping blocks and each block consists of 4 cells, i.e. 196 cells are defined in each window. Finally, each cell is described using 4 distance values, i.e. 784 descriptors are used (196×4).

The configuration $Dist_3$ is designed with the size of sliding window = 88×88, size of blocks = 22×22, size of cells = 11×11, horizontal block step size = 11. This configurations consists of 784 descriptors for one position of sliding window; each window consists of 49 overlapping blocks and each block consists of 4 cells, i.e. 196 cells are defined in each window. Finally, each cell is described using 4 distance values, i.e. 784 descriptors are used (196×4). The examples of visualization of distance function values of training images are shown in Fig. 5.

For comparison, we used the detectors that are based on the HOG features, LBP (Local Binary Patterns) features [15] and Haar features (Viola-Jones detection framework). For the HOG features, we used the classical parameters of HOG; the size of block = 16×16, size of cell = 8×8, horizontal step size = 8, number of bins = 9. The training images (for HOG) were resized to the size of 80×80 pixels (the size of sliding window was also set to this size) and the sliding window scanned 10 different resolutions of input image. This configuration of HOG consists of 2916 descriptors for one position of sliding window, and it is denoted as HOG. For the detectors that are based on the Haar and LBP features, the cascade classifiers was created and the training images were resized to the 19×19 pixels for these detectors; the detector based on Haar is denoted as $Haar$, the detector based on LBP is denoted as LBP.

We used the identical training set (2300 positive and 4300 negative samples) and testing set for all detectors. To test the detectors, we collected 300 images from the Faces in the Wild dataset [4]. Before the process of performance

Table 1. The face detection results.

	Precision	Sensitivity	F1 score
$Dist_1$	99.14 %	88.69 %	93.62 %
$Dist_2$	96.05 %	93.83 %	94.93 %
$Dist_3$	98.53 %	86.38 %	92.05 %
HOG	68.85 %	94.71 %	79.73 %
$Haar$	85.88 %	81.28 %	83.52 %
LBP	72.60 %	70.67 %	71.62 %

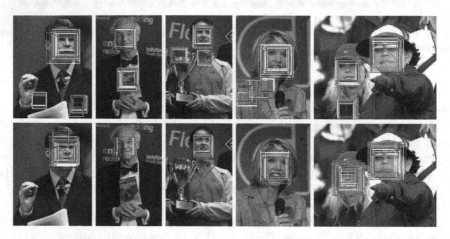

Fig. 6. The differences between the detection results. The first row: the detection results of HOG detector. The second row: the detection results of proposed detector based on the $Dist_2$ configuration. The results are without the postprocessing (the detection results are not merged).

calculation, the positive detections were merged to one if at least 5 positive detections hit approximately one place in the image. In Table 1, the detection results are shown.

The HOG based detector achieved the higher true positives rate (Sensitivity 94.71 %). It means that this detector achieved the large numbers of true positives and the detector had relatively small numbers of false negatives. On the other side, the positive predictive value (Precision 68.85 %) is quite low. It means that the number of false positives is rather large. This is caused by the large dimensionality of feature vector that is created by the 2916 values for one position of sliding window. Since we used the relatively small set of training data (2300 faces and 4300 non-faces) and dimensionality of feature vector of HOG is relatively large, the detector based on HOG detected the faces in the wrong places. Overall detection rate (F1 score) of HOG detector is 79.73 %. This problem also appeared in Haar (F1 score = 83.52 %) and LBP (F1 score = 71.62 %) based

Fig. 7. The face detection results of $Dist_2$ configuration. The results are without the postprocessing (the detection results are not merged).

detector. Although, the Haar based detector achieved the low number of false positives, in general, this detector needs a larger training set; similarly to LBP.

On the other side, the proposed method achieved very promising results. Since the detectors based on the $Dist_2$ and $Dist_3$ created the feature vector of a relatively small size (784), the selected training set (2300 faces and 4300 non-faces) was sufficiently large for them. Even the $Dist_2$ configuration (with 1296 values in feature vector) achieved better results than the state-of-the-art detectors (F1 score = 93.62 %). The best detection results achieved the detector based on the $Dist_2$ configuration (F1 score = 93.62 %). The sensitivity of this detector was lower than in the HOG detector (88.69 % vs. 94.71 %), nevertheless the proposed detector achieved considerably less false positives than the HOG, LBP, and Haar based detectors. The examples of differences between the detection results of the prosed detector and the HOG detector are shown in Fig. 6.

The achieved results confirmed the assumption that the distance function values can be effectively used to create the feature vector. Additionally, the values can be used to describe the information inside the sliding window better than the classical gradient-based approaches with a relatively small set of numbers. Finally, the calculation of geodesic distance and composition of the feature vector take approximately 2 ms for one position of sliding window in $Dist_2$ configuration on an Intel core i3 processor. The detection results of proposed detector based on the $Dist_2$ configuration are shown in Fig. 7.

5   Conclusion

In the paper, we presented an efficient way how the image information can be encoded into the feature vector, which can be used in sliding-window-based techniques of recognition. In essence, the method is based on the idea that the information contained in the window can be expressed by measuring distances along the image manifold. In the method, the sliding window is divided into the cells; inside each cell, a central point is selected. The distances are computed from the central point to all other points inside the cells. We also showed that the proposed method can be used for solving the problem of face detection with promising results and a relatively small size of feature vector. We will try to reduce the vector dimensionality using statistical methods for reducing the dimension of feature vector (e.g. PCA).

In the paper, we used the geodesic distance; however, it is worth mentioning that various distance metrics can be used. We leave experiments with various types of distances for future work.

Acknowledgments. This work was supported by the SGS in VSB Technical University of Ostrava, Czech Republic, under the grant No. SP2014/170.

References

1. Ahonen, T., Hadid, A., Pietikainen, M.: Face description with local binary patterns: application to face recognition. IEEE Trans. Pattern Anal. Mach. Intell. **28**(12), 2037–2041 (2006)
2. Ahonen, T., Hadid, A., Pietikäinen, M.: Face recognition with local binary patterns. In: Pajdla, T., Matas, J.G. (eds.) ECCV 2004. LNCS, vol. 3021, pp. 469–481. Springer, Heidelberg (2004)
3. Bai, X., Sapiro, G.: A geodesic framework for fast interactive image and video segmentation and matting. In: IEEE 11th International Conference on Computer Vision, ICCV 2007, pp. 1–8, October 2007
4. Berg, T.L., Berg, A.C., Edwards, J., Forsyth, D.: Who's in the picture. In: Saul, L.K., Weiss, Y., Bottou, L. (eds.) Advances in Neural Information Processing Systems 17, pp. 137–144. MIT Press, Cambridge (2005)
5. Bosch, A., Zisserman, A., Munoz, X.: Representing shape with a spatial pyramid kernel. In: Proceedings of the 6th ACM International Conference on Image and Video Retrieval, CIVR '07, pp. 401–408. ACM, New York (2007). http://doi.acm.org/10.1145/1282280.1282340

6. Criminisi, A., Sharp, T., Blake, A.: GeoS: geodesic image segmentation. In: Forsyth, D., Torr, P., Zisserman, A. (eds.) ECCV 2008, Part I. LNCS, vol. 5302, pp. 99–112. Springer, Heidelberg (2008)

7. Dalal, N., Triggs, B.: Histograms of oriented gradients for human detection. In: IEEE Computer Society Conference on Computer Vision and Pattern Recognition, CVPR 2005, vol. 1, pp. 886–893, June 2005

8. Economou, G., Pothos, V., Ifantis, A.: Geodesic distance and MST based image segmentation. In: European Signal Processing Conference, pp. 941–944 (2004)

9. Felzenszwalb, P.F., McAllester, D.A., Ramanan, D.: A discriminatively trained, multiscale, deformable part model. In: CVPR (2008)

10. Freund, Y., Schapire, R.E.: A decision-theoretic generalization of on-line learning and an application to boosting. In: Vitányi, P.M.B. (ed.) EuroCOLT 1995. LNCS, vol. 904, pp. 23–37. Springer, Heidelberg (1995). http://dl.acm.org/citation.cfm?id=646943.712093

11. Hadid, A., Pietikainen, M., Ahonen, T.: A discriminative feature space for detecting and recognizing faces. In: Proceedings of the 2004 IEEE Computer Society Conference on Computer Vision and Pattern Recognition, CVPR 2004, vol. 2, pp. II-797–II-804 (2004)

12. Kobayashi, T., Hidaka, A., Kurita, T.: Selection of histograms of oriented gradients features for pedestrian detection. In: Ishikawa, M., Doya, K., Miyamoto, H., Yamakawa, T. (eds.) ICONIP 2007, Part II. LNCS, vol. 4985, pp. 598–607. Springer, Heidelberg (2008)

13. Lazebnik, S., Schmid, C., Ponce, J.: Beyond bags of features: Spatial pyramid matching for recognizing natural scene categories. In: Proceedings of the 2006 IEEE Computer Society Conference on Computer Vision and Pattern Recognition - Volume 2, CVPR '06, pp. 2169–2178. IEEE Computer Society, Washington, DC, USA (2006). http://dx.doi.org/10.1109/CVPR.2006.68

14. Lee, K., Ho, J., Kriegman, D.: Acquiring linear subspaces for face recognition under variable lighting. IEEE Trans. Pattern Anal. Mach. Intell. 27(5), 684–698 (2005)

15. Liao, S.C., Zhu, X.X., Lei, Z., Zhang, L., Li, S.Z.: Learning multi-scale block local binary patterns for face recognition. In: Lee, S.-W., Li, S.Z. (eds.) ICB 2007. LNCS, vol. 4642, pp. 828–837. Springer, Heidelberg (2007)

16. Lienhart, R., Maydt, J.: An extended set of haar-like features for rapid object detection. In: 2002 International Conference on Image Processing, vol. 1, pp. I-900–I-903 (2002)

17. Lowe, D.G.: Distinctive image features from scale-invariant keypoints. Int. J. Comput. Vis. 60(2), 91–110 (2004). http://dx.doi.org/10.1023/B:VISI.0000029664.99615.94

18. Lowe, D.: Object recognition from local scale-invariant features. In: The Proceedings of the Seventh IEEE International Conference on Computer Vision, vol. 2, pp. 1150–1157 (1999)

19. Ojala, T., Pietikäinen, M., Harwood, D.: A comparative study of texture measures with classification based on featured distributions. Pattern Recogn. 29(1), 51–59 (1996). http://dx.doi.org/10.1016/0031-3203(95)00067-4

20. Papageorgiou, C., Poggio, T.: A trainable system for object detection. Int. J. Comput. Vision 38(1), 15–33 (2000). http://dx.doi.org/10.1023/A:1008162616689

21. Paragios, N., Deriche, R.: Geodesic active contours and level sets for the detection and tracking of moving objects. IEEE Trans. Pattern Anal. Mach. Intell. 22(3), 266–280 (2000)

22. Viola, P., Jones, M.: Rapid object detection using a boosted cascade of simple features. In: Proceedings of the 2001 IEEE Computer Society Conference on Computer Vision and Pattern Recognition, CVPR 2001, vol. 1, pp. I-511–I-518 (2001)
23. Wu, B., Ai, H., Huang, C., Lao, S.: Fast rotation invariant multi-view face detection based on real adaboost. In: Proceedings of Sixth IEEE International Conference on Automatic Face and Gesture Recognition, pp. 79–84 (2004)
24. Wu, C., Duan, L., Miao, J., Fang, F., Wang, X.: Detection of front-view vehicle with occlusions using adaboost. In: International Conference on Information Engineering and Computer Science, ICIECS 2009, pp. 1–4 (2009)
25. Zhang, L., Chu, R.F., Xiang, S., Liao, S.C., Li, S.Z.: Face detection based on multi-block LBP representation. In: Lee, S.-W., Li, S.Z. (eds.) ICB 2007. LNCS, vol. 4642, pp. 11–18. Springer, Heidelberg (2007). http://dl.acm.org/citation.cfm?id=2391659.2391662
26. Zhu, Q., Yeh, M.C., Cheng, K.T., Avidan, S.: Fast human detection using a cascade of histograms of oriented gradients. In: 2006 IEEE Computer Society Conference on Computer Vision and Pattern Recognition, vol. 2, pp. 1491–1498 (2006)

Hough Forests Revisited: An Approach to Multiple Instance Tracking from Multiple Cameras

Georg Poier[(⊠)], Samuel Schulter, Sabine Sternig, Peter M. Roth,
and Horst Bischof

Institute for Computer Graphics and Vision,
Graz University of Technology, Graz, Austria
poier@icg.tugraz.at, {schulter,bischof}@icg.tugraz.at,
sternig.sabine@gmail.com, p.m.roth@ieee.org

Abstract. Tracking multiple objects in parallel is a difficult task, especially if instances are interacting and occluding each other. To alleviate
the arising problems multiple camera views can be taken into account,
which, however, increases the computational effort. Evoking the need
for very efficient methods, often rather simple approaches such as background subtraction are applied, which tend to fail for more difficult
scenarios. Thus, in this work, we introduce a powerful multi-instance
tracking approach building on Hough Forests. By adequately refining the
time consuming building blocks, we can drastically reduce their computational complexity without a significant loss in accuracy. In fact, we show
that the test time can be reduced by one to two orders of magnitude,
allowing to efficiently process the large amount of image data coming
from multiple cameras. Furthermore, we adapt the pre-trained generic
forest model in an online manner to train an instance-specific model,
making it well suited for multi-instance tracking. Our experimental evaluations show the effectiveness of the proposed efficient Hough Forests for
object detection as well as for the actual task of multi-camera tracking.

1 Introduction

Visual object tracking is one of the most important tasks in computer vision,
building the basis for various applications such as surveillance, sports analysis,
and industrial (quality) inspection [24,26,32]. Even though there are considerable advances in single object tracking (*e.g.*, [21,22]), there are still open challenges for multi-object tracking [38,40], in particular if multiple instances of the
same class interact.

However, by conducting well elaborated reasoning relying on part-based
approaches (*e.g.*, [4,39]) even complex scenarios can be handled very well with
a single camera. Nevertheless, such methods fail if instances are fully occluding
each other. To overcome this problem, recent approaches additionally exploit
temporal information and analyze the tracks of individual instances over time

© Springer International Publishing Switzerland 2014
X. Jiang et al. (Eds.): GCPR 2014, LNCS 8753, pp. 499–510, 2014.
DOI: 10.1007/978-3-319-11752-2_41

[9,13]. However, such methods are condemned to fail if the assumptions (*e.g.*, constant velocity) are hurt. A natural way to resolve the problem of similar objects occluding each other is to exploit the additional information provided by multiple cameras [15,26].

The increased amount of data captured by multiple cameras, however, limits the computational complexity of the employed tracking algorithms. Thus, often simple techniques like background subtraction or color models are used in order to locate foreground objects [15,23,30]. However, the simplicity of these methods usually yields severe problems like mistaken instances or ghost detections. In contrast, computationally more complex object detection models, such as [35], are often lacking efficiency and cannot be applied for real-world scenarios.

Thus, the goal of this work is to provide stronger models for tracking multiple instances using multiple cameras, however, still ensuring short response times and thus real-time capabilities. In particular, we focus on Hough Forests (HFs) [19] as underlying object model in a tracking-by-detection setup. In general, HFs are a powerful and versatile extension of Random Forests (RFs) [8] and have been successfully applied to various tasks including object detection [27,36], tracking [17,33], and pose estimation [34,37].

For real-time tracking, however, their original formulation bears several drawbacks. For example, the runtime is strictly correlated with the number of training samples, as we will point out later. Hence, a major advantage of Random Forests, where the test time is independent from the amount of training data, is lost. This is especially a problem as the amount of training data is a crucial parameter for the accuracy of Random Forest based models [11,18,34]. Furthermore, as we also demonstrate, redundant and unnecessary image information is extracted and processed, indicating an *over-determined* object description.

Thus, the contribution of this work is twofold. First, we provide an efficient but still effective object detector based on Hough Forests. Second, this detector is adopted for online learning to allow real-time tracking from multiple cameras. We first, provide a theoretical as well as empirical analysis of Hough Forest based object detection, with special emphasis on computational efficiency. Investigating different aspects such as the classifier complexity, the underlying data complexity, and the final prediction reveals that the runtime can be drastically reduced, without a loss in accuracy. In fact, the performance of the proposed method is on par with the baseline approach, while being one to two orders of magnitude faster.

Once, having identified and eliminated the critical bottlenecks, Hough Forests can straightforwardly be applied for tracking. To this end, we first introduce a novel online learning strategy, where a pre-trained generic object classifier is adapted to discriminate specific instances on the fly. Second, this strategy – in contrast to similar approaches – allows integration of information from different views into one single classifier. This, on the one hand, reduces the complexity of the classifier as redundant information need not be modeled in parallel. On the other hand, it provides an elegant way to share features from different views in a single RF. Finally, the evaluations confirm that in this way an efficient and effective multi-camera tracking system can be built.

2 Hough Forests

Hough Forests [19,27] combine the flexibility of Implicit Shape Models (ISMs) [25] with the efficiency of Random Forests [1,8,11], by integrating part-based classification and regression into a single model. Given a small image patch, the goal is to classify if it originates from an object as well as to predict the possible object location. In the following, we will briefly review the main ideas of HFs [19]. For more details we would like to refer to, *e.g.*, [18,19].

The prediction process of HFs is based on the generalized Hough transform [3], where the Random Forest represents a codebook of local appearances (*i.e.*, small image patches). More specifically, the predictor model at the leafs of the trees includes a class histogram as well as offset vectors representing the relation of the observed image patches to a specified position on the object (*e.g.*, the object center). These offsets are used to cast votes for the object's location and scale. The codebooks are optimized such that the cast votes exhibit small uncertainty, which is crucial for accurate predictions [19]. In contrast to the codebook used in [25], the efficiency of the forest framework also permits dense sampling of local image patches, yielding an additional gain in accuracy.

In order to learn the codebook, a Hough Forest is given a labeled training set $\mathcal{L} = \{(\mathbf{x}_i, y_i, \mathbf{o}_i)\}$, where $\mathbf{x}_i \in \mathcal{X} = \mathbb{R}^D$ and $y_i \in \mathcal{Y} = \{0, \ldots, (C-1)\}^1$ represent the features and the corresponding class label, respectively, and $\mathbf{o}_i \in \mathcal{O} = \mathbb{R}^2$ is the offset vector. Each tree in the forest aims at splitting the given data into small subsets by simultaneously minimizing the uncertainty in the class labels and the offset vectors. For that purpose, a binary splitting test $\phi(\mathbf{x})$ is employed to split the data into two subsets $\mathcal{L}_R$ and $\mathcal{L}_L$. This test is found by generating a number of possible tests $\{\phi_k(\mathbf{x})\}$ at random for each node of a tree and selecting the test $\phi_*(\mathbf{x})$ which minimizes either a classification or a regression objective.

The classification objective $U_\mathcal{Y}(\mathcal{L}, \phi(\mathbf{x}))$ is designed to minimize the class uncertainty in the resulting sets $\mathcal{L}_R$ and $\mathcal{L}_L$. It is based on the entropy $H(\mathcal{L}) = -\sum_{y=0}^{C-1} p(y|\mathcal{L}) \ln p(y|\mathcal{L})$ computed for each set and weighted according to the number of samples contained in the respective set. Here $p(y|\mathcal{L})$ denotes the empirical class probability for class y, estimated from dataset $\mathcal{L}$. The regression objective $U_\mathcal{O}(\mathcal{L}, \phi(\mathbf{x}))$, on the other hand, aims at minimizing the uncertainty of the offset vectors and is based on their deviation from the mean $\bar{\mathbf{o}}$: $V(\mathcal{L}) = \sum_{y \in \mathcal{Y} \setminus \{0\}} \sum_{\mathbf{o} \in \mathcal{O}_p^{(y)}} \|\mathbf{o} - \bar{\mathbf{o}}\|_2^2$. The value of the objective function for a given test $\phi(\mathbf{x})$ is obtained by simply summing up the uncertainties in both sets, *i.e.*, $U_\mathcal{O}(\mathcal{L}, \phi(\mathbf{x})) = V(\mathcal{L}_L) + V(\mathcal{L}_R)$.

In terms of object detection with Hough Forests, a data sample $\mathbf{x}$ represents the appearance of a small image patch. To detect an object, all patches $\mathbf{x}_\mathbf{v}$ of a given image are associated to a leaf l_t of each tree by evaluating the splitting functions. The class probability $p_t(y|\mathbf{x}_\mathbf{v})$ and the offset vectors $\mathbf{o} \in \mathcal{O}_{l_t}^{(y)}$ of class y stored at the leaf are then used to cast votes for an object at position $\mathbf{u}$ and scale s. From the votes accumulated over a number of scales, finally the object positions and scales are identified.

---

[1] Note that $y = 0$ for the background class.

3   Efficient Hough Forests

In the following, we investigate the efficiency of Hough Forests for the task of object detection. For that purpose, we first determine the relevant parameters for the different parts of an ISM based detection approach, *i.e.*, feature computation, matching local image patches to the codebook, and casting votes into the Hough space. Second, we analyze the effectiveness of these parameters with respect to run-time and accuracy. We run different experiments on a standard benchmark dataset, namely *TUD-Pedestrian*[2] [2]. For performance evaluation we employ the standard PASCAL overlap criterion [16]. To ensure statistically valid results, each experiment was repeated 10 times and the averaged results are reported. As we are solely interested in the relative performance changes according to the specific parameters, the results are shown with respect to the original reference implementation.

3.1   Complexity Analysis

The first relevant parameter is the run-time for feature computation $C_{\mathcal{F}}$. As we use a dense sampling strategy, which has shown to be crucial for accurate detection [19], the features have to be computed for each image pixel. Thus, the feature computation is independent from n, the number of actually used patches $\mathbf{x}_i$. Thus, for a single scale, $C_{\mathcal{F}}$ only depends on the number of feature channels F and image pixels I: $C_{\mathcal{F}} = O(FI)$.

While for the original ISM [25] the cost $C_{\mathcal{M}}$ of matching a single data sample to the codebook linearly scales with the size of the codebook, for codebooks based on a Random Forest it logarithmically depends on the codebook size. In fact, $C_{\mathcal{M}}$ is estimated as the sum of the cost of matching a sample to a leaf of each tree within the forest: $C_{\mathcal{M}} = \sum_{t=1}^{T} O(\kappa \log(|l_t|))$, where $|l_t|$ specifies the number of leafs of tree $t \in \{1, \ldots, T\}$, and κ denotes the complexity of a single splitting test. Ignoring that $|l_t|$ may slightly differ from tree to tree, we get $C_{\mathcal{M}} = O(T\kappa \log(|l|))$.

Hence, assuming balanced trees, we finally derive the following overall detection costs:

$$C = C_{\mathcal{F}} + n\left(C_{\mathcal{M}} + \frac{T |\mathcal{O}_{\mathcal{L}}|}{|l|}\right) = O\left(FI + nT\left(\kappa \log(|l|) + \frac{|\mathcal{O}_{\mathcal{L}}|}{|l|}\right)\right), \quad (1)$$

where $|\mathcal{O}_{\mathcal{L}}|$ is the number of all offsets in the training set.

As our final target application is tracking, where the scale is roughly known, parameters regarding the scale space, which are essential for the efficiency of modern object detection systems (see, *e.g.*, [6,14]), are not considered in the following. On the other hand, Eq. (1) reveals that the test time directly depends on the amount of training data, as for ISM based object detection each training sample is assigned one offset vector, *i.e.*, $|\mathcal{O}_{\mathcal{L}}| = |\mathcal{L}|$.

In the following, we analyze the parameters identified in Eq. (1). We structure them into those related to the complexity of the data representation and those related to the complexity of the classifier.

---

[2] http://www.d2.mpi-inf.mpg.de/andriluka_cvpr08

3.2 Data Complexity

Data complexity is manifested in three parameters, which are discussed in the following: the number of data samples n extracted from a test image, the number of feature channels F, and the number of image pixels I at which the features need to be computed:

Amount of Data Samples: Gall *et al.* showed [19] that sampling patches densely results in more accurate detections compared to sampling from interest points only [25]. On the other hand, *nearly dense* sampling strategies have proven to be almost as effective in terms of accuracy, whereas the number of samples and, thus, the runtime is significantly reduced. Increasing the sampling distance from one (dense) to two, already results in a speed-up of 40 % at the same level of accuracy (see Fig. 1a).

Feature Channels: Originally, it was proposed to use 32 feature channels, however, different features may be highly correlated, representing similar information (see *e.g.*, [8]). To this end, evaluations using different subsets of feature channels reveal that more than half of the channels may be simply omitted without any loss in accuracy. Especially by omitting the min- and max-filtration [19] the runtime can be reduced by approx. 20 % (see next paragraph and Fig. 1b, respectively).

Object Size: The smaller the scale at which an object can be detected, the less pixels have to be considered for feature computation. Hence, size matters. An according evaluation, starting with a unit object height of 100 pixels [19] is given in Fig. 1b. Moreover, filtration of the feature channels was found to be unnecessary for smaller scales since noise suppression is implicitly achieved by down-scaling. In fact, an additional filter step even decreases the discriminability at smaller scales. Thus – similar to the experiments regarding the number of data samples and feature channels – this points out that the model is strong enough to work well on a more compact object description. Please note that all results in Fig. 1b were generated without filtration of the feature channels but are relative to the standard setting, *i.e.*, with filtration.

3.3 Classifier Complexity

The complexity of the Hough Forest classifier is given by the number of trees, their depth, the complexity of a split tests, and by the complexity of the offset distribution. Due to space limitations we omit discussions on the depth and split complexity since these parameters did not yield a speed up without sacrificing accuracy for our experiments:

Number of Trees: The number of trees T of a Random Forest provides a simple way to trade off accuracy versus runtime. In fact, the runtime scales linearly with the number of trees in the forest, while the accuracy usually levels up at

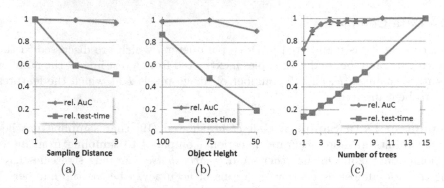

Fig. 1. Relative accuracy (in terms of *Area under the Curve* (AuC)) and relative test time per image for (a) different sampling distances during detection, (b) models trained on different object resolutions, and (c) using different forest sizes. For the results in (b) no min- and max-filtration were used, *i.e.*, 16 feature channels. The values are relative to the result obtained using the standard setting, *i.e.*, using dense sampling, object height: 100, patch size: 16 × 16, and using 32 channels (including min- and max-filtration). The error bars give the standard deviation for AuC.

some point (*c.f.* [28]). In our setup, the accuracy shows only a slight increase for $T > 4$ (see Fig. 1c).

Complexity of the offset distribution: A major drawback of Hough Forests, limiting their applicability, is that their test time depends on the amount of offset vectors stored at the leafs, and thus, on the amount of training data. While in [34] this drawback is addressed by summarizing voting information at the leaf nodes off-line, the authors of [21] employ a *grid-like* quantization of the offset distribution in order to update the distribution incrementally during online learning. Considering applications which require online learning as well as computational efficiency, we would like to take advantages of both ideas. Thus, we apply a grid-like approximation of the offset distribution on a coarse scale. The corresponding results over different grid resolutions compared to the results for vote compression using mean-shift as in [20] are shown in Fig. 2.

3.4 Discussion

The results presented in this section clearly show that we can enhance the efficiency of a Hough based object detector without negatively affecting the accuracy. In fact, integrating all findings gives an accuracy comparable to the original approach [19], while the computation is sped up by a factor of 40 (see Fig. 3). In particular, when combining the adaptions for all described parts the best results were obtained by using the following setup: two pixel sampling distance in x-/y-direction, an unit object height of 75 pixels, discarding the min-, and max-filtration of the feature channels, reducing the number of trees from 15 to 5 and summarizing the offset distributions by utilizing grids of 77 cells.

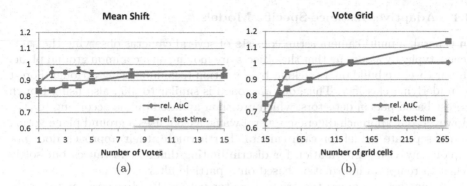

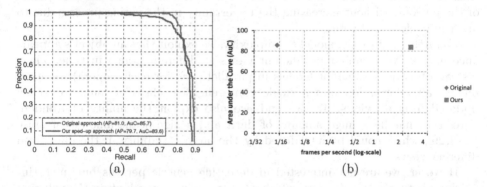

Fig. 2. Relative accuracy (in terms of *Area under the Curve* (AuC)) and relative test time per image: (a) As a function of the number of votes for vote compression by mean shift, and (b) as a function of the number of cells for vote compression by vote grid. Values are relative to voting with all offsets arrived at a leaf. The error bars give the standard deviation for AuC.

Fig. 3. Precision-Recall curves (a), and accuracy plotted vs. runtime (b) for the original [19], and our sped-up approach. It can be seen that the loss in accuracy is small, while the system is sped up by a factor of 40. (image size: 720×576)

Note that the findings of recent works regarding efficiency of scale space analysis [6,14] can be easily integrated, which bears the potential for an additional speed-up. Moreover, these adaptions are quite general and straightforwardly applicable to tracking in a tracking-by-detection manner.

4   Multi-camera Multi-instance Tracking

In the following, we first introduce a new forest-based multi-camera tracking method exploiting the findings from the previous section and then demonstrate the method in comparison to existing approaches for a realistic scenario.

4.1   Adaptive, Instance-Specific Models

In general, a multi-camera setup consists of several cameras observing the same scene, typically assuming that the objects are moving on a common ground plane. In our case, we build on a tracking-by-detection technique using a Hough Forest as underlying classifier. Thus, our approach is similar to [35], also using Hough Forest based person detectors, which are then updated online according to [17]. However, our approach differs in several ways, as [35] uses a ground plane voting and a separate model for each camera. Furthermore, their approach does not exploit any visual information for discriminating different instances but solely relies on temporal information based on a particle filter.

In contrast, we apply the original center voting [19]. However, at test time the votes are offset, such that they actually point to the ground plane. For that purpose, the offset from center to foot-point is computed based on the scale of the detected object. In this way, we limit the length of the learned vote vectors o_i, since it was shown that the performance of Hough voting schemes decrease with the length of those [10,12,34]. Hence, we can now vote for the foot-points of the persons, without increasing the uncertainty in the voting process due to artificially elongated votes.

To further reduce the model's complexity, we exploit that an object's appearance may be very similar in different views at different points in time. Thus, instead of learning different models in parallel (*i.e.*, for each camera) we accumulate all information into a single comprehensive classifier, simultaneously updated from all views. To this end, we also exploit the ability of Random Forests to handle a huge amount of data and a significant amount of noise [5,18,29], which could be introduced by the large variability in the data from different views.

However, we are not interested in detecting generic persons but in distinguishing between specific ones, which can be realized by adapting the splitting tests. Obviously, the splitting tests of a generic person detector are not very discriminative when aiming at separation of specific instances. For example, color information is not intensively used by a generic forest classifier, as it is not well suited for distinguishing between foreground and background. However, for instance separation this is apparently one of the most important cues. Hence, a generic person detector is not an appropriate choice for separation of specific instances. A Hough Forest – being a part-based detector – trained for person detection is able to segregate the individual parts of an object. For example, for person detection, parts like feet, the head, or the torso are clustered together. The color of each part is highly informative when it comes to discrimination between object instances. Thus, in order to differentiate between individual instances we extend the pre-trained model online by learning additional splitting tests, which are specifically optimized to separate the instances occurring in the scene.

To do so, the data samples extracted from the detected instances are traversed down the trees reaching the respective leafs. Similar to [33] we collect the samples together with their instance specific offsets at the leafs until a fixed number of samples n arrived. Subsequently, a split is optimized on a balanced sub-sample of $m\,|E|$ instance samples, where $|E|$ denotes the number of instances and $m < \frac{n}{|E|}$. In our experiments, this has shown to be preferable over sampling from the posterior distribution at the node. (For this work we set $m = 20$ and $n = 1.5m\,|E|$.) The process is repeated for three additional depth levels. Furthermore, in order to maintain the good generalization performance, we keep the samples from the pre-trained classifier, split them according to the selected splitting tests, and use them to build reliable statistics at the newly created leafs. Since only very few additional splits are generated, the samples need not be kept forever, which would obviously require infinite memory and hamper runtime. Instead, after a (pre-)defined number of splits, we keep the tree depth fixed and discard all collected samples. Subsequently, only the (class, instance, and offset) statistics are updated.

4.2 Experimental Results

To demonstrate the benefits of our approach, we pre-trained a Hough Forest on the publicly available INRIA person dataset[3] and compared our approach to three different methods. First, to a simple baseline based on mapping foreground masks obtained from a background subtraction onto the ground plane. Second, to Berlcaz et al. [7], which uses K-Shortest Patch (KSP) to link the detections obtained from probabilistic occupancy maps (POM). Third, to the most related approach of Sternig et al. [35], which also builds on Hough Forests.

All approaches were evaluated on *Set 1* from the publicly available dataset [31], which shows an indoor scene captured by three different cameras. Consisting of more than 2500 frames (where every tenth frame is annotated and used for evaluation), the scene shows three persons walking around, regularly occluding each other. The thus obtained results are presented in Table 1, where we show the averaged error (localization) on the ground plane. It can be seen that our single comprehensive classifier (*MultiInstanceHF*) outperforms not only the simple baseline but also the more sophisticated approaches of Berclaz et al. and Sternig et al.. The latter one is of particular interest, as we do not use an additional particle filter and ensure a much lower computational effort. Furthermore, illustrative results are given in Fig. 4.

Table 1. Mean pixel errors on ground plane for different approaches on the multi-camera sequence *Set 1* [31].

Method	Error (in pixel)
Background subtraction	75.7
Berclaz et al. [7]	106.3
Sternig et al. [35]	23.9
MultiInstanceHF	**18.8**

---

[3] http://pascal.inrialpes.fr/data/human/

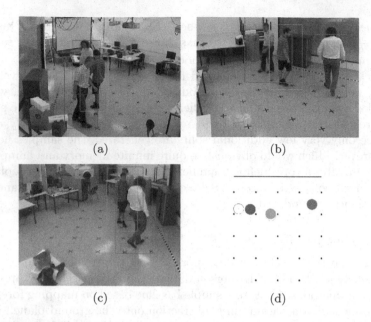

(a)   (b)

(c)   (d)

Fig. 4. Illustrative tracking results overlaid on the camera views (a-c) and the ground plane (d), where the filled circles represent the tracking result and the unfilled the corresponding ground truth annotations (best viewed in color and high definition).

5   Conclusion and Outlook

In this work, we revisited Hough Forests, a prominent approach to object detection, which has been successfully applied to numerous tasks within the field of computer vision. We pointed out that – using simple means – their runtime can be reduced by one to two orders of magnitude, while scoring in the same range of accuracy. This enables their use for applications with limited computational budget. The gathered insights were then exploited for tracking multiple instances from multiple cameras, where we showed that visual information can be as effective for instance discrimination as temporal information. This points out that instance discrimination should not be fully handed over to methods that only take temporal consistency into account. Instead, our work motivates an approach where both visual and temporal cues are incorporated.

Acknowledgment. This work was supported by the Austrian Science Foundation (FWF) project Advanced Learning for Tracking and Detection in Medical Workflow Analysis (I535-N23).

References

1. Amit, Y., Geman, D.: Randomized inquiries about shape; an application to hand-written digit recognition. Technical report 401, Department of Statistics, University of Chicago, IL (1994)
2. Andriluka, M., Roth, S., Schiele, B.: People-tracking-by-detection and people-detection-by-tracking. In: Proceedings of CVPR (2008)
3. Ballard, D.: Generalizing the hough transform to detect arbitrary shapes. Pattern Recognit. **13**(2), 111–122 (1981)
4. Barinova, O., Lempitsky, V.S., Kohli, P.: On detection of multiple object instances using hough transforms. IEEE Trans. PAMI **34**(9), 1773–1784 (2012)
5. Bauer, E., Kohavi, R.: An empirical comparison of voting classification algorithms: Bagging, boosting, and variants. Mach. Learn. **36**(1–2), 105–139 (1999)
6. Benenson, R., Mathias, M., Timofte, R., Van Gool, L.: Pedestrian detection at 100 frames per second. In: Proceedings of CVPR (2012)
7. Berclaz, J., Fleuret, F., Fua, P.: Multiple object tracking using k-shortest path optimization. IEEE Trans. PAMI **9**(33), 1806–1819 (2011)
8. Breiman, L.: Random forests. Mach. Learn. **45**(1), 5–32 (2001)
9. Butt, A.A., Collins, R.T.: Multi-target tracking by lagrangian relaxation to min-cost network flow. In: Proceedings of CVPR (2013)
10. Cootes, T.F., Ionita, M.C., Lindner, C., Sauer, P.: Robust and accurate shape model fitting using random forest regression voting. In: Fitzgibbon, A., Lazebnik, S., Perona, P., Sato, Y., Schmid, C. (eds.) ECCV 2012, Part VII. LNCS, vol. 7578, pp. 278–291. Springer, Heidelberg (2012)
11. Criminisi, A., Shotton, J., Konukoglu, E.: Decision forests: a unified framework for classification, regression, density estimation, manifold learning and semi-supervised learning. Found. Trends Comput. Graph. Vis. **7**(2–3), 81–227 (2012)
12. Dantone, M., Gall, J., Fanelli, G., Van Gool, L.: Real-time facial feature detection using conditional regression forests. In: Proc. CVPR (2012).
13. Dicle, C., Camps, O., Sznaier, M.: The way they move: tracking multiple targets with similar appearance. In: Proceedings of ICCV (2013)
14. Dollár, P., Belongie, S., Perona, P.: The fastest pedestrian detector in the west. In: Proceedings of BMVC (2010)
15. Eshel, R., Moses, Y.: Tracking in a dense crowd using multiple cameras. IJCV **88**(1), 129–143 (2010)
16. Everingham, M., Van Gool, L., Williams, C.K., Winn, J., Zisserman, A.: The pascal visual object classes (VOC) challenge. IJCV **88**(2), 303–338 (2010)
17. Gall, J., Razavi, N., Van Gool, L.: On-line adaption of class-specific codebooks for instance tracking. In: Proceedings of BMVC (2010)
18. Gall, J., Razavi, N., Van Gool, L.: An introduction to random forests for multi-class object detection. In: Dellaert, F., Frahm, J.-M., Pollefeys, M., Leal-Taixé, L., Rosenhahn, B. (eds.) Real-World Scene Analysis 2011. LNCS, vol. 7474, pp. 243–263. Springer, Heidelberg (2012)
19. Gall, J., Yao, A., Razavi, N., Van Gool, L., Lempitsky, V.S.: Hough forests for object detection, tracking, and action recognition. IEEE Trans. PAMI **33**(11), 2188–2202 (2011)
20. Girshick, R.B., Shotton, J., Kohli, P., Criminisi, A., Fitzgibbon, A.W.: Efficient regression of general-activity human poses from depth images. In: Proceedings of ICCV (2011)

21. Godec, M., Roth, P.M., Bischof, H.: Hough-based tracking of non-rigid objects. Comput. Vis. Image Underst. **117**(10), 1245–1256 (2013)
22. Kalal, Z., Mikolajczyk, K., Matas, J.: Tracking-learning-detection. IEEE Trans. PAMI **34**(7), 1409–1422 (2012)
23. Khan, S.M., Shah, M.: Tracking multiple occluding people by localizing on multiple scene planes. IEEE Trans. PAMI **31**(3), 505–519 (2009)
24. Küttel, D., Breitenstein, M.D., Van Gool, L., Ferrari, V.: What's going on? discovering spatio-temporal dependencies in dynamic scenes. In: Proceedings of CVPR (2010)
25. Leibe, B., Leonardis, A., Schiele, B.: Robust object detection with interleaved categorization and segmentation. IJCV **77**(1–3), 259–289 (2008)
26. Liu, J., Carr, P., Collins, R.T., Liu, Y.: Tracking sports players with context-conditioned motion models. In: Proceedings of CVPR (2013)
27. Okada, R.: Discriminative generalized hough transform for object dectection. In: Proceedings of ICCV (2009)
28. Özuysal, M., Calonder, M., Lepetit, V., Fua, P.: Fast keypoint recognition using random ferns. IEEE Trans. PAMI **32**(3), 448–461 (2010)
29. Perlich, C., Provost, F.J., Simonoff, J.S.: Tree induction vs. logistic regression: a learning-curve analysis. J. Mach. Learn. Res. **4**, 211–255 (2003)
30. Possegger, H., Sternig, S., Mauthner, T., Roth, P.M., Bischof, H.: Robust real-time tracking of multiple objects by volumetric mass densities. In: Proceedings of CVPR (2013)
31. Roth, P.M., Leistner, C., Berger, A., Bischof, H.: Multiple instance learning from multiple cameras. In: IEEE Workshop on Camera Networks (CVPR) (2010)
32. Schreiber, D., Cambrini, L., Biber, J., Sardy, B.: Online visual quality inspection for weld seams. Int. J. Adv. Manuf. Technol. **42**(5–6), 497–504 (2008)
33. Schulter, S., Leistner, C., Roth, P.M., Van Gool, L., Bischof, H.: On-line hough forests. In: Proceedings of BMVC (2011)
34. Shotton, J., Girshick, R.B., Fitzgibbon, A.W., Sharp, T., Cook, M., Finocchio, M., Moore, R., Kohli, P., Criminisi, A., Kipman, A., Blake, A.: Efficient human pose estimation from single depth images. IEEE Trans. PAMI **35**(12), 2821–2840 (2013)
35. Sternig, S., Mauthner, T., Irschara, A., Roth, P.M., Bischof, H.: Multi-camera multi-object tracking by robust hough-based homography projections. In: IEEE Workshop on Visual Surveillance (ICCV) (2011)
36. Tang, D., Liu, Y., Kim, T.K.: Fast pedestrian detection by cascaded random forest with dominant orientation templates. In: Proceedings of BMVC (2012)
37. Tang, D., Yu, T.H., Kim, T.K.: Real-time articulated hand pose estimation using semi-supervised transductive regression forests. In: Proceedings of ICCV (2013)
38. Tang, S., Andriluka, M., Milan, A., Schindler, K., Roth, S., Schiele, B.: Learning people detectors for tracking in crowded scenes. In: Proceedings of ICCV (2013)
39. Wohlhart, P., Donoser, M., Roth, P.M., Bischof, H.: Detecting partially occluded objects with an implicit shape model random field. In: Lee, K.M., Matsushita, Y., Rehg, J.M., Hu, Z. (eds.) ACCV 2012, Part I. LNCS, vol. 7724, pp. 302–315. Springer, Heidelberg (2013)
40. Roshan Zamir, A., Dehghan, A., Shah, M.: GMCP-tracker: global multi-object tracking using generalized minimum clique graphs. In: Fitzgibbon, A., Lazebnik, S., Perona, P., Sato, Y., Schmid, C. (eds.) ECCV 2012, Part II. LNCS, vol. 7573, pp. 343–356. Springer, Heidelberg (2012)

Graph-Based and Variational Minimization of Statistical Cost Functionals for 3D Segmentation of Aortic Dissections

Cosmin Adrian Morariu[1]([✉]), Tobias Terheiden[1], Daniel Sebastian Dohle[2], Konstantinos Tsagakis[2], and Josef Pauli[1]

[1] Intelligent Systems Group, Faculty of Engineering, University of Duisburg-Essen, Duisburg, Germany
adrian.morariu@uni-due.de
[2] Department of Thoracic and Cardiovascular Surgery, Universitatsklinikum Essen, Essen, Germany

Abstract. The objective of this contribution consists in segmenting dissected aortas in computed tomography angiography (CTA) data in order to obtain morphological specifics of each patient's vessel. Custom-designed stent-grafts represent the only possibility to enable minimally invasive endovascular techniques concerning Type A dissections, which emerge within the ascending aorta (AA). The localization of cross-sectional aortic boundaries within planes orthogonal to a rough aortic centerline relies on a multicriterial 3D graph-based method. In order to consider the often non-circular shape of the dissected aortic cross-sections, the initial circular contour detected in the localization step undergoes a deformation process in 2D, steered by either local or global statistical distribution metrics. The automatic segmentation provided by our novel approach, which widely applies for the delineation of tubular structures of variable shapes and heterogeneous intensities, is compared with ground truth provided by a vascular surgeon for 11 CTA datasets.

1 Introduction

Aortic dissection represents a pathology with imminently fatal outcome if not treated promptly. The pathophysiological formation process of an aortic dissection is being characterized by the occurrence of a false lumen caused by blood flow through a fissure in the inner lining of the aortic wall. A membrane, part of the original aortic wall, separates the false lumen from the true channel. Open surgical repair of aortic dissections is associated with perioperative mortality rates of up to 30 % and several long-term complications pertaining to elderly patients with multiple comorbidities [15, 22].

Minimally invasive techniques have proven successful in mitigating these negative effects of the classical procedure concerning Stanford Type B dissections as well as other pathologies affecting the descending aorta (e.g., aortic aneurysms) [15]. Stanford Type A dissections represent a higher challenge than Type B dissections for clinicians and equally for the computer vision recognition task, since

© Springer International Publishing Switzerland 2014
X. Jiang et al. (Eds.): GCPR 2014, LNCS 8753, pp. 511–522, 2014.
DOI: 10.1007/978-3-319-11752-2_42

Type B comprises only the descending aorta. The Type A pathology originates within the AA, but typically affects all parts of the aorta (ascending, arch and descending), sometimes also involving the branching vessels of the arch.

Bearing in mind that clinical datasets represent some of the most complex environments for pattern recognition, the universality of the presented graph-based and variational optimization strategies in regard to the segmentation of various tubular structures of heterogeneous shapes and intensities will be proven within the following sections by the successful handling of datasets containing extremely enlarged vessels, tortuous membranes or thrombosed subregions, upon which the intravenously injected contrast agent remains without influence.

2   Related Work

Increasing efforts have been directed in recent years towards segmenting pathological aortas (aneurysms, stenosis or dissections), as this task represents an imperative prerequisite for numerous clinical applications including surgical planning, catheter selection and stent-graft deployment. Dissected aortas are delineated in [13] under tedious manual interaction, where the user adjusts the orientation and the position of multiple slicing planes in order to obtain a tubular envelope with circular cross-sections as aortic model. A prototype-based approach for the segmentation of aneurysms is demonstrated in [7] by deriving a statistical model from given geometrical shapes of the targeted anatomy in the training set. The method described by [1] is applied to datasets without pathologies for the detection of thoracic aorta calcifications and implies a graph-based method relying on an entropy term. A segmentation scheme with focus on aortic dissections is discussed in [11]. By performing the Hough transformation (HT), circles are being extracted in several cross-sectional slices around the center of the aortic arch. They serve as an initial mesh for an elastically deformable model. The cross-sectional boundary of dissected aortas is represented in [16] by a piecewise constant curvature model and the segmentation process is performed within planes orthogonal to a rough aortic centerline. A homogeneity function in conjunction with radius constraints determines the update equation for the aortic region in circular sectors of equal degrees in polar coordinates. Reference [23] presents an important contribution towards endografting the aortic arch. An analytic intensity model serves to fit a 3D cylinder segment to the image intensities within a 3D ROI in order to estimate the vessel radius and centerline, the latter including position and orientation. Subsequently, the final segmentation result is projected onto a plane, which serves for the quantification of anatomical specifics essential to endoluminal repair. An intensity-based elastic image registration functional extends the method highlighted in [23], allowing for also non-circular aortic cross-sections [4].

3   Segmentation Technique for the Complete Aorta

Similarly to the initialization of the automated segmentation process in [16], the physician selects a subvolume of choice within the input dataset for further processing. The selection of two axial slices delimitating the desired subvolume is achieved by placing a point within the superior part of the aortic arch and two points within the ascending, respectively descending aorta, the lowest one determining the inferior limit of the subvolume. The three manually-placed points also serve for a cubic spline fitting as a coarse approximation of the aortic centerline. The spline is sampled at a constant interval of twice the slice spacing, wherein each sampled 3D point becomes the center of a planar region, extracted orthogonal to the spline (multiplanar reformat, MPR).

3.1   Localization Step: 3D Graph-Based Cost Minimization

The coarse localization of the aortic cross-section in each MPR-slice represents a core component of our segmentation framework. Due to the fact that our datasets (with up to 5 mm slice spacing) comprise vessels affected by severe pathologies and that elder patients often suffered prior surgical interventions leading to tortuous aortic trajectories, the fitted cubic spline only approximates the vessel's centerline in a harsh manner, as illustrated in Fig. 1 (a). Therefore, the selection of, e.g., the strongest HT accumulator peak in vicinity of the sampled spline point (MPR center) may be erroneous. Hence, we introduce a 3D multicriterial functional to select the optimal candidate for the vessel's cross-section in each MPR. In contrast to the approach proposed in [11], where the best circle in the HT accumulator was replaced by the second one depending on the probability to be an outlier, we establish $q = 10 + 4 \times 8 = 42$ (Fig. 1(c)) candidates in each MPR under utilization of the HT for cicles. A simple global search for the strongest 42 accumulator values unveils the inherent disadvantage that the relevant Hough peak may be omitted in case of weak edges, as depicted in Fig. 1(b). Dispensing the peak search on the cells of a 3×3-grid (Fig. 1(c)) ensures a uniform distribution of the peaks. For the marginal regions, 4 peaks are selected, and

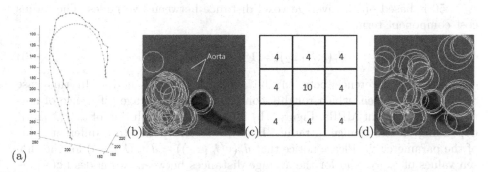

Fig. 1. (a) The aortic centerline resulting after the segmentation process (blue) may occasionally differ from the initial spline (green). (b) The $q = 42$ strongest Hough-Peaks. Lines indicate both lumina of the dissected aortic cross-section, which were not considered due to weak edges. (c) Local determination of the 42 peaks within a 3×3 grid. (d) Successful inclusion of the aortic cross-section within the candidates for the subsequent 3D optimal path search

from the center 10, since the probability to localize the aortic region within the MPR center is nevertheless greater. Figure 1(d) shows the result of this localized search.

Each of the q Hough peaks in one of the n MPRs represents a node $\mathcal{K}_{i,j}$ of the graph, with i the index of the corresponding MPR ($1 \leq i \leq n$) and $1 \leq j \leq q$. Each node $\mathcal{K}_{i,1..q}$ is unidirectionally connected to all nodes $\mathcal{K}_{i+1,1..q}$ belonging to the subsequent MPR $i+1$, resulting in a graph of size $n \times q$. Our goal consists in finding the most appropriate path $P = (\mathcal{K}_{1,j_1}, \mathcal{K}_{2,j_2}, ..., \mathcal{K}_{n,j_n})$ to approximate the vessel in 3D. A cost-transition matrix of $q \times q \times (n-1)$ elements contains for each edge the associated transition costs. Hence, we formulate the aortic localization as optimal path detection problem constrained by the cost functional

$$
\begin{aligned}
\mathcal{C}(\mathbf{c}(\mathcal{K}_{1 \rightarrow 2}), r(\mathcal{K}_{1 \rightarrow 2}), \bar{I}(\mathcal{K}_{1 \rightarrow 2}), v(\mathcal{K}_2)) &= w_{\mathbf{c}} d_{\mathbf{c}}(\mathbf{c}(\mathcal{K}_{1 \rightarrow 2})) + w_r d_r(r(\mathcal{K}_{1 \rightarrow 2})) \\
&\quad + w_{\bar{I}} d_{\bar{I}}(\bar{I}(\mathcal{K}_{1 \rightarrow 2})) + w_v d_v(v(\mathcal{K}_2)) \text{ , and } \mathbf{w} = (w_{\mathbf{c}}, \ w_r, \ w_{\bar{I}}, \ w_v),
\end{aligned}
\tag{1}
$$

with $\mathcal{K}_{1 \rightarrow 2}$ representing the transition from node $\mathcal{K}_1$ to $\mathcal{K}_2$ and $\mathbf{w} = (0.8, 0.5, 0.3, 0.15)$ the weighting vector for the cost components terms. We define the following symbols: the three-dimensional vector $\mathbf{c}(\mathcal{K}_i)$ comprises the coordinates of the circle center associated with node $\mathcal{K}_i$ and $r(\mathcal{K}_i)$ denotes the radius. Symbol $\bar{I}(\mathcal{K}_i)$ represents the mean intensity within the circular region, $v(\mathcal{K}_i)$ the corresponding variance of the intensity values. The distance cost component term

$$
d_{\mathbf{c}}(\mathbf{c}(\mathcal{K}_{1 \rightarrow 2})) = 1 - e^{\frac{-||\mathbf{c}(\mathcal{K}_1) - \mathbf{c}(\mathcal{K}_2)||_2}{\gamma_c}}
\tag{2}
$$

builds on the euclidean distance between the 3D coordinates of both circle centers. The exponential function, in conjunction with proper control parameters, ensures that all cost component terms are mapped to a similar range of values. Figure 2(a) shows values of $d_{\mathbf{c}}(\mathbf{c}(\mathcal{K}_{1 \rightarrow 2}))$ for different euclidean distances between centers of circular regions belonging to adjacent MPR. The choice of $\gamma_c = 50$ is based on the average voxel distance between two nodes. The radius cost component term

$$
d_r(r(\mathcal{K}_{1 \rightarrow 2})) = 1 - e^{\frac{-|r(\mathcal{K}_2) - r(\mathcal{K}_1)|}{\gamma_r}},
\tag{3}
$$

evaluates the difference between the radii of the two nodes. In this case $\gamma_r = 20$ was chosen, which roughly conforms to the average difference of aortic radii in adjacent MPRs. Figure 2(b) illustrates the evolution of $d_r(r(\mathcal{K}_{1 \rightarrow 2}))$ depending on the change in radius. Each plot corresponds to a different value of the parameter γ_r. Please notice that $d_{\mathbf{c}}(\mathbf{c}(\mathcal{K}_{1 \rightarrow 2})) \approx d_r(r(\mathcal{K}_{1 \rightarrow 2}))$ for the chosen values of γ_c, γ_r and for the average distances between two nodes $||\mathbf{c}(\mathcal{K}_1) - \mathbf{c}(\mathcal{K}_2)||_2 \approx 50$, $|r(\mathcal{K}_2) - r(\mathcal{K}_1)| \approx 20$. The third term in Eq. 1,

$$
d_{\bar{I}}(\bar{I}(\mathcal{K}_{1 \rightarrow 2})) = 1 - e^{\frac{-|\bar{I}(\mathcal{K}_1) - \bar{I}(\mathcal{K}_2)|}{\gamma_{\bar{I}}}}, \text{ with } \gamma_{\bar{I}} = \frac{1}{3},
\tag{4}
$$

restricts the permitted change in mean intensity $\bar{I}$ of the circular regions belonging to consecutive cross-sections. By sequentially restricting the change in mean

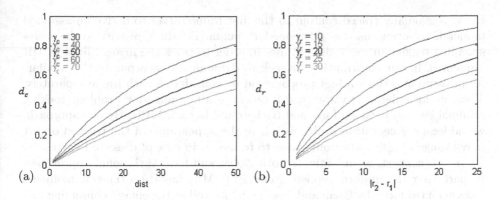

Fig. 2. (a) Distance costs for several γ_c values, (b) Radius costs for different γ_r values

intensity along the path, we take into account only one of the 3 nodes related to the input points provided by the clinician. By calculating the last cost component term

$$d_v(\mathcal{K}_2) = 1 - e^{\frac{-[\sum_{\mathbf{x}\in\mathcal{K}_2}(I(\mathbf{x})-\bar{I}_{exp})^2]}{\gamma_v}} \text{ , using } \gamma_v = 3, \tag{5}$$

two of the provided nodes are implicitly considered due to the expected mean intensity $\bar{I}_{exp}$. Expression $d_v(\mathcal{K}_2)$ demands a minimal variance of the intensity values within the circular region w.r.t. to $\bar{I}_{exp}$ for the currently regarded node. Depending on the position of this node, $\mathcal{K}_2$, along the path, the expected mean intensity $\bar{I}_{exp}$ for $\mathcal{K}_2$ is being calculated by linear interpolation between the mean intensities of starting and middle circular regions, respectively between the middle and the ending node.

We opt for Dijkstra's algorithm to minimize the multicriterial 3D cost functional, due to the fact that greedy algorithms, in contrast to dynamic programming methods, possess the capability to implicitly attribute higher priority to the starting nodes of a graph. Please note in this context that we build an unidirectional graph, in which, presuming several paths of equal costs, choosing the lower cost subpaths within the ending part of the graph, in detriment to the beginning nodes, would potentially select a completely erroneous route.

3.2 Embedding Distribution Metrics in a 2D Variational Framework

The results obtained after performing the localization step are used to initialize the dynamic embedment of statistical intensity distributions and moments in the segmentation process via level-set curve evolution, striving for an optimal separation of the aortic cross-section from the background. A statistical maximization of the distance between the kernel-based estimates of both class conditional densities [5], respectively the minimization of their mutual information, can be achieved by several criteria, such as the Kullback-Leibler divergence, the Maximum Likelihood principle or the Bhattacharyya coefficient. It has been shown that the latter measure performs better w.r.t. several use cases [10,14].

In [3], a manually traced contour in the first frame serves to derive an assumed model distribution, used as reference for maximizing the similarity to the corresponding regions in succeeding frames in order to track the myocardium in MR sequences. In our contribution we seek to dynamically incorporate this similarity measure into the level-set propagation by considering each frame separately. Thus, we formulate the variational problem as a minimization problem, seeking minimal overlap between fore- and background in each MPR. By our approach, we address segmentation problems where the appearance of the tracked object may change significantly from frame to frame, as in case of dissected aortas.

In this context, we investigate both global and local variational approaches. In particular the uniform modeling energy (UME) based on the raw moments (means) introduced by Chan and Vese in [9], as well as the energy combining histogram statistics (histogram separation energy, HSE) within the Bhattacharyya coefficient [14]

$$\mathfrak{B} = \int_n \sqrt{(P_f(n)P_b(n)}\, dn, \tag{6}$$

with $P_f(n)$ and $P_b(n)$ the intensity histograms of the foreground f and background region b, respectively. Variable n denotes the number of discrete intensity levels. Figure 3 (a) illustrates an aortic cross-section within an MPR-ROI, having the circular contour from the localization step overlayed. This contour represents the initial zero crossing of the implicit level-set function $\Phi(x)$, the zero-level-set $C = \{x|\Phi(x) = 0\}$ [18, 20].

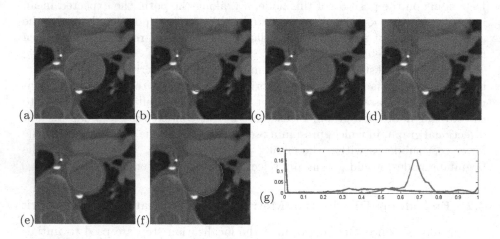

Fig. 3. (a)MPR-ROI with initial contour after localization step. (b), (c) Global UME with $\lambda = 0.75$ after 10, respectively 50 iterations. (d) Same energy with $\lambda = 10$ after 30 iterations. (e), (f) Global HSE flow after 120 iterations with $\lambda = 1$, respectively $\lambda = 0.01$. (g) Normalized statistical distributions of the foreground and background regions in preceding subfigure (red, resp. blue) enlighten the poor segmentation results yielded by the global flow

Global Flows. Figure 3 (b) and (c) depict the energy minimization by the *global UME* level-sets using a low weighting factor $\lambda = 0.75$ for the regularization term responsible for the contour smoothness. An extension of the curve to bone structures and other bright, adjoining tissue can be observed in the context of a background region revealing inhomogeneous intensities. This effect can be reduced, but not entirely prevented, by increasing the smoothness parameter λ (Fig. 3 (d)). However, in this case the contour only slightly deviates from the original circular shape.

The curve evolution related to the *global energy using the Bhattacharyya measure* in Eq. 6 in order to quantify the dissimilarity of statistical distributions is depicted in Fig. 3 (e), respectively (f), for different λ values (after 120 iterations). Obviously the value of the weighting factor for the regularization term plays a major role in steering the global HSE flow in this case. A closer look at the gray value histograms of the foreground (red) and background (blue) within Fig. 3 (g) sheds light on this behavior. For a broad range of the intensity spectrum the two normalized histograms are similar. Thus, computing a global HSE energy within medical CTA image data for the entire object and background fails to meet the crucial requirement of separable intensity profiles.

Local Flows. In order to improve the separability of the distribution metrics which guide the curve evolution process, we adopt the concept of calculating the statistical distributions only within a narrow-band around the contour [12]. Consequently, every contour pixel x_0 becomes the center of a circular region

$$\mathcal{R}(x_0, x) = \begin{cases} 1 & \text{for } ||x_0 - x|| < r \\ 0 & \text{otherwise} \end{cases} \tag{7}$$

of radius r, wherein only the pixels x included in $\mathcal{R}$ contribute to the energy term. With A_f and A_b the areas of the local foreground and local background regions, as well as $\mathfrak{B}$ the Bhattacharyya measure from Eq. 6 (computed locally), we derive the evolution equation of the *local HSE* flow as

$$\frac{\partial \Phi}{\partial t}(x_0) = \int_{\Omega_x} \frac{\mathcal{R}(x_0, x)\delta\Phi(x)}{2} \left[\mathfrak{B}\left(\frac{1}{A_b} - \frac{1}{A_f} \right) + \int_n \left(\frac{1}{A_f} \sqrt{\frac{P_b(n) + \varepsilon(n)}{P_f(n) + \varepsilon(n)}} \right. \right.$$
$$\left. \left. - \frac{1}{A_b} \sqrt{\frac{P_f(n) + \varepsilon(n)}{P_b(n) + \varepsilon(n)}} \right) dn \right] dx \cdot \delta\Phi(x_0) + \lambda \cdot \delta(\Phi(x_0)) \cdot div\left(\frac{\nabla\Phi(x_0)}{|\nabla\Phi(x_0)|} \right), \tag{8}$$

where $\delta\Phi$ represents the derivative of the Heaviside function. It assumes a nonzero value only for the zero-level-set (for the contour pixels), namely $\delta\Phi(x_0) = 1$. By not including a kernel density estimator in the update equation, but adding an infinitesimal value $\varepsilon(n)$ to all intensity bins, we achieve a greater separability of the local statistical distributions compared to [12] and the local flow accomplishes a satisfying segmentation result even for noisy image data (see Fig. 4 (a) and (b)). A kernel density estimator would have blended the local histogram samples to a broader spectrum, with the consequence of increasing the overall energy, a considerable disadvantage in our case. Experiments have shown that

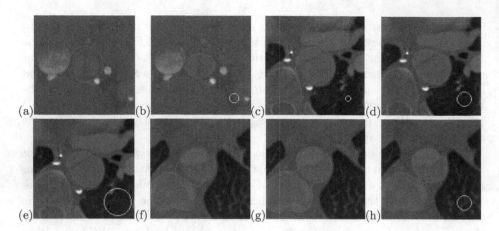

Fig. 4. (a) Initial contour in MPR from abdominal region. (b) Correct segmentation by the local HSE energy after 120 iterations with $\lambda = 0.01$ and $r = 5$. The cyan circle depicts the narrow-band's dimension around each contour pixel. (c),(d),(e) Local UME after 120 iterations and $\lambda = 0.75$ with the narrow-band radius $r = 3$, $r = 8$, respectively $r = 15$. (f) Shifted initial contour with the goal of assessing the robustness of the different flows under the circumstances of partially wrong initializations (g) Global UME ($\lambda = 0.75/$ 500 it.). (h) Local UME ($\lambda = 0.75/$ 500 it./ $r = 8$)

this energy provides best results for narrow-bands of relatively reduced dimension ($r \leq 7$) in order to prevent the boundary from spilling into the background.

The *local formulation of the UME* models foreground and background regions by the intensity averages of the local inner and outer regions in each contour point [12,21]. Figure 4 (c)–(e) illustrates the deformation of the initial contour from Fig. 3 (a) for different dimensions of the local mask $\mathcal{R}$. The number of iterations and the smoothing parameters are identical in all three cases. Excepting the case of very small radii of the local regions (e.g., $r = 3$), where insufficient image information leads to slight segmentation inaccuracies, this energy reveals an excellent convergence behaviour and robustness compared to the global formulation. Even partially wrong initial contours, i.e. the deliberately shifted initial contour in Fig. 4 (f), do not significantly affect the segmentation quality of the local flow (Fig. 4 (h)). The opposite effect occurs when employing the global energy: the contour expands to cover the spinal cross-section (Fig. 4 (g)). Due to its robustness against parameter changes and poor initializations, the local UME level set evolution scheme represents the method of choice for the accurate delineation of tubular structures in medical CT data. A grid search exposed $r = 8$, $\lambda = 0.75$ and 150 iterations as an appropriate parameter combination within this context.

Finally, the aortic boundaries delineated in cross-sectional MPRs are connected in 3D to a volume by a Multi-Level Partition of Unity Implicits (MPU) approach [6,17]. This method has proven more accurate than the quickhull algorithm [2] used in [16] when constructing a 3D aortic mesh based on cross-sections.

Table 1. Mean and standard deviation values (in percent) for the DSC, Precision and Recall, yielded by different segmentation approaches applied to aortic dissections

Segmentation method	DSC	Precision	Recall
Local UME	0.9422 ± 0.0085	0.9445 ± 0.0208	0.9404 ± 0.0149
Local HSE	0.8704 ± 0.0186	0.8304 ± 0.0266	0.9155 ± 0.0289
Global HSE	0.8382 ± 0.0714	0.7706 ± 0.1012	0.9263 ± 0.0567
Global UME	0.7697 ± 0.1038	0.6670 ± 0.1386	0.9280 ± 0.0484

4   Evaluation Results

The current section provides a quantitative assessment of the proposed graph-based and variational approaches targeting the minimization of statistical cost functionals for the segmentation and quantification of dissected aortas. In that perspective, a vascular surgeon produced ground truth data by performing manual annotations on the original axial slices of 11 CTA datasets from patients with aortic Type A dissection. The datasets from different Siemens CT scanners (Sensation 16, Emotion 6, Definition, Sensation 4) contain 87 to 1034 axial slices with a slice spacing ranging from 0.7 to 5 mm. The within-slice resolution of the axial slices (each of 512×512 voxels) varies between 0.445 and 0.863 mm.

The voxelwise comparison of the volumes resulting from automatic segmentation, respectively from ground truth serves to evaluate the accuracy of our proposed segmentation method in terms of the Dice Similarity Coefficient (DSC), Precision and Recall [19]. The 3D graph-based localization approach in conjunction with the 2D local UME curve evolution yields the highest DSC value of 94.2 % (Table 1). The DSC represents the harmonic mean of the 94.5 % Precision rate and the 94.0 % Recall rate.

Figure 5 shows several segmentation results for different CTA datasets obtained by intersecting the aortic volume, which has been generated from the 2D cross-sectional aortic boundaries by the MPU-method, with an axial, a coronal and a sagittal slice, respectively. The dataset containing the axial slice in Fig. 5(a) pertains to a patient with an extremely enlarged aorta (only the descending aorta depicted in the figure), wherein the thrombosed part of the false lumen exhibits a distinct contrast compared to the perfused areas. The successful segmentation of this dataset in spite of the abnormal aortic size and heterogeneous intensities demonstrates the capabilities of the proposed approach. The curve evolution based on local statistics also outperformes the global versions in quantitative terms, with both formulations having the optimal parameters adjusted. With Recall values not significantly lower, the global energies are noticeably surpassed by their local alternatives in terms of Precision, indicating that the lower convergence capabilities of the global methods occasionally lead to oversegmentations. In case that both lumina are perfused, as in Fig. 5(b) and (c), they are separated by a dissection membrane. Due to the embedment of *region-based* (and not edge-based) energies in the variational framework, the

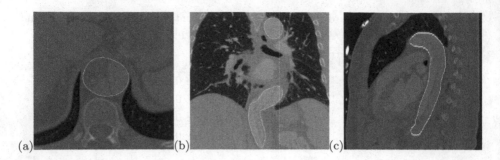

Fig. 5. Intersection of segmented 3D volumes with (a) an axial slice, respectively (b) with a coronal slice and (c) with a sagittal slice belonging to 3 different patients

thin membrane does not stop the contour evolution. The local flows accomplish this in conjunction with the sufficiently high radius $r = 8$. The robust initialization provided by the multicriterial 3D graph-based localization also plays a major part in the correct simultaneous segmentation of both lumina.

5   Discussion and Conclusions

By this contribution we introduce a genuine method for the segmentation of vessels affected by severe pathologies (e.g., dissections) in 3D CTA images. The approach has been evaluated on datasets comprising various pathological phenotypes, including extremely dilated vessels, highly folded dissection membranes or thrombosed subregions lacking contrast agent. Therefore, our framework would represent a viable alternative for the segmentation of tubular structures within other fields of application (e.g., defects in steel [8]). In the scope of general-purpose segmentation we embedded two types of distribution metrics, each of them in both global and local formulations, into a variational framework and demonstrated the superior capabilities of the local flows in regard to deliberately flawed initializations and to a broad range of parameter variations. Furthermore, the local analysis of image data emerges as advantageous in the context of delineating heterogeneous structures, where any assumed model distributions are deemed to fail. An advantage of our technique resides in not constraining the segmentation scheme to prior knowledge extracted from a finite training set, but using intrinsic statistical photometric data to recognize the targeted structures.

On application level, the highlighted segmentation approach serves as basis for contriving a lengths measurement scheme on the aortic surface, which will be subject of future publications. Aortic branch-vessel obstruction due to inadequate stent dimensioning or deployment could be efficiently avoided by predetermining key parameters related to the individual anatomy of each patient, paving the way to establishing minimally invasive endovascular techniques in the AA as a future common procedure in clinical practice.

References

1. Avila-Montes, O.C., Kukure, U., Kakadiaris, I.A.: Aorta segmentation in non-contrast cardiac CT images using an entropy-based cost function. In: SPIE Medical Imaging, pp. 76233J–76233J-8. International Society for Optics and Photonics (2010)
2. Barber, C.B., Dobkin, D.P., Huhdanpaa, H.: The quickhull algorithm for convex hulls. ACM Trans. Math. Softw. (TOMS) **22**(4), 469–483 (1996)
3. Ben Ayed, I., Li, S., Ross, I.: Embedding overlap priors in variational left ventricle tracking. IEEE Trans. Med. Imaging **28**(12), 1902–1913 (2009)
4. Biesdorf, A., Rohr, K., Feng, D., von Tengg-Kobligk, H., Rengier, F., Böckler, D., Kauczor, H.U., Wörz, S.: Segmentation and quantification of the aortic arch using joint 3D model-based segmentation and elastic image registration. Med. Image Anal. **16**(6), 1187–1201 (2012)
5. Bishop, C.M.: Neural networks for pattern recognition. Oxford University Press, Oxford (1995)
6. Braude, I., Marker, J., Museth, K., Nissanov, J., Breen, D.: Contour-based surface reconstruction using MPU implicit models. Graph. Models **69**(2), 139–157 (2007)
7. de Bruijne, M., van Ginneken, B., Viergever, M.A., Niessen, W.J.: Interactive segmentation of abdominal aortic aneurysms in CTA images. Med. Image Anal. **8**(2), 127–138 (2004)
8. Bürger, F., Buck, C., Luther, W., Pauli, J.: Image-based object classification of defects in steel using data-driven machine learning optimization. In: Proceedings of VISAPP 2014 - International Conference on Computer Vision Theory and Applications, Scitepress, pp. 143–152 (2014)
9. Chan, T.F., Vese, L.A.: Active contours without edges. IEEE Trans. Image Process. **10**(2), 266–277 (2001)
10. Freedman, D., Zhang, T.: Active contours for tracking distributions. IEEE Trans. Image Process. **13**(4), 518–526 (2004)
11. Kovács, T.: Automatic segmentation of the vessel lumen from 3D CTA images of aortic dissection. Diss., Eidgenössische Technische Hochschule ETH Zürich, Nr. 19167 (2010)
12. Lankton, S., Tannenbaum, A.: Localizing region-based active contours. IEEE Trans. Image Process. **17**(11), 2029–2039 (2008)
13. Lubniewski, P.J., Miguel, B., Sauvage, V., Lohou, C.: Interactive 3D segmentation by tubular envelope model for the aorta treatment. In: IS&T/SPIE Electronic Imaging, pp. 82901F–82901F-14. International Society for Optics and Photonics (2012)
14. Michailovich, O., Rathi, Y., Tannenbaum, A.: Image segmentation using active contours driven by the Bhattacharyya gradient flow. IEEE Trans. Image Process. **16**(11), 2787–2801 (2007)
15. Moon, M.C., Greenberg, R.K., Morales, J.P., Martin, Z., Lu, Q., Dowdall, J.F., Hernandez, A.V.: Computed tomography-based anatomic characterization of proximal aortic dissection with consideration for endovascular candidacy. J. Vasc. Surg. **53**(4), 942–949 (2011)
16. Morariu, C.A., Dohle, D.S., Terheiden, T., Tsagakis, K., Pauli, J.: Polar-based aortic segmentation in 3D CTA dissection data using a piecewise constant curvature model. Bildverarbeitung für die Medizin 2014, pp. 390–395. Springer, Heidelberg (2014)

17. Ohtake, Y., Belyaev, A., Alexa, M., Turk, G., Seidel, H.P.: Multi-level partition of unity implicits. ACM Trans. Graph. **22**(3), 463–470 (2003)
18. Osher, s, Fedkiw, R.: Level set methods and dynamic implicit surfaces. Applied Mathematical Sciences, vol. 153. Springer, New York (2003)
19. Powers, D.M.: Evaluation: from precision, recall and F-measure to ROC, informedness, markedness & correlation. J. Mach. Learn. Technol. **2**(1), 37–63 (2011)
20. Sethian, J.A.: Level set methods and fast marching methods: evolving interfaces in computational geometry, fluid mechanics, computer vision, and materials science, vol. 3. Cambridge University Press, Cambridge (1999)
21. Wang, X.F., Huang, D.S., Xu, H.: An efficient local Chan-Vese model for image segmentation. Pattern Recogn. **43**(3), 603–618 (2010)
22. Wendt, D., Thielmann, M., Melzer, A., Benedik, J., Droc, I., Tsagakis, K., Dohle, D.S., Jakob, H., Abele, J.E.: The past, present and future of minimally invasive therapy in endovascular interventions: A review and speculative outlook. Minim. Invasive Ther. Allied Technol. **22**(4), 242–253 (2013)
23. Wörz, S., von Tengg-Kobligk, H., Henninger, V., Rengier, F., Schumacher, H., Böckler, D., Kauczor, H.U., Rohr, K.: 3-D quantification of the aortic arch morphology in 3-D CTA data for endovascular aortic repair. IEEE Trans. Biomed. Eng. **57**(10), 2359–2368 (2010)

Mask-Specific Inpainting with Deep Neural Networks

Rolf Köhler$^{(\boxtimes)}$, Christian Schuler, Bernhard Schölkopf, and Stefan Harmeling

Max Planck Institute for Intelligent Systems, Tübingen, Germany
rolf.koehler@tuebingen.mpg.de

Abstract. Most *inpainting* approaches require a good image model to infer the unknown pixels. In this work, we directly learn a mapping from image patches, corrupted by missing pixels, onto complete image patches. This mapping is represented as a deep neural network that is automatically trained on a large image data set. In particular, we are interested in the question whether it is helpful to exploit the shape information of the missing regions, i.e. the masks, which is something commonly ignored by other approaches. In comprehensive experiments on various images, we demonstrate that our learning-based approach is able to use this extra information and can achieve state-of-the-art inpainting results. Furthermore, we show that training with such extra information is useful for *blind* inpainting, where the exact shape of the missing region might be uncertain, for instance due to aliasing effects.

Keywords: Inpainting · Deep learning · Neural-nets · Multi-layer-perceptrons

1 Introduction and Related Work

Image inpainting tries to fill-in missing parts of an image. Commonly, one can distinguish two settings, where pixels in an image are missing:

(i) In the first setting, the goal is to manipulate an existing image. Usually, some image details or larger regions of a given image should be removed. The resulting hole must be filled in to create a *plausible* complete image. For instance, consider an image with two persons, where one person should disappear. Then the task of image inpainting is to fill-in the resulting (possibly large) hole with some background textures or patterns. The goal is not to recover a *true* image but one that looks realistic. Many successful methods (some based on texture synthesis [8]) have been proposed in the past, for instance [5,7,13,17,26] and references therein. These ideas can be also generalized to video inpainting [23].

(ii) The second setting considers an image that is locally corrupted, for example by super-imposed text or scratches. In this case, the missing regions are small but possibly all over the image. The goal is to recover an image which is as close as possible to the *true* image. Also for this setting many good approaches exist which will be discussed in the following.

© Springer International Publishing Switzerland 2014
X. Jiang et al. (Eds.): GCPR 2014, LNCS 8753, pp. 523–534, 2014.
DOI: 10.1007/978-3-319-11752-2_43

Both settings are relevant for image processing, however, in this paper we solely consider the second setting. The existing methods for the second setting can be categorized into two groups:

(i) Many classical approaches are *diffusion-based* methods that propagate the local information, such as edges and gradients, from the boundary to the missing pixels, see e.g. [2,4,16].
(ii) The second class is based on *sparse representations* using dictionaries [9,11,14]. General purpose image priors based on Markov random fields (MRFs) can be learned on image databases [18,20]. These have been also successfully applied to inpainting.

While the existing methods lead to impressive image reconstruction results, none of them is exploiting the shape of the mask for inpainting. The (binary) mask has the same size as the image and is 1 for pixels which are missing in the corrupted image and 0 for all other pixels. In this paper we show how this additional information can be utilized to obtain better image reconstruction results. For this we choose a task-specific learning approach employing deep neural networks since they have recently been successfully applied to several other image restoration problems, e.g. to image denoising [3] or to image deblurring [21]. Closest to our work is [25] who apply a deep learning approach to both denoising and inpainting. They are also able to do blind inpainting (as we do in Sect. 3.4), but do not use the mask information.

In a nutshell, the contributions of the present paper are as follows:

- We show that a mask-specific inpainting method can be learned with neural networks, which leads to better results.
- We show that it is relevant to train the inpainting method with the correct masks.
- We show that by training an inpainting method for masks generated with certain fonts, it is possible to *blindly* inpaint an image, i.e. without knowing the locations of the missing pixels.

2   Learning Mask-Specific Inpainting Methods

The overall idea is to train a neural network to map corrupted image patches (=rectangular parts from the image) to their uncorrupted counterparts. This mapping is applied to all patches of a corrupted image. The recovered patches are averaged at their overlapping parts to obtain the reconstructed image. In case the true mask for the whole picture is given, the known pixels from the input image are directly copied to the reconstructed image. Note that the input patches are usually larger than the output patches, which matches the intuition that the missing pixels can be recovered by considering some large enough area around them. In the following we briefly recall multi-layer perceptrons (MLPs) and explain the training procedure.

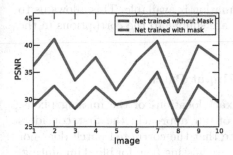
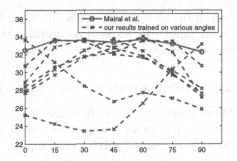

Fig. 1. PSNR for 10 test images for an MLP that takes only the corrupted patch as input (blue line) and an MLP that additionally has the masking patch as input (green line) (Color figure online).

Fig. 2. Performance of the MLP trained on different angles compared against the method by [14]. The comparison against [20] looks similar. The performance of the neural net depends on the orientation of the bars and is best for the orientation on which the neural net was trained on.

2.1 Multi-layer Perceptrons

A multi-layer perceptron is a nonlinear function f that maps input vectors x onto output vectors y. We follow the notation of [3]. An MLP with two hidden layers can be written as

$$f(x) = b_3 + W_3 \tanh(b_2 + W_2 \tanh(b_1 + W_1 x)) \tag{1}$$

(from Eq. (1) in [3]) where vectors b_1, b_2, b_3 and the matrices W_1, W_2, W_3 are the parameters of the MLP. More generally we denote the architecture of an MLP by some integer tuple: e.g. a $(256, 1024, 1024, 64)$-MLP has a 256-dimensional input layer, two 1024-dimensional hidden layers and a 64-dimensional output layer.

2.2 Mask-Specific Training

To adjust the parameters of the MLP, we need training data that consists of pairs of input patches x and output patches y. Using these we can automatically *learn* the mapping using the backpropagation algorithm [12,19]. To speed up convergence we use the ADADELTA method [27] that automatically adapts parameter-specific learning rates (fixing the decay rate $\rho = 0.95$, the conditioning constant $\epsilon = 10^{-6}$, and batch size to 128).

To generate large amounts of training pairs we randomly select image patches from an image database (we used imagenet [6] for training). These extracted patches are the uncorrupted output patches.

To corrupt those patches we utilize the knowledge about the masks. For instance for super-imposed text we corrupt the output patches by adding random text of the same font and size, if that information is available. This allows us to create input patches with corruptions that are similar to the corruptions in the test image.

2.3   Training with and w/o Mask as Input Patch

Many inpainting algorithms require the exact locations of the missing pixels. For this reason we consider two versions of our approach: the first considers only the corrupted patch as the input. Note that the corrupted patch does not always contain the information which pixels are missing (e.g. for blind inpainting, see Sect. 3.4). Version two of our approach requires the corrupted patch and additionally the masking patch as the input.

To show that feeding the masking patch additionally helps, we performed a comparison of the two approaches on 10 test images (randomly selected from the Berkeley segmentation dataset [15]) with super-imposed text. Figure 1 shows the results of two neural nets, both with architecture $(16^2, 512, 512, 512, 8^2)$ which were trained for the same amount of epochs. The first MLP was trained with the mask as an additional input, the second MLP was trained without the mask, but with correctly corrupted patches, i.e. using the correct font and text size. In all cases the MLP with the additional input was better. See also Fig. 3 for example images.

2.4   Training with the Correct and Wrong Masks

In this section we demonstrate the advantage that is gained through incorporating the correct mask into the learning process. For that purpose we generate

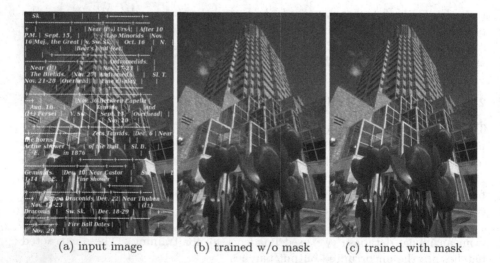

(a) input image   (b) trained w/o mask   (c) trained with mask

Fig. 3. The performance of the neural net improves if a mask is included in the training process. It can be seen that in image (b), artifacts, especially in the sky, are visible.

several masks which are similar. We start with a mask, showing vertical bars of size 14×3 pixels. We rotate those bars repeatedly by 15 degrees. We do not allow aliasing, but only binary masks, so each pixel in the mask which is greater than 0 after rotation is set to 1.

We trained several MLPs with the architecture $(16^2, 512, 512, 512, 8^2)$ using training images generated with these different masks. For each of those angles, also 10 test images are generated (10 randomly picked images from the Berkeley segmentation dataset [15]).

Figure 2 shows that the neural net performs best on the angle it was trained on, and that the performance deteriorates quickly for other angles. The methods by [14,20] perform about the same for all angles, no matter in what way the bars are orientated. Our approach is better if the correct angle has been considered during training time.

3 Experiments

We always trained the MLPs on gray-scale images. To apply them to color we applied the same MLP separately to the three color channels. To generate training patches we used 1.8 million color images from imagenet [6], which we converted to gray-scale. Note that in general the larger the architecture of the neural network, the better the results, but the longer the training time until convergence. In average the training time until convergence is 3 to 5 days.

Note that we did not have access to implementations for all competing methods. For Figs. 4, 5 and 6 we applied all algorithms that were available to us. Furthermore, we included for Fig. 5 those methods which published results on the "New Orleans" image in their papers. Figure 6 shows a comparison on images from [26].

3.1 Comparison with Horizontal/Vertical Lines

We perform an experiment with horizontal and vertical lines painted on an image. We used a $(16^2, 1024, 1024, 1024, 8^2)$ architecture. As can be seen in Fig. 4,

input, PSNR 17.36   [1], PSNR 32.91   [20], PSNR 35.31   [14], PSNR 35.32   ours, PSNR **37.34**

Fig. 4. Comparison on an image with line artifacts. Note that the chest-part of the shirt shows fewer artifacts with our approach than it does with the other methods.

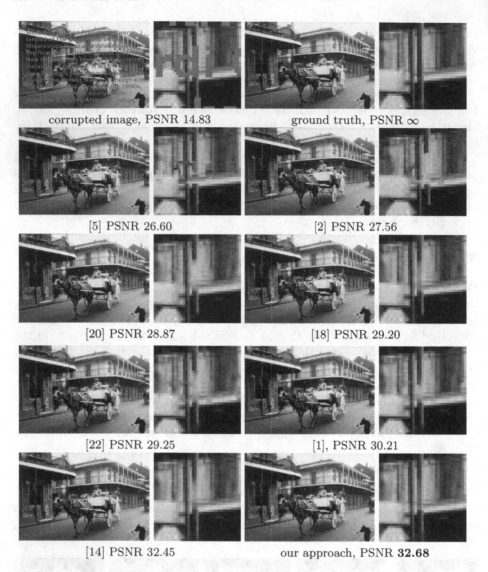

Fig. 5. Comparison on the image "New Orleans" from [2]. While [14] and our approach are close in terms of PSNR, the enlarged image detail (see green box in the corrupted image) shows that our approach is better able to recover the pole (Color figure online).

we achieve a better PSNR than the methods by [14, 20]. This is especially due to the fact that the MLP is able to inpaint the stripes on the shirt in a much better way than the other algorithms.

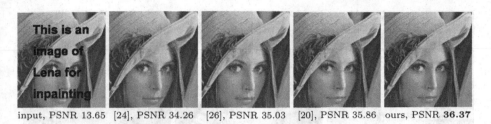

input, PSNR 13.65   [24], PSNR 34.26   [26], PSNR 35.03   [20], PSNR 35.86   ours, PSNR **36.37**

Fig. 6. Comparison on an image from [26].

3.2 Comparison on the New Orleans Image

We compare our method against other methods on the New Orleans image, used by [2]. For this comparison a $(39^2, 2047, 2047, 2047, 2047, 13^2)$ architecture was trained. Visually our approach performs better than the approaches by [1, 2, 5, 18, 20, 22]. Our result looks similar to the result of [14], but we are still able to recover more detailed information, e.g. the pole (marked with a green box in the corrupted image). One reason why we are better than the second best performing method [14] might be that we are able to use larger patch sizes. We used input patches of size 39×39 pixels, whereas [14] used a dictionary with patch size 9×9 pixels.

Note that for this example the original image was corrupted with some text of which the true mask is available. We used exactly this mask to corrupt images for the training set.

3.3 Comparison Against Images from Xu and Sun [26]

Figure 6 shows the results for one out of five images from [26] (Fig. 8 in their paper). In Sect. 3.5 we show the limitations of our proposed method on one of the other images. The neural net (with architecture $(39^2, 2047, 2047, 2047, 13^2)$) was trained with the given masks from [26]. On the Lena image we achieve better results (in terms of PSNR) than all other algorithms.

3.4 Comparison for Blind Inpainting

With an MLP we are also able to inpaint images without the exact knowledge of the mask while other inpainting methods do require a binary mask as an input. Figure 7 shows results for two images, on which random text was written with the same font and font-size as used for the training procedure of the MLP.

Note that the text written on the image is aliased, meaning it is not binary. Extracting a mask from such an image is rather tedious, as the optimal mask cannot be found by just thresholding the image. However, identifying the font and font-size is often possible and provides important information for the training process.

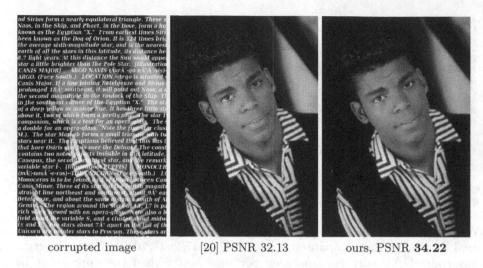

corrupted image   [20] PSNR 32.13   ours, PSNR **34.22**

Fig. 7. Inpainting an image without knowledge of the exact mask of each image. Only the font and font-size of the masks is known for training the neural net.

Optimally we would compare against [25] who also consider a deep learning approach to inpainting (but ignoring the mask based training). Unfortunately, we were not able to obtain their images nor their code, so we have to postpone this comparison for a later publication. For this reason we compare only against the method of [20] which seems to be our closest competitor in terms of PSNR. Since the latter method requires the mask we additionally provided it to their method. Although our blind approach did not have access to the mask, it was able to reconstruct a better result.

We see that our method can be used to automatically inpaint images with super-imposed text or for automatic removing of watermarks or logos. It can also be used to batch inpaint several images which were damaged by the same type of corruption, without creating a mask for each corrupted image. Though this method does not perform as well as with the mask as an input (see Sect. 2.3), it is a good way to inpaint several images at once without the necessity of the user to determine the pixels of an image which are corrupted.

3.5   Limitations

The proposed method performs well if the holes to be filled-in are small. If the holes are too large the inpainting result gets blurry, see the two left images of Fig. 8. While the sky is inpainted in an appropriate way (since it is smooth), the bridge pylons and the grass on the left part of the image are inpainted in a blurry way. Similarly, in the two right images of Fig. 8 we see that across the grass the inpainted regions appear blurry. This is similar to the phenomenon also present with diffusion-based inpainting methods. This is probably due to the fact that the nonlinear filter learned by the MLP can only propagate a few pixels into the

Fig. 8. Limitations of our approach: large holes (10×10 squares) result in blurry output on textured regions, e.g. at the poles of the bridge. Similar limitations can be seen on the horse image from [26] (line-width: 10 pixels).

mask, but fails to fill-in larger regions. It seems that the neural net used in this approach is only able to learn to continue image information along isophotes.

4   Towards Understanding the Trained Neural Network

A common criticism of methods based on neural networks is that they work like a black box, i.e. even though they are able to reach state-of-the-art performance it remains unclear how the task is solved. To gain insight into how the trained neural network achieves its performance, we study the feature generators for MLPs trained with different distortion types and look how the input feature depend on the shape of the masks.

4.1   Recognize and Play Back

Activation maximization [10] finds, for a given neuron in the last hidden layer, the respective input that maximizes that neuron's activation, while constraining the input's L_2-norm. The neurons in the last hidden layer are of interest since they are the *feature generators*, meaning the weights from these neurons to the output layer comprise the features that are linearly superimposed to reconstruct the image. We here analyze the neural net trained for Sect. 3.2. For Fig. 9, we additionally fixed the inpainting mask to be the letter 'e'. The bottom row depicts typical 13×13 outputs of the feature generators, the top row the maximizing 39×39 input. For the Gabor-like feature in the left column, the MLP tries to detect a continuation of the feature outside of the output size, which also motivates the importance of larger input than output patches. In the second column, the feature is not obstructed by the mask and is therefore just copied to the output. The feature in the third column is impossible to reconstruct with the given mask, the result of the activation maximization is only L_2-constrained noise. In the fourth column, the feature is partly visible, however, the input is similar to the previous column, indicating that either the MLP is not yet optimal or that the non-convex activation maximization converged to a local minimum.

Intuitively speaking, it seems that the neural network is able to detect certain basic features which are then generated without missing pixels in the last layers. So image features are detected even though pixels are missing, and played back including the missing information.

Fig. 9. Input patterns (top row) maximizing the activation of eight of the MLP's feature generators (bottom row) for a given inpainting mask (shown in red), the location of the output patch is marked with a green hairline. The MLP reconstructs features in the output image by trying to find a corresponding input pattern in regions not obstructed by the mask (Color figure online).

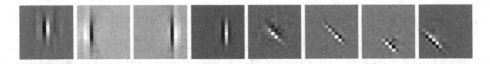

Fig. 10. Selection of weights for the first layer of two MLPs trained on images corrupted with vertical bars (left 4 images) or with 45 diagonal bars (right 4 images). It can be seen that the filters learned by the MLPs are dependent on the distortion type (shape of the mask) given to the MLP.

4.2 Input Features Depend on the Masks

In Sect. 2.4 we showed that the inpainting performance depends on the masks used for training. To gain additional insight, we can also look at the weights for the first layer, which are shown in Fig. 10. The first four images show weights of a neural network trained on vertical bars, while the second four images on the right were trained on diagonal bars. In general we see that the MLP learns mask specific feature detectors. It would be interesting to better understand the other features that appear; those currently have no obvious interpretation.

5 Conclusion

A purely learning-based inpainting approach has the advantage that it is easy to incorporate further information, like the shape of the mask: simply include that information into the input layer and let the training procedure figure out how to make best use of it. Implementing this idea, we are able to show that our inpainting approach based on deep neural networks is able to compete with the currently best inpainting methods that are based on other principles.

Clearly, the mask specific training makes the solution more specific, with the limitation that a trained network will not perform optimally if trained on the wrong mask.

Surprisingly, it is also possible to train the neural network to *blindly* inpaint an image. This is possible if we know the corruption process, e.g. font and font-size, because we can then generate characteristic data to train the neural network. This possibility goes beyond the capabilities of the usual approaches to inpainting which commonly require the exact location of the missing pixels.

References

1. Bertalmio, M., Bertozzi, A.L., Sapiro, G.: Navier-stokes, fluid dynamics, and image and video inpainting. In: 2001 IEEE Conference on Computer Vision and Pattern Recognition (CVPR), pp. I-355. IEEE (2001)
2. Bertalmio, M., Sapiro, G., Caselles, V., Ballester, C.: Image inpainting. In: Proceedings of the 27th Annual Conference on Computer Graphics and Interactive Techniques, pp. 417–424. ACM Press/Addison-Wesley Publishing Co. (2000)
3. Burger, H.C., Schuler, C.J., Harmeling, S.: Image denoising: can plain neural networks compete with bm3d? In: 2012 IEEE Conference on Computer Vision and Pattern Recognition (CVPR), pp. 2392–2399. IEEE (2012)
4. Chan, T.F., Shen, J.: Nontexture inpainting by curvature-driven diffusions. J. Vis. Commun. Image Represent. **12**(4), 436–449 (2001)
5. Criminisi, A., Pérez, P., Toyama, K.: Region filling and object removal by exemplar-based image inpainting. IEEE Trans. Image Process. **13**(9), 1200–1212 (2004)
6. Deng, J., Dong, W., Socher, R., Li, L.J., Li, K., Fei-Fei, L.: Imagenet: a large-scale hierarchical image database. In: 2009 IEEE Conference on Computer Vision and Pattern Recognition (CVPR), pp. 248–255. IEEE (2009)
7. Drori, I., Cohen-Or, D., Yeshurun, H.: Fragment-based image completion. ACM Trans. Graph. (TOG) - Proc. ACM SIGGRAPH **22**(3), 303–312 (2003)
8. Efros, A.A., Leung, T.K.: Texture synthesis by non-parametric sampling. In: The Proceedings of the Seventh IEEE International Conference on Computer Vision, vol. 2, pp. 1033–1038. IEEE (1999)
9. Elad, M., Starck, J.L., Querre, P., Donoho, D.L.: Simultaneous cartoon and texture image inpainting using morphological component analysis (MCA). Appl. Comput. Harmon. Anal. **19**(3), 340–358 (2005)
10. Erhan, D., Courville, A., Bengio, Y.: Understanding representations learned in deep architectures. Technical report, Technical Report 1355, Université de Montréal/DIRO (2010)
11. Fadili, M.J., Starck, J.L., Murtagh, F.: Inpainting and zooming using sparse representations. Comput. J. **52**(1), 64–79 (2009)
12. LeCun, Y.A., Bottou, L., Orr, G.B., Müller, K.-R.: Efficient BackProp. In: Orr, G.B., Müller, K.-R. (eds.) NIPS-WS 1996. LNCS, vol. 1524, pp. 9–50. Springer, Heidelberg (1998)
13. Liu, Y., Caselles, V.: Exemplar-based image inpainting using multiscale graph cuts. IEEE Trans. Image Process. **22**(5), 1699–1711 (2013)
14. Mairal, J., Elad, M., Sapiro, G.: Sparse representation for color image restoration. IEEE Trans. Image Process. **17**(1), 53–69 (2008)
15. Martin, D., Fowlkes, C., Tal, D., Malik, J.: A database of human segmented natural images and its application to evaluating segmentation algorithms and measuring ecological statistics. In: Proceedings of 8th International Conference on Computer Vision, vol. 2, pp. 416–423, July 2001

16. Masnou, S., Morel, J.M.: Level lines based disocclusion. In: Proceedings of the 1998 International Conference on Image Processing, ICIP 98, pp. 259–263. IEEE (1998)
17. Pérez, P., Gangnet, M., Blake, A.: Patchworks: example-based region tiling for image editing. Microsoft Research, Redmond, WA, Technical report MSR-TR-2004-04, pp. 1–8 (2004)
18. Roth, S., Black, M.J.: Fields of experts: a framework for learning image priors. In: IEEE Conference on Computer Vision and Pattern Recognition (CVPR), pp. 860–867. IEEE (2005)
19. Rumelhart, D.E., Hintont, G.E., Williams, R.J.: Learning representations by back-propagating errors. Nature **323**(6088), 533–536 (1986)
20. Schmidt, U., Gao, Q., Roth, S.: A generative perspective on MRFs in low-level vision. In: 2010 IEEE Conference on Computer Vision and Pattern Recognition (CVPR), pp. 1751–1758. IEEE (2010)
21. Schuler, C.J., Burger, H.C., Harmeling, S., Scholkopf, B.: A machine learning approach for non-blind image deconvolution. In: 2013 IEEE Conference on Computer Vision and Pattern Recognition (CVPR), pp. 1067–1074. IEEE (2013)
22. Telea, A.: An image inpainting technique based on the fast marching method. J. Graph. Tools **9**(1), 23–34 (2004)
23. Wexler, Y., Shechtman, E., Irani, M.: Space-time completion of video. IEEE Trans. Pattern Anal. Mach. Intell. **29**(3), 463–476 (2007)
24. Wong, A., Orchard, J.: A nonlocal-means approach to exemplar-based inpainting. In: 15th IEEE International Conference on Image Processing, ICIP 2008, pp. 2600–2603. IEEE (2008)
25. Xie, J., Xu, L., Chen, E.: Image denoising and inpainting with deep neural networks. In: NIPS, pp. 350–358 (2012)
26. Xu, Z., Sun, J.: Image inpainting by patch propagation using patch sparsity. IEEE Trans. Image Process. **19**(5), 1153–1165 (2010)
27. Zeiler, M.D.: Adadelta: an adaptive learning rate method. arXiv preprint arXiv:1212.5701 (2012)

Detection of Clustered Objects in Sparse Point Clouds Through 2D Classification and Quadric Filtering

Christopher Herbon[1]([✉]), Benjamin Otte[1], Klaus Tönnies[2], and Bernd Stock[1]

[1] HAWK Fakultät Naturwissenschaften und Technik, Von-Ossietzky-Straße 99,
37085 Göttingen, Germany
{herbon,otte,stock}@hawk-hhg.de
[2] Institut für Simulation und Graphik, Otto-von-Guericke-Universität Magdeburg,
Universitätsplatz 2, 39106 Magdeburg, Germany
klaus@isg.cs.uni-magdeburg.de

Abstract. A novel approach for detecting single objects in large clusters is presented. The proposed method is designed to work with structure from motion data, which typically includes a set of input images, a very sparse point cloud and camera poses. We use provided objects of interest from 2D classification, which are then projected to three dimensional space.

The main contribution of this paper is an algorithm, which accurately detects the objects of interest and approximates their locations in three dimensional space, by using 2D classification data and quadric filtering. Optionally, a partly dense reconstructed mesh, containing objects of interest only, is computed, without the need for applying patch based multiple view stereo algorithms first. Experiments are performed on a challenging database containing images of wood log piles with a known ground truth number of objects, provided by timber processing companies. The average true positive rate exceeds 98.0 % in every case, while it is shown how to reduce the false positive rate to less than 0.5 %.

1 Introduction

Image-based object detection is an important and very challenging task for many applications, such as automation of industrial processes, automatic vehicle navigation, pedestrian and vehicle detection, medical imaging, or facial recognition. In international forestry and wood processing organizations there is a strongly increasing demand for fully automatic wood log counting systems. Timber is usually stored in large piles, which must be counted, in order to determine the quantity which can be processed or sold. Manually counting wood logs is a very time consuming and error-prone task, with a high potential for automation.

For many purposes the field of view (FOV) of a regular digital camera is not sufficient to capture large objects. In such cases wide angle or fish eye lenses can be used to enlarge the FOV, e.g., [23]. Equivalently, panoramic images may be

© Springer International Publishing Switzerland 2014
X. Jiang et al. (Eds.): GCPR 2014, LNCS 8753, pp. 535–546, 2014.
DOI: 10.1007/978-3-319-11752-2_44

536   C. Herbon et al.

composed from multiple overlapping images [4,26]. These panoramas can either be captured from a single, fixed position regardless of the scene, or from multiple view points, when the scene can be constrained to one dominant plane [14]. In real world scenarios, scenes are often too large to be adequately captured by wide angle lenses or rotational panoramas without significant distortion towards the edges. Also, one may not be able to guarantee that the aforementioned single plane constraint for planar panoramas is always satisfied. In such cases, where general scene content must be captured, structure from motion (SfM) is used for camera calibration and sparse point cloud computation.

Several approaches have been presented to retrieve objects in digital images, which include techniques for feature description [3,6,19], template matching [17], classification [22,27], and bag-of-words (BoW) [8]. In addition, some techniques have been published, which can be used to detect three dimensional objects in dense point clouds (obtained through range scans or MRI data), or in stereo images, e.g., [1,7,24]. In dense point clouds, these methods require the object's 3D silhouette to be clearly distinguishable from its surroundings. For clustered objects, that were reconstructed by structure from motion, these constraints cannot be met and the aforementioned methods can therefore not be used for our purpose.

While much work has been done in the field of two dimensional object detection in images and three dimensional object detection in dense point clouds, little research addresses the use of 2D object detection combined with point clouds. In this paper we aim to fill this gap by performing a 2D object detection step on the input images of the sparse point cloud, which is representing a clustered object scene. To the best of our knowledge we are the first to use this approach. The detected 2D objects, in form of bounding circles, are projected back to 3D space using the camera pose of each input image. Each projection results in a virtual cone, which passes through the camera center. The vertices, which lie within the projection cone, are considered for the determination of the reconstructed 3D location of the object, which in turn is being used to establish correspondences between objects, that were captured across multiple frames. Through this proposed approach the total number objects may be accurately determined and their locations can be approximated.

2 Related Work

In this section we will focus on work that is relevant in the context of 2D and 3D object detection to establish the current state of the art, which this paper will build on. Feature based object detection methods have been widely popular for over two decades. Many object classification methods have since been published, that achieve high detection rates and short computation times. Local Binary Patterns (LBP) [22], Haar-like features [27], and Histograms of Oriented Gradients (HOG) [6] are among the most popular and widely used methods. Recently, advances have been made as to adapt two dimensional object detection methods for the use in three dimensional space. Multiple approaches have been suggested to apply HOG features to 3D data, e.g., [5,24]. LBPs have also been adapted for three dimensional usage, mainly for the purpose of analyzing medical MRI data [7]. As stated previously, these approaches require dense point clouds or layered images and are therefore not suitable for our purposes.

In this paper we are interested in combining 2D object detection methods with information about the 3D location and shape of the object cluster. In the field of facial recognition Yang et al. [29] propose the combination of 2D and 3D HOG features to improve the recognition of faces in depth images (RGBD data) captured by a Kinect device. A similar approach is followed by Wang et al. [28] who estimate the head pose through a combination of 2D SIFT and 3D HOG features. These setups differ significantly from our setup, since we do not obtain RGBD images, when the input data is constrained to structure from motion.

In [5] textures from traffic camera images are mapped onto 3D models of vehicles to improve vehicle classification. This approach requires a sparse occurrence of the object of interest and a strong silhouette, which is not the case for our clustered object scenario. References [18,25] transfer and adapt the 2D HOG descriptor to 3D volumetric images. Khan [16] uses images from multiple calibrated cameras to detect people in 2D and triangulate their position in 3D space using a planar homographic occupancy constraint. In our setup we do not wish to constrain our scenes in such a way and we will use a more general approach.

Recently, it has been proposed to extend the structure from motion approach by an *a contrario* model estimation, which has been shown to be able to compute high precision global camera calibration [21]. Some advances have been made in using high level cues in structure from motion pipelines. Reference [2] compares structure from motion results, which were obtained through either high level cues or low level features. The authors approach this problem with a maximum likelihood algorithm and show that using high level cues improves SfM. While this method may enhance SfM, it can only be regarded as a preprocessing step for object detection at the most, while special attention needs to be devoted to possible degeneracy in coplanar feature distribution cases.

The experiments we have conducted use a database of stacked wood log reconstructions (Fig. 2(c)). Some work has been published, which addresses the topic of detection and segmentation of wood logs in images, which our detection algorithm is built upon. Mainly [11,12] show graph cuts can be used for the segmentation of wood logs in large piles. Fink [9] successfully uses watersheds and active contours for segmentation in controlled environments. In [13] graph cut based segmentation is combined with automatic detection through Haar-like features. We utilize and adapt these methods to detect objects in structure from motion data, rather than in plain images.

3 Method

The contribution of this paper can be summarized as follows:

1. We are the first to perform detection of clustered objects based on 2D classification data and object reprojection to 3D.

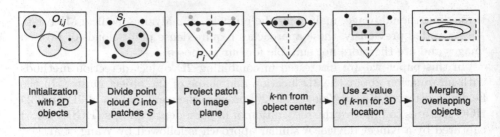

Initialization with 2D objects	→	Divide point cloud C into patches S	→	Project patch to image plane	→	k-nn from object center	→	Use z-value of k-nn for 3D location	→	Merging overlapping objects

Fig. 1. Method overview

2. We show how the 3D location of objects in a sparse point cloud can be recovered through quadric filtering and a k nearest neighbors (k-nn) approach.
3. The 3D location of individual objects is used to establish correspondences between objects over several input frames.

3.1   2D Object Detection

For 2D object detection we use the method provided in [15] as a preprocessing step for our input images. Reference [15] uses a combination of HOG and LBP detectors, which iteratively classify objects and maintain a Gaussian mixture model as a probability measure for possible candidate objects. The average true positive rate of [15] is reported to be 99.3 % while the false positive rate is below 0.4 %. The method is thus well suited for our purposes. Result for an example image can be seen in Fig. 2(b).

3.2   3D Object Detection

Our proposed method uses structure from motion input data and detects the objects of interest and their approximate locations in $\mathbb{R}^3$. The structure from motion data usually consists of the input images $I_0 \dots I_{n-1}$, the camera poses $P_0 \dots P_{n-1}$, camera intrinsic parameters $K_0 \dots K_{n-1}$, and the resulting point cloud C. The point cloud C consists of vertices in $\mathbb{R}^3$ and will thus be regarded as a set of vertices C_i with $i = 0...m - 1$. Figure 2 shows an example of an input image and the resulting point cloud.

Figure 1 provides an overview of our proposed method. For images $I_0 \dots I_{n-1}$ the 2D object set $O_{i,j}$ is given as input data, where i denotes the corresponding image index and j defines the object index within the image i. Each object is described by a 2D location (x, y) and a size parameter r, indicating the radius. Thus $O_{i,j} = \{x_{i,j}, y_{i,j}, r_{i,j}\}$. Secondly, we need to divide the input point cloud into patches S, where each patch $S_{i,j}$ is a subset of the point cloud C. Therefore every patch contains a specific number of vertices, which are regarded as locally planar, as suggested by Molton et al. [20].

We define the patch $S_{i,j}$, that corresponds to an object $O_{i,j}$, as all vertices which lie within a cone, defined by the camera center and the back projection

Fig. 2. (a) Input image with (b) detection result. (c) Sparse point cloud as a structure from motion result. (d) Partly dense reconstruction of detected objects.

of the 2D object to 3D space (see Fig. 3(a)). Assuming that the camera pose $P_i = [I]$, where $[I]$ is the 4×4 identity matrix, we define the circular cone through the quadric equation [30]

$$x^T A x + 2b^T x + c = 0 \tag{1}$$

where $A \in \mathbb{R}^{3 \times 3}$, $b \in \mathbb{R}^3$, $x \in \mathbb{R}^3$, and $c \in \mathbb{R}$ as

$$A = diag\left(\frac{1}{r_{i,j}^2}, \frac{1}{r_{i,j}^2}, -\frac{1}{f_i^2}\right) \tag{2}$$

$$b = \mathbf{0}, c = 0 \tag{3}$$

where f_i is the focal length of camera P_i and x is a vertex in $\mathbb{R}^3$. We can now use the quadric definition of the circular cone to retrieve all vertices from the point cloud, which lie inside of this cone and in front of the camera.

The vertices of the patch are then projected onto the image plane. Firstly, a vertex $v = \{v_x, v_y, v_z\}$ from a patch with its corresponding camera pose P_i is transformed in a way that the new camera pose $P_i' = P_i^{-1} \cdot P_i = [I]$. The transformed vertex is thus $v' = P_i^{-1} \cdot v$. The projection is then performed by

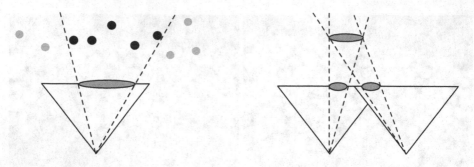

Fig. 3. (a) Definition of patch through quadric filtering. The dashed line defines the cone, vertices marked black are kept after the filtering. (b) Merging of two 3D objects.

defining $w_x = f_i \frac{v_x}{v_z}$ and $w_y = f_i \frac{v_y}{v_z}$, where $w = \{w_x, w_y\}$ denotes the projection of the vertex v in the image plane in $\mathbb{R}^2$. For each detected object $O_{i,j}$ the k neighbors with smallest Euclidean distance from the object center are selected. The median z-value of the k nearest vertices is then used to restore the approximate 3D location of the object, while the x- and y-value are known from their projection in the respective image.

When processing multiple overlapping image frames, most of the objects of interest will be detected several times. We now indicate the globally defined 3D objects by the set Q, where $Q_i = \{x_i, y_i, z_i, r_i\}$. In order to reduce multiple occurrences, a merging step is performed, where $d(Q_a, Q_b)$ is the Euclidean distance between two 3D objects Q_a and Q_b. Each object Q_a is combined with all remaining objects Q_j, with $j = 0 \ldots n$ and $j \neq a$, for which one of the following constraints holds true:

$$d(Q_a, Q_j) < r_a \quad \text{or} \quad d(Q_a, Q_j) < r_j \tag{4}$$

This ensures that all objects, whose centers lie within another object's boundaries, are combined and the final detection result is obtained. The output mesh, with objects of interest only, is then generated by projecting texture onto the 3D object from the camera whose optical axis aligns best with the object's surface normal. This way texture mapping distortions are minimized.

4 Experimental Results

4.1 Test Database

To the best of our knowledge, there does not exist any publicly available test database for our purposes. To evaluate our method, we have obtained 65 image sequences from international timber processing organizations, which will be used for our experiments. Each image was captured by a regular smartphone and was scaled to a resolution of 800×600 pixels. The ground truth number of wood logs has been manually determined and will be used for evaluation purposes.

Fig. 4. Subset of the images in datasets provided by timber processing organizations. Note the steep angles from which the images were captured and the poor wood quality.

The test database is highly challenging for several reason. As can be seen in Figs. 4 and 5 the angles, from which images are taken, tend to be extremely steep towards the edges, which makes object recognition quite difficult. The quality of the wood logs is also comparatively poor and image resolution is low, which again makes object detection challenging.

4.2   True / False Positive Rate

Our first experiment evaluates the overall true and false positive detection rate (tpr / fpr) of our proposed method. We analyze the influence of two important parameters on the outcome, which are the scale factor used in multi scale classification (s) and the minimum number of occurrences (ξ) for a single object, which is needed to validate the object. For practical purposes we are interested in high positive detection rates, since a user is able to dismiss false positives more easily than false negatives.

Scale factor. It is common to train classifiers with a fixed size of training images, e.g., 24×24 px. Classification is performed as a windowed search on an image (see Fig. 7(a)) and only objects with the approximate size of the classifier can be found. In order to detect objects which are larger or smaller, a Gaussian pyramid of the input image is constructed, which consists of the image at different resolutions. The scale factor s provides information about how much the image is scaled at each pyramid level. When the scale factor s is close to 1.0, it results in many different scales. For scale factors, which are much larger than 1.0, the image is scaled very coarsely and scale invariance of the classification result is poor. Figure 6 shows the tpr and fpr in relation to the scale factor for

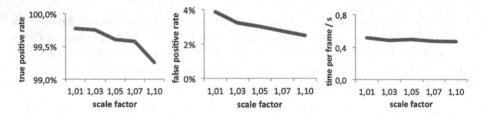

Fig. 5. Image samples in the evaluation database provided by timber processing organizations.

Fig. 6. (a) True positive rate, (b) false positive rate, and object detection time in relation to scale factor s

$\xi = 1$. It can be seen that there is a significant drop in the tpr for $s > 1.10$, while the fpr decreases almost linearly.

Minimum occurrences. In our second experiment we vary the parameter ξ. ξ is an integer, that represents the minimum number of occurrences of an object, which are necessary for the object to be validated. E.g., $\xi = 2$ would require an object to be detected in at least two input images. In Fig. 7 we show the relationship between ξ and tpr/fpr, with $s = 1.05$ as a good compromise between high tpr and low fpr. The edges of the scene are especially vulnerable to this parameter, because fewer images overlap at these points. We therefore choose a maximum of $\xi = 3$ in order to obtain meaningful results. It can be observed that there is a drop in the fpr for $\xi = 2$ / $\xi = 3$ down to less than 0.8% / 0.5%, since onetime outliers can be filtered effectively. The advantage of $\xi > 1$ is reduced by the lower tpr since many objects in our images are partly occluded and are only captured by a single image.

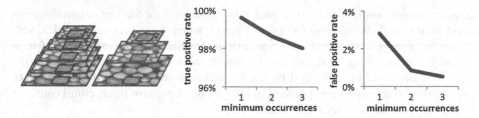

Fig. 7. (a) Multi scale classification (note the same absolute detector size marked green). The images are scaled at each level according to s. Left: small s, right: large s. (b) True positive and (c) false positive rate in relation to min. occurrences ξ.

Computational efficiency. The object detection time for each frame, in relation to the scale factor, is shown in Fig. 6. In all cases the processing time is around 500 ms on a Core i7 quad core computer with 3.5 GHz. The time per frame is only insignificantly influenced by the scale factor. For the overall processing time approximately 53 % is used for object detection, 41 % for global structure estimation and texture mapping, and 6 % for postprocessing.

4.3 Point Cloud Density

Patch definition. Our patch definition procedure filters a subset of vertices from the point cloud, which is then used to approximate the 3D location of a candidate object. The number of points within this radius is crucial to establish correspondences between objects in multiple images. If no vertices are found, the location must be interpolated from vertices outside the cone while assuming local planarity. We therefore analyze the average number of points in a patch in relation to the size of the detected object's radius in pixels. From Fig. 7(c) it becomes clear that a dense point cloud contains approximately five times as many vertices and that small objects in sparse point clouds are especially vulnerable to a lack of vertices for location determination.

Point cloud density. Our method aims to be lightweight and does not require a dense input point cloud. In order to investigate the impact of the point cloud

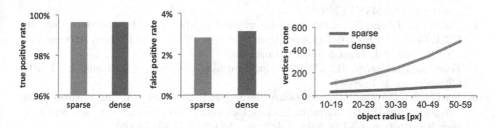

Fig. 8. (a) True positive and (b) false positive rate for sparse and dense point clouds. (c) Vertices in cone filter for sparse and dense point clouds.

density we now compare the true and false positive rate for the original sparse point clouds and the quasi-dense point clouds, which were processed by PMVS2 [10], with $\xi = 1$ and $s = 1.05$. Contrary to intuitive expectation, the fpr is actually higher for quasi-dense point clouds, while the tpr remains the same (Fig. 8). This could be explained by interpolation procedures and local planarity assumptions during patch expansion in [10], while the sparse point cloud contains high precision points only as shown by [21].

5   Conclusion and Future Work

We have presented a novel method for 3D object detection and location approximation. Our approach relies on 2D object classification and patch based k-nn detection in sparse point clouds, combined with quadric filtering. We have shown that it is indeed possible to achieve high detection rates with very little false positives. The true positive rate exceeded 98.0 % in any case with a false positive rate of less than 0.5 %. The effect of different parameters on the detection result have been studied. These include the scale factor, which is used in multi scale detection, the minimum number of occurrences, and the point cloud density. Our future work will address the optimization of the location estimation as well as the calculation of the partially dense reconstructed mesh.

References

1. Banerjee, J., Moelker, A., Niessen, W.J., van Walsum, T.: 3D LBP-based rotationally invariant region description. In: Park, J.-I., Kim, J. (eds.) ACCV Workshops 2012, Part I. LNCS, vol. 7728, pp. 26–37. Springer, Heidelberg (2013)
2. Bao, S.Y., Savarese, S.: Semantic structure from motion. In: IEEE Conference on Computer Vision and Pattern Recognition (CVPR), 2011, pp. 2025–2032. IEEE (2011)
3. Bay, H., Tuytelaars, T., Van Gool, L.: SURF: speeded up robust features. In: Leonardis, A., Bischof, H., Pinz, A. (eds.) ECCV 2006, Part I. LNCS, vol. 3951, pp. 404–417. Springer, Heidelberg (2006)
4. Brown, M., Lowe, D.G.: Automatic panoramic image stitching using invariant features. Int. J. Comput. Vis. **74**(1), 59–73 (2007)
5. Buch, N., Orwell, J., Velastin, S.A.: 3D extended histogram of oriented gradients (3DHOG) for classification of road users in urban scenes. In: British Machine Vision Association (BMVC) (2009)
6. Dalal, N., Triggs, B.: Histograms of oriented gradients for human detection. In: IEEE Computer Society Conference on Computer Vision and Pattern Recognition, CVPR 2005, vol. 1, pp. 886–893. June 2005
7. Fehr, J., Burkhardt, H.: 3D rotation invariant local binary patterns. In: ICPR, Citeseer, pp. 1–4 (2008)
8. Fei-Fei, L., Perona, P.: A Bayesian hierarchical model for learning natural scene categories. In: IEEE Computer Society Conference on Computer Vision and Pattern Recognition, CVPR 2005, vol. 2, pp. 524–531. IEEE (2005)
9. Fink, F.: Foto-optische Erfassung der Dimension von Nadelrundholzabschnitten unter Einsatz digitaler, bildverarbeitender Methoden. Ph.D. thesis, Albert-Ludwigs-Universitt (2004)

10. Furukawa, Y., Ponce, J.: Accurate, dense, and robust multiview stereopsis. IEEE Trans. Pattern Anal. Mach. Intel. **32**(8), 1362–1376 (2010)
11. Gutzeit, E., Ohl, S., Kuijper, A., Voskamp, J., Urban, B.: Setting graph cut weights for automatic foreground extraction in wood log images. In: VISAPP (2), pp. 60–67 (2010)
12. Gutzeit, E., Ohl, S., Voskamp, J., Kuijper, A., Urban, B.: Automatic wood log segmentation using graph cuts. In: Richard, P., Braz, J. (eds.) VISIGRAPP 2010. CCIS, vol. 229, pp. 96–109. Springer, Heidelberg (2011)
13. Gutzeit, E., Voskamp, J.: Automatic segmentation of wood logs by combining detection and segmentation. In: Bebis, G., Boyle, R., Parvin, B., Koracin, D., Fowlkes, C., Wang, S., Choi, M.-H., Mantler, S., Schulze, J., Acevedo, D., Mueller, K., Papka, M. (eds.) ISVC 2012, Part I. LNCS, vol. 7431, pp. 252–261. Springer, Heidelberg (2012)
14. Herbon, C., Tönnies, K., Stock, B.: Adaptive planar and rotational image stitching for mobile devices. In: Proceedings of the 5th ACM Multimedia Systems Conference, pp. 213–223. ACM (2014)
15. Herbon, C., Tönnies, K., Stock, B.: Detection and segmentation of clustered objects by using iterative classification, segmentation, and Gaussian mixture models and application to wood log detection. In: 36th German Conference on Pattern Recognition, Münster, 2–5 September 2014
16. Khan, S.M.: Multi-view Approaches to Tracking, 3D Reconstruction and Object Class Detection. Ph.D. thesis, University of Central Florida (2008)
17. Lewis, J.: Fast template matching. Vis. Interface **95**, 15–19 (1995)
18. Liu, K., Skibbe, H., Schmidt, T., Blein, T., Palme, K., Ronneberger, O.: 3D rotation-invariant description from tensor operation on spherical HOG field. NeuroImage **57**(2), 416–422 (2011)
19. Lowe, D.G.: Object recognition from local scale-invariant features. In: Proceedings of the Seventh IEEE International Conference on Computer Vision, 1999, vol. 2, pp. 1150–1157. IEEE (1999)
20. Molton, N., Davison, A.J., Reid, I.: Locally planar patch features for real-time structure from motion. In: BMVC, pp. 1–10 (2004)
21. Moulon, P., Monasse, P., Marlet, R., et al.: Global fusion of relative motions for robust, accurate and scalable structure from motion. In: Proceedings of IEEE International Conference on Computer Vision (2013)
22. Ojala, T., Pietikainen, M., Harwood, D.: Performance evaluation of texture measures with classification based on kullback discrimination of distributions. In: Proceedings of the 12th IAPR International Conference on Pattern Recognition, 1994, vol. 1, pp. 582–585 (1994)
23. Okamoto Jr., J., Grassi Jr., V.: Visual servo control of a mobile robot using omnidirectional vision. In: Proceedings of Mechatronics, pp. 413–422 (2002)
24. Scherer, M., Walter, M., Schreck, T.: Histograms of oriented gradients for 3d object retrieval. In: Proceedings of the WSCG, pp. 41–48 (2010)
25. Skibbe, H., Reisert, M., Burkhardt, H.: SHOG - spherical HOG descriptors for rotation invariant 3D object detection. In: Mester, R., Felsberg, M. (eds.) DAGM 2011. LNCS, vol. 6835, pp. 142–151. Springer, Heidelberg (2011)
26. Szeliski, R.: Image alignment and stitching: a tutorial. Found. Trends Comput. Graph. Vis. **2**(1), 1–104 (2006)
27. Viola, P., Jones, M.: Rapid object detection using a boosted cascade of simple features. In: Proceedings of the 2001 IEEE Computer Society Conference on Computer Vision and Pattern Recognition, CVPR 2001. vol. 1, pp. I-511-I-518 (2001)

28. Wang, B., Liang, W., Wang, Y., Liang, Y.: Head pose estimation with combined 2D sift and 3D HOG features. In: Seventh International Conference on Image and Graphics (ICIG 2013), pp. 650–655 July 2013
29. Yang, J., Liang, W., Jia, Y.: Face pose estimation with combined 2D and 3D HOG features. In: 21st International Conference on Pattern Recognition (ICPR), 2012, pp. 2492–2495 (2012)
30. Zwillinger, D.: CRC Standard Mathematical Tables and Formulae, 32nd Edn., Discrete Mathematics and Its Applications, CRC Press (2011)

On the Second Order Statistics of Essential Matrix Elements

M. Hossein Mirabdollah$^{(\boxtimes)}$ and Bärbel Mertsching

GET Lab, University of Paderborn, Pohlwegstr 47-49, 33098 Paderborn, Germany
mirabdollah@get.upb.de

Abstract. In this paper, we investigate the second order statistics of essential matrix elements. Using the Taylor expansion for a rotation matrix up to second order terms and considering relatively high uncertainties for the rotation angles and translation parameters, a covariance matrix is obtained which includes the second order statistics of essential matrix elements. The covariance matrix is utilized along with the coplanarity equations and acts as a regularization term. Using the regularization term brings considerable improvements in the recovery of camera motion which will be proven based on simulation and different real image sequences.

Keywords: Essential matrix · Second order statistics

1 Introduction

Relative monocular camera motion (ego-motion) estimation based on the coplanarity constraint for calibrated cameras was initially addressed by Higgins with the well-known 8-point method [12]. In this method, eight matched points between two frames are used to obtain a 3×3 matrix known as *essential matrix*. Consequently, based on an essential matrix, the rotation matrix and the translation vector which define the motion of the camera (up to a scale) can be obtained. For uncalibrated cameras, in [8] an 8-point method is proposed to extract a 3×3 matrix known as *fundamental matrix* to recover camera motion and camera focal length simultaneously, if the principal point of the camera is known and the horizontal and vertical focal lengths are the same. Two main applications for the ego-motion estimation can be named: visual odometry [16,19] and optical flow calculation [18].

Although the 8-point methods proposed by Higgins and Hartley were simple, they had poor performances in the presence of measurement noise, especially for the uncalibrated cameras. Therefore, different nonlinear optimization methods were proposed to estimate the essential and fundamental matrices iteratively [13]. Clearly, the iterative methods require good initial guesses to converge to correct solutions; otherwise, they would get stuck in local minima. On the other hand, Hartley in [9] claimed that if matched points are transformed such that their centroids become zero and their average distances from the centroids become $\sqrt{2}$, the performance of the 8-point method would improve noticeably. Nevertheless,

© Springer International Publishing Switzerland 2014
X. Jiang et al. (Eds.): GCPR 2014, LNCS 8753, pp. 547–557, 2014.
DOI: 10.1007/978-3-319-11752-2_45

by comparing the results presented in the mentioned work, we can see that this method also performs poorly for measurement noise greater than 0.15 pixel.

One reason for the poor performance of the 8-point method is ignoring the dependencies of essential matrix elements. Hartley et. al proposed a 7-point method which used rank deficiency of fundamental matrices leading to enhancement in the estimation of camera motions [10]. Considering the fact that ego-motion is defined up to a scale, it can be defined with three rotation angles and two translation parameters. Thus, an essential matrix can also be determined with only five points. In this regard, David Nister in [15] introduced an algebraic solution using five matched points. He used rank deficiency and trace equations which hold for any essential matrix and obtained a polynomial equation of the order of ten, of which real roots yielded different valid solutions for the essential matrix. In [11], the authors proposed a more direct solution to reach the polynomial of the order of ten, which is the determinant of a 10×10 matrix of a variable. Obviously, implementations of both algorithms require symbolic processing, which makes their implementations inconvenient. Additionally, they would be very slow. Both 5-point methods are ad-hoc as there are 3 unknowns (x, y and z) but 10 equations; thus, none of them can guarantee that the solutions minimize all 10 equations. In Nister's method four equations are ignored and in the second method, we may confront contradictory elementary equations for an unknown in the forms of first, second or third order equations. In [14], the authors assumed large uncertainties for the motion parameters and obtained a variance for each coplanarity equation depending on the coordinates of the matching point. Consequently, the 8-point method was modified and solved based on a Mahalanobis criterion and the variance of each coplanarity equation to come up with a method which is more robust against the measurement noise.

In this paper, we demonstrate that in addition to the variance for each coplanarity equation, a covariance matrix for the nine elements of an essential matrix can be found, which encodes the second order statistics and dependencies of the elements. Then, in the essential matrix recovery, the covariance matrix will be augmented as a regularization term to direct the final solution in a feasible region defined by the physical constraints of cameras.

This paper is structured as follows: in Sect. 2, the 8-point, 7-point and 5-point methods are briefly introduced. The derivation of our proposed method is discussed in Sect. 3. Through the simulation in Sect. 4 the proposed method will be evaluated in two different cases. The evaluation of the proposed method in comparison with the 8-point, 7-point and 5-point methods based on the KITTI benchmark sequences is done in Sect. 5. Section 6 concludes this paper.

2   Essential Matrix Estimation

It is known that for a calibrated camera for each pair of matched points between two frames such as (x_1, y_1) and (x_2, y_2), the following equation (coplanarity constraint) is valid:

$$\mathbf{p}_2^T E \mathbf{p}_1 = 0 \tag{1}$$

where $\mathbf{p}_1 = [x_1\, y_1\, 1]^T$, $\mathbf{p}_2 = [x_2\, y_2\, 1]^T$ and E is an essential matrix. An essential matrix is a 3×3 matrix with nine elements:

$$E = \begin{bmatrix} e_1 & e_2 & e_3 \\ e_4 & e_5 & e_6 \\ e_7 & e_8 & e_9 \end{bmatrix} \tag{2}$$

If an essential matrix can be determined using a set of matched points between two frames, the relative camera motion can be estimated up to a scale factor. In the 8-point method, given eight matched points, a homogeneous equation system of nine unknown elements, consisting of eight equations, is formed. Thus, the essential matrix elements will be the null space of the matrix containing the coefficients of the equation system. Since the dependencies of essential matrix elements are ignored in this method, the solution could be sensitive to the measurement noise. Based on [4,17], the dependencies of essential matrix elements can be formulated with the following equations:

$$det(E) = 0 \tag{3}$$

$$(EE^T)E - \frac{1}{2}trace(EE^T)E = 0 \tag{4}$$

Using Eq. 3, the 7-point method is obtained. To utilize Eq. 3, a homogeneous equation system including seven coplanarity equations is formed and the coefficients of the equation system are stacked in a matrix. The matrix has a null space spanned by two vectors such as $\mathbf{y}$ and $\mathbf{z}$. Thus, we have:

$$\mathbf{e} = y\mathbf{y} + \mathbf{z} \tag{5}$$

where $\mathbf{e} = [e_1 \ldots e_9]^T$. Plugging Eq. 5 in Eq. 3 results in a third order polynomial equation of y, which may have up to three real roots. It means that three valid essential matrices could explain the camera motion. However, if Eq. 4 is used, the solution which minimizes Eq. 4 can be selected as the best essential matrix. For the 5-point method, five coplanarity equations are used and therefore, we have $\mathbf{e} = \mathbf{w} + x\mathbf{x} + y\mathbf{y} + z\mathbf{z}$. Using the 10 equations obtained from Eqs. 3 and 4, a polynomial equation of the order of ten of the variable z is obtained. The real roots of the equation lead to different valid essential matrices. Generally, the true solution can be found based on the multiple observations of the matched points in multiple frames.

3   Second Order Statistics of an Essential Matrix

In this section, we find the mean and the covariance matrix for essential matrix elements and use them as regularization terms in different original N-point methods. In this regard, we first find the means and variances of motion parameters based on physical or equation constraints, and then the mean and covariance matrix of essential matrix elements will be calculated.

Concerning the translation vector, it can be verified that, if we find the null space vectors under the condition: $\mathbf{e}_1^2 + \ldots + \mathbf{e}_9^2 = 1$ (for instance using singular value decomposition (SVD)), the following equation will hold:

$$t_x^2 + t_y^2 + t_z^2 = \frac{1}{2} \tag{6}$$

It means that $\mathbf{t} = [t_x, t_y, t_z]^T$ has a uniform distribution over a sphere with the radius $\frac{\sqrt{2}}{2}$. To obtain means, variances and correlations of the translation elements, we may need to marginalize out one and two variables from their joint probability distribution function: $p([t_x, t_y, t_z]^T)$. However, due to the symmetry of the distribution, it can be simply proven that mean values are zero: $\mu_\mathbf{t} = \mathcal{E}(\mathbf{t}) = [0 \ 0 \ 0]^T$ ($\mathcal{E}$ is expectation operator) and also covariance between each two translation elements are zero: $\sigma_{t_x t_y} = \sigma_{t_x t_z} = \sigma_{t_y t_z} = 0$. Consequently, Eq. 6 results in $\sigma_{t_x}^2 = \sigma_{t_y}^2 = \sigma_{t_z}^2 = \frac{1}{6}$.

On the other hand, by encoding a rotation matrix using the three elementary rotations about the X, Y and Z axis: $R = R_Z(\psi)R_Y(-\theta)R_X(\phi)$ and using the Taylor expansion for the $\sin(.)$ and $\cos(.)$ up to desired orders, the rotation matrix can be written in the form of polynomials of ϕ, θ and ψ which makes the expectation operations much simpler. It should be mentioned that physical constraints of cameras and limitation of feature tracking methods force the angles to hardly exceed $30°$. As a result, the terms with more than second order can be ignored. Thus, the following approximation for the rotation matrix holds:

$$R = \begin{bmatrix} 1 - \frac{\theta^2}{2} - \frac{\psi^2}{2} & -\psi - \phi\theta & -\theta + \phi\psi \\ \psi & 1 - \frac{\phi^2}{2} - \frac{\psi^2}{2} & -\phi - \theta\psi \\ \theta & \phi & 1 - \frac{\phi^2}{2} - \frac{\theta^2}{2} \end{bmatrix} \tag{7}$$

Obviously, for larger rotation angles higher order terms should be included, but the rest of the algorithm will be similar. Now, by considering maximum possible deviations from zero for the angles such as σ_ϕ, σ_θ and σ_ψ, we can model the angles as Gaussian random variables as: $\phi \sim \mathcal{N}(0, \sigma_\phi^2)$, $\theta \sim \mathcal{N}(0, \sigma_\theta^2)$ and $\psi \sim \mathcal{N}(0, \sigma_\psi^2)$. Additionally, we know that:

$$E = RT; \quad T = \begin{bmatrix} 0 & -t_z & t_y \\ t_z & 0 & -t_x \\ -t_y & t_x & 0 \end{bmatrix} \tag{8}$$

Using Eqs. 7 and 8, we can obtain the vector of essential matrix elements $\mathbf{e} = [e_1, e_2, \ldots, e_9]^T$ as nonlinear functions of the motion parameters:

$e_1 = -t_z(\psi + \phi\theta) + t_y(\theta - \phi\psi)$, $e_2 = t_x(\phi\psi - \theta) + t_z(\frac{\psi^2 + \theta^2}{2} - 1)$,
$e_3 = t_y(1 - \frac{\psi^2 + \theta^2}{2}) + t_x(\psi + \phi\theta)$, $e_4 = -t_z(\frac{\phi^2 + \psi^2}{2} - 1) + t_y(\phi + \psi\theta)$,
$e_5 = -t_z\psi - t_x(\phi + \psi\theta)$, $e_6 = \psi t_y + t_x(\frac{\phi^2 + \psi^2}{2} - 1)$, $e_7 = t_z\phi + t_y(\frac{\phi^2 + \theta^2}{2} - 1)$,
$e_8 = -\theta t_z - t_x(\frac{\phi^2 + \theta^2}{2} - 1)$ and $e_9 = -\phi t_x + \theta t_y$.

Using the predefined Gaussian distributions for the motion parameters, we can now calculate the first and second order statistics of $\mathbf{e}$. For the first order statistics (mean vector), we can simply verify that all the nine elements include first order terms of the motion parameter and therefore we have:

$$\mu_{\mathbf{e}} = \mathcal{E}(\mathbf{e}) = \mathbf{0} \tag{9}$$

Since $\mu_{\mathbf{e}} = \mathbf{0}$, the second order statistics can be calculated as follows:

$$P_{\mathbf{e}} = \mathcal{E}(\mathbf{e}\mathbf{e}^T) \tag{10}$$

The calculation of the above covariance matrix is tedious and should be done by using a symbolic math package once. The covariance matrix looks as follows:

$$P_{\mathbf{e}} = \begin{bmatrix}
\sigma_{e_1}^2 & 0 & 0 & 0 & \sigma_{e_1,e_5} & 0 & 0 & 0 & \sigma_{e_1,e_9} \\
0 & \sigma_{e_2}^2 & 0 & \sigma_{e_2,e_4} & 0 & 0 & 0 & 0 & 0 \\
0 & 0 & \sigma_{e_3}^2 & 0 & 0 & 0 & \sigma_{e_3,e_7} & 0 & 0 \\
0 & \sigma_{e_2,e_4} & 0 & \sigma_{e_4}^2 & 0 & 0 & 0 & 0 & 0 \\
\sigma_{e_1,e_5} & 0 & 0 & 0 & \sigma_{e_5}^2 & 0 & 0 & 0 & \sigma_{e_5,e_9} \\
0 & 0 & 0 & 0 & 0 & \sigma_{e_6}^2 & 0 & \sigma_{e_6,e_8} & 0 \\
0 & 0 & \sigma_{e_3e_7} & 0 & 0 & 0 & \sigma_{e_7}^2 & 0 & 0 \\
0 & 0 & 0 & 0 & 0 & \sigma_{e_6e_8} & 0 & \sigma_{e_8}^2 & 0 \\
\sigma_{e_1e_9} & 0 & 0 & 0 & \sigma_{e_5,e_9} & 0 & 0 & 0 & \sigma_{e_9}^2
\end{bmatrix}$$

where

$$\sigma_{e_1}^2 = \sigma_\phi^2 \sigma_\psi^2 \sigma_{t_y}^2 + \sigma_\phi^2 \sigma_\theta^2 \sigma_{t_z}^2 + \sigma_\psi^2 \sigma_{t_z}^2 + \sigma_\theta^2 \sigma_{t_y}^2,$$

$$\sigma_{e_1 e_5} = \sigma_\psi^2 \sigma_{t_z}^2, \quad \sigma_{e_1 e_9} = \sigma_\theta^2 \sigma_{t_y}^2,$$

$$\sigma_{e_2}^2 = \sigma_\phi^2 \sigma_\psi^2 \sigma_{t_x}^2 + \tfrac{3}{4}\sigma_\psi^4 \sigma_{t_z}^2 + \tfrac{1}{2}\sigma_\psi^2 \sigma_\theta^2 \sigma_{t_z}^2 - \sigma_\psi^2 \sigma_{t_z}^2 + \tfrac{3}{4}\sigma_\theta^4 \sigma_{t_z}^2 + \sigma_\theta^2 \sigma_{t_x}^2 - \sigma_\theta^2 \sigma_{t_z}^2 + \sigma_{t_z}^2,$$

$$\sigma_{e_2 e_4} = -\tfrac{1}{4}\sigma_\phi^2 \sigma_\psi^2 \sigma_{t_z}^2 - \tfrac{1}{4}\sigma_\phi^2 \sigma_\theta^2 \sigma_{t_z}^2 + \tfrac{1}{2}\sigma_\phi^2 \sigma_{t_z}^2 - \tfrac{3}{4}\sigma_\psi^4 \sigma_{t_z}^2 - \tfrac{1}{4}\sigma_\psi^2 \sigma_\theta^2 \sigma_{t_z}^2 +$$
$$\sigma_\psi^2 \sigma_{t_z}^2 + \tfrac{1}{2}\sigma_\theta^2 \sigma_{t_z}^2 - \sigma_{t_z}^2,$$

$$\sigma_{e_3}^2 = \sigma_\phi^2 \sigma_\theta^2 \sigma_{t_x}^2 + \tfrac{3}{4}\sigma_\psi^4 \sigma_{t_y}^2 + \tfrac{1}{2}\sigma_\psi^2 \sigma_\theta^2 \sigma_{t_y}^2 + \sigma_\psi^2 \sigma_{t_x}^2 - \sigma_\psi^2 \sigma_{t_y}^2 + \tfrac{3}{4}\sigma_\theta^4 \sigma_{t_y}^2 - \sigma_\theta^2 \sigma_{t_y}^2 + \sigma_{t_y}^2,$$

$$\sigma_{e_3 e_7} = -\tfrac{1}{4}\sigma_\phi^2 \sigma_\psi^2 \sigma_{t_y}^2 - \tfrac{1}{4}\sigma_\phi^2 \sigma_\theta^2 \sigma_{t_y}^2 + \tfrac{1}{2}\sigma_\phi^2 \sigma_{t_y}^2 - \tfrac{1}{4}\sigma_\psi^2 \sigma_\theta^2 \sigma_{t_y}^2 + \tfrac{1}{2}\sigma_\psi^2 \sigma_{t_y}^2 - \tfrac{3}{4}\sigma_\theta^4 \sigma_{t_y}^2 +$$
$$\sigma_\theta^2 \sigma_{t_y}^2 - \sigma_{t_y}^2,$$

$$\sigma_{e_4}^2 = \tfrac{3}{4}\sigma_\psi^4 \sigma_{t_z}^2 + \tfrac{1}{2}\sigma_\phi^2 \sigma_\psi^2 \sigma_{t_z}^2 + \sigma_\phi^2 \sigma_{t_y}^2 - \sigma_\phi^2 \sigma_{t_z}^2 + \tfrac{3}{4}\sigma_\psi^4 \sigma_{t_z}^2 + \sigma_\psi^2 \sigma_\theta^2 \sigma_{t_y}^2 - \sigma_\psi^2 \sigma_{t_z}^2 + \sigma_{t_z}^2,$$

$$\sigma_{e_5}^2 = \sigma_\theta^2 \sigma_{t_x}^2 + \sigma_\psi^2 \sigma_\theta^2 \sigma_{t_x}^2 + \sigma_\psi^2 \sigma_{t_x}^2, \quad \sigma_{e_5 e_9} = \sigma_\theta^2 \sigma_{t_x}^2,$$

$$\sigma_{e_6}^2 = \tfrac{3}{4}\sigma_\phi^4 \sigma_{t_x}^2 + \tfrac{1}{2}\sigma_\phi^2 \sigma_\psi^2 \sigma_{t_x}^2 - \sigma_\phi^2 \sigma_{t_x}^2 + \tfrac{3}{4}\sigma_\psi^4 \sigma_{t_x}^2 - \sigma_\psi^2 \sigma_{t_x}^2 + \sigma_\psi^2 \sigma_{t_y}^2 + \sigma_{t_x}^2,$$

$$\sigma_{e_6,e_8} = -\tfrac{3}{4}\sigma_\phi^4 \sigma_{t_x}^2 - \tfrac{1}{4}\sigma_\phi^2 \sigma_\psi^2 \sigma_{t_x}^2 - \tfrac{1}{4}\sigma_\phi^2 \sigma_\theta^2 \sigma_{t_x}^2 + \sigma_\phi^2 \sigma_{t_x}^2 - \tfrac{1}{4}\sigma_\psi^2 \sigma_\theta^2 \sigma_{t_x}^2 + \tfrac{1}{2}\sigma_\psi^2 \sigma_{t_x}^2 +$$
$$\tfrac{1}{2}\sigma_\theta^2 \sigma_{t_x}^2 - \sigma_{t_x}^2,$$

$$\sigma_{e_7}^2 = \tfrac{3}{4}\sigma_\phi^4 \sigma_{t_y}^2 + \tfrac{1}{2}\sigma_\phi^2 \sigma_\theta^2 \sigma_{t_y}^2 - \sigma_\phi^2 \sigma_{t_y}^2 + \sigma_\phi^2 \sigma_{t_z}^2 + \tfrac{3}{4}\sigma_\theta^4 \sigma_{t_y}^2 - \sigma_\theta^2 \sigma_{t_y}^2 + \sigma_{t_y}^2,$$

$$\sigma_{e_8}^2 = \tfrac{3}{4}\sigma_\phi^4 \sigma_{t_x}^2 + \tfrac{1}{2}\sigma_\phi^2 \sigma_\theta^2 \sigma_{t_x}^2 - \sigma_\phi^2 \sigma_{t_x}^2 + \tfrac{3}{4}\sigma_\theta^4 \sigma_{t_x}^2 - \sigma_\theta^2 \sigma_{t_x}^2 + \sigma_\theta^2 \sigma_{t_z}^2 + \sigma_{t_x}^2 \quad \text{and}$$

$$\sigma_{e_9}^2 = \sigma_\phi^2 \sigma_{t_x}^2 + \sigma_\theta^2 \sigma_{t_y}^2.$$

Interestingly, the dependencies of the essential elements appeared as non-zero off-diagonal elements. Inspired by the work of [3], we applied a smoothing method for the essential matrix estimation by minimizing a cost function consisting of the coplanarity equations as data terms and the covariance matrix as a smoothness term. The cost function will be:

$$C = \sum_{i=0}^{N} \frac{1}{\sigma_{c_i}} (A_i \mathbf{e})^T (A_i \mathbf{e}) + \mathbf{e}^T P_e^{-1} \mathbf{e} \tag{11}$$

where N is the number of matched points, A_i is a row vector of ith coplanarity coefficients and σ_{c_i} is the standard deviation for the ith coplanarity equation.

In the cost function, σ_{c_i} s play important roles. Assuming that the true motion of a camera and consequently the essential matrix is given, we can see in the presence of the measurement noise, the coplanarity equations will not hold exactly. In this case, each coplanarity equation has an almost zero mean Gaussian distribution if the measurement noises are also Gaussian. The variance of the distributions can be calculated as a function of the matched points and measurement noise variances. However, since the essential matrix is not known initially, considering large uncertainties for the motion parameters, we can obtain a relative measure which shows how much each coplanarity equation is allowed to deviate from zero. For the calculation of this measure, we assume matched points (x_1, y_1) and (x_2, y_2) have Gaussian distributions as:

$$x_1 = \mathcal{N}(\bar{x}_1, \sigma_p^2),\ y_1 = \mathcal{N}(\bar{y}_1, \sigma_p^2),\ x_2 = \mathcal{N}(\bar{x}_2, \sigma_p^2), y_2 = \mathcal{N}(\bar{y}_2, \sigma_p^2) \tag{12}$$

where $\bar{x}_1, \bar{y}_1, \bar{x}_2, \bar{y}_2$ are the measured values and σ_p is the standard deviation for the measurement noise. Using a similar procedure proposed in [14], and since we can always say $\sigma_{t_x}^2 = \sigma_{t_y}^2 = \sigma_{t_z}^2 = \frac{1}{6}$ (as already discussed), we have:

$$\sigma_c^2 = \frac{\sigma_p^2}{6} \left(4 + \sigma_\phi^2 + \sigma_\theta^2 + 2\sigma_\psi^2 + (1 + \sigma_\phi^2 + \sigma_\theta^2 + \sigma_\psi^2)(\bar{x}_1^2 + \bar{y}_1^2 + \bar{x}_2^2 + \bar{y}_2^2)\right) \tag{13}$$

The inverse of σ_p weights the importance of the data term: the larger σ_p, the more estimations of essential matrix elements are affected by the regularization term. Therefore, σ_p should be selected to achieve a balance between the data and regularization terms such that the method can work well for a wide range of measurement noise. Based on simulation and experimental results, we found $\sigma_p = 0.5$ pixel a proper selection.

For the minimization of the cost function Eq. 11, we reform it as follows:

$$C = (B\mathbf{e})^T P^{-1} (B\mathbf{e}) \tag{14}$$

where

$$B = \begin{bmatrix} A_1 \\ \vdots \\ A_N \\ I_{9 \times 9} \end{bmatrix} \quad \text{and} \quad P = \begin{bmatrix} \sigma_{c,1}^2 & 0 & 0 & 0 \\ \vdots & \ddots & \cdots & 0 \\ 0 & 0 & \sigma_{c,N}^2 & 0 \\ 0 & 0 & 0 & P_e \end{bmatrix} \tag{15}$$

Now we should find the vectors which minimize the cost function. It can be done using the SVD. We only need to calculate the SVD of the matrix $P^{-1/2}B = U\Sigma V^T$. Then the columns of V^T which are associated to the smallest eigenvalues can form the basis for different N-point methods.

4   Simulation

We have simulated a camera with the resolution 2000×2000 [pixels^2] and a focal length of 1000 pixels. The camera was moved based on random translation and rotation parameters (ϕ, θ and ψ had the means 0.3 rad and standard deviations 0.1 rad) for two different types of motions: dominant forward and dominant side translations. During the motion, the camera could observe spatial points randomly distributed at depths from 10 to 20 m. The projection of the spatial points on the camera screen were added by zero mean Gaussian noises with varying standard deviation as measurement noise. The original 8-, 7- and 5-point methods were compared with the smoothed versions (named as 8-, 7- and 5- point-S). The standard deviations of the rotation angles (σ_ϕ, σ_θ and σ_ψ) for the regularization constraint were set to 0.5 rad. For the evaluation, two different measures were used: the mean of magnitudes of errors between the estimated translation and true translation vectors (MME_t) and the mean of magnitudes of errors between the estimated rotation angles and the true angles (MME_a). Assuming the estimated translation vector at time k is $\hat{t}_k$ and the ground truth is t_k, the error between two vectors will be $\epsilon_{t,k} = \hat{t}_k - t_k$. Then, we have: $MME_t = \frac{1}{K} \sum_{k=1}^{K} \sqrt{\epsilon_{t,k}^T \epsilon_{t,k}}$, where K is the number of frames. If the estimated rotation matrix at time k is $\hat{R}_k$ and the ground truth is R_k the rotation matrix error will be $R_{e,k} = R_k^T \hat{R}_k$. Consequently, using Eq. 8, we can calculate the error angles as : $\epsilon_{a,k} = \frac{180}{\pi} \sqrt{3 - trace(R_{e,k})}$. The mean square error for angles will be: $MME_a = \frac{1}{K} \sum_{k=1}^{K} \epsilon_{a,k}$. It is good to mention that since the regularization term is centered on the origin; not surprisingly, the proposed method could perform well if the rotation parameters were also distributed about the origin. Therefore, we selected random rotation angles with non-zero means to evaluate the method in a more challenging case.

The results can be seen in Tables 1 and 2. We can see that the regularization constraint improved the translation estimation slightly; however, it resulted in noticeable improvements for the rotation estimation in case of forward translation and good improvements for side translation. Surprisingly, we see that the 5-point method had a poor performance in rotation estimation (possibly due to

Table 1. Mean of magnitudes of errors for dominant forward translations.

σ_p[pixel]	0.1	0.5	1.0	1.5	2.0	0.1	0.5	1.0	1.5	2.0
	$MME_t[m/frame]$					$MME_a[deg/frame]$				
5-point-S	0.03	0.11	0.14	0.17	0.18	0.13	0.62	0.92	1.23	1.39
5-point	0.03	0.13	0.16	0.19	0.20	0.15	0.80	1.42	1.53	1.57
7-point-S	0.03	0.14	0.15	0.17	0.17	0.11	0.66	0.98	1.28	1.40
7-point	0.03	0.2	0.22	0.22	0.23	0.14	0.99	1.49	2.03	2.48
8-point-S	0.03	0.14	0.17	0.22	0.26	0.12	0.67	0.99	1.32	1.74
8-point[14]	0.03	0.20	0.26	0.32	0.38	0.14	1.10	1.48	2.04	2.38
8-point[9]	0.03	0.20	0.26	0.33	0.40	0.15	1.10	1.62	2.38	3.10

Table 2. Mean of magnitudes of errors for dominant side translations.

σ_p[pixel]	0.1	0.5	1.0	1.5	2.0	0.1	0.5	1.0	1.5	2.0
	$MME_t[m/frame]$					$MME_a[deg/frame]$				
5-point-S	0.05	0.16	0.24	0.29	0.32	1.86	5.89	9.73	10.92	14.19
5-point	0.06	0.17	0.25	0.29	0.33	2.06	6.09	10.28	11.86	14.54
7-point-S	0.10	0.35	0.55	0.63	0.68	0.70	2.15	3.38	4.44	5.10
7-point	0.11	0.36	0.59	0.67	0.72	0.75	2.81	4.51	5.92	6.77
8-point-S	0.11	0.33	0.53	0.66	0.74	0.71	2.19	3.93	5.76	6.88
8-point[14]	0.11	0.35	0.58	0.71	0.79	0.76	2.84	4.82	6.32	7.23
8-point[9]	0.11	0.35	0.57	0.71	0.79	0.76	2.84	5.04	6.34	7.27

numerical errors) in case of side motion. As a result, we can conclude that the 7-point-S method provides more reliable estimations if the camera experiences different types of motions.

5 Experimental Results

We used the KITTI dataset for visual odometry [6] to evaluate the performance of the proposed method. We applied the proposed regularization constraint for the 8-point, 7-point and 5-point methods and compared them with the original methods. The methods were implemented using C++ and ran on a computer with an Intel(R) Core(TM)2 Duo 3.33GHz CPU. All the methods were run in the context of a random sample consensus algorithm (RANSAC) [5], in which several essential matrices based on randomly selected N-matched points were calculated and then the best essential matrix which defined the flow of all points in the best way was selected.

As monocular ego-motion can generally be calculated only up to a scale factor, and the scale drift detection is not within the scope of this work, we removed the scale factor from the provided ground truths of the first eleven sequences. For the submitted data (for 7-point-S [1]), we have only used a fixed scale factor and obviously the submitted results were not meant for the ranking evaluations due to the large scale drifts in some of the sequences (for instance sequence 13) and also most of the other methods have used stereo or laser data which make the comparison unfair. What could be compared is the mean error angle. It can be seen that the proposed method (named as RMCPE) outperforms even some of the stereo based algorithms and also obviously has an error 2.7 times less than the 8-point based method (VISO2-M [7]). In this benchmark, the errors are calculated only based on a few frames, while in case of all frames, the differences will be much more as shown in Tables 3 and 4.

Since apparently instantaneous motion parameters of ground truth had drifts, we used two types of accumulating errors as evaluation measures: first, the mean of magnitudes of errors between all estimated and ground truth camera poses: MME_p, and second, the mean of magnitudes of errors between the estimated

Table 3. Mean of magnitudes of errors for estimated camera poses.

sequence	00	01	02	03	04	05	06	07	08	09	10
	MME_p[m/frame]										
7-point-S	23.4	25.0	12.8	2.4	0.9	22.9	5.5	12.2	58.7	11.2	8.3
7-point	32.3	139.3	30.0	4.7	1.5	27.4	10.5	14.9	61.5	25.6	12.0
5-point-S	9.2	53.5	24.0	3.0	1.1	30.1	8.3	10.5	58.0	12.0	9.2
5-point	9.4	78.7	36.3	3.6	1.3	30.1	8.8	10.8	67.3	16.6	9.4
8-point-S	25.3	40.7	47.3	6.2	1.6	28.6	13.00	13.8	67.6	16.2	14.8
8-point [14]	70.3	126.2	93.3	11.2	1.6	67.3	10.2	22.8	82.9	44.0	24.3
8-point[9]	70.0	443.5	97.4	23.6	5.3	66.5	27.3	22.7	124.9	42.0	38.6

Table 4. Mean of magnitudes of errors for estimated camera angles.

sequence	00	01	02	03	04	05	06	07	08	09	10
	MME_a[deg/frame]										
7-point-S	0.53	0.59	0.42	0.22	0.03	0.18	0.27	0.17	0.40	0.34	0.22
7-point	1.50	4.28	1.37	0.20	0.14	0.86	1.08	1.05	0.64	0.93	1.26
5-point-S	0.31	1.19	0.80	0.23	0.07	0.94	0.53	0.60	0.31	0.43	0.30
5-point	0.33	1.75	1.37	0.24	0.1	1.03	0.54	0.72	0.30	0.47	0.30
8-point-S	1.51	1.63	1.69	0.37	0.07	0.64	1.10	0.86	0.73	0.71	1.09
8-point [14]	1.72	4.12	2.86	0.57	0.07	2.77	0.89	1.65	1.31	2.00	0.74
8-point [9]	2.09	10.92	2.98	0.59	1.23	2.58	1.14	1.63	2.12	1.90	1.61

and ground truth angles at all poses: MME_a. For corner feature tracking, we used pyramid Lucas-Kanade optical flow (implemented in opencv) [2], which worked very well in real-time. Nevertheless, the percentage of outliers and the amount of measurement noise were relatively high. The comparison results can be seen in Table 3 (MME_p), Table 4 (MME_a) and Table 5 (elapsed time). The regularization constraint brings noticeable improvement to the original 8- and 7- point methods and in some sequences sound improvement for the 5-point method. In average, the 7-point-S method gave the best result. To understand the results better, it should be mentioned that in the estimation of essential matrices the most challenging case is when the base line is small and the rotations are high. In this case, measurement noise can destructively affect the recovery of camera motion, especially the rotation matrix. This situation occurs very often in the KITTI dataset when the car is driven trough sharp bends. Consequently, the 8-point methods had very poor performance almost in all of the sequences. On the contrary, applying the regularization constraint made the 8-point-S method have almost the same performance as the 7-point method at a much lower elapsed time which shows that it could be a proper option for real-time applications. The 5-point(-S) method, as expected, had a better performance than the 7-point method and even for the sequence 0, it outperformed 7-point-S (Fig. 1-top) but for the other sequences, it performed either similar to or worse than the 7-point-S (for instance Fig. 1-bottom). We analyzed the

Table 5. Average elapsed time (E. T.) per frame for different methods.

Method	5-point-S	5-point	7-point-S	7-point	8-point-S	8-point [14]	8-point [9]
E. T. [s/frame]	4.82	4.12	0.39	0.32	0.21	0.12	0.13

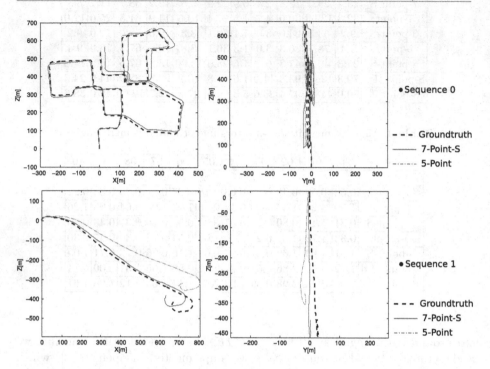

Fig. 1. The estimated and ground truth paths for Seq. 0 (top) and Seq. 1 (bottom). To avoid visual confusion for the Seq. 0, the path was plotted by the frame 3650.

sequences more precisely and noticed that in the sequences in which the outlier ratio is high, the 5-point method works more robust than other methods. The reason lies in RANSAC part, in which N-matched points are selected randomly. Clearly, the less N matched points are selected, the more it is possible that the matched points do not contain any outliers. On the other hand, based on simulation results, the 5-point(-S) method does not perform well to estimate rotation matrices in case of dominant side motions (occurred in sharp bends) which explains why it cannot outperform the 7-point-S method.

6   Conclusion

A new method was developed to include prior knowledge about the ranges of the motion parameter in the form of a covariance matrix for the estimation of essential matrices. By applying the method for the original 8- and 7-point method, great improvements were obtained which have been proven based on simulated and real datasets. Future work could be applying the same analysis to the case of uncalibrated cameras.

References

1. Relative Monocular Camera Pose Estimation (RMCPE). http://www.cvlibs.net/datasets/kitti/eval_odometry.php
2. Bouguet, J.Y.: Pyramidal implementation of the Lucas Kanade feature tracker description of the algorithm (2000). http://robots.stanford.edu/cs223b04/algo_tracking.pdf
3. Dellaert, F., Kaess, M.: Square root SAM: simultaneous localization and mapping via square root information smoothing. Int. J. Rob. Res. **25**(12), 1181–1203 (2006)
4. Faugeras, O.: Three-Dimensional Computer Vision: A Geometric Viewpoint. MIT Press, Cambridge (1993)
5. Fischler, M.A., Bolles, R.C.: Random sample consensus: a paradigm for model fitting with applications to image analysis and automated cartography. Commun. ACM **24**(6), 381–395 (1981)
6. Geiger, A., Lenz, P., Urtasun, R.: Are we ready for autonomous driving? the kitti vision benchmark suite. In: Conference on Computer Vision and Pattern Recognition (2012)
7. Geiger, A., Ziegler, J., Stiller, C.: StereoScan: dense 3D reconstruction in real-time. In: Intelligent Vehicles Symposium, pp. 963–968 (2011)
8. Hartley, R.I.: Estimation of relative camera positions for uncalibrated cameras. In: Sandini, G. (ed.) ECCV 1992. LNCS, vol. 588, pp. 579–587. Springer, Heidelberg (1992)
9. Hartley, R.I.: In defense of the eight-point algorithm. IEEE Trans. Pattern Anal. Mach. Intell. **19**(6), 580–593 (1997)
10. Hartley, R.I., Zisserman, A.: Multiple View Geometry in Computer Vision. Cambridge University Press, Cambridge (2004). ISBN: 0521540518
11. Li, H., Hartley, R.I.: Five-point motion estimation made easy. In: 18th International Conference on Pattern Recognition, pp. 630–633 (2006)
12. Longuet-Higgins, H.C.: A computer algorithm for reconstructing a scene from two projections. Nature **293**, 133–135 (1981)
13. Luong, Q., Deriche, R., Faugeras, O., Papadopoulo, T.: On determining the fundamental matrix: analysis of different methods and experimental results. Technical report 1894, INRIA (1993)
14. Mirabdollah, M.H., Mertsching, B.: Single camera motion estimation: modification of the 8-point method. In: Lee, J., Lee, M.C., Liu, H., Ryu, J.-H. (eds.) ICIRA 2013, Part I. LNCS, vol. 8102, pp. 117–128. Springer, Heidelberg (2013)
15. Nistér, D.: An efficient solution to the five-point relative pose problem. IEEE Trans. Pattern Anal. Mach. Intell. **26**(6), 756–777 (2004)
16. Strasdat, H., Montiel, J.M.M., Davison, A.J.: Scale drift-aware large scale monocular SLAM. In: Proceedings of Robotics: Science and Systems (2010)
17. Tsai, R.Y., Huang, T.S.: Uniqueness and estimation of three-dimensional motion parameters of rigid objects with curved surfaces. IEEE Trans. Pattern Anal. Mach. Intell **6**(1), 13–27 (1984)
18. Yamaguchi, K., McAllester, D., Urtasun, R.: Robust monocular epipolar flow estimation. In: CVPR, pp. 1862–1869 (2013)
19. Zhao, L., Huang, S., Yan, L., Dissanayake, G.: Parallax angle parametrization for monocular SLAM. In: IEEE International Conference on Robotics and Automation, pp. 3117–3124 (2011)

Geometric Reasoning for Uncertain Observations of Man-Made Structures

Jochen Meidow[✉]

Fraunhofer Institute of Optronics, System Technologies and Image Exploitation,
Ettlingen, Germany
jochen.meidow@iosb.fraunhofer.de

Abstract. Observations of man-made structures in terms of digital images, laser scans or sketches are inherently uncertain due to the acquisition process. Thus reverse engineering has to be applied to obtain topologically consistent and geometrically correct model instances by feature aggregation. The corresponding spatial reasoning process usually implies the detection of adjacencies, the generation and testing of hypotheses, and finally the enforcement of the detected relations. We present a complete and general work-flow for geometric reasoning that takes the uncertainty of the observations and of the derived low-level features into account. Thereby we exploit algebraic projective geometry to ease the formulation of geometric constraints. As this comes at the expense of an over-parametrization, we introduce an adjustment model which stringently incorporates uncertainty and copes with singular covariance matrices. The size of the resulting normal equation system depends only on the number of established constraints which paves the way to efficient solutions. We demonstrate the usefulness and the feasibility of the approach with results for the automatic analysis of a sketch and for a building reconstruction based on an airborne laser scan.

1 Introduction

Motivation. Observations of man-made structures are available in various forms. Examples are digital images, laser scans, or line drawings showing technical objects or diagrams. What is common to these kinds of representation is the inherent uncertainty due to the acquisition process. Although such representations can reflect scenes of arbitrary complexity, there is no explicit representation of structure. Therefore data analysis by spatial reasoning must be applied in order to get as-built or "as-drawn" model instances.

This instantiation can be done by looking for patterns corresponding to predefined templates stored in a library. Alternatively, data-driven approaches can be applied starting with the grouping of observations to elementary entities such as straight lines or planes. As a result, instances of generic representations can be inferred which allow to describe easily arbitrary complex structures in a flexible way, for example by boundary representations. To detect man-made structures, a feature adjacency graph can be traversed in order to find all cliques which constitute potential geometric relations such as parallelism or orthogonality.

© Springer International Publishing Switzerland 2014
X. Jiang et al. (Eds.): GCPR 2014, LNCS 8753, pp. 558–568, 2014.
DOI: 10.1007/978-3-319-11752-2_46

After the verification of geometric relations by hypothesis testing, these constraints can be introduced in a joint adjustment for all entities. During this process, the creation of redundant or even contradicting constraints has to be prevented. As a result, model instances with enforced regular structures are produced, a convenient "beautification" for many visualization tasks. However, more important is the fact that the established model instances feature topological consistency, completeness and integrity [14]—a prerequisite for many applications with geometric questions, such as volume or flooding calculations, occlusion analysis, or simulation of wave propagation.

Related Work. Spatial reasoning with uncertain geometric entities has been explored intensively in [10,11] for instance. An adjustment model for optimal parameter estimation with homogeneous entities and arbitrary constraints can be found in [17] and in [16] especially for geometric reasoning with 2D entities. Compact representations for uncertain geometric entities can be found in [5] for algebraic projective geometry, later used for testing and estimation in [8].

Major approaches for Geometric Constraint Solving include graph-based, logic-based, and algebraic methods, cf. [9] for instance. Among the algebraic methods, the computation of Gröbner bases for a given system of equations is convenient. The methods can solve nonlinear systems of algebraic equations, but require long running times and large memory requirements in general. However, for the task at hand moderate requirements are expected. Algebraic methods to deduce redundant or implicit constraint automatically are discussed in [13] in the context of 3D city modeling. Numerical methods for constrained fitting by sequential constraint satisfaction can be found in [1].

Contribution and Approach. We present a work-flow for geometric reasoning based on uncertain point observations. After low-level feature extraction by region growing or segmentation, we estimate the corresponding entities (or features) such as straight lines or planes by minimizing the squared point-to-entity distances. Subsequently, the adjacency information is established by determining the distance between point groups. This is done by simply considering the minimum distance of any two points p_i and p_j where p_i and p_j belong to different sets.

The resulting feature adjacency graph is used to generate hypotheses for geometric relations at hand. By checking the corresponding test statistics, we avoid non-interpretable thresholds. The set of verified and therefore recognized geometric relations is subsequently examined by a greedy algorithm to obtain independent constraints for the simultaneous adjustment of all geometric entities. Hence we perform a constrained optimization for the parameters of the features only.

We restrict out investigations on planes in 3D and straight lines in 2D respectively. For convenience and a consistent representation of different geometric entities, we exploit algebraic projective geometry, leading to constraints which are bilinear in the unknown parameters. The over-parametrization comes at the

expense of singular covariance matrices for the estimated parameters. Therefore, we show the proper incorporation of such an uncertainty representation
contemplated in [18] within the framework of an adjustment model.

2   Theoretical Background

Notation. For the representation of geometric entities we use homogeneous vectors, denoted by upright boldface letters, e.g., $\mathbf{l}$ or $\mathbf{A}$, while Euclidean vectors
and matrices are denoted by slanted boldface letters, e.g., $\boldsymbol{x}$ or $\boldsymbol{R}$. For homogeneous coordinates '=' means an assignment or an equivalence up to a scaling
factor $\lambda \neq 0$. With the skew-symmetric matrix $\boldsymbol{S}(\boldsymbol{x})$ we present the cross product
$\boldsymbol{S}(\boldsymbol{x})\boldsymbol{y} = \boldsymbol{x} \times \boldsymbol{y}$ of two vectors $\boldsymbol{x}$ and $\boldsymbol{y}$.

2.1   Uncertain Geometric Entities and Constraints

Uncertain Entities. In our applications we consider straight lines in 2D and
planes only, both represented in homogeneous coordinates, i.e., $\mathbf{l}$, $\mathbf{m}$, $\mathbf{n}$ and $\mathbf{A}$, $\mathbf{B}$,
$\mathbf{C}$, $\mathbf{D}$. The natural spaces for of homogeneous coordinates are the unit spheres.
Thus eventually a spherical normalization has to be performed for an entity
represented by a vector $\mathbf{x}$ accompanied by variance propagation. With the given
vector $\mathbf{x}$ representing a straight 2D line or a plane and its covariance matrix $\boldsymbol{\Sigma}_{\mathrm{xx}}$
we obtain

$$\mathbf{x}^{(s)} = \frac{\mathbf{x}}{||\mathbf{x}||}, \qquad \boldsymbol{\Sigma}_{\mathrm{x^sx^s}} = \boldsymbol{J}_s \boldsymbol{\Sigma}_{\mathrm{xx}} \boldsymbol{J}_s^\mathsf{T} \quad \text{with} \quad \boldsymbol{J}_s = \frac{1}{||\mathbf{x}||}\left(\boldsymbol{I} - \frac{\mathbf{x}\mathbf{x}^\mathsf{T}}{\mathbf{x}^\mathsf{T}\mathbf{x}}\right) \quad (1)$$

with the Jacobian $\boldsymbol{J}_s$ for the spherically normalized vector $\mathbf{x}^{(s)}$. The covariance
matrix has a rank deficiency due to over-parametrization and the vector $\mathbf{x}^{(s)}$
constitutes the null space of the covariance matrix, thus $\boldsymbol{\Sigma}_{\mathrm{xx}}\mathbf{x}^{(s)} = \mathbf{0}$ holds.

In the following we will use spherically normalized vectors only and drop the
superscript and subscript s for clarity.

Geometric Constraints. We restrict our investigation to orthogonality ($\perp$),
parallelism ($\parallel$), and concurrence ($\circ$) of multiple entities in one point. For further
potential relations such as identity or incidence of different types of entities we
refer to [8]. Table 1 summarizes the used constraints and their degrees of freedom
for three straight 2D lines and four planes respectively.

For the formulation of the constraints we use the dual conic $\boldsymbol{C}_\infty^* = \mathrm{Diag}(1,1,0)$,
the canonical form of absolute dual quadric $\boldsymbol{Q}_\infty^* = \mathrm{Diag}(1,1,1,0)$, cf. [6], and the
special rotation matrix

$$\boldsymbol{R}_\perp = \begin{bmatrix} 0 & -1 & 0 \\ 1 & 0 & 0 \\ 0 & 0 & 1 \end{bmatrix} \qquad (2)$$

to construct parallelism in 2D. The 3×4 projection matrix $\boldsymbol{P} = [\boldsymbol{I}_3, \mathbf{0}]$ extracts
the homogeneous parts of the plane representations. Note that one has to select
two independent equations to check and enforce parallelism in 3D.

Table 1. Distance measures for planes $\{A, B, C, D\}$ and straight lines $\{l, m, n\}$ together with the number of degrees of freedom.

dim. constraint	2D		3D	
	distance measure	dof	distance measure	dof
orthogonality	$d_\perp^{(2)} = l^\mathsf{T} C_\infty^* m$	1	$d_\perp^{(3)} = A^\mathsf{T} Q_\infty^* B$	1
parallelism	$d_\parallel^{(2)} = -l^\mathsf{T} R_\perp^\mathsf{T} C_\infty^* m$ $= +m^\mathsf{T} R_\perp C_\infty^* l$	1	$d_\parallel^{(3)} = +S(PA)PB$ $= -S(PB)PA$	2
concurrence	$d_\circ^{(2)} = \det([l, m, n])$	1	$d_\circ^{(3)} = \det([A, B, C, D])$	1

2.2 Hypothesis Generation and Testing

Once the adjacencies have been determined, a feature adjacency graph can be constructed. The cliques of this graph constitute the potential geometric relations which can to be checked by conventional hypothesis testing. To do so, we formulate the null-hypotheses that the assumed relations are valid, i.e., the true distances are zeros. The test statistics for the hypotheses based on the distances listed in Table 1 read

$$T = d^\mathsf{T} \Sigma_{dd}^{-1} d \qquad \text{or} \qquad T = \frac{d^2}{\sigma_d^2} \qquad (3)$$

respectively. For the relation *parallelism of two planes* the vector d in (3) has length two since two independent constraints have to be selected. The test statistics are χ^2-distributed with $\nu = 1$ or $\nu = 2$ degrees of freedom. Such a hypothesis will not be rejected in favor of the alternative hypothesis if $T < \chi^2_{\nu;q}$ holds for the test statistic T and the q-quantile of the corresponding χ^2_ν-distribution.

The covariance matrix and the variance σ_d^2 for the distances are obtained by variance propagation $\Sigma_{dd} = J_c \Sigma_{xx} J_c^\mathsf{T}$ whereby Σ_{xx} is the block-diagonal, singular covariance matrix of the involved spherical normalized geometric entities, e.g. $x = [l^\mathsf{T}, m^\mathsf{T}]^\mathsf{T}$. The Jacobians J_c can easily be read from the equations listed in Table 1, for example $\partial d_\perp^{(2)} / \partial m = l^\mathsf{T} C_\infty^*$. For the relation *concurrence* the Jacobian of the determinant is simply $J_c = \mathrm{adj}(M)^\mathsf{T}$ with the adjoint of the matrix $M = [l, m, n]$ or $M = [A, B, C, D]$ respectively.

2.3 Constrained Optimization

Adjustment Model and Objective Function. To enforce the postulated hard constraints, we adjust the parameter values of the geometric entities—in our case straight 2D lines or planes—in a least squares manner. In principle the solution to this problem is given by an adaption of Sequential Quadratic Programming [21]. Considering the features' parameters to be uncertain observations, the solution is equivalent to the adjustment with constraints for observations only [12, 15]. However, these models assume regular covariance matrices, as their inverses serve as weight matrices. For the correct treatment of singular covariance matrices, we modify the general adjustment model introduced in [17].

We compile the unconstrained, spherically normalized parameters of E given geometric entities in a parameter vector $\boldsymbol{x} = [\mathbf{x}_1^\mathsf{T}, \ldots, \mathbf{x}_E^\mathsf{T}]^\mathsf{T}$ whose singular covariance matrix $\boldsymbol{\Sigma}_{xx}$ features a block-diagonal shape. The adjusted or constrained parameters $\widehat{\boldsymbol{x}}$ are related to the unconstrained parameters $\boldsymbol{x}$ by additive unknown corrections $\boldsymbol{\epsilon}$ which have to be estimated, thus $\widehat{\boldsymbol{x}} = \boldsymbol{x} + \widehat{\boldsymbol{\epsilon}}$ holds.

The functional model contains the constraints to be enforced by the adjustment. We distinguish two types of constraints:

1. K restrictions $\boldsymbol{k}(\widehat{\boldsymbol{x}}) = \mathbf{0}$ on the parameters reflecting the *intrinsic constraints*, i.e., the spherical normalizations (1) of the homogeneous entities needed to obtain unique results.
2. H *geometric constraints* $\boldsymbol{h}(\widehat{\boldsymbol{x}}) = \mathbf{0}$ on the parameters of the geometric entities, e.g., orthogonality or parallelism.

The stochastic model reflects the uncertainty of the unconstrained parameters, described by an initial covariance matrix $\boldsymbol{\Sigma}_{xx}^{(0)}$ which has block-diagonal structure and results from the feature extraction stage. The covariance matrix $\boldsymbol{\Sigma}_{xx}$ is singular due to over-parametrization. But its null space is properly handled by the intrinsic constraints: If the null space of the covariance matrix $\boldsymbol{\Sigma}_{xx}$ reflects the constraints on the parameters the result using homogeneous coordinates and the pseudo inverse is the same as when minimizing the corresponding Euclidean version [16]. The explicit constraints, i.e., the normalization of the homogeneous entities, and the implicit constraints comprised in the covariance matrix $\boldsymbol{\Sigma}_{xx}$ must be consistent.

Thus finding the optimal estimates $\widehat{\boldsymbol{x}}$ can be done by minimizing the Lagrangian

$$L = \frac{1}{2}\widehat{\boldsymbol{\epsilon}}^\mathsf{T}\boldsymbol{\Sigma}_{xx}^{+}\widehat{\boldsymbol{\epsilon}} + \boldsymbol{\lambda}^\mathsf{T}\boldsymbol{k}(\boldsymbol{x} + \widehat{\boldsymbol{\epsilon}}) + \boldsymbol{\mu}^\mathsf{T}\boldsymbol{h}(\boldsymbol{x} + \widehat{\boldsymbol{\epsilon}}) \tag{4}$$

with the Lagrangian multipliers $\boldsymbol{\lambda}$ and $\boldsymbol{\mu}$ and the pseudo inverse $\boldsymbol{\Sigma}_{xx}^{+}$.

Normal Equation System. For the derivation of the normal equation system we linearize the constraints. The sought constrained parameters $\widehat{\boldsymbol{x}}$ are given by the unconstrained parameters adjusted by estimated corrections $\widehat{\boldsymbol{\epsilon}}$, or by approximate values $\boldsymbol{x}_0$ and estimated updates $\Delta\boldsymbol{x}$. Thus

$$\widehat{\boldsymbol{x}} = \boldsymbol{x} + \widehat{\boldsymbol{\epsilon}} = \boldsymbol{x}_0 + \widehat{\Delta\boldsymbol{x}} \tag{5}$$

holds and the linearization of the constraints reads

$$\boldsymbol{k}(\widehat{\boldsymbol{x}}) = \boldsymbol{k}(\boldsymbol{x}_0) + \boldsymbol{K}\widehat{\Delta\boldsymbol{x}} = \boldsymbol{K}\widehat{\boldsymbol{\epsilon}} + \boldsymbol{k}_0 \tag{6}$$

$$\boldsymbol{h}(\widehat{\boldsymbol{x}}) = \boldsymbol{h}(\boldsymbol{x}_0) + \boldsymbol{H}\widehat{\Delta\boldsymbol{x}} = \boldsymbol{H}\widehat{\boldsymbol{\epsilon}} + \boldsymbol{h}_0 \tag{7}$$

with the Jacobians $\boldsymbol{K}$ and $\boldsymbol{H}$ and the auxiliary variables

$$\boldsymbol{k}_0 = \boldsymbol{k}(\boldsymbol{x}_0) + \boldsymbol{K}(\boldsymbol{x} - \boldsymbol{x}_0) \tag{8}$$

$$\boldsymbol{h}_0 = \boldsymbol{h}(\boldsymbol{x}_0) + \boldsymbol{H}(\boldsymbol{x} - \boldsymbol{x}_0). \tag{9}$$

For the intrinsic constraints $K = \text{Diag}(\mathbf{x}_1, \mathbf{x}_2, \ldots, \mathbf{x}_E)$ and $KK^\mathsf{T} = I$ holds due to the constraints $(\mathbf{x}_e^\mathsf{T}\mathbf{x}_e - 1)/2 = 0$ for each entity indexed with e. Again, the Jacobian H can easily be extracted from the equations listed in Table 1.

Minimizing the sum of weighted squared residuals yields the Lagrangian

$$L = \frac{1}{2}\widehat{\boldsymbol{\epsilon}}^\mathsf{T} \Sigma_{xx}^+ \widehat{\boldsymbol{\epsilon}} + \boldsymbol{\lambda}^\mathsf{T}(K\widehat{\boldsymbol{\epsilon}} + k_0) + \boldsymbol{\mu}^\mathsf{T}(H\widehat{\boldsymbol{\epsilon}} + h_0) \tag{10}$$

with the Lagrangian vectors $\boldsymbol{\lambda}$ and $\boldsymbol{\mu}$. The necessary conditions for a minimum

$$\frac{\partial L}{\partial \widehat{\boldsymbol{\epsilon}}^\mathsf{T}} = \Sigma_{xx}^+ \widehat{\boldsymbol{\epsilon}} + K^\mathsf{T}\boldsymbol{\lambda} + H^\mathsf{T}\boldsymbol{\mu} = 0 \tag{11}$$

yield the normal equation system

$$\begin{bmatrix} \Sigma_{xx}^+ & K^\mathsf{T} & H^\mathsf{T} \\ K & O & O \\ H & O & O \end{bmatrix} \begin{bmatrix} \widehat{\boldsymbol{\epsilon}} \\ \boldsymbol{\lambda} \\ \boldsymbol{\mu} \end{bmatrix} = \begin{bmatrix} 0 \\ -k_0 \\ -h_0 \end{bmatrix}. \tag{12}$$

The estimates are $\widehat{\mathbf{x}} = \mathbf{x} + \widehat{\boldsymbol{\epsilon}}$ and one has to iterate until convergence. Note that the covariance Σ_{xx} must be adjusted during the iterations since the linearization points for K change and $\Sigma_{xx}K^\mathsf{T} = O$ must be fulfilled. This can be achieved by error propagation for the spherical normalization (1).

Further Simplification. Solving (12) constitutes a solution to the problem at hand. Unfortunately this implies the computation of the pseudo inverse for Σ_{xx}, a matrix whose size depends on the number of parameters, i.e., the number of entities and their representation. Thus usually one has to face numerical problems leading to unreliable results.

However, the equation system can be further simplified. Rewriting (12) leads to the reduced system

$$\begin{bmatrix} \Sigma_{xx}^+ & K^\mathsf{T} \\ K & O \end{bmatrix} \begin{bmatrix} \widehat{\boldsymbol{\epsilon}} \\ \boldsymbol{\lambda} \end{bmatrix} = \begin{bmatrix} -H^\mathsf{T}\boldsymbol{\mu} \\ -k_0 \end{bmatrix}. \tag{13}$$

Then, with $\Sigma_{xx}K^\mathsf{T} = O$ and $KK^\mathsf{T} = I$ the relation

$$\begin{bmatrix} \Sigma_{xx}^+ & K^\mathsf{T} \\ K & O \end{bmatrix} \begin{bmatrix} \Sigma_{xx} & K^\mathsf{T} \\ K & O \end{bmatrix} = \begin{bmatrix} I & O \\ O & I \end{bmatrix} \tag{14}$$

holds due to the definition of pseudo inverses and the normal equation matrix can be inverted explicitly:

$$\begin{bmatrix} \widehat{\boldsymbol{\epsilon}} \\ \boldsymbol{\lambda} \end{bmatrix} = \begin{bmatrix} \Sigma_{xx} & K^\mathsf{T} \\ K & O \end{bmatrix} \begin{bmatrix} -H^\mathsf{T}\boldsymbol{\mu} \\ -k_0 \end{bmatrix} \tag{15}$$

Thus the substitution of the estimate $\widehat{\boldsymbol{\epsilon}} = -\Sigma_{xx}H^\mathsf{T}\boldsymbol{\mu} - K^\mathsf{T}k_0$ in (7) yields the Lagrangian vector $\boldsymbol{\mu} = \Sigma_{hh}^{-1}(h_0 - HK^\mathsf{T}k_0)$ with the regular covariance matrix

$\Sigma_{hh} = H\Sigma_{xx}H^\top$ of the contradictions (9) due to variance propagation. Finally we obtain the estimate

$$\widehat{\epsilon} = -\Sigma_{xx}H^\top\Sigma_{hh}^{-1}(h_0 - HK^\top k_0) - K^\top k_0 \tag{16}$$

for the corrections. Note that the size of the matrix to be inverted depends only on the number of constraints and that the pseudo inverse of Σ_{xx} has not to be computed explicitly. Furthermore, no special treatment of entities not being affected by any extrinsic constraint is necessary. In this case the corresponding estimates for the corrections (16) will simply be zeros.

2.4   Selection of Independent Constraints

The hypotheses for geometric constraints are obtained by checking all cliques of the feature adjacency graph. These cliques are obtained by the Bron-Kerbosch algorithm, cf. [2,3] and the set of potential constraints is then fed into a greedy algorithm, as proposed in [4] for instance. Criteria for success are the estimates for the rank of the Jacobian H and for the condition number of Σ_{hh}.

To certify the simplicity of the approach, we list a pseudo code for the adjustment procedure in Fig. 1.

Require: set of E unconstrained entities $\{x_e, \Sigma_{ee}\}$
Ensure: E constrained entities $\{x_e, \Sigma_{ee}\}$

```
 1:  x ← [x₁ᵀ, x₂ᵀ, ..., x_Eᵀ]ᵀ
 2:  x₀ ← x {approx. values, to be improved}
 3:  while (not converged) do
 4:      compute contradictions k₀(x, x₀) and h₀(x, x₀)
 5:      compute Jacobians K(x₀) and H(x₀)
 6:      Σxx ← Diag(Σ₁₁, Σ₂₂, ..., Σ_EE)
 7:      ε̂ ← Σxx Hᵀ(HΣxxHᵀ)⁻¹(HKᵀk₀ - h₀) - Kᵀk₀
 8:      x₀ ← x + ε̂
 9:      for all entities do
10:          {spherical normalization}
11:          Je ← (I - x₀e x₀eᵀ/(x₀eᵀx₀e)) /‖x₀e‖
12:          Σee ← Je Σee Jeᵀ
13:          x₀e ← x₀e/‖x₀e‖
14:      end for
15:      x₀ ← [x₀₁ᵀ, x₀₂ᵀ, ..., x₀Eᵀ]ᵀ
16:  end while
17:  return {x₀e, Σee}
```

Fig. 1. Pseudo code for the adjustment. The contradictions and Jacobians are obtained by the equations listed in Table 1.

3   Results

We demonstrate the feasibility of the proposed approach by explicating two examples—the analysis of a line drawing (Sect. 3.1) and a building reconstruction based on a given laser scan (Sect. 3.2). While for our 2D example the groups are already given by the user interaction, we apply a region growing for the 3D reconstruction using the method explained in [19] as provided by the Point Cloud Library [20]. For the built-up of the feature adjacency graph we determine adjacent groups of points. Candidates for neighbors are given by point groups having bounding boxes with an overlap. These hypotheses are then checked by the consideration of the pairwise Euclidean distances of the points of the two groups. If the minimal found distance is below a threshold, the groups are considered to be adjacent. The magnitude of the threshold should correspond to the average point spacing at hand.

3.1   2D Application: Line Drawing of a Flow Chart

Figures 2 and 3 illustrate the fully automatic reasoning for a hand drawing, here a simple—not necessarily meaningful—part of a flow chart. The sketch had been captured interactively by a user interface, that's why groups of points are already at hand. In case of scanned documents, a grouping has to be carried out, e.g., by region growing or random sample consensus.

After point acquisition, grouping, and adjacency determination, the feature extraction has been performed, i.e., the estimation of the parameter values and corresponding covariance matrices for the 22 straight lines. Then for neighboring point groups all hypotheses for geometric relations of the corresponding straight lines have been tested—here a set of 33 constraints usable for the reasoning process.

During the optimization the greedy algorithm discovered a set of 23 linearly independent constraints which were enforced. The computing time was 2 s in Matlab. The results for the adjusted entities are shown in Fig. 3, too.

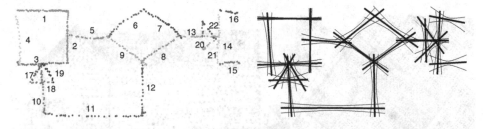

Fig. 2. Left: Interactively captured sketch with 22 point groups. Right: Result of the feature extraction for the line drawing. The uncertainty regions feature hyperbolic shapes, plotted somewhat exaggerated.

566   J. Meidow

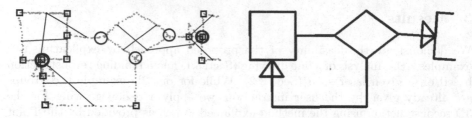

Fig. 3. Left: Geometric relations, automatically found. Orthogonality (□, 13 times), parallelism (–, 11 times), and concurrent straight lines (○, 9 times). The relation *concurrence* can appear multiple times at one locations for more than three intersecting lines. Right: Final result of the reasoning process with 23 enforced constraints.

3.2   3D Application: Building Reconstruction

The "Abenberg 2009" data set is an accumulated point cloud captured by airborne laser scanning [7]. The test site has been approached by a helicopter in a cross pattern, which results in overlapping point clouds. The co-registration and accumulation of the stripes led to a reference data set which features points lying on the building's façades and an average point density of 21 pts/m². Figure 4 shows a cut of this data set containing the point cloud of an L-shaped building and its surrounding area. Region growing delivered 16 planar regions.

After the estimation of the unconstrained plane parameters and their covariances, the adjacencies have been determined and the geometric relations checked. The set of 26 found constraints includes: 14 times orthogonality, five times parallelism, and three times four planes concurring in one 3D point. Please note that not all adjacencies and geometric relations have been found due to noisy and missing data.

The greedy algorithm found 21 independent constraints. Introducing these constraints into the adjustment leads to the result depicted in Fig. 5, right. Observe that in the roof's valley and on its ridge, multiple planes intersect in one point in each case.

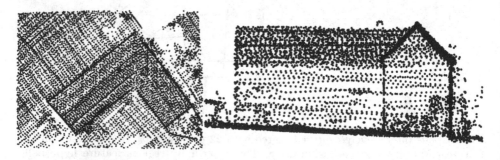

Fig. 4. Airborne captured point cloud, showing an L-shaped gable roof building and its surrounding.

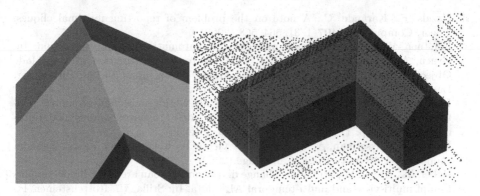

Fig. 5. Left: four unconstrained planes do not intersect in one point. Right: The result of the reasoning process featuring 18 enforced constraints. Among them the fact that 4 planes intersect exactly in one point.

4 Conclusions and Outlook

We presented a complete work-flow for the geometric reasoning based on uncertain observations. The approach exploits the power of algebraic projective geometry to represent geometric entities and to formulate constraints between them. The general representation comes at the expense of an over-parametrization and corresponding singular covariance matrices. Therefore we adopted an adjustment model capable to deal with singular covariance matrices by considering the normalization of homogeneous vectors as intrinsic constraints. The enforcement of the geometric constraints yields consistent model instances needed for applications with geometric questions.

The conducted experiments indicate the usability and feasibility of the approach for applications in 2D and 3D. The essential thresholds have a sound interpretation, e.g., the significance level for testing of hypothetical constraints. The size of the equations systems to be solved depends only on the number of constraints at hand which paves the way for large-scale applications.

The proposed framework allows for the seamless incorporation of further geometric entities such as points or conics in 2D, and points or straight lines in 3D. Further types of topological constraints, e.g. incidence or coplanarity, equal slopes etc., can be introduced easily as well as metric constraints, such as fixed angles or distances. For the future, more sophisticated methods to determine sets of independent constraints are envisaged, especially algebraic methods using Gröbner bases [13].

References

1. Benkő, P., Kós, G., Várady, T., Andor, L., Martin, R.: Constrained fitting in reverse engineering. Comput. Aided Geom. Des. **19**(3), 173–205 (2002)
2. Bron, C., Kerbosch, J.: Algorithm 457: finding all cliques of an undirected graph. Commun. ACM **16**(9), 575–577 (1973)

3. Cazals, F., Karande, C.: A note on the problem of reporting maximal cliques. Theor. Comput. Sci. **407**(1–3), 564–568 (2008)
4. Förstner, W.: Mid-level vision processes for automatic building extraction. In: Gruen, A., Kuebler, O., Agouris, P. (eds.) Automatic Extraction of Man-Made Objects from Aerial and Space Images. Monte Verita, pp. 179–188. Birkhäuser, Basel (1995)
5. Förstner, W., Brunn, A., Heuel, S.: Statistically testing uncertain geometric relations. In: Sommer, G., Krüger, N., Perwass, C. (eds.) Mustererkennung 2000. Informatik aktuell, pp. 17–26. Springer, Heidelberg (2000)
6. Hartley, R., Zisserman, A.: Multiple View Geometry in Computer Vision, 2nd edn. Cambridge University Press, Cambridge (2000)
7. Hebel, M., Arens, M., Stilla, U.: Change detection in urban areas by direct comparison of multi-view and multi-temporal ALS data. In: Stilla, U., Rottensteiner, F., Mayer, H., Jutzi, B., Butenuth, M. (eds.) PIA 2011. LNCS, vol. 6952, pp. 185–196. Springer, Heidelberg (2011)
8. Heuel, S.: Uncertain Projective Geometry. LNCS, vol. 3008. Springer, Heidelberg (2004)
9. Hoffmann, C.M., Joan-Arinyo, R.: A brief on constraint solving. Comput. Aided Des. Appl. **2**(5), 655–663 (2005)
10. Kanatani, K.: Statistical analysis of geometric computation. CVGIP: Image Underst. **59**(3), 286–306 (1994)
11. Kanatani, K.: Statistical Optimization for Geometric Computation: Theory and Practice, 2nd edn. Artificial Intelligence Laboratory, Department of Computer Science, Gumma University, Japan (1995)
12. Koch, K.R.: Parameter Estimation and Hypothesis Testing in Linear Models, 2nd edn. Springer, Berlin (1999)
13. Loch-Dehbi, S., Plümer, L.: Automatic reasoning for geometric constraints in 3D city models with uncertain observations. ISPRS J. Photogrammetry Remote Sens. **66**, 177–187 (2011)
14. Mäntylä, M.: An Introduction to Solid Modeling. Computer Science Press, Inc., New York (1987)
15. McGlone, J.C., Mikhail, E.M., Bethel, J. (eds.): Manual of Photogrammetry, 5th edn. American Society of Photogrammetry and Remote Sensing, Bethesda (2004)
16. Meidow, J., Beder, C., Förstner, W.: Reasoning with uncertain points, straight lines, and straight line segments in 2D. ISPRS J. Photogrammetry Remote Sens. **64**(2), 125–139 (2009)
17. Meidow, J., Förstner, W., Beder, C.: Optimal parameter estimation with homogeneous entities and arbitrary constraints. In: Denzler, J., Notni, G., Süße, H. (eds.) Pattern Recognition. LNCS, vol. 5748, pp. 292–301. Springer, Heidelberg (2009)
18. Pohl, M., Meidow, J., Bulatov, D.: Extraction and refinement of building faces in 3D point clouds. In: Image and Signal Processing for Remote Sensing XIX, p. 88920V. Society of Photo-Optical Instrumentation Engineers (2013)
19. Rabbani, T., van den Heuvel, F.A., Vosselmann, G.: Segmentation of point clouds using smoothness constraint. In: Proceedings of the ISPRS Commission V Symposium 'Image Engineering and Vision Metrology'. ISPRS Archives, vol. XXXVI, Part 5, pp. 248–253 (2006)
20. Rusu, R.B., Cousins, S.: 3D is here: Point Cloud Library (PCL). In: IEEE International Conference on Robotics and Automation (ICRA), p. 14 (2011)
21. Triggs, B., McLauchlan, P.F., Hartley, R.I., Fitzgibbon, A.W.: Bundle adjustment – a modern synthesis. In: Triggs, B., Zisserman, A., Szeliski, R. (eds.) Vision Algorithms 1999. LNCS, vol. 1883, pp. 298–372. Springer, Heidelberg (2000)

Locality Sensitive Hashing Using GMM

Fabian Schmieder[(✉)] and Bin Yang

ISS, University of Stuttgart, Stuttgart, Germany
`fabian.schmieder@iss.uni-stuttgart.de`

Abstract. We propose a new approach for locality sensitive hashes (LSH) solving the approximate nearest neighbor problem. A well known LSH family uses linear projections to place the samples of a dataset into different buckets. We extend this idea and, instead of using equally spaced buckets, use a Gaussian mixture model to build a data dependent mapping.

1 Introduction

Nearest neighbor search is a basic building block in many machine learning tasks and retrieval systems. A direct linear search for the k nearest neighbors (kNN) is often not an option for current applications as the datasets are very large. Therefore, a sub-linear search strategy is needed. For low dimensional datasets of size N, the well known k-d trees can be used to find the nearest neighbors in $O(\log N)$ time [5]. However, for high dimensional datasets, no general solution has yet been found to solve the exact kNN problem in a sub-linear time. So instead of searching for the exact nearest neighbors a search for the approximate nearest neighbors (ANN) may be considered. This can be achieved much more efficiently. In recent years the locality sensitive hashes (LSH) have gained a lot of attention to solve this task [3,6,7,10].

The basic idea of LSH is to use a family of locality sensitive hash functions to hash samples into buckets. The LSH function is designed in such a way that with a high probability, samples which are close together will be mapped into the same bucket. To find the closest samples to a query sample, the LSH function is used to find the bucket of the query. The samples of the same bucket are considered as the set of possible nearest neighbors and ranked with the used distance measure to return the approximate kNN.

The used LSH family has a very large influence in the performance of the system and should be chosen according to the used distance measure D. For the l_p norm Datar et al. [3] used p-stable distributions to create a number of random projections and divided the projection space into buckets of a fixed section length, see Fig. 1(a). Due to this fixed length, buckets in a region with a high density of samples will contain a large amount of samples, whereas buckets in low density regions will be very sparsely populated. This is not a problem for their primary use case, namely the approximate r nearest neighbor problem. In this case, all samples with a distance smaller or equal to r should be returned by

X. Jiang et al. (Eds.): GCPR 2014, LNCS 8753, pp. 569–581, 2014.
DOI: 10.1007/978-3-319-11752-2_47

(a) Basic   (b) GMM   (c) K-Means

Fig. 1. The bucket regions for different LSH families.

the system. Therefore, the number of samples the system returns is not provided by the user. It depends on how dense the region of the query sample is populated. However, when searching for the kNN this creates the problem of choosing a good section length for the buckets, as the number of returned samples should now be of equal length independent of the query itself. By using a big section length, the candidate set of possible nearest neighbors is very large for samples in dense regions. This will result in a very expensive search for the approximate neighbors in the candidate set. However, for a small section length, the candidate set is to small or empty for samples in sparsely populated regions and the system may return less than the defined number of nearest neighbors. A common solution to this problem is to use multiple Hashes with a varying section length covering different r values.

In [10] Paulevé et al. argued that a data dependent hash function will provide a better mapping and proposed to use a k-means vector quantizer, see Fig. 1(c). However, this creates the problem to find an efficient way to select the closest mean for a query as the number of means is very large and should grow linearly with the number of samples. They proposed to use a hierarchical k-means structure to find the corresponding mean more efficiently. At the end their method has a close relationship to other space-partitioning search strategies like multiple random k-d trees, see [8, 11].

Instead of using a totally new hash function like [10] we propose an extension of the well known family of hash functions based on the l_p norm from [3]. Instead of using a fixed size for the buckets we use a new strategy to partition the projections. We approximate the distribution of the samples for each projection with a very simple one-dimensional Gaussian mixture model (GMM) and create an adaptive mapping using the resulting cumulative distribution function (CDF), see Fig. 1(b). Even for very large datasets the estimation of the GMM can be achieved very efficiently for this simple one-dimensional case by using a basic stemming technique.

2   Locality Sensitive Hashing

LSH was first introduced in [6] and is based on the LSH functions. They have the property that with a high probability samples which are close together will land in the same bucket, whereas samples which are far apart will land in the same buckets with a very low probability. The formal definition of LSH is

Definition 1. *A function family* $\mathcal{H} = \{h : \mathcal{S} \rightarrow \mathcal{U}\}$ *is called* (r, cr, p_1, p_2) *sensitive for D if* $\forall p, q \in \mathcal{S}$

- *if* $D(p, q) \leq r$ *then* $Pr_{\mathcal{H}}[h(p) = h(q)] \geq p_1$,
- *if* $D(p, q) \geq cr$ *then* $Pr_{\mathcal{H}}[h(p) = h(q)] \leq p_2$.

For an approximate nearest neighbor search the family has to further satisfy $r > 0$, $c > 1$ and $p_1 > p_2$. Now the problem is to find a function family $\mathcal{H}$ which has a very low collision probability p_2 and still retains a large retrieval probability p_1. In [3] Datar *et al.* described a function family for the l_p norm using p-stable distributions of the form

$$h_{n,b}(p) = \left\lfloor \frac{<n, p> + b}{W} \right\rfloor, \tag{1}$$

where the elements of n are drawn independently from a p-stable distribution and b is equally distributed between 0 and W. For example, in the case of the euclidean norm ($p = 2$) the elements of n are independently drawn from a standard normal distribution and n is then normalized to unit length. The length of the individual sections can be controlled with W and has to be fitted to the dataset.

This family only has a very small gap between p_1 and p_2. This means, close samples are only slightly more likely to be mapped to the same bucket than far away samples. To increase the gap, M functions of $\mathcal{H}$ are concatenated to create a new family of functions $\mathcal{G} = \{g : \mathcal{S} \rightarrow \mathcal{U}^M\}$, with $g(p) = [h_1(p), \dots, h_M(p)]$ and $h_m \in \mathcal{H}$. $\mathcal{G}$ decreases the collision probability p_2 because all M scalar functions h_m have to be equal for a collision. This is visualized in Fig. 1(a) where three scalar functions h_m are used to divide the space into different buckets $g(p)$. While this strategy is able to decrease p_2, it will also decrease the retrieval probability p_1. To still ensure a large retrieval probability p_1, Datar uses L different hash functions of $g_l \in \mathcal{G}, l = 1, \dots, L$ and returns the union of the L corresponding buckets for each query. This makes it possible to adjust p_1 and p_2 by choosing M and L accordingly. However, for practical problems L can increase to large values of 500 and above. This is a significant disadvantage because the memory requirements for the index structure increase linearly with L, as does the computational complexity. A few modifications have been proposed to decrease the required L by querying multiple buckets of each hash function g_l, see [7,9].

3   GMM Mapping

Instead of simply dividing the projection of (1) into equally long sections, we propose to use a data dependent approach to divide the projection into sections with a roughly equal amount of samples per section. We use the same projections as in [3] given by

$$f_n(p) = <n, p>. \tag{2}$$

Instead of dividing f_n into sections of length W, we use the dataset $\boldsymbol{X} = \{\boldsymbol{x}^{(1)}, \ldots, \boldsymbol{x}^{(N)}\}$, with $\boldsymbol{x}^{(i)} \in \mathcal{S} \subset \mathbb{R}^d$ to estimate the distribution of $f_n^{(i)} = f_n(\boldsymbol{x}^{(i)})$ for $i = 1, \ldots, N$ by assuming a one-dimensional GMM of the form

$$p_n(f_n) = \sum_{j=1}^{J} w_j \mathrm{N}(f_n; \mu_j, \sigma_j^2). \tag{3}$$

$\mathrm{N}(f_n; \mu_i, \sigma_j^2)$ is the normal distribution with mean μ_j and variance σ_j^2. We use an adaption of the algorithm proposed by Figueiredo et al. in [4] to estimate the model parameters w_j, μ_j, σ_j^2 and model order J. They proposed an extension to the well known Expectation Maximization (EM) algorithm incorporating the estimation of the model order. Starting with a model using a large number of modes, they successively drop modes with a low support by the data using a sparsity constraint on the model weights w_j.

We applied a few changes to their proposed algorithm. First, as we do not require a perfect fit, we apply a stemming technique to implement the estimation of the GMM efficiently. Instead of using all N $f_n^{(i)}$ directly, we compute a histogram of $f_n^{(i)}$ with N_H bins and use the bin centers $\tilde{f}_n^{(j)}$ and counts $c_n^{(j)}$, $j = 1, \ldots, N_H$ in the EM updates. This ensures that the running time of the algorithm is independent of the number of samples N. Stemming allows us to estimate the GMM parameters rapidly even for very large datasets. A further modification is to increase the sparsity constraint of [4] to create a very compact GMM model containing only a few modes.

Using the cumulative distribution function (CDF) P_n of the estimated GMM, we can now define a new family of hash functions $\mathcal{H}$ with

$$\tilde{h}_{n,b}(\boldsymbol{p}) = \lfloor a \, P_n(<\boldsymbol{n}, \boldsymbol{p}>) + b \rfloor, \tag{4}$$

where b is uniformly distributed between 0 and 1 and a is a positive scale factor. An advantage of this definition is, that the number of unique hash values depends only on a and b. It is given by $S = \lceil a+b \rceil$ and even queries with a much larger or smaller $f_n^{(i)}$ than the samples of the dataset will still be mapped to one of these S values. Furthermore, because we use the CDF to compute the hash values, excluding the first and last value, each hash value will be assigned to roughly the same number of samples, i.e.

$$\mathrm{Pr}_{\boldsymbol{X}}[\tilde{h}_{n,b}(\boldsymbol{p}) = u] \approx 1/(S - 1),$$

for $u = 1, \ldots, S - 2$. Under the assumption that different f_n are independent given different $\boldsymbol{n}$, this is also true for the number of samples per bucket, i.e.

$$\mathrm{Pr}_{\boldsymbol{X}}[g(\boldsymbol{p}) = \boldsymbol{u}] = \prod_{m=1}^{M} \mathrm{Pr}_{\boldsymbol{X}}[\tilde{h}_m(\boldsymbol{p}) = u_m] \approx \mathrm{const.},$$

with $\boldsymbol{u} = [u_1, \ldots, u_M]$. In practice, this assumption will not be fulfilled for most problems. However, even if different f_n for different $\boldsymbol{n}$ are not truly independent, the number of samples per bucket will vary by a much smaller degree than for the original hash functions $\mathcal{H}$ using a constant section length W.

The scale factor a replaces the section length W of the original formulation. There are two possibilities to select a,

- a constant value $a = S - 1$, creating S different values for each $\tilde{h}_{n,b}$.
- compute a adaptively for each $\tilde{h}_{n,b}$ using a minimal section length W_{min}, see Sect. 5.3.

4   Experimental Setup

We used three different datasets in our experiments, all of which are freely available at the UCIML Repositories [2]:

Corel contains 67033 images from the Corel image collection. A HSV color space histogram containing 32 colors was extracted for each image in the dataset.

MNIST contains 60000 handwritten digits from 0 to 9. Each digit is described by a 28×28 grayscale image creating a 784 dimensional feature vector. We reduced the number of dimensions by selecting the 128 principal components of the dataset.

Year Prediction MSD contains 463715 audio features extracted from songs. 12 timbre features were extracted from the Echo Nest API and the mean and covariance over the local timbre descriptors of a song create a 90 dimensional feature vector for each song in the dataset.

4.1   Performance Metrics

The performance of a LSH scheme can be described with a number of different metrics, each of which focusing on different requirements for the ANN search. To find suitable measures, we first examine what the results of a query to a LSH look like. For each query q, the LSH will produce a set of possible nearest neighbors $A(q)$. The size of this set can vary widely and depends on the query q itself, the length of the dataset N and the parameters used for the LSH. In a final step, the approximate k nearest neighbors to q are returned by the ANN system using a simple linear search for the nearest samples in $A(q)$.

Based on this observation, we define the relative average set size for a set $\mathcal{Q}$ of query samples

$$S_{\mathcal{Q}} = \frac{1}{N|Q|} \sum_{q \in \mathcal{Q}} |A(q)|,$$

$S_{\mathcal{Q}}$ describes how many samples in the dataset have to be examined on average, relative to the total number of samples in the dataset. This indicates how efficiently a LSH scheme finds the nearest neighbors. However, this is not enough to describe the performance of the system as we still do not know how accurate the returned nearest neighbors are.

In the strictest sense, an ANN search can be considered as accurate, if it returns as many of the true nearest neighbors as possible. To describe this property, we compute the recall for the set Q of query samples

$$R_Q(k) = \frac{1}{k|Q|} \sum_{q \in Q} |A(q) \cap I_k(q)|,$$

where $I_k(q)$ is the set of the true k nearest neighbors of q in the dataset. However, as the recall computes the number of exact matches, it fails to describe an important detail. For an ANN search we do not require to find the true nearest neighbors. We also accept samples which are nearly as close as the true neighbors.

To capture how close the found samples are compared to the true nearest samples, we first sort the samples in the sets $A(q)$ and $I_k(q)$ by their distance to q. This creates the sequences $(d_{A,1}(q), \ldots, d_{A,|A|}(q))$ and $(d_{I,1}(q), \ldots, d_{I,k}(q))$ of increasing distance values. Using these sequences we define the relative distance accuracy

$$D_Q(k) = \frac{1}{k|Q|} \sum_{q \in Q} \sum_{i=1}^{k} \frac{d_{I,i}(q)}{d_{A,i}(q)}.$$

We use the ratio $d_{I,i}(q)/d_{A,i}(q)$ instead of the more conventional ratio $d_{A,i}(q)/d_{I,i}(q)$ to cover the special case of $|A(q)| < k$, i.e. the LSH returned less possible nearest neighbors than we required. In this case we will set $d_{A,i}(q)$ to ∞ for $i > |A(q)|$ resulting in the ratio $d_{I,i}(q)/d_{A,i}(q) = 0$.

For each dataset 1000, previously unseen test samples form the query set Q which is used to compute the performance measures. Furthermore, as the projections are created randomly, each experiment is run 3 times and the average result over the different runs is reported.

5   Implementation

To compare the performance of the proposed GMM mapping with the original mapping, the multi-probe LSH indexing approach of Lv et al. was implemented in python. First we will give a short introduction to multi-probe LSH indexing and remark on the changes made to adapt it to the proposed mapping.

5.1   Multi-probe LSH Indexing

In [7] multi-probe LSH was presented to reduce the number of required hashes L by using multiple buckets of each hash. The large number of L is needed because for G the retrieval probability p_1 decreases fast with increasing M. This means, for a given query q a nearest neighbor p in the dataset will often not be placed in the same bucket, i.e. $g(q) \neq g(p)$. Using the union of L different hash functions g to build the set $A(q)$ mitigates this effect. However, as Lv et al. observed, for hash functions where the nearest neighbor was not placed in the same bucket, only a small number of individual h_m are different. Furthermore, in nearly all

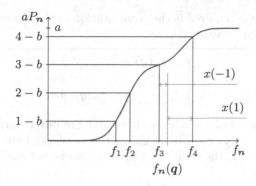

Fig. 2. An example of an estimated CDF $P_n(f_n)$, the resulting buckets with the borders f_i based on a scale $a = 4$ creating $S = 5$ buckets and the projection $f_n(q)$ for a query q with the distances $x(\delta)$.

cases the buckets containing the nearest neighbors are direct neighbors of the query bucket, i.e. the h_m are only one section apart.

Based on these observations, Lv *et al.* proposed to not only query the bucket $g(q)$, but also its direct neighbors where some of the individual h_m where shifted by a section, i.e. $\bar{h}_m(q) = h_m(q) + \delta_m$ with $\delta_m \in \{-1, 0, 1\}$. As the number of neighboring buckets grows exponentially with M, it is not feasible to query all the neighboring buckets. They derived a query-directed probing sequence to only query the T most likely buckets. To this end, each bucket is assigned a score given by

$$s(\delta) = \sum_{m=1}^{M} (x_m(\delta_m))^2,$$

where $\delta \in \{-1, 0, 1\}^M$ describes a bucket by its offsets relative to the true values $g(q) = [h_1(q), \ldots, h_M(q)]$ and $x_m(\delta)$ defines the distance of q to the corresponding section, see Fig. 2. Lv *et al.* described an efficient algorithm to find the probing sequence for the T buckets with the lowest scores in [7]. Per default Lv *et al.* derived a probing sequence for each of the L hash function independently. However, as they suggested, we will use only one probing sequence using all L hash functions together. This means that in our case T directly provides us with the number of buckets used to build the resulting set $A(q)$. So changing L will not change how many buckets are used, it only changes the number of buckets the algorithm has to choose from.

5.2 Changes for GMM Mapping

We implemented the algorithm of [7] without the optimization to buffer the probing sequences. For each query we maintain a heap containing the buckets with lowest score to create the probing sequence. The proposed query scheme can be directly applied to the new hash family $\tilde{H}$ based on GMM mapping.

The only modification required is the computation of the distance $x_m(\delta)$. In our case the distance can be expressed by

$$x_{n,b}(\delta) = \begin{cases} f_n(q) - f_{\tilde{h}_{n,b}(q)}, & \delta = -1 \\ 0, & \delta = 0 \\ f_{\tilde{h}_{n,b}(q)+1} - f_n(q), & \delta = 1 \end{cases},$$

where f_i defines the border of the buckets and the outer borders are given by $f_0 \rightarrow -\infty$ and $f_S \rightarrow \infty$, see Fig. 2. Using the borders of the buckets and the unit step function $u(f)$, the hash function can also be reformulated to

$$\tilde{h}_{n,b}(q) = \sum_{i=1}^{S-1} u(<n, q> -f_i). \qquad (5)$$

Because the f_i can be precomputed, we now only need S additional operations compared to the basic mapping of (1) to compute the hash value $\tilde{h}_{n,b}(q)$ and the distances $x(\delta)$. As S is typically small, i.e. $S < 6$, the computational complexity increases only slightly as $< n, q >$ will dominate the computation for problems in high dimensions.

5.3   Computation of the Scale Factor

The scale factor a for the hash function $\tilde{h}_{n,b}(p)$ of (4) can be defined implicitly using a minimal section length W_{min}, allowing a finer control of the size of the bucket regions. However, an exact computation of a is difficult because the CDF P_n of the GMM would have to be inverted. Therefore, we use a lower bound $\tilde{a}$ ensuring that the resulting section lengths $W_i = f_i - f_{i-1}$ $(i = 1, \ldots, S)$ are not smaller than W_{min}. This is fulfilled for

$$\tilde{a} = \min_f \frac{1}{P_n(f) - P_n(f - W_{min})}. \qquad (6)$$

In practice, we sample equation (6) at $N_t = 250$ discrete positions between the minimal and maximal value of $f_n^{(i)}$ to find the an approximation for a.

6   Experiments

For each dataset, we computed the performance of the original family $\mathcal{H}$ and the proposed GMM mapping $\tilde{\mathcal{H}}$ for various parameter settings. The euclidean distance metric was used to find the exact $k = 1, 5, 10, 20$ nearest neighbors for all datasets. For each choice of (L, M), the original mapping and the GMM mapping used the same projections $f_{l,m}$, with $l = 1, \ldots, L$ and $m = 1, \ldots, M$. The only difference was how the hash families mapped these projections to integer values. Furthermore, to allow for a better comparison between $\mathcal{H}$ and $\tilde{\mathcal{H}}$, we did not define a directly. Instead, we defined the minimal section length

W_{min} and computed the required scale factor a for each $f_{l,m}$ as described in Sect. 5.3. The used common parameters for all sets are $L = 10, \ldots, 50$ and $T = 100, \ldots, 600$. The remaining parameters were chosen based on the dataset. For Corel $M = 8, \ldots, 18$ and $W/W_{min} = 0.002, \ldots, 0.01$, for MNIST $M = 14, \ldots, 24$ and $W/W_{min} = 100, \ldots, 400$ and for Year Prediction MSD $M = 12, \ldots, 20$ and $W/W_{min} = 100, \ldots, 400$. Changing the number of nearest neighbors K did not change the behavior of the GMM mapping compared the original mapping. Therefore, all reported results are for the case of $K = 10$ nearest neighbors.

Table 1 provides the results for the best choice of L, M and T. W/W_{min} was chosen to achieve a length S_Q close to 1 % and 2 %. The results show that GMM mapping is superior for all datasets in both the recall $R_Q(k)$ and the distance $D_Q(k)$. Furthermore, the percentiles over the set length $|A(q)|$ show that, while the average lengths S_Q are close together, the percentiles differ between the original and the GMM mapping. The Interquartile range (IQR) is constantly smaller for the GMM mapping. This supports our original statement that by using GMM mapping, the length of the created buckets varies less. The most significant change is the increase of the 25 percentile for GMM mapping. This means, queries in sparsely populated regions return larger and most importantly non-empty candidate sets $A(q)$.

Figure 3 provides us with a closer look at the performance by plotting the recall $R_Q(k)$ and the distance accuracy $D_Q(k)$ over the average length S_Q when slowly increasing W or W_{min}. Each plot includes the results for the best parameter choice given in Table 1 and a further parameter setting to highlight the influence of the parameters.

The results show that GMM mapping is not as useful for the Corel dataset. One possible explanation is given by the used histogram features which are positive integers counting the pixels for each color. This type of data can be described by a multinomial distribution and the used GMM may not approximate the

Table 1. Results for a specific choice of L, M, T and W/W_{min}. W/W_{min} was chosen to achieve a length S_Q close to 1 % and 2 %. In addition to the average length S_Q, the 25 and 75 percentiles for the set length $|A(q)|$ are provided, too.

		W/W_{min}	S_Q in % (25 %, 75 %)	$R_Q(10)$	$R_Q(20)$	$D_Q(10)$	$D_Q(20)$
Corel $L = 50, M = 16, T = 600$	Orig.	$8.67e - 02$	1.15 (0.47, 1.63)	0.974	0.959	0.994	0.989
		$1.00e - 01$	2.22 (0.90, 3.10)	0.990	0.983	0.997	0.996
	GMM	$6.67e - 02$	1.12 (0.50, 1.59)	0.971	0.951	0.996	0.993
		$8.00e - 02$	2.17 (1.25, 3.01)	0.992	0.986	0.999	0.999
MNIST $L = 50, M = 18, T = 600$	Orig.	$3.25e + 02$	1.25 (0.28, 1.45)	0.736	0.675	0.971	0.957
		$3.50e + 02$	2.02 (0.55, 2.57)	0.817	0.764	0.985	0.976
	GMM	$2.75e + 02$	1.15 (0.41, 1.20)	0.807	0.752	0.985	0.977
		$3.12e + 02$	2.08 (0.96, 2.42)	0.893	0.852	0.994	0.990
Year prediction MSD $L = 50, M = 14, T = 600$	Orig.	$2.12e + 02$	1.06 (0.00, 0.84)	0.427	0.401	0.677	0.647
		$2.50e + 02$	2.41 (0.00, 2.81)	0.542	0.518	0.766	0.741
	GMM	$1.88e + 02$	1.21 (0.01, 0.66)	0.444	0.413	0.888	0.843
		$2.12e + 02$	2.03 (0.02, 1.76)	0.561	0.528	0.945	0.926

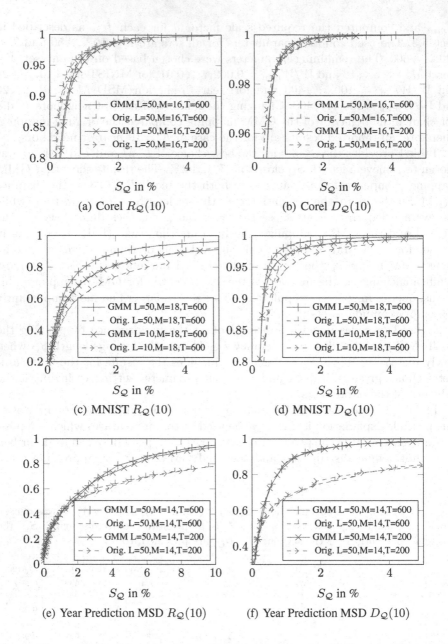

(a) Corel $R_Q(10)$

(b) Corel $D_Q(10)$

(c) MNIST $R_Q(10)$

(d) MNIST $D_Q(10)$

(e) Year Prediction MSD $R_Q(10)$

(f) Year Prediction MSD $D_Q(10)$

Fig. 3. The recall $R_Q(k)$ and relative distance accuracy $D_Q(k)$ for $k = 10$ over the average length S_Q when W/W_{min} is slowly increased.

projections as closely as for the other datasets. This worse approximation is most obvious for a small section length W, where the CDF of the GMM is sampled more often.

6.1   Recall Vs. Relative Distance Accuracy

It can be seen in Fig. 3 that the recall $R_Q(k)$ and the distance accuracy $D_Q(k)$ show a slightly different behavior. First, the relative distance accuracy $D_Q(k)$ is greater than the corresponding recall $R_Q(k)$. This is easily explained, as $D_Q(k)$ can be compared to a soft decision recall allowing inexact matches. Second, for short set lengths S_Q the improvements for GMM mapping of the relative distance accuracy $D_Q(k)$ are larger. This behavior can be explained by the previous observation that GMM is better able to find possible matches in sparsely populated regions. Using a short section lengths W, the original mapping will return very small or even empty candidate sets $A(q)$ for these queries. On the other hand, GMM mapping will use a longer section length in these regions. Because of this, it is able to return closer matches for these queries, even if they are not exact nearest neighbors.

6.2   Parameter Influence

It can be observed in Fig. 3 that both the original and the GMM mapping behave the same when the parameters change. For example, Fig. 3(c) and (d) show that the performance increases for larger values of L. This is easily explained, because if L increases, the query algorithm has more buckets with a smaller score to choose from. The likelihood of finding a close neighbor increases.

In Fig. 3(a) and (b), the influence of T is shown. It can be seen that by increasing the number of buckets T used to build the set $A(q)$, the size $|A(q)|$ of the set increases if W remains the same. The distance accuracy $D_Q(k)$ shows a similar behavior when increasing the number of buckets T or simply increasing the section size W. However, the recall $R_Q(k)$ shows that while the accuracy $D_Q(k)$ is similar in this case, the number of exact matches increases faster if T is larger. This can be explained by the observation that, while increasing T or W will both return larger sets $A(q)$, an increase in T will cover the space for queries close to the borders of a bucket better. Therefore, the likelihood of finding an exact match increases.

The parameters influence the original and GMM mapping in a similar way. This means, a good choice of L, M and T for the original mapping will remain a good choice for the GMM mapping. Only the choice for the section length W and the minimal section length W_{min} are different. In most cases, W_{min} should be chosen slightly smaller than the original W to create a hash with the same average length S_Q, see Table 1.

6.3   Space and Time Requirements

The space requirements do not change for the GMM mapping and still grow linearly with the number of samples and the number of hashes L. The time required to build a LSH still grows linearly with the size of the dataset, too. Furthermore, for the proposed GMM mapping only the time required to compute the histogram used for the stemming grows with the dataset size. Therefore, the time required to build a LSH index for the GMM mapping is still closely related to the original formulation. For our implementation in python, the GMM mapping was $\approx 1.5\ldots 2$ times slower for building a LSH index. However, this observation is not very informative, as the required times can change significantly for highly optimized implementations in C.

The time required to solve a query can also be compared to the original formulation. In our implementation, the query time was dominated by merging the sets, building the querying sequence and searching for the nearest neighbors in the candidate set $A(q)$. The computation of the mapping h based on the projections f contributed only partially to the total query time. As the parameters L, M and T have a large influence on the query time, it is difficult to compare the resulting query times of the original and the GMM mapping.

7   Conclusion

In this paper we presented a new approach using a GMM mapping to extend the LSH family of p-stable distributions. The provided results show that this new mapping can improve the performance of LSH when searching for the kNN significantly. For a given retrieval quality, GMM mapping is able to reduce the size of the set of possible nearest neighbors $A(q)$ by a half. Furthermore, for the original mapping we are required to build multiple hashes with different, carefully selected choices of W for a satisfactory result in a real life application. Using the GMM mapping this selection becomes less difficult, as it automatically uses a larger section length in sparse regions.

Instead of using M scaler hash functions $h_m(q)$ to identify the bucket index $g(q)$, more recent LSH approaches have studied lattice structures to use multiple projections $f_m(q)$ jointly to map to a bucket index, see [1]. If and how GMM mapping can be applied to this new approach is an open question and we intend to examine it further.

References

1. Andoni, A., Indyk, P.: Near-optimal hashing algorithms for approximate nearest neighbor in high dimensions. In: 47th Annual IEEE Symposium on Foundations of Computer Science, 2006. FOCS '06, pp. 459–468 (2006)
2. Bache, K., Lichman, M.: UCI machine learning repository (2013). http://archive.ics.uci.edu/ml
3. Datar, M., Immorlica, N., Indyk, P., Mirrokni, V.S.: Locality-sensitive hashing scheme based on p-stable distributions. In: Proceedings of the Twentieth Annual Symposium on Computational Geometry, SCG '04, pp. 253–262. ACM, New York (2004). http://doi.acm.org/10.1145/997817.997857
4. Figueiredo, M., Jain, A.: Unsupervised learning of finite mixture models. IEEE Trans. Pattern Anal. Mach. Intell. **24**(3), 381–396 (2002)
5. Friedman, J.H., Bentley, J.L., Finkel, R.A.: An algorithm for finding best matches in logarithmic expected time. ACM Trans. Math. Softw. **3**(3), 209–226 (1977). http://doi.acm.org/10.1145/355744.355745

6. Indyk, P., Motwani, R.: Approximate nearest neighbors: towards removing the curse of dimensionality. In: Proceedings of the Thirtieth Annual ACM Symposium on Theory of Computing, STOC '98 pp. 604–613. ACM, New York (1998). http://doi.acm.org/10.1145/276698.276876
7. Lv, Q., Josephson, W., Wang, Z., Charikar, M., Li, K.: Multi-probe LSH: efficient indexing for high-dimensional similarity search. In: Proceedings of the 33rd International Conference on Very Large Data Bases, VLDB '07, VLDB Endowment, pp. 950–961 (2007). http://dl.acm.org/citation.cfm?id=1325851.1325958
8. Muja, M., Lowe, D.G.: Fast approximate nearest neighbors with automatic algorithm configuration. In: VISAPP International Conference on Computer Vision Theory and Applications, pp. 331–340 (2009)
9. Panigrahy, R.: Entropy based nearest neighbor search in high dimensions. In: Proceedings of the Seventeenth Annual ACM-SIAM Symposium on Discrete Algorithm, SODA '06, pp. 1186–1195. ACM, New York (2006). http://doi.acm.org/10.1145/1109557.1109688
10. Paulevé, L., Jégou, H., Amsaleg, L.: Locality sensitive hashing: a comparison of hash function types and querying mechanisms. Pattern Recogn. Lett. **31**(11), 1348–1358 (2010). http://hal.inria.fr/inria-00567191, qUAERO
11. Silpa-Anan, C., Hartley, R.: Optimised KD-trees for fast image descriptor matching. In: IEEE Conference on Computer Vision and Pattern Recognition, 2008, CVPR 2008, June 2008, pp. 1–8 (2008)

Coded Aperture Flow

Anita Sellent$^{(\boxtimes)}$ and Paolo Favaro

Institut für Informatik und angewandte Mathematik,
Universität Bern, Bern, Switzerland
sellent@iam.unibe.ch

Abstract. Real cameras have a limited depth of field. The resulting defocus blur is a valuable cue for estimating the depth structure of a scene. Using coded apertures, depth can be estimated from a single frame. For optical flow estimation between frames, however, the depth dependent degradation can introduce errors. These errors are most prominent when objects move relative to the focal plane of the camera. We incorporate coded aperture defocus blur into optical flow estimation and allow for piecewise smooth 3D motion of objects. With coded aperture flow, we can establish dense correspondences between pixels in succeeding coded aperture frames. We compare several approaches to compute accurate correspondences for coded aperture images showing objects with arbitrary 3D motion.

Keywords: Coded aperture · Optical flow

1 Introduction

Optical flow algorithms estimate the apparent motion between succeeding frames of a video sequence [6] by comparing the brightness values of pixels. Optical flow is an approximation of the projection of 3D motion to the image plane. Traditionally, optical flow algorithms consider pinpoint sharp images, without any degradations other than moderate levels of noise [3]. Real recording conditions, however, rarely allow to capture pinpoint sharp images. When the amount of light in a scene is limited, real cameras require a finite aperture to capture images with a usable signal to noise ratio. Finite aperture sizes introduce defocus blur into images of non fronto-planar scenes. We found that this depth dependent image degradation can lead to erroneous optical flow estimates. However, the size of the blur provides depth information. In fact, defocus blur is a frequently exploited depth cue [18].

Conventional depth from defocus approaches acquire several images of a static scene to estimate a depth map and reconstruct sharp textures. By introducing a coded mask into the aperture of a conventional camera, depth estimates as well as texture restoration can be obtained from a single input image [9,21].

Funded by the Deutsche Forschungsgemeinschaft, project Se-2134/1.

X. Jiang et al. (Eds.): GCPR 2014, LNCS 8753, pp. 582–592, 2014.
DOI: 10.1007/978-3-319-11752-2_48

This single image method is highly suited to provide monocular depth cues in dynamic scenes, where 3D location and shape of every object in a scene changes independently from frame to frame.

For the estimation of pixel trajectories over time, coded aperture frames are a challenging input. The appearance of objects changes dramatically whenever they move relative to the focal plane. Conventional optical flow algorithms do not take this change into account. In this work we consider several approaches to model the effect of defocus blur in optical flow estimation. We evaluate these formulations in the particular setup of high frequency aperture masks that are optimized for the estimation of depth from a single input frame.

2   Related Work

The estimation of optical flow from image sequences is a challenging problem. For a summary and evaluation of modern approaches we refer the reader to the work of Sun et al. [19] and Baker et al. [3]. In our work we build upon the TV-L^1 optical flow approach of Zach et al. [23] and its anisotropic extension by Werlberger et al. [22]. These approaches estimate dense optical flow with a robust L^1 norm for comparing brightness values in two frames and (anisotropic) total variation regularization. The algorithms use a dual optimization scheme that, as a GPU implementation, allows for real time dense flow estimation with state-of-the-art accuracy. While most optical flow algorithms ignore depth altogether, some approaches assign each pixel to a layer and can thus achieve improved regularization and high accuracy, see e.g. Ref. [20]. Still, the layers do not incorporate a model for defocus blur that may change from frame to frame. In contrast, the filter flow by Seitz and Baker [14] models *relative blur* between two images, and allows to compute accurate correspondences also in the presence of defocus blur. For the high frequency apertures that enable single frame depth estimation, relative blurring is not applicable. Thus for coded aperture flow deblurring is necessary for the comparison of points that move in depth. In Sect. 3 we adapt the filter flow to coded apertures to evaluate its performance in our application.

Other approaches to consider defocus effects in dynamic scenes have been introduced by Kubota et al. [8] and Shroff et al. [17]. Both build on the assumption that objects do not move in depth. Shroff et al. acquire focal stacks of moving scenes. They initialize optical flow estimation on images with the same focus settings. Then they refine this flow by considering images with different focal settings, re-blurring deblurred images according to the current, constant depth estimate. Kubota et al. avoid deblurring by applying blur to both images. They evaluate all possible combinations of blur for the best correspondences, before a common depth map is estimated with a depth-from-defocus approach. In contrast to this approach, coded apertures allow to estimate depth from a single frame. In our setup we can therefore simplify correspondence estimation by applying the estimated depth map directly.

Finite apertures improve signal to noise ratio by admitting more light than the ideal pinhole camera. When instead the exposure time is extended, the images are affected by motion blur. The modeling of motion blur can improve the performance of tracking [7] or dense optical flow estimation [13]. However, extended exposure time does not introduce additional information on the depth structure of a dynamic scene such as provided by the coded apertures. To obtain depth information Xu and Jia consider stereo images and then remove depth dependent motion blur. However, even for the stereo camera setup, scene motion is restricted to mainly translational camera motion. Independent object motion is not allowed.

In our work we profit from defocus blur as a depth cue. There are some recent advances in single image depth estimation using a conventional aperture, e.g. Ref. [10]. However, using a conventional aperture sacrifices high frequency image content in the low-pass property of the full-aperture blur. Coded aperture images preserve high frequency content more faithfully. This preservation property can be used for improved deblurring results. By evaluating the quality of the images deblurred with different depth hypotheses, a depth map can be estimated, [9,21]. In contrast, filter-based coded aperture depth estimation [5,11] evaluates the blurred images themselves for depth estimation. More accurate results can be obtained with less computational burden. To obtain smoother and temporally consistent depth maps, Martinello and Favaro [12] use succeeding frames in a coded aperture image sequence for regularization of the depth map. However, they do not compute explicit correspondences between frames and exploit only objects that move parallel to the image plane. We focus on scenes where objects can move arbitrarily and estimate their motion from frame to frame.

3   Brightness Constancy Assumptions

The basic assumption of optical flow estimation is that the brightness of a pixel does not change through the motion [6]. Given two focused images $I_1, I_2 : \Omega \to [0, 1]$ of a moving scene, this brightness constancy assumption can be written as $I_1(x) = I_2(x + u)$ for $x \in \Omega \subset \mathbb{R}^2$ and a displacement vector $u \in \mathbb{R}^2$. To solve this equation, usually the Taylor linearization $I_2(x + u) \approx I_2(x) + u^\top \nabla I_2(x)$ is used. For focused images, the estimation of the displacement u is then based on the data-term

$$D_F(x, u) = I_2(x) + u^\top \nabla I_2(x) - I_1(x). \tag{1}$$

When a planar scene is out of focus, we measure the defocused image $B_1 = k_1 * I_1$ were the defocus blur is expressed as the convolution with a depth dependent point spread function (PSF) k_1. In this case, the brightness of a pixel x depends on the PSF and the brightness of the neighboring pixels. The PSF is depth dependent, so in sum the brightness of a pixel depends on the neighborhood and its depth. When a surface point moves towards the camera, its brightness in every frame is different.

In coded aperture photography, aperture masks are designed to effect a texture highly differently for different depths [15]. For optical flow estimation based on the brightness constancy assumption, highly depth dependent brightness is a hinderance. On the other hand, the coded aperture masks allow for single frame depth estimation. We can profit from the estimated depth to improve optical flow estimation.

In our evaluation, we concentrate on the estimation of optical flow on coded aperture images. For depth estimation we use the state-of-the-art algorithm of Martinello and Favaro [11]. In the following we assume that a spatially variant depth map $d_i : \Omega \to \mathbb{R}$ is given for all measured frames B_i.

We compare several approaches to obtain optical flow estimates for coded aperture images. The first approach is based on the idea that high frequency aperture masks conserve high image frequencies better than conventional apertures. They therefore provide better deblurring results [9]. The deblurred images $\hat{I}_1, \hat{I}_2$ are all-in-focus representation of the scene, for which the linearized brightness constancy, Eq. (1), can be directly applied

$$D_D(x, u) = \hat{I}_2(x) + u^\top \nabla \hat{I}_2(x) - \hat{I}_1(x) \tag{2}$$

Thereby we compute the images $\hat{I}_i$ by the conjugate gradient based, spatially variant deblurring that has been applied successfully for coded aperture images with multiple objects [16]. We use the optimized smoothness weight of 0.01 and 50 iterations. The advantage of operating optical flow estimation on the deblurred image is that any state-of-the-art optical flow implementation can be used out-of-the-box. The disadvantage is that we need to perform the ill-posed procedure of deblurring twice. Apart from disturbance by deblurring artifacts, we also have to deal with the computational burden of the deblurring.

Under the assumption of fronto-parallel scene patches we also consider a second approach by adapting the idea of Refs. [14,17] to coded aperture images. We compare the measured image B_2 to the re-blurred image $k_{d_2} * \hat{I}_1$ where k_{d_2} is the PSF corresponding to the estimated depth in $B_2(x + u)$. Under the assumption of local planarity, we can apply Taylor linearization and obtain the brightness term

$$D_S(x, u) = B_2(x) + u^\top \nabla B_2(x) - \left(k_{d_2(x)} * \hat{I}_1 \right)(x). \tag{3}$$

In comparison to the therm D_D in Eq. (2) we now only have to estimate the deblurred image $\hat{I}_1$. Still, this procedure can introduce deblurring artifacts that might not be compensated by the convolution with k_{d_2}.

Under the same assumptions as above, we also consider a third approach. Based upon the idea of mutual blurring of Refs. [7,8,13] we compare the images $k_{d_1} * B_2$ and $k_{d_2} * B_1$. Linearization leads to the brightness term

$$D_M(x, u) = \left(k_{d_1(x)} * B_2 \right)(x) + u^\top \nabla \left(k_{d_1(x)} * B_2 \right)(x) - \left(k_{d_2(x)} * B_1 \right)(x). \tag{4}$$

This approach has the advantage that deblurring is not required. However, relying on mutual blurring of defocused images potentially sacrifices those image frequencies that are required for accurate optical flow estimation.

From the three above formulations of depth dependent brightness constancy we want to evaluate which provides us with the most accurate flow estimates.

Generally, scenes consist of multiple objects and therefore incorporate depth discontinuities. The brightness of pixels at depth discontinuities is determined by objects in the foreground as well as the background [2]. In this case convolution with a single, depth dependent PSF is not an accurate description. Instead of introducing more elaborate defocus models, we decided to disable the brightness constancy assumption for pixels close to discontinuities. We introduce a weight function $\phi_{d_1}(x) = \exp(\frac{\int_{N_x} \|\nabla d_1(x)\|^2 \, dx}{\sigma_d})$ where N_x is a small neighborhood around pixel x and σ_d a constant. This weight function disables the brightness constancy assumption at known depth discontinuities, i.e., when ∇d_1 is large. We set N_x to $\frac{1}{4}$ the maximally considered blur size and fix $\sigma_d = 3$.

A further cue to depth discontinuities are occlusion boundaries of moving objects. As proposed by Alvarez et al. [1] we therefore compare forward motion estimate w and backward motion estimate $v : \Omega \to \mathbb{R}^2$. When the difference is large a point is most probable occluded. We let $\phi_s(x) = \exp(\frac{\|w(x)-v(x+w)\|^2}{\sigma_s})$ with the parameter σ_s set to 2% of the smallest image dimension. Our final confidence in the brightness constancy is $\phi_w = \phi_s \phi_d$.

4   Estimating Correspondence Fields

All brightness constancy assumptions introduced in the previous section provide only one equation for the two unknown components of the displacement vector. To solve for a dense optical flow field $w : \Omega \to \mathbb{R}^2, x \mapsto u = \binom{w_1}{w_2}$, we additionally assume piecewise smoothness of the flow. A typical difficulty is the determination of the pieces to impose smoothness on. In conventional optical flow estimation, only the images are available to determine regions. In coded aperture flow we also have the depth map available. We expect the flow to be discontinuous at the same locations where the depth of the scene changes rapidly.

To study the effect of coded aperture blur in optical flow estimation on a comparable basis, we incorporate the modified brightness constancies and the depth dependent regularization in the state-of-the-art optical flow of Werlberger et al. [22]. For completeness, we here give a short summary of the approach highlighting our modifications. For more details on the original optical flow algorithm, we refer the reader to Ref. [22].

The first modification in our implementation is the data term. Instead of the conventional brightness constancy, Eq. (1), we consider alternative expressions, Eqs. (2)–(4). Additionally, we include an occlusion weight ϕ_w to circumvent false brightness comparisons at object boundaries. The second modification is to consider depth gradients for regularization. For the depth map normal $n_i = \frac{\nabla d_i}{\|\nabla d_i\|}$ and its perpendicular vector $n^\perp$ we consider the diffusion tensor $T = \exp(a\|\nabla d_i\|)nn^\top + n^\perp(n^\perp)^\top$. Thus, our variational formulation of the problem takes the form

$$\min_{w:\Omega \to \mathbb{R}^2} \sum_{x \in \Omega} \lambda \, \phi_w |D_q| + \sum_{i=1}^{2} \psi_\epsilon \left((\nabla w_i)^\top T \nabla w_i \right) \tag{5}$$

with D_q either of our brightness constancy formulations D_D, D_S or D_M, $\lambda > 0$ a constant and ψ_ϵ the Huber-norm from Ref. [22]. Given the linearizations in Sect. 3 we can apply the solution scheme of Ref. [22]. For comparability we picked suitable parameters of the algorithm for conventional optical flow estimation and kept them fixed for all experiments. In detail, for normalized images we set $a = 0.20$, $\lambda = 50$, and, from Ref. [22] $\epsilon = 0.1$, $\theta = 1$ in the solution scheme.

The implementation of Werlberger et al. works on a Gaussian image pyramid to increase speed and obtain robust estimates for large displacements. To compute D_S and D_M for downscaled images, we require corresponding PSFs and depth maps. We downscale the PSF from the camera calibration, Sect. 5 to obtain PSFs for each level of the image pyramid. To obtain a down-sampled depth map, we consider all depth levels that contribute to a pixel on a coarser level and pick the depth level that is closest to the camera. This heuristic is motivated by the fact that for constant motion the projection of foreground motion spans larger 2D displacements. In our experiments we use 6 levels of an image pyramid with a down-sampling factor of 2.

5 Experiments

We evaluate the different approaches to calculate optical flow on coded aperture frames in several experiments. First we perform evaluation on synthetic images with known ground truth. Then we show results on real images.

5.1 Synthetic Experiments

We render several synthetic scenes with blur size between 4 and 11 pixels. The rendered frames for the 5×5 optimized coded aperture from Ref. [15] are shown in Fig. 1. The scenes contain different challenges ranging from a simple plane moving away from the focal plane, Fig. 1a, to a complex object moving in space, Fig. 1d. Note that for all experiments we keep all parameter of the algorithm fixed. All flow fields in this work are visualized with the color scale in Fig. 4c using black for points with $\phi_w(x) < 0.5$ that are rejected as occluded.

Accuracy Evaluation. In our first experiment, we evaluate the different approaches to coded aperture flow for their accuracy. First we observe that optical flow estimation on defocused images with a conventional algorithm leads to noisy results, Fig. 2a and b. We also find that on synthetic images with known PSF and estimated depth map the results of the deblurring is visually very pleasing, Fig. 2c. Still, optical flow estimation between two deblurred images is noisy, Fig. 2d. Better results can be obtained by using re-blurred images or mutually blurred images, Fig. 2e and f. By evaluating the endpoint error of the unoccluded optical flow, Table 1, we observe that any formulation of depth dependent brightness improves the agnostic approach. The improvement is clearly visible, even though the estimated depth maps have a remaining depth estimation error. Over all our synthetic data-sets we observe best performance by the data-term

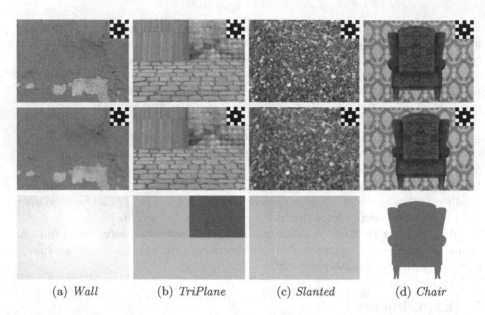

(a) *Wall*   (b) *TriPlane*   (c) *Slanted*   (d) *Chair*

Fig. 1. For the evaluation of coded aperture flow we render coded aperture frames for scenes of which the 3D motion, i.e. depth maps for each frame and the 2D projection of the motion, is known. From top to bottom: input frames B_1 and B_2 and ground truth 2D motion, color coded with the map in Fig. 4c (Color figure online).

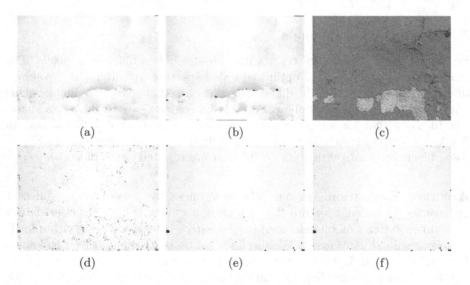

(a)   (b)   (c)

(d)   (e)   (f)

Fig. 2. Estimating conventional optical flow on defocused images with objects moving in depth (scene *Wall*, Fig. 1) leads to erroneous flow estimation (a), (b). Deblurring the defocused input image with the estimated depth map provides visually pleasing images (c). Still, optical flow estimation between two deblurred images is noisy (d). Better results can be obtained when only one image is deblurred (e) or images are mutually blurred (f) (color coding with Fig. 4c) (Color figure online).

Table 1. We compare the average endpoint error of different formulations of brightness constancy. Computing optical flow (OF) on images with a conventional, full aperture in most cases results in a smaller error than optical flow on coded aperture images. Better results can be obtained when the estimated depth map is incorporated in the brightness constancy assumption by using D_D, D_S or D_M although the estimated depth has a certain mean squared error.

	OF, full	OF, coded	D_D	D_S	D_M	Depth
Wall	0.46 px	0.67 px	0.22 px	0.09 px	**0.08 px**	0.17 px^2
TriPlane	0.28 px	0.30 px	0.23 px	0.21 px	**0.15 px**	0.55 px^2
Slanted	0.68 px	0.85 px	0.49 px	0.10 px	**0.06 px**	0.20 px^2
Chair	0.58 px	0.61 px	0.36 px	0.38 px	**0.28 px**	0.65 px^2

based on mutual blurring. Although the deblurred images are visually pleasing, deblurring artifacts seem to deteriorate the accuracy of other approaches to coded aperture flow.

Similar results can be obtained for a variety of coded aperture masks proposed in literature, see supplementary material.

In our second experiment we evaluate the robustness of the coded aperture flow towards errors in the depth estimation. For our synthetic scenes, ground truth depth maps are known. For additional comparison we also use the point-wise depth estimates returned by the algorithm [11]. We compute flow fields with these depth maps as input and observe that deblurring both input images still gives the worst coded aperture flows, Table 2.

In the next experiment we evaluate the influence of the occlusion term. The effect is most prominent in the *Chair* sequence. E.g. for data-term D_M the average endpoint error for setting $\psi_w = 1$ is 1.43 px. By setting $\psi_w = \psi_d$, i.e. considering only the depth dependent cue, we can reduce the error to 0.97 px. Setting $\psi_w = \psi_s$ the error is reduced to 0.56 px. By combination of the terms with $\psi_w = \psi_s \psi_d$ a further reduction of the error to 0.28 px can be obtained (see supplement for the other data-sets).

Table 2. We evaluate the robustness of the different approaches to coded aperture flow towards the estimated depth map. Due to deblurring artifacts, D_D has the highest endpoint error even when ground truth depth is known (a). The point-wise estimated depth map is slightly less accurate than its smoothed version, but allows for comparable flow estimation, see (b) and Table 1.

(a)				(b)				
GT depth	D_D	D_S	D_M	Pointwise	D_D	D_S	D_M	Depth
Wall	0.14 px	**0.06 px**	0.07 px	Wall	0.20 px	0.09 px	**0.08 px**	0.18 px^2
TriPlane	0.10 px	0.06 px	**0.04 px**	TriPlane	0.25 px	0.21 px	**0.15 px**	0.58 px^2
Slanted	0.37 px	0.08 px	**0.05 px**	Slanted	0.47 px	0.10 px	**0.06 px**	0.21 px^2
Chair	0.12 px	0.12 px	**0.11 px**	Chair	0.34 px	0.34 px	**0.26 px**	0.68 px^2

Runtime Evaluation. We implemented the coded aperture flow estimation using MATLAB. We use the same basic framework for each of the data-terms. Deblurring two images and estimating optical flow with data-term D_D takes 81 s an a 3.2 GHz Mac Pro. Deblurring one image and employing data-term D_S takes 72 s. The deblurring free data-term D_M allows for optical flow estimation in 61 s.

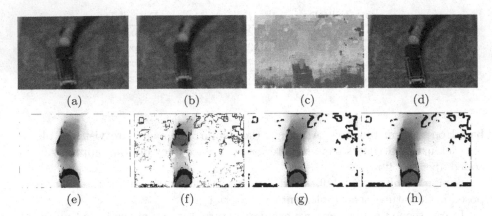

Fig. 3. A toy train backs away from the focal plane, (a) and (b). Coded apertures allow to estimate depth independently for each frame (c) and eases to deblur the images (d). Ignoring coded defocus effects in optical flow estimation (e) leads to inaccurate flow. Deblurring the images before conventional optical flow estimation (f) is susceptible to deblurring effects. Better results can be obtained by a combination of deblurring and re-blurring (g) or the application of mutual blur (h) (Color figure online).

5.2   Real Images

We acquire real image sequences by introducing the binary 5×5 mask from Ref. [15] into a Canon EF f/1.8 II lens [4]. We attach the lens to a Canon EOS 5D, Mark II camera that we set to continuous shooting mode. The camera is calibrated by acquiring a single point spread function (PSF) from a calibration point light source. The blur kernels for all other scales are generated synthetically from the measured image by downscaling to adjust to different depth levels.

Figure 3 shows the scene *train*, and the optical flow we obtain with conventional algorithms and with coded aperture flow. Note how only the data-terms in Eqs. (3) and (4) can estimate the motion of the whole train correctly, even for the weakly textured locomotive. In the scene *walking* (Fig. 4) a person approaches the camera. Here all coded aperture approaches provide a good flow estimate in spite of noisy depth maps. However, deblurring both images introduces more noise on the background stones to the right than the other two approaches.

6   Conclusion

We consider dense optical flow between images that are acquired with a coded aperture. Unlike the ideal sharp image usually assumed for optical flow estimation,

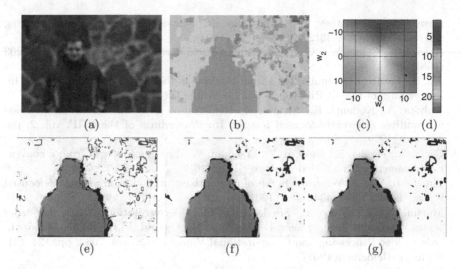

Fig. 4. A person approaches the camera, (a) Although the depth map is noisy (color coded with (d)) coded aperture flow estimation provides reasonable flow estimates (e) by deblurring both input images, (f) by re-blurring a deblurred image and (g) by applying mutual blur (color coded with (c)) (Color figure online).

coded aperture defocus allows for single frames depth estimation. We show that conventional optical flow estimation is unsuitable to estimate accurate motion for objects moving relative to the focus plane. Instead, we evaluate three different formulations that take defocus maps into consideration for flow estimation. We find that the most accurate results can be obtained by comparing a measured image to a reblurred deconvolved image or by comparing mutually blurred images. As the latter approach is faster, we plan to use this approach in our future work on coded aperture video. Generally, the high accuracy that can be obtained with all evaluated methods also shows that coded aperture defocus blur preserves a sufficient amount of high frequency texture for dense optical flow estimation.

References

1. Alvarez, L., Deriche, R., Papadopoulo, T., Sánchez, J.: Symmetrical dense optical flow estimation with occlusions detection. In: Heyden, A., Sparr, G., Nielsen, M., Johansen, P. (eds.) ECCV 2002, Part I. LNCS, vol. 2350, pp. 721–735. Springer, Heidelberg (2002)
2. Asada, N., Fujiwara, H., Matsuyama, T.: Analysis of photometric properties of occluding edges by the reversed projection blurring model. T-PAMI **20**(2), 155–167 (1998)
3. Baker, S., Scharstein, D., Lewis, J., Roth, S., Black, M., Szeliski, R.: A database and evaluation methodology for optical flow. IJCV **92**(1), 1–31 (2011)
4. Bando, Y.: How to disassemble the canon EF 50mm f/1.8 II lens (2013). http://web.media.mit.edu/~bandy/rgb/disassembly.pdf

5. Dowski Jr., E., Cathey, W.: Single-lens single-image incoherent passive-ranging systems. Appl. Opt. **33**(29), 6762–6773 (1994)
6. Horn, B.K., Schunck, B.G.: Determining optical flow. Artif. Intell. **17**(1), 185–203 (1981)
7. Jin, H., Favaro, P., Cipolla, R.: Visual tracking in the presence of motion blur. In: Proceedings of the CVPR, vol. 2, pp. 18–25. IEEE (2005)
8. Kubota, A., Kodama, K., Aizawa, K.: Registration and blur estimation methods for multiple differently focused images. In: Proceedings of the ICIP, vol. 2, pp. 447–451 (1999)
9. Levin, A., Fergus, R., Durand, F., Freeman, W.: Image and depth from a conventional camera with a coded aperture. TOG **26**(3), 70 (2007)
10. Lin, J., Ji, X., Xu, W., Dai, Q.: Absolute depth estimation from a single defocused image. T-IP **22**(11), 4545–4550 (2013)
11. Martinello, M., Favaro, P.: Single image blind deconvolution with higher-order texture statistics. In: Cremers, D., Magnor, M., Oswald, M.R., Zelnik-Manor, L. (eds.) Video Processing and Computational Video. LNCS, vol. 7082, pp. 124–151. Springer, Heidelberg (2011)
12. Martinello, M., Favaro, P.: Depth estimation from a video sequence with moving and deformable objects. In: Proceedings of the Image Processing Conference (2012)
13. Portz, T., Zhang, L., Jiang, H.: Optical flow in the presence of spatially-varying motion blur. In: Proceedings of the CVPR, pp. 1752–1759. IEEE (2012)
14. Seitz, S., Baker, S.: Filter flow. In: Proceedings of the ICCV, pp. 143–150. IEEE (2009)
15. Sellent, A., Favaro, P.: Optimized aperture shapes for depth estimation. Pattern Recogn. Lett. **40**, 96–103 (2014)
16. Sellent, A., Favaro, P.: Which side of the focal plane are you on? In: Proceedings of the ICCP, pp. 1–8. IEEE (2014)
17. Shroff, N., Veeraraghavan, A., Taguchi, Y., Tuzel, O., Agrawal, A., Chellappa, R.: Variable focus video: reconstructing depth and video for dynamic scenes. In: Proceedings of the ICCP, pp. 1–9. IEEE (2012)
18. Subbarao, M., Surya, G.: Depth from defocus: a spatial domain approach. IJCV **13**(3), 271–294 (1994)
19. Sun, D., Roth, S., Black, M.: Secrets of optical flow estimation and their principles. In: Proceedings of the CVPR, pp. 2432–2439. IEEE (2010)
20. Sun, D., Wulff, J., Sudderth, E., Pfister, H., Black, M.: A fully-connected layered model of foreground and background flow. In: Proceedings of the CVPR, pp. 2451–2458 (2013)
21. Veeraraghavan, A., Raskar, R., Agrawal, A., Mohan, A., Tumblin, J.: Dappled photography: mask enhanced cameras for heterodyned light fields and coded aperture refocusing. TOG **26**(3), 69 (2007)
22. Werlberger, M., Trobin, W., Pock, T., Wedel, A., Cremers, D., Bischof, H.: Anisotropic Huber-L1 optical flow. In: Proceedings of the BMVC, pp. 1–11 (2009)
23. Zach, C., Pock, T., Bischof, H.: A duality based approach for realtime TV-L^1 optical flow. In: Hamprecht, F.A., Schnörr, C., Jähne, B. (eds.) DAGM 2007. LNCS, vol. 4713, pp. 214–223. Springer, Heidelberg (2007)

Kernel Density Estimation for Post Recognition Score Analysis

Sebastian Sudholt[✉], Leonard Rothacker, and Gernot A. Fink

Technische Universität Dortmund, Dortmund, Germany
{sebastian.sudholt,leonard.rothacker,gernot.fink}@tu-dortmund.de

Abstract. Post processing pattern recognition results has long been an effective way to reduce the false recognitions by rejecting results that are deemed wrong by a verification system. Recent work laid down a theoretical foundation for a specific post recognition approach. This approach was termed Meta Recognition by its inventors and is based on a statistical outlier detection that makes use of the Weibull distribution. Using distance or similarity scores that are generated at recognition time, Meta Recognition automatically classifies a recognition result to be correct or incorrect. In this paper we present a novel approach to Meta Recognition using a kernel density estimation. We show this approach to be able to outperform the aforementioned post processing technique in different scenarios.

Keywords: Meta Recognition · Kernel density estimation · Post processing · Cohort Analysis · Recognition score analysis

1 Introduction

The field of pattern recognition aims at reproducing the human ability to perceive information from an environment. As with all recognition processes, one is interested in keeping the error rate as low as possible. One possible way to reduce false classifications is to only forward the recognition system's result to a user if a verification system approves of the result. This adds another layer of checking to the recognition process and makes the results more reliable. Building on Cox' [6] work on metacognition, Scheirer et al. [19] term this recognition-verification-process *Meta Recognition*. It is inspired by the human capability to infer information about its own cognitive processes, referred to as "knowing about knowing" [7]. Scheirer et al. define the Meta Recognition process to be made up of two components, a recognition system and a Meta Recognition system. The recognition system sends data from recognition processes to the Meta Recognition component, while the latter analyzes the recognition performance and sends back control information to the recognition system [19]. In this paper we introduce a novel approach to post processing pattern recognition results that is able to outperform Meta Recognition in different scenarios. Furthermore, we introduce the use of Meta Recognition in Information Retrieval and demonstrate its usefulness.

© Springer International Publishing Switzerland 2014
X. Jiang et al. (Eds.): GCPR 2014, LNCS 8753, pp. 593–603, 2014.
DOI: 10.1007/978-3-319-11752-2_49

2   Related Work

Meta Recognition has its root in biometrics. Here, a classification or matching system has to decide whether a supplied probe is known to the system or is an imposter. One concept is to use the distance or similarity scores produced at recognition time (e.g., distances in a nearest neighbor classifier or votings in a multi class SVM) to predict whether the recognition system succeeded or failed. A naiv approach, known as standard thresholding, is to select a threshold that the best score produced has to overcome in order to be accepted in a verification step [19]. However, there exists a more sophisticated technique called *Cohort Analysis* [4,9,19,20]. Here, the best score is compared to a cohort of neighbor scores and normalized based on this cohort. This can be done calculating the T-norm of the best score w.r.t. the cohort selected [14]. Other techniques use a multilayer perceptron to learn the individual scaling factors [20] or normalize the best score by dividing by the second best score [4].

Meta Recognition pursues a different direction. The hypothesized match score (the best score) is compared to the distribution of non-match scores [19] which is produced at recognition time. The claim is that if the best score is the match score it should be an outlier with respect to the non-match distribution. Rather than modelling the overall non-match distribution, as is done in [10], Scheirer et al. choose to use the non-match distribution that is generated at each recognition instance. They argue that modelling an overall non-match and match distribution has problems in the overlapping tails.

Furthermore, of the per instance non-match distribution, Scheirer et al. decide to only model the tail. They prove, that any non-match distribution's tail is best modelled by a Weibull distribution. Dropping the best score, they estimate a Weibull distribution using the next m best scores. The cumulative distribution function (CDF) of the estimated distribution is then evaluated at the best score's value. If the CDF value is above a previously selected area threshold δ then the best score is interpreted to be an outlier with respect to the non-match distribution and the recognition result is accepted. Otherwise the recognition result is rejected.

Compared to the standard thresholding approach, Meta Recognition's use of a model assumption makes it more robust. While the standard thresholding method always calls for a threshold specific to the application's context, Meta Recognition's area threshold is always in the interval $[0; 1]$ as this is any CDF's range. Choosing a small area threshold relaxes the acceptance condition for a recognition result but increases the probability of falsely accepted incorrect results. A high area threshold implies a strict acceptance policy while increasing the probability of correct recognition results being falsely rejected.

The proof showing the family of Weibull distributions to be best to model the non-match distribution's tail is based on a number of assumptions, one being that the number of available scores is big enough [19]. This implies that there are sufficiently many classes as the number of scores directly depends on the number of classes. While Scheirer et al. do not further discuss the influence of the number of classes on the Weibull distribution, it is possible that a different distribution might model that tail better if only few classes are available. Another challenge is the choice of scores selected for the fitting of the Weibull distribution. Nandaku-mar et al. [12] notice that the non-match distribution might be multimodal. If the number of scores selected for fitting is chosen unfavorably, the selected score might not represent the true tail of the distribution anymore. In this scenario, the selected scores' distribution might be multimodal and the unimodal Weibull distribution might also not be best suited to model the distribution of the selected scores.

Additionally, Fragoso and Turk [8] notice that Meta Recognition's outlier detections has a somewhat rapid drop off. They manage to conquer this deficiency by replacing the Weibull distribution with a Rayleigh distribution, which has a smoother drop off. However, the Rayleigh distribution is unimodal as the Weibull distribution and thus depends on an accurate selection of tail size.

3   Kernel Density Estimation for Meta Recognition

In order to overcome the mentioned limitations we propose the use of a *Kernel Density Estimation* (KDE) [5,13]. KDE resembles a non-parametric estimate of an unknown and possibly multi modal distribution. Given a kernel k and a bandwidth h a KDE can be expressed as

$$\hat{f}(x) = \frac{1}{nv} \sum_{i=1}^{n} k\left(\frac{x - x_i}{h}\right),$$ (1)

where n is the number of samples used and x_i is the i-th sample. Typically a normal distribution is chosen as the kernel leaving only the bandwidth to be selected. The latter serves as a smoothing parameter. Small bandwidths lead to sharp peaks in the distribution while high bandwidths yield a smoother function. Selecting an appropriate bandwith is pivotal in determining a good estimate. Besides empiric ways to determine the bandwidth, the standard approach is to minimize the asymptotic mean integrated squared error (AMISE) of the estimate $\hat{f}(x)$ and the true distribution $f(x)$ by using a Taylor-series [11,16].

Using a KDE in Meta Recognition, it is now possible to model multimodal parts of the distribution as well as the entire non-match distribution. Modelling the entire non-match distribution was already proposed by Grother and Phillips [10] but Scheirer et al. point out that the binomial model used does not model the distribution's tail well and is thus not a good model. Nandukamar et al. [12] state that score distributions usually have a long tail which is one possible explanation why binomial models are not suitable. KDE has much more flexibility in this regard, as different kernels or bandwidths can be chosen thus enabling KDE to adept to this trait of a score distribution. As we will show, even the standard Gaussian kernel produces encouraging results.

Algorithm 1.1 shows the steps of how to use a KDE in Meta Recognition. As we explained before, selecting a number of best scores is optional as all scores can be used for fitting. After fitting, the CDF of the KDE is evaluated at the best score. As with Meta Recognition, the recognition result is accepted if the CDF value calculated is above the area threshold and rejected otherwise.

Algorithm 1.1. Meta Recognition with kernel density estimation

---

Require: a set of scores S, an area threshold $\delta \in [0, 1]$, a bandwidth h
1: sort S and take the k best scores $s_1, \ldots, s_m \in S$
2: use a kernel density estimation to generate a model $\mathcal{P}$ from $s_2, \ldots, s_m$ using bandwidth h
3: **if** Cumulative distribution function of $\mathcal{P}$ in $s_1 > \delta$ **then**
4:      **return** correct
5: **else**
6:      **return** incorrect

Using a Gaussian kernel, the probability density function for the KDE can be expressed as

$$\hat{g}(x) = \frac{1}{n} \sum_{i=1}^{n} \frac{1}{\sqrt{(2\pi h^2)}} e^{-\frac{(x - x_i)^2}{2h^2}}. \tag{2}$$

In this scenario, the bandwidth represents the standard deviation for the kernel used in the fitting. Calculating the CDF for $\hat{g}$ is straight forward. Each Gaussian's CDF value is computed and the results are summed up to form the CDF value:

$$\hat{G}(x) = \frac{1}{2n} \sum_{i=1}^{n} 1 + \operatorname{erf}\left(-\frac{x - x_i}{\sqrt{2h^2}}\right). \tag{3}$$

4   Experiments

We conduct two experiments on Meta Recognition in pattern classification and one on Meta Recognition in Information Retrieval.

The result of a Meta Recognition system in combination with a classifier is one of four possible outcomes: true accept (Meta Recognition accepts the correct classification result), true reject (Meta Recognition rejects the wrong classification result), false accept (Meta Recognition accepts the wrong classification result), or false reject (Meta Recognition rejects the correct classification result). Scheirer et al. report their results in terms of *Meta Recognition Miss Detection Rate* (MRMDR) and *Meta Recognition False Accept Rate* (MRFAR). They are defined as follows:

$$\text{MRMDR} = \frac{|\text{false rej.}|}{|\text{false rej.}| + |\text{true acc.}|}, \text{MRFAR} = \frac{|\text{false acc.}|}{|\text{false acc.}| + |\text{true rej.}|}. \tag{4}$$

The curve of plotting MRMDR against MRFAR is called *Meta Recognition Error Tradeoff* (MRET) curve. While Scheirer et al. chose to use the *Equal Error Rate* as a measure to compare two MRET curves, we use the area under the curve (AUC) as a performance measure. The rational is that we are interested in minimizing both error rates rather than looking for a point where the rates are identical. Inspired by precision recall curves in Information Retrieval, this measure represents a plausible way to numerically express the result of a tradeoff curve in a single value. Note that while in Information Retrieval the goal is to maximize the AUC, our goal is to minimize this value as the MRET curve is made up of two error rates. Hence, minimizing the AUC leads to small errors in both MRMDR and MRFAR.

In order to put both Meta Recognition and Meta Recognition with KDE in perspective, we conduct a standard thresholding in all experiments as a third form of post recognition, just like the one in [19]. Additionally, we compare both paradigms to the Meta Recognition with Rayleigh as proposed in [8].

In the following experiments we choose to use the standard Gaussian kernel for Meta Recognition with KDE. We will abbreviate Meta Recognition with MR-WBL, Meta Recognition with Rayleigh with MR-RL, and Meta Recognition with kernel density estimation with MR-KDE where convenient.

4.1 BSSR1

For the first experiment we examine the NIST BSSR1 [2]. It contains three databases, each made up of at least two datasets of biometric scores. In order to compare our approach to the one of Scheirer et al. [19], we chose to use the four datasets *Face C*, *Face G*, *Left Index V* and *Right Index V* from BSSR1's $fing \times face$ database, each containing 517 samples. Scheirer et al. choose a tail size of 75 to estimate the Weibull distribution. This tail size will be used throughout this experiment as well.

Our experiments show that selecting the bandwidth automatically by means of minimizing the AMISE always generates an unsuitable bandwidth that was not even close to being competitive. Thus we decided to learn a suitable bandwidth.

As there exist no dedicated training sets and the number of samples is relatively small, we chose to learn the bandwidth in a leave one out approach: For each sample in a set all other samples were used to minimize the AUC. The bandwidth corresponding to the smallest AUC value was chosen to classify the scores from the remaining sample. Please note that the other BSSR1 databases besides $fing \times face$ can not be used as training sets as their intersections with $fing \times face$ are non-zero.

Figure 1 displays the MRET curves for standard thresholding, Meta Recognition, and two curves for Meta Recognition with KDE. Minimization of the AUC yielded almost the same value for each sample of 0.034. Of these two, one is generated using the minimal AMISE for bandwidth selection while the other is obtained by using the best bandwidth as explained before. Table 1 shows the different AUC values for the aforementioned curves. In the table, Meta Recognition, Meta Recognition with KDE and Meta Recognition with Rayleigh are represented by their respectively used density functions. As one can see from the results, Meta Recognition with KDE clearly outperforms all other post recognition methods investigated.

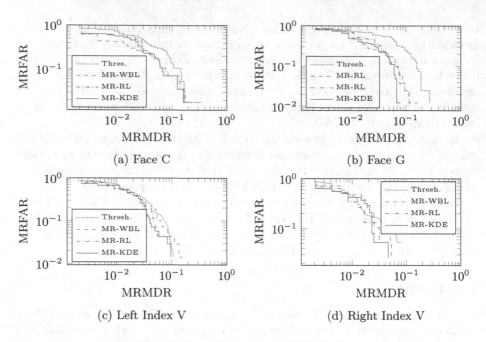

Fig. 1. MRET curves for the four BSSR1 datasets

4.2 MNIST

On order to examine the behavior of Meta Recognition and Meta Recognition with KDE in the case of a small number of classes, we conduct an experiment on the well known MNIST dataset [3]. This dataset serves well for this purpose as it only contains ten different classes. We use the distances generated by a nearest neighbour classifier to obtain a plausible set of scores. As a smaller score is better here, the scores are transformed in order to be used in Meta Recognition. Let $S = \{s_1, \cdots, s_n\}$ be a set of n scores then a transformed set $S' = \{s'_1, \cdots, s'_n\}$ is generated as follows:

$$s'_i = \max(S) - s_i. \tag{5}$$

This way the numerical distance between scores is maintained while their order is transformed to a higher score being better than a lower score.

Table 1. AUC values for the BSSR1

Method	Face C	Face G	Left Index V	Right Index V
Thresholding	$4.95 \cdot 10^{-2}$	$6.65 \cdot 10^{-2}$	$2.75 \cdot 10^{-2}$	$1.58 \cdot 10^{-2}$
Weibull	$1.02 \cdot 10^{-1}$	$2.52 \cdot 10^{-2}$	$2.65 \cdot 10^{-2}$	$1.47 \cdot 10^{-2}$
Rayleigh	$3.21 \cdot 10^{-1}$	$2.72 \cdot 10^{-2}$	$3.99 \cdot 10^{-2}$	$3.55 \cdot 10^{-2}$
KDE	$2.88 \cdot 10^{-2}$	$2.15 \cdot 10^{-2}$	$1.83 \cdot 10^{-2}$	$1.16 \cdot 10^{-2}$

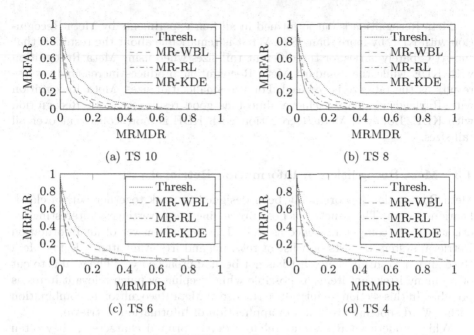

Fig. 2. The image shows the MRET curves for the four post recognition techniques.

For this experiment we chose to use tail sizes of ten, eight, six and four. It is important to notice that a tail size of ten means dropping the first score and estimating the respective distributions with the remaining nine scores. The different experiments are labeled TS10, TS8, TS6, and TS4 respectively.

As with the first experiment, a bandwidth needs to be selected for Meta Recognition with KDE. In order to achieve this, we randomly split the MNIST training set into a 50 000 sample training set and a 10 000 sample validation set. The best bandwidth was calculated by minimizing the AUC for the validation set. Figure 2 shows the MRET curves for the four tail sizes. Table 2 lists the AUC values for the different post processing techniques and tail sizes. The AUC is constant for standard thresholding for all tail sizes as it only depends on the single best score. This score doesn't change if the tail size is altered.

Table 2. AUC values for the MNIST experiment

Method	TS10	TS8	TS6	TS4
Thresholding	$12.55 \cdot 10^{-2}$			
Weibull	$8.51 \cdot 10^{-2}$	$9.02 \cdot 10^{-2}$	$10.65 \cdot 10^{-2}$	$16.92 \cdot 10^{-2}$
Rayleigh	$6.53 \cdot 10^{-2}$	$5.93 \cdot 10^{-2}$	$5.30 \cdot 10^{-2}$	$4.93 \cdot 10^{-2}$
KDE	$4.39 \cdot 10^{-2}$	$4.38 \cdot 10^{-2}$	$4.29 \cdot 10^{-2}$	$4.30 \cdot 10^{-2}$

Meta Recognition is outperformed in all four experiments by Meta Recognition with KDE by more than 48 %. What is remarkable about the results is that the AUC barely varies for the different tail sizes when using Meta Recognition with KDE while the standard Meta Recognition produces increasingly worse results as the tail size is reduced. For the smaller tail sizes, Meta Recognition with Rayleigh is able to achieve almost as good results as Meta Recognition with KDE. However, Meta Recognition with KDE is more consistent over all tail sizes.

4.3   Meta Recognition in Information Retrieval

Meta Recognition has originally been designed to work together with a classification system. The benefits of post processing are nevertheless interesting for other fields in computer science as well. The ranked answer of an Information Retrieval system consists of a list of relevant and irrelevant items. Using Meta Recognition to shrink such a list has not been approached yet. The goal is to cut off as many irrelevant items as possible while keeping as much relevant items as possible. In this section we introduce the use of Meta Recognition in combination with a word spotting problem, an application of Information Retrieval.

While modern OCR systems reliably detect printed characters, they often fail on images of handwritten documents, especially when they are old [15]. This is largely due to the cursive handwriting, noise in the images or fading ink. Thus, the concept of word spotting is to manually label a single word in the document and use techniques from the field of pattern recognition to identify similar regions in the document images.

The *Bag-of-Features Hidden Markov Models* [17,18] have been a promising method for the word spotting problem. Rothacker et al. examine this paradigm on the *George Washington Database* [1], a collection of handwritten documents from George Washington. Their approach is a *query-by-example* method, in which each word in the document is used as a query for all pages. The best results for each of the 20 pages are returned up to a maximum of 200 per page. In contrast to using Meta Recognition in combination with a classifier, it is infeasible to drop a number of scores for the fitting of a probability density function as this represents an assumption about how many elements in the list will be relevant. Assuming the number of relevant documents is small, the irrelevant documents will dominate if all scores are used for fitting though. For the George Washington Database this assumption holds as the average occurrence of a word is 52 compared to the average size of the returned list of 2604 elements. Thus we use all scores for fitting. Evaluating the use of Meta Recognition for Information Retrieval problems is no longer possible by means of MRET curves. Similar to the F-score when looking at precision and recall, we evaluate the harmonic mean of Mean Average Precision (MAP) and Mean Recall (MR) for each area threshold.

We found all bandwidths between 0.01 and 10 to perform almost equally well with minor advantages for a bandwidth of 1. Hence, we set the bandwidth to 1 for the following experiment.

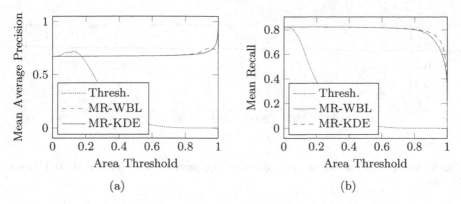

Fig. 3. Mean average precision and mean recall for the word spotting experiment.

Figure 3 shows the MAP and MR with respect to the area threshold δ (see Algorithm 1.1). As can be seen, Meta Recognition with KDE allows for an almost arbitrary selection of MAP while Meta Recognition only allows for a maximum MAP of 0.756. Table 3 lists the MAP, MR and the average number of retrieved elements for the best harmonic mean value. While all techniques generate the same harmonic mean to roughly the same MAP and MR values, Meta Recognition is able shrink the list to 742 elements on average. This means that a user has to inspect 71 % less items on average. However, using Meta Recognition with KDE allows for an almost arbitrary MAP. Applications, that value MAP over MR, can benefit from the proposed concept.

5   Discussion

As can be seen in the experiments, Meta Recognition with KDE outperforms Meta Recognition as proposed by Scheirer et al. in a pattern classification scenario. The reason for this lies in its ability to model the tail of the non-match-distribution more accurately. While this claim can be visually verified in a Q-Q-Plot, we express this behavior numerically: for each probe in the experiments we calculate the average sum of squared errors in the respective Q-Q-Plots. Figure 4 visualizes this measure for the *face G* dataset from the first experiment. As the results of all other experiments lead to the same conclusion in this regard, we chose to only display the results for one experiment. The index of each probe

Table 3. MAP, MR and average number of retrieved elements (ANR) for the best harmonic mean

Method	Harm. mean	MAP	MR	ANR
KDE	0.741	0.677	0.817	1048
Weibull	0.741	0.681	0.811	742
Thresholding	0.741	0.689	0.802	1337

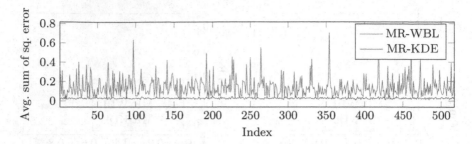

Fig. 4. The picture shows the average sum of squared errors for the BSSR1 face G dataset.

in its respective dataset is displayed on the x-axis while the average sum of squared errors for each input sample is shown on the y-axis. For the majority of the probes the kernel density estimation generates a smaller error in the Q-Q-Plot and thus better models the non-match-distribution's tail.

6   Conclusion

In this paper we proposed an extension for Meta Recognition that overcomes certain limitations of this paradigm. Building on the work of Scheirer et al. [19], we used a kernel density estimation for Meta Recognition. Our approach proved to outperform the standard Meta Recognition on test sets from the BSSR1 and on the MNIST dataset. Our approach especially excelled in classification scenarios where only a limited number of classes were available. Meta Recognition with Rayleigh as proposed in [8] was able to produce similar results for the experiments with very small tail sizes. However, Meta Recognition with kernel density estimation was more consistent over all tail sizes thus demonstrating its independence of a tail size.

The reason for KDE's success over the Weibull distribution is credited to its better modelling of the tail of the non-match-distribution. This could already be achieved by using the standard Gaussian kernel.

When applied to the George Washington dataset Meta Recognition was able to shrink the returned list to almost one third of its previous size, outperforming Meta Recognition with kernel density estimation in this regard. However, Meta Recognition with KDE was able to increase the MAP to almost arbitrary values while Meta Recognition wasn't able to produce MAP values above 0.756. In an application that puts a high value on MAP, this value can be optimized at the cost of a decreasing MR.

References

1. George   Washington   Database.   http://www.iam.unibe.ch/fki/databases/iam-historical-document-database/washington-database. Accessed 2 May 2014
2. NIST Biometric Score Set - Release 1. http://www.nist.gov/itl/iad/ig/. Accessed 2 May 2014

3. The MNIST Database. http://yann.lecun.com/exdb/mnist/. Accessed 2 May 2014
4. Aggarwal, G., Ratha, N.K., Bolle, R.M., Chellappa, R.: Multi-biometric cohort analysis for biometric fusion. In: IEEE International Conference on Acoustics, Speech and Signal Processing, pp. 5224–5227. IEEE (2008)
5. Bishop, C.M.: Pattern Recognition and Machine Learning, 9th edn. Springer, New York (2006)
6. Cox, M.T.: Metacognition in computation: a selected research review. Artif. Intell. **169**(2), 104–141 (2005)
7. Flavell, J.H., Wellmann, H.M.: Metamemory. In: Kail Jr., R.V., Hagen, J.W. (eds.) Perspectives on the Development of Memory and Cognition. Hillsdale, NJ: Erlbaum (1977)
8. Fragoso., V., Turk, M.: Swigs: a swift guided sampling method. In: Proceedings of the 2013 IEEE Conference on Computer Vision and Pattern Recognition (CVPR), pp. 2770–2777 (2013)
9. Furui, S.: Recent advances in speaker recognition. Pattern Recogn. Lett. **18**(9), 859–872 (1997)
10. Grother, P., Phillips, P.J.: Models of large population recognition performance. In: Proceedings of the 2004 IEEE Computer Society Conference on Computer Vision and Pattern Recognition, vol. 2, pp. 68–75 (2004)
11. Jones, M.C., Marron, J.S., Sheather, S.J.: A brief survey of bandwidth selection for density estimation. J. Am. Stat. Assoc. **91**, 401–407 (1996)
12. Nandakumar, K., Chen, Y., Dass, S.C., Jain, A.: Likelihood ratio-based biometric score fusion. IEEE Trans. Pattern Anal. Mach. Intell. **30**(2), 342–347 (2008)
13. Parzen, E.: On estimation of a probability density function and mode. Ann. Math. Stat. **33**(3), 1065–1076 (1962)
14. Poh, N., Merati, A., Kittler, J.: Making better biometric decisions with quality and cohort information: a case study in fingerprint verification. In: 17th European Signal Processing Conference (EUSIPCO 2009), pp. 70–74 (2009)
15. Rath, T.M., Manmatha, M.: Features for word spotting in historical manuscripts. In: Proceedings of the 7th International Conference on Document Analysis and Recognition (ICDAR), vol. 1, pp. 218–222 (2003)
16. Raykar, V.C., Duraiswami, R.: Fast optimal bandwidth selection for kernel density estimation. In: Ghosh, J., Lambert, D., Skillicorn, D., Srivastava, J. (eds.) Proceedings of the Sixth SIAM International Conference on Data Mining, pp. 524–528 (2006)
17. Rothacker, L., Fink, G.A., Banerjee, P., Bhattacharya, U., Chaudhuri, B.B.: Bag-of-features HMMS for segmentation-free bangla word spotting. In: MOCR '13 Proceedings of the 4th International Workshop on Multilingual OCR, pp. 5:1–5:5 (2013)
18. Rothacker, L., Rusiñol, M., Fink, G.A.: Bag-of-features HMMS for segmentation-free word spotting in handwritten documents, pp. 1305–1309 (2013)
19. Scheirer, W.J., Rocha, A., Micheals, R.J., Boult, T.E.: Meta-recognition: the theory and practice of recognition score analysis. IEEE Trans. Pattern Anal. Mach. Intell. **33**(8), 1689–1695 (2011)
20. Tulyakov, S., Zhang, Z., Govindaraju, V.: Comparison of combination methods utilizing t-normalization and second best score models. In: Proceedings of the IEEE Workshop on Biometrics (2008)

Recognizing Scene Categories of Historical Postcards

Rene Grzeszick[(✉)] and Gernot A. Fink

Department of Computer Science, TU Dortmund, Dortmund, Germany
{rene.grzeszick,gernot.fink}@tu-dortmund.de

Abstract. The recognition of visual scene categories is a challenging issue in computer vision. It has many applications like organizing and tagging private or public photo collections. While most approaches are focused on web image collections, some of the largest unorganized image collections are historical images from archives and museums. In this paper the problem of recognizing categories in historical images is considered. More specifically, a new dataset is presented that addresses the analysis of a challenging collection of postcards from the period of World War I delivered by the German military postal service. The categorization of these postcards is of greater interest for historians in order to gain insights about the society during these years. For computer vision research the postcards pose various new challenges such as high degradations, varying visual domains like sketches, photographs or colorization and incorrect orientations due to an image in the image problem. The incorrect orientation is addressed by a pre-processing step that classifies the images into portrait or landscapes. In order to cope with the different visual domains an ensemble that incorporates global feature representations and features that are derived from detection results is used. The experiments on a development set and a large unexplored test set show that the proposed methods allow for improving the recognition on the historical postcards compared to a Bag-of-Features based scene categorization.

1 Introduction

Archives, museums and libraries have access to tremendous amounts of historical documents and images, which provide valuable insight into the past not only for historians but also for the greater public. Most of these samples are only interesting in the context of a whole data collection, referred to as a mass source. Despite the availability of such mass sources, they are very difficult to interpret manually. The amount of data makes their digitization difficult, but a crucial part of the work is assigning machine readable labels to the data samples.

In this paper the automatic recognition of categories within a challenging collection of postcards from the period of World War I is considered. The postcards were delivered by the German military postal service, so called *feldpost* [2]. More specifically, thematic categories on the images of the postcards, e.g. *landscapes, portraits, love* or *patriotic* themed images, are recognized. The images

© Springer International Publishing Switzerland 2014
X. Jiang et al. (Eds.): GCPR 2014, LNCS 8753, pp. 604–615, 2014.
DOI: 10.1007/978-3-319-11752-2_50

Fig. 1. Example of German feldpost postcards from World War I. From top left to bottom right showing the following categories: *cartoon, destruction, frontline, landscape, love & poem, patriotism, portrait* and *weapons*. The samples show degradations, destructions of the images, different color schemes and different orientations. The images are taken from [1].

in some of these categories provide insights about the society by expressing political opinions or representing the everyday life or typical role models and the changes of these over time (cf. [2]). Examples of the postcards and their categories are shown in Fig. 1. For a detailed analysis an automatic recognition of certain categories is desirable. Here, scene recognition methods are applied that use global image representations like Bag-of-Features and spatial tiling, GIST or color histograms in order to classify natural scene images [19]. In contrast to scene recognition problems on modern benchmarks that are often based on web images some additional challenges arise on the historical postcard images.

The first challenge is their visual appearance. The postcards are often degraded or the images are faded. In addition, some of the postcards are black and white photographs, others are painted color images or colorized images and some of them are only drawn sketches. These color schemes create different visual domains and thus increased intra class variations. An ambitious approach for dealing with these visual domains is cross-domain image matching, as described in [14]. Features are re-weighted based on their uniqueness. In order to learn the uniqueness of features, the features of a query are compared with features from thousands of images. Thus, the unique features in a visual domain are found and used for matching a query with a labeled dataset. However, the application is still quite impractical since the uniqueness of features cannot be efficiently computed. Also problems might arise from the fact that common images are typically obtained from web databases which may not reflect the domain of the historical images appropriately. A more common approach toward these variations is the adaption between different domains using SVM margin based approaches for adapting from a training set to a test set in a different domain, as described in [6]. However, in contrast to [6] the visual domains are not known beforehand. This makes adaption between them impractical without further manual effort. Therefore, multiple visual cues are used in order to capture information from different domains. Global image feature representations and features that are derived from detections are combined in an ensemble. The idea is to infer information about a category based on entities that occur in the image. Useful detections could be buildings, persons or even text. In this paper the focus is on faces which are the most promising entity for distinguishing the categories and can be computed in an unsupervised manner. A face detector can be pretrained on existing data and then be adapted to the historical image domain. The second and less obvious challenge is the orientation of the images. Natural scene images are typically in the proper orientation, which allows for context constraints that are widely used in scene recognition. For example, the tiling in the Spatial Pyramid approach implicitly learns that the sky is on the top and people are rarely upside down (cf. [9,19]). In contrast, the images of historical postcards can best be described as an image in an image whose orientation is not necessarily known beforehand.

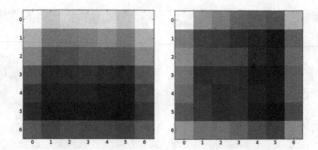

Fig. 2. Example of the average 7×7 tiny image for landscape (left) and portrait (right) images. The landscape images show a strong horizontal structure (e.g. buildings), whereas the portrait images show a similar but vertical structure.

2 Method

For identifying categories in the historic postcards a two stage approach is proposed. Before classifying the categories the postcards are automatically put in the proper orientation by a pre-processing step that is described in Sect. 2.1. Then, an ensemble approach is used that addresses the high intra class variations by using different representations including global image feature representations and a face detection based feature, which are discussed in Sects. 2.2 and 2.3.

2.1 Postcard Orientation

Most feature representations, e.g. Spatial Pyramids [9], GIST [11], Tiny Images [16], build upon the assumption that the image is correctly orientated. For example, the Spatial Pyramid subdivides the image in tiles and SIFT descriptors for a Bag-of-Features representation are computed on a grid with a fixed orientation. Thus, when rotating an image, it will have a completely different representation in the feature space. Typically, the assumption of a correct spatial configuration holds. Web images, which are often used for creating image datasets, are mostly provided in the proper orientation and photographs from private collections are typically rotated by the user or simply by using the EXIF data from the camera.

However, for images of postcards this assumption does not hold. While the text side of a postcard is always landscape the front side image can either be a landscape or a portrait image. Therefore, we propose a pre-processing step in which the orientation of the postcards is determined automatically. Even if putting the images in the proper orientation does not improve the recognition rates, it is desirable for archives to do so without further manual effort. A 7×7 pixel gray scale thumbnail is computed that covers the typical structures appearing in landscape or portrait images. The average images for both categories are shown in Fig. 2. In addition, a gradient histogram with 3×3 tiles and 8 orientations, as well as the mean and standard deviation of the gradients along the X and Y-axis are computed. This results in a 125 dimensional feature representation. An SVM is trained in order to classify images as portrait or landscape oriented. For further processing, the postcards in an incorrect format are rotated.

 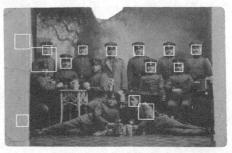

Fig. 3. Qualitative example showing the difficulty to recognize faces in the postcard images. Top left: an example of a group portrait with degradations and destructions of the image. Top right: detection results using a haar cascade [18]. A few false positives are shown due to the sensitivity to noise. Lower left: detection after applying a global threshold on the detections. Due to the large amount of clutter in the whole collection most positive detections are removed. Lower right: using a face classification trained on the best detections instead of thresholding. Most faces are recognized and the noise is removed.

2.2 Face Detection

It is known that scene level information can be leveraged for improving detection tasks in images (cf. [3]). Here, this relation is reversed by deriving features from detection results. In general such relations could be learned from training data (cf. [20]). In case of the historical postcards considered in this paper, basic knowledge about the problem domain can be applied. Useful detections could be buildings, persons, their faces or texts like poems (cf. Fig. 1). From these detections, faces appear to be the most promising ones. On postcards, historical or today's, the appearance of persons can be a very meaningful feature for several categories. For example, some images show countrysides or buildings while others show close-up or group portraits.

The most common method for face detection uses haar cascades as first introduced in [18]. The detections are computed in different scales. Detections at the same position at different scales are then returned as one result and assigned a weight w_i based on the number of detections at this position. Typically, a threshold ϵ is used so that only results with a weight $w_i > \epsilon$ are kept and the others removed in order to reduce false positive detections. However, there are additional challenges that arise on historical images. On modern images the detection is often enhanced by skin color estimation in order to reduce false positives (cf. [8]) which is not possible on gray scale images. Also, the postcards are degraded and show noises toward which the detector is very sensitive. Especially geometric normalizations that mostly build on top of eye detection [15] are almost impossible. Thus, the detection results contain a lot of clutter and simple thresholding does not yield satisfactory results.

Therefore, the face detector is adapted to the domain of historical postcard images. Note that no ground truth annotations are available that could be used for training a detector on historical images. First, faces are detected in the training set without any thresholding and the detections are ordered by their weights w_i. A gradient orientation feature with three scales and $8, 8$ and 4 orientations is computed on each of the detected regions. An SVM is trained on the k best detections and the $3k$ worst detections. This allows for learning typical faces and typical background clutter. It also covers the characteristics of the persons in that time, e.g. facial hair or hats that can only rarely be found in modern images, which is why it is not possible to train a classifier on large available datasets like FERET [12] or LFW [7]. Figure 3 illustrates the issues and shows a qualitative result for the proposed detection approach. The detected faces are then used for deriving features that help identifying the image category. A quantitative evaluation of the proposed face detection is shown in Sect. 3.3.

2.3   Classification Ensemble

In order to address the variations in the data an ensemble of different features is computed. For each feature representation j a random forest classifier is trained. Then, an ensemble decision is made by combining the probabilities:

$$\Omega_{max} = argmax_i \prod_j P(\Omega_i | c_j) \tag{1}$$

Random Forests are used in the ensemble since they outperform SVMs on this task and are able to represent classes with only a few training samples well. Also the computation of class-wise probabilities is not obvious for SVMs. Typical scene recognition representations are Bag-of-Features, LBP histograms and GIST [19]. Color histograms are also widely used, but not as powerful on diverse image datasets. Here, the color histogram may account for the bias of various categories toward different visual domains, e.g. *portraits* are mostly black and white photograph, while *cartoons* are mostly drawn color images (cf. [13]). In addition faces are detected, as described in Sect. 2.2, and features are derived from the detections. In the following the features are discussed in detail:

Bag-of-Features: SIFT features are extracted on a dense grid with a step size of 5 pixels and bin sizes of $4, 6, 8$ and 10 pixels. In order to incorporate spatial information the concept of Spatial Visual Words is applied [5]. Here, a spatialtiling of 2×2 areas is considered. The descriptors are quantized into a codebook of 1,000 Spatial Visual Words resulting in a Bag-of-Features representation. The histogram is then represented by square rooted frequencies [17].

LBP: A histogram of rotation invariant Local Binary Patterns is computed [10]. At each pixel 12 comparison points are chosen on a circle with a radius of one. The interpolated pixel values are compared with the center pixel in order to derive the rotation invariant LBP code. The compressed rotation invariant code has 352 dimensions. A pyramid scheme is built that computes an LBP histogram for 3×3 subregions and a histogram of the complete image is derived using max pooling.

GIST: For a global image description the GIST of a scene is computed. Here, the Spatial Envelope representation is computed as introduced in [11] using the color GIST implementation described in [4]. Three scales with 12, 12 and 4 orientations are computed, which gives a 1344 dimensional feature representation.

Color Histogram: For each color channel of an RGB image 16 equally sized bins are computed. Additionally for each color channel a mean and standard deviation is computed, resulting in a 54 dimensional color feature.

Faces: Faces are detected as described in Sect. 2.2. Here, the $k = 30$ best detections, which can safely be assumed as a lower bound for the number of faces that can be found in the dataset, are used for learning typical faces. A seven dimensional feature is computed that consists of the number of faces and the mean and standard deviation of the position (x, y) and size of the faces.

3   Evaluation

The proposed method has been evaluated on two sets of historical postcards that are described in more detail in Sect. 3.1. The experiments concerning the orientation and categorization are described in Sects. 3.2 and 3.3.

3.1   Dataset of Historical Postcards

The dataset of German feldpost postcards considered in this work is part of a private collection of postcards from World War I [1]. In total 1,346 postcards from nine different acquisition campaigns have already been digitized[1]. All postcards were photographed at approximately 600 dpi with a total resolution of 4288×2484 pixels. The photographs were taken in front of a red background that is removed by color space thresholding. In the following, images from the postcards that are approximately 800×525 pixels in size are considered. 256 of these postcards have been annotated. From the typical categories described in

---

[1] The dataset will be made available for research purposes in the resources section at http://patrec.cs.tu-dortmund.de.

Table 1. Number of correctly oriented postcards after the orientation classification. Note that 65.8 % of the images in the validation set are landscapes and thus already correctly oriented without any processing.

Features	Dim.	Correctly oriented images	Landscape	Portrait	Class average
Nothing	-	65.8 %	100 %	0 %	50 %
Tiny Image	125	83.5 ± 1.0 %	81.1 ± 6.8 %	88.2 ± 7.1 %	84.6 ± 4.1 %
GIST	960	71.2 ± 5.4 %	69.2 ± 13.3 %	75.2 ± 10.7 %	72.2 ± 2.0 %
Bag-of-Features	100	81.2 ± 2.7 %	79.6 ± 6.5 %	84.5 ± 8.3 %	82.0 ± 2.5 %
Bag-of-Features	1,000	84.8 ± 2.8 %	82.9 ± 5.6 %	88.5 ± 3.7 %	85.7 ± 1.6 %

[2] the following eight could clearly be identified: *cartoon, destruction, frontline, landscape, love & poem, patriotism, portrait* and *weapons*. Images that could not be clearly associated with one of these categories are assigned to a background class. In addition to these annotations a few sub categories (e.g. portrait and group portraits) and the correct orientation of the images have been annotated. This small 256 postcard dataset is used for validating the proposed method. The remaining images are a large, so far unexplored, collection from three different acquisitions that can be used as a realistic test set. For these images the recognition quality can only be estimated as no ground truth is available.

For all of the following experiments the 256 postcard dataset was randomly split into a training and a validation set. A major issue is that the dataset is highly unbalanced. Therefore, for each category 50 % of the images, but no more than 30, were used for training and the remaining ones for validation. All experiments were repeated five times using different training samples.

3.2 Orientation

For the orientation classification the 256 postcard dataset was split into a training and validation set of portraits and landscapes. The results are shown in Table 1. The proposed approach is compared to global image representations: a GIST and a Bag-of-Features representation using SIFT descriptors. Note that without any classification already 65.8 % of the images are correctly oriented since all postcards were digitized as landscapes. The GIST representation yields a classification rate of 71.2 %. The proposed approach can be computed more efficiently and also allows for correctly estimating the orientation of 83.5 % of the images. The Bag-of-Features representation achieves competitive recognition rates in comparison with the proposed approach but at much higher computational costs due to the clustering and quantization of the local image features.

In the following the effect of rotating the images correctly with respect to the classification performance has been evaluated. A categorization experiment was performed using a single feature type. The postcards were used without changing the orientation and after changing it by using the proposed orientation

Table 2. Classification results of a random forest on the 256 postcard dataset. Left: the postcards were used as photographed. Middle: the orientation was corrected using the classification as proposed in Sect. 2.1 (*). Right: the images are correctly oriented using the ground truth annotations (GT).

Features	Unoriented	Oriented*	Oriented (GT)
GIST	$43 \pm 2.2\%$	$46.3 \pm 1.3\%$	$46.5 \pm 1.1\%$
Bag-of-Features	$59.1 \pm 0.9\%$	$59.2 \pm 1.5\%$	$60.5 \pm 1.1\%$
LBP	$53.6 \pm 2.9\%$	$54.2 \pm 2.0\%$	$56.6 \pm 1.6\%$
Color Histogram	$45.9 \pm 3.1\%$	-	-
Faces	-	$51.3 \pm 1.2\%$	$53.2 \pm 1.5\%$

classification approach. In addition, an oracle experiment has been performed on the ground truth orientation as an upper baseline. Table 2 shows the results for all feature representations described in Sect. 2.3. Please note that the color histogram is rotation invariant and that the face features do not work properly on incorrectly oriented images. Using the ground truth orientations improves the classification rate significantly in all three cases which clearly states the usefulness of having the images in the proper orientation. Correcting the orientation using the proposed method also yields an improvement on the classification rate. The Bag-of-Features representation shows the best recognition results and appears to be relatively robust against changes in the orientation. Most likely because it covers finer structures that are typical for the categories instead of global orientation information.

3.3   Ensemble Recognition

In the following, the ensemble approach is evaluated using the features described in Sect. 2.3. An extensive study of all combinations has been performed. Some of the results are shown in Table 3. The best classification rate of $62.5 \pm 1.1\%$ is achieved with a combination of the Bag-of-Features, GIST and the face detection based features. Since the dataset is highly unbalanced, the categories *landscape* and *portrait* are highly overrepresented while some of the other categories do not have more than a few samples. Therefore, the overall recognition rate as well as the average class-wise recognition rate over all nine classes is shown. While the classification rate is $62.5 \pm 1.1\%$ the respective class average is only around $22 \pm 0.7\%$ due to the limited amount of samples. Note that the ensemble improves the overall classification rate but on the other hand tends to slightly reduce the class average. The results also show that it is not beneficial to use all possible features. This is more clearly shown by the ensemble combination using the color histogram which reduces the classification rate. Although it might help to identify the visual domain, e.g., distinguish photographs from color sketches, this information does not appear to pose valuable information for the categorization. The face features on the other hand add some very specific information about the

Table 3. Classification results on both sets of historical postcards. In all cases a random forest has been used for classification. For the unexplored set the recognition rates could only be estimated as no ground truth annotations are available.

Dataset	Method	Features	Classification rate	Class avg.
256 Postcards	Bag-of-Features	SIFT	$59.2 \pm 1.5\%$	$23.0 \pm 1.3\%$
	Ensemble	SIFT GIST Faces	$62.5 \pm 1.1\%$	$22.0 \pm 0.7\%$
	Ensemble	SIFT GIST Color	$50.1 \pm 2.7\%$	$17.8 \pm 0.9\%$
	Ensemble	All	$60.4 \pm 1.6\%$	$21.8 \pm 0.8\%$
Unexplored set	Ensemble	SIFT GIST Faces	47.4%	25.2%

Table 4. Face detection on the 256 postcard dataset (GT oriented). The detector that is adapted to the historical postcards is compared to a baseline detection cascade using a fixed threshold $\epsilon = 6$ and a scaling factor of 1.3, which showed the best results.

Method	Precision	Recall	F_1 score
Cascade	89.2%	22.4%	35.8%
Cascade & adaption	94.7%	31.0%	46.7%

postcards that helps recognizing them correctly. It especially helps to recognize the categories *portrait* and *love & poem* very well.

The face detection method that is adapted to the historical images yields a precision of 94.7 % and an estimated recall of approximately 31 %. Due to fading and clutter in some of the images the exact number of faces is not known. The results in Table 4 also show that the proposed method outperforms a cascade detection without any adaption.

3.4   Unexplored Set

The best performing method has then been evaluated on the so far unexplored set. While this set poses a realistic test scenario, no annotations are available for this set. Therefore, the results are estimated by manually verifying the results of 50 samples in each category. The results are shown in Table 3. Here, a recognition rate of 47.4 % is achieved. This estimation is rather a lower bound as a majority of the images belong to the categories *landscape* and *portrait* which are recognized with 80 % and 70 % respectively. The estimated class average of 25.2 % is comparable to the 256 postcard dataset and slightly improved due to the larger number of training samples for the rare categories. On the negative side no *weapon, patriotism* or *frontline* themed images were recognized as these categories are very rare and can easily be mistaken.

4   Conclusion

In this paper a new task for scene categorization has been introduced: a challenging set of historical postcard images from the period of World War I. First approaches to address the challenges that arise on this image collection have been proposed.

The orientation of postcards as an image in the image has been evaluated and an ensemble approach has been applied for categorization. Also features that are based on detection results, in this case based on face detections, have been incorporated in the ensemble. It could be shown that all three steps show improvements for recognizing thematic categories in the historical postcards. Recognition rates of more than 62.5 % have been shown on a validation set and 47.4 % have been estimated on a large unexplored test set.

References

1. Bley, B.: Feldpostkarten im 1. Weltkrieg (Feldpost Postcards of World War I). Private Collection
2. Brocks, C.: Die bunte Welt des Krieges: Bildpostkarten aus dem Ersten Weltkrieg 1914–1918 (The Colorful World of the War: Picture Postcards from the First World War 1914–1918). Klartext-Verlag, Essen (2008, in German)
3. Divvala, S., Hoiem, D., Hays, J., Efros, A., Hebert, M.: An empirical study of context in object detection. In: IEEE Conference on Computer Vision and Pattern Recognition, pp. 1271–1278 (2009)
4. Douze, M., Jégou, H., Sandhawalia, H., Amsaleg, L., Schmid, C.: Evaluation of gist descriptors for web-scale image search. In: Proceedings of the ACM International Conference on Image and Video Retrieval. ACM (2009)
5. Grzeszick, R., Rothacker, L., Fink, G.A.: Bag-of-features representations using spatial visual vocabularies for object classification. In: Proceedings of the IEEE International Conference on Image Processing (2013)
6. Hoffman, J., Rodner, E., Donahue, J., Darrell, T., Saenko, K.: Efficient learning of domain-invariant image representations. In: Proceedings of the International Conference on Learning Representations (ICLS) (2013)
7. Huang, G.B., Ramesh, M., Berg, T., Learned-Miller, E.: Labeled faces in the wild: a database for studying face recognition in unconstrained environments. Technical report 07-49, University of Massachusetts, Amherst, October 2007
8. Kakumanu, P., Makrogiannis, S., Bourbakis, N.: A survey of skin-color modeling and detection methods. Pattern Recogn. **40**(3), 1106–1122 (2007)
9. Lazebnik, S., Schmid, C., Ponce, J.: Beyond bags of features: spatial pyramid matching for recognizing natural scene categories. In: IEEE Conference on Computer Vision and Pattern Recognition (CVPR), pp. 2169–2178 (2006)
10. Ojala, T., Pietikäinen, M., Mäenpää, T.: Gray scale and rotation invariant texture classification with local binary patterns. In: Vernon, D. (ed.) ECCV 2000. LNCS, vol. 1842, pp. 404–420. Springer, Heidelberg (2000)
11. Oliva, A., Torralba, A.: Building the gist of a scene: the role of global image features in recognition. Prog. Brain Res. **155**, 23–36 (2006)
12. Phillips, P.J., Moon, H., Rizvi, S.A., Rauss, P.J.: The feret evaluation methodology for face-recognition algorithms. IEEE Trans. Pattern Anal. Mach. Intell. **22**(10), 1090–1104 (2000)
13. Schroff, F., Criminisi, A., Zisserman, A.: Harvesting image databases from the Web. IEEE Trans. Pattern Anal. Mach. Intell. **33**(4), 754–766 (2011)

14. Shrivastava, A., Malisiewicz, T., Gupta, A., Efros, A.A.: Data-driven visual similarity for cross-domain image matching. In: ACM Transactions on Graphics (TOG), vol. 30, p. 154. ACM (2011)
15. Talele, K., Kadam, S.: Face detection and geometric face normalization. In: TENCON 2009–2009 IEEE Region 10 Conference, pp. 1–6. IEEE (2009)
16. Torralba, A., Fergus, R., Freeman, W.T.: 80 million tiny images: a large data set for nonparametric object and scene recognition. IEEE Trans. Pattern Anal. Mach. Intell. **30**(11), 1958–1970 (2008)
17. Vedaldi, A., Gulshan, V., Varma, M., Zisserman, A.: Multiple kernels for object detection. In: IEEE 12th International Conference on Computer Vision, pp. 606–613. IEEE (2009)
18. Viola, P., Jones, M.: Rapid object detection using a boosted cascade of simple features. In: Proceedings of the 2001 IEEE Computer Society Conference on Computer Vision and Pattern Recognition (CVPR), vol. 1, pp. 511–518. IEEE (2001)
19. Xiao, J., Hays, J., Ehinger, K.A., Oliva, A., Torralba, A.: Sun database: large-scale scene recognition from abbey to zoo. In: 2010 IEEE Conference on Computer Vision and Pattern Recognition (CVPR), pp. 3485–3492. IEEE (2010)
20. Zhu, S., Yung, N.H.: Improve scene categorization via sub-scene recognition. Mach. Vis. Appl. **25**(6), 1561–1572 (2014)

A Stochastic Late Fusion Approach to Human Action Recognition in Unconstrained Images and Videos

Muhammad Shahzad Cheema[1](✉), Abdalrahman Eweiwi[1], and Christian Bauckhage[2]

[1] Bonn-Aachen International Center for IT, University of Bonn, Bonn, Germany
cheema@bit.uni-bonn.de
[2] Multimedia Pattern Recognition Group, Fraunhofer IAIS, Sankt Augustin, Germany

Abstract. Recognizing human actions in unconstrained videos and still images has attracted considerable interest in recent research. An increasingly popular trend is to use ensembles of multiple features and classifiers in order to cope with different aspects such as motion, scene, pose and context. It has been observed that *late fusion* of predictions from individual classifiers offers more robustness than the *early fusion* of feature descriptors. In this paper, we present a novel framework for the late fusion of probabilistic predictions of different classifiers which is based on formulating and solving constrained quadratic optimization problems. In contrast to late fusion methods such as the sum-rule and the linear weighting, our approach binds constraints on mixture coefficients such that they represent the posterior of every participating classifier for each class. Further, unlike fusion by Bayesian inference, the proposed approach minimizes an error function that also considers correlations among different models. Experiments on three video and image action datasets show that our approach outperforms other late fusion techniques. In particular we report 6 %–8 % improvement compared to previously published results on two benchmark datasets.

1 Introduction

The problem of recognizing human activities from realistic images or videos has received considerable interest over last decade. Accordingly, existing research has achieved promising advances in terms of informative features and efficient classification models. Despite this progress, human action recognition in unconstrained scenarios is largely an unsolved problem – mainly due to the challenges arising from substantial inter-and intra-class variations. To handle such variations, a practical system must incorporate representations based on a range of cues such as human motion, appearance, scene, and body pose. Most existing methods rely on individual representations based on motion [19], pose [22] or appearance features [5,12]. Recently, Wang et al. [25] obtained state-of-the-art results on several benchmark video datasets by encoding both motion and

X. Jiang et al. (Eds.): GCPR 2014, LNCS 8753, pp. 616–628, 2014.
DOI: 10.1007/978-3-319-11752-2_51

appearance features in a Bag-of-Words (BoW) method, affirming the benefits of combining several action representations.

The tangible performance enhancement achieved by using multiple features motivated several other attempts to combine various action representations [3,5,7,18,23,26]. Often, these approaches rely on feature level fusion to achieve robust recognition. For instance, [5,18,23] combine a variety of heterogeneous representation by concatenating feature descriptors. This, however, may undermine the discriminative potential of each individual representation for particular classes. Wang et al. [26] follows a rather principled approach to combine a set of mined action features called *actionlets* using Multiple Kernel Learning (MKL) [3], which assigns different weights to the feature kernels in order to obtain better similarity measures. Yet, recent evaluation [7] shows that the simple kernel averaging, a much faster method, can achieve similar results as MKL.

Classifier level fusion, often called *late fusion*, has been thoroughly investigated and it has shown certain key-advantages over other fusion schemes. For example, it is faster than feature level fusion, especially as the trained system grows to adapt new features. In this case, classifier level fusion requires only to re-train the fusion part in contrast to feature level fusion where the whole system needs to be retrained. Also, it abstracts away details of the underlying classifiers, giving the freedom of selecting classification models that best suit a given feature. Baseline approaches for classifier level fusion such as the sum-rule or the SVM-rule [11] have been extensively evaluated for several application [11,28]. These baselines assume that individual classifier outputs are normalized to an estimate of posterior probabilities so that they can be combined homogeneously [10]. Despite their good performance, these approaches are criticized for neglecting the discriminative power of features w.r.t. particular classes, thus remaining subject to suboptimal fusion performance. To remedy this limitation, alternative methods have been suggested that use shared latent spaces [6] or learn weights for classifier scores [21], clustering results [15], or even for data samples [14].

In this paper, we purpose a novel late fusion strategy that determines stochastic weights of *models* for each class through a quadratic optimization formulation, where a model (with a slight abuse of term) is a combination of a feature and a classifier. Unlike common linear weighting scheme for late fusion, our approach constraints on the semantics of the mixture coefficients (weights) i.e. to represent posterior of a model for each class. Our experimental results show that the proposed late fusion approach outperforms other late fusion techniques – often providing state-of-the-art classification accuracies on the benchmark image and video datasets.

2   Related Work

Fusing complementary modalities and feature representations has become popular in computer vision. Conventional approaches such as kernel averaging and the sum-rule have been widely adopted for their simplicity [11,28]. Alternatively, a principled early fusion strategy consists in *Multiple Kernel Learning*

(MKL) [3] which aims at optimized combinations of kernels. For instance, [8] formulated a quadratic optimization approach to learn optimal discriminative kernel weights for classification. However, recent work [7] show that even baseline approaches such as kernel averaging can be as effective as MKL. In contrast to these approaches, our efforts focus on late fusion which builds on the confidence scores of different models of different features.

A comprehensive study of baseline strategies [11] for late fusion conclude that the sum-rule with uniform linear weighting performs best in almost all situations. Most existing linear weighting approaches for late fusion use equal (sum), static, or classifier level fusion weights [2]. Other approaches [1,20] employ domain heuristics or classifiers histories to assign weights to individual classifiers. A Gaussian mixture was used in [16] to fit the scores of different features and then use a likelihood ratio test to fuse classifier scores. Reference [21] optimize for the best linear combination in terms of the misclassification rate for multiple binary classifiers. In comparison, our approach considers the individual results of each model for each particular class in an optimization setup and determines a joint stochastic linear weighting of individual models for each class.

Another late fusion scheme was recently presented in [31] late fusion is approached via rank minimization on the pairwise relation matrices of the learned models. This approach ignores model confidence scores; which, however, are of great interest for indexing and retrieval. Recent work in [14] presents a promising approach that adopts a sample-specific late fusion scheme by propagating the learned fusion weights of labeled samples to unlabeled samples. Again, our approach differs from these ideas as we determine class level weights in a supervised fashion.

In the context of action recognition, fusion of multiple modalities is of particular interest since different actions are often best characterized in terms of different representations pertaining to motion, appearance, scene, and body pose. The issue of efficiently combining these modalities in a principled manner is often overlooked. Approaches in [5,23] utilize different spatial or spatio-temporal representations and combine them by averaging their kernels. Reference [26] learns a linear combination of mined actionlets for classification which may not outperform kernel averaging [7]. Below, we address these limitation and propose a stochastic late fusion technique based on quadratic optimization that jointly learns the best linear combination of models in a multi-class classification scenario. The learned weights can reveal the significance of the utilized features as well as the discriminating potential of each model for their respective classes.

3   Late Fusion by Quadratic Programming

Let $\mathbf{D}$, $\mathbf{V}$, and $\mathbf{T}$ be three independent sets of data and let M be the number of constituting models trained on the training set $\mathbf{D}$ which contains C classes or categories. Further let N be the number of samples in the validation set $\mathbf{V}$ that will be used for learning the fusion model. Given that each model provides a probabilistic predictions, let $\mathbf{V}^{(m)}$ be an $N \times C$ matrix of predictions of all the N instances according to model m, i.e. each row of $\mathbf{V}^{(m)}$ is a stochastic vector. Let $\mathbf{Y}$ be an $N \times C$ binary indicator matrix based on true labels such that $y_{ic} = 1$ only if sample i belongs to category c. Then our objective is to find stochastic mixture coefficients for each class and each model that minimize the sum of squared errors over all in the training data. Let $\mathbf{w}_m$ denote the C-dimensional column vector of the target mixture coefficients for model m and let $\odot$ represents the Hadamard (element-wise) product of each row of a matrix with a row vector, then our objective is to solve

$$\min_{\mathbf{w}_m} \left\| \mathbf{Y} - \sum_m \mathbf{w}_m^T \odot \mathbf{V}^{(m)} \right\|_F$$

$$\text{s.t.} \quad \sum_{m=1}^{M} w_{mc} = 1, \quad w_{mc} \geq 0 \;\; \forall m, c \tag{1}$$

where $\|.\|_F$ denotes the Frobenius norm. Solving this system yields weights w_{mc} that encode the belief of a model m regarding its performance for class c. The sum-to-one constraint in the above formulation ensures that weights of different models are normalized for each class and hence that beliefs are measured relative to each class.

In the remainder of this section, we give discuss further details regarding the formulation of the above quadratic optimization problem and present a strategy for its solution.

First of all, note that the objective function in (1) is equivalent to

$$\min_{w_{mc}} \sum_{i=1}^{N} \sum_{c=1}^{C} \left(y_{ic} - \sum_{m=1}^{M} w_{mc} v_{ic}^{(m)} \right)^2 \tag{2}$$

where $v_{ic}^{(m)}$ represents the probability that sample i belongs to class c according to model m. Expanding this expression yields the following coefficients of the unknowns

$$coeff\left(w_{mc}^2\right) = \sum_{i=1}^{N} \left(v_{ic}^{(m)}\right)^2 \tag{3}$$

$$coeff\left(w_{mc}w_{kc}\right) = \frac{1}{2} \sum_{i=1}^{N} v_{ic}^{(m)} v_{ic}^{(k)} \tag{4}$$

$$coeff\left(w_{mc}\right) = -2 \sum_{i=1}^{N} y_{ic} v_{ic}^{(m)} \tag{5}$$

Next let $\mathbf{w}$ be a MC-dimensional column vector obtained by stacking the $\mathbf{w}_m$ such that

$$\mathbf{w} = [w_{11} w_{12} \ldots w_{1C} w_{21} w_{22} \ldots w_{MC}]^T . \tag{6}$$

Further consider $\mathbf{P}$ to be a $MC \times MC$ matrix whose $C \times C$ blocks contain coefficients corresponding to the quadratic terms in (2). Specifically, $\mathbf{P}$ has the following shape:

$$\mathbf{P} = \begin{bmatrix} \mathbf{P}_{1,1} & \mathbf{P}_{1,2} & \cdots & \mathbf{P}_{1,M} \\ \mathbf{P}_{2,1} & \mathbf{P}_{2,2} & \cdots & \mathbf{P}_{2,M} \\ \vdots & \vdots & \ddots & \vdots \\ \mathbf{P}_{M,1} & \mathbf{P}_{M,2} & \cdots & \mathbf{P}_{M,M} \end{bmatrix} \tag{7}$$

where each $\mathbf{P}_{m,k}$ is $C \times C$ diagonal (sub)matrix and contains the coefficients $coeff\left(w_{mk}^2\right)$ if $m = k$ and the coefficients $coeff\left(w_{mc}w_{kc}\right)$ otherwise.

Also let $\mathbf{q}$ be a $1 \times MC$ stacked vector containing coefficients of the linear terms in (2). That is

$$\mathbf{q} = [\mathbf{q}_1 \mathbf{q}_2 \cdots \mathbf{q}_M] \tag{8}$$

where each $\mathbf{q}_i$ is a C-dimensional row vector.

Accordingly, the problem defined in (1) is equivalent to the following standard quadratic program

$$\min_{\mathbf{w}} \quad \frac{1}{2}\mathbf{w}^T\mathbf{P}\mathbf{w} + \mathbf{q}\mathbf{w}$$

$$\text{s.t.} \quad \mathbf{I}_{C \times MC}\mathbf{w} = \mathbf{1}$$

$$\mathbf{I}_{MC \times MC}\mathbf{w} \geq \mathbf{0} \tag{9}$$

where $\mathbf{I}_{C \times MC}$ is a matrix containing stacked identity matrices of dimension C and $\mathbf{0}$ and $\mathbf{1}$ are MC dimensional column vectors containing zeros and ones, respectively.

The optimal solution of this convex problem determines the mixture coefficients w_{mc} such that each such coefficient can be interpreted as $P(c|m)$. A given query instance x is then classified accordingly, i.e.

$$\arg\max_{c} P(c|x) = \arg\max_{c} \sum_{m=1}^{M} P(c|m,x)P(c|m)P(m). \tag{10}$$

where $P(c|m,x)$ is the probabilistic prediction by model m, $P(c|m)$ is the stochastic weight learned by our approach and $P(m)$ is the prior probability of a model. The prior $P(m)$ can be considered uniform or can be estimated in terms of average accuracies through cross-validation in the training/validation phase [1].

3.1   Regularization and Normalization

Formulating the energy function for optimization problems such as the one in (1) allows us to penalize the weight vector and to reduce effects due to unbalanced data.

Handling unbalanced data: Unbalanced data may cause a bias in the objective function in (2). This, however, is easily overcome by reformulating the objective function as

$$\min_{w_{mc}} \sum_{i=1}^{N} \sum_{c=1}^{C} \frac{1}{\sum_{i=1}^{N} y_{ic}} \left(y_{ic} - \sum_{m=1}^{M} w_{mc} v_{ic}^{(m)} \right)^2 . \tag{11}$$

This formulation normalizes the artifacts of having different sizes for different categories in the training data. Readers may verify that this formulation effects only the diagonal entries in the **P** matrix in (7).

Regularizing and penalizing the weight vector: The energy function formulation permits further parametrization to introduce certain properties of optimal solution e.g. sparsity (L_1-normalization) or smoothness (L_2-normalization). Note that L_1 regularization is embedded in our framework as a constraint, i.e. weight vectors must be stochastic. The L_2 regularization can be added to the objective function which will become

$$\min_{w_{mc}} \sum_{i=1}^{N} \sum_{c=1}^{C} \frac{1}{\sum_{i=1}^{N} y_{ic}} \left(y_{ic} - \sum_{m=1}^{M} w_{mc} v_{ic}^{(m)} \right)^2 + \lambda \sum_{c,m} w_{mc}^2. \tag{12}$$

where λ is regularization constant and can be evaluated through cross-validation.

4   Datasets and Feature Descriptors

Here, we briefly discuss 3 well known action datasets containing challenging videos and images and provide details as to which features are extracted from each dataset.

4.1   HMDB Video Dataset

HMDB [12] is one of the largest and most versatile datasets for action recognition in videos. It contains 6,766 video sequences of 51 action categories such as facial actions, body movements, and human interactions. In our experiments on this dataset, we considered following feature descriptors which are known to show good performance.

Action Bank [19]. Action bank consists of a set of high level action detectors sampled broadly in semantic space and viewpoint spaces. Action bank feature extraction is based on spotting different motion templates in the multiple scale spatio-temporal cuboids. We used the same settings as in [19] to extract $14,965$ dimensional features. They have shown good performance in combination with linear SVM classification.

HOG/HOF Around Harris3d Corners [13]. Histogram of oriented gradient (HOG) and histogram of oriented flow (HOF) features determined around space time interest points (STIP) are often considered as baseline. We used the binaries provided by [13] to extract HOG/HOF features along 3D Harris STIPs. We adopted the Bag-of-Words(BoW) method by sampling 100,000 descriptors from the training data, building a vocabulary of 2000 words using k-means, and representing each video as a unit-sum histograms. The best baseline classifier is an SVM with a Gaussian kernel.

Motion Boundary Histograms and HOG/HOF Along Dense Trajectories [24]. Most 3D STIP detectors are extensions of their 2D counter-parts, they may fail to keep track of interesting spatio-temporal regions. To this end, Wang et al. [24] proposed an efficient way to track densely sampled points using optical flow fields. They also proposed a novel feature descriptor based on motion boundary histograms which is robust to camera motion. Recently [25] used multiple features along dense trajectories and achieved state-of-the-art results on a number of action recognition datasets including HMDB. In particular, they used the BoW approach for 5 different types of descriptors with 6 different types of spatio-temporal griding schemes ending up in using 480,000 dimension features. The results of 30 channels were combined in a multi-channel chi-square kernel setting. Obviously, the application of spatio-temmporal griding in feature extraction and use of many BoW channels can improve recognition accuracy. In our experiments, however, we focus on expressing power of late fusion and use only two dense trajectories channels namely motion boundary histograms xy (mbhxy) and hog/hof along dense trajectories with BoW scheme using code books of size 4000.

4.2   PPMI and Web Actions Datasets

The Web actions dataset [9] contains 2458 images of 5 human actions. We follow the experimental setup proposed in [9]. People playing musical instruments (PPMI) [30] is another popular action recognition dataset in still images. It contains images with many challenges such as cluttered background, partial occlusion, and viewpoint variation. Our experimental evaluation considers the 7 class classification task for the evaluation.

HOG and SIFT-BoW Image Features. For these image datasets, we harness HOG features [4] with *Convolved Trilinear Interpolation(CTI)* to distribute the effect of each pixel over its neighborhood. Our motivation for using this descriptor is that certain activities (e.g. "walk") show limited variations of pose and appearance. As a more flexible action representation, we use BoW with SIFT local features. We gather SIFT local features at different scales and construct a codebook of size 512. Local features are then encoded using Locality-constrained Linear Encoding (LLC) [27] and pooled into a three level spatial pyramid representation. For both features, we build our models using SVM classifiers with Gaussian kernels [17].

Table 1. Recognition accuracies(%) of different approaches

	Model	Datasets		
		HMDB [12]	Web-actions [9]	PPMI7 [30]
Individual	Action Bank	25.90	-	-
	STIP HOG/HOF	18.45	-	-
	Dense Trajectories MBH	34.31	-	-
	Dense Trajectories HOG/HOF	30.13	-	-
	HOG	-	57.4	68.7
	BoW-Sift	-	56.8	63.6
Late Fusion	Bayes	40.76	65.6	72.4
	Sum Rule	40.70	66.5	72.9
	SVM	40.63	64.2	70.32
	Our Approach	**41.83**	**67.15**	**74.0**

5   Experimental Results

In our experiments, we use the models (feature descriptors and classifiers) described earlier and compare our late fusion approach to three different baselines: (i) the sum-rule (ii) the Bayes method and (iii) SVM fusion. The sum-rule is a simple fusion strategy that sums the confidence posteriors of a test sample across multiple modalities and assigns it the class label with the highest response. The Bayesian fusion, as in [1], uses class-wise accuracies as probabilistic weights in the classification i.e. $P(c|m)$ in (10). The SVM based fusion scheme builds a classifier on the prediction score space of different models. Note that, for all experiments, we assume that the models' confidence scores are normalized and transformed into posterior probabilities [10].

In order to train our classifiers, we divide the data into their standard train- and test-splits following standard guidelines or conventions for each dataset. We further divide the training data into a training set $\mathbf{D}$ and a validation set $\mathbf{V}$ to learn the model parameters along with the fusion weights w_{ij} or $P(c|m)$ discussed in Sect. 3. In case of the video dataset (HMDB) where different clips may belong to the scenes of a longer video, we use the train-test and train-validation splits which ensure that train, validation and test sets do not share clips of the same video scene. Specifically, we use the 3 train-test splits provided by [12] and divide the training data in 3 train-validation splits following the same guidelines.

5.1   Recognition Performance

Table 1 shows classification accuracies obtained from using our proposed fusion scheme on each dataset and compares them to three different late fusion strategies (sum-rule, Bayes and SVM-based fusion). For the HMDB dataset, features along dense trajectories show superior performance as compared to Action Bank

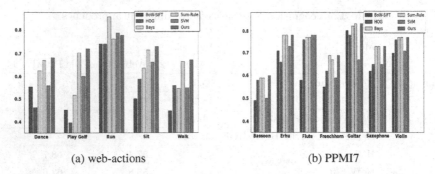

(a) web-actions   (b) PPMI7

Fig. 1. Class-wise accuracies of different individual and ensemble classifiers for (a) web-actions and (b) PPMI7 action images data sets

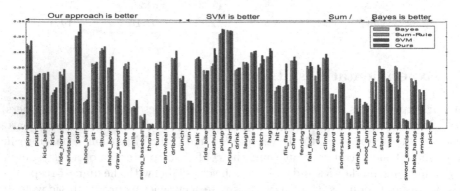

Fig. 2. Class-wise accuracies of individual and ensemble classifiers for HMDB video dataset

features which show better results in comparison to STIP HOG/HOF. Recall that we are using only 2 channels of dense trajectories features each represented as a 4000 BoW vector. This is in contrast to [25] which uses 30 channels with BoW representations of dimensionality 480,000. Adding further channels may enhance the classification accuracy of our approach, yet its results are still superior to other methods that use moderate numbers of features. Notice further that our fusion scheme of these modalities yields the best performance of 41.83 % as compared with the sum-rule (40.70 %) and the SVM-rule (40.63).

Results from PPMI7 and Web-action image dataset shows that our weighting scheme significantly boosts classification performance (by 3 %–4 %) as compared to SVM fusion. Our results also express power of the naive sum-rules towards late fusion. While SVM and Bayes rule suffer from degradation of overall performance, our method of computing class-wise probabilistic fusion weights ensures no degradation in results. Note also that our results in Table 1 better state-of-the-art recognition accuracies on both datasets by 6 %–8 %. In particular, for web-action dataset, we achieve 67.15 % accuracy compared to the earlier best result of 61.07 % [29]. For PPMI7, we report 74.0 % compared to 65.7 % of the

Table 2. Example images from PPMI7 and the recognition results using individual models and different fusion methods.

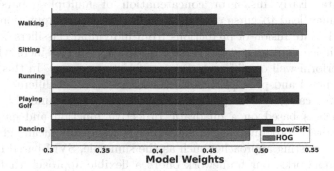

BoW-SIFT	✓	✗	✓	✗	✓	✗	✓
HOG	✗	✗	✓	✓	✗	✓	✗
SVM	✓	✗	✓	✗	✗	✗	✗
Sum-rule	✓	✗	✓	✗	✓	✗	✓
Bayes method	✓	✗	✗	✓	✓	✗	✓
Ours	✓	✓	✓	✓	✓	✗	✓

Fig. 3. The models weights of each action for the web-action dataset.

Grouplets features of [30]. Figures 1 and 2 plots class-wise accuracies of different models and the fusion methods. Our approach consistently outperforms other fusion schemes on all datasets where the sum-rule ranks second.

5.2   Distribution and Impact of Fusion Weights

As discussed above, our method for learning the stochastic weight vectors for each modality addresses the limitation of assigning fixed weights across classes for fusing the predicted scores. This can be seen, for instance, by looking at the individual class-wise classification accuracies of each modality on the web-actions dataset, Fig. 1(a). Note that for actions that are characterized by a limited set of poses (e.g. "walk" and "sit"), employing HOG templates achieve good performance. For other actions that stand for a broad set of body poses (e.g. "dance" and "play golf"), using the flexible representation of BoW-SIFT proves a more appropriate choice for recognition. In this sense, it is more intuitive to assign greater fusion weights to the features that best suit certain classes.

Figure 3 shows the fusion weights learned for each class on the web-actions dataset. These weights are stochastic vectors and therefore can be interpreted as the posterior probabilities $P(model|class)$. Notice that for the actions "walk" and "sit", the weights obtained for the HOG model are greater than those for the BOW-SIFT model. While for the actions "dance" and "playgolf", the BOW-SIFT weights dominates. However, the two models are assigned comparable weights for the action "run". Table 2 shows challenging images from the PPMI7 dataset and their recognition results using HOG, BOW-SIFT and various fusion methods. In comparison to other methods, our approach is consistent and it correctly classifies the example images in most cases.

6 Conclusion

Fusing multiple modalities to incorporate different sources of inter- and intra-class variations has become a paramount to human action recognition in unconstrained data. Early fusion or concatenation of multiple (sparse) feature descriptors may lead to curse of dimensionality and is computation-intensive. Therefore, the late fusion of predictions from individual classifiers is becoming a popular choice as it provides a robust integration of classifiers which (individually) perform well on different regions of instance space. In this paper, we presented a novel and principled method for late fusion of different modalities that estimates category-wise probabilistic weights for the underlying models. Our approach is based on a quadratic objective function and employs constrained quadratic programming to determine semantically meaningful weights. Compared to existing approaches such as the sum-rule, SVM-based fusion, and Bayesian frameworks, our framework offers a flexible approach that combines favorable characteristics of these earlier methods – it considers error minimization like in SVM-based methods and computes probabilistic weighting factors just as Bayesian approaches do. Experimental results on 3 challenging video and image action datasets show the prevalence and consistency of our approach. Moreover, we report 6%–8% improvement compared to previously published results on the two image action datasets.

Acknowledgement. This work was carried out in the project "automatic activity recognition in large image databases" which is funded by the German Research Foundation (DFG).

References

1. Atrey, P., Kankanhalli, S., Jain, R.: Information assimilation framework for event detection in multimedia surveillance systems. ACM Multimedia Syst. J. **12**(3), 239–253 (2006)
2. Atrey, P.K., Hossain, M.A., Saddik, A.E., Kankanhalli, M.S.: Multimodal fusion for multimedia analysis: a survey. Multimedia Syst. **16**, 345–379 (2010)
3. Bach, F., Lanckriet, G.: Multiple kernel learning, conic duality, and the smo algorithm. In: ICML (2004)
4. Dalal, N., Triggs, B.: Histograms of oriented gradients for human detection. In: CVPR (2005)

5. Deltaire, V., Laptev, I., Sivic, J.: Recognizing human actions in still images: a study of bag-of-features and part-based representations. In: BMVC (2010)
6. Eweiwi, A., Cheema, M.S., Bauckhage, C.: Discriminative joint non-negative matrix factorization for human action classification. In: Weickert, J., Hein, M., Schiele, B. (eds.) GCPR 2013. LNCS, vol. 8142, pp. 61–70. Springer, Heidelberg (2013)
7. Gehler, P., Nowozin, S.: On feature combination for multiclass object classification. In: CVPR. pp. 221–228 (2009)
8. He, J., Chang, S., Xie, L.: Fast kernel learning for spatial pyramid matching. In: CVPR (2008)
9. Ikizler-Cinbis, N., Cinbis, R., Sclaroff, S.: Learning actions from the web. In: ICCV (2009)
10. Jain, A., Duin, R., Mao, J.: Statistical pattern recognition: a review. TPAMI **22**, 4–37 (2000)
11. Kittler, J., Hatef, M., Duin, R., Matas, J.: On combining classifiers. TPAMI **20**, 226–239 (1998)
12. Kuehne, H., Jhaung, H., Garrote, E., Poggio, T., Serre, T.: Hmdb: A large video database for human motion recognition. In: ICCV (2011)
13. Laptev, I., Marszalek, M., Schmid, C., Rozenfeld, B.: Learning realistic human actions from movies. In: CVPR (2008)
14. Liu, D., Lai, K., Ye, G., Chen, M., Chang, S.: Sample-specific late fusion for visual category recognition. In: CVPR (2013)
15. Liu, J., Yang, Y., Saleemi, I., Shah, M.: Learning semantic features for action recognition via diffusion maps. CVIU **116**, 361–377 (2012)
16. Nandakumar, K., Chen, Y., Dass, S., Jain, A.: Likelihood ratio-based biometric score fusion. TPAMI **30**, 342–347 (2008)
17. Pedregosa, F., Varoquaux, G., Gramfort, A., Michel, V., Thirion, B., Grisel, O., Blondel, M., Prettenhofer, P., Weiss, R., Dubourg, V., Vanderplas, J., Passos, A., Cournapeau, D., Brucher, M., Perrot, M., Duchesnay, E.: Scikit-learn: machine learning in python. J. Mach. Learn. Res. **12**, 2825–2830 (2011)
18. Rohrbach, M., Amin, S., Andriluka, M., Schiele, B.: A database for fine grained activity detection of cooking activities. In: CVPR (2012)
19. Sadanand, S., Corso, J.J.: Action bank: a high-level representation of activity in video. In: CVPR (2012)
20. Tavakoli, A., Zhang, J., Son, S.H.: Group-based event detection in undersea sensor networks. In: International Workshop on Networked Sensing Systems (2005)
21. Terrades, O., Valveny, E., Tabbone, S.: Optimal classifier fusion in a non-bayesian probabilistic framework. TPAMI **31**, 1630–1644 (2009)
22. Thurau, C., Hlavac, V.: Pose primitive based human action recognition in videos or still images. In: CVPR (2008)
23. Wang, H., Klaeser, A., Schmid, C., Cheng-Lin, L.: Action recognition by dense trajectories. In: CVPR (2011)
24. Wang, H., Kläser, A., Schmid, C., Liu, C.L.: Action recognition by dense trajectories. In: IEEE Conference on Computer Vision and Pattern Recognition, June 2011
25. Wang, H., Kläser, A., Schmid, C., Liu, C.L.: Dense trajectories and motion boundary descriptors for action recognition. IJCV **103**(1), 60–79 (2013). http://hal.inria.fr/hal-00803241
26. Wang, J., Liu, Z., Wu, Y., Yuan, J.: Mining actionlet ensemble for action recognition with depth cameras. In: CVPR (2012)

27. Wang, J., Yang, J., Yu, K., Lv, F., Huang, T., Gong, Y.: Locality-constrained linear coding for image classification. In: CVPR (2010)
28. Xu, L., Krzyzak, A., Suen, C.: Methods of combining multiple classifiers and their applications to handwriting recognition. IEEE Trans. Syst. Man Cybern. **22**, 418–435 (1992)
29. Yang, W., Wang, Y., Mori, G.: Recognizing human actions from still images with latent poses. In: CVPR (2010)
30. Yao, B., Fei-Fei, L.: Grouplet: a structured image representation for recognizing human and object interactions. In: CVPR (2010)
31. Ye, G., D.Liu, Chang, I.J.S.: Robust late fusion with rank minimization. In: CVPR (2012)

A Dense Pipeline for 3D Reconstruction from Image Sequences

Timm Schneevoigt(✉), Christopher Schroers, and Joachim Weickert

Mathematical Image Analysis Group, Faculty of Mathematics and Computer Science,
Saarland University, Campus E1.7, 66041 Saarbrücken, Germany
{schneevoigt,schroers,weickert}@mia.uni-saarland.de

Abstract. We propose a novel pipeline for 3D reconstruction from image sequences that solely relies on dense methods. At no point sparse features are required. As input we only need a sequence of color images capturing a static scene while following a continuous path. Furthermore, we assume that an intrinsic camera calibration is known. Our pipeline comprises three steps: (1) First, we jointly estimate correspondences and stereo geometry for each two consecutive images. (2) Subsequently, we connect the individual pairwise estimates and globally refine them through bundle adjustment. As a result, all camera poses are merged into a consistent global model. This allows us to create accurate depth maps. (3) Finally, these depth maps are merged using variational range image integration techniques. Experiments show that our dense pipeline is an interesting alternative to sparse approaches. It yields accurate camera poses as well as 3D reconstructions.

1 Introduction

Recovering high-quality 3D models of a static scene from images of a moving camera is an important task in computer vision. Many of the existing algorithms for this problem rely on *sparse* features. Such methods have to select the most appropriate data carefully and eliminate outliers. On the other hand, *dense* computer vision methods have made enormous progress in the last decade. For optical flow computation they belong to the leading approaches; see e.g. [1]. Moreover, dense strategies can also be on par with sparse methods for other problems such as the estimation of the fundamental matrix [16]. Dense methods do not have to put effort into selecting the best data but rather draw their robustness from using all data.

Our Goals. Motivated by these achievements, the goal of our paper is to present a pipeline for 3D reconstruction that consistently relies on dense methods: It does not require sparse features at any point. The pipeline can be divided into three stages. First we compute dense correspondences and the fundamental matrix for each consecutive image pair in a joint approach. Every pairwise estimate generally has its own scale. To estimate a consistent motion sequence these scales

© Springer International Publishing Switzerland 2014
X. Jiang et al. (Eds.): GCPR 2014, LNCS 8753, pp. 629–640, 2014.
DOI: 10.1007/978-3-319-11752-2_52

have to be unified. Thus, in the second step, we connect the pairwise estimates and perform a global refinement with bundle adjustment. Finally, we compute accurate depth maps and merge them into a 3D model using anisotropic range image integration.

In contrast to many existing approaches for camera motion estimation that rely on sophisticated feature descriptors, our novel pipeline uses state-of-the-art dense variational methods in every part of the process of reconstructing a 3D object from unregistered cameras. This is especially advantageous in sequences with less texture and many similar structures, where unordered feature matching is difficult.

Organization of the Paper. Our paper is organized as follows: After discussing related work in Sect. 2, Sect. 3 presents our dense reconstruction pipeline. Section 4 then evaluates its performance before the paper is concluded with a summary.

2 Related Work

Monocular tracking and range image integration are most closely related to our work. A prominent tracking approach is the PTAM algorithm by Klein and Murray [7], which tracks sparse features of the scene and creates a map of the environment. With this map the new camera pose is estimated depending on the matched features. Newcombe et al. [11] follow a similar tracking approach, but use a dense scene model instead of a sparse feature map. This way, they achieve a higher robustness to rapid motion. However, they still need standard point features to initialize their tracking algorithm. In order to describe such point features one can employ the well-known SIFT descriptor by Lowe [8], or one of the many modifications such as SURF [2] or GLOH [9].

Instead of using RGB images for tracking, also range images can be successfully employed for this task. Newcombe et al. [10] present a tracking algorithm based on dense depth frame alignment, and Izadi et al. [6] investigate a similar approach that focuses on reconstruction. Zhou et al. use points of interest to reconstruct a dense model from range images [21]. They employ a global optimization scheme which protects parts of the scene that have been scanned already. This leads to more consistent and detailed reconstructions. The disadvantage of depth sensors is their typically lower resolution compared to RGB images. Moreover, these algorithms are not applicable in certain situations such as the reconstruction from old image sequences or benchmarks where no depth data is available.

3 Our 3D Reconstruction Pipeline

This section describes the three steps of our 3D reconstruction pipeline and explains how they can be connected. First, we discuss how to obtain correspondences and epipolar geometry for all consecutive image pairs following a joint approach. Then we describe how we connect this pairwise information to compute a globally consistent camera motion with the help of bundle adjustment. Finally we explain how this enables us to construct a high quality 3D model using anisotropic range image integration.

3.1   Joint Estimation of Correspondences and Epipolar Geometry

As we assume a static scene, the only scene element that moves is the camera itself. Therefore, the moving camera can equivalently be understood as multiple identical cameras that capture the scene from different positions at the same time. In this paper, we assume that there is no knowledge about the camera movement other than that contained in the images. In order to cope with this, we propose an anisotropic modification of the joint method of Valgaerts et al. [16]. It is capable of estimating dense point correspondences and the associated fundamental matrix at the same time. Besides obtaining the fundamental matrix, this has the inherent advantage that the correspondence search is simplified as it is guided by the evolving epipolar constraint. The flow field $w : \Omega \to \mathbb{R}^2$ over the image domain Ω between two images f_i, f_{i+1} and the fundamental matrix $F \in \mathbb{R}^{3\times3}$ are found as a minimizer of a suitable energy in this joint approach [16]:

$$E(w, F) = E_D + \alpha E_S + \beta E_E. \tag{1}$$

Here E_D is the data term, E_S the smoothness term, and E_E denotes the epipolar term. The weights $\alpha, \beta > 0$ balance the individual terms. In the data term we assume brightness and gradient constancy for all image points, expressed by the first and second line of the following data term, respectively:

$$E_D(w) = \int_\Omega \Psi\big(|f_{i+1}(x + w) - f_i(x)|^2$$
$$+ \gamma|\nabla f_{i+1}(x + w) - \nabla f_i(x)|^2 \big)\mathrm{d}x, \tag{2}$$

with a sub-quadratic penalizer function $\Psi(s^2) = \sqrt{s^2 + \epsilon^2}$ and a positive weight parameter $\gamma \in \mathbb{R}$ that balances brightness and gradient constancy assumptions.

To obtain a smooth flow field $w = (u, v)^T$, Valgaerts et al. [16] employed an isotropic flow-driven regularizer with sub-quadratic penalization. To obtain better performance, we use a flow-driven anisotropic regularizer [18]

$$E_S(\nabla w) = \int_\Omega \mathrm{tr}\big(\Psi \big(\nabla u \nabla u^T + \nabla v \nabla v^T\big)\big) \mathrm{d}x, \tag{3}$$

where tr is the matrix trace operator, and Ψ is an extension of the scalar valued function that acts on the eigenvalues of the matrix. The epipolar term directly couples the flow field w with the fundamental matrix F:

$$E_E(w, F) = \int_\Omega \Psi\left(\begin{pmatrix} x + w \\ 1 \end{pmatrix}^T F \begin{pmatrix} x \\ 1 \end{pmatrix} \right) \mathrm{d}x. \tag{4}$$

In this way it softly constrains the correspondences to fulfill an epipolar constraint. Note that the trivial solution of $F = 0$ has to be excluded by imposing a constraint on the Frobenius norm $\|F\|_F^2 = 1$.

As in [16], we employ the method of Lagrange multipliers to solve the constrained optimization problem (1) subject to $\|F\|_F^2 = 1$. Thus we have to find critical points for which the functional derivatives of the Lagrangian vanish. This comes down to solving a system of equations, which in this case can be achieved by iterating between optical flow computation and fundamental matrix estimation. In the first iteration, we can simply use a zero matrix as initialisation for the fundamental matrix. This corresponds to switching off the epipolar constraint such that the fundamental matrix is recovered from pure optical flow in the first iteration.

3.2 Enforcing Scale Consistency

After processing all subsequent image pairs of the sequence as discussed in the previous subsection, we obtain fundamental matrices that describe the epipolar geometry for each image pair along with dense point correspondences (flow field). From each estimated fundamental matrix the relative pose of a canonical camera pair can be extracted [5]. Since we assume that the cameras are calibrated, i.e. their intrinsic parameters are known, the main problem at this stage is merely the remaining scale ambiguity for each pairwise estimate.

To find a mutually consistent scale we have to enforce a set of constraints that connects the whole image sequence. Each constraint has to span at least three images to contribute to a consistent scale. To build such constraints, we construct point trajectories by simply adding up flow vectors. Displacement vectors between the flow field's grid positions are estimated by bilinear interpolation. We test the validity of the resulting trajectories with a forward-backward flow consistency check. Regions of large flow gradients can originate from flow boundaries, where the estimates are less reliable. They are excluded from our computation, as proposed by Sundaram et al. [14].

However, unlike Sundaram et al., we do not limit the sampling of trajectory starting points to well textured regions. Since the idea of dense optical flow is to provide meaningful displacements even in less textured regions, we propose to spare this additional filtering effort. In our experiments, we show that a simple uniform subsampling can be used to decrease the computational effort and that this can even lead to slightly more accurate results when compared to the texture based sampling. The trajectory points should just lie on the scene object. Background regions are filtered out by given masks, which typically can be obtained by image segmentation approaches.

Even with highly accurate optical flow and despite the mentioned consistency checks, estimation and interpolation errors add up at some point. Therefore, we propose to limit the length of trajectories to a fixed value. However, short trajectories that connect only two frames may still help to improve or preserve the initially estimated relative scene geometry, even if it does not contribute to obtaining a unified scale directly.

After processing all subsequent image pairs of the sequence, we concatenate the pair-wise poses one by one to form an initial scene model. Furthermore, we triangulate the depth of one 3D point for each point trajectory to serve as an initialization. This initial scene model is then refined by enforcing the constraints imposed by the point trajectories. To this end, we minimize the reprojection error at each trajectory point by adjusting the triangulated scene point coordinates and the camera poses. This optimization procedure is known as bundle adjustment. For a detailed discussion of the theoretical concept we refer to the work of Triggs et al. [15]. We employ the implementation by Zach [20], which uses Levenberg-Marquardt steps to update the camera poses and initially estimated scene points. Since the intrinsic camera parameters are assumed to be fixed, we leave them unchanged during the optimization. We stop when either a gradient threshold or a fixed number of maximum iterations is reached. In all of our experiments we used 60 as the maximum number of iterations.

To estimate initial 3D scene points, we use the first two frames containing a trajectory thus adopting their local scale. Since badly triangulated scene points resulting from very small baselines or difficult object geometry may harm the optimization, we propose to exclude points that are triangulated to implausible depths by using a threshold. However, it is also possible to choose a different camera pair instead, in order to perform the triangulation.

3.3 Anisotropic Range Image Integration

From the optimization with bundle adjustment explained in the previous section, we have obtained a globally consistent estimation of camera motion. In return, we can now use this knowledge to compute fundamental matrices that are in accordance with the global estimation for each image pair. Then we minimize the energy from Eq. 1 again, but this time we keep the fundamental matrix fixed, i.e. we only minimize w.r.t. the flow w. By putting a high weight on the epipolar term, we can elegantly make use of the fact that stereo matching constitutes a 1D search problem only. This high weight allows to constrain the correspondences to closely obey the epipolar constraint. These correspondences define registered depth maps, which can be merged using variational range image integration techniques. To this end we define a 3D bounding box of interest Ω_3 and compute a signed distance field g within it for each input depth map. In the second step we find a cumulative signed distance function $u : \Omega_3 \rightarrow \mathbb{R}$ as a minimizer of the energy

$$E(u) = \int_{\Omega_3} \left(\sum_{i=1}^{N} c_i \, \Psi((u - g_i)^2) + \alpha \, S(\boldsymbol{\nabla} u) \right) \mathrm{d}\boldsymbol{x}, \qquad (5)$$

where c_i are confidence weights. The unknown surface is then given by the zero level set of u. We employ the anisotropic range image integration algorithm of Schroers et al. [12]. It incorporates a direction dependent smoothing behaviour that is capable of creating smooth surfaces while preserving ridges and corners.

Fig. 1. Sequences used for the experiments, from left to right: **(a)** Monkey. **(b)** Indy. **(c)** Plant. **(d)** Middlebury DinoRing. The Indy and Plant models are available on the *blendswap.com* platform.

4   Experiments

To evaluate our approach, we compare the accuracy and stability of our estimated camera motion to a sparse feature tracker. We have selected the popular Voodoo Camera (VCT) for comparison [17]. It uses the Kanade-Lucas-Tomasi (KLT) tracker [13] and monitors the tracked image windows to detect outliers. Furthermore, it allows to perform a global refinement via bundle adjustment. Again we treat the intrinsic camera calibration as a fixed parameter for all cameras.

We test both approaches on four image sequences, which are shown in Fig. 1. Three of the sequences are synthetic ones, where the camera moves around an object. The Monkey and Indy dataset consist of 112 images. The plant dataset contains 91 images. The baseline lengths between two poses vary by a factor of 2 to 3. The last sequence is the *DinoRing* dataset of the Middlebury multi-view stereo benchmark, containing 48 images.

To judge the quality of the final model one needs to overcome the gauge freedom that is basically inherent to all 3D reconstructions. Therefore, the cameras are aligned with the ground truth model before computing the individual pose errors. Each pose error consists of the Euclidean distance of the estimated camera center to its ground truth counterpart and also the difference in orientation in radians. For the orientation we do not only consider the heading but also pitch and roll.

Accuracy. Table 1 shows the errors in position and orientation of the final globally aligned model estimate for the evaluation sequences. They are being compared with the errors of the VCT. Our dense method turns out to yield competitive results compared to VCT which relies on sparse features. Figure 2 shows the resulting 3D reconstructions for the used datasets with our dense pipeline. To obtain the textured reconstructions, we estimated the color of a point on the surface by projecting it into all cameras and then averaging the individual color values from all cameras where the corresponding surface point is visible.

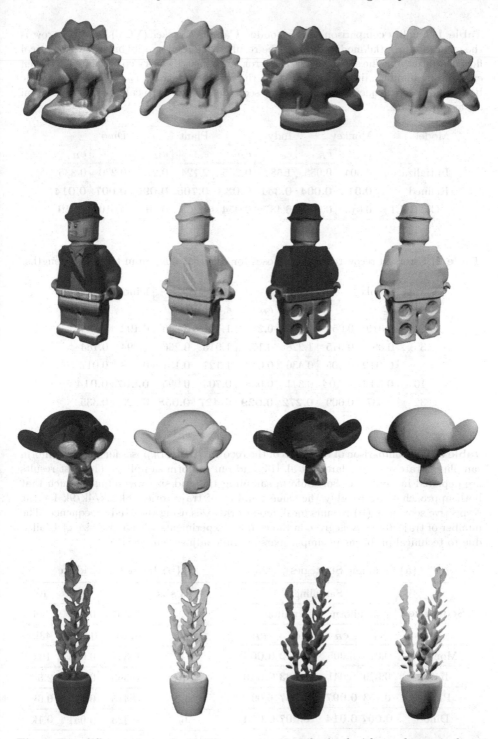

Fig. 2. Two different views of each 3D reconstruction obtained with our dense pipeline. Each view is shown with and without texture

Table 1. Quality comparison to the Voodoo Camera Tracker (VCT). The first row is the error of our initial model, showing the result when relying only on pair-wise optical flow computation. The second row is the error after the trajectory constraints have been enforced by bundle adjustment. Note that the errors in position e_c are not comparable between different sequences because every sequence has its own scale. Errors in rotation e_R are in radians.

Model	Monkey		Indy		Plant		Dino	
	e_c	e_R	e_c	e_R	e_c	e_R	e_c	e_R
Initialization	0.304	0.055	1.582	0.255	2.124	0.278	0.230	0.335
Refined	**0.012**	**0.004**	**0.341**	**0.038**	**0.705**	**0.095**	**0.007**	**0.014**
VCT (KLT)	0.013	0.005	0.437	0.054	0.809	0.105	0.010	0.019

Table 2. Errors of recovered camera poses for different maximum trajectory lengths.

Max	Monkey		Indy		Plant		Dino	
	e_c	e_R	e_c	e_R	e_c	e_R	e_c	e_R
2	0.310	0.055	1.501	0.285	1.701	0.246	0.194	0.344
3	0.055	0.015	1.228	0.179	1.915	0.266	0.094	0.184
8	**0.012**	0.005	0.436	0.052	1.121	0.151	0.008	0.017
16	**0.012**	**0.004**	0.341	0.038	0.705	0.095	**0.007**	**0.014**
32	0.037	0.009	**0.272**	**0.029**	**0.427**	**0.058**	0.229	0.335

Table 3. (a) Comparison of the errors of the recovered camera poses for the nonuniform sampling strategy of Sundaram et al. [14] and our uniform sampling. The best results are depicted in bold face. For uniform sampling the grid size was adjusted such that both approaches use roughly the same number of trajectories: 8k/2k/4k/9k for the respective sequence. (b) Results for different grid sizes using the "Indy" sequence. The number of trajectories K is given in thousands. Experiments with a grid size of 1 failed due to technical problems in simple sparse bundle adjustment (SSBA).

(a) Sampling Strategies

	Sampling			
Sequence	nonuniform		uniform	
	e_c	e_R	e_c	e_R
Monkey	0.015	0.005	**0.012**	**0.004**
Indy	0.379	0.041	**0.363**	**0.040**
Plant	0.753	**0.097**	**0.737**	0.098
Dino	**0.007**	**0.014**	**0.007**	**0.014**

(b) Impact of Gridsize

Grid size	e_c	e_R	K
2	0.307	0.035	209k
4	**0.306**	**0.034**	43k
8	0.341	0.038	10k
16	0.363	0.040	2.5k
32	0.415	0.050	0.6k
64	0.423	0.051	0.1k

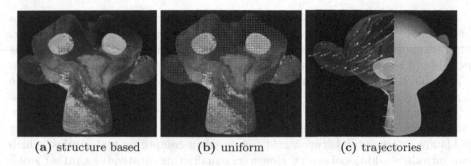

| (a) structure based | (b) uniform | (c) trajectories |

Fig. 3. (a) Trajectory starting points located where λ_2 of the structure tensor is above half of the average of all λ_2. (b) Trajectory starting points uniformly spaced on the grid. (c) Subset of trajectories depicted over many frames blended with the current dense flow field. Our dense flow field is accurate everywhere, which is the reason why we could show that a simple uniform sampling for trajectory starting points is preferable.

4.1 Maximum Length

To find the optimal limit for trajectory lengths we list the errors of the recovered camera poses for different maximum lengths in Table 2. Basically we see that the additional information in longer trajectories improves the result, but at some point the accumulated errors can start degrading quality.

4.2 Sampling

As mentioned previously, Sundaram et al. [14] use a texture based sampling strategy for determining trajectory starting points, cf. Figure 3(a). We employ a uniform sampling strategy instead (Fig. 3(b)). In Table 3(a) we compare their approach to ours. The results show that ignoring the local texture measure does not decrease the accuracy of the recovered camera poses. Our more dense approach gives even slightly better results. This also holds if the grid size is adjusted such that both methods use roughly the same amount of trajectories. Table 3(b) compares the results for different sampling grid sizes. It shows that our approach is very robust with respect to the chosen degree of density. Therefore, one may choose a coarser grid to speed up the computations. A subsampling by a factor 4 even improved the accuracy very slightly.

5   Limitations and Discussion

Bundle adjustment can lead to an undesired local minimum for an erroneous initialization. Since the main source of error in our initial model is the varying scaling between scene points triangulated from different camera pairs, it might be beneficial to only optimize for the baseline scales in a first step. This can help guiding a subsequent full bundle adjustment into a preferable minimum. A similar approach is taken in the odometry method of Fraundorfer et al. [4].

Loop closure is not directly included into our pipeline since the trajectories are initially computed over neighboring frames of an image sequence. However, after computing camera poses and the 3D reconstruction with our pipeline, it is possible to identify new frame correspondences to extend existing trajectories. This could help to solve the problem of closing loops when iterating the pipeline a second time.

Errors in the optical flow are not corrected by the bundle adjustment step, since the trajectory constraints build upon the flow field. Furthermore, the flow fields are interpolated between grid points when computing trajectories, which can introduce additional errors. However, usually this constitutes a rather small error and it is possible to limit the maximum length of trajectories. In terms of variational methods one could estimate the flow over multiple frames simultaneously. It is well-known that one can increase the flow quality by using temporal regularization; see e.g. [19].

6   Conclusions and Future Work

We have presented a novel pipeline for 3D reconstruction that completely relies on dense methods. From an image sequence of a static scene, optical flow fields and stereo geometry are jointly estimated for each consecutive image pair with a variational approach. Subsequently, the pairwise estimates are connected and globally refined through bundle adjustment. After this step, we obtain a refined model of the scene camera positions. Using these, depth maps are computed and fused by anisotropic range image integration.

Even for bad initializations our optical flow based trajectories prove to be sufficiently robust constraints, enabling the bundle adjustment to recover the correct global model. This and our comparisons to VCT show that dense approaches are an interesting alternative to sparse methods.

To be able to integrate our approach into a tracking framework for navigation purposes, future work could be based on an incremental optical flow approach, as for example discussed by Black [3]. This may speed up the estimation due to good initializations for new frames and may additionally improve the accuracy of the flow by temporal regularization.

Acknowledgements. We gratefully acknowledge partial funding by the Cluster of Excellence for Multimodal Computing and Interaction.

References

1. Baker, S., Scharstein, D., Lewis, J.P., Roth, S., Black, M.J., Szeliski, R.: A database and evaluation methodology for optical flow. Int. J. Comput. Vis. **92**(1), 1–31 (2011)
2. Bay, H., Tuytelaars, T., Van Gool, L.: SURF: speeded up robust features. In: Leonardis, A., Bischof, H., Pinz, A. (eds.) ECCV 2006, Part I. LNCS, vol. 3951, pp. 404–417. Springer, Heidelberg (2006)

3. Black, M.J.: Robust incremental optical flow. Ph.D. thesis, Yale University, Department of Computer Science, New Haven, CT (1992)
4. Fraundorfer, F., Scaramuzza, D., Pollefeys, M.: A constricted bundle adjustment parameterization for relative scale estimation in visual odometry. In: Proceedings of IEEE International Conference on Robotics and Automation, pp. 1899–1904 (2010)
5. Hartley, R.I., Zisserman, A.: Multiple View Geometry in Computer Vision, 2nd edn. Cambridge University Press, Cambridge (2004)
6. Izadi, S., Kim, D., Hilliges, O., Molyneaux, D., Newcombe, R., Kohli, P., Shotton, J., Hodges, S., Freeman, D., Davison, A., Fitzgibbon, A.: Kinectfusion: real-time 3d reconstruction and interaction using a moving depth camera. In: Proceedings of 24th Annual ACM Symposium on User Interface Software and Technology, UIST '11, pp. 559–568. ACM, New York (2011)
7. Klein, G., Murray, D.: Parallel tracking and mapping for small AR workspaces. In: Proceedings of Sixth IEEE and ACM International Symposium on Mixed and Augmented Reality, ISMAR '07, pp. 1–10. IEEE Computer Society, Washington, DC (2007)
8. Lowe, D.G.: Distinctive image features from scale-invariant keypoints. Int. J. Comput. Vis. **60**(2), 91–110 (2004)
9. Mikolajczyk, K., Schmid, C.: A performance evaluation of local descriptors. IEEE Trans. Pattern Anal. Mach. Intell. **27**(10), 1615–1630 (2005)
10. Newcombe, R.A., Davison, A.J., Izadi, S., Kohli, P., Hilliges, O., Shotton, J., Molyneaux, D., Hodges, S., Kim, D., Fitzgibbon, A.: Kinectfusion: real-time dense surface mapping and tracking. In: Proceedings of Tenth IEEE International Symposium on Mixed and Augmented Reality, ISMAR '11, pp. 127–136. IEEE Computer Society, Washington, DC (2011)
11. Newcombe, R.A., Lovegrove, S.J., Davison, A.J.: DTAM: dense tracking and mapping in real-time. In: Proceedings of IEEE International Conference on Computer Vision, pp. 2320–2327. IEEE Computer Society, Washington, DC (2011)
12. Schroers, C., Zimmer, H., Valgaerts, L., Bruhn, A., Demetz, O., Weickert, J.: Anisotropic range image integration. In: Pinz, A., Pock, T., Bischof, H., Leberl, F. (eds.) DAGM/OAGM 2012. LNCS, vol. 7476, pp. 73–82. Springer, Heidelberg (2012)
13. Shi, J., Tomasi, C.: Good features to track. In: Proceedings of IEEE Conference on Computer Vision and Pattern Recognition, pp. 593–600. IEEE Computer Society, Seattle, WA (1994)
14. Sundaram, N., Brox, T., Keutzer, K.: Dense point trajectories by GPU-accelerated large displacement optical flow. In: Daniilidis, K., Maragos, P., Paragios, N. (eds.) ECCV 2010, Part I. LNCS, vol. 6311, pp. 438–451. Springer, Heidelberg (2010)
15. Triggs, B., McLauchlan, P.F., Hartley, R.I., Fitzgibbon, A.W.: Bundle adjustment – a modern synthesis. In: Triggs, B., Zisserman, A., Szeliski, R. (eds.) Vision Algorithms'99. LNCS, vol. 1883, pp. 298–372. Springer, Heidelberg (2000)
16. Valgaerts, L., Bruhn, A., Mainberger, M., Weickert, J.: Dense versus sparse approaches for estimating the fundamental matrix. Int. J. Comput. Vis. **96**(2), 212–234 (2012)
17. Viscoda: Voodoo camera tracker 1.2.0 (2012). http://www.viscoda.com/index. php/en/products/non-commercial/voodoo-camera-tracker. Accessed 30 Oct 2013
18. Weickert, J., Schnörr, C.: A theoretical framework for convex regularizers in PDE-based computation of image motion. Int. J. Comput. Vis. **45**(3), 245–264 (2001)
19. Weickert, J., Schnörr, C.: Variational optic flow computation with a spatio-temporal smoothness constraint. J. Math. Imaging Vis. **14**(3), 245–255 (2001)

20. Zach, C.: Simple sparse bundle adjustment 3.0 source code (2011). http://www. inf.ethz.ch/personal/chzach/opensource.html. Accessed 30 Oct 2013
21. Zhou, Q.Y., Koltun, V.: Dense scene reconstruction with points of interest. ACM Trans. Graph. **32**(4), 112:1–112:8 (2013)

Active Online Learning for Interactive Segmentation Using Sparse Gaussian Processes

Rudolph Triebel$^{(\boxtimes)}$, Jan Stühmer, Mohamed Souiai, and Daniel Cremers

Computer Vision Group, Department of Computer Science, TU Munich,
Boltzmannstrasse 3, 85748 Garching, Germany
{rudolph.triebel,jan.stuehmer,mohamed.souiai}@in.tum.de, cremers@tum.de

Abstract. We present an active learning framework for image segmentation with user interaction. Our system uses a sparse Gaussian Process classifier (GPC) trained on manually labeled image pixels (user scribbles) and refined in every active learning round. As a special feature, our method uses a very efficient online update rule to compute the class predictions in every round. The final segmentation of the image is computed via convex optimization. Results on a standard benchmark data set show that our algorithm is better than a recent state-of-the-art method. We also show that the queries made by the algorithm are more informative compared to randomly increasing the training data, and that our online version is much faster than the standard offline GPC inference.

1 Introduction

Automatic image segmentation is one of the most important problems in computer vision. Its attractiveness stems from its very large range of applications, including medical imaging and robotics. However, in general the image segmentation problem is ill-posed, because a correct segmentation depends strongly on the application. Therefore, we focus on the *interactive* segmentation problem, where the user provides information about the regions to be segmented, e.g. by manually sampling image pixels and assigning them to a predefined region class. These *user scribbles* are used as ground truth information, and the aim is to infer a good segmentation using these scribbles as constraints on the labelling. To do this, many approaches have been presented in the literature with impressive results. However, current methods can reach high classification rates only by requiring comparably many user scribbles, and the number of user scribbles needed usually grows very fast as the segmentation quality approaches 100 %.

In this paper, we present a method that asks for user input more intelligently by actively querying pixels to be labeled where the classification was made with high uncertainty. This way, only a few user scribbles are needed to obtain a high quality segmentation. Our method uses an efficient sparse Gaussian Process classifier (GPC) to learn background and foreground models, providing an accurate estimation of the classification uncertainty. We also present a very efficient way to compute the class predictions on every round using an online update rule.

© Springer International Publishing Switzerland 2014
X. Jiang et al. (Eds.): GCPR 2014, LNCS 8753, pp. 641–652, 2014.
DOI: 10.1007/978-3-319-11752-2_53

1.1   Related Work

Many previous works use energy minimization for image segmentation, and since the work of Boykov *et al.* [1], intensive research, e.g. [7,13], has been done on embedding the input image onto a discrete lattice and computing a segmentation using the min-cut framework. Another line of work [8,16] models segmentation in the continuous domain and is based on the convex relaxation technique of Nikolova *et al.* [9]. Both discrete and continuous approaches impose spatial consistency as a prior on the image labelling. Our work is related to [8] where the data term describing the pixel class probabilities includes spatial information while estimating the colour distribution using a Parzen window estimator. However, we use an Informative Vector Machine (IVM) [5], a sparse version of the Gaussian Process Classifier, and employ active learning, which improves the segmentation result quickly after only a few training rounds (see Fig. 1). In contrast to the sparse GP algorithm of Csató and Opper [2], the IVM has advantages in the context of active learning, mainly due to the information-theoretic criterion used to select the subset of the training points.

In the field of active learning, Kapoor *et al.* [4] address object categorization using a GP classifier (GPC) where data points possessing large uncertainty (using posterior mean and variance) are queried for labels and used to improve classification. Triebel *et al.* [15] use an IVM to actively learn traffic lights in urban traffic images. Here, we use a similar approach, but with a very efficient online update method for the classification step of the GPC. Vezhnevets *et al.* [17], as well as

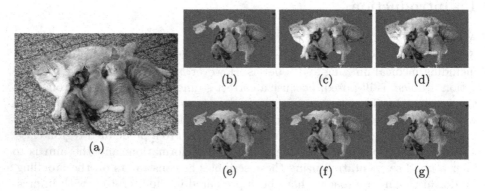

Fig. 1. Comparison between the Parzen window estimator [8] and our sparse GP classifier for foreground classification. From the initial scribble image (a) both approaches learn a model for the foreground. As none of the scribble pixels for the foreground class is white, both approaches fail to classify the white neck of the cat correctly (b, e). However, in the next active learning round, the GP manages to query this part from the user based on its accurate estimation of the predictive uncertainty (c). In contrast, the Parzen window estimator does not query this part, because its uncertainty is low despite its incorrect classification, i.e. it is over-confident (f). After 6 rounds the GP achieves a very good segmentation (d), while the Parzen window estimator still gives a lower-quality segmentation (g).

Wang *et al.* [18] also use active learning for interactive image segmentation, but either with a CRF+NaiveBayes [17] or a Gaussian Mixture Model (GMM) [18] as an underlying classifier. We use a GPC, because it is non-parametric, i.e. it does not assume a functional model for the data, and it was shown to provide very accurate uncertainty estimates, which is crucial in active learning.

2   Algorithm Overview

Figure 2 shows an example sequence of our active learning framework for interactive image segmentation. From a set of initial user scribbles from both foreground and background regions (Fig. 2a), our algorithm learns a sparse Gaussian Process Classifier (GPC) and classifies the remaining pixels. Then, a segmentation is obtained using regularization (Fig. 2b), and an uncertainty measure is computed from the predictive variance returned by the GPC. We use a GPC, because its uncertainty estimates are more reliable than those produced by other learning methods such as Support Vector Machines, where reliable refers to a strong correlation between uncertain and incorrectly classified samples (see, e.g., [10]). Then, we perform an over-segmentation of the original image based on super-pixels [3] and compute the average classification uncertainty (entropy) for each segment (see Fig. 2c). In the next step, the algorithm selects the segment with the highest uncertainty to query a ground truth label from the user, samples pixels uniformly from the segment, and adds the samples with the obtained labels to the training data set (see Fig. 2d). Note that, due to imperfections in the segmentation, some segments can contain both foreground and background pixels.

(a)   (b)   (c)   (d)   (e)   (f)

Fig. 2. Example sequence of our proposed active learning framework. The algorithm starts with initial user scribbles as shown in (a). It then learns a sparse GP classifier and segments the image using the GP prediction and a regularization term (b). Then, candidate regions for new, informative user scribbles are computed (c). These are based on the normalized entropy of the GP prediction, i.e. bright regions represent a higher classification uncertainty than darker regions. In this case, a segment at the upper right border is chosen. A label is queried for these pixels (here it is background), and a sub-set of uniformly sampled pixels together with the class labels is added to the training data (d). In the next round, the classification is improved and the result is refined (e). After a few rounds (here 4 in total), the final segmentation is obtained (f).

In that case, the user can select a "don't know" option, and the next segment is chosen in the order of decreasing entropies. This however, occurs only rarely when the segmentation is done sufficiently fine-grained. The whole learning and classification process is then repeated for a fixed number of times or until an appropriate stopping criterion is met (Fig. 2e and f).

3   Gaussian Process Classification

Every round of our active learning algorithm starts by training a Gaussian Process Classifier (GPC) on the current set of user scribbles. If we denote the scribbles as pairs $(\mathbf{x}_1, y_1), \ldots, (\mathbf{x}_N, y_N)$, where $\mathbf{x}_i$ are feature vectors[1] and $y_i \in \{-1, 1\}$ are binary labels denoting background or foreground, then the task is to compute a *predictive distribution* $p(y_* = 1 \mid \mathcal{X}, \mathbf{y}, \mathbf{x}_*)$. Here, $(\mathbf{x}_*, y_*)$ is an unseen pixel/label pair, $\mathcal{X}$ the set of all training pixels, and $\mathbf{y}$ the training labels. To compute the predictive distribution, the GPC first estimates a distribution $p(\mathbf{f} \mid \mathcal{X}, \mathbf{y})$ over the *latent variables* $\mathbf{f} \in \mathbb{R}^N$, approximating it with a multivariate normal distribution with mean $\boldsymbol{\mu}$ and covariance matrix Σ, i.e.: $p(\mathbf{f} \mid \mathcal{X}, \mathbf{y}) \approx \mathcal{N}(\mathbf{f} \mid \boldsymbol{\mu}, \Sigma)$. This is done using Bayes' rule:

$$p(\mathbf{f} \mid \mathcal{X}, \mathbf{y}) = \frac{p(\mathbf{y} \mid \mathbf{f}) p(\mathbf{f} \mid \mathcal{X})}{\int p(\mathbf{y} \mid \mathbf{f}) p(\mathbf{f} \mid \mathcal{X}) d\mathbf{f}}, \tag{3.1}$$

where $p(\mathbf{f} \mid \mathcal{X}) = \mathcal{N}(\mathbf{f} \mid \mathbf{0}, K)$ is the prior of the latent variables, and

$$p(\mathbf{y} \mid \mathbf{f}) = \prod_i p(y_i \mid f_i) \tag{3.2}$$

are the likelihoods, which are conditionally independent. These likelihoods are determined using a *sigmoid function* Φ, i.e. $p(y_i \mid f_i) = \Phi(y_i f_i)$, which has the effect that Eq. (3.1) cannot be computed in closed form. Here, Expectation Propagation (EP) and Assumed Density Filtering (ADF) are commonly used approximations based on a Gaussian $q(y_i \mid f_i)$ that minimises the Kullback-Leibler (KL) divergence between $q(\mathbf{y} \mid \mathbf{f}) p(\mathbf{f} \mid \mathcal{X})$ and the numerator of Eq. (3.1).

Then, for a given new test data point $\mathbf{x}_*$, the GP classifier computes the mean μ_* and the variance σ_*^2 of the latent variable distribution

$$p(f_* \mid \mathcal{X}, \mathbf{y}, \mathbf{x}_*) = \int p(f_* \mid \mathcal{X}, \mathbf{x}_*, \mathbf{f}) p(\mathbf{f} \mid \mathcal{X}, \mathbf{y}) d\mathbf{f} \tag{3.3}$$

and uses that to compute the predictive distribution

$$p(y_* = 1 \mid \mathcal{X}, \mathbf{y}, \mathbf{x}_*) = \int \Phi(f_*) p(f_* \mid \mathcal{X}, \mathbf{y}, \mathbf{x}_*) df_*. \tag{3.4}$$

---

[1] These can be either RGB pixel values or a combination of image coordinates and RGB values of the pixels. In our implementation, we use the latter, because it also provides locality information about background and foreground.

If Φ is the cumulative Gaussian function this can be done in closed form using

$$p(y_* = 1 \mid \mathcal{X}, \mathbf{y}, \mathbf{x}_*) = \Phi\left(\frac{\mu_*}{\sqrt{1 + \sigma_*^2}}\right). \tag{3.5}$$

3.1 Information-Theoretic Sparsification

One problem with the GPC is its huge demand of memory and run time, because it maintains an $N \times N$ covariance matrix, and the number of training samples N can be very large. Therefore, we use a sparsification known as the Informative Vector Machine (IVM) [6]. The main idea here is to only use a sub-set of training points denoted the *active set* $\mathcal{I}_D$, from which an approximation q of the posterior is computed. As above, q is Gaussian, i.e. $q(\mathbf{f} \mid \mathcal{X}, \mathbf{y}) = \mathcal{N}(\mathbf{f} \mid \boldsymbol{\mu}, \Sigma)$. The IVM computes $\boldsymbol{\mu}$ and Σ incrementally, i.e. in step j a new $\boldsymbol{\mu}_j$ and Σ_j are computed:

$$\boldsymbol{\mu}_j = \boldsymbol{\mu}_{j-1} + \Sigma_{j-1}\mathbf{g}_j \tag{3.6}$$
$$\Sigma_j = \Sigma_{j-1} - \Sigma_{j-1}(\mathbf{g}_j\mathbf{g}_j^T - 2\Gamma_j)\Sigma_{j-1} \tag{3.7}$$

where

$$\mathbf{g}_j = \frac{\partial \log Z_j}{\partial \boldsymbol{\mu}_{j-1}}, \qquad \Gamma_j = \frac{\partial \log Z_j}{\partial \Sigma_{j-1}}, \tag{3.8}$$

and Z_j is the approximation to the normalizer in Eq. (3.1) using the estimate q_j. Initially, $\boldsymbol{\mu}_0 = \mathbf{0}$, and $\Sigma_0 = K$, where K is the prior GP covariance matrix. Then, at iteration j the training point $(\mathbf{x}_k, y_k)$ that maximizes the entropy difference between q_{j-1} and q_j is selected into the active set. The algorithm stops when the active set has reached a desired size D. In our implementation, we choose D as a fixed fraction of N.

Due to a circular dependence between $\mathcal{I}_D$ and the kernel hyper parameters θ, the IVM training algorithm loops a given number of times over two steps: estimation of $\mathcal{I}_D$ from θ and minimizing the *marginal likelihood* Z_D using $\partial Z_D/\partial\theta$, thereby keeping $\mathcal{I}_D$ fixed. Although there are no convergence guarantees, in practice a few iterations are sufficient to find good kernel hyper-parameters.

4 Online Update of the IVM

In addition to its sparsity, the IVM differs from the standard GP also by its ability to compute the posterior distribution $p(\mathbf{f} \mid \mathcal{X}, \mathbf{y})$ *incrementally*. Thus, the algorithm loops over all active points and updates mean vector $\boldsymbol{\mu}$ and covariance matrix Σ by increasing their lengths in every iteration. In particular, it keeps the lower triangular matrix L_d of a Cholesky decomposition in memory and updates it using rank-1 Cholesky updates, where L_d is of size $d \times d$ and $d = 1, \ldots, D$. Further details of this procedure are given in Algorithm 1 of Lawrence *et al.* [6]. For our purpose, this incremental scheme is particularly useful, because it avoids the complete re-computation of the GP parameters in every training round and adds only a fixed number of rows and columns to L_d. This decreases the training time substantially, as we show below. For an efficient *class prediction*, we furthermore propose a novel online update rule, as described next.

4.1   Online Computation of the Class Prediction

To predict a class label y_* for a new test data point $\mathbf{x}_*$, the IVM computes the mean μ_* and the covariance σ_* of the approximation to the predictive distribution given in Eq. (3.3), and uses them to obtain the class probability (Eq. (3.5)). With the notation of Rasmussen and Williams [12], this can be expressed as

$$\mu_* = \mathbf{k}_*^T(K + \tilde{\Sigma})^{-1}\tilde{\mu} \tag{4.1}$$

$$\sigma_* = k(\mathbf{x}_*, \mathbf{x}_*) - \mathbf{k}_*^T(K + \tilde{\Sigma})^{-1}\mathbf{k}_*, \tag{4.2}$$

where $\tilde{\mu}$ and $\tilde{\Sigma}$ are the *site parameters* of the approximate Gaussian likelihood $q(\mathbf{y} \mid \mathbf{f})$, K is the prior covariance matrix, i.e. the kernel function k applied to all pairs of training points $\mathbf{x}_1, \ldots, \mathbf{x}_N$, and $\mathbf{k}_* = (k(\mathbf{x}_*, \mathbf{x}_1), \ldots, k(\mathbf{x}_*, \mathbf{x}_N))$. Note that $\mathbf{k}_*$, $\tilde{\mu}$, and $B := K + \tilde{\Sigma}$ are only computed for D active points with $D < N$.

In general, Eqs. (4.1) and (4.2) have to be computed completely anew for every new test point $\mathbf{x}_*$, and it is usually unlikely to observe the same test point again. In active learning, this means that the complexity of making predictions increases quadratically with the training rounds, because in every training round the matrix B is larger due to the additional active points in the training data. However, for interactive image segmentation, we can use the fact that class predictions are made on the same pixels (i.e. test points) in every round. This means that $\mathbf{k}_{*,t}$ from round t can be obtained from $\mathbf{k}_{*,t-1}$ of the previous round by appending the covariances $k(\mathbf{x}_*, \mathbf{x}_{D_{t-1}+1}), \ldots, k(\mathbf{x}_*, \mathbf{x}_{D_t})$ between $\mathbf{x}_*$ and the new active points, where D_t is the total number of active points in round t. This can be used to compute $\mu_{*,t}$ and $\sigma_{*,t}$ incrementally from $\mu_{*,t-1}$ and $\sigma_{*,t-1}$. To do this, we note that B_t is given by its Cholesky decomposition $L_t L_t^T$, and

$$L_t := \begin{pmatrix} L_{t-1} & 0 \\ A & L_+ \end{pmatrix}, \tag{4.3}$$

where L_+ is lower-triangular. To compute B_t^{-1}, we use

$$B_t^{-1} = \begin{pmatrix} L_{t-1}L_{t-1}^T & L_{t-1}A^T \\ AL_{t-1}^T & AA^T + L_+L_+^T \end{pmatrix}^{-1}, \tag{4.4}$$

and compute the Schur complement as

$$S = AA^T + L_+L_+^T - AL_{t-1}^T(L_{t-1}L_{t-1}^T)^{-1}L_{t-1}A^T = L_+L_+^T.$$

With this, we obtain

$$B_t^{-1} = \begin{pmatrix} C & -L_{t-1}^{-T}A^T S^{-1} \\ -S^{-1}AL_{t-1}^{-1} & S^{-1} \end{pmatrix}, \tag{4.5}$$

where $C = (L_{t-1}L_{t-1}^T)^{-1} + L_{t-1}^{-T}A^T S^{-1} A L_{t-1}^{-1}$. We now formulate Eq. (4.2) as:

$$\sigma_* = k(\mathbf{x}_*, \mathbf{x}_*) - \begin{pmatrix} \mathbf{k}_{*,t-1} & \mathbf{k}_{*,+} \end{pmatrix} B_t^{-1} \begin{pmatrix} \mathbf{k}_{*,t-1} \\ \mathbf{k}_{*,+} \end{pmatrix}, \tag{4.6}$$

where $\mathbf{k}_{*,+}$ is the vector of newly added covariances in round t. Plugging Eq. (4.5) into Eq. (4.6) we obtain for the rightmost term r of Eq. (4.6):

$$r = \hat{\mathbf{k}}_{t-1}^T \hat{\mathbf{k}}_{t-1} + \hat{\mathbf{k}}_{t-1}^T A^T S^{-1} A \hat{\mathbf{k}}_{t-1} - 2\hat{\mathbf{k}}_{t-1}^T A^T S^{-1} \mathbf{k}_{*,+} + \mathbf{k}_{*,+}^T S^{-1} \mathbf{k}_{*,+} \quad (4.7)$$

where $\hat{\mathbf{k}}_{t-1} = L_{t-1}^{-1}\mathbf{k}_{*,t-1}$. It follows that the first term of r in Eq. (4.7) and the first term in Eq. (4.6) define the predictive variance of the previous round $\sigma_{*,t-1}$

$$\sigma_{*,t-1} = k(\mathbf{x}_*, \mathbf{x}_*) - \hat{\mathbf{k}}_{t-1}^T \hat{\mathbf{k}}_{t-1}, \quad (4.8)$$

whereas the remaining terms of r can be subsumed into

$$(L_+^{-1}\mathbf{k}_{*,+} - L_+^{-1}A\hat{\mathbf{k}}_{t-1})^T (L_+^{-1}\mathbf{k}_{*,+} - L_+^{-1}A\hat{\mathbf{k}}_{t-1}), \quad (4.9)$$

which simplifies into

$$(L_+^{-1}\Delta\mathbf{k})^T (L_+^{-1}\Delta\mathbf{k}) \quad (4.10)$$

where $\Delta\mathbf{k} = \mathbf{k}_{*,+} - A\hat{\mathbf{k}}_{t-1}$. This results in an efficient way to compute $\sigma_{*,t}$: We store $\hat{\mathbf{k}}_{t-1}$ from the previous round and compute $\Delta\mathbf{k}$ and $L_+^{-1}\Delta\mathbf{k}$. Then we multiply the result with itself (Eq. (4.10)) and subtract it from $\sigma_{*,t-1}$. Similarly, we can compute $\mu_{*,t}$ from $\mu_{*,t-1}$ of the previous round using the difference vector $\Delta\mu := \mu_{*,+} - A\hat{\mu}_{t-1}$, where $\hat{\mu}_{t-1} = L_{t-1}^{-1}\tilde{\mu}_{t-1}$. To summarize, we have

$$\mu_{*,t} = \mu_{*,t-1} + (L_+^{-1}\Delta\mu)^T (L_+^{-1}\Delta\mathbf{k}) \quad (4.11)$$
$$\sigma_{*,t} = \sigma_{*,t-1} - (L_+^{-1}\Delta\mathbf{k})^T (L_+^{-1}\Delta\mathbf{k}). \quad (4.12)$$

4.2   The Kernel Hyper-Parameters

As mentioned before, finding optimal hyper parameters for the kernel function involves several iterations over active set determination and gradient-descent on the marginal likelihood. However, doing this in every training round has several disadvantages: first, it requires a large computational effort, and second it makes the formulation of the online computation developed in the previous section invalid. The reason for the latter is that the online formulation relies on the fact that the active set does not change across the learning rounds, because otherwise $\mathbf{k}_*$ would have to be recomputed completely in every round. Fortunately, it turns out that the kernel hyper parameters do not change significantly across the training rounds, and, even when they do change, they only have a minor impact on the classification results of the GPC. This is another strength of the GPC framework, because essentially it represents a non-parametric model. In our implementation, we obtain the kernel hyper-parameters using cross-validation on a hold-out set. Compared to the usual gradient-descent based maximization of the log marginal, this has the advantage that the kernel parameters are optimized *across* a number of images, and not for each individual image. Especially, as we use locality and color features in combination with an Automatic Relevance Determination (ARD) kernel, the obtained length scales represent a general weighting between position and color. This turned out to achieve much better results than a per-image training of the ARD kernel parameters.

5   Segmentation

The class predictions from the IVM are only local estimates, and they disregard global properties of the image I. Therefore, we formulate the segmentation problem on the image domain $\Omega \subset \mathbb{R}^2$ as finding a *foreground region* $\hat{\Omega}_F$ so that

$$\hat{\Omega}_F = \arg\min_{\Omega_F} \lambda \int_{\Omega_F} -\log p(y_* = 1 \mid \mathcal{X}, \mathbf{y}, \mathbf{x}_*) \, d\mathbf{x}$$

$$+ \lambda \int_{\Omega \setminus \Omega_F} -\log p(y_* = -1 \mid \mathcal{X}, \mathbf{y}, \mathbf{x}_*) \, d\mathbf{x} + \mathrm{Per}_\alpha(\Omega_F), \qquad (5.1)$$

where Per_α is the perimeter of Ω_F, weighted by a local metric $\alpha(\mathbf{x}) = e^{-\gamma|\nabla I|}$ that depends on the image gradient, and λ is the weight of the dataterm. This functional favours spatial regularity by penalizing the boundary length of the foreground region. First, we define an indicator function $u(\mathbf{x})$ that is 1 for $\mathbf{x} \in \Omega_F$ and 0 otherwise. Then, the segmentation problem can be written in a variational formulation:

$$\min_{u \in [0,1]} \int_\Omega \varrho(\mathbf{x}) u(\mathbf{x}) \, d\mathbf{x} + \frac{1}{\lambda} \int_\Omega \alpha(\mathbf{x}) |\nabla u(\mathbf{x})| \, d\mathbf{x}, \qquad (5.2)$$

where the first term encodes the cost of a pixel to belong to the foreground and $\varrho(\mathbf{x}) = \log p\,(u\,(\mathbf{x}) = 0) - \log p\,(u\,(\mathbf{x}) = 1)$. The second term of Eq. (5.2) is the total variation (TV) of the indicator function u which penalizes the perimeter of the foreground region. Since the TV is not differentiable everywhere, we rewrite Eq. (5.2) as a saddle point problem:

$$\min_{u \in [0,1]} \max_{|v| \leq \alpha(\mathbf{x})} \int_\Omega \varrho(\mathbf{x}) u(\mathbf{x}) + \int_\Omega u(\mathbf{x}) \, \mathrm{div}\, v(\mathbf{x}). \qquad (5.3)$$

This can be efficiently minimized using a first-order primal-dual method [11].

6   Experimental Results

We evaluate our active learning approach on the benchmark data set from the University of Graz [14]. It consists of images with ground truth segmentations and user scribbles. As our method applies for foreground and background segmentation we chose a subset of 44 images from the dataset which contain only two object classes. As performance measure for this benchmark we use the f_1 measure, which is defined as the harmonic mean of precision and recall.

6.1   Benefits of the GP Classifier

We compare our approach with the method of Nieuwenhuis and Cremers [8]. There, the data term is computed using a Parzen window (PW) estimator, and the training data consists of color information and positions of user scribbles.

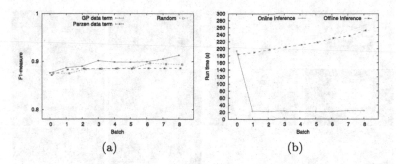

Fig. 3. (a) Average f-measure over 8 active learning rounds. The GPC steadily improves the segmentation, because its label queries are more informative for classification. In contrast, the Parzen window only improves slightly and then remains at a lower performance level. We also show GPC results where new user scribbles are chosen randomly and not based on the entropy. This also improves the segmentation, as it increases the training data, but it is worse than the entropy-based method. (b) Run time of online and offline inference, averaged over all images. Note that in batch 0, the online and the offline method take the same time, because they both build up the initial covariance matrix. However, in later steps the online computation time drops down significantly.

We use the same idea, but employ a GPC instead of the PW. Our benefit is the ability to detect misclassifications using the predictive uncertainty, which is more strongly correlated to incorrect classifications than for the PW. As a result, in active learning the GPC generates more informed questions (see Fig. 1). For a quantitative evaluation, we ran active learning with the GPC and the PW on the Graz data set (Fig. 3a). Both approaches perform equally well in the first rounds, but then the GPC (red curve) outperforms the PW (blue curve), because it asks more informed label queries, while the PW tends to be overconfident. We also show the results for randomly selected scribbles (magenta curve) instead of those with the highest uncertainty. We see that random sampling also improves the classification, as it provides more training data in every round, but the improvement is smaller compared to selecting the most uncertain segments. This is because the GP requests the more informative user scribbles.

Some results from the Graz data set are shown in Fig. 4. The left column shows the images with the initial user scribbles. Columns two and three show the uncertainties of the GPC (brighter is more uncertain) and the segmentation after the first learning round. The general segmentation is good, but small misclassifications occur. However, these often correspond to locations of high uncertainty, e.g. the lower right corner of the helicopter image or the third peg on the wardrobe: here the classification is incorrect, but the uncertainty is also high. This enables the classifier to correct the error in subsequent training rounds.

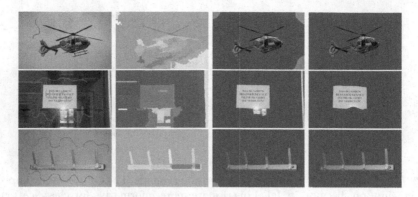

Fig. 4. Examples from the Graz benchmark. **First column:** original images with initial user scribbles. **Second column:** classification uncertainties after the first learning round. **Third column:** resulting segmentation after the first round. Note how the algorithm misclassifies some small areas, but the classification in those same areas is often very uncertain (see, e.g., the third peg on the wardrobe). Thus, the errors can be corrected by querying more useful, i.e. informative user scribbles. **Last column:** final results, obtained after a few further active learning rounds (between 1 and 5). Here, a high-quality segmentation is obtained.

6.2 Advantage of the Online Inference Algorithm

As mentioned in Sect. 4.1, we use a very efficient online class prediction step. Note that this is different from an online *training* step: while the latter is inherently provided by the IVM approach, the former is a novel contribution. In Fig. 3b, we show its benefit over the standard offline technique in every active learning round. Observe that for all but the first learning round the average run time drops from the order of minutes to the order of seconds. Also note that the increase in run time over the learning rounds is super-linear in the offline case, where for the online method it is roughly linear. In the first round, the online and the offline method perform the same steps, because every pixel is compared to all training points. Currently, we compute this in parallel on 8 CPU threads, but we expect a substantial speed-up when using a GPU implementation.

7 Conclusions

We present an efficient active learning approach and show its application to inter-active segmentation. Our method learns models for background and foreground adaptively by informed questions based on the classification uncertainty and uses a regularizer that favors regions with smooth contours. To make the classification process efficient, we use an online update method that incrementally estimates the class posteriors. This reduces computation time substantially, without reducing the high segmentation performance of the active learning method.

Acknowledgment. This work was partly funded by the EU project SPENCER (ICT-2011-600877).

References

1. Boykov, Y., Veksler, O., Zabih, R.: Fast approximate energy minimization via graph cuts. Trans. Patt. Anal. Mach. Intell. **23**(11), 1222–1239 (2001)
2. Csató, L., Opper, M.: Sparse on-line gaussian processes. Neural Comput. **14**(3), 641–668 (2002)
3. Felzenszwalb, P.F., Huttenlocher, D.P.: Efficient graph-based image segmentation. Int. J. Comput. Vis. **59**(2), 167–181 (2004)
4. Kapoor, A., Grauman, K., Urtasun, R., Darrell, T.: Gaussian processes for object categorization. Int. J. Comput. Vis. **88**(2), 169–188 (2010)
5. Lawrence, N., Seeger, M., Herbrich, R.: Fast sparse gaussian process methods: the informative vector machine. Adv. Neural Inf. Process. Syst. **15**, 609–616 (2002)
6. Lawrence, N.D., Platt, J.C., Jordan, M.I.: Extensions of the informative vector machine. In: Winkler, J.R., Niranjan, M., Lawrence, N.D. (eds.) Deterministic and Statistical Methods in Machine Learning. LNCS (LNAI), vol. 3635, pp. 56–87. Springer, Heidelberg (2005)
7. Lombaert, H., Sun, Y., Grady, L., Xu, C.: A multilevel banded graph cuts method for fast image segmentation. In: International Conference on Computer Vision (ICCV), pp. 259–265 (2005)
8. Nieuwenhuis, C., Cremers, D.: Spatially varying color distributions for interactive multi-label segmentation. Trans. Patt. Anal. Mach. Intell. **35**(5), 1234–1247 (2013)
9. Nikolova, M., Esedoglu, S., Chan, T.F.: Algorithms for finding global minimizers of image segmentation and denoising models. SIAM J. Appl. Math. **66**(5), 1632–1648 (2006)
10. Paul, R., Triebel, R., Rus, D., Newman, P.: Semantic categorization of outdoor scenes with uncertainty estimates using multi-class Gaussian process classification. In: Proceedings of the International Conference on Intelligent Robots and Systems (IROS) (2012)
11. Pock, T., Cremers, D., Bischof, H., Chambolle, A.: An algorithm for minimizing the piecewise smooth mumford-shah functional. In: International Conference on Computer Vision (ICCV) (2009)
12. Rasmussen, C.E., Williams, C.K.I.: Gaussian Processes for Machine Learning. MIT Press, Cambridge (2006)
13. Rother, C., Kolmogorov, V., Blake, A.: Grabcut: Interactive foreground extraction using iterated graph cuts. ACM Trans. Graph. (TOG) **23**(3), 309–314 (2004)
14. Santner, J., Pock, T., Bischof, H.: Interactive multi-label segmentation. In: Kimmel, R., Klette, R., Sugimoto, A. (eds.) ACCV 2010, Part I. LNCS, vol. 6492, pp. 397–410. Springer, Heidelberg (2011)
15. Triebel, R., Grimmett, H., Paul, R., Posner, I.: Driven learning for driving: how introspection improves semantic mapping. In: Proceedings of International Symposium on Robotics Research (ISRR) (2013)
16. Unger, M., Pock, T., Trobin, W., Cremers, D., Bischof, H.: TVSeg - interactive total variation based image segmentation. In: British Machine Vision Conference (2008)
17. Vezhnevets, A., Buhmann, E.J., Ferrari, V.: Active learning for semantic segmentation with expected change. In: Conference on Computer Vision and Pattern Recognition (2012)

652 R. Triebel et al.

Multi-view Tracking of Multiple Targets with Dynamic Cameras

Till Kroeger[1]([✉]), Ralf Dragon[1], and Luc Van Gool[1,2]

[1] Computer Vision Laboratory, ETH Zurich, Zürich, Switzerland
{kroegert,dragonr,vangool}@vision.ee.ethz.ch
[2] ESAT - PSI / IBBT, K.U. Leuven, Leuven, Belgium

Abstract. We propose a new tracking-by-detection algorithm for multiple targets from multiple dynamic, unlocalized and unconstrained cameras. In the past tracking has either been done with multiple static cameras, or single and stereo dynamic cameras. We register several moving cameras using a given 3D model from Structure from Motion (SfM), and initialize the tracking given the registration. The camera uncertainty estimate can be efficiently incorporated into a flow-network formulation for tracking. As this is a novel task in the tracking domain, we evaluate our method on a new challenging dataset for tracking with multiple moving cameras and show that our tracking method can effectively deal with independently moving cameras and camera registration noise.

1 Introduction

Simultaneous object tracking across multiple views is commonly solved with strong restrictive assumptions of known static cameras and planar movement constraints. It is easy to see that both constraints generally do not hold for many tracking tasks, such as simultaneous tracking using cameras on unmanned aerial vehicles, or tracking in synchronized dynamic videos from hand-held cameras. The knowledge about camera configurations will be unreliable or nonexistent. The movement on ground planes (GP) is an important special case, but even in standard tracking scenarios, e.g. pedestrians within cities, often too restrictive. We exploit connections between tracking and the Structure from Motion (SfM) domain, which can help making generic tracking scenarios solvable. Knowledge about camera localization and 3D structure can help with this task. Structure models built by today's SfM methods generally show good quality, are easy to create and widely available. Because of the ubiquity of available 3D models we propose to merge established methods for multi-view tracking-by-detection and localization methods developed for SfM to enable more generic tracking tasks (see Fig. 1).

Our contribution is a method for tracking-by-detection for multiple dynamic cameras, with known but noisy 6-DoF camera motion. We propose a novel method for linking data across time and views incorporating the motion uncertainty of the cameras.

© Springer International Publishing Switzerland 2014
X. Jiang et al. (Eds.): GCPR 2014, LNCS 8753, pp. 653–665, 2014.
DOI: 10.1007/978-3-319-11752-2_54

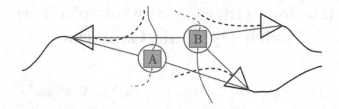

Fig. 1. Multi-view tracking from three dynamic cameras (colored in black, past: solid, future: dashed) of two moving objects (red). Object A is tracked in two, and B in three cameras. Camera locations are unknown and approximately determined by registration to SfM models (Color figure online).

We are the first to present a generalization of the strongly constrained 2D multi-view tracking scenario to unconstrained 3D scenarios. We do not make common restrictive assumptions, such as planar motion, known number of objects, constrained camera configurations, background modeling and explicit introduction of knowledge about the tracked objects movement characteristics in motion models.

We describe the tracking framework in Sect. 2. In Sect. 3 we explain unary and pairwise cost terms for the flow-network framework. In Sect. 4 an experimental analysis is given. In Sect. 5 we conclude and discuss future work. We use *view* and *camera* to refer to image data, orientation and position of one camera at a given time. We use *frame* to refer to image data from all views at a given time. A *dynamic camera* means a moving camera.

Related Work. Multi-object tracking has been studied extensively. The most successful methods define tracking as a global optimization over a complete sequence, with given frame-wise object hypotheses. This is usually called *tracking-by-detection.*

Optimal single view data association for many detections per frame has been formulated in [11,27]. Reference [26] tracks without explicit detections on probabilistic occupancy grids. Reference [2] extends this with iterations of discrete associations and continuous refinement.

Data association across multiple views adds the challenge of associating over time and views simultaneously. Reference [15] solves one tracking graph for each view, multi-view couplings are solved in an additional graph. Reference [9] solves associations across time using pre-computed merging candidates across views. Reference [13] uses a homographic constraint to associate across views. References [1,3,8] propose a probabilistic occupancy grid. Tracking is solved over discretized grid positions. These methods rely on static cameras. References [6,16] propose multi-view tracking with dynamic cameras in a stereo setup.

Data is usually collected as independent detections from a given detector [1,2,9,17,20,22,23,27] or by background subtraction [11,15]. References [1,3,8, 24,26] work directly on a discretized grid or volume representing the space of all possible locations.

Several common constraints are used to facilitate tracking. Motion is often limited to a known GP [2,3,8,9,13,17,22,24,26]. References [6,16] assume planar motion, but infer the GP automatically. Reference [26] uses global appearance constraints. Reference [22] uses social grouping cues. References [3,8,11,15] need static background and cameras. Specific camera configurations are needed: head-level cameras [8], top-down views [9], visible feet locations [13].

Solutions for association networks are found by Linear Programming (LP) [1,9,11,15,20,26], Dynamic Programming (DP) [3,8,27], Energy minimization in MRF/CRFs [2,22]. To reduce computational demands greedy approximations are available [23].

Camera registration/localization has a variety of methods and distinct applications. In SLAM [5], methods estimate precise 6-DoF poses to 3D models, given location priors and feature matches. Reference [19] solves robot self-localization given large-scale 3D models. References [10,25] propose methods for registering image collections for large scale models. Reference [14] achieves high-precision registration for Videos to SfM models.

Our work is related to [9], with important differences. We perform the data association in 3D, while [9] uses a given GP. Reference [9] has static entry/exit regions and hence the number of views to be used is known a priori. We incorporate camera and detector uncertainties into linking (16) and reconstruction probabilities (12), while [9] only uses these for error bounds while pruning reconstructions. Reference [9] needs a form of enumeration for all possible sets of reconstructions, which grows factorially with number of cameras. Our work uses LP to optimally select the top reconstructions without enumeration, with a lower dependency on number of cameras, thus removing a serious bottlenecks of [9].

2   Tracking

We adopt a similar notation to [9]. We extract person detections using the Deformable Part Model detector [7]. Each 2D detection in camera j at time t is described by $\mathbf{x}_i^{(j,t)} = (x, s)$, with position x and scale s. All detections at time t from all cameras are denoted $\mathbf{X}^{(t)}$. One set of coupled detections from multiple cameras at time t is denoted as reconstruction $\mathcal{R}_k^{(t)} \in \mathcal{R}^{(t)} \subset \mathcal{R}$. From each camera at most one detection can be included in each $\mathcal{R}_k^{(t)}$:

$$\mathcal{R}_k^{(t)} \subset \{\mathbf{X}^{(t)} | \forall \mathbf{x}_i^{(j,t)}, \mathbf{x}_{i'}^{(j',t)} \in \mathcal{R}_k^{(t)} : j \neq j' \vee i = i'\} . \tag{1}$$

Ideally one reconstruction $\mathcal{R}_k^{(t)}$ corresponds to one real world object which caused the detections in each view. We denote the number of detections in one reconstruction as its *cardinality*. The set $\mathcal{R}^{(t)}$ denotes all reconstructions at time t. A trajectory hypothesis is defined as $\mathcal{T}_u = \{\mathcal{R}_{u_1}^{(t)}, \ldots, \mathcal{R}_{u_n}^{(t+n)}\}$. This formulation allows missing detections for a real world object track in several views at different times. These missing detections can be due to detector errors or occlusions in a subset of views.

Similar in spirit to [9,27] we formulate multi-target tracking as a MAP problem and attempt to find an optimal set of tracks $\mathcal{T}$ which can be written as

$$\mathcal{T}^* = \arg\max_{\mathcal{T}} P(\mathcal{T}|\mathcal{R}) \quad = \quad \arg\max_{\mathcal{T}} P(\mathcal{R}|\mathcal{T})P(\mathcal{T}) \tag{2}$$

$$= \arg\max_{\mathcal{T}} \prod_i P(\mathcal{R}_i|\mathcal{T}) \prod_{\mathcal{T}_k \in \mathcal{T}} P(\mathcal{T}_k) . \tag{3}$$

This assumes conditional independence of likelihoods given $\mathcal{T}$ and independent object motion in (3). We further assume non-overlap of tracks, i.e. $\mathcal{T}_k \cap \mathcal{T}_j = \emptyset, \forall k \neq j$. The terms in (3) are defined as follows:

$$P(\mathcal{T}_k) = P_{en}(\mathcal{R}_{u_1}^{(t)}) \; P_{li}(\mathcal{R}_{u_2}^{(t+1)}|\mathcal{R}_{u_1}^{(t)}) \; \ldots \; P_{ex}(\mathcal{R}_{u_n}^{(t+n)}) , \tag{4}$$

$$P(\mathcal{R}_i|\mathcal{T}) = \begin{cases} P_{rec}, & \text{if } \exists \mathcal{T}_k \in \mathcal{T}, \mathcal{R}_i \in \mathcal{T}_k \\ 1 - P_{rec}, & \text{otherwise} \end{cases} . \tag{5}$$

This problem can be turned into a cost-flow network and solved *exactly* in an LP as in [1,9,11,15,26], or using DP [3,8,27], or *approximately* and fast as in [23]. We use an LP, similar to [9]. In order to solve the LP we have to compute a set of reconstruction hypothesis per frame $\mathcal{R}^{(t)}$ and provide probabilities P_{en}, P_{ex}, P_{rec} for all reconstructions, and P_{li} for all pairs of reconstructions. We turn the probabilities into costs W_{en}, W_{ex}, W_{rec} and W_{li} for the flow-network as in [27] (Eq. 11), or [9] (Eq. 13–16).

3   Modeling of Probabilities

The flow-network method links reconstructions across views and time given unary and binary costs. Similar to [9,27] we model these costs as probabilities P_{rec} and P_{li}.

3.1   Extraction of Reconstruction Candidates

To start the tracking we need a set $\mathcal{R}^{(t)}$ of object hypotheses for all times t containing the true solution. Because of this we focus on extracting the sets $\mathcal{R}^{(t)}$ large enough to ensure maximum recall. However, the combinatorial explosion prohibits inclusion of all feasible reconstructions in $\mathcal{R}^{(t)}$. We extract a set of L top-ranking reconstructions for a each cardinality m as follows: We apply a distance function $D_1(\mathbf{x}_i^{(j,t)}, \mathbf{x}_{i'}^{(j',t)})$ in 3D world coordinates between all pairs of two detections originating from different views, which we define in (9). Using this distance function we define the compactness of a reconstruction $\mathcal{R}_k^{(t)}$ as the sum of all pairwise detection distances $\sum_{x,x' \in \mathcal{R}_k^{(t)}, x \neq x'} D_1(\mathbf{x}, \mathbf{x}')$. This allows solving for the best (i.e. most compact) reconstruction of a given cardinality m at time t as an LP. We extract the L top ranking solution using CPLEX for each cardinality and insert all extracted reconstructions at time t into $\mathcal{R}^{(t)}$.

3.2   Localization of Detections and Reconstructions

Most previous multi-view multi-target tracking methods rely on the assumption of a given or automatically inferred ground plane (GP). However, GPs may not exist in some scenes, may not be visible (i.e. occluded in head-level cameras), or non-planar tracking is required. Localization of 2D object detections in 3D from head-level cameras using a GP assumption will be unreliable, even if a perfect GP is available, due to the small viewing angle. We side-step these problems by using the height of a detection as depth cue. Even if a bounding box is partly occluded, given the width and expected aspect ratio of a walking person, approximation of the height is possible. We use a distribution of heights with large uncertainty: mean $\lambda = 1.74\,\text{m}$ and standard deviation $\sigma = \lambda/4$.

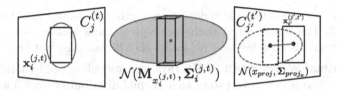

Fig. 2. Detection $\mathbf{x}_i^{(j,t)}$ from camera $C_j^{(t)}$ is transformed into 3D with center $\mathbf{M}_{x_i^{(j,t)}}$ and uncertainty $\Sigma_i^{(j,t)}$. We propagate both into view j' at time t' and compute $D_3(x_{proj}, \mathbf{x}_{i'}^{(j',t')})$ (11) as the Mahalanobis distance in the image (colored in red) (Color figure online).

We assume a calibrated camera j at time t is given with approximately known pose $C_j^{(t)} = (R, tr)$, where tr and R denote the position and orientation given from some arbitrary noisy image-based registration method, and known internal calibration matrix K_j. Let $\Sigma_{C_j^{(t)}}$ be the positional covariance of the camera. Given a detection $\mathbf{x}_i^{(j,t)} = (x, s)$ with center x and scale s in view j at time t we compute a 3D location and uncertainty in world coordinates for the detection. On undistorted images, the detection direction $\hat{x}$ is given by $\hat{x} = K_j^{-1}\tilde{x}$, where $\tilde{x}$ is homogeneous x and $\|\hat{x}\| = 1$.

We compute the detection distance from the camera center as $d = \lambda \cdot f_j/s$, where s is the detection's pixel height, and f_j the focal length. This is a reasonable distance approximation if the object is close to the optical axis. Knowing d, we model the location in world coordinates as a normal distribution with mean $\mathbf{M}_{x_i^{(j,t)}}$ and covariance $\Sigma_i^{(j,t)}$:

$$\mathbf{M}_{x_i^{(j,t)}} = R^T \cdot \left[(\hat{x} \cdot d) - tr \right], \tag{6}$$

$$\Sigma_i^{(j,t)} = R^T \cdot \text{diag}\left(\lambda/2, \; \lambda/2, \; \lambda \cdot f_j/(4\,s) \right) \cdot R + \Sigma_{C_j^{(t)}}. \tag{7}$$

The covariance is composed of two terms: Orthogonal to the optical axis, the detection 3D uncertainty is only dependent on the object scale λ. The detection

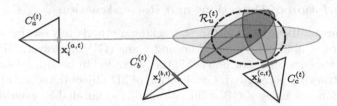

Fig. 3. A reconstruction $\mathcal{R}_u^{(t)}$ consisting of 3 detections (red, blue, green), seen from 3 cameras at time t is localized in 3D. Detection uncertainties in 2D become depth uncertainties in 3D, which increase with distance to the camera. Equation (8) gives center and uncertainty of $\mathcal{R}_u^{(t)}$ (black, dashed) (Color figure online).

height uncertainty in 2D becomes detection depth uncertainty in 3D as shown in Figs. 2 and 3. The second term is the isotropic camera localization uncertainty $\Sigma_{C_j^{(t)}}$. A normal distribution over height in 2D will generally not translate into a normal distribution over depth in 3D. However, modeling (and thereby approximating) the true depth uncertainty as a normal distribution in 3D allows us to easily propagate the resulting uncertainty back into other 2D views and obtain a normal distributions again. See Fig. 2 and (10) for an explanation of the back-propagation.

To model the 3D reconstruction $\mathcal{R}_k^{(t)}$ consisting of many detections, we fit a normally distributed uncertainty with mean $\mathbf{M}_{R_k^{(t)}}$ and covariance $\Sigma_{R_k^{(t)}}$ such that:

$$x \sim \mathcal{N}(\mathbf{M}_{R_k^{(t)}}, \Sigma_{R_k^{(t)}}) \sim \frac{1}{Z} \sum_{x \in R_k^{(t)}} \mathcal{N}(\mathbf{M}_x, \Sigma_x) \,. \qquad (8)$$

Figure 3 shows an example of 3 detections combined into one reconstruction. We use only one normal distribution for $\mathcal{R}_k$ and not the mixture of all distribution from included detections to allow quick propagation of a single normal distribution to different views.

3.3   Detection and Reconstruction Distances

In order to model P_{rec} and P_{li}, or the reconstruction unary and pairwise costs, respectively, we need a set of geometric and appearance-based comparison measures between detections from different views and reconstructions. Given the localization and uncertainties of detections and reconstructions in 3D we will define four geometric and one appearance-based distance measure to be used for P_{rec} and P_{li}, in Sects. 3.5 and 3.6.

– D_1: Mahalanobis distance between detections in 3D:

$$D_1(x, y) = D_{mah}(\mathbf{M}_y; \mathbf{M}_x, \Sigma_x) + D_{mah}(\mathbf{M}_x; \mathbf{M}_y, \Sigma_y) \,. \qquad (9)$$

- D_2: Similar to D_1, Mahalanobis distance between two reconstructions $\mathcal{R}_k^{(t)}, \mathcal{R}_{k'}^{(t')}$ using mean $\mathbf{M}_{R_k^{(t)}}, \mathbf{M}_{R_{k'}^{(t')}}$ and covariance $\boldsymbol{\Sigma}_{R_k^{(t)}}, \boldsymbol{\Sigma}_{R_{k'}^{(t')}}$.

- D_3: Defined between a 2D detection $\mathbf{x}_i^{(j,t)}$, which has been projected into 3D as $\mathbf{M}_{x_i^{(j,t)}}$ with uncertainty $\boldsymbol{\Sigma}_i^{(j,t)}$, and a 2D detection $\mathbf{x}_{i'}^{(j',t')}$ (see Fig. 2). We project $\mathbf{M}_{x_i^{(j,t)}}$ into view j' at time t' with camera $C_{j'}^{(t')} = (R, tr)$ and produce $x_{proj} = K_{j'}[R\ tr] \cdot \mathbf{M}_{x_i^{(j,t)}}$. To propagate the uncertainty, the 3D covariance $\boldsymbol{\Sigma}_i^{(j,t)}$ is projected into view j' at time t' using the projection's Jacobian matrix $J^C(x)$ of $C_{j'}^{(t')}$. We also include $C_{j'}^{(t')}$'s localization uncertainty to handle camera errors:

$$\boldsymbol{\Sigma}_{proj_x} = J^C(x)^T \cdot (\boldsymbol{\Sigma}_i^{(j,t)} + \boldsymbol{\Sigma}_{C_{j'}^{(t')}}) \cdot J^C(x) . \tag{10}$$

Using projected mean x_{proj} and covariance $\boldsymbol{\Sigma}_{proj_x}$ we define D_3 as the Mahalanobis distance

$$D_3(x_{proj}, \mathbf{x}_{i'}^{(j',t')}) = D_{mah}(\mathbf{x}_{i'}^{(j',t')}; x_{proj}, \boldsymbol{\Sigma}_{proj_x}) . \tag{11}$$

- D_4: Defined between a reconstruction and a 2D detection equivalently to D_3 using the reconstruction's mean and covariance for reprojection.
- D_5: Between two 2D detections we define D_5 as the Earth mover's distance over RGB color histograms. We experimented with adding a HoG-based distance measure, but found it not to be helpful, due to large baselines between different views.

3.4   Entry and Exit Probability

Another often exploited advantage of static cameras is the possibility of defining explicit *entry* and *exit zones* in the images. Only in those areas are tracks allowed to start and end. This is easily modeled by setting P_{en} and P_{ex} to zero for all detections occurring outside these zones. For dynamic cameras this option does not exist. We estimate $P_{en} = P_{ex}$ for all reconstructions uniformly as the average missdetection probability of the detector over all sequences and cameras. These probabilities are transformed into costs for the flow-network as $W_{en} = W_{ex} = -\log(P_{ex}) = -\log(P_{en})$.

3.5   Reconstruction Probability

The probability P_{rec} measures the pairwise similarity in appearance and spatial proximity of all detections in a reconstruction. For a reconstruction with detections of the same object, all detections should be similar in appearance and spatially close together. We set

$$P_{rec} = P(d_{rep}) \cdot P(d_{col}) . \tag{12}$$

$P(d_{rep})$ is computed as the probability that each detection reprojects well to all detections in other views. We average the distance D_3 for all pairs of included detections:

$$d_{rep} = \frac{1}{|\mathcal{R}_k^{(t)}|^2} \sum_{x_i, x_j \in \mathcal{R}_k^{(t)}} D_3(x_i, x_j) . \tag{13}$$

Average bounding box color dissimilarities are computed:

$$d_{col} = \frac{1}{|\mathcal{R}_k^{(t)}|^2} \sum_{x_i, x_j \in \mathcal{R}_k^{(t)}} D_5(x_i, x_j) . \tag{14}$$

The distributions of d_{rep}, d_{col} are trained from ground truth and turned into matching probabilities $P(d_{rep})$, $P(d_{col})$. We transform the probability P_{rec} into the node's flow cost W_{rec} and add the detector confidence for every detection:

$$W_{rec} = \log\left(\frac{1 - P_{rec}}{P_{rec}}\right) + \sum_{x_i \in \mathcal{R}_k^{(t)}} \log\left(\frac{\beta(x_i)}{1 - \beta(x_i)}\right) , \tag{15}$$

where $\beta(x_i)$ describes the detection's false positive probability based on the detector score. Note that this cost neither penalizes nor prefers reconstructions with large or small cardinality. Unlike [9] we cannot assume to know the number of cameras in which an object is visible, and prefer to keep this score invariant to the cardinality.

3.6   Linking Probability

For two given reconstruction candidates $\mathcal{R}_k^{(t)}$, $\mathcal{R}_{k'}^{(t+1)}$ the transition probability P_{li} indicates the probability of $\mathcal{R}_{k'}^{(t+1)}$ following $\mathcal{R}_k^{(t)}$ in a track. We set

$$P_{li} = P(d_{pro}) \cdot P(d_{col}) \cdot P(d_{cen}) . \tag{16}$$

To establish these probabilities we will compute projection distances d_{pro}, pairwise color dissimilarities d_{col}, and the distance between the reprojections' center points d_{cen}. Let m, n be the cardinalities of $\mathcal{R}_k^{(t)}$, $\mathcal{R}_{k'}^{(t+1)}$, respectively. We calculate how close the reconstruction $\mathcal{R}_k^{(t)}$'s 3D center projects to all detections $x_j \in \mathcal{R}_{k'}^{(t+1)}$ and vice versa:

$$d_{pro} = \frac{1}{n} \sum_{j=1}^{n} D_4(\mathbf{M}_{R_k^{(t)}}, x_j) + \frac{1}{m} \sum_{i=1}^{m} D_4(\mathbf{M}_{R_{k'}^{(t+1)}}, x_i) . \tag{17}$$

d_{col} is computed as the average color dissimilarity between all detection pairs $x_i \in \mathcal{R}_k^{(t)}$, $x_j \in \mathcal{R}_{k'}^{(t+1)}$, similar to (14). We compute the distance between reconstructions centers $d_{cen} = D_2(\mathbf{M}_{R_k^{(t)}}, \mathbf{M}_{R_{k'}^{(t+1)}})$. This implicitly assumes maximum probability for stationary objects, which is generally not correct. However, this

Fig. 4. Tracking result in 4 dynamic cameras of seq. 3. Static cameras $1+2$ show current (red) and past (blue) estimated camera positions. Anonymized for publication purposes (Color figure online).

is an acceptable approximation because the frame-to-frame motion (1.5 m/s avg. walking speed) is 1-2 magnitudes smaller than the camera localization error. The distributions of d_{pro}, d_{col} and d_{cen} are trained from ground truth and turned into matching probabilities for (16). The probability is transformed into a cost as $W_{li} = -\log(P_{li})$.

4 Experiments

Dataset for Evaluation. We used three sets of videos for tracking, each set is called a *sequence*. Each sequence consists of 301 frames, seen from 7 synchronized cameras. Two of the 7 cameras are static, show the scene from a top-down view, are manually registered to the SfM model, and only used to create ground truth 3D locations for pedestrians and the 5 videographers. The remaining 5 cameras are dynamic, always seen from the static cameras, and used for tracking. In Figs. 4 and 5, 4 out of 5 dynamic and 2 static cameras are shown. Due to space constraints the 5th camera was omitted. Pedestrian tracks are annotated in all 7 views and identities across views are given. A manually determined GP is calculated for visualization and evaluation, but is not used in the algorithm. We manually collect 3D locations on this GP for all person tracks, including the videographers. The scene is not perfectly planar as can be seen in Figs. 4 and 5. Nevertheless, we expect the 3D localization error for all tracks to be below 30 cm.

Video Registration. For each sequence, we independently register all 5 dynamic videos to the SfM model using [21]. The SfM model covering the scene was manually created using high-resolution DSLR images. Because we have ground truth 3D position on a GP for each videographer track, we can evaluate the localization of each camera. The camera registration results in a mean and median error of 0.71 and 0.62 m, respectively. Column *C.Err1* in Table 1 show this positional

Table 1. Tracking and video registration result for three sequences. Explanation in Sect. 4.

	MOTA (%)	MOTP io(%)	MOTP 3d(m)	MT [%]	ML [%]	PT [%]	ids	Frag	C.Err1 (m)	C.Err2 (m)
S.1	1.01	0.88/0.93	2.65/1.36	7.14	28.57	64.29	21	91	0.73/0.65	2.16/1.79
S.2	0.64	0.91/0.95	1.25/0.96	9.09	54.55	36.36	3	104	0.78/0.67	3.55/1.71
S.3	1.44	0.89/0.94	1.32/0.88	0	54.55	45.45	4	119	0.62/0.48	2.82/1.21

error for all sequences separately. Column *C.Err2* shows the mean and median error between GT camera location and the estimated position of the corresponding videographer.

Tracking. We use an independently trained Deformable Part model detector [7] for person detection. Given registered videos we start the tracking by computing 3D positions and uncertainties for all detections, in all views and frames. We only have tracking annotation for persons which are seen from the two static cameras in at least one frame. We cannot evaluate the tracking for persons outside of this area. Therefore, we manually eliminate all detections, based on their 3D projection, which cannot be seen from the static cameras. This avoids creating tracks outside of the valid tracking region, which would result in a misleading FP number. Following Sect. 3.1 we extract reconstruction candidates for all frames. We set $L = 100$ as sufficiently large set of hypotheses. We create the flow graph exactly as in [9], using our defined probabilities, and solve it using CPLEX. As evaluation criterion we use the metric presented in [18]: Identity switches (ids), Number of interruptions of a ground truth track (Frag), Mostly tracked (MT), Mostly lost (ML), Partly tracked (PT). Additionally, we use the CLEAR metric [4,12]: Multiple Object Tracking Accuracy (MOTA) and Multiple Object Tracking Precision (MOTP). The hit/miss decision is computed in image space. A hit has an intersection over union score of over 0.4. We compute MOTP in image space as intersection over union (MOTP io, in percent, mean/median) and as the distance between estimated and true 3D track location (MOTP 3D, in meters, mean/median).

Discussion. We achieve consistently good precision in image space (MOTP io). In addition, we achieve good precision in 3D localization (MOTP 3d) considering that detection height was used as initial depth estimate and no GP was available. Additionally, we only have low confusion, as indicated by low identity switches (ids). The low ids score is partly explained by frequent interruptions of the ground truth track, resulting in high fragmentation (frag). Tracks break, rather than incorrectly switching targets. Often detections are incorrectly left out from a track, resulting in many false negatives and a low MOTA. This is due to the fact that the scene setup does not allow to specify a known cardinality for an object detection. Furthermore, several difficult tracks (severely occluded persons, far away persons) are missed altogether, resulting in a low MT, and high ML.

Fig. 5. Tracking result in 4 dynamic cameras of seq. 1. Static cameras $1+2$ show current (red) and past (blue) estimated camera positions. Anonymized for publication purposes (Color figure online).

Figures 4 and 5 show the tracking result from 4 of 5 dynamic cameras in two sequences. Identical colors and numbers in different views indicates the same track. The two static cameras show current estimated camera positions (red) and past videographer position (blue) on the GP. Typical failure cases include track 143 and 153 in Fig. 4: The object is correctly tracked but the tracks are not merged over views. Track 114 and 131 in Fig. 5 show a confusions between tracked objects within each track. In Fig. 5 some far-away persons are missed, due to unreliable detections.

5 Conclusion

With this work we contribute to the solution of multi-target tracking in multiple moving, approximately localized cameras. We extend an established tracking-by-detection framework using flow-networks, and show that even without many common constraints, such as availability of GP, static cameras or background subtraction, a satisfying tracking solution can be found nevertheless. This work presents the first generalization of multi-view multi-target tracking of objects on a GP to moving objects and cameras in 3D. Our tracking method is not limited to person detections. All objects with approximately constant size can be tracked given an appropriate detector. We consider several directions for future work. The MOTA scores we obtain are significantly lower than scores usually obtained on established datasets with static and known cameras, such as PETS2009. Primarily, this is due to the additional unknown and varying camera locations and the lack of available track entry and exit regions. We aim to improve this by using the 3D model visibility information to infer likely entry and exit regions. Another reason is the strong influence of incorrect and noisy camera poses. We aim to improve this by including pairwise essential matrix constraints and estimation of the relative poses directly within the tracking framework. We also explore possibilities of creating reconstruction candidates build from short 2D object tracklets, instead of single-frame detections to reduce computational demands.

Acknowledgments. This work was supported by the European Research Council (ERC) under the project VarCity (#273940).

References

1. Andriyenko, A., Schindler, K.: Globally optimal multi-target tracking on a hexagonal lattice. In: Daniilidis, K., Maragos, P., Paragios, N. (eds.) ECCV 2010, Part I. LNCS, vol. 6311, pp. 466–479. Springer, Heidelberg (2010)
2. Andriyenko, A., Schindler, K., Roth, S.: Discrete-continuous optimization for multi-target tracking. In: CVPR (2012)
3. Berclaz, J., Fleuret, F., Türetken, E., Fua, P.: Multiple object tracking using k-shortest paths optimization. PAMI **33**(9), 1806–1819 (2011)
4. Bernardin, K., Elbs, A., Stiefelhagen, R.: Multiple object tracking performance metrics and evaluation in a smart room environment performance metrics for multiple object tracking. In: EURASIP (2008)
5. Davison, A.J., Reid, I.D., Molton, N.D., Stasse, O.: MonoSLAM: real-time single camera SLAM. PAMI **29**(6), 1052–1067 (2007)
6. Ess, A., Leibe, B., Schindler, K., Van Gool, L.: Robust multi-person tracking from a mobile platform. PAMI **31**(10), 1831–1846 (2009)
7. Felzenszwalb, P.F., Girshick, R.B., McAllester, D., Ramanan, D.: Object detection with discriminatively trained part-based models. PAMI **32**(9), 1627–1645 (2010)
8. Fleuret, F., Berclaz, J., Lengagne, R., Fua, P.: Multi-camera people tracking with a probabilistic occupancy map. PAMI **30**(2), 267–282 (2008)
9. Hofmann, M., Wolf, D., Rigoll, G.: Hypergraphs for joint multi-view reconstruction and multi-object tracking. In: CVPR (2013)
10. Irschara, A., Zach, C., Frahm, J.M., Bischof, H.: From structure-from-motion point clouds to fast location recognition. In: CVPR (2009)
11. Jiang, H., Fels, S., Little, J.J.: A linear programming approach for multiple object tracking. In: CVPR (2007)
12. Kasturi, R., Goldgof, D., Soundararajan, P., Manohar, V., Garofolo, J., Bowers, R., Boonstra, M., Korzhova, V., Zhang, J.: Framework for performance evaluation of face, text, and vehicle detection and tracking in video: data, metrics, and protocol. PAMI **31**(2), 319–336 (2009)
13. Khan, S.M., Shah, M.: A multiview approach to tracking people in crowded scenes using a planar homography constraint. In: Leonardis, A., Bischof, H., Pinz, A. (eds.) ECCV 2006. LNCS, vol. 3954, pp. 133–146. Springer, Heidelberg (2006)
14. Kroeger, T., Van Gool, L.: Video registration to SfM models. In: Fleet, D., Pajdla, T., Schiele, B., Tuytelaars, T. (eds.) ECCV 2014, Part V. LNCS, vol. 8693, pp. 1–16. Springer, Heidelberg (2014)
15. Leal-Taixé, L., Pons-Moll, G., Rosenhahn, B.: Branch-and-price global optimization for multi-view multi-target tracking. In: CVPR (2012)
16. Leibe, B., Cornelis, N., Cornelis, K., Van Gool, L.: Dynamic 3D scene analysis from a moving vehicle. In: CVPR (2007)
17. Leibe, B., Schindler, K., Van Gool, L.: Coupled detection and trajectory estimation for multi-object tracking. In: CVPR (2007)

18. Li, Y., Huang, C., Nevatia, R.: Learning to associate: HybridBoosted multi-target tracker for crowded scene. In: CVPR (2009)
19. Lim, H., Sinha, S.N., Cohen, M.F., Uyttendaele, M.: Real-time image-based 6-dof localization in large-scale environments. In: CVPR (2012)
20. Morefield, C.L.: Application of 0–1 integer programming to multitarget tracking problems. Autom. Control **22**(3), 302–312 (1977)
21. Moreno-Noguer, F., Lepetit, V., Fua, P.: Accurate non-iterative O(n) solution to the PnP problem. In: ICCV (2007)
22. Pellegrini, S., Ess, A., Van Gool, L.: Improving data association by joint modeling of pedestrian trajectories and groupings. In: Daniilidis, K., Maragos, P., Paragios, N. (eds.) ECCV 2010, Part I. LNCS, vol. 6311, pp. 452–465. Springer, Heidelberg (2010)
23. Pirsiavash, H., Ramanan, D., Fowlkes, C.C.: Globally-optimal greedy algorithms for tracking a variable number of objects. In: CVPR (2011)
24. Possegger, H., Sternig, S., Mauthner, T., Roth, P.M., Bischof, H.: Robust real-time tracking of multiple objects by volumetric mass densities. In: CVPR (2013)
25. Sattler, T., Leibe, B., Kobbelt, L.: Fast image-based localization using direct 2D-to-3D matching. In: ICCV (2011)
26. Shitrit, H.B., Berclaz, J., Fleuret, F., Fua, P.: Tracking multiple people under global appearance constraints. In: ICCV (2011)
27. Zhang, L., Li, Y., Nevatia, R.: Global data association for multi-object tracking using network flows. In: CVPR (2008)

Quality Based Information Fusion in Fully Automatized Celiac Disease Diagnosis

Michael Gadermayr[1]([⊠]), Andreas Uhl[1], and Andreas Vécsei[2]

[1] Department of Computer Sciences, University of Salzburg, Salzburg, Austria
mgadermayr@cosy.sbg.ac.at
[2] Department of Pediatrics, St. Anna Children's Hospital,
Medical University Vienna, Vienna, Austria

Abstract. Up to now, for most endoscopical computer aided celiac disease diagnosis approaches, image regions showing discriminative features have to be manually extracted by the physicians, prior to their automatized classification. This is obligatory to get idealistic and reliable data which is free from strong image degradations. On the one hand such a human interaction during endoscopy is subjective, expensive and tedious, but on the other hand state-of-the-art fully automatized selection corresponds to decreased classification accuracies compared to experienced human experts. In this work, a fully automatized approach is introduced which exploits the availability of a significant number of subimages within one original endoscopic image. A weighted decision-level and a weighted feature-level fusion method are introduced and investigated with respect to the achieved classification accuracies. The outcomes are compared with simple decision-level and feature-level fusion methods and the manual and the automatized patch selection. Finally, we show that the proposed feature-level fusion method outperforms all other automatized methods and comes close to manual patch selection.

1 Introduction

Celiac disease [15] is a disorder affecting the small bowel. After introduction of gluten containing food, the disease leads to an inflammatory reaction in the mucosa of the small intestine caused by a dysregulated immune response triggered by ingested gluten proteins of certain cereals. During the course the disease, the mucosa loses its absorptive villi completely and hyperplasia of the enteric crypts occurs, leading to a strongly diminished ability to absorb nutrients. The overall prevalence [5] of the disease in the USA is about 1:133. Figure 1 shows example images, captured during endoscopy.

For most computer aided celiac disease diagnosis approaches [3,4,11,21], reliable image regions (e.g. patches with a size of 128×128 pixels) showing discriminative features have to be identified prior to the automatized classification. This must be done to get idealistic data which is free from strong image degradations as in case of strong degradations the classification accuracy of the decision support system decreases [9,10]. The identification of reliable regions could be done

© Springer International Publishing Switzerland 2014
X. Jiang et al. (Eds.): GCPR 2014, LNCS 8753, pp. 666–677, 2014.
DOI: 10.1007/978-3-319-11752-2_55

manually [11,21] on the one hand or by means of a computer based method [9] on the other hand. Other approach for detection of "informative" frames [1,2], do not directly focus on a succeeding computer aided diagnosis and are certainly not optimized for celiac disease diagnosis. Although the manual method seems to be beneficial if done by experienced medical doctors [9], there are two incentives to use a computer based selection method: Firstly, a human interaction during endoscopy is time consuming and tedious, which probably leads to a diminished acceptance of the physicians. Apart from that, especially in case of physicians which are inexperienced, inattentive or just unfamiliar with the (new) decision support system, a weak selection automatically leads to weak classification accuracies [9].

The reason for the decreased classification accuracies in case of randomly or weakly selected patches (or if using the complete images) is the vulnerability of image classification methods to various types of degradations which are prevalent in endoscopic images [10]. It could be shown that image degradations definitely affect the feature extraction and consequently lead to a reduced classification accuracy. Such degradations are blur, noise, a lack of contrast and reflections caused by the light of the endoscope.

In this work we exploit the availability of numerous small subimages (patches) in each original endoscopic image. This availability of large data not just allows to select the best patch per image, as done in previous work [9], but also facilitates a redundant processing (i.e. feature extraction and classification) of these multiple images aiming at improving the overall accuracy. In order to generate one final decision for each image, these redundant threads have to be fused. This can be done on varying levels [19], such as feature-, score- or decision-level. Whereas in celiac disease diagnosis, information fusion has not been investigated so far, in biometric systems considerable improvements with these techniques could be obtained by considering multiple modalities [19], multiple instances of one modality [20] or multiple processing techniques [17]. As the simple fusion methods do not lead to improved accuracies, we utilize patch quality measures [9] to introduce a weighting. Based on this weighting, we propose a weighted decision-level as well as a weighted feature-level fusion method. The training set in each scenario investigated in this work consists of manually extracted idealistic patches. It should be mentioned that this manual stage can be done beforehand

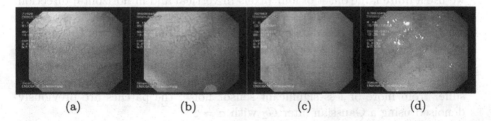

(a)   (b)   (c)   (d)

Fig. 1. Endoscopic images of healthy mucosa (1a and 1b) clearly showing the villi structure and images of diseased mucosa (1c and 1d).

668   M. Gadermayr et al.

by experts and does not require any interaction during medical treatment. The ground-truth, which has been determined by histological examination of biopsies, is available for each original image and can be directly taken for all patches extracted from the respective image.

The paper is organized as follows: In Sect. 2 the two fusion approaches are introduced and the quality measures are outlined. In Sect. 3 the experimental results are analyzed and discussed. Finally, Sect. 4 concludes this paper.

2   Fusion Methods

Before introducing the two weighted fusion methods, sensible quality measures have to be defined. In the following we use the same metrics as utilized in previous work [9] to identify one single "best" patch, as these methods seem to be appropriate for our problem definition.

2.1   Quality Measures

- The first measure addresses the problem of a too low illumination. As such a weak illumination generally corresponds to images with a low average gray value, we propose a quality measure being based on the mean of the pixel intensities

$$q_A(P) = \frac{1}{|Z|} \cdot \sum_{z \in Z} P(z) \,, \tag{1}$$

where Z comprises the coordinates of the image patch P.
- The next measure is utilized to detect image regions lacking from any significant gray value differences. Such image patches can be identified by measuring the contrast which is defined by

$$q_C(P) = \sum_{i,j \in K} |i - j| \cdot p(i,j) \,, \tag{2}$$

where K comprises all gray values in P and $p(i,j)$ stands for the probability of these two gray values to be present in a certain image neighborhood in P. In order to focus on real contrast rather than on noise, for this neighborhood we use a quite large offset of four pixels in vertical and in horizontal direction and average these two values.
- The next measure is based on a blur metric b [14]. For computing this metric, first in one direction the edges are identified by extracting all local minima and maxima. Finally the ratio between the lengths and the pixel differences of the edges is computed which indicates the blur level. As all of our images suffer from more or less significant sensor noise, the patches are previously denoised using a Gaussian filter G_2 with $\sigma = 2$.

$$q_B(P) = -b(P * G_2) \,. \tag{3}$$

- To detect noisy image patches, we sum up the differences between the original image and a denoised version of the same image

$$q_N(P) = \sum_{z \in Z} |P - G_1 * P| . \tag{4}$$

The denoised image is achieved by filtering the original image with a Gaussian G_1 with $\sigma = 1$.
- Finally, we need a measure to address the problem of reflections and extremely high illuminations. These regions can be detected quite easily by considering the maximum gray values.

$$q_I(P) = \begin{cases} 1, & \text{if } max(P) < T \\ 0, & \text{otherwise .} \end{cases} \tag{5}$$

T is set to 245 (eight bit gray scale), which turned out to be appropriate for separating extremely bright regions (by manual inspection of a set of training images).

   As shown in [9], one single quality measure is unable to represent the "quality" of a patch with respect to the classification performance. Therefore, we do not focus on single measures but instead utilize a weighted combination of the proposed quality measures.

2.2   Weighted Fusion

Let Q be a matrix containing the row vectors $(q_A, q_C, q_B, q_N, q_I)$ of each patch of one original image and let W be a properly chosen column vector containing a weight for each quality measure. Then the column vector $Q \cdot W$ is the weighted summed overall quality measure as used in the original work on patch selection [9]. In this previous work, the row with the maximum value of this product is evaluated and the corresponding image is used for feature extraction and classification. Classification in this context refers to the discrimination between images showing healthy and diseased mucosa. In the current work, by computing the element-wise exponentiation of $Q \cdot W$ with the properly chosen exponent k, the ratio between the impact of high and low quality patches can be adjusted ($^\circ$ denotes the element-wise matrix exponentiation which corresponds to the repeated Hadamard matrix product). In case of setting k to zero, the quality measures and the weights are ignored and each image finally has the same impact. The thereby achieved fusion methods (unweighted decision-level (DLF) and unweighted feature-level fusion (FLF)) are compared with the weight based methods in the experimental section. If assigning a large value to k, the methods converge to the patch selection strategy as small values are thereby suppressed. In the following two subsections we show how the quality vector $(Q \cdot W)^{\circ k}$ can be used in patch fusion. For the experiments, W and k are evaluated during exhaustive search based on a separate data set.

2.3   Weighted Decision-Level Fusion (W-DLF)

The first method based on the computed quality vector $(Q \cdot W)^{\circ k}$ operates on the decision level. That is, for each patch in an original image, first the classifier's decision is computed by means of traditional feature extraction and classification. All decisions for one original image are stored in the row vector D, where 1 stands for a positive and -1 stands for a negative decision. By computing

$$D_f = sgn(D \cdot (Q \cdot W)^{\circ k}),\tag{6}$$

the single decisions are multiplied with the corresponding weights (image qualities), summed up and finally thresholded using the sign function sgn. We have to content with the rather simple sum rule, as more elaborate decision-level fusion approaches like the Behavior-Knowledge Space [18] or Decision Templates [12] are developed for fusing different classifiers and not different input data.

2.4   Weighted Feature-Level Fusion (W-FLF)

In opposite to W-DLF, W-FLF operates on the feature level. This implies that the features are fused prior to the classification step. In this approach the classification step that corresponds to a loss of information is postponed and applied to the fused features, which could be a benefit compared to the simpler decision-level fusion. The fused feature vector F_f which is used for classification is calculated by

$$F_f = F \cdot \frac{(Q \cdot W)^{\circ k}}{||(Q \cdot W)^{\circ k}||},\tag{7}$$

where F is a matrix containing the feature vectors (columns) for each patch. The quality vector $(Q \cdot W)^{\circ k}$ is normalized to ensure that the sum of all contributions is one. The column vector F_f contains the element-wise weighted sum of all feature vectors and can be directly given to the classifier. F_f could be interpreted as a weighted average feature vector. We pursue this strategy, as it intuitively allows a weighting of the individual features, which cannot be achieved easily in case of a feature concatenation. The averaging theoretically requires that the decision boundaries are linear as otherwise the averaging of two features of one class could lead to an averaged descriptor located in the subspace of the other class. However, in the experiments we do not restrict to linear classification. To investigate the impact of the decision boundary on our approach, the utilized features are individually analyzed with respect to this problem in Sect. 3 with variable classifier adjustments.

2.5   Runtime Analysis

The major steps, as far as computational effort is concerned, consist of quality measurement (consisting of five single measures) and feature extraction. Whereas in the fused approach the quality measures as well as the features must be computed for each patch, in case of patch selection [9] the feature must be computed only for the best patch. The overall computation time[1] for all quality measures on 128×128 pixel gray value patches is 37 milliseconds (ms) (q_A: 1 ms, q_C: 16 ms, q_B: 1 ms, q_N: 1 ms, q_I: 18 ms). The computation time for the features ranges from 6 to 142 ms (6 ms (LBP), 6 ms (ELBP), 13 ms (SCH), 142 ms (MFS), 2 ms (FPS)). For example in case of fusion based classification with LBP or ELBP and extracting 16 patches per original image, for each original image the computation time would be about 688 ms where 592 of them are consumed for quality measurement and only 96 are used for feature extraction. In case of patch selection based classification, it would take $598 \, (592 + 6)$ ms which is not significantly faster. Thus, we claim that the small additional computational effort is justified if the fusion leads to increased accuracies.

3   Experiments

3.1   Experimental Setup

The image test set used contains images of the duodenal bulb and the pars descendens taken during duodenoscopies at the St. Anna Children's hospital using pediatric gastroscopes (with a resolution of 768×576 (Olympus GIF Q165) and 528×522 pixels (GIF N180), respectively).

To generate the ground-truth, the condition of the mucosal areas covered by the images was determined by histological examination of biopsies from the corresponding regions. Severity of villous atrophy was classified according to the modified Marsh classification as proposed in [15]. Although it is possible to distinguish between the several stages of the disease, we only aim in distinguishing between images of patients with (Marsh-3) and without the disease (Marsh-0), because this 2-classes case is more relevant in practice [21]. Another incentive for preferring the 2-classes case is that the distinction between the different stages of the disease is considerably subjective even as far as the histological examination is concerned [22]. Thereby, the ground-truth and furthermore the evaluation in a multi-classes case would be less reliable.

Our experiments are based on a balanced database containing 612 idealistic patches (i.e. patches 306 per class) which are used for classifier training and 172 original images that are used for evaluation. From each original image, 16 non-overlapping 128×128 pixel patches are automatically extracted and furthermore used for fused classification. The patch size is chosen in order to be able to compare the results with the manual extraction that is done by a highly experienced endoscopist. The original images are captured during endoscopies from 72 different patients. To allow an efficient parameter estimation, this database (consisting of 612 idealistic and 172 original images) is divided into two equally sized sets (DB 1 and DB 2). Each of them contains 306 idealistic images (patches) which are used for training as well as 86 original images. In case of multiple images of

---

[1] Runtime tests are executed on an Intel i5 architecture with 3.1 MHz. All functions are implemented in MATLAB 2013a.

one patient, we had to ensure that they end up in the same set. The weight vector W as well as the exponent k are evaluated during exhaustive search, based on the opposing data set as follows: In order to evaluate the accuracies based on the original images of DB 1, the idealistic images of DB 2 are utilized for training and the original images of DB 2 are used for parameter estimation. The same procedure is applied (vice versa) to evaluate DB 2. Thereby a strict separation between training set, test set and evaluation set is achieved. The search space for each element of W is between 0.0 and 1.0 with a step-size of 0.33 and k is within $\{2^{-1}, 2^0, 2^1, \ldots, 2^6, 2^7\}$. The parameters of the quality measures are taken from the previous work on patch selection [9] as described in Sect. 2.

We perform two different experiments. Experiment A corresponds to the natural fusion of patches extracted from one distinct original image. Experiment B should show if the accuracy improvements are limited by the correlations within one original image. Such correlations are quite natural, as degradations like blur or noise often do not occur only in a small region, but sometimes even compromise a whole image. Therefore, in this experiment the patches of each patient are randomly interchanged across the images leading to virtual images consisting of patches from the same patient, but from different original endoscopic images. This is done as the patches from the new virtual images are supposed to be less correlated and the used database does not contain enough patients to fuse all patches from one patient.

For classification the k-nearest neighbor classifier is used. We utilize this simple classifier in order to focus on the effect of different settings rather than on achieving the highest overall classification rates. For the first experiments (A and B), the rates achieved with odd k values reaching from 1 to 31 are averaged, to get highly stable results rather than to get the highest possible rates. In an analysis (Fig. 4) we investigate the impact of different k values.

3.2   Feature Extraction Techniques

For the experimental analysis we deploy the following feature extraction techniques which proved to the adequate for celiac disease classification in previous work [6]:

- Local Binary Patterns [16] (LBP):
  The commonly used Local Binary Patterns describe a texture by computing the joint distribution of binarized intensity differences within a certain neighborhood. This widely used feature extraction technique is used with eight neighbors and a radius (i.e. the distance to the neighbors) of two pixels.
- Extended Local Binary Patterns [13] (ELBP):
  ELBP is an edge based derivative of Local Binary Patterns. As LBP it is used with eight neighbors and a radius of two pixels.
- Fourier Power Spectra Rings [8] (FPSR):
  To get this descriptor, first the Fourier power spectra of the image patches are computed, in a way that the low frequencies are in the image center. Afterwards, a ring with a fixed inner and outer radius is extracted and the

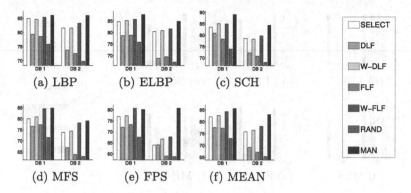

(a) LBP   (b) ELBP   (c) SCH

(d) MFS   (e) FPS   (f) MEAN

Fig. 2. Experiment A: these plots show the overall classification accuracies (in percent) achieved with patch selection (SELECT), decision-level fusion (DLF and W-DLF), feature-level fusion (FLF and W-FLF), a random patch selection (RAND) and the manual patch selection (MAN).

median of the values in this ring are calculated. For our experiments we use an inner radius of seven and an outer radius of eight pixels, which turned out to be suitable in previous work [8].
– Shape Curvature Histogram [7] (SCH):
  SCH is a shape feature, especially developed for celiac disease diagnosis. After detection of significant locations, a histogram collects the occurrences of the contour curvature values in these regions. As in the original work, we consider a histogram bin count of eight.
– Multi-Fractal Spectrum [23] (MFS):
  The local fractal dimension is computed for each pixel using three different types of measures for computing the local density. The feature vector is built by concatenation of these fractal dimensions.

3.3 Results and Discussion

In Fig. 2 the overall classification accuracies achieved with patch selection (SELECT), unweighted and weighted decision-level fusion (DLF and W-DLF), unweighted and weighted feature-level fusion (FLF and W-FLF), a random patch selection (RAND) and the patch selection based on the human experts (MAN) are shown for experiment A. It can be seen that the unweighted feature-level fusion method FLF as well as the unweighted decision-level method DLF are unable to compete with the single patch selection in case of any feature. The rates obtained with the manual selection are totally out of reach. However, these methods, which are not based on any (obviously highly important) quality measure, are at least able to outreach the random selection. Considering the weight based methods we recognize that especially the weighted feature-level based method W-FLF is able to outperform the single patch selection method SELECT in case

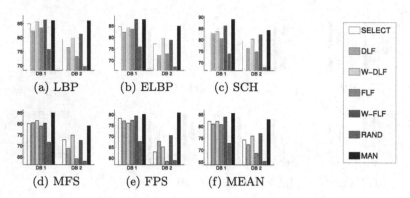

Fig. 3. Experiment B: These plots show the accuracies (in percent) with the same strategies as in Fig. 2. In opposite to experiment A, the patches of one patient are randomly interchanged.

of all features and all databases with differing extent. Considering MFS, LBP and ELBP, the accuracies of the manual patch selection can be virtually reached. Quite high differences are observed in case of SCH which does not strongly benefit from fusion. A quite interesting aspect is the difference between the two weighted fusion techniques. In almost each case, the feature based W-FLF corresponds to the higher accuracy compared to W-DLF. Obviously the early fusion prior to the (information reducing) classification has a positive impact on the final discriminative power.

In Fig. 3 the results of experiment B, which is based on randomly interchanged patches across images of the same patient, are shown. As in this experiment not only patches extracted from the same image are fused, but patches from different images, we expected that in this scenario more significant improvement could be obtained in case of information fusion. Actually, on average (see Fig. 3f) the rates with the weight based fusion methods are similar. Interestingly, it can be observed that the expected accuracy increase holds off in case of the methods DLF and FLF, not being based on weighting. However, as they are unable to outperform the weight based ones, this observation has no practical relevance.

As especially the weight based feature-level fusion method in both experiments leads to increased accuracies, we expect that a fusion on the one hand across all patches in an image (derived from experiment A) and on the other hand across all images, captured during endoscopy of one distinct patient (derived from experiment B) could improve the rates from our experiments once again. Unfortunately, the data currently available is not large enough for such an experiment.

So far, we experimentally showed that the W-FLF approach is able to improve the classification accuracies of state-of-the-art patch selection, without regarding the theoretical issues in case of non-linear decision boundaries. Finally we investigate the impact of the classifier's decision boundary on the effectiveness of W-FLF. As stated in Sect. 2.4, the feature averaging theoretically requires that the decision boundaries are linear as otherwise the averaging of two features of one class could lead to an averaged descriptor located in the subspace of the other class. To investigate how often an averaged feature of two correctly classified images would be incorrectly classified, now we consider all correctly classified images (from the idealistic patches data set). For each pair of these images, the average feature is computed and classified with varying settings (different k values). This is done as especially small k values correspond to highly non-linear decision boundaries, whereas with higher k values this effect is softened. Figure 4a shows that especially in combination with low dimensional features (FPS, SCH, MFS) and small k values (majorly for k = 1), the feature averaging leads to decreased classification accuracies (as 100 % accuracy is expected in case of linear classification).

In Fig. 4b, the impact of a small k value (k = 1) compared to the averaging (with k reaching from 3 to 31) on the improvement achieved with W-FWF compared to patch selection is shown. Apart from the classifier settings, the same setup as in experiment A is used. As expected, if k is set to one, the improvements of W-FLF are (especially in combination with the low dimensional features) considerably smaller or even negative, which is expected to be due to the highly non-linear decision boundaries. Therefore, we recommend to take care about the classifier choice using the proposed method for weighted feature-level fusion.

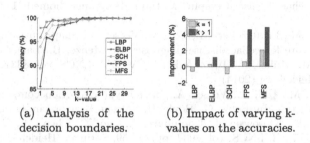

(a) Analysis of the decision boundaries.

(b) Impact of varying k-values on the accuracies.

Fig. 4. Analysis of effects in weighted feature-level fusion.

4   Conclusion

We have shown that information fusion in celiac disease classification is able to increase the classification accuracies with small additional computation costs compared to single patch selection. Whereas the simple decision-level and feature-level methods are not able to improve the performance, the introduced weight-based methods definitely are. Especially the feature-level fusion approach turned out to be most appropriate on average. Thereby, this paper showed that the measurement of image quality has a major impact not only in case of a single patch selection, but also in case of information fusion. Getting nearer to the classification rates of manual patch selection, this work brings us one step closer to fully automatized non-interactive celiac disease diagnosis.

References

1. Atasoy, S., Mateus, D., Lallemand, J., Meining, A., Yang, G.-Z., Navab, N.: Endoscopic video manifolds. In: Jiang, T., Navab, N., Pluim, J.P.W., Viergever, M.A. (eds.) MICCAI 2010, Part II. LNCS, vol. 6362, pp. 437–445. Springer, Heidelberg (2010)
2. Bashar, M., Kitasaka, T., Suenaga, Y., Mekada, Y., Mori, K.: Automatic detection of informative frames from wireless capsule endoscopy images. Med. Image Anal. **14**(3), 449–470 (2010)
3. Ciaccio, E.J., Tennyson, C.A., Bhagat, G., Lewis, S.K., Green, P.H.R.: Classification of videocapsule endoscopy image patterns: comparative analysis between patients with celiac disease and normal individuals. BioMed. Eng. Online **9**(1), 1–12 (2010)
4. Ciaccio, E.J., Tennyson, C.A., Lewis, S.K., Krishnareddy, S., Bhagat, G., Green, P.: Distinguishing patients with celiac disease by quantitative analysis of videocapsule endoscopy images. Comput. Methods Programs Biomed. **100**(1), 39–48 (2010)
5. Fasano, A., Berti, I., Gerarduzzi, T., Not, T., Colletti, R.B., Drago, S., Elitsur, Y., Green, P.H.R., Guandalini, S., Hill, I.D., Pietzak, M., Ventura, A., Thorpe, M., Kryszak, D., Fornaroli, F., Wasserman, S.S., Murray, J.A., Horvath, K.: Prevalence of celiac disease in at-risk and not-at-risk groups in the united states: a large multicenter study. Arch. Intern. Med. **163**, 286–292 (2003)
6. Gadermayr, M., Liedlgruber, M., Uhl, A., Vécsei, A.: Evaluation of different distortion correction methods and interpolation techniques for an automated classification of celiac disease. Comput. Methods Programs Biomed. **112**(3), 694–712 (2013)
7. Gadermayr, M., Liedlgruber, M., Uhl, A., Vécsei, A.: Shape curvature histogram: a shape feature for celiac disease diagnosis. In: Menze, B., Langs, G., Montillo, A., Kelm, M., Müller, H., Tu, Z. (eds.) MCV 2013. LNCS, vol. 8331, pp. 175–184. Springer, Heidelberg (2014)
8. Gadermayr, M., Uhl, A., Vécsei, A.: Barrel-type distortion compensated fourier feature extraction. In: Bebis, G., Boyle, R., Parvin, B., Koracin, D., Li, B., Porikli, F., Zordan, V., Klosowski, J., Coquillart, S., Luo, X., Chen, M., Gotz, D. (eds.) ISVC 2013, Part I. LNCS, vol. 8033, pp. 50–59. Springer, Heidelberg (2013)
9. Gadermayr, M., Uhl, A., Vécsei, A.: Getting one step closer to fully automatized celiac disease diagnosis. In: IEEE International Conference on Image Processing Theory, Tools and Applications 2014 (IPTA'14), October 2014
10. Hegenbart, S., Uhl, A., Vécsei, A.: Impact of endoscopic image degradations on LBP based features using one-class SVM for classification of celiac disease. In: Proceedings of the 7th International Symposium on Image and Signal Processing and Analysis (ISPA'11), Dubrovnik, Croatia, pp. 715–720, September 2011
11. Hegenbart, S., Uhl, A., Vécsei, A.: Impact of histogram subset selection on classification using multiscale LBP. In: Handels, H., Ehrhardt, J., Deserno, T.M., Meinzer, H.-P., Tolxdorff, T. (eds.) BVM'11. Informatik Aktuell, pp. 359–363. Springer, Heidelberg (2011)
12. Kuncheva, L.I., Bezdek, J.C., Duin, R.P.: Decision templates for multiple classifier fusion: an experimental comparison. Pattern Recogn. **34**(2), 299–314 (2001)
13. Liao, S.C., Zhu, X.X., Lei, Z., Zhang, L., Li, S.Z.: Learning multi-scale block local binary patterns for face recognition. In: Lee, S.-W., Li, S.Z. (eds.) ICB 2007. LNCS, vol. 4642, pp. 828–837. Springer, Heidelberg (2007)

14. Marziliano, P., Dufaux, F., Winkler, S., Ebrahimi, T., Sa, G.: A no-reference perceptual blur metric. In: IEEE International Conference on Image Processing (ICIP'02), pp. 57–60 (2002)
15. Oberhuber, G., Granditsch, G., Vogelsang, H.: The histopathology of coeliac disease: time for a standardized report scheme for pathologists. Eur. J. Gastroenterol. Hepatol. **11**, 1185–1194 (1999)
16. Ojala, T., Pietikäinen, M., Harwood, D.: A comparative study of texture measures with classification based on feature distributions. Pattern Recogn. **29**(1), 51–59 (1996)
17. Prabhakar, S., Jain, A.K.: Decision-level fusion in fingerprint verification. Pattern Recogn. **35**(4), 861–874 (2002)
18. Raudys, Š., Roli, F.: The behavior knowledge space fusion method: analysis of generalization error and strategies for performance improvement. In: Windeatt, T., Roli, F. (eds.) MCS 2003. LNCS, vol. 2709, pp. 55–64. Springer, Heidelberg (2003)
19. Ross, A., Jain, A.: Information fusion in biometrics. Pattern Recogn. Lett. **24**(13), 2115–2125 (2003)
20. Uhl, A., Wild, P.: Single-sensor multi-instance fingerprint and eigenfinger recognition using (weighted) score combination methods. Int. J. Biometrics **1**(4), 442–462 (2009). Special Issue on Multimodal Biometric and Biometric Fusion
21. Vécsei, A., Amann, G., Hegenbart, S., Liedlgruber, M., Uhl, A.: Automated marsh-like classification of celiac disease in children using an optimized local texture operator. Comput. Biol. Med. **41**(6), 313–325 (2011)
22. Weile, B., Hansen, B.F., Hägerstrand, I., Hansen, J.P.H., Krasilnikoff, P.A.: Inter-observer variation in diagnosing coeliac disease, a joint study by danish and swedish pathologists. APMIS **108**(5), 380–384 (2000)
23. Xu, Y., Ji, H., Fermüller, C.: Viewpoint invariant texture description using fractal analysis. Int. J. Comput. Vis. **83**(1), 85–100 (2009)

Fine-Grained Activity Recognition
with Holistic and Pose Based Features

Leonid Pishchulin[1]([✉]), Mykhaylo Andriluka[1,2], and Bernt Schiele[1]

[1] Max Planck Institute for Informatics, Saarbrücken, Germany
leonid@mpi-inf.mpg.de
[2] Stanford University, Stanford, USA

Abstract. Holistic methods based on dense trajectories [29,30] are currently the de facto standard for recognition of human activities in video. Whether holistic representations will sustain or will be superseded by higher level video encoding in terms of body pose and motion is the subject of an ongoing debate [12]. In this paper we aim to clarify the underlying factors responsible for good performance of holistic and pose-based representations. To that end we build on our recent dataset [2] leveraging the existing taxonomy of human activities. This dataset includes 24,920 video snippets covering 410 human activities in total. Our analysis reveals that holistic and pose-based methods are highly complementary, and their performance varies significantly depending on the activity. We find that holistic methods are mostly affected by the number and speed of trajectories, whereas pose-based methods are mostly influenced by viewpoint of the person. We observe striking performance differences across activities: for certain activities results with pose-based features are more than twice as accurate compared to holistic features, and vice versa. The best performing approach in our comparison is based on the combination of holistic and pose-based approaches, which again underlines their complementarity.

1 Introduction

In this paper we consider the task of human activity recognition in realistic videos such as feature movies or videos from YouTube. We specifically focus on how to represent activities for the purpose of recognition. Various representations were proposed in the literature, ranging from low level encoding using point trajectories [29,30] to higher level representations using body pose trajectories [12] and collection of action detectors [23]. At the high level human activities can often be accurately characterized in terms of body pose, motion, and interaction with scene objects. Representations based on such high level attributes are appealing as they allow to abstract the recognition process from nuisances such as camera viewpoint or person clothing, and facilitate sharing of training data across activities. However, articulated pose estimation is a challenging and non-trivial task that is subject of ongoing research [8,18,19,24,32]. Therefore, state-of-the-art methods in activity recognition rely on holistic representations [9,15,29,30]

© Springer International Publishing Switzerland 2014
X. Jiang et al. (Eds.): GCPR 2014, LNCS 8753, pp. 678–689, 2014.
DOI: 10.1007/978-3-319-11752-2_56

that extract appearance and motion features from the entire video and leverage discriminative learning techniques to identify information relevant for the task.

Recent results on the JHMDB dataset [12] suggest that state-of-the-art pose estimation methods might have reached sufficient accuracy to be effective for activity recognition. Motivated by these results, we further explore holistic and pose based representations aiming for much broader scale and coverage of activity classes. To that end we employ our recent "MPI Human Pose" dataset [2]. Compared to 21 activity classes considered in [12] the "MPI Human Pose" dataset includes 410 activities and more than an order of magnitude more images ($\sim$32 K in JHMDB vs. over 1M images in "MPI Human Pose"). "MPI Human Pose" aims to systematically cover a range of activities using an existing taxonomy [1]. This is in contrast to existing datasets [13, 26] that typically include ad-hoc selections of activity classes. Using the rich labelling of people provided with "MPI Human Pose" we evaluate the robustness of holistic and pose based representations to factors such as body pose, viewpoint, and body-part occlusion, as well as to the number and speed of dense trajectories covering the person.

This paper makes the following contributions. First, we perform a large-scale comparison of holistic and pose based features on the "MPI Human Pose" dataset. Our results complement the findings in [12], indicating that pose based features indeed outperform holistic features for certain cases. However we also find that both types of features are complementary and their combination performs best. Second, we analyze factors responsible for success and failure, including number and speed of trajectories, occlusion, viewpoint and pose complexity.

Related work. There is a large body of work on human activity recognition and its review is out of the scope of this paper. Instead we point to the respective surveys [4, 28] and concentrate on the methods most relevant for this work.

Holistic appearance based features combined with the Bag-of-Words representation [9, 12, 15, 29, 30] are considered the de facto standard for human activity recognition in video. Many methods create discriminative feature representations of a video by first detecting spatiotemporal interest points [5, 14] or sampling them densely [31] and then extracting various feature descriptors in the space-time volume. Most commonly used feature descriptors are histograms of oriented gradients (HOG) [6], histograms of flow (HOF) [7] or Harris 3D interest points [14]. In this paper we examine the recent holistic approach [29, 30] which tracks dense feature points and extracts strong appearance based features along the trajectories. This method achieves state-of-the-art results on several datasets. Other holistic approaches include template based [20] or graph based methods constructing graphs from a spatiotemporal segmentation of the video [3].

Another line of research explores ways of higher level video encoding in terms of body pose and motion [11, 12, 21, 25]. The intuition there is that many activities exhibit characteristic body motions and thus can reliably be described using human body pose based features. Pose based activity recognition was shown to work particularly well in images with little clutter and fully visible people [25]. However, in more challenging scenarios with frequent occlusions, truncations and complex poses, body features significantly under-perform holistic appearance based representations [21]. Recently, it was shown that body features extracted from detected joint positions outperform holistic methods, and their combination did not improve over using body features only [12]. However, these conclusions were made on a subset of the HMDB dataset [13], where actions with global body motion are performed by isolated and fully visible individuals – a setting that seems well suited for pose estimation methods. In this work we examine a wide range of underlying factors responsible for good performance of body based and holistic methods. In contrast to [12] our analysis on hundreds of activity classes reveals that holistic and pose based methods are highly complementary, and their performance varies significantly depending on the activity.

We build our analysis on our recent "MPI Human Pose" dataset collected by leveraging an existing taxonomy of every day human activities and thus aiming for a fair coverage. This is in contrast to existing activity recognition datasets [13,16,17,20,26] that typically include ad-hoc selections of activity classes. A large number of activity classes (410) and more than an order of magnitude more images compared to [12] facilitate less biased evaluations and conclusions.

2   Dataset

In order to analyze the challenges for fine-grained human activity recognition, we build on our recent publicly available "MPI Human Pose" dataset [2]. The dataset was collected from YouTube videos using an established two-level hierarchy of over 800 every day human activities. The activities at the first level of the hierarchy correspond to thematic categories, such as "Home repair", "Occupation", "Music playing", etc., while the activities at the second level correspond to individual activities, e.g. "Painting inside the house", "Hairstylist" and "Playing woodwind". In total the dataset contains 20 categories and 410 individual activities covering a wider variety of activities than other datasets, while its systematic data collection aims for a fair activity coverage. Overall the dataset contains 24, 920 video snippets and each snippet is at least 41 frames long. Altogether the dataset contains over a 1M frames. Each video snippet has a key frame containing at least one person with a sufficient portion of the body visible and annotated body joints. There are 40, 522 annotated people in total. In addition, for a subset of key frames richer labels are available, including full 3D torso and head orientation and occlusion labels for joints and body parts.

Static pose estimation complexity measures. In addition to the dataset, in [2] a set of quantitative complexity measures aiming to asses the difficulty of pose estimation in each particular image was proposed. These measures map body image annotations to a real value that relates the complexity of each image w.r.t. each factor. These complexity measures are listed below.

1. *Pose*: deviation from the mean pose on the entire dataset.
2. *Occlusion*: number of occluded body parts.
3. *Viewpoint*: deviation of 3D torso rotation from the frontal viewpoint.
4. *Part length*: deviation of body part lengths from the mean part lengths.
5. *Truncation*: number of truncated body parts.

Novel motion specific complexity measures. We augment the above set with the measures assessing the amount of motion present in the scene.

1. *# dense trajectories (# DT)*: total number of DT computed by [30].
2. *# dense trajectories on body (# DT body)*: number of DT on body mask.
3. *Motion speed (MS)*: mean over all trajectory displacements in the video.
4. *Motion speed on body (MS body)*: MS extracted on body mask.
5. *# people*: number of people in the video.

3   Methods

In order to analyze the performance on the challenging task of fine-grained human activity recognition, we explore two lines of methods that extract relevant features. The first line of methods extracts holistic appearance based features and is represented by the recent "Dense Trajectories" method [29] which achieves state-of-the-art performance on several datasets. The second line of methods computes features from locations of human body joints following the intuition that body part configurations and motion should provide strong cues for activity recognition. We now describe both types of methods and their combinations.

3.1   Dense Trajectories (DT)

DT computes histograms of oriented gradients (HOG) [6], histograms of flow (HOF) [15], and motion boundary histograms (MBH) [7] around densely sampled points that are tracked for 15 frames using median filtering in a dense optical flow field. In addition, x and y displacements in a trajectory are used as a fourth feature. We use a publicly available implementation of the improved DT method [30], where additional estimation removes some of the trajectories consistent with camera motion. Following [30] we extract all features on our data and generate a codebook for each of the four features of 4 K words using k-means from a million of sampled features, and stack L_2–normalized histograms for learning.

3.2   Pose-Based Methods

It has recently been shown that body features provide a strong signal for recognition of human activities on a rather limited set of 21 distinctive full body actions in monocular rgb video sequences [12]. We thus investigate the usefulness of body features for our task where the variability of poses and granularity of activities is much higher. We explore different ways of obtaining body joint locations and extract the body features using the code kindly provided by [12]. We use the same trajectory length of 7 frames with a step size of 3, generate a codebook of 20 words for each descriptor type and finally stack the L_2–normalized histograms for learning. We now present different ways of obtaining body joint locations.

GT single pose (GT). We directly use the ground truth locations of body joints in the key frame to compute single pose based features. As some of the body parts may be truncated, we compute features only for the present parts.

GT single pose + track (GT-T). As the ground truth information is not available for the rest of the frames in a sequence, we approximate the positions of body joints in the neighboring frames by tracking the joints using sift-based tracker [21]. The tracker is initialized with correct positions of body joints, and thus provides reliable tracks of joints in the local temporal neighborhood.

PS single pose + track (PS-T). It is not realistic to expect the ground truth information to be available at test time in real world scenarios. We thus replace the given body joint locations by automatically estimated ones using the publicly available implementation [32]. This efficient method is based on pictorial structures (PS) and obtained good performance on the "MPI Human Pose" [2].

PS multi-pose (PS-M). Using the method [32] also allows to obtain joint locations independently in each frame of a sequence without using the sift tracker. Notably, the same method was shown by [12] to outperform the holistic approach.

3.3   Combinations of Holistic and Body Based Methods

As the holistic DT approach does not extract any pose information, and pose based methods do not compute any appearance features, both are potentially complementary. Thus we expect that an activity recognition system will profit from their combinations. We investigate two ways of combining the methods.

PS-M + DT (features). We perform a *feature* level fusion of both DT and PS-M by matching both types of features independently to the respective codebooks and then stacking the normalized histograms into a single representation.

PS-M + DT (classifiers). We also investigate a *classifier* level fusion. To do so we first run pre-trained DT and PS-M classifiers (see Sect. 4) independently on each sequence and stack the scores together into a single feature vector.

PS-M filter DT. Another way of combining both types of methods is using estimated joint locations to filter the trajectories computed by DT. We first estimate poses in all video frames and generate a binary mask using the union of rectangles around detected body parts for all single top detections per frame. We then only preserve the trajectories overlapping with the mask in all frames.

4   Analysis of Activity Recognition Performance

In this section we analyze the performance of holistic and pose based methods and their combinations on the challenging task of fine-grained human activity recognition with hundreds of activity classes.

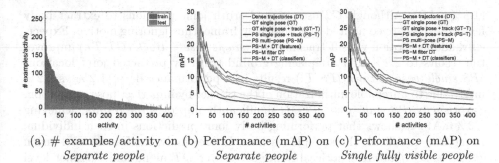

(a) # examples/activity on   (b) Performance (mAP) on   (c) Performance (mAP) on
    Separate people   *Separate people*   *Single fully visible people*

Fig. 1. Dataset statistics and performance (mAP) as a function training set size. Shown are (a) number of training/testing examples/activity in *Separate people* subset; performance on (b) *Separate people* and (c) *Single fully visible people*. Best viewed in color (Color figure online).

Data splits. As main test bed for our analysis, a split with videos containing sufficiently separated individuals [2] is used. This restriction is necessarily for using the pose estimation method [32]. This *Separate people* split contains 411 activities with 15244 training and 5699 testing video snippets. Figure 1(a) shows statistics of the training and testing videos per activity. Notably, the videos may still contain multiple people and some body parts may be truncated by a frame border. To rule out the confusion potentially caused by the presence of multiple truncated people, we define a subset of the test set from *Separate people*. This subset contains 2622 videos with exactly one fully visible person per video. This *Single fully visible people* setup is inspired by [12] and is favorable for the pose estimation method [32] designed to predict body joints of fully visible people.

Training and evaluation. We train activity classifiers using feature representations described in Sect. 3 and ground truth activity labels. In particular, we train one-vs-all SVMs using mean stochastic gradient descent (SGD) [22] with a χ^2 kernel approximation [27]. At test time we perform one-vs-all prediction per each class independently and report the results using mean Average Precision (AP) [10]. When evaluating on a subset, we always report the results on the top N activity classes arranged w.r.t training set sizes.

4.1 Overall Performance

We start the evaluation by analyzing the performance on all activity classes.

Separate people. It can be seen from Fig. 1(b) that performance is reasonable for a relatively small number of classes (the typical case for many activity recognition datasets), but quickly degrades for a large number of classes, clearly leaving room for improvement of activity recognition methods.

We observe that *Dense trajectories (DT)* alone outperforms all pose based methods achieving 5.1 % mAP. Expectedly, *GT single pose (GT)* performs worst

(1.8 % mAP). Although *GT* uses ground truth joint positions to extract body features, they are computed in a single key frame, thus ignoring motion. Expectedly, adding motion via sift tracking (*GT single pose + track (GT-T).*) improves the results to 2.2 % mAP. Replacing ground truth by predicted joint locations (*PS single pose + track (PS-T)*) results in a performance drop (1.2 % vs 2.2 % mAP) due to unreliable initialization of the tracker by imperfect pose estimation. Surprisingly, *PS multi-pose (PS-M)* significantly improves the results, achieving 4.2 % mAP. It shows that performing body joint predictions in each individual frame is more reliable than simple tracking. Interestingly, the feature level fusion *PS-M + DT (features)* noticeably improves over *DT* alone and classifier level fusion *PS-M + DT (classifiers)*, achieving 5.5 % mAP. This shows that both holistic *DT* and pose based *PS-M* methods are complementary. We analyze whether the complementarity of *DT* comes from the holistic features extracted on the person or elsewhere in the scene. By restricting the extraction to the body mask (*PS-M filter DT*), we observe a drop of performance w.r.t. *DT* (4.3 % mAP vs. 5.1 % mAP). It shows that the features extracted outside of the body mask do contain additional information which helps to better discriminate between activities in a fine-grained recognition setting. This intuition is additionally supported by the similar performance of *PS-M filter DT* w.r.t. *PS-M*. Overall, we conclude that holistic and pose based methods are complementary and should be used in a combination for better activity recognition.

Single fully visible people. We now analyze the results in Fig. 1(c). Although the absolute performances are higher, which is explained by an easier setting, the ranking is similar to Fig. 1(b). Two differences are: (1) *GT-T* achieves similar performance to *PS-M* on many activity classes, but looses in total (3.4 % mAP vs. 4.2 % mAP); and (2) *PS-M filter DT* is better than both *DT* and *PS-M* on a small set of classes, probably because the trajectory features on the background mostly contribute to confusion on this set of activities.

Differences to [12]. Our analysis in a fine-grained activity recognition setting on hundreds of classes leads to conclusions which go beyond the results of [12] obtained from much smaller number of classes from HMDB dataset [13]. First, we compare the performance of *DT* to a larger number of pose based methods and show the superior performance of *DT*, when evaluated on hundreds of activities. This is in contrast to [12] showing that the pose based *PS-M* is better. Second, we discover that holistic *DT* and pose based *PS-M* are complementary and their combination outperforms each of the approaches alone. This contradicts the conclusions of [12] which does not show any improvement when combining *DT* and *PS-M*. Finally, we showed that using the trajectories restricted to body only degrades the performance, which suggests that the context adds to the discrimination between activity classes.

4.2   Analysis of Activity Recognition Challenges

After analyzing the overall recognition performance on all classes, we explore which factors affect the performance of best performing holistic *DT*, pose based

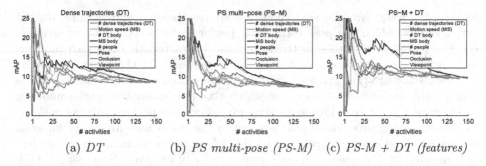

(a) *DT*   (b) *PS multi-pose (PS-M)*   (c) *PS-M + DT (features)*

Fig. 2. Performance (mAP) on a subset of 150 activities from *Separate people* as a function of the complexity measures. Best viewed in color and with additional zooming (Color figure online).

PS-M and combination *PS-M + DT (features)* of both methods. We use the complexity measures 1–3 specific for static pose estimation and our novel 1–5 motion specific complexity measures described in Sect. 2. To make the evaluation consistent with the rest of the experiments, we compute the average complexities for the whole activity class and use the obtained real values to rank the classes. This is in contrast to [2] which computes the measures per single pose and thus operates on individual instance level. To visualize the performance, we sort the activities using the pose related complexity measures in *increased* complexity order, and motion related complexity measures in the *decreased* order. As performance may still be dominated by the training set size when only few examples are available, we restrict the evaluation to the 150 largest activity classes. This corresponds to a slice at 150 in Fig. 1(b). The results are shown in Fig. 2.

Dense trajectories (DT). Analyzing the results in Fig. 2(a) we observe that a high number of dense trajectories everywhere in the video (*# DT*) and on human body (*# DT body*) leads to the best performance of the *DT* method. Also, we notice that high motion speed (*MS, MS body*) is an indicative factor for good recognition results. Surprisingly, *DT* performs better on activities with a high number of people (*# people*). This is explained by the fact that more people potentially produce more motion, which is a positive factor for *DT*. On the other hand, being close to the average pose (*Pose*) and having little occlusion (*Occlusion*) hurts performance. The former is not very surprising, as the average pose is common to many activities, which makes it more difficult for *DT* to capture distinctive features. We discover that activities with little occlusion often contain either little motion (e.g. "sitting, talking in person") or fine-grained motion (e.g. "wash dishes") and thus are hard for *DT*.

PS multi-pose (PS-M). We now analyze Fig. 2(b) and observe several distinctive differences w.r.t. which factors mostly affect the performance of *PS-M*. It can be seen that *MS* and *MS body* have stronger effect on *PS-M* compared to *DT*, and the higher the speed, the better the performance. In order to better understand this nontrivial trend, we analyze which activities happen to produce highest *MS body*. We note that those are sports, dancing and running related activities, for which the pose estimation performance of [32] is above average (cf. Fig. 7 in [2]). Also, these activities exhibit characteristic body part motions and can successfully be encoded using body features. At the other end of the *MS body* scale are the activities with low fine-grained motion, related to home repair, self care and occupation, for which the pose estimation performance is much worse. *Pose* and *Viewpoint* strongly affect the performance as well, as frontal upright people whose pose is close to the mean pose are easier for pose estimation. This is again in contrast to *DT*, where the performance is not noticeably affected by *Viewpoint* and even drops in case of low *Pose*. Surprisingly, high # *people* positively affects *PS-M*. Looking at top ranked activities, we notice that many of them are related to active group exercises or team sports, such as "aerobic" and "frisbee", or to simple standing postures, such as "standing, talking in person". Body features can again be successfully used to encode the corresponding motions. On the other hand, we observe that the high # *DT* and # *DT body* hurts performance, which is in contrast to the *DT* method. We observe that for high # *DT body* many activities correspond to water related activities, such as "fishing in stream", "swimming, general", "canoeing, kayaking". Interestingly, the presence of water leads to high # *DT* and characteristic motions captured by *DT*. At the same time *PS-M* fails due to unreliable pose estimation caused by complex appearance and occlusions.

PS-M + DT (features). The differences for *DT* and *PS-M* methods suggest that both methods are complementary. We analyze in Fig. 2(c) which factors affect the performance of *PS-M + DT (features)*. It can be seen that positively affecting factors are either positive for both *DT* and *PS-M* (*MS*, *MS body*, # *people*), or positive for *PS-M* only (*Pose*, *Viewpoint*). In contrast to *PS-M* the high # *DT* slightly improves the performance, while high # *DT body* does not hurt as much. Expectedly, *Viewpoint* hurts performance as it does for both *DT* and *PS-M*. This shows the complementarity of both *DT* and *PS-M* and leaves room for improvement in finding better ways of combining both methods.

4.3   Detailed Analysis on a Subset of Activities

After analyzing the factors affecting the results by different methods, we conduct a detailed analysis on a smaller set of the top 15 activities from *Separate people*.

The results are shown in Table 1. None of the methods outperforms all others on all activities and different approaches are better on different activities. On average methods perform well on activities with simple poses and motions e.g. "rope skipping", "skiing, downhill" and "golf" - typical cases in most of the current activity recognition benchmarks. However, the performance of all methods is low for activities with more variability in motion and poses, e.g. "cooking", "carpentry, general" and "forestry". This leaves room for improvement of current methods. Analyzing the performance on individual activities, we observe that for "yoga, power" activity *GT* outperforms holistic *DT* and *PS-M filter DT* methods (22.3 % vs. 10.6 % and 16.1 % mAP, respectively) and is better than the

Table 1. Activity recognition results (mAP) on 15 largest classes from *Separate people*.

	yoga, power	bicycl., mount.	skiing, downh.	cooking or food	skate-board.	rope skip.	softball, general	forestry
Dense trajectories (DT)	10.6	14.5	51.9	0.5	11.4	36.0	**12.7**	8.4
GT single pose (GT)	22.3	26.5	7.5	1.8	3.4	51.2	2.2	1.4
GT single pose + track (GT-T)	**37.0**	28.0	10.9	**2.6**	4.6	69.2	3.6	1.2
PS single pose + track (PS-T)	8.8	6.6	6.0	1.3	1.7	63.1	1.6	1.8
PS multi-pose (PS-M)	18.3	34.0	27.3	2.6	17.2	**90.5**	3.0	5.2
PS-M + DT (features)	19.6	**40.7**	32.9	2.2	**19.5**	88.7	3.9	7.2
PS-M filter DT	16.1	20.4	**52.2**	0.8	13.5	55.7	4.2	**10.6**
	carpentry, general	bicycl., racing	golf	rock climb.	ballet, modern	aerobic step	resist. train.	total
Dense trajectories (DT)	5.5	5.5	33.0	**41.5**	12.7	24.5	**16.5**	19.0
GT single pose (GT)	2.7	7.1	36.1	2.3	1.0	1.1	1.4	11.2
GT single pose + track (GT-T)	2.8	8.7	25.3	8.9	1.7	3.3	1.3	13.9
PS single pose + track (PS-T)	5.3	0.5	14.7	1.2	2.8	11.1	1.6	8.5
PS multi-pose (PS-M)	3.4	8.6	47.9	4.7	22.9	10.4	7.2	20.2
PS-M + DT (features)	5.0	12.1	**51.9**	14.4	**23.7**	17.1	14.4	**23.5**
PS-M filter DT	**6.1**	**15.5**	15.9	38.6	7.1	**25.8**	9.6	19.5

Fig. 3. Successful and failure cases on several activity classes. Shown are the most confident prediction per class. False positives are highlighted in red (Color figure online).

pose based *PS-M* (22.3 % vs. 18.3 % mAP). It is interesting, as *GT* does not use any motion and relies on static body features only. The explanation is that the "yoga, power" activity contains distinctive body poses and thus can be reliably captured by *GT*, while *PS-M* fails due to unreliable pose estimations. It can be seen that in many cases the combination *PS-M + DT (features)* noticeably outperforms both *PS-M* and *DT* alone. The differences are most pronounced for "bicycling, mountain", "bicycling, racing", "skateboarding" exhibiting characteristic motions, and "golf" activity having distinctive body motion and poses. Overall *PS-M + DT (features)* achieves the best performance of 23.5 % mAP. We visualize several successful and failure cases of the methods in Fig. 3.

5   Conclusion

In this work we address the challenging task of fine-grained human activity recognition on a recent comprehensive dataset with hundreds of activity classes. We study holistic and pose based representations and analyze the factors responsible for their performance. We reveal that holistic and pose based methods are complementary, and their performance varies significantly depending on the activity. We found that both methods are strongly affected by the speed of trajectories. While the holistic method is also strongly influenced by the number of trajectories, pose based methods are strongly affected by human pose and viewpoint. We observe striking performance differences across activities and experimentally show that the combination of both methods performs best.

Acknowledgements. The authors would like to thank Marcus Rohrbach and Sikandar Amin for helpful discussions. This work has been supported by the Max Planck Center for Visual Computing & Communication.

References

1. Ainsworth, B., Haskell, W., Herrmann, S., Meckes, N., Bassett, D., Tudor-Locke, C., Greer, J., Vezina, J., Whitt-Glover, M., Leon, A.: 2011 compendium of physical activities: a second update of codes and MET values. MSSE **43**(8), 1575–1581 (2011)
2. Andriluka, M., Pishchulin, L., Gehler, P., Schiele, B.: 2D human poseestimation: new benchmark and state of the art analysis. In: CVPR'14
3. Brendel, W., Todorovic, S.: Learning spatiotemporal graphs of human activities. In: ICCV'11
4. Cardinaux, F., Bhowmik, D., Abhayaratne, C., Hawley, M.S.: Video based technology for ambient assisted living: a review of the literature. J. Ambient Intell. Smart Environ. **3**(3), 253–269 (2011)
5. Chakraborty, B., Holte, M.B., Moeslund, T.B., Gonzalez, J., Xavier Roca, F.: A selective spatio-temporal interest point detector for human action recognition in complex scenes. In: ICCV'11
6. Dalal, N., Triggs, B.: Histograms of oriented gradients for human detection.In: CVPR'05

7. Dalal, N., Triggs, B., Schmid, C.: Human detection using oriented histograms of flow and appearance. In: ECCV'06
8. Dantone, M., Gall, J., Leistner, C., Gool., L.V.: Human pose estimation usingbody parts dependent joint regressors. In: CVPR'13
9. Duchenne, O., Laptev, I., Sivic, J., Bach, F., Ponce, J.: Automatic annotation of human actions in video. In: ICCV'09
10. Everingham, M., Van Gool, L., Williams, C.K.I., Winn, J., Zisserman, A.: The PASCAL Visual Object Classes Challenge 2007 (VOC2007) Results. http://www.pascal-network.org/challenges/VOC/voc2007/workshop/index.html
11. Ferrari, V., Marin, M., Zisserman, A.: Progressive search space reduction for human pose estimation. In: CVPR'08
12. Jhuang, H., Gall, J., Zuffi, S., Schmid, C., Black, M.J.: Towards understanding action recognition. In: ICCV'13
13. Kuehne, H., Jhuang, H., Garrote, E., Poggio, T., Serre, T.: HMDB: a large video database for human motion recognition. In: Proceedings of the International Conference on Computer Vision (ICCV) (2011)
14. Laptev, I.: On space-time interest points. IJCV 64(2/3), 107–123 (2005)
15. Laptev, I., Marszałek, M., Schmid, C., Rozenfeld, B.: Learning realistichuman actions from movies. In: CVPR'08
16. Liu, J., Luo, J., Shah, M.: Recognizing realistic actions from videos in thewild. In: CVPR'09
17. Marszałek, M., Laptev, I., Schmid, C.: Actions in context. In: CVPR'09
18. Pishchulin, L., Andriluka, M., Gehler, P., Schiele, B.: Poselet conditionedpictorial structures. In: CVPR'13
19. Pishchulin, L., Andriluka, M., Gehler, P., Schiele, B.: Strong appearance and expressive spatial models for human pose estimation. In: ICCV'13
20. Rodriguez, M.D., Ahmed, J., Shah, M.: Action mach: a spatio-temporal maximum average correlation height filter for action recognition. In: CVPR'08
21. Rohrbach, M., Amin, S., Andriluka, M., Schiele, B.: A database for fine grained activity detection of cooking activities. In: CVPR'12
22. Rohrbach, M., Stark, M., Schiele, B.: Evaluating knowledge transfer andzero shot learning in a large-scale setting. In: CVPR'11
23. Sadanand, S., J., C.J.: Action bank: a high-level representation of activity in video. In: ECCV'12
24. Sapp, B., Taskar, B.: Multimodal decomposable models for human pose estimation. In: CVPR'13
25. Singh, V.K., Nevatia, R.: Action recognition in cluttered dynamic scenes usingpose-specific part models. In: ICCV'11
26. Soomro, K., Zamir, A.R., Shah, M.: Ucf101: a dataset of 101 human action classes from videos in the wild. Technical report CRCV-TR-12-01, UCF (2012)
27. Vedaldi, A., Zisserman, A.: Efficient additive kernels via explicit feature maps. In: CVPR'10
28. Vishwakarma, S., Agrawal, A.: A survey on activity recognition and behavior understanding in video surveillance. VC 29(10), 983–1009 (2013)
29. Wang, H., Kläser, A., Schmid, C., Liu, C.L.: Dense trajectories and motion boundary descriptors for action recognition. IJCV 103(1), 60–79 (2013)
30. Wang, H., Schmid, C.: Action recognition with improved trajectories. In:ICCV'13
31. Wang, H., Ullah, M.M., Kläser, A., Laptev, I., Schmid, C.: Evaluation oflocal spatio-temporal features for action recognition. In: BMVC'09
32. Yang, Y., Ramanan, D.: Articulated human detection with flexible mixtures of parts. PAMI 61(1), 55–79 (2013)

Obtaining 2D Surface Characteristics from Specular Surfaces

Mathias Ziebarth[1]([✉]), Markus Vogelbacher[1],
Sabine Olawsky[1], and Jürgen Beyerer[2]

[1] Vision and Fusion Laboratory, Karlsruhe Institute of Technology,
Adenauerring 4, 76131 Karlsruhe, Germany
`mathias.ziebarth@kit.edu`
[2] Fraunhofer Institute of Optronics, System Technologies and Image Exploitation,
Fraunhoferstr. 1, 76131 Karlsruhe, Germany

Abstract. Today's surface appearance measures often ignore the inherent two-dimensionality. This paper proposes a method to acquire and assess the appearance of larger specular surfaces in 2D. First, we describe a deflectometric setup to obtain a gradient field of the surface microstructure. Hence, we propose an areal measure based on the angular power spectrum, as defined in ISO 25178, to characterize the waviness of coated surfaces in relevant scales. To verify the validity of this measure, we compare it with an 1D industry standard appearance measurement system (wave-scan). While our method shows the same characteristics when mapped to the wave-scan values, we observed differences between both systems. These are mainly caused by the different measurement principles and the resulting information of the surface.

1 Introduction

Surface coatings on automobiles, furniture or other objects have two functions: first, they seal the surface of the underlying material and second, they define the object's appearance. Especially on specular coatings, this appearance is highly defined by the surface roughness and waviness. To ensure a steady quality in a production process, the homogeneity of this appearance has to be monitored. These requirements are met when the whole surface area is acquired with sufficient accuracy and assessed using a descriptive measure. The relevant surface waviness on coated surfaces is called *orange peel* and typically extends between 0.1 mm and 10 mm in longitudinal direction. It is caused by an inappropriate leveling process during the drying of the sprayed paint and is influenced e.g. by the spraying technique, the paint condition and surrounding conditions [5,7,15].

After looking at related work in Sect. 2, we describe a set of coated metal sheets used for the comparison of our work with the wave-scan in Sect. 3. Then the two measurement principles appropriate to obtain the surface microstructure, deflectometry and the wave-scan, are introduced in Sects. 3.1 and 3.2. Specifically for the deflectometric measurement, we propose a measure for the surface

X. Jiang et al. (Eds.): GCPR 2014, LNCS 8753, pp. 690–700, 2014.
DOI: 10.1007/978-3-319-11752-2_57

appearance in Sect. 3.3. Finally, we compare our results with the results of the wave-scan in Sect. 4.

2   Related Work

The appearance of specular surfaces is caused by its waviness [18]. The waviness can be characterized by appearance or technical measures. Widely used appearance measures are loss of gloss (GLOSS, [4]), distortion of reflected images (DOI, [9,16]) and loss of sharpness (HAZE, [4]). These measures describe the clarity of the surface reflection and are based on the analysis of the spreading of a light point around the direction of the specular reflection [4,16]. An overview of light-scattering and structured light methods to characterize surfaces is given by Tian et al. [14]. A method measuring the 2D surface slope of $153 \times 153\,\text{mm}$ large specular tiles using a photo diode array as screen was proposed by Lu and Forrest [12]. Miranda-Medina et al. [6] use 2D-interferometry and ellipsometry to measure orange peel on very small parts ($1270 \times 950\,\mu\text{m}$) of highly polished surfaces. Chen et al. [2] used specular highlights on glossy and wavy surfaces to reconstruct the surface mesostructure for rendering purposes. Byk-Gardner developed an instrument (wave-scan, [1]) to describe the appearance of wavy surfaces by means of their point spread function, as described in Sect. 3.2. Pietschmann [11] gives a general overview for appearance measurements of coated surfaces. Some technical 3D-surface measures are defined in ISO 25178-2:2012 and are used by Rebbegiani et al. [13]. Osterhold [8] derived a measure similar to ours, but based on 1D-profile measurements. Fletcher [3] studied the correlation between surface roughness and gloss and found that gloss measured at $60°$ correlates with the slope of the roughness.

3   Surface Acquisition

We created a set of six $30 \times 3\,\text{cm}$ flat metal sheets with black basecoat and a clearcoat with 3 different degrees of gloss, ranging from low gloss (sheet A), high gloss (sheet B) to very high gloss (sheet C). For each degree of gloss there is one sheet with minimal waviness (A, B, C) and one sheet with a defined one-directional gradient of waviness (AO, BO, CO) as shown in Fig. 1. The gradient of the waviness ranges from the minimal waviness at the borders to a maximal waviness in the center of the sheet.

3.1   Deflectometry

Deflectometry allows us to obtain the surface slope of specular surfaces with high accuracy [10,17]. The measurement system (Fig. 2) consists of an AVT Prosilica-GX2750 camera I (pixel size $4.45\,\mu\text{m}$, resolution 2752×2200) and a white ceiling as screen L. A projector (pixel size $2.28\,\text{mm}$, resolution 1280×800) projects sinusoidal stripe patterns onto the screen. With a 60 mm camera lens and $d_{\text{cam}} = 435\,\text{mm}$ we observe an area of $81 \times 65\,\text{mm}$ on the surface, leading to a

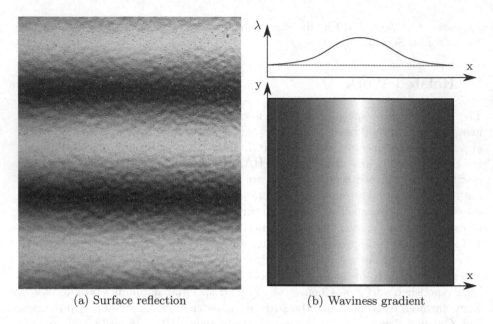

(a) Surface reflection   (b) Waviness gradient

Fig. 1. The reflection of a sinusoidal stripe pattern as seen by the camera over the surface of sheet BO (a). Waviness λ over the surface in x-direction (top of b) for sheets A, B and C (dotted line) and for sheets AO, BO and CO (solid line). Gradient of AO, BO and CO in x-direction (bottom of b).

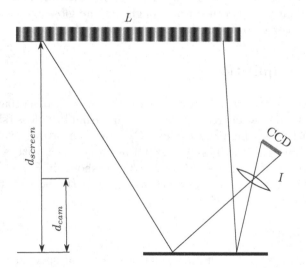

Fig. 2. Deflectometric system setup consisting of camera I, specular surface S and screen showing sinusoidal stripe patterns L.

pixel size of $2\,\mu$m on the surface. The large screen distance of $d_{\text{screen}} = 1983\,$mm leads to a high angular resolution. A sequence of phase shifted sinusoidal stripe patterns codes each point on the screen uniquely. The sequence is observed by the camera over the specular surface, where the viewing rays are reflected depending on the surface normals. By decoding this information we get the deflectometric registration. Hence, using the system setup we get the gradient fields $\frac{\partial g(x,y)}{\partial x}$ and $\frac{\partial g(x,y)}{\partial y}$ of the surface $g(x,y)$.

3.2 Wave-Scan

In order to objectively assess the quality of coated surfaces and to assess the occurrence of orange peel, Byk-Gardner developed an orange peel measuring instrument that is often used in the industry. Two measurement principles of this device enable the estimation of waviness and light scattering. For the evaluation of waviness, according to the reflection law, the reflected light intensity of a laser point source is measured. Thus, the optical profile of the surface is recorded. Using mathematical filters, this profile is divided into different wavelength ranges and from these parameters an assessment is derived. The wavelength ranges are grouped into $\lambda_a = 0.1 \ldots 0.3\,$mm, $\lambda_b = 0.3 \ldots 1\,$mm, $\lambda_c = 1 \ldots 3\,$mm, $\lambda_d = 3 \ldots 10\,$mm and $\lambda_e = 10 \ldots 30\,$mm. For assessing the dullness ($\lambda_{du} \leq 0.1\,$mm) caused by surface roughness the reflection of an LED is observed with a very low incident angle with respect to the surface normal. The observed image is divided into a focus and scattering area. A perfectly specular surface would reflect all light into the focus area. Rough surfaces reflect light into the scattering area. The ratio of scattering and focal intensity delivers a characteristic measure for the surface dullness. Based on these methods the wave-scan calculates the appearance measures *dullness* (du) and *waviness* in the corresponding wavelengths $(\text{W}_a \ldots \text{W}_e)$.

3.3 Obtaining Surface Characteristics

At first, local defects in the measurement are detected and suppressed. Then the Fourier transform $\mathscr{F}\{\cdot\}$ of the gradient fields $\frac{\partial g(x,y)}{\partial x}$ and $\frac{\partial g(x,y)}{\partial y}$ is calculated (examples in Fig. 3). The resulting frequency spectrum is divided using bandpass filters with the wavelength ranges λ_i with $i \in \{du, a, b, c, d, e\}$. As described in ISO 25178-2:2012, the angular power spectrum (APS) can be calculated between the cutoff frequencies for the frequency bands $f_i = 1/\lambda_i$ by

$$\text{APS}_i(\vartheta) = \int\limits_{f_i} r \left| \mathscr{F}\{g(r \sin(\vartheta), r \cos(\vartheta))\} \right|^2 \, dr. \tag{1}$$

r and ϑ describe the polar coordinates corresponding to the Cartesian coordinates f_x and f_y of the frequency spectrum. By using the differentiation law of the Fourier transform

$$\mathscr{F}\left\{ \frac{\partial g(x,y)}{\partial x} \right\} = \mathrm{j} 2\pi f_x \mathscr{F}\{g(x,y)\}, \tag{2}$$

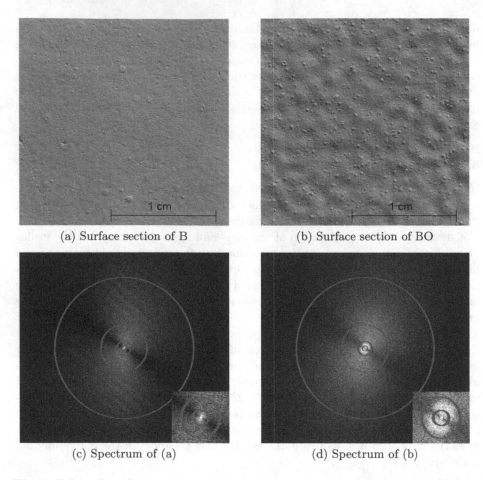

(a) Surface section of B   (b) Surface section of BO

(c) Spectrum of (a)   (d) Spectrum of (b)

Fig. 3. Enlarged gradient images of a section of surfaces B and BO without (a) and with waviness (b) and radial cutoff frequencies in red in corresponding spectra (c,d). Cutoff frequency rings are visible from $1/0.1$ mm to $1/3$ mm. Lower right image in (c) and (d) shows an enlarged spectrum with cutoff frequencies $1/1\,mm$ and $1/3$ mm (Color figure online).

and analogously for $\frac{\partial g(x,y)}{\partial y}$, we get APS_i without knowing the surface topography $g(x,y)$:

$$\mathrm{APS}_i(\vartheta) = \int\limits_{f_i} \frac{4\pi^2}{r} \left(\left| \mathscr{F}\left\{ \frac{\partial g(r\sin(\vartheta), r\cos(\vartheta))}{\partial(r\sin(\vartheta))} \right\} \right|^2 \right.$$

$$\left. + \left| \mathscr{F}\left\{ \frac{\partial g(r\sin(\vartheta), r\cos(\vartheta))}{\partial(r\cos(\vartheta))} \right\} \right|^2 \right) dr. \tag{3}$$

As we are not interested in the preferred direction of the surface microstructure, but in its existence in different frequency bands f_i, we integrate (1) over all angles ϑ:

$$\text{BPS}_i = \int_0^{2\pi} \text{APS}_i(\vartheta) \, d\vartheta. \tag{4}$$

Thus, we obtain an areal measure BPS_i for the corresponding wavelength range that allows us to estimate the surface waviness. In our experiments we used a Hamming window function to reduce boundary effects.

4 Results and Discussion

To compare the results of our areal measurements with the 1D measurements of the wave-scan, we measured the sample sheets in sections, as shown in Fig. 4. Additionally, each section captured by the camera (Fig. 4 (b)) was subdivided into subsections of 30 mm width and 65 mm height to limit the size to the homogeneous regions on the sheets. Furthermore, we had to sample the surface in similar distances as the wave-scan.

In Fig. 5 our results are plotted on a logarithmic scale. In the first column we observe, that the BPS measure remains constant over the sheets A, B and C in all scales, which meets our expectations that these sheets have no change in waviness. The measures $\text{BPS}_{du} \ldots \text{BPS}_c$ decrease starting with higher waviness values for the low gloss sheet to minimum waviness values for the very high gloss sheet. This meets our expectation that gloss and waviness are linked, because a higher gloss was obtained by polishing the sheet for a longer time which effects the waviness, too. One deviation from our expectation is a local peak between $20 \ldots 25$ cm that is caused by a large and only partially suppressed defect in this area. In the second column we observe, next to the observations made in the first

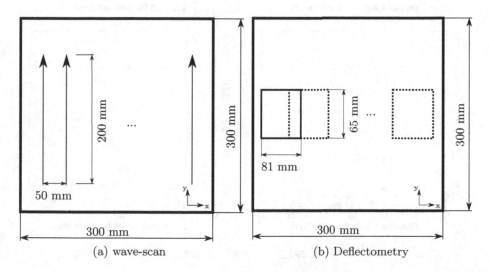

(a) wave-scan   (b) Deflectometry

Fig. 4. Measuring of sample sheets with 1D-wave-scan profile scan (a) and 2D-Deflectometry camera acquisition (b).

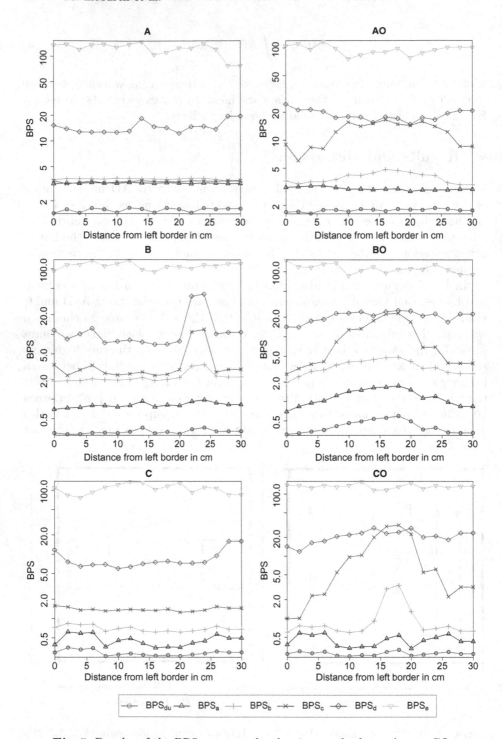

Fig. 5. Results of the BPS_i measure for the six sample sheets A, ..., CO.

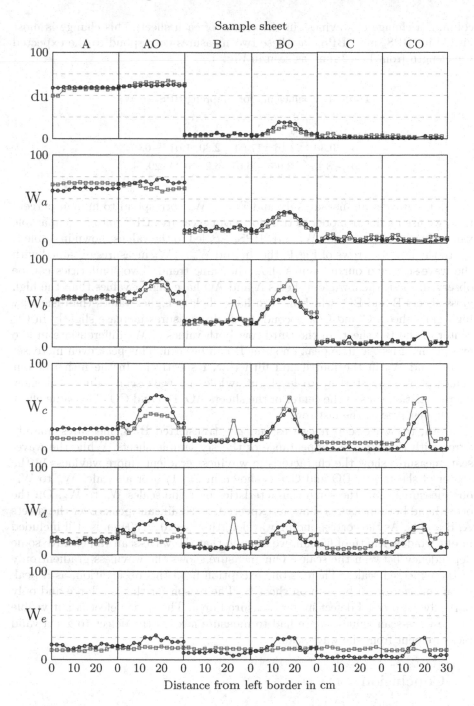

Fig. 6. Comparison of the wave-scan measures (blue line with circles) and our transferred BPS_i measure (red line with squares) for the six sample sheets A, ..., CO. (Color figure online)

column, a change of waviness in the center of each sheet. This change is most visible in BPS_b and BPS_c, as these two measures correspond to the expected wavelength from $1\ldots2$ mm, as seen in Fig. 4.

Table 1. Coefficients for mapping BPS_i to W_i.

	λ_{du}	λ_a	λ_b	λ_c	λ_d	λ_e
α	40.30	22.16	17.04	2.30	1.01	−0.07
β	−8.13	−3.95	−6.07	6.29	0.44	20.44

As the wave-scan measures du and $W_a\ldots W_e$ correspond to human perception, we used a linear model to transform our measures BPS_i into a comparable value range. The regression $W_i = \alpha_i\,BPS_i + \beta_i$ led to the values shown in Table 1.

In the first four rows of Fig. 6, the du and $W_a\ldots W_c$ measures obtained with the wave-scan and our measures show the same trend. Two similarities can be observed: First, the low gloss sheets A and AO have higher values than the high gloss sheets B and BO, which in turn have slightly higher values than the very high gloss sheets C and CO. Second, both du measures increase slightly in the center of the BO sheet. In the third row both values of W_b differentiate in two exceptions: First, a little peak on sheet B can be seen. This peak even increases in W_c and W_d in the fourth and fifth row, respectively. In the fifth row our values of W_d are starting to level out, while the wave-scan values still show characteristic peaks in the center of the sheets AO, BO and CO. The same effect can be observed in the last row.

Both, the wave-scan measures and the transferred BPS_i measures, mainly agree with the expected appearance of the six sample sheets. While the wave-scan measures show the characteristic waviness gradient (more waviness in the center of sheets AO, BO and CO as shown in Fig. 1) over all scales W_a to W_e, our measures show the same characteristics over the scales W_a to W_c. On the other hand both measures show, as expected, no significant peaks over the sheets A, B and C. As the corresponding wavelength $\lambda_e = 10\ldots30$ mm is still included in our window length of 65 mm, we assume that the wave-scan values have some dependency between the scales. Our measure shows the waviness gradient only in the expected scales. There is one exception from this observation, as a peak in our measures can be seen on sheet B. The reason for this is a large and only partially suppressed defect in our measured area. The same defect is not visible in the wave-scan values, as we had to measure next to the defect to get a valid measurement from the wave-scan.

5   Conclusion

We proposed a method to acquire and assess the appearance of specular surfaces in 2D and applied it to representative sample sheets. We described the setup used for the acquisition of the surface gradient and derived a measure from the angular power spectrum as defined in ISO 25178-2:2012. Using a linear transformation of the derived measure, we observed that our measure can be mapped to the values of the appearance measurement system wave-scan. Therefore our proposed method is suitable to assess the appearance of specular surfaces in 2D. This has several advantages over the 1D wave-scan method. First, our setup is more appropriate for an automated inspection, due to larger measurement areas and faster inspection times. Second, defect areas in the measurement can be handled better, because after suppressing those areas there is still enough information left to assess the surface part. On the other hand the wave-scan rejects to return a valid measurement for those areas.

In future works it should be investigated to which extent the measurement area can be maximized and the screen distance can be minimized, provided that the necessary resolution limits are met. Hence, it should be possible to assess if the whole surface appearance is homogeneous. Furthermore, our measure didn't exploit all available information in the deflectometric measurement. It could be interesting to extend our measure to also consider directional information. As seen in the results the defect detection should be optimized.

This work was financed by Baden-Württemberg Stiftung.

References

1. BYK-Gardner GmbH, Orange Peel and 2014, DOI Meters. https://www.byk.com/en/instruments/products/appearance-measurement/orange-peel-doi-meter.html, Accessed 13 Feb 2014
2. Chen, T., Goesele, M., Seidel, H.P.: Mesostructure from specularity. In: Proceedings of IEEE Conference on Computer Vision and Pattern Recognition (2006)
3. Fletcher, T.: A simple model to describe relationships between gloss behaviour, matting agent concentration and the rheology of matted paints and coatings. Prog. Org. Coat. **44**, 25–36 (2002)
4. Kigle-Boeckler, G.: Measurement of gloss and reflection properties of surfaces. Met. Finish. **93**, 28–31 (1995)
5. Koleske, J.: Paint and Coating Testing Manual: Fourteenth Edition of the Gardner-Sward Handbook. ASTM manual series, ASTM (1995)
6. Miranda-Medina, M., Wagner, T., Bhm, J., Vernes, A., Hingerl, K.: Optical analysis of orange peel on metallic surfaces. In: Proceedings of SPIE Optical Micro- and Nanometrology IV (2012)
7. Mischke, P.: Filmbildung in modernen Lacksystemen, chap. 3. Vincentz Network GmbH & Co KG (2007)
8. Osterhold, M.: Characterization of surface structures by mechanical and optical fourier spectra. Prog. Org. Coat. **27**, 195–200 (1996)
9. Osterhold, M.: Characterizing the surface structure of plastics coatings. Prog. Org. Coat. **57**, 165–169 (2006)
10. Pérard, D.: Automated visual inspection of specular surfaces with structured-lightning reflection techniques. Ph.D. Thesis, University of Karlsruhe (2000)
11. Pietschmann, J.: Industrielle Pulverbeschichtung, vol. 3. Vieweg+Teubner (2010)
12. Lu, R.-S., Forrest, A.K.: 3D surface topography from the specular lobe of scattered light. Opt. Lasers Eng. **45**, 1018–1027 (2007)

13. Rebeggiani, S., Rosn, B.G., Sandberg, A.: A quantitative method to estimate high gloss polished tool steel surfaces. J. Phys. Conf. Ser. **311**(1) (2011)
14. Tian, G., Lu, R., Gledhill, D.: Surface measurement using active vision and light scattering. Opt. Lasers Eng. **45**(1), 131–139 (2007)
15. of Transportation, U.D. (ed.): Aviation Maintenance Technician Handbook - Airframe, vol. 2, chap. 8. Federal Aviation Administration (2012)
16. Tse, M.K., Forrest, D., Hong, E.: An improved method for distinctness of image (doi) measurements. Technical report, Quality Engineering Associates (QEA) Inc, Burlington, MA, USA (2005)
17. Werling, S., Mai, M., Heizmann, M., Beyerer, J.: Inspection of specular and partially specular surfaces. Metrol. Meas. Syst. **16**, 415–431 (2009)
18. Ho, Y.-X., Maloney, L.: Conjoint measurement of gloss and surface texture. Psychol. Sci. **19**, 196–204 (2008)

Learning Must-Link Constraints for Video Segmentation Based on Spectral Clustering

Anna Khoreva[1](✉), Fabio Galasso[1], Matthias Hein[2], and Bernt Schiele[1]

[1] Max Planck Institute for Informatics, Saarbrücken, Germany
{khoreva,galasso,schiele}@mpi-inf.mpg.de
[2] Saarland University, Saarbrücken, Germany
hein@cs.uni-saarland.de

Abstract. In recent years it has been shown that clustering and segmentation methods can greatly benefit from the integration of prior information in terms of must-link constraints. Very recently the use of such constraints has been integrated in a rigorous manner also in graph-based methods such as normalized cut. On the other hand spectral clustering as relaxation of the normalized cut has been shown to be among the best methods for video segmentation. In this paper we merge these two developments and propose to learn must-link constraints for video segmentation with spectral clustering. We show that the integration of learned must-link constraints not only improves the segmentation result but also significantly reduces the required runtime, making the use of costly spectral methods possible for today's high quality video.

1 Introduction

Video segmentation is an open problem in computer vision, which has recently attracted increasing attention. The problem is of high interest due to its potential applications in action recognition, scene classification, 3D reconstruction and video indexing, among others. The literature on the topic has become prolific [2,4,7,10,11,19,27–29,43] and a number of techniques have become available, e.g. generative layered models [25,26], graph-based models [20,36,46] and spectral techniques [8,15,16,18,32,35,39].

Spectral methods, stemming from the seminal work of [34,39], have received much attention from the theoretical viewpoint [9,21,31], and currently provide state-of-the-art segmentation performance [3,16,18,32,35,40–42]. Spectral clustering, as a relaxation of the NP-hard normalized cut problem, is suitable due to its ability to include long-range affinities [18,40] and its global view on the problem [14], providing balanced solutions.

In this paper, we focus on two important limitations of spectral techniques: the *excessive resource requirements* and the *lack of exploiting available training data*. The large demands of spectral techniques [18,40] are particularly clear in the case of high-quality video datasets [17], limiting their current large-scale applicability. While often a labeled dataset is available, a systematic learning

© Springer International Publishing Switzerland 2014
X. Jiang et al. (Eds.): GCPR 2014, LNCS 8753, pp. 701–712, 2014.
DOI: 10.1007/978-3-319-11752-2_58

a. Video sequence   b. SPX   c. Proposed $\mathcal{M}$ SPX   d. Video segm.

Fig. 1. Video segmentation [18] employs fine superpixels (b), resulting in large resource requirements, *esp.* when using spectral methods. We propose learned must-links to merge superpixels into fewer must-link-constrained $\mathcal{M}$ superpixels (c). This reduces runtime and memory consumption and maintains or improves the segmentation (d).

of the affinities used to build the graph for spectral clustering is very difficult. In particular, as the normalized cut itself is a NP-hard problem and even the spectral relaxation is non-convex, the optimization of the minimizer which yields the segmentation is out of reach. Thus in practice one typically validates a few model parameters [8,18,32], refraining spectral methods to make use of recently available large training data [17].

We propose to *learn must-link constraints* to overcome both limitations. Recent spectral theory [16,38] has shown that the integration of must-links (i.e. forcing two vertices to be in the same cluster) allows to reduce the size of the problem, while preserving the original optimization objective for all partitions satisfying the must-links. On the other hand by learning must-link constraints we can leverage the available training data in order to guide spectral clustering towards a desired segmentation. Figure 1 illustrates the advantages of learning must-links: superpixel-based techniques [18] build spectral graphs on fine superpixels, Fig. 1(b); by contrast, we propose to build graphs merging superpixels based on learned must-link constraints, Fig. 1(c). In particular, specifically training a classifier to minimize the number of false positives allows conservative superpixel merging, which: **i.** reduces the problem size significantly; **ii.** preserves the original optimization problem; and **iii.** improves the video segmentation, Fig. 1(d), because *correct* must-links avoid undesired solutions (cf. Sect. 3).

In the following, we present the integration and learning of must-link constraints in Sect. 3 and validate them experimentally under various setups in Sect. 4 on two recent video segmentation datasets [8,17].

2   Related Work

The usage of must-link constraints, first introduced in [44], is an active area of research in machine learning known as *constrained clustering* (see [5] for an overview). The goal of integrating must-link constraints into spectral clustering has been tried via: **i.** modifying the value of affinities (cf. [24], which first considered constrained spectral clustering); **ii.** modifying the spectral embedding

[30]; or **iii.** adding constraints in a post-processing step [13,33,45,48,49]. Interestingly, none of these methods can guarantee that the must-link constraints are actually satisfied in the final clustering. By contrast, we employ must-link constraints to reduce the original graph to one of smaller size, thus enforcing the constraints while additionally benefiting runtime and memory consumption.

In particular, [16,38] have shown that must-link constraints can be used to reduce the graph, based on the corresponding point groupings, and proved equivalence between the reduced and the original graph, respectively in terms of NCut [38] and SC [16], for any clustering satisfying the must-link constraints. We employ these recent advances and propose to learn the must-link constraints in a data-driven discriminative fashion for video segmentation.

Other related work in segmentation have looked at merging superpixels with equivalence [1], but using hand-designed affinities, or learned pair-wise relations between superpixels [23], disregarding equivalence in the agglomerative merging process. This work brings together learning affinities and merging with equivalence guarantees for the first time.

3   Learning Spectral Must-Link Constraints

We provide here the steps of a video segmentation framework based on the normalized cut [22,34,39] and review the integration of must-link constraints by graph reductions as proposed in [16,38]. While the idea of learning must-link constraints applies to any segmentation problem, we discuss in detail learning and inference in the specific case of the video segmentation features of [18].

3.1   Segmentation and Must-Link Constraints

We represent a video sequence as a graph $\mathcal{G} = (\mathcal{V}, \mathcal{E})$: nodes $i \in \mathcal{V}$ represent superpixels, extracted at each frame of the video sequence with an image segmentation algorithm [3]; edges $e_{ij} \in \mathcal{E}$ between superpixels i and j take non-negative weights w_{ij} and express the similarity (*affinity*) between the superpixels.

A video segmentation can be defined as a partition $S = \{S_1, S_2, \ldots, S_K\}$ of the (superpixel) vertex set $\mathcal{V}$, i.e. $\cup_k S_k = \mathcal{V}$, $S_k \cap S_m = \varnothing \ \forall \ k \neq m$. Given $\mathcal{S}$ the set of all partitions, we look for an optimal video segmentation $S^* = \{S_1^*, S_2^*, \ldots, S_N^*\} \in \mathcal{S}$ (where N is the number of visual objects), minimizer of an objective function, implicit [20,37,47] or explicit [10,34,39,43].

Must-link constraints alter the video segmentation by reducing the set of feasible partitions $\mathcal{S}$. Given *correct*[1] must-links, a video segmentation algorithm generally improves in performance, since the solver is constrained to disregard non-optimal segmentations *wrt* S^*. Moreover, the integration of must-links leads to reduced runtime and memory load as the recent work [16,38] suggests.

---

[1] correct refers to the desired ground truth segmentation, which ideally corresponds with the optimal segmentation S^*.

We are interested in learning a *must-link grouping function* $\mathcal{M}$, which groups *certain*[2] superpixels in the graph, while respecting S^*. $\mathcal{M}$ should *conservatively* associate each node i with a point grouping $I_k \subseteq S_l^*$ (in most uncertain cases a point grouping may only include a single node). More formally:

$$\mathcal{M} : \mathcal{V} \mapsto \mathcal{P}, \quad i \mapsto I_k \tag{1}$$
$$s.t. \ I_k \subseteq S_l^* \subseteq \mathcal{V}, \ \cup_k I_k = \mathcal{V}, \ I_k \cap I_m = \varnothing \ \forall \ k \neq m,$$

where $\mathcal{P}$ is the set of possible partitions of $\mathcal{V}$.

3.2   Framework

Here we tailor the general theory to a video segmentation framework based on the normalized cut, solved either via the spectral [34,39] or 1-spectral [9,21] relaxation. Further, we discuss the integration of learned must-link constraints via graph reduction techniques [16,38] and learning and inference strategies.

Video Segmentation Setup. We build upon Galasso et al. [18]. Their constructed graph $\mathcal{G} = (\mathcal{V}, \mathcal{E})$ uses superpixels extracted from the lowest level (level 1) of a hierarchical image segmentation [3]. Edges connect superpixels from spatial and temporal neighbors and are weighted by their pair-wise affinities, computed from motion, appearance and shape features.

We consider six pairwise affinities: spatio-temporal appearance (STA), based on the median CIE Lab color distance; spatio-temporal motion (STM), based on median optical flow distance; across boundary appearance (ABA) and motion (ABM), computed across the common boundary of superpixels; short-term-temporal (STT), measuring shape similarity by the spatial overlap of optical flow-propagated superpixels; long-term-temporal (LTT), given by the fraction of common trajectories between the superpixels. Additionally we consider the number of common intersecting trajectories (IT). We distinguish four types of affinities, depending on whether the related superpixels: **i.** lie within the same frame (STA,STM,ABA,ABM); **ii.** lie on adjacent frames (STA,STM,STT); **iii-iv.** lie on frames at a distance of 2 (STT,LTT,IT) or more frames (LTT,IT) respectively.

Video Segmentation Objective Function. Given a partition of $\mathcal{V}$ into N sets $S_1, \ldots, S_N$, the normalized cut (NCut) is defined [31] as:

$$\text{NCut}(S_1, \ldots, S_N) = \sum_{k=1}^{N} \frac{\text{cut}(S_k, \mathcal{V} \backslash S_k)}{\text{vol}(S_k)}, \tag{2}$$

where $\text{cut}(S_k, \mathcal{V} \backslash S_k) = \sum_{i \in S_k, j \in \mathcal{V} \backslash S_k} w_{ij}$ and $\text{vol}(S_k) = \sum_{i \in S_k, j \in \mathcal{V}} w_{ij}$. The balancing factor prevents trivial solutions and is ideal when unary terms cannot be defined, but is also the reason why minimization of the NCut is NP-Hard.

---

[2] certain groupings are the conservative grouping decisions which we propose to learn.

Spectral Relaxations. The most widely adopted relaxation of NCut is spectral clustering (SC) [31,34,39], where the solution of the relaxed problem is given by representing the data points with the first few eigenvectors and then clustering them with k-means.

While widely adopted [3,8,16,18,32,40,41], the SC relaxation is known to be *loose*. We therefore additionally consider the 1-spectral clustering (1-SC) [21,22] - a tight relaxation based on the 1-Laplacian. However, the relaxation is only tight for bi-partitioning, for multi-way partitioning recursive splitting is used as greedy heuristic.

Reducing the original graph size with learned must-link constraints allows to experiment with 1-SC on state-of-the-art video segmentation benchmarks [8,17], notwithstanding the increased computational costs.

Graph Reduction Schemes. Given must-link constraints provided as point groupings $\{I_1, I_2, \ldots, I_q\}$ on the original vertex set $I_k \subseteq \mathcal{V}$, recent work [16,38] shows how to integrate such constraints into the original problem with respectively preserving the NCut and the spectral clustering objective function.

In more detail, integration proceeds by reducing the original graph $\mathcal{G}$ to one of smaller size $\mathcal{G}^M = (\mathcal{V}^M, \mathcal{E}^M)$, whereby the vertex set is given by the point grouping $\mathcal{V}^M = \{I_1, I_2, \ldots, I_q\}$, the edge set $\mathcal{E}^M$ preserves the original node connectivity and weights w_{IJ}^M are estimated so as to preserve the original video segmentation problem in terms of the NCut or spectral clustering objective. In particular, the NCut reduction is given by

$$w_{IJ}^M = \sum_{i \in I} \sum_{j \in J} w_{ij} \tag{3}$$

while the spectral clustering reduction is defined as

$$w_{IJ}^M = \begin{cases} \displaystyle\sum_{i \in I} \sum_{j \in J} w_{ij} & \text{if } I \neq J \\ \displaystyle\frac{1}{|I|} \sum_{i \in I} \sum_{j \in J} w_{ij} - \frac{(|I| - 1)}{|I|} \sum_{i \in I} \sum_{j \in \mathcal{V} \setminus I} w_{ij} & \text{if } I = J, \end{cases} \tag{4}$$

provided equal affinities of elements of $\mathcal{G}$ constrained in $\mathcal{G}^M$, cf. [16].

3.3 Learning

An ideal must-link constraining function $\mathcal{M}$ (Eq. 1) should only merge superpixels which are *correct*, i.e. belong to the same set in the optimal segmentation. From an implementation viewpoint, it is convenient to consider instead $\mathcal{M}_{pw}$, defined over the set of edges $\mathcal{E}$ of the graph $\mathcal{G}$ representing the video sequence:

$$\mathcal{M}_{pw} : \mathcal{E} \mapsto \{0, 1\} \tag{5}$$

$\mathcal{M}_{pw}$ casts the must-link constraining problem as a binary classification one, where a TRUE output for an input edge e_{ij} means that i and j belong to the same point grouping, in the must-link constrained graph $\mathcal{G}^M$.

We learn $\mathcal{M}_{pw}$ with Random Forests [6,12] using as features the affinities of [18] (STA,STM,ABA,ABM,STT,LTT) and the additional IT which we described in Sect. 3.2. Since different sets of affinities are available depending on whether two superpixels lie on the same or on different frames, we learn 4 different classifiers to match the 4 types of affinities.

We train a set of independent trees by estimating optimal parameters θ_p for the split functions $h(x, \theta_p)$ at each tree node p, as a function of the computed features x. Given a training set $T_p \subset X \times Y$, with X the vector of computed features and $Y = \{0, 1\}$ the corresponding ground truth video annotations, we seek to maximize the information gain I_p:

$$I_p(T_p, T_p^L, T_p^R) = H(T_p) - \frac{|T_p^L|}{|T_p|} H(T_p^L) - \frac{|T_p^R|}{|T_p|} H(T_p^R), \tag{6}$$

with $T_p^L = \{(x, y) \in T_p | h(x, \theta_p) = 0\}$, $T_p^R = T_p \backslash T_p^L$, the Shannon entropy $H(T) = -\sum_{y \in \{0,1\}} p_y \log(p_y)$ and p_y is the pdf of outcome y.

We extend the formulation of (6) to allow for learning must-link constraints on pre-grouped nodes. Reference [16] uses superpixel groupings (larger superpixel named *level 2*, cf. 4). It is important, as we found out, to consider the node multiplicity. We define therefore $|T_p| = \sum_{k \in T_p} m_k$, where $m_k = |I_k| \cdot |J_k|$ is the multiplicity of the edge between superpixel groupings I_k and J_k, thus $p_y = \frac{\sum_y m_y}{\sum_{y \in \{0,1\}} m_y}$.

Must-link constraints have a transitive nature: $\mathcal{M}_{pw}(e_{ij}) = 1$ and $\mathcal{M}_{pw}(e_{ik}) = 1$ imply $\mathcal{M}_{pw}(e_{jk}) = 1$. It is therefore crucial that all decided constraints are correct, as a few wrong ones may result in a larger set of incorrect decisions by transitive closure and potentially spoil the segmentation. Thus we define the hyperparameters (threshold of the classifier and tree depth) such that $\mathcal{M}_{pw}$ provides the largest number of positive predictions (the must-link decisions), while making zero false positives on the validation set. In such a conservative way we ensure that the resulting classifier makes only a very small number of false positives on unseen data. Although this conservative classifier might imply that in the worst case, no must-link constraints are predicted, it turns out our classifier actually predicts for a large fraction of the edges to be linked and thus leads to a significant reduction in size, while making a few false positives on the unseen data (overall, 1 false positive per 242 k true predictions).

3.4   Inference

The learned must-link constraining function $\mathcal{M}_{pw}$ provides must-link decisions for each edge of graph $\mathcal{G} = (\mathcal{V}, \mathcal{E})$. A further propagation of merge decisions in the graph accounts for the transitivity closure of $\mathcal{M}_{pw}$, consistently with the validation procedure (cf. Sect. 3.3). Based on the must-link decisions, we use the graph reduction techniques of Sect. 3.2, which integrate must-link decisions into graph $\mathcal{G}$ by reducing it to the smaller one $\mathcal{G}^M = (\mathcal{V}^M, \mathcal{E}^M)$ based on the determined groupings.

The described framework allows for evaluating different reduction schemes (equivalence in terms of NCut [38] and SC [16]) and various spectral partitioning functions (1-SC [22] and SC [34,39]). It further allows to include spatial must-link constraints and use larger superpixels, as done in [16]. We report experimental results on all these combinations in the following section.

4   Experimental Validation

We conduct two sets of experiments to analyze performance and efficiency of must-link constrained graphs $\mathcal{G}^M$. In both cases we adopt the recently proposed benchmark metrics of [17]: the boundary precision-recall (BPR) from [3] and the volume precision-recall (VPR) metric. Besides the PR curves, we report aggregate performance for BPR and VPR: optimal dataset scale [ODS], optimal segmentation scale [OSS], average precision [AP].

In the first set of experiments, we consider the *Berkeley Motion Segmentation Dataset* (BMDS) [8], which consists of 26 VGA-quality video sequences, representing mainly humans and cars, which we arrange into training, validation and test sets (6+4+16). We restrict sequences to the first 30 frames. The ground truth is provided for the 1st, 10th, 20th, 30th frame. We further annotate the 2nd, 9th, 11th frame to learn must-links across 1 and 2 frames (we release the extra annotations).

We compare the baseline of [18] with the proposed variants, $[\mathcal{M}(\mathcal{G})]^{\text{NCut}}$ - SC and $[\mathcal{M}(\mathcal{G})]^{\text{SC}}$ - SC, reducing the original graph $\mathcal{G}$ of [18] with learned must-links to $\mathcal{G}^M$ by using respectively the normalized cut (NCut) and spectral clustering (SC) reductions, and then performing SC. Figure 2 (*plots*) shows that both proposed variants outperform the baseline algorithm [18] both on BPR and VPR. The table shows improvement by 4.7% in BPR and 9% in VPR. Since the average number of superpixels is reduced by 66.7%, the better performance is accompanied by a reduction of 60% in runtime and 90% in memory load.

In Fig. 2, we further experiment by adopting 1-spectral clustering (1-SC) [22] for the NCut within the baseline algorithm (Galasso et al. [18] - 1-SC), and we compare this with our proposed variants, $[\mathcal{M}(\mathcal{G})]^{\text{NCut}}$ - 1-SC and $[\mathcal{M}(\mathcal{G})]^{\text{SC}}$ - 1-SC, where we have grouped superpixels according to learned must-links prior to processing (here with 1-SC). Since 1-SC is more costly, the provided computational reduction is even more desirable here. Again, our proposed variants improve in performance, as it appears both in the plots and the tables (average improvement of 12.3% in BPR and 9% in VPR), while significantly reducing runtime (improved by 80%) and memory load (improved by 90%). We note the similar performance of 1-SC for both reduction variants, $[\mathcal{M}(\mathcal{G})]^{\text{NCut}}$ and $[\mathcal{M}(\mathcal{G})]^{\text{SC}}$, which surprises because only the NCut reduction is theoretically justified in combination with 1-SC. Moreover, we observe the better performance of SC over 1-SC. This may indicate that the affinities of [18], designed for SC, do not fit as well the original (but different) NCut problem.

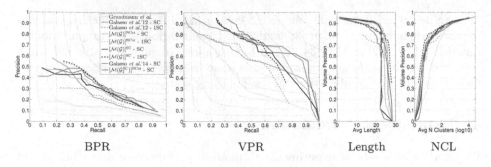

	BPR			VPR			Length	NCL
Algorithm	ODS	OSS	AP	ODS	OSS	AP	$\mu(\delta)$	μ
Grundmann et al. [20]	0.22	0.25	0.12	0.42	0.44	0.39	26.06(6.34)	13.81
Galasso et al. [18] - SC	0.37	0.39	0.24	0.57	0.72	0.59	25.75(6.46)	4.00
$[\mathcal{M}(\mathcal{G})]^{NCut}$ - SC	0.40	0.45	0.26	**0.69**	**0.77**	**0.69**	24.17(8.57)	6.00
$[\mathcal{M}(\mathcal{G})]^{SC}$ - SC	**0.41**	**0.46**	**0.27**	0.64	0.75	0.67	22.66(9.55)	6.00
Galasso et al. [18] - 1SC	0.34	0.36	0.19	0.56	0.62	0.49	25.99(6.61)	5.00
$[\mathcal{M}(\mathcal{G})]^{NCut}$ - 1SC	**0.44**	**0.48**	**0.34**	**0.64**	0.70	**0.60**	26.62(5.80)	5.00
$[\mathcal{M}(\mathcal{G})]^{SC}$ - 1SC	0.43	**0.48**	**0.34**	**0.64**	**0.71**	**0.60**	26.41(5.95)	5.00
Galasso et al.'14 [16] - SC	**0.43**	**0.48**	**0.29**	0.71	0.79	0.71	22.04(8.92)	7.00
$[\mathcal{M}(\mathcal{G}_2^{SC})]^{NCut}$ - SC	**0.43**	**0.48**	0.28	**0.71**	**0.80**	**0.75**	24.77(7.49)	5.00

Fig. 2. Comparison of state-of-the-art video segmentation algorithms with the learned must-links, on BMDS (restricted to first 30 frames) [8]. The plots and table show BPR and VPR, aggregate measures ODS, OSS and AP, and length statistics (mean μ, std. δ, no. clusters NCL) [17].

Additionally, we consider the recent work of [16], which uses superpixels extracted from a higher hierarchical level of an image segmentation algorithm [3] (superpixels at level 2), computes affinities between them and re-weights them according to SC, to take the finest superpixels at level 1 into account. Our proposed method based on must-links also allows learning constraints on the larger superpixel graph $\mathcal{G}_2$ (the multiplicity of point groupings plays a role in this case, cf. Sect. 3.3). Figure 2 shows that the reduction $[\mathcal{M}(\mathcal{G}_2^{SC})]^{NCut}$ - SC leads to the same performance as the original algorithm [16] on BPR and improves on VPR, while reducing the problem size *wrt* [16] (runtime by 30 % and memory load by 70 %).

Figure 3 qualitatively supports the positive results. Note that the learned must-links respect the GT objects while reducing the number of employed superpixels, $\mathcal{M}$ SPX. Improvements in the video segmentation output ($\mathcal{M}$ Segm Vs. (SPX) Segm.) are more evident for 1-SC. The proposed learned must-links determine merging both in the spatial and temporal dimension. It is interesting to note that for the BMDS [8] most merging comes from the first: it seems easier to make conservative merging assumptions within the frame.

In the second set of experiments we consider the novel benchmark VSB100 [17], which includes 100 HD quality videos [41] arranged into train and test sets (40+60) (we split test – 25 – and validation set – 15). In Fig. 4 we compare the

Spectral Clustering

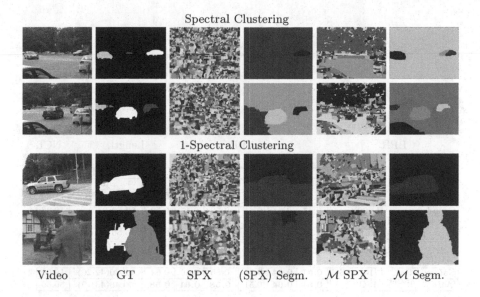

1-Spectral Clustering

Video   GT   SPX   (SPX) Segm. $\mathcal{M}$ SPX $\mathcal{M}$ Segm.

Fig. 3. Sample superpixels (SPX) and segmentation results of [18], compared with the proposed learned must-link variants, both when employing SC and 1-SC (cf. Sect. 4 for details). The proposed superpixels ($\mathcal{M}$ SPX) respect the video segmentation output while reducing the problem size. Additionally, $\mathcal{M}$ SPX improve results, *esp.* for 1-SC.

proposed method $[\mathcal{M}(\mathcal{G}_2^{SC})]^{\text{NCut}}$ - SC to the baseline [16] and state-of-the-art video segmentation algorithms. Our method maintains the performance of [16] on BPR and slightly improves on VPR. This shows that [16], by jointly leveraging large powerful superpixels [3], *saturates* the few affinities of [18], which we also use here. Thus learned must-links closely follow the spectral clustering optimization and our proposed method only provides further reduction of the problem size. With similar arguments, as also maintained in [16], the segmentation propagation method of [17] is only partially outperformed, due to its more complex image features e.g. textures. Both observations suggest to use more complex features for learning. With respect to the efficient reduction of [16], we further reduce runtime by 30 % and memory load by 65 %, while we reduce runtime by 97 % and memory load by 87 % *wrt* [18].

In addition, we adopt 1-spectral clustering [22] within the baseline (Galasso et al. [16] - 1-SC), and compare this with our proposed method ($[\mathcal{M}(\mathcal{G}_2)]^{\text{NCut}}$ - 1SC). Figure 4 shows that $[\mathcal{M}(\mathcal{G}_2^{SC})]^{\text{NCut}}$ - 1SC results in the same performance on BPR and minor improvement on VPR, while significantly reducing runtime (by 70 %) and memory load (by 65 %) *wrt* [16].

IMPLEMENTATION DETAILS. We use the Random Forests implementation of [12]. The number of features to sample for each node split is set to $\sqrt{F}$, where F is the dimensionality of the feature space. The averaged prediction of the individual trees is taken for prediction of the ensemble. As weak learners we use

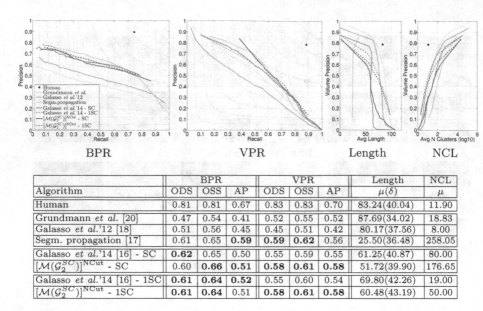

	BPR			VPR			Length	NCL
Algorithm	ODS	OSS	AP	ODS	OSS	AP	$\mu(\delta)$	μ
Human	0.81	0.81	0.67	0.83	0.83	0.70	83.24(40.04)	11.90
Grundmann *et al.* [20]	0.47	0.54	0.41	0.52	0.55	0.52	87.69(34.02)	18.83
Galasso *et al.*'12 [18]	0.51	0.56	0.45	0.45	0.51	0.42	80.17(37.56)	8.00
Segm. propagation [17]	0.61	0.65	**0.59**	**0.59**	0.62	0.56	25.50(36.48)	258.05
Galasso *et al.*'14 [16] - SC	**0.62**	0.65	0.50	0.55	0.59	0.55	61.25(40.87)	80.00
$[\mathcal{M}(\mathcal{G}_2^{SC})]^{NCut}$ - SC	0.60	**0.66**	**0.51**	0.58	0.61	0.58	51.72(39.90)	176.65
Galasso *et al.*'14 [16] - 1SC	**0.61**	**0.64**	**0.52**	0.55	0.60	0.54	69.80(42.26)	19.00
$[\mathcal{M}(\mathcal{G}_2^{SC})]^{NCut}$ - 1SC	**0.61**	**0.64**	0.51	**0.58**	**0.61**	**0.58**	60.48(43.19)	50.00

Fig. 4. Comparison of state-of-the-art video segmentation algorithms with our proposed method based on the learned must-links, on VSB100 [17]

linear binary split functions and conic sections, and the forest size is set to 100 trees. The tree depth is varied in the range $[2, 12]$ and validated along with the threshold, which yields the largest number of must-links with zero false positives. Following [18], we extract the first 6 eigenvectors.

5   Conclusions

We have formalized must-link constraints and proposed the relevant learning and inference algorithms. While this theory is applicable to general clustering and segmentation problems, we have particularly shown the use of learned must-link constraints in conjunction with spectral techniques, whereby recent theoretical advances employ these to reduce the original problem size, hence the runtime and memory requirements. Experimentally, we have shown that learned must-link constraints improve efficiency and, in most cases, performance, as these allow discriminatively training on GT data.

Acknowledgments. The authors would like to thank Syama Sundar Rangapuram for his support on the use of the 1-spectral clustering code.

References

1. Alpert, S., Galun, M., Brandt, A., Basri, R.: Image segmentation by probabilistic bottom-up aggregation and cue integration. TPAMI **34**(2), 315–326 (2012)
2. Andres, B., Kappes, J.H., Beier, T., Köthe, U., Hamprecht, F.A.: Probabilistic image segmentation with closedness constraints. In: ICCV (2011)
3. Arbeláez, P., Maire, M., Fowlkes, C.C., Malik, J.: Contour detection and hierarchical image segmentation. TPAMI **33**(5), 898–916 (2011)
4. Banica, D., Agape, A., Ion, A., Sminchisescu, C.: Video object segmentation by salient segment chain composition. In: ICCV, IPGM Workshop (2013)
5. Basu, S., Davidson, I., Wagstaff, K.: Constrained Clustering: Advances in Algorithms, Theory, and Applications. Chapman & Hall, Boca Raton (2008)
6. Breiman, L.: Random forests. Mach. Learn. **45**(1), 5–32 (2001)
7. Brendel, W., Todorovic, S.: Video object segmentation by tracking regions. In: ICCV (2009)
8. Brox, T., Malik, J.: Object segmentation by long term analysis of point trajectories. In: Daniilidis, K., Maragos, P., Paragios, N. (eds.) ECCV 2010, Part V. LNCS, vol. 6315, pp. 282–295. Springer, Heidelberg (2010)
9. Bühler, T., Hein, M.: Spectral clustering based on the graph p-Laplacian. In: ICML (2009)
10. Chang, J., Wei, D., Fisher, J.W.: A video representation using temporal superpixels. In: CVPR (2013)
11. Cheng, H.T., Ahuja, N.: Exploiting nonlocal spatiotemporal structure for video segmentation. In: CVPR (2012)
12. Criminisi, A., Shotton, J., Konukoglu, E.: Decision forests: a unified framework for classification, regression, density estimation, manifold learning and semi-supervised learning. In: Foundations and Trends in Computer Graphics and Vision (2012)
13. Eriksson, A.P., Olsson, C., Kahl, F.: Normalized cuts revisited: a reformulation for segmentation with linear grouping constraints. In: ICCV (2007)
14. Fowlkes, C., Malik, J.: How much does globalization help segmentation?. Technical report, EECS - UC Berkeley (2004)
15. Fragkiadaki, K., Shi, J.: Video segmentation by tracing discontinuities in a trajectory embedding. In: CVPR (2012)
16. Galasso, F., Keuper, M., Brox, T., Schiele, B.: Spectral graph reduction for efficient image and streaming video segmentation. In: CVPR (2014)
17. Galasso, F., Nagaraja, N.S., Cardenas, T.Z., Brox, T., Schiele, B.: A unified video segmentation benchmark: annotation, metrics and analysis. In: ICCV (2013)
18. Galasso, F., Cipolla, R., Schiele, B.: Video segmentation with superpixels. In: Lee, K.M., Matsushita, Y., Rehg, J.M., Hu, Z. (eds.) ACCV 2012, Part I. LNCS, vol. 7724, pp. 760–774. Springer, Heidelberg (2013)
19. Galasso, F., Iwasaki, M., Nobori, K., Cipolla, R.: Spatio-temporal clustering of probabilistic region trajectories. In: ICCV (2011)
20. Grundmann, M., Kwatra, V., Han, M., Essa, I.: Efficient hierarchical graph-based video segmentation. In: CVPR (2010)
21. Hein, M., Bühler, T.: An inverse power method for nonlinear eigenproblems with applications in 1spectral clustering and sparse pca. In: NIPS (2010)
22. Hein, M., Setzer, S.: Beyond spectral clustering - tight relaxations of balanced graph cuts. In: NIPS (2011)
23. Jain, V., Turaga, S.C., Briggman, K.L., Helmstaedter, M., Denk, W., Seung, H.S.: Learning to agglomerate superpixel hierarchies. In: NIPS (2011)

24. Kamvar, S.D., Klein, D., Manning, C.D.: Spectral learning. In: IJCAI (2003)
25. Kannan, A., Jojic, N., Frey, B.J.: Generative model for layers of appearance and deformation. In: AISTATS (2005)
26. Kumar, M.P., Torr, P., Zisserman, A.: Learning layered motion segmentations of video. IJCV **76**, 301–319 (2008)
27. Lee, Y.J., Kim, J., Grauman, K.: Key-segments for video object segmentation. In: ICCV (2011)
28. Lezama, J., Alahari, K., Sivic, J., Laptev, I.: Track to the future: Spatio-temporal video segmentation with long-range motion cues. In: CVPR (2011)
29. Li, F., Kim, T., Humayun, A., Tsai, D., Rehg, J.M.: Video segmentation by tracking many figure-ground segments. In: ICCV (2013)
30. Li, Z., Liu, J., Tang, X.: Constrained clustering via spectral regularization. In: CVPR (2009)
31. von Luxburg, U.: A tutorial on spectral clustering. Stat. Comput. **17**(4), 395–416 (2007)
32. Maire, M., Yu, S.X.: Progressive multigrid eigensolvers for multiscale spectral segmentation. In: ICCV (2013)
33. Maji, S., Vishnoi, N.K., Malik, J.: Biased normalized cuts. In: CVPR (2011)
34. Ng, A.Y., Jordan, M., Weiss, Y.: On spectral clustering: Analysis and an algorithm. In: NIPS (2001)
35. Ochs, P., Malik, J., Brox, T.: Segmentation of moving objects by long term video analysis. TPAMI **36**, 1187–1200 (2013)
36. Palou, G., Salembier, P.: Hierarchical video representation with trajectory binary partition tree. In: CVPR (2013)
37. Paris, S.: Edge-preserving smoothing and mean-shift segmentation of video streams. In: Forsyth, D., Torr, P., Zisserman, A. (eds.) ECCV 2008, Part II. LNCS, vol. 5303, pp. 460–473. Springer, Heidelberg (2008)
38. Rangapuram, S., Hein, M.: Constrained 1-spectral clustering. In: AISTATS (2012)
39. Shi, J., Malik, J.: Normalized cuts and image segmentation. TPAMI **22**, 1–8 (2000)
40. Sundaram, N., Keutzer, K.: Long term video segmentation through pixel level spectral clustering on gpus. In: ICCV Workshops (2011)
41. Sundberg, P., T.Brox, Maire, M., Arbelaez, P., Malik, J.: Occlusion boundary detection and figure/ground assignment from optical flow. In: CVPR (2011)
42. Taylor, C.: Towards fast and accurate segmentation. In: CVPR (2013)
43. Vazquez-Reina, A., Avidan, S., Pfister, H., Miller, E.: Multiple hypothesis video segmentation from superpixel flows. In: Daniilidis, K., Maragos, P., Paragios, N. (eds.) ECCV 2010, Part V. LNCS, vol. 6315, pp. 268–281. Springer, Heidelberg (2010)
44. Wagstaff, K., Cardie, C., Rogers, S., Schrödl, S.: Constrained k-means clustering with background knowledge. In: ICML (2001)
45. Wang, X., Davidson, I.: Flexible constrained spectral clustering. In: KDD (2010)
46. Xu, C., Corso, J.J.: Evaluation of super-voxel methods for early video processing. In: CVPR (2012)
47. Xu, C., Xiong, C., Corso, J.J.: Streaming hierarchical video segmentation. In: Fitzgibbon, A., Lazebnik, S., Perona, P., Sato, Y., Schmid, C. (eds.) ECCV 2012, Part VI. LNCS, vol. 7577, pp. 626–639. Springer, Heidelberg (2012)
48. Xu, L., Li, W., Schuurmans, D.: Fast normalized cut with linear constraints. In: CVPR (2009)
49. Yu, S.X., Shi, J.: Grouping with bias. In: NIPS (2001)

Young Researcher Forum

Automatic 3D Reconstruction of Indoor Manhattan World Scenes Using Kinect Depth Data

Dominik Wolters[(✉)]

Institute of Computer Science, Kiel University, Kiel, Germany
dwol@informatik.uni-kiel.de

Abstract. This paper discusses a system to reconstruct indoor scenes automatically and evaluates its accuracy and applicability. The focus is on the realization of a simple, quick and inexpensive way to map empty or slightly furnished rooms. The data is acquired with a Kinect sensor mounted onto a pan-tilt head. The Manhattan world assumption is used to approximate the environment. The approach for determining the wall, floor and ceiling planes of the rooms is based on a plane sweep method. The floor plan is reconstructed from the detected planes using an iterative flood fill algorithm. Furthermore, the developed method allows to detect doors and windows, generate 3D models of the measured rooms and to merge multiple scans.

1 Introduction

Today there are no simple and cost-effective systems available to map interiors. Acquisition is usual performed either manually or with 3D laser scanners. Laser scanner achieve a high level of accuracy, however, the acquisition is costly.

The purpose of this research is the development of a low-cost solution for automatic mapping of empty or slightly furnished interiors. For a few years, with the Kinect sensor, a new, light and cost-effective sensor is available that can provide both color and depth images of the environment in real time. The used active structured light method ensures the acquisition of depth data also in poorly textured and dark areas. An automatic movement of the sensor performed by a pan-tilt unit guarantees a full coverage of the environment.

Related Work. Since floor plans are often needed for robotic navigation tasks, much research has been done on mapping of building interiors. The maps are usually captured with laser scanners during the movement of the robots. Example approaches are region growing using surface normals [6], plane sweeping [2,7] or extracting walls using a Hough transform [1,10].

Geometry estimation of indoor environments from single Kinect images has been studied in several works [8,12]. In the last years 3D reconstruction from multiple images with a handheld Kinect has been a popular task [4,9].

Recommended for submission to YRF2014 by Prof. Dr.-Ing. Reinhard Koch.

X. Jiang et al. (Eds.): GCPR 2014, LNCS 8753, pp. 715–721, 2014.
DOI: 10.1007/978-3-319-11752-2_59

Contribution. The developed method allows an automatic acquisition and generation of floor plans. It is based on an approach proposed by Budroni and Böhm [2], which was adapted and extended in this work for use with data from the Kinect sensor. The described plane sweep approach for the detection of walls has been extended by an edge-based method that allows the detection of smaller and partially covered wall surfaces. The floor plan is determined from the detected planes using a novel iterative flood fill algorithm. Furthermore, a method for the detection of doors, including their opening direction, and windows is introduced. A three-dimensional reconstruction allows the detection of block-like objects in the room. To enable the reconstruction of larger rooms a histogram-based merging of multiple scans is presented.

2   Method

2.1   Equipment Overview

The equipment consists of a tripod on which a pan-tilt unit is mounted with the Kinect sensor on the head. The system is intended to capture the entire area from a single point. For this purpose, a gradual horizontal rotational movement is performed and at each position images with different angles of inclination are recorded. The data processing is performed on a connected laptop.

Based on the known rotation angle of the pan-tilt unit and Kinect sensor as well as the transformation between the camera coordinate system and the pan-tilt coordinate system (determined by a hand-eye calibration), a point cloud of the environment is created.

2.2   Modeling of Floor Plans

Plane-Sweep. The point cloud is used as input for the plane sweep method according to Budroni and Böhm [2]. First the alignment of the point cloud is determined using the Manhattan world assumption [3], which states that the scene contains three orthogonal, dominant directions. Then the main wall planes are detected by plane sweeping along this directions.

The rotational sweep used for the alignment proved to be error-prone in case of the noisy data of the Kinect sensor and additional spatial structures that do not correspond to the Manhattan world. Therefore, this step was replaced by an entropy based method similar to the one used by Gallup et al. [5].

Edge-Based Plane Detection. The developed edge-based plane detection allows to detect planes that can not be detected with the plane sweep approach due to a small size or because of occlusion (Fig. 1(a)).

To determine the edges, histograms of the point coordinates distributions along the detected planes are created. For each histogram bin, the gradient is determined by discrete differences. Areas with high gradients represent potentially the beginning or end of a wall surface, which indicates that orthogonal walls are linked.

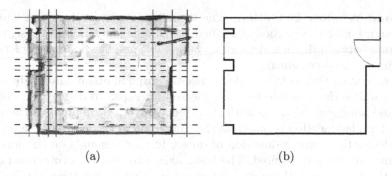

(a)   (b)

Fig. 1. (a) Projection of the point cloud with planes detected by the plane sweep (red/solid) and edge-based plane detection (blue/dashed). (b) Determined floor plan (Color figure online)

Orthogonal planes are determined for all significant minima and maxima of the gradient. The planes are added to the previously detected planes, sorted by the strength of the gradient. However, to prevent duplicate or closely spaced planes, they are only added if there is not already a plane present in a given neighborhood.

Iterative Interior Detection. For the determination of the interior, an iterative flood fill algorithm has been developed. The algorithm employs the rectangular cells, caused by cutting the detected planes. In the first step, the cell which contains the recording system is labeled as interior. Then the edges to the 4-neighbors are examined. If the number of points around the edges is below a threshold, the neighboring cell is considered as part of the interior. Similarly, the edges of the newly added cells are examined (Fig. 2). The edges to neighboring cells are only examined if they have a greater distance to the starting cell than the currently considered cell. The purpose is that only edges are examined, which can be seen from the recording location.

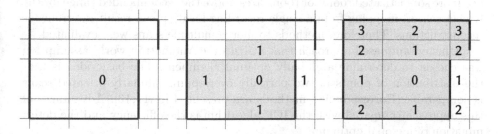

Fig. 2. Process flow of the iterative interior detection. The current interior cells are highlighted in yellow. The numbers indicate the distance to the starting cell (Color figure online)

Door and Window Detection. The technique to detect door and windows is an extension of the method described by Budroni und Böhm [2], so that a distinction between doors and windows is possible and the opening direction of the doors can be determined.

In order to do this a 50 cm wide horizontal stripe in about 1 m height on the detected wall surfaces is selected and checked for gaps. To distinguish between doors and windows the areas with detected gaps are examined on the entire height. The classification is based on the dimension of the gap.

To detect the opening direction of doors, further demands on the measurement environment are required. The basic idea is to provide a rotational sweep around the door frames. If the door is opened and visible from the recording location, it generates an additional peak, which does not correspond to the direction of the wall.

Floor Plan. The final result is shown in Fig. 1(b). The presented method is generally suitable to map unfurnished and geometrically simple interiors. In order to map slightly furnished or more sophisticated rooms an extension of the proposed algorithm is necessary.

2.3 3D Reconstruction

The 3D reconstruction allows the detection of block-like objects in the room. Furthermore a virtual 3D model of the room can be created. The colored images captured by the Kinect sensor can be used to texture the model.

The basic algorithm substantially remains unchanged for the 3D reconstruction. Additionally vertical planes are determined, so that the cells conform to cuboids. These cuboids are used for the iterative interior detection. Analogous to the 4-connected neighborhood in the 2D reconstruction, the 6-connected neighborhood of the cuboid is considered here.

2.4 Merging of Multiple Scans

For more sophisticated rooms or room sizes above the recommended range for the Kinect sensor, mapping from a single recording location can result in erroneous reconstructions. Therefore, methods to merge multiple scans were evaluated.

A histogram-based approach that utilizes the Manhattan world assumption was chosen to determine a globally optimal alignment. The basic idea is that the distribution of points in two partially overlapping similarly oriented scans is comparable. Corresponding methods are presented in [11,13]. Advantages of this approach compared with the ICP algorithm is the efficiency and the determination of a global optimum.

3 Results

3.1 Experiments with Synthetic Data

The evaluation is performed first on synthetic data. Point clouds with random size between $4\,\text{m} \times 3\,\text{m} \times 2.5\,\text{m}$ and $5\,\text{m} \times 4\,\text{m} \times 3\,\text{m}$ were generated. To simulate the noise of the Kinect sensor, different levels of Gaussian noise are added. With a low noise level ($\sigma \leq 20\,\text{mm}$), the deviations for the plane sweep and the edge-based plane detection are less than $10\,\text{mm}$. The deviation for the rotation determination is less than $0.1°$.

3.2 Experiments with Real Data

Distance Experiments. To evaluate the influence of the distance to the sensor, planar surfaces of walls were measured at various distances from $0.3\,\text{m}$ to $6.2\,\text{m}$. The depth resolution decreases quadratically with increasing distance from the sensor. Thus, the detection of an unambiguous wall plane is more difficult with an increasing distance. In the study, the relative deviation between measured and real wall distance is usually less than $1\,\%$. The ground truth was manually measured with a laser distance meter. Truly orthogonal walls are assumed.

Determination of Floor Plans. As a first example a reconstruction of an empty room with a size of about $3\,\text{m} \times 4\,\text{m}$ is considered. Figure 3 shows the projected point cloud (left) and the determined floor plan (right). The door and the two windows were correctly detected. The comparison between real and determined room sizes for the main planes, i.e. length, width and height, results in only small deviations of less than $1\,\%$. These accuracies were confirmed also in the evaluation of other rooms.

Merging of Multiple Scans. For large rooms merging of multiple scans can be useful to achieve a higher accuracy. Figure 4 shows a room with a size of about $8.5\,\text{m} \times 5\,\text{m}$. Clearly visible is the high noise level at the farther wall, i. e. the right-hand wall shown in Fig. 4(a). By merging two scans (Fig. 4(b)) the deviation of the room length was reduced from $1.1\,\%$ to $0.8\,\%$.

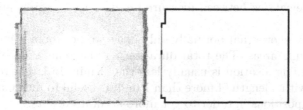

Fig. 3. Point cloud and determined floor plan of an empty room. Solid walls: thick lines. Door and windows: thin lines

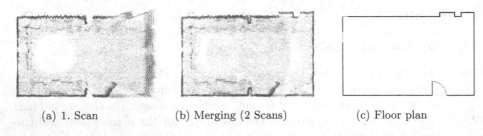

(a) 1. Scan   (b) Merging (2 Scans)   (c) Floor plan

Fig. 4. Merging from two scans to determine the floor plan

3D Reconstruction. Figure 5 shows the point cloud of a slightly furnished room, the determined floor plan and the created 3D models. The size of the room is about 6 m×3 m. Door and window as well as the objects in the room were correctly recognized. A comparison of the real dimensions of the objects with the determined measures results in only minor deviations of up to 25 mm.

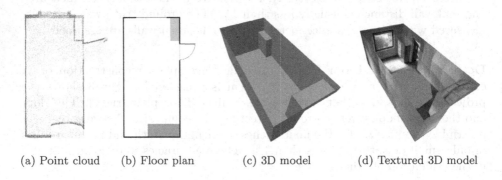

(a) Point cloud   (b) Floor plan   (c) 3D model   (d) Textured 3D model

Fig. 5. 3D reconstruction of a slightly furnished room

4   Conclusion

The developed algorithm provides an automatic acquisition and generation of floor plans that reaches an accuracy which is already sufficient in many fields. The relative deviations between measured and real room sizes are usually less than 1 %.

The analysis shows that normally only one scan per room is required for the mapping of living areas. The total duration for the generation of a floor plan from one recording location is usually less than 3 min. In larger rooms, such as living rooms, with a length of more than 7 m, it is useful to merge multiple scans from different positions in order to get more accurate results.

References

1. Adan, A., Huber, D.: 3D reconstruction of interior wall surfaces under occlusion and clutter. In: International Conference on 3D Imaging, Modeling, Processing, Visualization and Transmission (3DIMPVT), pp. 275–281 (2011)
2. Budroni, A., Böhm, J.: Automatic 3D modelling of indoor manhattan-world scenes from laser data. In: Proceedings of the ISPRS Commission V Mid-Term Symposium 'Close Range Image Measurement Techniques', Newcastle upon Tyne, UK, vol. XXXVIII, pp. 115–120 (2010)
3. Coughlan, J.M., Yuille, A.L.: Manhattan world: compass direction from a single image by bayesian inference. In: Proceedings of the Seventh IEEE International Conference on Computer Vision, vol. 2, pp. 941–947 (1999)
4. Du, H., Henry, P., Ren, X., Cheng, M., Goldman, D.B., Seitz, S.M., Fox, D.: Interactive 3D modeling of indoor environments with a consumer depth camera. In: Proceedings of the 13th International Conference on Ubiquitous Computing, pp. 75–84 (2011)
5. Gallup, D., Frahm, J.M., Mordohai, P., Yang, Q., Pollefeys, M.: Real-time plane-sweeping stereo with multiple sweeping directions. In: IEEE Conference on Computer Vision and Pattern Recognition (CVPR '07), Minneapolis, USA, pp. 1–8, June 2007
6. Hähnel, D., Burgard, W., Thrun, S.: Learning compact 3D models of indoor and outdoor environments with a mobile robot. Robot. Auton. Syst. **44**(1), 15–27 (2003)
7. Johnston, M., Zakhor, A.: Estimating building floor-plans from exterior using laser scanners. In: SPIE Electronic Imaging Conference, 3D Image Capture and Applications, vol. 3 (2008)
8. Neverova, N., Muselet, D., Trémeau, A.: $2^1/_2$D scene reconstruction of indoor scenes from single RGB-D images. In: Tominaga, S., Schettini, R., Trémeau, A. (eds.) CCIW 2013. LNCS, vol. 7786, pp. 281–295. Springer, Heidelberg (2013)
9. Newcombe, R.A., Davison, A.J., Izadi, S., Kohli, P., Hilliges, O., Shotton, J., Molyneaux, D., Hodges, S., Kim, D., Fitzgibbon, A.: KinectFusion: real-time dense surface mapping and tracking. In: 10th IEEE International Symposium on Mixed and Augmented Reality (ISMAR), pp. 127–136 (2011)
10. Okorn, B., Xiong, X., Akinci, B., Huber, D.: Toward automated modeling of floor plans. In: Proceedings of the Symposium on 3D Data Processing, Visualization and Transmission, vol. 2 (2010)
11. Rofer, T.: Using histogram correlation to create consistent laser scan maps. In: Proceedings of the IEEE International Conference on Robotics Systems (IROS-2002), Lausanne, pp. 625–630 (2002)
12. Taylor, C., Cowley, A.: Parsing indoor scenes using RGB-D imagery. In: Proceedings of Robotics: Science and Systems. Sydney, July 2012
13. Weiß, G., Wetzler, C., von Puttkamer, E.: Keeping track of position and orientation of moving indoor systems by correlation of range-finder scans. In: Proceedings of the IEEE/RSJ/GI International Conference on Intelligent Robots and Systems '94. 'Advanced Robotic Systems and the Real World', IROS '94. vol. 1, pp. 595–601 (1994)

Can Cosegmentation Improve
the Object Detection Quality?

Timo Lüddecke[✉]

Faculty of Computer Science, Image Processing and Image Understanding Group,
Otto von Guericke University Magdeburg, Magdeburg, Germany
tluedde@gwdg.de

Abstract. In order to train an object detector usually a large annotated dataset is needed, which is expensive and cumbersome to acquire. In this paper the task of collecting these annotations is automated to a large extent by cosegmentation, i.e. the simultaneous segmentation of multiple images. This way only weak requirements on the input must be obeyed: The respective object must occur in every image exactly once and has to be at least slightly salient. Obviously, this facilitates the collection of an appropriate training set. On the cosegmentation's result a straightforward object detector is trained for the underlying object. Both steps, cosegmentation and detection, share the representation of regions. Results show competitive results on cosegmentation datasets and indicate that detection actually benefits from a prior cosegmentation.

1 Introduction

The segmentation of objects in natural images can be considered a very sophisticated task since its region appearance is in general not uniform, especially not when it is projected onto a two-dimensional camera sensor. In addition, it is often not clear which one, out of multiple possible objects, is central to an image, i.e. represents its topic. At least the latter problem of ambiguity can be addressed by utilizing the massive amounts of (image) data available today, e.g. on the internet. Instead of considering every image isolated, sets of images are formed that share an object to be segmented. Now, enforcing constraints on the similarity of some segments in the set turns out to be greatly beneficial for solving the problem, because – given an appropriately large number of images – the object of interest remains relatively constant in appearance while other segments vary. This method is called cosegmentation [8] and has been used in several works before [3,10,12]. In this work, the number of labels is set to two, so the problem reduces to a differentiation between foreground and background in an image set and an object instance is expected in every image. Having obtained a proper cosegmentation, it can be leveraged as pre-training to object detection: Based on the extracted segments, an object detector can be trained – either directly

---

Recommended for submission to YRF2014 by Prof. Klaus Tönnies.

© Springer International Publishing Switzerland 2014
X. Jiang et al. (Eds.): GCPR 2014, LNCS 8753, pp. 722–728, 2014.
DOI: 10.1007/978-3-319-11752-2_60

using the cosegmentation's model (which is done in this work) or indirectly by training a new model.

In the following, first a novel cosegmentation algorithm is proposed being designed with the purpose of detection in mind. It is compared with other methods in an object cosegmentation setting and finally evaluated for its intended use in a detection task.

2   A Simple Cosegmentation Algorithm

The algorithm to be proposed here follows the idea of [12] as depicted in Fig. 1: First a set of hypotheses is generated for each image, then the most plausible hypotheses are selected in order to obtain a cosegmentation. The latter is then used to train a detector and to search images for object instances by a sliding-window method.

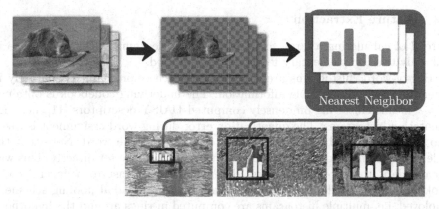

Fig. 1. The idea of the algorithm: First cosegment an image set, then use the obtained region features to train a nearest neighbor detector. (The depicted images in the top row are part of the iCoSeg dataset [1]. Those in the bottom row originate from http://www.flickr.com and have been licensed under creative commons. Attribution from left to right: Timo Heuer (CC BY-NC-SA 2.0), Linda Tanner (CC BY 2.0), M Kuhn (CC BY-NC-SA 2.0). The red boxes indicate detections of our algorithm.)

While common cosegmentation algorithms are tuned to segment entire objects, this is not necessarily desirable when detection is the goal. If a constituent is more stable than the object as a whole, it appears to be a good idea to rely just on this part during detection since it represents the most reliable trait of the object.

2.1   Hypothesis Generation

The objective of the first step of the algorithm is to generate a set of hypotheses, each being a binary image. It is crucial to have the correct hypothesis or at least

724   T. Lüddecke

a close one included in this set, since this cannot be corrected later on. On the other hand, the number of hypotheses is limited as it dramatically increases the computational demands.

The hypothesis generation is based on GrabCut [9] with multiple initializations. The latter has shown very good results in a setting relying on minimal user interaction. Given a set of images with a certain degree of saliency it seems natural to automate this process. This idea is not new and a respective algorithm has been proposed by [4]. However, the method here is mostly inspired by [2] but varies in some details, e.g. GrabCut initializations are used and saliency is taken into account. Basically, three different types of initialization can be separated: Rectangles of decreasing size around the entire image, rectangles of increasing size around the salient region thresholds and the salient regions as foreground seeds with different parts of the image border as background seeds.

Usually, there are some duplicates or very similar hypotheses in the set. As this would lead to a longer runtime they are removed.

2.2 Feature Extraction

Instead of calculating a large set of features and learning relevant ones like in [12], only two complementary feature types are used.

Bag-of-words histograms are computed on each color channel separately in order to exploit all available information. The underlying codebook is obtained by a k-means clustering on densely computed DAISY descriptors [11], i.e. edge orientation histograms. The quantization error during word assignment is taken into account to weight the contribution of the respective word. Note that the codebook is computed on the cosegmentation input image set directly. This way it can adapt to content and stress differences in contrast to a general codebook. As bag-of-words neglects spatial information, a radial pooling scheme is employed, i.e. multiple histograms are computed in rings around the hypothesis center and subsequently concatenated.

However, as edge histograms are computed on the first-derivative of the image, some information is dropped in order to gain invariance. In particular, pixel intensities are substantial to represent certain types of image regions (e.g. sky), hence color histograms can be employed as a second feature in addition to the edge orientation histogram based bag-of-words representation – depending on the task.

2.3 Probabilistic Model

The goal is to find the cosegmentation $\mathbf{s}^*$ that is most probable given the images $\mathbf{i}$, i.e $\mathbf{s}^* = \arg\max_{\mathbf{s}} P(\mathbf{s} \mid \mathbf{i})$. with the probability $P(\mathbf{s} \mid \mathbf{i})$ being modeled as a Markov random field.

$$P(\mathbf{s} \mid \mathbf{i}) = \frac{1}{Z} \cdot \exp\left(E_{unary} + E_{pairwise}\right) = \frac{1}{Z} \cdot \exp\left(0 - \sum_{i,j<I} f(\mathbf{s}_i, \mathbf{s}_j)\right) \quad (1)$$

Unary energy E_{unary} can be seen as a measure of the segmentation quality. The term is dropped, however, as the available information does not suffice to derive it reasonably. The labeling, i.e. the segmentation, of an image is modeled to be dependent on the labels of the direct neighbors, so the energy can be expressed by pairwise potentials. Note, contrary to MRFs at the pixel level, the graph is fully connected, thus the set of neighbors involves all other nodes.

Since only the maximal probability state is of interest, the MRF can be simplified. Z is a constant and exp is monotonic, i.e. removing them does not alter the maximal state.

$$\mathbf{s}^* = \arg\max_{\mathbf{s}} \frac{1}{Z} \exp\left(-\sum_{i,j<I} f(\mathbf{s}_i, \mathbf{s}_j)\right) = \arg\min_{\mathbf{s}} \sum_{i,j<I} f(\mathbf{s}_i, \mathbf{s}_j) \qquad (2)$$

Our bag-of-word and color histogram representations allow to express similarity by comparing related histograms. The intersection measure has proven to be quite useful in such a case. Since it is possible to combine multiple features, there can also be multiple histograms associated with each hypothesis. The different histogram types (e.g. color histogram for the red channel) are defined by the set T. The function $h_i^t(\mathbf{s})$ can be considered the feature extractor as it maps the segmentation $\mathbf{s}$ to a histogram based on the feature type t in the i-th image. As the goal is to minimize the energy, f is defined as the reciprocal of the sum of all histogram intersections in T.

$$f(\mathbf{s}_i, \mathbf{s}_j) = \left(\sum_{t \in T} intersection(h_i^t(\mathbf{s}_i), h_j^t(\mathbf{s}_j))\right)^{-1} \qquad (3)$$

The histogram intersection is the sum of the overlaps between each pair of bins in the histograms, formally $intersection(a,b) = \sum_i min(a_i, b_i)$. The quantization error, i.e. the distance to cluster center in the descriptor space, is incorporated for bag-of-word histograms by weighting the contribution of each word by its error during feature extraction.

2.4 Optimization

Finding the minimal energy state is not a trivial task, as the search space is very large. Even for a small number of hypotheses and images, it prohibits to precompute all pairwise potentials. In this work, optimization is carried out by the simulated annealing algorithm that was introduced by [5] for image restoration using an MRF. Due to its simple structure, it can easily be extended to compute pairwise potentials only when they are required (lazy evaluation).

2.5 Experiments

An evaluation has been carried out on two common datasets for cosegmentation. The first is a selection of 16 categories from the iCoSeg dataset [1] assembled by [12], the experiment was carried out using color histograms and bag-of-word

iCoseg subset

	This	[3]	[10]	[12]
P	81.4%	**94.4 %**	89.6 %	85.4 %
J	0.53	**0.79**	0.68	0.62

Internet dataset

	This	This*	[10]	[6]	[7]
P	72.81%	77.37%	**85.4 %**	74.74%	56.96%
J	0.299	0.453	**0.573**	0.046	0.255

Fig. 2. Average precision (P) and jaccard index (J) for iCoSeg and Internet dataset. The results marked with * were obtained on the Internet dataset after removing non-object images.

features. The second dataset has been compiled by [10] from the internet, here bag-of-words were used as the only feature type.

The results depicted in Fig. 2 indicate a lower performance compared to competitive algorithms on iCoSeg, although it is in the same magnitude. At least a part of the inferior performance can be attributed to the segmentation of object parts. On the Internet dataset of [10] two out of three other algorithms are outperformed. A possible reason for inferior performance compared to [10] might be due to the dataset comprising centered objects, which apply smaller force onto the deformation field. We decided not to implement a feature of this kind as it would interfere with the intended purpose since objects might also occur non-centered in a detection task.

3 Incorporating Object Detection

A straightforward approach to object detection could be carried out by adding the detection image to the cosegmentation's training set. However, the additional image would bias the object model and detection would have fairly high computational demands as every image in the detection set requires its own cosegmentation. Therefore the two processes are separated and a discrete detector is trained on the cosegmentation's result. Contrary to common object detection settings, the set of positive examples is quite small so detection becomes more challenging. Hence typical methods that rely on large training sets cannot be applied in this case and we used a fairly straightforward approach: The score for a detection window is its maximal histogram intersection with all models from the cosegmentation set, i.e. only the accordance with the nearest neighbor matters. This is different from the cosegmentation procedure where the intersection with all other images is maximized. The latter is motivated by the need for stability in cosegmentation, as both, object location and model, need to be found while the model is already available for detection. The degree of variation in the cosegmentation set determines the maximal diversity of objects that can be detected. So it is assumed that the cosegmentation image set is versatile enough to ensure a robust detection.

3.1   Evaluation

In order to evaluate the detection quality, bounding boxes computed from entire image features are compared to boxes where only the cosegmented region was considered for feature computation. Both are compared to ground truth bounding boxes through dividing intersection by union.

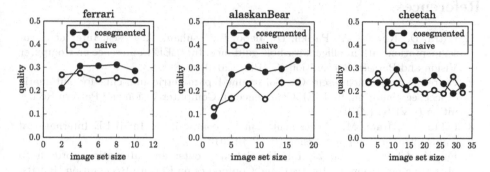

Fig. 3. The detection quality for different training image set sizes and three categories.

The detection quality reported in Fig. 3 clearly shows a superior performance of the cosegmentation-trained detector in almost all categories and for all set sizes. However, the overall performance is quite low, given that it does not exceed 0.3 where 1.0 indicates an optimal detection. A possible explanation is that constituents rather than entire objects have been detected as discussed above. Also the algorithm is quite simple and can be easily biased by both, a bad cosegmentation and a bad training image, whose foreground object strongly resembles the background of another image. Due to the nearest neighbor principle detection can easily fail in such a case. In addition, consider that the number of training examples is even for the largest sets of 30 images still fairly small. This might not be sufficient to ensure a stable nearest neighbor matching for all possible transformations in terms of perspective and lighting, especially if there are redundancies inside the training set.

Despite these minor issues it can be concluded that cosegmentation actually improves detection. Admittedly, this has been shown only for a fairly simple case, but these observations clearly give rise to conduct more elaborate investigations.

4   Conclusion

The main idea of this paper is to train an object detector in an unsupervised setting through cosegmentation. Along the way several contributions have been made: First, a custom-tailored cosegmentation algorithm is proposed which relies only on expressive histogram features, namely color and bag-of-word histograms. Despite being tuned for detection, competitive results on common cosegmentation datasets are attained. Regarding detection, results are very promising:

Cosegmentation has improved the detection quality in almost all cases, although the degree of improvement seems to be category dependent.

Future work might involve an alternation between hypothesis generation and cosegmentation to capture objects with a lower degree of saliency. Also more sophisticated representations could be employed.

References

1. Batra, D., Kowdle, A., Parikh, D., Jiebo, L., Tsuhan, C.: icoseg: Interactive co-segmentation with intelligent scribble guidance. In: IEEE Conference on Computer Vision and Pattern Recognition (CVPR) (2010)
2. Carreira, J., Sminchisescu, C.: Constrained parametric min-cuts for automatic object segmentation. In: IEEE Conference on Computer Vision and Pattern Recognition (CVPR) (2010)
3. Faktor, A., Irani, M.: Co-segmentation by composition. In: IEEE International Conference on Computer Vision (ICCV) (2013)
4. Fu, Y., Cheng, J., Li, Z., Lu, H.: Saliency cuts: an automatic approach to object segmentation. In: International Conference on Pattern Recognition (ICPR), pp. 1–4 (2008)
5. Geman, S., Geman, D.: Stochastic relaxation, gibbs distributions, and the bayesian restoration of images. IEEE Trans. Pattern Anal. Mach. Intell. (PAMI) 6(6), 721–741 (1984)
6. Gunhee, K., Xing, E.: On multiple foreground cosegmentation. In: IEEE Conference on Computer Vision and Pattern Recognition (CVPR) (2012)
7. Joulin, A., Bach, F., Ponce, J.: Multi-class cosegmentation. In: IEEE Conference on Computer Vision and Pattern Recognition (CVPR) (2012)
8. Rother, C., Minka, T., Blake, A., Kolmogorov, V.: Cosegmentation of image pairs by histogram matching - incorporating a global constraint into MRFs. In: IEEE Conference on Computer Vision and Pattern Recognition (CVPR), vol. 1 (2006)
9. Rother, C., Kolmogorov, V., Blake, A.: Grabcut: interactive foreground extraction using iterated graph cuts. ACM Trans. Graph. (TOG) - SIGGRAPH 23, 309–314 (2004)
10. Rubinstein, M., Joulin, A., Kopf, J., Liu, C.: Unsupervised joint object discovery and segmentation in internet images. In: IEEE Conference on Computer Vision and Pattern Recognition (CVPR) (2013)
11. Tola, E., Lepetit, V., Fua, P.: Daisy: an efficient dense descriptor applied to wide-baseline stereo. IEEE Trans. Pattern Anal. Mach. Intell. (PAMI) 32(5), 815–830 (2010)
12. Vicente, S., Rother, C., Kolmogorov, V.: Object cosegmentation. In: IEEE Conference on Computer Vision and Pattern Recognition (CVPR) (2011)

Multi-atlas Based Segmentation of Corpus Callosum on MRIs of Multiple Sclerosis Patients

Anneke Meyer[✉]

Otto-von-Guericke University of Magdeburg, Magdeburg, Germany
anneke.meyer@st.ovgu.de

Abstract. In this work, a supervised automatic multi-atlas based segmentation method for corpus callosum (CC) in magnetic resonance images (MRIs) of MS patients is presented. Due to atrophy, the shape of disease affected CC differs distinctively from healthy ones. Therefore, atlases are used that are built from the underlying dataset and do not originate from atlas datasets of healthy brains. The atlas construction is done by clustering the patient images into subgroups of similar images and building a mean image from each cluster. During this work, the optimal number of atlases and the best label fusion method are analyzed. The method is evaluated on 100 T1-weighted brain MRI images from MS patients. Accuracy is assessed by comparing the overlap of the segmentations from the developed method against manual segmentations obtained by a medical student.

Keywords: Multi-atlas based segmentation · Atlas construction · Corpus callosum · Label fusion

1 Introduction

Research in Multiple Sclerosis often relies on morphometry of multiple internal brain structures. Therefore, these structures need to be segmented. Since manual segmentation is time-consuming, expensive and subject to various errors including inter-and intra-rater errors, accurate automatic segmentation methods are in high demand.

A popular segmentation method for cerebral structures is atlas based segmentation. An *atlas* is described as an anatomical template image in which relevant structures are already segmented (*labels*). For segmentation purposes, the atlas is aligned to the target image via registration. The obtained transformation is used to align the manually segmented labels to the target image, resulting in segmentation of the query image.

More accurate results can be achieved when multiple atlases are used in order to reduce bias in the segmentation result [1]. Multiple atlases produce multiple candidate segmentations that are then fused to a final consensus segmentation (*label fusion*).

Recommended for submission to YRF2014 by Prof. Dr.-Ing. Klaus Tönnies, Otto-von-Guericke University of Magdeburg.

© Springer International Publishing Switzerland 2014
X. Jiang et al. (Eds.): GCPR 2014, LNCS 8753, pp. 729–735, 2014.
DOI: 10.1007/978-3-319-11752-2_61

Besides the registration algorithm, the results in multi-atlas based segmentation are highly dependent on the chosen atlases and the label fusion technique. If the atlas differs distinctly from the target image, accurate segmentation can not be achieved. Thus, a lot of research is done in the construction of an unbiased atlas. For example, Blezek et al. [2] showed that large datasets have different modes and should be therefore better represented by multiple atlases. The modes are discovered in population using mean shift algorithm where mean square distance and mutual information define the similarity measure.

A major cause for inaccurate automatic segmentation results of CC from MS patients is the thinning of the CC due to atrophy in the course of the disease [3–5]. Thus, in this work multiple atlases are built from the underlying dataset in order to represent the variations of shape of the CC. The approach is to cluster the dataset according to the estimated length and average width of the CC. From each cluster a mean image is computed that represents the atlas for this cluster.

In addition to the atlas construction method, the focus in this work is set to analyze the optimal number of atlases and the best performing label fusion algorithm (STAPLE, majority voting, similarity weighted majority voting).

2   Methods and Materials

In this section the pipeline for the supervised automatic multi-atlas based segmentation is presented.

2.1   Preprocessing

Improved segmentation results can be expected if the registration process is concentrated just on the important area (region around the CC). Thus, a region of interest (ROI) is computed that contains the whole CC in each image of the dataset.

Region of Interest. For ROI computation, the patient images are spatially normalized via affine registration to the MNI152 template, provided by the FSL toolbox.[1] Then, atlas based segmentation is performed by registering the patient images to the MNI template using symmetric diffeomorphic image registration[2] (SyN) [6]. According to Klein et al. [7], SyN is one of the best performing registration algorithms for brain MRI. After registration, all segmentations are accumulated to a sum image. This image is trimmed to the smallest rectangular region that contains all of the non-background voxels. This region is expanded by 10 voxels in each dimension for ensuring that each CC is contained completely and no registration errors impact the size of the ROI. Finally, the computed ROI is applied to all spatially normalized images.

---

[1] http://fsl.fmrib.ox.ac.uk/fsl/fslwiki

[2] Provided by ANTS toolbox (https://www.nitrc.org/projects/ants).

For subsequent computation of length and average width of the CC, a special label based on the MNI152 template was created that contains control points on the centerline of the MNI152 label. The control points are drawn manually on the centerline in the midsagittal slice of the label and have increasing intensity values as identification. This label is propagated to the patient images via the transformation from the non-rigid registration of MNI152 template to the patient images. The length L is then the accumulated Euclidean distances between consecutive control point pairs. The average width W is defined as $W = {}^{A_{CC}}/_L$, where A_{CC} is the area of CC as defined by the number of pixels on the midsagittal slice.

2.2   Atlas Construction and Registration

In order to represent the variation of shape of the CC, the dataset is clustered according to the length and average width of the CC using the K-means algorithm. This clustering algorithm allows the user to determine the number of clusters and therefore the number of atlases.

For each cluster a mean intensity image is constructed that is later used as the anatomical atlas. Two algorithms are tested for this purpose: atlas building by self-organized registration and bundling (ABSORB [8]) and ANTS parallel template building (APTB) that is based on symmetric diffeomorphic groupwise registration [9, 10]. APTB turned out to result in much better anatomical atlases than ABSORB and is therefore used in this pipeline.

After building the anatomical atlas, an atlas label needs to be created. Five slices around the midsagittal slice and the midsagittal slice itself are manually segmented. Then, the ROI is applied to both the anatomical atlas and the atlas label.

Finally, each atlas needs to be aligned to the individual patient images. As in the single-atlas based segmentation from Sect. 2.1, registration of the multiple atlases is done by symmetric diffeomorphic normalization (SyN) [6].

2.3   Label Fusion

After aligning multiple anatomical atlases and their corresponding manual labels to the patient images, the labels need to be fused to obtain final consensus segmentation. Three different label fusion methods are studied in this work: majority voting, similarity weighted voting and STAPLE.

Majority voting is the simplest approach to combine labels. It assigns to each voxel the label that the majority of segmentations vote for.

Since majority voting weights each candidate segmentation equally, it assumes different atlases represent the subject equally well. But as described in [2] and due to the high variety in shape of CC, the dataset is expected to have different modes that are best represented by different atlases. Thus it is reasonable to take into account the similarity between atlas and query image and weight each segmentation accordingly.

There are three similarity measures tested in this work: Cross correlation, mutual information and the distance of the image to the cluster centers (regarding length and average width of the CC), that is defined as:

$$m = \frac{min_i \sqrt{(l_p - l_{C_i})^2 + (w_p - w_{C_i})^2}}{\sqrt{(l_p - l_{C_i})^2 + (w_p - w_{C_i})^2}} \qquad (1)$$

where l_p is the length of the roughly segmented patient image, l_{C_i} the center of the i-th cluster and w_p the average width of the segmented patient image, respectively.

STAPLE (simultaneous truth and performance level estimation), developed by Warfield et al. [11], estimates the true segmentation of different classifiers as well as the performance quality of different atlases (sensitivity and specificity) using expectation maximization. The candidate segmentations are fused by weighting the atlases according to their performance level.

2.4   Image Data and Evaluation Tools

The dataset which was used for evaluation consists of 100 T1-weighted 3D MRI images from MS patients. The images in the dataset are from different sites and are therefore acquired by different MRI scanners.

Manual segmentations of the CC on the midsagittal slice obtained by a medical student were used as ground truth. Since 15 manual segmentations were unusable, only 85 images could be used for comparing the manual segmentations with the automatic segmentations.

For comparison purposes the Dice coefficient (DC) between manual and automatic segmentations is computed as described in [12]. In addition, the fraction of overestimation (FOE) and fraction of underestimation (FUE) are computed as described in [13]. They give a better idea whether an algorithm tends to oversegment or undersegment the CC and help to understand the DC score.

3   Results

In the following section the impact of different numbers of atlases, different label fusion methods as well as the impact of the ROI is studied. Results of the average evaluation measures DC, FOE and FUE as well as the standard deviation (SD) of DC can be found in Table 1.

3.1   Experiment I - Best Label Fusion Strategy

The following label fusion strategies were analyzed: majority voting (MAJ), weighted majority voting and STAPLE. For weighted majority voting three similarity measures (mutual information (WMI), cross correlation (WCC) and distance of the image to the cluster center (WDIS)) were investigated. Five atlases were used in this experiment.

Table 1. Results of the methods evaluated in the experiment.

Experiment	Method	FOE	FUE	DC	SD
Experiment I	MAJ	0.137	0.03	0.922	0,029
	WCC	0.134	0.031	0.922	0,029
	WMI	0.139	0.027	0.922	0,029
	WDIS	0.131	0.08	0.897	0,061
	STAPLE	0.043	0.133	0.908	0,022
Experiment II	Single atlas	0.161	0.066	0.895	0,088
	WCC_2	0.118	0.044	0.921	0,053
	WCC_3	0.095	0.05	0.929	0,035
	WCC_4	0.105	0.049	0.925	0,036
	WCC_5	0.134	0.031	0.922	0,029
	WCC_6	0.093	0.031	0.929	0,041
Experiment IV	MNI ROI	0.161	0.066	0.895	0,088
	MNI whole brain	0.295	0.015	0.867	0,052

MAJ, WCC and WMI achieve best segmentation results. The average DC for each of these three methods is 0.92. The range of DC values for STAPLE is smaller than the ones from MAJ, WCC and WMI. But the average DC for STAPLE is 0.91 and therefore slightly worse than the ones for the three best performing fusion techniques. WDIS performs worst by achieving only a mean DC of 0.90.

Most of the techniques tend to oversegment. Only STAPLE tends to undersegment (FOE = 0.04 and FUE = 0.13).

Since there is no performance difference between MAJ, WCC and WMI, a label fusion strategy can be chosen arbitrarily in order to determine the optimal number of atlases. The decision was made for cross correlation driven weighted majority voting.

3.2 Experiment II – Optimal Number of Atlases

For this experiment different number of atlases (1–6) were tested for the atlas based segmentation. For each number of atlases the images in the dataset were clustered and processed separately. Single-atlas segmentation is performed by using the MNI152 template as atlas.

Single-atlas segmentation obtains an average DC of 0.9. The results with more than one atlas are similar to each other and better than with one atlas: the DC is increased up to 0.93. The standard deviation for single atlas is 0.09 and higher than the ones from multiple atlases.

It is worth noting that the accuracy differences may result from manual segmentation variability. Therefore, small variation may result from differences in manual segmentations. But in the case of single atlas segmentation, the difference is clear enough to conclude that the use of multiple atlases improved the accuracy.

Examples for segmentation results can be found in Figs. 1 and 2. Figure 1 depicts an example for an inaccurate segmentation result whereas Fig. 2 depicts a very good

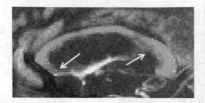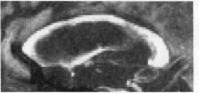

Fig. 1. Example for a segmentation that achieved a DC of 0.86 in WCC with 6 atlases. Left: midsagittal slice of original patient data with multi-atlas based CC segmentation (red) and the manual segmentation as overlay (green). Right: the same midsagittal slice of this brain volume without any segmentation overlays (Color figure online).

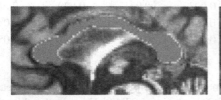

Fig. 2. Example for a very good automatic segmentation by 6 atlases and WCC. The achieved DC for this image is 0.96.

segmentation result. Both segmentations were achieved by use of 6 atlases and cross correlation driven weighted voting.

3.3   Experiment III - Influence of Region of Interest

For analyzing the impact of the ROI, single-atlas based segmentation with the MNI152 template as atlas is conducted on the whole image and on the ROI. The segmentation accuracy was improved from an average DC of 0.87 to a value of 0.90 by applying a ROI.

4   Discussion

The best segmentation accuracy of the CC on MRI images of MS patients was obtained by multi-atlas based segmentation with majority voting and weighted voting (similarity measure: cross correlation and mutual information) as label fusion. The ROI plays an important role in improving the accuracy.

Considering inter- and intra-rater variability of experts, the achieved average DC of 0.93 of this implemented pipeline is on the level of manual segmentation accuracy.

The FOE value indicates that the segmentation results tend to oversegmentation (average FOE of WCC with 6 atlases is 0.09). The oversegmentation is caused to a large part by partial volume effect or adjacent structures with similar intensity values (see Fig. 1). Additionally, the CC has not always clear boundaries at its tips.

This makes it difficult to automatically segment this part of the CC accurately without having any prior knowledge of comparing the segmentations of neighboring slices where the boundary of the CC at this position might be represented more clearly.

Acknowledgements. This work arose during my internship at the MSMRI Research Group (University of British Columbia in Vancouver, Canada). The dataset that was used for development and evaluation was kindly provided by MSMRI.

References

1. Rohlfing, T., Brandt, R., Menzel, R., Maurer Jr., C.R.: Evaluation of atlas selection strategies for atlas-based image segmentation with application to confocal microscopy images of bee brains. NeuroImage **21**(4), 1428–1442 (2004)
2. Blezek, D.J., Miller, J.V.: Atlas stratification. Med. Image Anal. **11**(5), 443–457 (2007)
3. Edwards, S., Liu, C., Blumhardt, L.: Cognitive correlates of supratentorial atrophy on MRI in multiple sclerosis. Acta Neurol. Scand. **104**(4), 214–223 (2001)
4. Martola, J., Stawiarz, L., Frederikson, S., Hillert, J., Bergström, J., Flodmark, O., Kristoffersen Wilberg, M.: Progression of non-age-related callosal brain atrophy in multiple sclerosis: a 9-year longitudinal MRI study representing four decades of disease development. J. Neurol. Neurosurg. Psychiatry **78**(4), 375–380 (2007)
5. Llufriu, S., Blanco, Y., Martinez-Heras, E., Casanova-Molla, J., Gabilondo, I., Sepulveda, M., Falcon, C., Berenguer, J., Bargallo, N., Villoslada, P.: Influence of corpus callosum damage on cognition and physical disability in multiple sclerosis: a multimodal study. PloS one **7**(5), e37167 (2012)
6. Avants, B.B., Epstein, C.L., Grossman, M., Gee, J.C.: Symmetric diffeomorphic image registration with cross-correlation: evaluating automated labeling of elderly and neurodegenerative brain. Med. Image Anal. **12**(1), 26–41 (2008)
7. Klein, A., Andersson, J., Ardckani, B.A., Ashburner, J., Avants, B., Chiang, M.-C., Christensen, G.E., Collins, D.L., Gee, J., Hellier, P.: Evaluation of 14 nonlinear deformation algorithms applied to human brain MRI registration. Neuroimage **46**(3), 786–802 (2009)
8. Jia, H., Wu, G., Wang, Q., Shen, D.: ABSORB: atlas building by self-organized registration and bundling. NeuroImage **51**(3), 1057–1070 (2010)
9. Avants, B.B., Yushkevich, P., Pluta, J., Minkoff, D., Korczykowski, M., Detre, J., Gee, J.C.: The optimal template effect in hippocampus studies of diseased populations. Neuroimage **49** (3), 2457–2466 (2010)
10. Yushkevich, P.A., Avants, B.B., Pluta, J., Das, S., Minkoff, D., Mechanic-Hamilton, D., Glynn, S., Pickup, S., Liu, W., Gee, J.C.: A high-resolution computational atlas of the human hippocampus from postmortem magnetic resonance imaging at 9.4 T. Neuroimage **442**, 385–398 (2009)
11. Warfield, S.K., Zou, K.H., Wells, W.M.: Simultaneous truth and performance level estimation (STAPLE): an algorithm for the validation of image segmentation. IEEE Trans. Med. Imaging **23**(7), 903–921 (2004)
12. Dice, L.R.: Measures of the amount of ecologic association between species. Ecology **26**(3), 297–302 (1945)
13. McAusland, J., Tam, R.C., Wong, E., Riddehough, A., Li, D.K.: Optimizing the use of radiologist seed points for improved multiple sclerosis lesion segmentation. IEEE Trans. Biomed. Eng. **57**(11), 2689–2698 (2010)

Committees of Deep Feedforward Networks Trained with Few Data

Bogdan Miclut[(✉)]

Institut für Neuro- und Bioinformatik, Universität zu Lübeck, Lübeck, Germany
miclutbogdan@gmail.com

Abstract. Deep convolutional neural networks are known to give good results on image classification tasks. In this paper we present a method to improve the classification result by combining multiple such networks in a committee. We adopt the STL-10 dataset which has very few training examples and show that our method can achieve results that are better than the state of the art. The networks are trained layer-wise and no backpropagation is used. We also explore the effects of dataset augmentation by mirroring, rotation, and scaling.

1 Introduction

Recently, deep convolutional neural networks have been shown to perform very well on various challenging pattern-recognition benchmarks. Such networks trained in a supervised way via backpropagation achieve state of the art performance on the Caltech-101 [10], Caltech-256 [10], PASCAL VOC [7], MNIST [11], and ImageNet [5] datasets. However, the drawback of this approach is that the required vast amounts of labeled data are not always available.

This paper regards unsupervised training of deep neural networks and investigates whether a voting scheme (by a committee of networks) can improve the classification result. We test our method on the STL-10 dataset [2], because it is challenging and has only a small number of labeled training data.

For filter training, we use k-means as in [2]. After the convolutional step, we found that the local normalization presented by [9] improves the classification result. For the connection between layers, we adopt the random grouping of [4].

Recent state of the art results [6] on the STL-10 dataset prove that methods from supervised training can be adapted for unsupervised training of networks. Their method creates virtual classes by largely augmenting single images, then training networks on each of these virtual classes using backpropagation. In our paper, we show that better results can be obtained without using back propagation.

Another important result on STL-10 is presented in [12] where filters are trained layer-wise: in the first layer, filters are learned via k-means, while in the second layer, filters are being supervised trained via Fisher weight maps for maximizing the between-class distance of descriptors obtained after the second

---

Recommended for submission to YRF2014 by Erhardt Barth and Thomas Martinetz.

© Springer International Publishing Switzerland 2014
X. Jiang et al. (Eds.): GCPR 2014, LNCS 8753, pp. 736–742, 2014.
DOI: 10.1007/978-3-319-11752-2_62

layer. In our work, we only perform layer-wise unsupervised learning of filters. Finally, we show that a committee of networks improves the classification result.

2 Network

We view the network as a chaining of two stages: a feature extractor and a classifier. The term unsupervised refers to the first stage, which is blind to image labels. The output of the feature extractor is a set of descriptors (one descriptor for each input image). The descriptors of the training image set are used to train the classifier, which will, in the end, assign a label to a descriptor corresponding to a test image.

The feature extractor consists of one or more almost identical layers. In the following we will present the operations that are being done by such a layer. We define a feature map as a 2D array given as output by a layer of the network, when presented with one image as input. Thus, an input image is characterized by a set of feature maps given as output by any of the layers of the network. The main goal is to make these representations invariant to certain transformations, such as translation, scaling, rotation.

2.1 Preprocessing and Filter Training

Let $F = \{f_i | f_i \in \mathbb{R}^{m_k \times n_k \times l_k}\}$ be the input set of layer k. Here m_k, n_k represent the 2D dimensions of one feature map, l_k is the number of feature maps in layer k and $i = 1, ..., N_{train}$ where N_{train} is the number of training images considered. (in the case of the input of the first layer, $l_1 = 1$ and m_1, n_1 are the dimensions of the images).

From the set F, we extract a set of patches $X = \{x_k | x_k \in \mathbb{R}^{p \times p \times l_2}\}$; for simplicity we extract volume patches consisting of $p \times p$ squares across l_2 feature maps (for the first layer, $l_2 = 1$, therefore each patch is a standard $p \times p$ square). The elements of x_k are unrolled, thus forming the matrix $X \in \mathbb{R}^{(p \cdot p \cdot l_2) \times N_{patches}}$, where $N_{patches}$ is the number of patches extracted. Each column of X is a x_k.

We employ patch-wise normalization as follows: we scale each patch by dividing by the maximum of the absolute value of its elements, then we center each patch by subtracting its mean. After this, we do ZCA whitening.

Filters are trained using k-means clustering on the preprocessed patches. Thus, we obtain $U = \{u_k | u_k \in \mathbb{R}^{(p \cdot p \cdot l_2) \times K}\}$, where K is the number of trained filters.

2.2 Convolutions

After learning the filters, we do a dense feature extraction: for each patch in the input feature maps, we apply all filters via dot product: $y_{ik} = < x_i, u_k > = \sum_j x_{ij} \cdot u_{kj}$. Up to this point, the network is doing pattern matching.

2.3 Rectification and ON-OFF Separation

We use absolute value as the simplest nonlinearity function. In addition to simple rectification (taking the absolute value) we use ON-OFF separation, where the values $max(0, x)$ and $max(0, -x)$ are fed separately into the next stage.

2.4 Local Contrast Normalization

This step was adopted from [9]. It performs local subtractive and divisive operations. Let x_{ijk} be the set of feature maps obtained after the *rectification* stage for one input image. Then, we have: $v_{ijk} = x_{ijk} - \sum_{ipq} w_{pq} \cdot x_{i,j+p,k+q}$, where w_{pq} is a Gaussian weighting window (of size $S \times S$, S depending on the size of the input) normalized so that $\sum_{ipq} w_{pq} = 1$. This subtractive operation is a form of edge detection. For the divisive normalization, we have: $y_{ijk} = v_{ijk}/max(c, \sigma_{jk})$, where $\sigma_{jk} = (\sum_{ipq} w_{pq} \cdot v_{i,j+p,k+q}^2)^{1/2}$. This divisive operation is a form of automatic gain control.

2.5 Pooling

Pooling is performed only within a feature map (across a 2D domain). Pooling is done within patches of size $p \times p$ with stride s (typically $s = p$, meaning pooling is done on disjoint neighboring patches).

Let x be a 2D patch. As a pooling function we implemented: $y = (\sum_i x_i^\alpha)^{1/\alpha}$. Such we can move from average pooling by setting $\alpha = 1$ to max pooling, for a large α.

The output of the pooling stage is also the output of a layer of the network. We can view a layer of the network as a black box, having a set of feature maps as input, and yielding K (the number of trained filters) smaller feature maps as output.

2.6 Connections Between Layers

This section addresses the issue of layer interconnectivity. The first layer gives for each input images a set of K_1 feature maps; therefore, we will have a total number of $K_1 \times N_{training}$ feature maps (K_1 is the number of filters trained in the first layer and $N_{training}$ is the number of images in the training set).

We implemented the *random grouping* explored by [4]: the K_1 feature maps are divided into $\frac{K_1}{n_K}$ groups of n_K. Each group will be treated separately, namely, each group will give K_2 feature maps as output (where K_2 is the number of filters trained for each group; we chose the same number of filters for each group for simplicity). So, as an example, after the second layer, we will have $K_2 \times \frac{K_1}{n_K} \times N_{training}$ feature maps, given we present all the training data to the input of the network.

2.7   Classifier

As a classifier we use a multiclass one-vs-all linear L2 SVM. When presented with a test image, the classifier will give a set of C values (scores). Here C is the number of classes. The assigned class will be the position of the maximum among these scores.

3   Committee of Networks

Here, we ignore the inner workings of the network and consider it as a black box that, when presented with an input image, will predict correct labels with a certain probability. Now, we can ask whether there is a strategy of combining multiple such black boxes in order to get a higher success rate. In order to investigate this, we will have to look at the output of the classifiers.

Consider an abstract classifier that, when presented with an input, gives a set of C scores between some arbitrary C_{min} and C_{max}. How does one pick the most likely label? A straight forward answer is to pick the label corresponding to the highest score. Now, two cases can occur: (i) one score is very large in comparison to all the others; (ii) one score is largest, but has some other scores very close to it.

In case (ii), and knowing that the classifier is not always right, some of the true labels are hidden in the scores close to the maximum one. Therefore, we must solve the following problem: the predicted label is false, but the true label has a score close to that of the predicted one.

One way of looking at this is to consider the process of assigning labels as a noisy stochastic process. The goal is to filter the noise out. In order to be able to do this, we need multiple realizations of this process for each input image.

Training a network with the same parameters does not necessarily give the same descriptors as output. There are three reasons for this: the clustering process of k-means, the ordering of filters obtained by k-means, and the random grouping of feature maps.

In order to get a wider variety, we use the augmented dataset to train multiple networks; we also vary the parameters, such as pooling size and stride, to obtain even more networks. If we are to combine values from the classifiers, these values have to be comparable. Thus, we disregard the original meaning of the classifier output and scale the set of values to the range $[0, 1]$ in order to obtain some scores that will be later combined.

Now, for an input test image, we have N sets of scores in $[0, 1]$. Each set of scores corresponds to the output of one network.

We have chosen the simplest method to combine the scores, which is to sum them up. Let $S_i = [s_{i1}, s_{i2}, ..., s_{iC}]$ be one set of scores, where $i = 1, ..., N$ and C is the number of classes.

Then, $S_{committee} \equiv S_c = \sum_{j=1}^{N} S_j$. The decision of the committee is taken in the same way as before: the highest score gives the class label.

4   Experimental Results

In this section we will describe the specific parameters chosen for the network architecture and present the results obtained.

All experiments were done on the STL-10 dataset with networks having 2 layers of feature extractors. We use this dataset to prove that good results can be obtained even with few training examples. The STL-10 dataset is comprised of a large collection of unlabeled data, which we do not use in our experiments, 5000 labeled training images and 8000 test images. There are 10 predefined folds, each fold containing 1000 training images. We test in the following way: we train on each fold of 1000 examples and test on the full set of 8000 images; we then report the average success rate over the 10 folds and the standard deviation. One of the reasons for choosing this dataset is the ratio of training vs test images, which is 1 to 8.

Images are first converted to grayscale. For each layer, the input goes through this chain: patch-wise preprocessing (as described above) and filter training, convolution, absolute value rectification or ON-OFF separation, local contrast normalization, and average pooling.

For the first layer, we worked with 300 filters of size 16×16. Pooling was done only across the spatial domain. Various pooling sizes and strides have been tried. In our experiments, average pooling ($\alpha = 1$) was performing better than other values of α.

For the connection between the first layer and the second one, we employed random grouping: feature maps were stacked together in groups of 4. We experimented other sizes of grouping, and the main finding was that it does not matter so much how many feature maps are grouped together, but it does matter that the dictionary trained for them is overcomplete. In the second layer, filters were of size $3 \times 3 \times 4$: 3×3 in spatial dimension, across 4 feature maps. Hence, the dimensionality is 36. We typically trained 70–80 filters for each group. So the choice for grouping of 4 was a practical one in order to keep the dimensionality low and for k-means to be able to converge rapidly.

Pooling in the second layer (again average pooling) was done only in the spatial domain with a fixed 3×3 pixels pooling window and a stride of 3 pixels. Other pooling sizes have been tried, but no significant differences were observed.

To further improve generalization performance, we considered a few transformations used to augment the data set: left-right mirroring, rotations of $\pm 10°$ and scaling by a factor of $1/3$.

Once several networks were trained, we obtained the classification results of each one and normalized them in order to create the scores. Then, the scores were added up with the position of the maximum giving the class label.

In Table 1 we present the results obtained by varying parameters of individual networks, and the effects of dataset augmentation.

Using these base networks, we build the committee consisting of: N_1 (mirroring + rotations), N_2 (mirroring + rotations), N_3 (mirroring + rotations), N_4 (mirroring), N_5 (ON-OFF + mirroring). With this committee, by summing up the scores, we obtain a classification rate of 68.0 %. With a reduced committee

Table 1. Classification results on STL-10 with single networks and varying network parameters

Network	Scaling	Rectification	Pooling, Stride	No augmentation	Mirroring Mirroring	Mirroring and rotations
N_1	no	absolute value	$12 \times 12, 12$	60.57	63.27	63.60
N_2	no	absolute value	$12 \times 12, 8$	60.61	63.41	64.59
N_3	no	absolute value	$9 \times 9, 9$	59.89	62.90	64.27
N_4	1/3	absolute value	$4 \times 4, 4$	58.9	60.52	not tested
N_5	no	ON-OFF	$12 \times 12, 12$	61.1	64.73	not tested

Table 2. Classification accuracy on STL-10 dataset

Paper	Result
Unsupervised feature learning by augmenting single images [6]	67.4 ± 0.6
Efficient Discriminative Convolution Using Fisher Weight Map [12]	66.0 ± 0.7
Unsupervised Feature Learning for RGB-D Based Object Recognition [1]	64.5 ± 1
Discriminative Learning of Sum-Product Networks [8]	62.3 ± 1
Selecting Receptive Fields in Deep Networks [3]	60.1 ± 1
This paper	68.0 ± 0.55

with only mirroring as dataset augmentation we obtain 67.39 %. Table 2 shows a comparison of the results of this paper and the state of the art.

5 Conclusions

In this paper we showed that combining different networks and employing a voting scheme improves the classification result. The committee can be constructed from any base network by varying its parameters or presenting as input different transformations applied on the training set. When building the committee, one has to bare in mind that the output of one network is not independent from the output of the others, thus the performance will not always increase with the number of networks, but will eventually saturate or even decrease if the choice of networks is poor (for example, adding a very bad performing network). In our experiments, the committee always performed better than any of the individual networks. Note that we achieve results that are better than state of the art by using rather simple two-layer networks for feature extraction and linear SVMs for classification.

References

1. Bo, L., Ren, X., Fox, D.: Unsupervised Feature Learning for RGB-D Based Object Recognition. In: ISER, June 2012
2. Coates, A., Lee, H., Ng, A.Y.: An analysis of single-layer networks in unsupervised feature learning. AISTATS (2011). http://cs.stanford.edu/~acoates/stl10
3. Coates, A., Ng, A.Y.: Selecting receptive fields in deep networks. In: Shawe-Taylor, J., Zemel, R., Bartlett, P., Pereira, F., Weinberger, K. (eds.) Advances in Neural Information Processing Systems 24, pp. 2528–2536. Curran Associates, Inc. (2011). http://papers.nips.cc/paper/4293-selecting-receptive-fields-in-deep-net-works.pdf
4. Culurciello, E., Jin, J., Dundar, A., Bates, J.: An analysis of the connections between layers of deep neural networks. CoRR abs/1306.0152 (2013)
5. Deng, J., Dong, W., Socher, R., Li, L.J., Li, K., Fei-Fei, L.: ImageNet: a large-scale hierarchical image database. In: CVPR09 (2009)
6. Dosovitskiy, A., Springenberg, J.T., Brox, T.: Unsupervised feature learning by augmenting single images. CoRR abs/1312.5242 (2013)
7. Everingham, M., Van Gool, L., Williams, C.K.I., Winn, J., Zisserman, A.: The pascal visual object classes (voc) challenge. Int. J. Comput. Vision **88**(2), 303–338 (2010)
8. Gens, R., Domingos, P.: Discriminative learning of sum-product networks. In: Pereira, F., Burges, C., Bottou, L., Weinberger, K. (eds.) Advances in Neural Information Processing Systems 25, pp. 3239–3247. Curran Associates, Inc. (2012). http://papers.nips.cc/paper/4516-discriminative-learning-of-sum-product-networks.pdf
9. Jarrett, K., Kavukcuoglu, K., Ranzato, M., LeCun, Y.: What is the best multi-stage architecture for object recognition? In: Proceedings International Conference on Computer Vision (ICCV 2009). IEEE (2009)
10. Fei-Fei, L., Fergus, R., Perona, P.: Learning generative visual models from few training examples: an incremental bayesian approach tested on 101 object categories. IEEE CVPR (2004)
11. LeCun, Y., Bottou, L., Bengio, Y., Haffner, P.: Gradient-based learning applied to document recognition. Proc. IEEE **86**(11), 2278–2324 (1998)
12. Nakayama, H.: Efficient discriminative convolution using fisher weight map. In: Proceedings of the British Machine Vision Conference. BMVA Press (2013)

Gas Bubble Shape Measurement and Analysis

Claudius Zelenka[(✉)]

Institute of Computer Science, Kiel University, Kiel, Germany
`cze@informatik.uni-kiel.de`

Abstract. This work focuses on the precise quantification of bubble streams from underwater gas seeps. The performance of the snake based method and of ellipse fitting with the CMA-ES non-linear optimization algorithm is evaluated. A novel improved snake based method is presented and the optimal choice of snake parameters is studied. A Kalman filter is used for bubble tracking. The deviation between the measured flux and a calibrated flux meter is 4% for small and 9% for larger bubbles. This work will allow a better data gathering on marine gas seeps for future climatology and marine research.

1 Introduction

Underwater gas seeps have essential influence on marine life [6] and global climate [12]. Modeling and understanding of their complex impacts on the biosphere requires exact knowledge of the composition and volume of the emitted gases and their absorption in the surrounding water [13].

Gas bubbles emerging from a seep in the sea floor are rising with high speed, and may change form, contour and volume during their ascent [4]. The goal of this research is the development of robust methods for automatic image processing, which can be employed to extract the shape, motion and volume of bubbles from image sequences.

The first part of this paper is dedicated to the detection of gas bubbles. Several methods are compared, the snake [8] approach and ellipse fitting with the CMA-ES [5] non-linear optimization algorithm. Diffuse back light illumination is used, in a similar configuration as in previous work by [10,11], or [15].

For the detection of bubbles different detection methods are established in previous work. In [2] active contours are initialized using the boundaries of the binarized input image. Another approach, presented in [3] uses a specialized Hough transformation on the input image to locate the circular or ellipsoid shape of bubbles. In [14] the application of a Canny edge detector [1] for contour extraction is presented. These contours are fitted and smoothed with an ellipse. In this work the Canny edge detection according to this publication [14] is used as a baseline method. On this basis, methods and algorithms based on the standard and improved snake algorithm are developed and evaluated. The CMA-ES based ellipse fitting method with diverse novel fitness functions is in

---

Recommended for submission to YRF2014 by Prof. Dr.-Ing. Reinhard Koch, Kiel University.

© Springer International Publishing Switzerland 2014
X. Jiang et al. (Eds.): GCPR 2014, LNCS 8753, pp. 743–749, 2014.
DOI: 10.1007/978-3-319-11752-2_63

this work extensively studied. The results are compared with manual ground truth evaluation of the bubble size.

In the second part of this work, methods for the tracking of bubbles and measuring of their velocity are developed. The accuracy of the detection and tracking methods in combination are verified by comparison with manually obtained flux data. Figure 1 illustrates the typical irregular bubble shapes and the bubble matching.

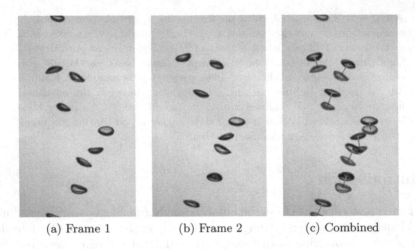

(a) Frame 1   (b) Frame 2   (c) Combined

Fig. 1. Movement of bubbles in between two frames at 100 fps. The combination shows frame 1 and 2 with the bubble detection in frame 2 in red and matching bubbles connected in green (Color figure online).

2   Bubble Detection

In this section methods for the reliable detection of bubble contour in image sequences are developed. The Canny Edge detector based method by [14] is used as a baseline. It sporadically suffers from false detections, which motivates the search for more stable algorithms. These false detections are caused by the light area inside the bubble. As an improvement a bounding box is formed around these detections and expanded. This bounding box is used as an initialization for the snake [8] and CMA-ES [5] based methods, which are introduced, tested and evaluated in the following.

2.1   Snake Method

The term snake was coined by Kass et al. in [8] and describes a spline, which is adjusted by the snake method to fit a prior defined criterion. For every contour, an energy function is declared. It can be described in two parts, the internal

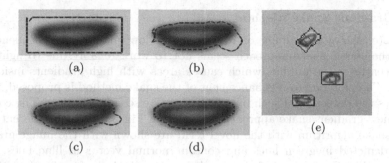

(a)   (b)

(c)   (d)   (e)

Fig. 2. Subfigures (a)–(d) show steps in the optimization process of the classic snake algorithm, from initialization to termination. The control points are depicted in green, while the linear interpolation in between them is shown in blue. Subfigure (e) shows an imperfect detection with the classic snake algorithm on an image with low image quality (Color figure online).

energy E_{int}, weighted with α and β, which is dependent purely on the snake's shape, and the external energy E_{ext}. Let $S : M \to P$ be a snake, with $M = [0,1]$ and $P \subset \mathbb{R}^2$ for all points on the snake. Let S_{final} be the optimal configuration of the snake:

$$S_{final} = arg\,min(E_{int}(S) + E_{ext}(S)). \qquad (1)$$

$$E_{int}(S) = \int_0^1 \alpha \left\| S'(c) \right\|^2 + \beta \left\| S''(c) \right\|^2 dc. \qquad (2)$$

The external energy is used to describe the part of the energy, which depends on the image signal. It is weighted with parameter γ. Given image $I : [0,x] \times [0,y] \to \mathbb{R}^+$ with width $x \in \mathbb{Z}^+$ and height $y \in \mathbb{Z}^+$, it holds that:

$$E_{ext}(S) = \int_0^1 -\gamma \left\| \nabla I(S(c)) \right\|^2 dc. \qquad (3)$$

The snake contour is initialized with the bounding box from the baseline localization. Figure 2 shows different iterations of a succesful localization of a bubble with a snake, from initialization to convergence.

The choice of snake parameters α, β and γ, which control the snakes affinity to continuity, low curvature and high gradients have significant impact on the convergence behaviour of snakes. An important question, which is not sufficiently adressed in previous work, concerns the optimum choice of snake parameters. In this work extensive investigation is done to find out, whether the snake parameters can stay constant, or must be adapted to each bubble stream, or perhaps to each bubble. The entire parameter space of α, β and γ has been tested with different bubble sequences. It was shown, that if α is high enough, then the parameters choice is not critical and the same optimized set of parameters can be used for different bubble streams, even from different cameras.

2.2   Gradient Snake Method

Figure 2(e) shows an example of a bubble in an low quality image sequence, on which the classic snake approach seems not to work reliably due to light area in the centre of the bubble, which causes areas with high gradients inside the bubble. Therefore, a novel enhancement of the snake method is proposed, which improves the detection of such bubbles, as can be seen in Fig. 3(e). The concept of the new gradient snake approach is shown on Fig. 3, in which different steps of the snake algorithm with the novel term are shown with the image gradient vector depicted in green hues and contour normal vector in lilac hues. They illustrate the effectiveness of the novel term particularly for the detection of bubbles.

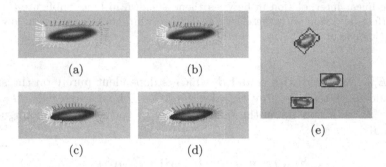

(a)   (b)

(c)   (d)

(e)

Fig. 3. Subfigures (a)–(d) show every third step in the optimization process of the snake algorithm with directional gradient, from initialization to termination. Lilac hues show the normal of the contour, green hues show the direction and magnitude of the gradient indicated by the length of the greenish indicators. Subfigure (e) shows the detection with the gradient snake algorithm on the same input image as Fig. 2(e) (Color figure online).

At position c, the dot product is applied to the image gradient and the outwards normal $\widehat{n}$ of the contour. If they point in the same direction, the result is positive and if they point in different directions it is negative. The result is scaled with the squared absolute value of the image gradient and is similar to E_{ext} weighted with parameter γ. For the following applies:

$$E_{gd}(S) = \int_0^1 -\gamma \left\langle \frac{\nabla I(S(c))}{\|\nabla I(S(c))\|}, \widehat{n}(S(c)) \right\rangle \|\nabla I(S(c))\|^2 \, dc. \qquad (4)$$

Using the sum of the normal and gradient direction vectors, instead of the dot product, has the advantage of gaining linear sensitivity to an angular difference between $\widehat{n}$ and $\nabla S(c)$ over the *cosine* and lower computational complexity. The difference is then scaled by 0.5 to achieve values, which are in the same scale as the original snake algorithm, for coinciding vectors:

$$E_{gd}(S) = \int_0^1 -0.5 \, \gamma \left\| \frac{\nabla I(S(c))}{\|\nabla I(S(c))\|} + \widehat{n}(S(c)) \right\| \|\nabla I(S(c))\|^2 dc. \qquad (5)$$

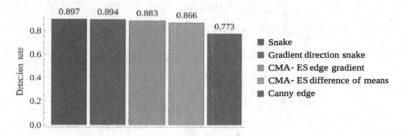

Fig. 4. Comparison of detection rates for several selected methods on an high quality GoPro image sequence with manual ground truth on 20 images.

2.3   CMA-ES Based Ellipse Fitting

Covariance Matrix Adaption - Evolution Strategy (CMA-ES) is a derivative free non-linear optimization algorithm [5]. For the detection of gas bubbles the algorithm is initialized with the expanded bounding box of the Canny edge detection, as described in Sect. 2. The CMA-ES algorithm then adjusts an ellipse to the bubble, according to a prior defined fitness function. The ellipse used in the optimization is defined by five parameters, the horizontal and vertical position of the bubble, its minor and major axis and a rotation parameter. Different fitness functions have been developed and tested. Two designs showed the best results, firstly a fitness function aiming for a high gradient on the edge of the ellipse and secondly an area based fitness function evaluating the difference between the mean intensities of the ellipse area and its surroundings.

2.4   Comparison of Detection Methods

In Fig. 4 the detection rates of different methods are compared with manual ground truth for an image sequence of 20 images with ca. 10 bubbles visible per frame. All tested methods show good results and a clear improvement compared to the baseline Canny edge detector, however both snake based methods showed the best performance. The fastest method is the classic snake algorithm with timings of about 30 ms per frame. Even though it is slower, being most reliable on images with low image quality (see Fig. 3(e)), the gradient direction snake has been determined as the bubble detection method of choice.

3   Bubble Tracking

Bubble tracking is employed to determine the movement of bubbles in between two frames and from this data over the entire image sequence. To effectively find the corresponding bubble detections between two frames, the bubble positions are detected in every frame. The bubble motion since the last frame is predicted using a Kalman filter [7]. These predictions and the new detections of bubble positions are matched with minimum weighted matching between prediction and

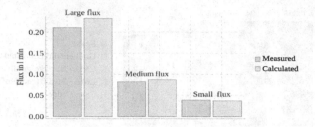

Fig. 5. Comparison between measured and calculated flux in liter per minute.

measurement using the hungarian algorithm [9], with the distances between these as edge weights. This resulting matching is the mapping of bubble movement between the last frame and this frame. A resulting matching is shown in Fig. 1 on the right image. In the tested 20 sequences, with a bubble detection rate according Fig. 4, the tracking is highly reliable. For a sequence such as Fig. 1, the framerate can be lowered by 50 % without deteriorating the tracking quality.

4 Conclusion

With the established methods for detecting and tracking bubbles, an accurate measurement of the flow volume from images of bubble stream can be done with high accuracy for wide range of flux volume, as shown in Fig. 5. The computed flux data has been compared with flux volume obtained by independent measurement with a calibrated volume flux meter. With small bubbles a very good accuracy of 4 % deviation can be achieved, for medium bubbles the deviation is 5 % and for larger bubbles the deviation is higher with 9 %. This can be explained by the rather irregular shape of larger bubbles.

It has been shown, that the snake method and the CMA-ES method are well suited for flux measurement of gas bubble seeps. The stability and performance of both methods was explored and improved.

Acknowledgments. This work was supported by the University of Kiel and GEOMAR Helmholtz Centre for Ocean Research Kiel. I am also grateful for support from Prof. Dr.-Ing. Reinhard Koch and Dr. Anne Jordt for my work.

References

1. Canny, J.: A computational approach to edge detection. IEEE Trans. Pattern Anal. Mach. Intell. PAMI **8**(6), 679–698 (1986)
2. Cheng, D.c., Burkhardt, H.: Bubble recognition from image sequences. In: Proceedings of Eurotherm Seminar 68: Inverse Problems and Experimental Design in Thermal and Mechanical Engineering (2001)
3. Goulermas, J.Y., Liatsis, P.: Novel combinatorial probabilistic hough transform technique for detection of underwater bubbles. In: Machine Vision Applications in Industrial Inspection V, vol. 3029, pp. 147–156 (1997)

4. Greinert, J., Artemov, Y., Egorov, V., Debatist, M., Mcginnis, D.: 1300-m-high rising bubbles from mud volcanoes at 2080m in the black sea: hydroacoustic characteristics and temporal variability. Earth Planet. Sci. Lett. **244**(1–2), 1–15 (2006)
5. Hansen, N., Kern, S.: Evaluating the CMA evolution strategy on multimodal test functions. In: Yao, X., Burke, E.K., Lozano, J.A., Smith, J., Merelo-Guervós, J.J., Bullinaria, J.A., Rowe, J.E., Tiňo, P., Kabán, A., Schwefel, H.-P. (eds.) PPSN 2004. LNCS, vol. 3242, pp. 282–291. Springer, Heidelberg (2004)
6. Ishimatsu, A., Kikkawa, T., Hayashi, M., Lee, K.S., Kita, J.: Effects of CO_2 on marine fish: larvae and adults. J. Oceanogr. **60**(4), 731–741 (2004)
7. Kalman, R.: A new approach to linear filtering and prediction problems. Trans. ASME J. Basic Eng. **82**(D), 35–45 (1960)
8. Kass, M., Witkin, A., Terzopoulos, D.: Snakes: active contour models. Int. J. Comput. Vis. **1**(4), 321–331 (1988)
9. Kuhn, H.W.: The hungarian method for the assignment problem. Naval Res. Logistics Q. **2**(1–2), 83–97 (1955)
10. Leifer, I., De Leeuw, G.: Optical measurement of bubbles: system design and application. J. Atmos. Oceanic Technol. **20**(9), 1317–1332 (2003)
11. Leifer, I., de Leeuw, G., Kunz, G., Cohen, L.: Calibrating optical bubble size by the displaced-mass method. Chem. Eng. Sci. **58**(23–24), 5211–5216 (2003)
12. Lelieveld, J., Crutzen, P.J., Dentener, F.J.: Changing concentration, lifetime and climate forcing of atmospheric methane. Tellus B **50**(2), 128–150 (1998)
13. McGinnis, D.F., Greinert, J., Artemov, Y., Beaubien, S.E., West, A.: Fate of rising methane bubbles in stratified waters: how much methane reaches the atmosphere? J. Geophys. Res. **111**(C9), C09007 (2006). doi:10.1029/2005JC003183
14. Thomanek, K., Zielinski, O., Sahling, H., Bohrmann, G.: Automated gas bubble imaging at sea floor; a new method of in situ gas flux quantification. Ocean Sci. **6**(2), 549–562 (2010)
15. Wang, H., Dong, F.: Image features extraction of gas/liquid two-phase flow in horizontal pipeline by GLCM and GLGCM. In: Proceedins of 2009 9th International Conference on Electronic Measurement & Instruments (ICEMI' 09), pp. 2-135–2-139. IEEE, August 2009

Scene Segmentation in Adverse Vision Conditions

Evgeny Levinkov[(✉)]

Max Planck Institute for Informatics, Saarbrücken, Germany
`levinkov@mpi-inf.mpg.de`

Abstract. Semantic road labeling is a key component of systems that aim at assisted or even autonomous driving. Considering that such systems continuously operate in the real-world, unforeseen conditions not represented in any conceivable training procedure are likely to occur on a regular basis. In order to equip systems with the ability to cope with such situations, we would like to enable adaptation to such new situations and conditions at runtime. We study the effect of changing test conditions on scene labeling methods based on a new diverse street scene dataset. We propose a novel approach that can operate in such conditions and is based on a sequential Bayesian model update in order to robustly integrate the arriving images into the adapting procedure.

1   Introduction

Driving assistance systems have been rapidly evolving lately due to a constantly increasing interest in real-world application as well as studies conducted in the field of computer vision. An important task of such systems is road scene labeling in order to derive the semantic structure of the observed scenes. One of the big challenges is making such systems robust so that they can reliably operate in a wide range of conditions. However, capturing and training every possible condition a car can encounter throughout years of driving seems to be an impossible task.

Recently, there has been an increased interest in approaches of domain adaptation [7,8] in computer vision that are able to adapt existing classifiers to new domains and conditions. These require supervision from the target domain, that can not be provided by the envisioned systems that continuously operate in the real-world. Existing adaptive methods [1] allow the use of machine generated labels in order to refine the classifier and help it to adapt to changing conditions. However, they perform only global adaptation, for which they require access to the whole test set. Again, this is against the idea of a continuously operating system.

In contrast, we aim at an adaptive algorithm that is able to perform adaptation on the fly. Therefore, this paper proposes a sequential Bayesian update strategy that pursues multiple model hypothesis for semantic scene labeling.

Recommended for submission to YRF2014 by Dr. Mario Fritz.

© Springer International Publishing Switzerland 2014
X. Jiang et al. (Eds.): GCPR 2014, LNCS 8753, pp. 750–756, 2014.
DOI: 10.1007/978-3-319-11752-2_64

In order to circumvent typical problems of online learning by a "self-training" procedure, we perform model updates under the assumption of a stationary label distribution.

2   Naïve Model Update

Typical self-training approaches are based on a two step procedure. First, a lately arrived batch of images is labeled using the current model. Second, after an optional threshold on a confidence rating, these samples are used to update/ re-train the model. In more detail, we get an output probability distribution $P(x_{(i,j)})$ from our classifier for each pixel (i,j) and the predicted class-label for it $c^* = \operatorname{argmax}_{c \in \mathcal{Y}} P(x_{(i,j)} = c)$. Then, samples are taken for which $P(x_{(i,j)} = c^*) > \lambda$ holds, where λ is a acceptance threshold parameter. High probability $P(x_{(i,j)} = c^*)$ should indicate high confidence of the classifier in the predicted label. This is a completely heuristic approach, as the classification of the test data is only an approximation to the un-accessible groundtruth.

Taking new samples with the predicted labels which have high confidence is not necessarily a reliable way of updating the model due to inaccuracies in the intermediate models. While we want to be robust w.r.t. changes in the feature distribution, stationarity of the label distribution is a milder assumptions in many scenarios. We adopt ideas from Alvarez et al. [1] who employ a pixel-wise, normalized class-histogram on the off-line data as a prior distribution to weight the output probability distribution of the classifier at testing time.

In detail, we compute histogram for each pixel and after per-pixel L_1-normalization we get a prior $P_{pr}^{(i,j)}$ for each pixel $(i,j), i = 1, \ldots, W_{pr}, j = 1, \ldots, H_{pr}$. In our experiments images in the testing dataset all have various dimensions, so we perform nearest-neighbor sampling from the prior distribution $P_{pr}^{(i,j)}$. Then at testing time output probability distribution $P(x_{(i,j)})$ for all pixels $(i,j), i = 1, \ldots, W, j = 1, \ldots, H$ from our classifier for an image with dimensions $W \times H$ is element-wised multiplied with the corresponding prior

$$\tilde{P}(x_{(i,j)}) \propto P(x_{(i,j)}) P_{pr}^{(\lfloor i\frac{H_{pr}}{H}\rfloor, \lfloor j\frac{W_{pr}}{W}\rfloor)}. \tag{1}$$

This is used for accepting or rejecting new training examples on a per-pixel basis $\tilde{P}(x_{(i,j)} = c^*) > \lambda$.

3   Sequential Bayesian Model Update Under Structured Scene Prior

We propose a new model to leverage unlabeled data for a sequential model update for scene labeling. Our approach is based on a Bayesian model update. We maintain a population of models (particles) that approximate the distribution over the model-space $p(h_t|L_t)$, instead of relying on a single model, as in the previous formulations. The required integration over the model-space is solved by a Monte-Carlo method – just like in Condensation and Particle Filters that are well known from tracking applications [4,5]. Consequently, scene labeling at test time will be performed by marginalization over the model distribution

$$p(X|L_t) = \int p(X|h_t)p(h_t|L_t)\,\mathrm{d}h_t, \tag{2}$$

where X is the labeling of a test image for which we want to do prediction.

While the above-mentioned tracking formulations have a measurement step that evaluates image evidence, we measure the compatibility with the scene prior S. This is again based on the assumption of a stationary label distribution $P_{pr}^{(i,j)}$ as for the previous method.

Bayesian Model Update. We are interested in modeling an evolving target distribution over models in order to account for the uncertainty in the unobserved scene labels. Therefore, we model the unobserved scene labels l_t of the unlabeled data u_t at time step t as a latent variable (Fig. 1). Rather than sticking to a single model hypothesis, we seek to model a distribution over model hypothesis h_t. Therefore we update a distribution over model hypothesis given labels $p(h_t|L_t)$. Here $L_t = \{l_0, l_1, \dots, l_{t-1}, l_t\}$.

We describe the incorporation of the unlabeled examples in a Bayesian framework by integrating over all model hypothesis

$$p(h_t|L_{t-1}) = \int p(h_t|h_{t-1}, u_t)p(h_{t-1}|L_{t-1})\,\mathrm{d}h_{t-1}. \tag{3}$$

In the measurement step, we apply the Bayes' rule in order to get the updated distribution over model hypothesis

$$p(h_t|L_t) = \frac{p(l_t|h_{t-1}, S)p(h_t|L_{t-1})}{p(l_t|L_{t-1})}, \tag{4}$$

with

$$p(l_t|h_{t-1}, S) = p(l_t|h_{t-1})p(l_t|S), \tag{5}$$

where $p(l_t|h_{t-1})$ is the probability of a certain scene labeling prediction given a model hypothesis h_{t-1} and $p(l_t|S)$ is a scene labeling prior $P_{pr}^{(i,j)}$.

Sampling. We perform inference with a Monte-Carlo approach [5]. At each time step the model distribution $p(h_t|L_t)$ is represented by a set of particles $s_t^{(N)}$ with weights $\pi_t^{(N)}$. Next, the particles are propagated to the next time step via $p(h_t|h_{t-1}, u_t)$ that takes into account the existing models and the unlabeled data. In traditional tracking application this transition is modeled with a deterministic part and a stochastic component. In our setting, we propose to do model propagation by randomly choosing a subset of images which are provided to a particular particle to retrain as well as picking a randomized acceptance threshold λ per particle. The benefits are twofold. First, a diverse set of models is generated for the next iteration. Second, parameters like the acceptance thresholds are dealt with within the model and no hard choices have to be made.

In summary, our particle filter over model space works as follows. For each particle i out of N:

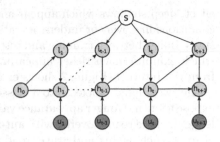

1. Pick a particle s_t^i from $s_t^{(N)}$, which represents $p(h_t|L_t)$, according to the weights $\pi_t^{(N)}$.
2. Sub-sample set of unlabeled images u_t to $\hat{u}_t$.
3. Predict labels $\hat{l}_t = \text{argmax}_l p(l|h_t)$ for subset $\hat{u}_t$.
4. Accept or reject samples based on some threshold λ.
5. Retrain model using $(\hat{u}_t, \hat{l}_t)$ and L_{t-1}.

Fig. 1. Bayesian network for the proposed Sequential Bayesian model update.

Fig. 2. First column shows examples of road scene dataset from [9]. Other columns show examples of the new diverse road scene dataset exhibiting very different appearances and a wider range of conditions.

Traditional tracking approaches would now follow up with a measurement in order to update the weights $\pi_t^{(N)}$. Similarly, we update the weight $\pi_t^{(N)}$ of each sample (model hypothesis) according to (4). In this equation $p(h_t|L_{t-1})$ is the distribution represented by our particles after the propagation step from above and $p(l_t|h_{t-1}, S)$ is the product of the likelihood of the labeling times the likelihood of the labeling given the scene labeling prior. We don't compute the denominator - but rather directly normalize the weights of the particles $\pi_t^{(N)}$ to sum to 1.

4 New Diverse Road Scenes Dataset

In order to study the problem of adaptation we need a dataset, which exhibits considerable amount of appearance variation between the training and test set. Typical road scene datasets like [2,9] (Fig. 2, first column) already exhibit some visually difficult situations like changes in object appearances due to motion blur effect, deep shadows which appear and disappear suddenly, changes in lightning conditions like over- or under-saturated regions, but the overall feature statistics stays similar between training and test. Therefore, we have collected a new dataset which exhibits much richer appearance variation, using freely available images from the Internet. Figure 2 shows examples from 220 images in our dataset. All images were hand-labeled into 3 classes: road, sky, and background. The dataset expose a much stronger appearance variation than previous datasets. Typical challenges include roads covered with autumn leaves or snow as well as different types of roads such as dirt and gravel roads and even images taken at night, although we leave out such issues like bad lighting, low contrast, or rain.

5   Experimental Results

In our implementations we employed a Random Forest [3] classifier (consisting of 10 trees each having maximal depth 15 with 20 % bagging) and features from [9]. We used the training set from [9] for training (Fig. 2, first column) and performed testing or adaptation on the new diverse road test set. Groundtruth annotation of the test set is not used in any way, other than for computing error rates.

Non-adaptive methods. In order to show that non-adaptive methods have a limited capability of generalizing to a different and strongly varying feature distribution as presented in our new dataset, we took one of the state-of-the-art methods for semantic image labeling of Krähenbühl et al. [6], and trained it on the training set and tested on both the old and the new test set (Table 1). The old test set has a similar appearance as the training set (Fig. 2, first column), so the resulting numbers are very good. But when we test on the new test set, the method shows strong accuracy degradations caused by the changed feature distribution. Particularly, the road recognition rate gets more than 50 times worse, because background and sky have more or less similar appearance as in the training set, while appearance of the road usually does not resemble the one in the training set.

Adaptive methods. Global adaptive methods consider the whole test set at once and try to adapt to it. The main restriction of such methods is that they require access to the whole test set. In the real world setting, when new images constantly arrive, global algorithms would have to deal with a constantly increasing test set. Alvarez et al. [1] proposed such an globally adaptive scheme for road

Table 1. Comparison of Krähenbühl et al. [6] semantic image labeling algorithm on the old and the new test test.

Test set	Fully connected CRF error, %			
	Road	Background	Sky	Average
Old	0.7	2.2	2.7	1.9
New	52.7	6.5	35	31.4

Table 2. Comparison of different adaptive approaches after processing the whole test set (mean plus std). Bold font highlights the best numbers.

Update type	Method	Error, %			
		Road	Background	Sky	Average
global	Alvarez *et al.* [1]	76.2	**12.7**	25.5	38.2
sequential	Naïve	26 ± 1.4	15.4 ± 0.4	9.3 ± 1.4	17 ± 0.7
	Naïve + Scene Prior	21 ± 2.7	18.5 ± 0.6	6.5 ± 0.9	15.5 ± 1.4
	Bayesian Model	19 ± 0.6	18.3 0.6	4.5 ± 0.4	13.9 ± 0.3

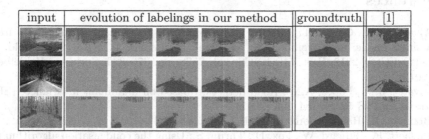

| input | evolution of labelings in our method | groundtruth | [1] |

Fig. 3. Example results showing the input image and evolution of the labelings through the proposed Sequential Bayesian Update method. The last two columns show the corresponding ground truth annotation and the output of the global adaptive method of Alvarez *et al.* [1]. Green color denotes background, red - sky, and blue - road.

scene segmentation. Table 2 (first row) presents resulting numbers for their original method, which the authors kindly agreed to run on our test set.

Table 2 shows resulting numbers for adaptive methods after they have processed the whole test set. Our algorithms were run 3 times and the results were averaged over. The numbers show that sequential adaptive methods have a larger capability in adapting to changing feature distribution in the setting, when new images arrive constantly during (possibly infinite) test time. Our proposed Sequential Bayesian Model Update shows the best average performance and the lowest variance.

Figure 3 shows some examples of how labelings for certain images evolve as our Bayesian Model Update method processes one batch of consequent images from the test set after another. It is remarkable how our approach can recover from initially poor segmentation results and adapts to the new conditions. We also show the results of the method of Alvarez *et al.* [1], over which we show quantitative as well as qualitative improvements.

6   Conclusion

Today's semantic scene labeling methods show good performance if the training distribution is representative for the test scenario. But when this feature distribution does change, as we showed, such techniques deteriorate in performance quickly. We collected a challenging dataset of images which has very different appearance statistic compared to the established scene segmentation datasets.

We proposed a Bayesian Model Update that sequentially updates the segmentation model as new data arrives, allowing to benefit from new information at test time and providing a possible application in scenarios when the new data is not available all at once, but rather arrives constantly in small batches.

References

1. Alvarez, J.M., Gevers, T., LeCun, Y., Lopez, A.M.: Road scene segmentation from a single image. In: Fitzgibbon, A., Lazebnik, S., Perona, P., Sato, Y., Schmid, C. (eds.) ECCV 2012, Part VII. LNCS, vol. 7578, pp. 376–389. Springer, Heidelberg (2012)
2. Álvarez, J.M., López, A.M.: Road detection based on illuminant invariance. IEEE Trans. on ITS **12**(1), 184–193 (2011)
3. Breiman, L.: Random forests. Mach. Learn. **45**, 5–32 (2001)
4. Dellaert, F., Burgard, W., Fox, D., Thrun, S.: Using the condensation algorithm for robust, vision-based mobile robot localization. In: CVPR (1999)
5. Isard, M., Blake, A.: Condensation - conditional density propagation for visual tracking. Int. J. Comput. Vis. **29**(1), 5–28 (1998)
6. Krähenbühl, P., Koltun, V.: Efficient inference in fully connected CRFs with gaussian edge potentials. In: NIPS (2011)
7. Kulis, B., Saenko, K., Darrell, T.: What you saw is not what you get: domain adaptation using asymmetric kernel transforms. In: CVPR (2011)
8. Saenko, K., Kulis, B., Fritz, M., Darrell, T.: Adapting visual category models to new domains. In: Daniilidis, K., Maragos, P., Paragios, N. (eds.) ECCV 2010, Part IV. LNCS, vol. 6314, pp. 213–226. Springer, Heidelberg (2010)
9. Wojek, C., Schiele, B.: A dynamic conditional random field model for joint labeling of object and scene classes. In: Forsyth, D., Torr, P., Zisserman, A. (eds.) ECCV 2008, Part IV. LNCS, vol. 5305, pp. 733–747. Springer, Heidelberg (2008)

Learning Multi-scale Representations for Material Classification

Wenbin Li[✉]

Max Planck Institute for Informatics, Saarbrücken, Germany
Wenbinli@mpi-inf.mpg.de

Abstract. The recent progress in sparse coding and deep learning has made unsupervised feature learning methods a strong competitor to hand-crafted descriptors. In computer vision, success stories of learned features have been predominantly reported for object recognition tasks. In this paper, we investigate if and how feature learning can be used for material recognition. We propose two strategies to incorporate scale information into the learning procedure resulting in a novel multi-scale coding procedure. Our results show that our learned features for material recognition outperform hand-crafted descriptors on the FMD and the KTH-TIPS2 material classification benchmarks.

1 Introduction

Perceiving and recognizing material is a fundamental aspect of visual perception. In contrast to texture recognition it requires generalization over large variations between material instances and discriminance between visually similar materials. Studies have shown that material recognition in real-world scenarios is far from solved [6,18]. While well established manually designed features [20,21,23] have been shown to be most powerful on material recognition tasks. It is non-trivial to come up with a good design of visual features and efforts are clearly needed to explore the question how we can automatically learn features for this challenging and relevant problem. Further, it is known that multi-scale representations are key for competitive performance on this task [6,16,20]. However, current feature learning techniques do not include multi-scale representations. Therefore, we present the first study of applying unsupervised feature discovery algorithms for material recognition and show improved performance over hand-crafted feature descriptors. Further, we investigate different ways to incorporate multi-scale information in the feature learning process and propose the first multi-scale coding procedure that results in a joint representation of multi-scale patches (see Fig. 1 for examples of multi-scale codes).

2 Related Work

Material Recognition. Curet database [9] was first proposed to address the recognition of single material instance, which motivated progress on texture

---

Recommended for submission to YRF2014 by Dr. Mario Fritz.

© Springer International Publishing Switzerland 2014
X. Jiang et al. (Eds.): GCPR 2014, LNCS 8753, pp. 757–764, 2014.
DOI: 10.1007/978-3-319-11752-2_65

(a) Filters learned on the KTH-TIPS2.   (b) Filters learned on the FMD.

Fig. 1. Examples of multi-scale filters learned on KTH-TIPS2a (a) and FMD (b) from the proposed MS4C model.

recognition [25,27]. Later research [6,12] shifted the focus towards whole material class, emphasizing challenges like scale and intra-class variations. Liu et al. [18] presented the Flickr material dataset using images captured under unknown real-world conditions and has been evaluated in [14,17,23].

Feature Learning. In machine learning literature, a rich set of models have been proposed for feature learning aimed to find a better representation for data. Examples include sparse coding [24], restricted Boltzmann machines [8,13] and various autoencoder-based models [4]. The Spike-and-Slab Sparse coding (S3C) [11] has recently been proposed to combine the advantages of SC and RBMs and has shown superior performance. Our representation is based on the S3C model and extended to multi-scale representation.

Multi-scale Representation. Early texton work included multi-scale filters to enrich the representation. Although the clustering step can be seen as a form of feature learning, the filters are hand-crafted. Also the LBP work has multi-scale extension [20] that has substantially improved the performance. Recently, a multi-scale convolutional neural network (CNN) [10] has been proposed. It differs from our multi-scale feature learning approach as we learn a representation jointly across scales. The image codes derived from our representation directly encode the multi-scaled information. Figure 1 illustrates some examples of multi-scale codes learned by our model.

3   Feature Learning

In a typical patch-based feature learning setting, random patches $\{v_n\}$ are firstly extracted from training images and a feature mapping f is learned (dictionary learning). Then one can encode the patches covering the input image and pool the codes together in order to form the final feature representation (feature extraction). By altering the model used for feature mapping, we can get different feature representations.

Sparse Coding (SC). Sparse coding for visual feature coding was originally proposed [22] as an unsupervised learning model of low-level sensory processing in humans. The dictionary W is obtained by optimizing $\underset{W,s_i}{\text{minimize}} \sum_i \|v_i - W s_i\|_2^2 + \beta \|s_i\|_1, w.r.t. \|W_j\|_2 \leq 1, \forall j$, then feature s_i for each input v_i is obtained by solving the same form of optimization problem but with the learned dictionary.

Auto-Encoder (AE). Auto-encoder is another popular model for learning feature representation. v is firstly mapped into a latent representation s (encoding) with a nonlinear function f such as the sigmoid function: $s = f(Wv + b)$. Then it is mapped back into a reconstruction $\widetilde{v}$ through a similar transformation $\widetilde{v} = f(\widetilde{W}s + \widetilde{b})$, and the dictionary (or weights) W is obtained by optimizing the reconstruction error $W = \underset{W}{\operatorname{argmin}} \sum_i L(v_i, \widetilde{v}_i)$, where $L(v_i, \widetilde{v})$ is a loss function such as the squared error $L(v, \widetilde{v}) = \|v - \widetilde{v}\|_2^2$. During encoding phase, the features are computed by applying the forward-pass only in order to obtain s.

Spike-and-Slab Sparse Coding (S3C). The Spike-and-Slab Sparse Coding (S3C) [11] has been recently proposed to combine the merits of both sparse coding and RBMs: the first layer is a real-valued D-dimensional visible vector $v \in R^D$, where v_d corresponding to the pixel value at position d; the second layer consists of two different kinds of latent variables, the binary *spike* variables $h \in \{0, 1\}^N$ and the real-valued *slab* variables $s \in R^N$. The spike variable h_i gates the slab variable s_i, and those two jointly define the ith hidden unit as $h_i s_i$. The process can be more formally described as follows:

$$\forall i \in \{1, ..., N\}, d \in \{1, ..., D\}$$
$$p(h_i = 1) = g(b_i)$$
$$p(s_i|h_i) = N(s_i|h_i\mu_i, \alpha_{ii}^{-1})$$
$$p(v_d|s, h) = N(v_d|W_{d:}(h \circ s), \beta_{dd}^{-1})$$

where g is the logistic sigmoid function, b is the biases on the spike variables, μ and W govern the linear dependence of s on h and v on s respectively, α and β are diagonal precision matrices of their respective conditionals, and $h \circ s$ denotes the element-wise product of h and s. Columns of W are constrained to have unit norm, α is restricted to be a diagonal matrix and β to be a diagonal matrix or a scalar. The model has shown to outperform previous feature learning techniques [11] and is the best performer on a recent transfer learning challenge [1].

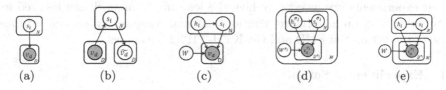

(a)   (b)   (c)   (d)   (e)

Fig. 2. Graphical model for: (a) SC, (b) AE, (c) S3C, (d) S4C, (e) MS4C.

4   Multi-scale Feature Learning

As shown in [6,17,19,20], encoding scale information is important for material recognition task. Hence we propose two different strategies to include multi-scale information in feature learning:

Stacked Spike-and-Slab Sparse Coding (S4C). We perform the encoding at multiple scales and stack the obtained codes, then use this code for classification. We convolve the patch with different sized Gaussians before encoding in order to represent scale information. While there is a common dictionary, the representation already encodes how the patch evolves in scale-space and therefore multi-scale information is captured:

$$\forall i \in \{1, ..., N\}, j \in \{1, ..., M\}, d \in \{1, ..., D\}$$
$$p(h_i^{\sigma_j} = 1) = g(b_i^{\sigma_j})$$
$$p(s_i^{\sigma_j}|h_i^{\sigma_j}) = N(s_i^{\sigma_j}|hi^{\sigma_j}\mu_i^{\sigma_j}, (\alpha_{ii}^{\sigma_j})^{-1})$$
$$p(v_d^{\sigma_j}|s^{\sigma_j}, h^{\sigma_j}) = N(v_d^{\sigma_j}|W_{d:}^{\sigma_j}(h^{\sigma_j} \circ s^{\sigma_j}), (\beta_{dd}^{\sigma_j})^{-1})$$

where M denotes the number of scales and σ_j indexes units and parameters at specific scale.

Multi-scale Spike-and-Slab Sparse Coding (MS4C). We construct a multi-scale pyramid for each image, apply the feature learning directly on the pyramid and then use the obtained codes for classification. In contrast to the S4C approach, the MS4C approach yields codes that model each patch jointly across scales:

$$\forall i \in \{1, ..., N\}, j \in \{1, ..., M\}, d \in \{1, ..., D\}$$
$$p(h_i = 1) = g(b_i)$$
$$p(s_i|h_i) = N(s_i|hi\mu_i, \alpha_{ii}^{-1})$$
$$p(v_d^{\sigma_j}|s, h) = N(v_d^{\sigma_j}|W_{d:}(h \circ s), \beta_{dd}^{-1})$$

where $v_d^{\sigma_j}$ denotes the joint representation of visible units at specific scale σ_j. Inference is carried out as in the S3C model as the different scales can be seen as a decomposition of a larger multi-scale patch that includes all the scales. Figure 2 shows the graphical models for both single-scale and multi-scale models.

5   Experiments

In our experiments, we investigate how the learning framework can be used for feature discovery on material recognition task and compare our approach to the state-of-the-art on the FMD and the KTH-TIPS2 databases.

5.1   Experimental Setup

We use KTH-TIPS2 [6] and FMD [18] in our experiments. For KTH-TIPS2 database, we use two instances for training and the other two for test per category. For FMD, we randomly split half for training and the other half for testing as suggested in [18]. We compare the learned features with hand-crafted features on the two databases with standard SVM classifier [7]. For single scale, we compare to the LBP [21] and its variants. For multi-scale approaches, we consider: Texton [16], MLBP [20]. For the learned representation, we compare to vector quantization, sparse coding, auto encoders and the spike-and-slab approach. In all our experiments we fix the size of dictionary at 1600 for consistency.

5.2   Single-Scale

For learned features, we apply the K-means, AE, SC and the S3C on the patch data where we vary the patch size; for hand-crafted features, we examine the original LBP and its variants as described in [20]. We use Theano [5] and Pylearn2 [2] for the auto-encoder and the S3C, the SPAMS [3] package for SC.

Results are shown in Figs. 3 and 4. On both datasets, the S3C model in combination with the linear kernel outperforms all other hand-crafted and learned features. With a performance of 71.3 % and 48.4 % for the KTH-TIPS2a and the FMD respectively it improves by 4.1 % (over LBP_u with the $exp - \chi^2$ kernel) and 9 % (over LBP with the linear kernel) respectively. The best performance is achieved for a patch size of 12. We verified that this parameter can be found via cross-validation on the training set. We attribute the decrease in the performance for the patch size of 24 to a lack of data to learn the required number of parameters. Best performance for feature learning technique is typically obtained in combination with linear kernel, while the hand-crafted features have to rely on the non-linear $exp - \chi^2$ kernel. This is another appealing property of the learned features from a computational point of view.

Overall, we found that the S3C feature performs better than other learning approaches and the hand-crafted features for the single-scale setting, and hence we further developed the S3C model to multi-scale approaches in the following experiments.

5.3   Multi-scale

Here we examine multi-scale feature representation for the task. Also we investigate the combination of color information for the learned representation. For hand-crafted features, we include the MLBP and also the texton with the MR8 filter [26]. As shown in Fig. 3 (a), (c) and Fig. 4 (a),(c). MLBP shows better performance than textons. While the S4C produces slightly worse performance than the MLBP on KTH-TIPS2, we see an improvement of 1.4 % for the MS4C. Further including color information improves the performance to 70.5 % which is an overall improvement of 3.8 % over the best hand-crafted descriptor. From the numbers on the FMD database, we see S4C and MS4C beat the best hand-crafted feature (MLBP) by 7.2 % and 8 % respectively. On this database, inclusion of color information does not yield additional improvements. The new joint multi-scale coding of of the MS4C consistently improves over the stacked approach of S4C.

Further we reproduced two additional settings in order to provide more points of comparison to the state-of-the-art. We follow the protocol in [15] on the KTH-TIPS2-a data, and then report results via a 3-NN classifier, feature learned by single scale S3C at patch size of 12 x 12 achieved 70.2 %, which is significantly better than the reported results of 64.2 % for LQP. Also we did additional experiments on the FMD database, following the settings in [14], with multi-scale collaborated representation, we got average recognition rate of 48.3 % and standard deviation of 1.8 %, which is comparable to the best single kernel descriptor with 49 %.

ClassificationRate(%)					
Single-Scale				Multi-Scale	
LBP	LBP_u	LBP_{ri}	$LBP_{ri,u}$	Texton	MLBP
58.7/64.8	60.3/67.2	55.0/53.6	50.9/51.4	54.0/58.9	66.7/66.1

(a) Hand-crafted Feature.

PatchSize	ClassificationRate(%)			
	KM	AE	SC	S3C
6	60.6/64.8	54.3/48.6	60.8/64.8	63.8/57.5
12	58.4/65.5	49.6/44.2	66.0/64.8	71.3/66.0
24	58.3/65.0	48.9/39.1	*	55.9/60.8

(b) Standard Feature Learning.

ClassificationRate(%)		
S4C	MS4C	MS4C+Color
65.6/58.6	68.1/66.6	70.5/69.3

(c) Multi-scale Feature Learning.

Fig. 3. Results on KTH-TIPS2a with linear kernel and the $exp - \chi^2$ kernel.

ClassificationRate(%)					
Single-Scale				Multi-Scale	
LBP	LBP_u	LBP_{ri}	$LBP_{ri,u}$	Texton	MLBP
39.4/36	38.2/36.2	34.2/35.6	27.8/31.8	29.4/35.6	41.4/42.0

(a) Hand-crafted Feature.

PatchSize	ClassificationRate(%)			
	KM	AE	SC	S3C
6	29.2/38.0	37.6/25.0	34.8/30.8	42.6/39.2
12	26.0/39.6	32.4/25.0	39.4/26.4	48.4/41.8
24	26.8/37.2	29.2/22.0	*	40.8/44.0

(b) Standard Feature Learning.

ClassificationRate(%)		
S4C	MS4C	MS4C+Color
49.2/42.2	50.0/41.0	48.8/43.2

(c) Multi-scale Feature Learning.

Fig. 4. Results on FMD with linear kernel and $exp - \chi^2$ kernel.

Figure 1 visualizes the MS4C model where we see how each filter has a multi-scale response. Together with the strong performance in our experiments, we conclude that a multi-scale code indeed captures additional information about how edge structures propagate through scales.

6   Conclusions

We have investigated different feature learning strategies for the task of material classification. Our results match and even surpass standard hand-crafted descriptors. Furthermore, we extend feature learning techniques to incorporate scale information. We propose the first coding procedure that learns and encodes features with a joint multi-scale representation. The comparison of our learned features with state-of-the-art descriptors shows improved performance on standard material recognition benchmarks.

References

1. Challenges in learning hierarchical models: Transfer learning and optimization. https://sites.google.com/site/nips2011workshop/transfer-learning-challenge
2. Pylearn2 vision, a python library for machine learning. http://deeplearning.net/ software/pylearn2/
3. Sparse modeling software, an optimization toolbox for solving various sparse estimation problems. http://spams-devel.gforge.inria.fr/
4. Bengio, Y., Lamblin, P., Popovici, D., Larochelle, H.: Greedy layer-wise training of deep networks. In: NIPS (2007)
5. Bergstra, J., Breuleux, O., Bastien, F., Lamblin, P., Pascanu, R., Desjardins, G., Turian, J., Warde-Farley, D., Bengio, Y.: Theano: a CPU and GPU math expression compiler. In: Proceedings of the Python for Scientific Computing Conference (SciPy) (2010)
6. Caputo, B., Hayman, E., Mallikarjuna, P.: Class-specific material categorisation. In: ICCV (2005)
7. Chang, C.C., Lin, C.J.: LIBSVM: a library for support vector machines. ACM Trans. Intell. Syst. Technol. (TIST) $2(3)$, 1–27 (2011)
8. Courville, A., Bergstra, J., Bengio, Y.: A spike and slab restricted boltzmann machine. In: JMLR (2011)
9. Dana, K.J., van Ginneken, B., Nayar, S.K., Koenderink, J.J.: Reflectance and texture of real-world surfaces. ACM Trans. Graph. $18(1)$, 1–34 (1999)
10. Farabet, C., Couprie, C., Najman, L., LeCun., Y.: Scene Parsing with Multiscale Feature Learning, Purity Trees, and Optimal Covers. In: ICML (2012)
11. Goodfellow, I., Couville, A., Bengio, Y.: Large-scale feature learning with spike-and-slab sparse coding. In: ICML (2012)
12. Hayman, E., Caputo, B., Fritz, M., Eklundh, J.-O.: On the significance of real-world conditions for material classification. In: Pajdla, T., Matas, J.G. (eds.) ECCV 2004. LNCS, vol. 3024, pp. 253–266. Springer, Heidelberg (2004)
13. Hinton, G., Osindero, S., Teh, Y.: A fast learning algorithm for deep belief nets. Neural Comput. 18, 1527–1554 (2006)
14. Hu, D., Bo, L., Ren, X.: Toward robust material recognition for everyday objects. In: BMVC (2011)
15. Hussain, S., Triggs, B.: Visual recognition using local quantized patterns. In: Fitzgibbon, A., Lazebnik, S., Perona, P., Sato, Y., Schmid, C. (eds.) ECCV 2012, Part II. LNCS, vol. 7573, pp. 716–729. Springer, Heidelberg (2012)
16. Leung, T., Malik, J.: Representing and recognizing the visual appearance of materials using three-dimensional textons. IJCV $43(1)$, 29–44 (2001)
17. Li, W., Fritz, M.: Recognizing materials from virtual examples. In: Fitzgibbon, A., Lazebnik, S., Perona, P., Sato, Y., Schmid, C. (eds.) ECCV 2012, Part IV. LNCS, vol. 7575, pp. 345–358. Springer, Heidelberg (2012)
18. Liu, C., Sharan, L., Adelson, E.H., Rosenholtz, R.: Exploring features in a bayesian framework for material recognition. In: CVPR (2010)
19. Mäenpää, T., Pietikäinen, M.: Multi-scale binary patterns for texture analysis. In: Bigun, J., Gustavsson, T. (eds.) SCIA 2003. LNCS, vol. 2749, pp. 885–892. Springer, Heidelberg (2003)
20. IEEE Trans. Pattern Anal. Mach. Intell. Multiresolution gray-scale and rotation invariant texture classification with local binary patterns. $24(7)$, 971–987 (2002)
21. Ojala, T., Pietikäinen, M., Harwood, D.: A comparative study of texture measures with classification based on featured distributions. Pattern Recogn. $29(1)$, 51–59 (1996)

22. Olshausen, B.A., et al.: Emergence of simple-cell receptive field properties by learning a sparse code for natural images. Nature **381**, 607–609 (1996)
23. Qi, X., Xiao, R., Guo, J., Zhang, L.: Pairwise rotation invariant co-occurrence local binary pattern. In: Fitzgibbon, A., Lazebnik, S., Perona, P., Sato, Y., Schmid, C. (eds.) ECCV 2012, Part VI. LNCS, vol. 7577, pp. 158–171. Springer, Heidelberg (2012)
24. Raina, R., Battle, A., Lee, H., Packer, B., Ng, A.: Self-taught learning: transfer learning from unlabeled data. In: ICML (2007)
25. Varma, M., Zisserman, A.: A statistical approach to material classification using image patch exemplars. IEEE Trans. Pattern Anal. Mach. Intell. **31**(11), 2032–2047 (2009)
26. Varma, M., Zisserman, A.: Classifying images of materials: achieving viewpoint and illumination independence. In: Heyden, A., Sparr, G., Nielsen, M., Johansen, P. (eds.) ECCV 2002, Part III. LNCS, vol. 2352, pp. 255–271. Springer, Heidelberg (2002)
27. Varma, M., Zisserman, A.: A statistical approach to texture classification from single images. Int. J. Comput. Vis. - Special Issue on Texture Anal. Synth. **62**(1–2), 61–81 (2005)

Casting Random Forests as Artificial Neural Networks (and Profiting from It)

Johannes Welbl[✉]

Heidelberg Collaboratory for Image Processing, Heidelberg, Germany
johannes.welbl@gmx.de

Abstract. While Artificial Neural Networks (ANNs) are highly expressive models, they are hard to train from limited data. Formalizing a connection between Random Forests (RFs) and ANNs allows exploiting the former to initialize the latter. Further parameter optimization within the ANN framework yields models that are intermediate between RF and ANN, and achieve performance better than RF and ANN on the majority of the UCI datasets used for benchmarking.

1 Introduction

In supervised machine learning, both RF [1] classifiers and feedforward ANNs [2] are in widespread use for tackling classification problems. In this work, it will be demonstrated that the predictive behaviour of RFs can equally be transferred into the framework of two-layer ANNs, where it can serve as an initialization for training. This transferability allows for a new functional interpretation of the RF model as special limit case of the general sigmoid ANN model.

ANNs are universal approximators [3], and with their many free parameters they are expressively very rich. However, their expressive power comes with the downside of increased overfitting risk, especially on small datasets. Conversely, RFs generalize well, but with their greedy tree construction process they yield fine, but often suboptimal classification performance.

Harnessing the novel RF-reformulation, a major practical intention is to exploit the benefits of RF and ANN models to overcome the shortcomings of the other. The novel network initialization allows for an optimization of RF predictions within the ANN framework and can possibly help ANNs to reduce their overfitting risk. The approach is developed in theoretical detail, later different training schemes are experimentally evaluated on various classification datasets.

Random Forests. In this section some notation will briefly be introduced. RFs are ensembles of decision trees [4] T_i, $i = 1, \ldots, N_{Trees}$, (with N_{Trees} being the number of trees) which consist of both *inner nodes* $\mathcal{N}_i^{Inner}$ and *leaf nodes*

Recommended for submission to the YRF2014 by Prof. Dr. Fred Hamprecht and Dr. Ullrich Köthe.

© Springer International Publishing Switzerland 2014
X. Jiang et al. (Eds.): GCPR 2014, LNCS 8753, pp. 765–771, 2014.
DOI: 10.1007/978-3-319-11752-2_66

$\mathcal{N}_i^{Leaf}$. For RFs using paraxial node splits, the split rule for $n \in \mathcal{N}_i^{Inner}$ and samples $x \in n$ is

$$x \in cl(n) \iff x_{f_n} < \theta_n \quad \text{and} \quad x \in cr(n) \iff x_{f_n} \geq \theta_n \qquad (1)$$

with f_n being the split feature of n, θ_n the threshold value, and $cl(n)$ and $cr(n)$ being the left and right child node of n.

For a tree leaf $L \in \mathcal{N}_i^{Leaf}$ the *ratio vote* is $y^L = (y_1^L, \ldots, y_{N_{Labels}}^L)$ with $y_l^L \in [0, 1] \forall l = 1, \ldots, N_{Labels}$, and $\sum_{l=1}^{N_{Labels}} y_l^L = 1$, where N_{Labels} is the number of labels.

Finally, the *path* $\mathcal{P}(L)$ to a leaf L will be the sequence of inner split nodes that has to be passed on the way from the tree root node n_0 to L:

$$\mathcal{P}(L) = (n_0, \ldots, n_d) \quad \text{with} \quad n_0, \ldots, n_d \in \mathcal{N}_i^{Inner} \quad \text{and} \quad L \subset n_d \subset \cdots \subset n_0 \qquad (2)$$

Thus, leaf membership is expressed in terms of conditions satisfied along $\mathcal{P}(L)$:

$$x \in L \iff \forall n_j \in \mathcal{P}(L) : \begin{cases} x_{f_{n_j}} < \theta_{n_j} & \text{if } L \subseteq cl(n_j) \\ x_{f_{n_j}} \geq \theta_{n_j} & \text{if } L \subseteq cr(n_j) \end{cases} \qquad (3)$$

Artificial Neural Networks. The ANNs chosen here are feedforward networks with two hidden layers (HLs). They use a tanh activation function in the HLs, and softmax activation for the output neurons, which allows for a probabilistic interpretation of the network output. During training, wrong predictions are backpropagated [5], and stochastic gradient descent (SGD) [6] is applied to maximize the likelihood of predicting true labels on randomized minibatches of training data.

2 Reformulating the RF as ANN

In this chapter, it will be described how a given pretrained RF can be reformulated as a two-layer ANN with the same predictive behaviour.

Both the RF and the ANN model rely on linear separation. In the RF any $n \in \mathcal{N}_i^{Inner}$ is *linearly* divided into $cl(n)$ and $cr(n)$. In ANNs with sigmoid activity the neurons *linearly* split their input space into two halfspaces of either positive or negative activity. This common characteristic of the two classifiers enables the transfer from RF to ANN, which will first be described for a single decision tree T_i and later be extended to a whole RF. There will be three functionally different parts in the resulting ANN architecture (also visualized in Fig. 1):

1. In the first hidden layer (HL_1) the neurons compute all tree split decisions $\forall n \in \mathcal{N}_i^{Inner}$ and indicate the split directions for a given input sample.
2. In the second hidden layer (HL_2) the information from HL_1 is combined to indicate if $x \in L$, $\forall L \in \mathcal{N}_i^{Leaf}$.
3. The weights from HL_2 to the output layer are set proportional to the ratio votes, effectively mimicking the RF voting system.

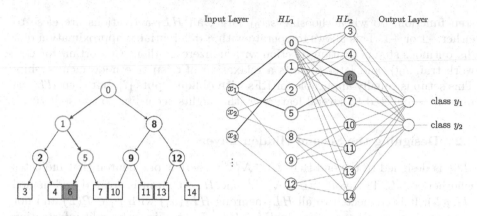

Fig. 1. A single decision tree (left) and its corresponding network structure (right). The circle nodes in the tree belong to $\mathcal{N}_i^{Inner}$ and the square nodes to $\mathcal{N}_i^{Leaf}$. The path to the red shaded leaf (6) consists of all light green nodes (0, 1, 5). Numbers in neurons correspond to numbers in tree nodes. The highlighted connections in the network are those relevant for the activity of the red neuron and its output vote.

2.1   Designing the First Hidden Layer

The first HL is designed to indicate tree split directions of given network inputs x, i.e. $\forall n \in \mathcal{N}_i^{Inner}$ one dedicated HL_1-neuron will encode if $x_{f_n} \geq \theta_n$ or not. Therefore, $\forall n \in \mathcal{N}_i^{Inner}$ one HL_1-neuron is created, named $HL_1(n)$, and then connected to input f_n. The corresponding weight and bias of $HL_1(n)$ are set to

$$w_{f_n, HL_1(n)} = str_{01} \quad \text{and} \quad b_{HL_1(n)} = -str_{01} \cdot \theta_n \qquad (4)$$

with $str_{01} \gg 0$ a global linear scaling hyperparameter. No other connections between input layer and HL_1 are established, so the connectivity is sparse. With this choice of weights and biases, the following statements are equivalent

$$x \in cr(n) \overset{(1)}{\Longleftrightarrow} x_{f_n} \geq \theta_n \overset{(4)}{\Longleftrightarrow} w_{f_n, HL_1(n)} \cdot x_{f_n} \geq -b_{HL_1(n)}$$
$$\Longleftrightarrow tanh(w_{f_n, HL_1(n)} \cdot x_{f_n} + b_{HL_1(n)}) \geq 0$$
$$\Longleftrightarrow activity(HL_1(n)) \geq 0$$

and likewise for $cl(n)$. Hence, for a given input x positive activities of HL_1-neurons encode tree splits to the right, and negative activities tree splits to the left. This allows to rewrite leaf membership from (3) in terms of HL_1-activations:

$$x \in L \iff \forall n_j \in \mathcal{P}(L) : \begin{cases} activity(HL_1(n_j)) < 0 & \text{if } L \subseteq cl(n_j) \\ activity(HL_1(n_j)) \geq 0 & \text{if } L \subseteq cr(n_j) \end{cases} \qquad (5)$$

The *strength*-hyperparameter str_{01} determines the contrast of the tanh activations: the larger str_{01} the sharper the transition from -1 to $+1$, and as str_{01} approaches infinity, the continuous tanh activation function converges to a binary step function. So when choosing str_{01} large, all HL_1-activations are close to either -1 or $+1$. This allows to operate with a differentiable approximation of a discontinous step activation function with nonzero gradient (important for network training), and later permits a relaxation of crisp tree node membership. This setup is equally applicable for RFs with oblique splits [7], but then $HL_1(n)$ must be connected to several features with weights according to the split tilt.

2.2   Designing the Second Hidden Layer

HL_2 is designed to the end that $\forall L \in \mathcal{N}_i^{Leaf}$ there is one neuron that indicates whether $x \in L$. Therefore, $\forall L \in \mathcal{N}_i^{Leaf}$ one HL_2-neuron is created and named $HL_2(L)$. It is connected to all HL_1-neurons $HL_1(n_j)$ with $n_j \in \mathcal{P}(L)$ and not connected to the other neurons in HL_1. $HL_2(L)$ combines the split information from the activations of all $HL_1(n_j)$, and for determining whether all conditions from (5) are satisfied, $HL_2(L)$ computes a softAND function of the $HL_1(n_j)$ activities. For this the incoming weights of $HL_2(L)$ are set according to the direction that $\mathcal{P}(L)$ takes at each n_j:

$$w_{HL_1(n_j),HL_2(L)} = \begin{cases} -str_{12} & \text{if } L \subseteq cl(n_j) \\ +str_{12} & \text{if } L \subseteq cr(n_j) \end{cases}, \qquad str_{12} > 0 \qquad (6)$$

Assuming str_{01} large $\Rightarrow HL_1(n_j)$-activations approximate either -1 or $+1$, so the incoming weights (6) act as a matched filter for the specific $HL_1(n_j)$-activation pattern (5) which appears iff $x \in L$. For fixed str_{12} this maximizes the activation response of $HL_2(L)$ if the HL_1-pattern (5) is present.

Note that for large str_{01} all datapoints x are mapped into the very corners of the hypercube of all HL_1-activations. Also the datapoints in the orthant defined by (5) (case $x \in L$) are located in one hypercube corner and are linearly separable from the other datapoints ($x \notin L$) in other hypercube corners. The bias determines the threshold of the softAND function to separate the cases:

$$b_{HL_2(L)} = -str_{12} \cdot [len(\mathcal{P}(L)) - 1] \qquad (7)$$

with $len(\mathcal{P}(L))$ the number of tree nodes n_j on $\mathcal{P}(L)$. In case of infinite str_{01} where $\forall n_j : activity(HL_1(n_j)) \in \{-1,1\}$ this bias choice maximizes the margin between the different HL_1-activation patterns appearing in the two cases $x \in L$ and $x \notin L$. Thus for large str_{01} the activity of $HL_2(L)$ effectively distinguishes the two cases and indicates if $x \in L$. Again, choosing a large global scaling hyperparameter ($str_{12} \gg 0$) makes the activities in HL_2 more contrastive.

2.3   Transferring the RF Voting System

So now, $\forall L \in \mathcal{N}_i^{Leaf}$ there is one HL_2-neuron which indicates whether $x \in L$. Next, all activities of HL_2 are linearly rescaled from $[-1,1]$ to $[0,1]$ and $HL_2(L)$ is connected to all outputs l, setting the weights proportional to the ratio vote:

$$w_{HL_2(L),l} = y_l^L \cdot str_{23} \tag{8}$$

with y_l^L the vote of L for label l, and $str_{23} > 0$ another scaling hyperparameter.

This concludes the ANN-reformulation of a single decision tree. For a complete RF, the HLs of several network structures are concatenated, so that the votes of all trees are expressed in the connections to the output neurons, effectively averaging the votes of the whole tree ensemble.

3   Training Schemes and Relaxations

The resulting network is now ready for training for which three approaches will be investigated. The *sparse* training scheme consists of the application of a standard network optimization algorithm (here: SGD) on the sparsely connected network. The *relaxed* training scheme has two differences to the *sparse*: (i) Allow for full connectivity between subsequent layers. While still initializing the previously non-existent connections close to 0, they will be free to change during network optimization, and hence establish cross-connections between functionally unrelated RF modules. Possibly more complex interactions between different RF elements can be captured this way. (ii) Decrease the global *strength* hyperparameters str_{01} and str_{12} to make the sigmoid activations less contrastive. This can be interpreted as a relaxation of crisp to fuzzy tree node membership. Samples not merely fall into one direction per split and one final leaf but simultaneously into several tree branches and leaves. Finally, the *vote* training scheme restricts network optimization to only weights of the last layer. While maintaining pure tree structure in the layers before, merely the RF votes will collectively be optimized. After training, this scheme enables a retransfer of the network into a classical RF with an altered (and possibly enhanced) voting system.

4   Experimental Evaluation

Next, several experiments are performed to investigate the novel initialization and different training schemes on classification datasets from UCI Machine Learning Repository [8] (*Breast Cancer Wisconsin (Diagnostic), Ionosphere, Sonar, Landsat, Pima Indians Diabetes, Heart Disease* and *German Credit*). The continuous feature dimensions of these datasets are in a generic preprocessing step linearly scaled to $[-1, 1]$. For each of the datasets the novel ANN initializations are examined with the three introduced training schemes, and for comparison also RFs and randomly initialized ANNs are tested. Each experimental setup is repeated 30 times on a randomly permuted separation into 3/4 training data and 1/4 test data (if training/test data not prespecified). Samples with missing values are discarded. The trees in the RF are grown applying the entropy criterium until a minimum number of 10 samples per node is reached. The best split feature is chosen among $\sqrt{M}$ random features, with M the input dimensionality. All RFs possess $N_{Trees} = 200$ trees (50 for Landsat). To account

Table 1. Predictive performance on test data for different classifiers and training schemes across seven datasets. Table entries are the empirical mean test error with standard deviation in parentheses. Bold writing highlights the winning classifier.

	RF	ANN	Sparse	Relaxed	Vote
Wisconsin	4.8 (1.7)	3.4 (1.7)	**2.9 (1.3)**	3.5 (1.9)	4.1 (1.8)
Ionosphere	6.5 (2.2)	11.0 (3.3)	**6.2 (2.0)**	6.9 (2.7)	6.7 (1.9)
Sonar	21.5 (5.1)	18.0 (5.5)	**14.4 (4.8)**	16.0 (5.2)	14.8 (4.1)
Landsat	10.9 (0.3)	10.1 (0.4)	9.1 (0.3)	9.1 (0.4)	**9.0 (0.4)**
Pima	**24.5 (2.1)**	28.9 (3.0)	26.8 (3.1)	26.4 (2.3)	26.6 (2.6)
Heart	**16.3 (4.4)**	21.8 (4.3)	19.5 (4.1)	19.5 (4.3)	19.0 (4.1)
Credit	25.6 (2.2)	28.3 (2.7)	**24.6 (1.8)**	24.7 (2.9)	25.3 (2.7)

for the random number of nodes in RFs and transcoded RFs, randomly initialized 2-layer networks with different sizes of $40 - 3500$ neurons per HL are trained, and the best result is chosen. All randomly initialized ANNs are initialized with biases 0 and weights following a narrow Gaussian $\mathcal{N}(0, \sigma^2)$, $\sigma = 0.01$.

Hyperparameters are set to $str_{01} = 1000$, $str_{12} = 1000$ and $str_{23} = 0.1$. For the relaxation str_{12} is decreased to 10, which is chosen as the lowest feasible value while evaluating the congruence of predictions of RFs and transcoded RFs on an exponentially scaled hyperparameter grid. SGD network training is applied using a batchsize of 100 and a momentum of 0.99. Learning rate is manually selected for each dataset so that training convergence is guaranteed. All networks are trained until convergence. To harness the power of parallelized GPU-computation the python package *theano* [9] is used. For all datasets RF construction and network training to convergence happens within seconds to minutes on a custom GPU. Table 1 shows the final test perfomances for all experimental setups.

5   Discussion of Results and Conclusion

A trivial result is that RF-initialized networks yield better test error at the start of training than randomly initialized networks, effectively giving the network a warmstart at the prediction level of the RF.

Comparing the performances of RFs with randomly initialized ANNs, it becomes apparent that the latter at times overfit (where RF performance is better than ANN), and that overfitting is mitigated by the novel network initialization as transcoded RF.

On the other hand, RF predictions are often improved by SGD-optimization within the ANN framework. The *sparse* scheme outperforms *both* ANN and RF in most cases. The *vote* optimization training (which allows for a retransfer into a classical RF) is not as competitive but still improves the RF predictions for several datasets. *Relaxed* training yields similar results to *sparse* training, hence the benefits of RF-initialization are not limited to sparsely connected networks but are equally available for the more widespreadly used fully connected ANNs.

In conclusion, the newfound theoretical link between RFs and ANNs can fruitfully be exploited to enhance both classifiers in practice. Optimizing RFs within the ANN framework often improves the predictive performance of the two original models. The benefits of both can be united in one novel hybrid classifier.

References

1. Breiman, L.: Random forests. Mach. Learn. **45**(1), 5–32 (2001)
2. Cybenko, G.: Approximation by superpositions of a sigmoidal function. Math. Control Signals Syst. **2**, 303–314 (1989)
3. Hornik, K., Stinchcombe, M., White, H.: Multilayer feedforward networks are universal approximators. Neural Netw. **2**(5), 359–366 (1989)
4. Breiman, L., Friedman, J., Olshen, R., Stone, C.: Classification and Regression Trees. Wadsworth and Brooks, Monterey (1984)
5. Rumelhart, D.E., Hinton, G.E., Williams, R.J.: Parallel Distributed Processing: Explorations in the Microstructure of Cognition. MIT Press, Cambridge (1986)
6. Bottou, L.: Online learning and stochastic approximations. In On-line Learning in Neural Networks, pp. 9–42. Cambridge University Press, Cambridge (1998)
7. Murthy, S.K., Kasif, S., Salzberg, S.: A system for induction of oblique decision trees. J. Artif. Intell. Res. **2**, 1–32 (1994)
8. Bache, K., Lichman, M.: UCI Machine Learning Repository. University of California, School of Information and Computer Sciences, Irvine, CA (2013). http://archive.ics.uci.edu/ml
9. Bergstra, J., Breuleux, O., Bastien, F., Lamblin, P., Pascanu, R., Desjardins, G., Turian, J., Warde-Farley, D., Bengio, Y.: Theano: a CPU and GPU math expression compiler. In: Proceedings of the Python for Scientific Computing Conference (SciPy), Oral Presentation, June 2010

Author Index

Printed in the United States
By Bookmasters